WORLD HEALTH ORGANIZATION

INTERNATIONAL AGENCY FOR RESEARCH ON CANCER

ONCOGENESIS AND HERPESVIRUSES

*Proceedings of a Symposium held at Christ's College, Cambridge, England
20 to 25 June 1971*

ORGANIZING COMMITTEE AND EDITORS

P. M. BIGGS (HPRS)   G. DE-THÉ (IARC)

L. N. PAYNE (HPRS)

IARC Scientific Publications No. 2

INTERNATIONAL AGENCY FOR RESEARCH ON CANCER
LYON

1972

The International Agency for Research on Cancer (IARC) was established in 1965 by the World Health Assembly as an independently financed organization within the framework of the World Health Organization. The headquarters of the Agency are at Lyon, France, and it has Regional Centres in Iran, Jamaica, Kenya and Singapore.

The Agency conducts a programme of research concentrating particularly on the epidemiology of cancer and the study of potential carcinogens in the human environment. Its field studies are supplemented by biological and chemical research carried out in the Agency's laboratories in Lyon and, through collaborative research agreements, in national research institutions in many countries. The Agency also conducts a programme for the education and training of personnel for cancer research.

The publications of the Agency are intended to contribute to the dissemination of authoritative information on different aspects of cancer research.

ACKNOWLEDGEMENT

The organizers of the Symposium wish to thank the Special Virus Cancer Program, National Cancer Institute, National Institutes of Health, USA, for its generous financial support.

CONTENTS

SCIENTIFIC PAPERS

Introductory Lecture
Chairman — P. M. Biggs

Pathology of Marek's disease
Chairman — P. M. Biggs
Rapporteur — J. G. Campbell

Lucké frog renal carcinoma

Chairman — A. Granoff

*Pathology of Burkitt's lymphoma, infectious
mononucleosis and nasopharyngeal carcinoma*

Chairman — J. B. Moloney
Rapporteur — C. S. Muir

*Epidemiology of Burkitt's lymphoma, infectious
mononucleosis and nasopharyngeal carcinoma*

Chairman — W. Henle

Rapporteur — R. H. Morrow, Jr

Other herpesviruses associated with neoplasms

Chairman — G. de-Thé
Rapporteur — R. J. C. Harris

FOREWORD

Since its inception, the International Agency for Research on Cancer has been engaged in studying the rôle of environmental factors in the causation of human cancer. Oncogenic viruses may prove to be amongst the most important of these factors.

With financial support from the National Cancer Institute, National Institutes of Health, Bethesda, USA, the Agency has developed programmes in several countries to investigate the aetiological rôle of herpes-type viruses in human cancer, especially cancer of the nasopharynx and Burkitt's lymphoma. The pioneer studies already undertaken by Epstein, Henle and Klein in this field provided a starting point for the programme.

Concurrently with the development of these studies in man, studies on herpesvirus infection in animal neoplasia suggested wider possibilities for investigating the biological action of these viruses.

Thus, when the Agency was approached by the team from the Houghton Poultry Research Station to discuss the possibility of organizing an international symposium on oncogenesis and herpesviruses, agreement to collaborate was quickly reached since such a symposium was clearly most pertinent to the Agency's programme. The Symposium was arranged at the earliest possible date and the publication of the proceedings was planned to follow soon after.

Although I was not present at the Symposium, it is clear from this volume that the programme had been very well designed and I hope it will contribute something towards the solution of the human cancer problem.

I would therefore like to convey my thanks to Dr Biggs, Dr Payne and Dr de-Thé for the immense amount of work that they have put into the preparations for this conference, and to thank also all those who were able to participate and contribute to the discussions formal and informal.

I wish to express my special thanks to the Special Virus Cancer Program, National Cancer Institute, National Institutes of Health, USA, and the Houghton Poultry Research Station, UK, for their support.

John HIGGINSON, M.D.
Director
International Agency
for Research on Cancer
Lyon, France

INTRODUCTION

The rapidly accumulating evidence for an association between herpesviruses and neoplasia suggested to us the need for a symposium on the subject, particularly as herpesviruses are the only DNA viruses which have been associated with spontaneous malignancies in man and other animals. A herpesvirus has been shown to be the cause of Marek's disease of fowl, and other herpesviruses have been closely associated with Burkitt's lymphoma, infectious mononucleosis and nasopharyngeal carcinoma in man, and with the Lucké renal carcinoma in frogs. We believed that a comparative approach would be useful and that the knowledge and techniques of workers in each disease would be of mutual value.

When we discussed the problem together, the proposal that workers at Houghton Poultry Research Station and at the International Agency for Research on Cancer should collaborate in organizing a Symposium followed inevitably. The aim was to bring together those engaged in the study of oncogenic herpesviruses and associated diseases to review the current state of knowledge and to present new findings. Rapporteurs were enlisted to prepare summaries of the discussions, and the proceedings of the entire Symposium were brilliantly summarized by Professor George Klein, to whom we are most indebted.

We have now collected together the papers and discussions to form a publication which we hope will provide an account of contemporary knowledge on oncogenic herpesviruses in all species.

P. M. BIGGS
G. DE-THÉ
L. N. PAYNE

PARTICIPANTS

H. K. ADLDINGER — Department of Microbiology, Albert Einstein Medical Center, York and Tabor Roads, Philadelphia, Pa. 19141, USA. Present address: Department of Veterinary Microbiology, School of Veterinary Medicine, University of Missouri — Columbia, Connaway Hall — Connaway Hall Annex, Columbia, Mo. 65 201, USA

J. C. AMBROSIONI — International Agency for Research on Cancer, 16 avenue Maréchal Foch, 69 Lyon (6e), France

D. P. ANDERSON — Poultry Disease Research Center, University of Georgia, 953 College Station Road, Athens, Georgia 30601, USA

S. L. BACHENHEIMER — Department of Microbiology, The University of Chicago, 939 East 57th Street, Chicago, Illinois 60637, USA

M. BAILEY — Standards Laboratory for Serological Reagents, Central Public Health Laboratory, Colindale Avenue, London, NW9 5HT, UK

J. E. BANATVALA — Department of Clinical Virology, St. Thomas' Hospital and Medical School, London, SWl, UK

S. M. BARD — University Health Service, University of Hong Kong, Hong Kong

O. BASARAB — Wellcome Research Laboratories, Langley Park, Beckenham, Kent, UK

W. BAXENDALE — Poultry Biologicals Limited, The Elms, Houghton, Huntingdon, UK

J. M. BÉCHET — The Wallenberg Laboratory, University of Uppsala, Uppsala, Sweden

Y. BECKER — Department of Virology, Hebrew University/Hadassah Medical School, Jerusalem, Israel

P. M. BIGGS — Houghton Poultry Research Station, Houghton, Huntingdon, UK

C. M. P. BRADSTREET — Standards Laboratory for Serological Reagents, Central Public Health Laboratory, Colindale Avenue, London, NW9 5HT, UK

V. VON BÜLOW — Bundesforschungsanstalt für Viruskrankheiten der Tiere, PO Box 1149, 74 Tübingen, Federal Republic of Germany

D. P. BURKITT — Medical Research Council External Staff, 172 Tottenham Court Road, London, W1, UK

B. R. BURMESTER — Regional Poultry Research Laboratory, 3606 East Mount Hope Road, East Lansing, Michigan 48823, USA

B. W. CALNEK — Department of Avian Diseases, Cornell University, Ithaca, New York 14850, USA

J. G. CAMPBELL — Cancer Campaign Research Unit, Bush House, Milton Bridge, Midlothian, UK

R. L. CARTER — Chester Beatty Research Institute, Fulham Road, London, SW3, UK

R. P. S. Carvalho — Instituto de Medicina Tropical de São Paulo, Av. Eneas De Carvalho Aguiar 470, Caixa Postal 2921, São Paulo, Brazil

R. Cassingena — Unité de Virologie, Institut de Recherches Scientifiques sur le Cancer, B. P. N° 8, 94 Villejuif, France

A. E. Churchill — Poultry Biologicals Limited, The Elms, Houghton, Huntingdon, UK

R. K. Cole — Poultry Science Department, Rice Hall, Cornell University, Ithaca, New York 14850, USA

M. D. Daniel — New England Regional Primate Research Center, Harvard Medical School, Southborough, Massachusetts 01772, USA

R. J. L. Davidson — The Laboratory, City Hospital, PO Box 42, Aberdeen, UK

W. Davis — International Agency for Research on Cancer, 16 avenue Maréchal Foch, 69 Lyon (6e), France

M. L. Didier-Fichet — International Agency for Research on Cancer, 16 avenue Maréchal Foch, 69 Lyon (6e), France

R. G. Duff — Department of Microbiology, The Milton S. Hershey Medical Center, The Pennsylvania State University, Hershey, Pennsylvania 17033, USA

V. C. Dunkel — National Cancer Institute, National Institutes of Health, Bethesda, Maryland 20014, USA

C. S. Eidson — Poultry Disease Research Center, University of Georgia, 953 College Station Road, Athens, Georgia 30601, USA

A. Y. Elliott — Division of Urology, Department of Surgery, University of Minnesota, A595 Mayo Memorial Building, Minneapolis, Minnesota 55455, USA

R. W. Else — Department of Veterinary Medicine, University of Bristol, Langford House, Langford, Bristol, BS18 7DU, UK

M. A. Epstein — Department of Pathology, Medical School, University of Bristol, University Walk, Bristol, BS8 1TD, UK

S. Eridani — Chester Beatty Research Institute, Fulham Road, London, SW3, UK

A. S. Evans — Department of Epidemiology, Yale University School of Medicine, New Haven, Connecticut 06510, USA

A. M. Field — Virus Reference Laboratory, Central Public Health Laboratory, Colindale Avenue, London, NW9 5HT, UK

W. E. H. Field — Department of Pathology, Medical School, University of Bristol, University Walk, Bristol, BS8 1TD, UK

M. A. Fink — National Cancer Institute, National Institutes of Health, Bethesda, Maryland 20014, USA

J. A. Frazier — Houghton Poultry Research Station, Houghton, Huntingdon, UK

I. G. S. Furminger — Evans Medical Limited, Speke, Liverpool, L24 9JD, UK

W. M. Gallmeier — Innere Klinik und Poliklinik (Tumorforschung), Essen, der Ruhr-universität, Bochum, Federal Republic of Germany

D. Gaudry — Institut Merieux/IFFA, 254 rue Marcel Merieux, 69 Lyon (7e), France

L. Geder	Department of Microbiology, Medical School, The University of Debrecen, Debrecen 12, Hungary
E. F. Gelenczei	Babcock Poultry Farm, Inc., PO Box 280, Ithaca, New York 14850, USA
P. Gerber	Division of Biologics Standards, National Institutes of Health, Bethesda, Maryland 20014, USA
A. Geser	International Agency for Research on Cancer, 16 avenue Maréchal Foch, 69 Lyon (6e), France
K. D. Gibson	Roche Institute of Molecular Biology, Nutley, New Jersey 07110, USA
R. F. Gordon	Houghton Poultry Research Station, Houghton, Huntingdon, UK
A. Granoff	St Jude Children's Research Hospital, 332 North Lauderdale, PO Box 318, Memphis, Tennessee 38101, USA
T. B. Greenland	International Agency for Research on Cancer, 16 avenue Maréchal Foch, 69 Lyon (6e), France
I. W. Halliburton	M. R. C. Virology Unit, Institute of Virology, Church Street, Glasgow, W1, UK
R. J. C. Harris	Microbiological Research Establishment, Porton Down, Salisbury, Wiltshire, UK
H. zur Hausen	Virologisches Institut der Universität, 8700 Würzburg, Versbacher Landstr. 7, Federal Republic of Germany
G. Henle	Research Department, The Children's Hospital of Philadelphia, 1740 Bainbridge Street, Philadelphia, Pa. 19164, USA
W. Henle	Research Department, The Children's Hospital of Philadelphia, 1740 Bainbridge Street, Philadelphia, Pa. 19164, USA
P. N. R. Heseltine	The Cottage, 6 Raglan Road, Dublin 4, Ireland
I. Hlozanek	Institute of Experimental Biology and Genetics, Czechoslovak Academy of Sciences, Flemingovo nám. 2, Prague 6, Czechoslovakia
H. C. Ho	M. & H. D. Institute of Radiology, Queen Elizabeth Hospital, Kowloon, Hong Kong
Y. Ito	Aichi Cancer Centre, Research Institute, Tashiro-Cho, Chikusa-Ku, Nagoya, Japan
I. Jack	Virus Laboratory, Royal Children's Hospital, Flemington Road, Parkville, Victoria 3052, Australia
J. H. Joncas	Institut de Microbiologie et d'Hygiène de l'Université de Montréal, PO Box 100, Laval, Quebec, Canada
H. M. Jones	The Institute of Laryngology and Otology, University of London, 330/332 Gray's Inn Road, London, UK
P. W. Jones	Department of Virology, Wellcome Research Laboratories, Beckenham, Kent, UK
O. R. Kaaden	Bundesforschungsanstalt für Viruskrankheiten der Tiere, PO Box 1149, 74 Tübingen, Federal Republic of Germany

S. S. Kalter — Division of Microbiology and Infectious Diseases, Southwest Foundation for Research and Education, PO Box 28147, 7480 West Commerce Street, San Antonio, Texas 78228, USA

A. Karpas — Department of Medicine, University of Cambridge, Hills Road, Cambridge, CB2 1QT, UK

S. Kato — Department of Pathology, Research Institute for Microbial Diseases, Osaka University, Yamada-kami, Suita, Osaka, Japan

N. H. Kemp — The Clinical Research Laboratories, St George's Hospital Medical School, Blackshaw Road, Tooting, London, SW17, UK

S. G. Kenzy — Department of Veterinary Microbiology, Washington State University, Pullman, Washington 99163, USA

E. D. Kieff — Department of Medicine, University of Chicago Hospitals and Clinics, 950 East 59th Street, Chicago, Illinois 60637, USA

J. Kirkwood — Yale University School of Medicine, 1 South Street, New Haven, Connecticut 06510, USA

E. Klein — Department of Tumor Biology, Karolinska Institutet, S-104 01, Stockholm 60, Sweden

G. Klein — Department of Tumor Biology, Karolinska Institutet, S-104 01, Stockholm 60, Sweden

S. H. Kleven — Poultry Disease Research Center, University of Georgia, 953 College Station Road, Athens, Georgia 30601, USA

P. H. Levine — National Cancer Institute, National Institutes of Health, Bethesda, Maryland 20014, USA

J. T. Lewin — National Cancer Institute, National Institutes of Health, Bethesda, Maryland 20014, USA

V. M. Lucké — Department of Pathology, Medical School, University of Bristol, University Walk, Bristol, BS8 1TD, UK

R. E. Luginbuhl — The University of Connecticut, College of Agriculture and Natural Resources, Department of Animal Diseases, Storrs, Connecticut 06268, USA

H. J. L. Maas — TNO, Division of Veterinary Science, Centraal Diergeneeskundig Instituut, Afdeling Rotterdam, Prof. Poelslaan 35, Rotterdam 7, The Netherlands

J. M. K. Mackay — Animal Diseases Research Association, Moredun Institute, 403 Gilmerton Road, Edinburgh, EH17 7JH, UK

C. R. Madeley — Institute of Virology, Church Street, Glasgow, W1, UK

B. Mahy — Department of Pathology, University of Cambridge, Tennis Court Road, Cambridge, CB2 1QF, UK

R. Manaker — National Cancer Institute, National Institutes of Health, Bethesda, Maryland 20014, USA

P. W. A. Mansell — Department of Pathology, University of Bristol, The Medical School, University Walk, Bristol BS8 1TD, UK

L. Martos — Flow Laboratories Inc., Research Building 4, 1710 Chapman Avenue, Rockville, Maryland 20852, USA

R. G. McKinnell   Department of Zoology, University of Minnesota, Minneapolis, Minnesota 55455, USA

L. V. Melendez   Division of Microbiology, Harvard Medical School, Southborough, Massachusetts 01772, USA

M. Mizell   Tulane University, New Orleans, La. 70118, USA

J. B. Moloney   National Cancer Institute, National Institutes of Health, Bethesda, Maryland 20014, USA

D. G. Morgan   Department of Pathology, University of Bristol, The Medical School, University Walk, Bristol, BS8 1TD, UK

R. H. Morrow, Jr   Department of Tropical Public Health, Harvard University, 665 Huntingdon Avenue, Boston, Massachusetts 02115, USA

C. Muir   International Agency for Research on Cancer, 16 avenue Maréchal Foch, 69 Lyon (6e), France

K. Munk   Institut für Virusforschung, Deutsches Krebsforschungszentrum, Thibautstrasse 3, 69 Heidelberg, Federal Republic of Germany

N. Muñoz   International Agency for Research on Cancer, 16 avenue Maréchal Foch, 69 Lyon (6e), France

G. M. R. Munube   East African Community, E. A. Virus Research Institute, PO Box No. 49, Entebbe, Uganda

A. J. Nahmias   Department of Pediatrics and Preventive Medicine, Emory University School of Medicine, Atlanta, Georgia 30322, USA

K. Nazerian   Regional Poultry Research Laboratory, 3606 East Mount Hope Road, East Lansing, Michigan 48823, USA

A. Newton   Department of Biochemistry, Tennis Court Road, Cambridge, CB2 1QF, UK

J. C. Niederman   Department of Epidemiology and Public Health, Yale University School of Medicine, 60 College Street, New Haven, Connecticut 06510, USA

K. N. Nilsson   The Wallenberg Laboratory, University of Uppsala, Uppsala, Sweden

D. I. Nisbet   Moredun Institute, 408 Gilmerton Road, Edinburgh, UK

K. Nishioka   National Cancer Centre, Research Institute, Tsukiji 5-Chome, Chuo-Ku, Tokyo, Japan

T. E. O'Connor   National Cancer Institute, National Institutes of Health, Bethesda, Maryland 20014, USA

T. Osato   Department of Virology, Cancer Institute, Hokkaido University School of Medicine, Sapporo N15, W7, Japan

P. K. Pani   Houghton Poultry Research Station, Houghton, Huntingdon, UK

L. N. Payne   Houghton Poultry Research Station, Houghton, Huntingdon, UK

M. S. Pereira   Virus Reference Laboratory, Central Public Health Laboratory, Colindale Avenue, London, NW9 5HT, UK

J. F. Peutherer   Bacteriology Department, University of Edinburgh, Medical School, Teviot Place, Edinburgh, UK

P. A. PHILIPS   Houghton Poultry Research Station, Houghton, Huntingdon, UK

M. PIKE   University of Oxford, Department of the Regius Professor of Medicine, Radcliffe Infirmary, Oxford, OX2 6HE, UK

T. M. POLLOCK   Epidemiological Research Laboratory, Central Public Health Laboratory, Colindale Avenue, London, NW9 5HT, UK

M. PROBERT   Department of Pathology, University of Bristol, The Medical School, University Walk, Bristol, BS8 1TD, UK

H. G. PURCHASE   Regional Poultry Research Laboratory, 3606 East Mount Hope Road, East Lansing, Michigan 48823, USA

K. A. RAFFERTY   Department of Anatomy, University of Illinois at the Medical Center, 1853 West Polk Street, PO Box 6998, Illinois 60690, USA

W. L. RAGLAND, III   Poultry Disease Research Center, The University of Georgia, Athens, Georgia 30601, USA

F. RAPP   Department of Microbiology, The Milton S. Hershey Medical Center, The Pennsylvania State University, Hershey, Pennsylvania 17033, USA

W. E. RAWLS   Department of Virology and Epidemiology, Baylor College of Medicine, Texas Medical Center, Houston, Texas 77025, USA

B. ROIZMAN   Virology Laboratory, The University of Chicago, 939 East 57th Street, Chicago, Illinois 60637, USA

L. J. N. ROSS   Houghton Poultry Research Station, Houghton, Huntingdon, UK

K. E. K. ROWSON   Institute of Laryngology and Otology, 330 Gray's Inn Road, London, WC1, UK

E. D. RUBERY   Department of Biochemistry, Tennis Court Road, Cambridge, CB2 1QF, UK

O. P. SETTNES   Institute of Medical Microbiology, University of Copenhagen, 22 Juliana Maries Vej, 2100 Copenhagen Ø, Denmark

K. SHANMUGARATNAM Department of Pathology, University of Singapore, Outram Road, Singapore 3

M. J. SIMONS   WHO Immunology Centre, Faculty of Medicine, University of Singapore, Sepoy Lines, Singapore 3

I. W. SMITH   Department of Bacteriology, University of Edinburgh, University Medical School, Teviot Place, Edinburgh, UK

R. SOHIER   Unité de Virologie, Institut National de la Santé et de la Recherche Médicale (INSERM), 1 place Professeur Joseph Renaut, 69 Lyon (8e), France

N. SOTIROV   Chester Beatty Research Institute, Fulham Road, London, SW3, UK

J. LLOYD SPENCER   Animal Diseases Research Institute, PO Box 1400, Hull, Quebec, Canada

P. G. STANSLY   National Cancer Institute, National Institutes of Health, Bethesda, Maryland 20014, USA

P. B. STONES   Pfizer Limited, Sandwich, Kent, UK

G. DE-THÉ	International Agency for Research on Cancer, 16 avenue Maréchal Foch, 69 Lyon (6e), France
M. C. TIMBURY	Institute of Virology, University of Glasgow, Church Street, Glasgow, W1, UK
J. O'H. TOBIN	Public Health Laboratory, Withington Hospital, Manchester, M20 8LR, UK
I. TOPLIN	Electro-Nucleonics Laboratories, 4921 Auburn Avenue, Bethesda, Maryland 20014, USA
K. S. TWEEDELL	Department of Biology, University of Notre Dame, College of Science, Notre Dame, Indiana 46556, USA
V. VONKA	Department of Experimental Virology, Institute of Sera and Vaccines, Srobarova 48, Prague 10, Czechoslovakia
B. WAHREN	Department of Virology, National Bacteriological Laboratories, Stockholm, Sweden
A. S. WALLIS	Beedal Laboratories, Poulton-le-Fylde, Blackpool, Lancashire, UK
K. W. WASHBURN	Department of Poultry Science, The University of Georgia, College of Agriculture, Athens, Georgia 30601, USA
A. WEINBERG	Department of Virology, Hebrew University/Hadassah Medical School, Jerusalem, Israel
C. N. WIBLIN	Imperial Cancer Research Fund, PO Box 123, Lincoln's Inn Fields, London, WC2A 3PX, UK
P. WILDY	Department of Virology, The Medical School, University of Birmingham, Birmingham 15, UK
R. L. WITTER	Regional Poultry Research Laboratory, 3606 East Mount Hope Road, East Lansing, Michigan 48823, USA
G. N. WOODE	Agricultural Research Council, Institute for Research on Animal Diseases, Compton, Nr. Newbury, Berkshire, UK
T. O. YOSHIDA	Aichi Cancer Centre, Research Institute, Tashiro-Cho, Chikusa-ku, Nagoya, Japan
N. ZYGRAICH	Biologics Department, RIT, 1330 Rixensart, Belgium
Mrs H. D. RICHARDS	Symposium Secretary, Houghton Poultry Research Station, Houghton, Huntingdon, UK

SCIENTIFIC PAPERS

The Biochemical Features
of Herpesvirus-Infected Cells, Particularly as they
Relate to their Potential Oncogenicity—A Review

B. ROIZMAN [1]

INTRODUCTION

The problems

In the last few years herpesviruses have been associated with neoplasia in man, monkey, fowl and frogs. The number of herpesviruses found in association with neoplasia is increasing. Paramount among the questions raised by the discovery of the viruses is whether they are the causative agents of the neoplasia with which they are associated. Clearly this is the theme of this Symposium and the problem before us is not whether a simple answer exists but rather what sort of experiments must be done to obtain this answer. Two questions, however, can be answered at least in part and have been the subject of study in our laboratory.

The first question relates to the uncertainty that existed for many years as to whether these viruses were in fact herpesviruses. This is reflected in the fact that some of them are still called herpes-type viruses rather than herpesviruses. The question arises therefore whether the herpesviruses associated with neoplasia are physically, genetically and in other characteristics similar to other members of the group or whether they have arisen by convergent evolution from an entirely different parentage. The implications of the latter alternative are immense and, regardless of how improbable the hypothesis of convergent evolution may appear, it must be considered. This question is dealt with by Bachenheimer *et al.* in another paper to this Symposium.[2] I shall anticipate that paper by indicating that so far as we can tell none of the genetic or physical properties studied to date indicate that the viruses associated with neoplasia differ from those that are not.

The second question assumes that herpesviruses associated with neoplasia do not form a unique group within the family *Herpesviridae* and asks which, if any, biochemical property of herpesviruses most closely reflects their putative oncogenic function. The present paper is concerned with this topic.

Definitions

The functions specified by the viruses of interest to us are expressed in the infected cell. For heuristic reasons it is convenient to differentiate the functions specified by the virus according to the outcome of infection. We currently recognize two states of the infected cell: productive and nonproductive. The consequence of *productive infection* is viral progeny. Herpesviruses can also infect and be perpetuated in the infected cell without production of progeny. By definition the *nonproductively infected cell* makes no viral progeny, infectious or noninfectious. Parenthetically, nonproductive infection is different from *abortive* and *restrictive* infections, in which small amounts of viral progeny and structural components of the virus are produced. In accordance with what we observe, it is convenient to differentiate between *viral productive functions,* which lead to the biosynthesis of viral structural components and progeny, and *viral nonproductive functions,* such as those that could be expressed in the nonproductively infected cell. In addition, we shall refer to "*early*" functions as those that, in productively infected cells, are independent of the synthesis of viral DNA, and "*late*" functions as those that are dependent for their expression on the prior synthesis of viral DNA. Also for heuristic reasons it is convenient to differentiate

[1] Departments of Microbiology and Biophysics, The University of Chicago, Chicago, Illinois, USA.

[2] See p. 74 of this publication.

between viral functions concerned with the production of progeny and a possible subset of these functions—*the oncogenic functions*—that might lead to malignant transformation.

The thesis

To facilitate the presentation of the data it is convenient to present the thesis of this paper as follows:

(i) Viral productive functions of all herpesviruses appear to be expressed in a similar fashion. Death of the infected cell, probably due to the irreversible inhibition of host macromolecular synthesis, invariably accompanies production of progeny virus.

(ii) Nonproductive infection is a common characteristic of all herpesviruses. Potentially oncogenic functions may be expressed in both productively and nonproductively infected cells. However, malignant transformation can occur only in nonproductively infected cells since cell death is a consequence of productive infection.

(iii) Of the known biochemical functions expressed by herpesviruses in the infected cell, the most likely candidate for a potentially oncogenic function is the modification of cellular membranes by virus-specific products.

THE EXPRESSION OF VIRAL FUNCTIONS IN
PRODUCTIVE INFECTION

The functions uniformly expressed by all herpesviruses in productively infected cells are: (i) the biosynthesis of viral macromolecules; (ii) the assembly of the virus; and (iii) the modification of the host. Expression of viral productive functions has been reviewed elsewhere in detail (Roizman, 1969, 1971a; Roizman & Spear, 1971a, b).

Biosynthesis and assembly of structural components

The biosynthesis and assembly of structural components have now been characterized quite well in cells infected with herpes simplex and pseudorabies viruses. The steps involved have been described elsewhere in greater detail (Roizman, 1969) and are as follows: (i) Virions enter the cell and become partially uncoated in the cytoplasm. Upon entry of the DNA-protein complex into the nucleus, the DNA becomes dissociated from the protein and is transcribed. The product—virus-specific RNA—is processed and transported into the cytoplasm (Roizman *et al.,*

1970). (ii) In the cytoplasm, virus-specific RNA enters into free and membrane-bound polyribosomes (Sydiskis & Roizman 1966, 1968; Wagner & Roizman 1969b) and directs the synthesis of structural and nonstructural proteins of the virus. (iii) Most of the proteins specified by the virus migrate to the nucleus. A few remain in the cytoplasm and bind to cellular membranes (Olshevsky, Levitt & Becker, 1967; Spear & Roizman, 1968; Ben-Porat, Shimono & Kaplan, 1969). (iv) The number and functions of the nonstructural proteins are uncertain. At least two enzymes (Keir, 1968) have been clearly identified as virus-specific. These are thymidine kinase and DNA polymerase. There are grounds for suspecting that the exonuclease found in infected cells is virus-specific (Keir, 1968). These and other nonstructural proteins of the virus are involved in the synthesis of herpesvirus nucleic acids, which takes place in the nucleus. DNA, soon after its synthesis, has numerous single-stranded breaks that are in part repaired and/or ligated afterwards (Kieff, Bachenheimer & Roizman, 1971). The structural proteins entering the nucleus aggregate with the DNA to form a capsid. The assembly of the capsid does not appear to take place at random, but is rather associated with structures, undefined in origin and composition, that appear to aggregate near the nuclear membrane (Schwartz & Roizman 1969b; Roizman, 1969). The process of assembly is not efficient. Only a small fraction of the DNA and protein aggregate to form virions. In particular, the structural components of some herpesviruses form aberrant aggregates resembling microtubules in both the nucleus and cytoplasm (Couch & Nahmias, 1969; Schwartz & Roizman, 1969b). (v) Electron microscopic studies indicate that there are at least three different kinds of capsids in the nucleus. These are: (*a*) capsids lacking a DNA core; (*b*) capsids with a DNA core, dispersed or in paracrystalline aggregates; and (*c*) capsids surrounded by some amorphous material and adhering to the nuclear membrane. Empty capsids rarely become enveloped in herpes simplex infected cells. Envelopment of empty capsids of the virus associated with Burkitt's lymphoma (Toplin & Schidlovsky, 1966) is more common.

Envelopment of nucleocapsids

There is considerable controversy regarding the envelopment of the herpesvirus nucleocapsid. Our knowledge of this process is as follows:

(i) Early in infection most herpesviruses become enveloped at the nuclear membrane (Fig. 1A) (Darlington & Moss, 1968; Schwartz & Roizman, 1969a, b). Capsids surrounded by an amorphous material undergo envelopment by the inner lamella of the nuclear membrane (Roizman, 1969). At the site of attachment, the nuclear membrane becomes thickened and surrounds the capsid. Envelopment occurs as the particle buds through the membrane to the perinuclear space (Fig. 1B). The composition of the enveloped nucleocapsid is shown in Fig. 2.

(ii) In cells infected with some herpesviruses, envelopment seems to occur not only at the nuclear membrane but also at the Golgi membranes, the endoplasmic reticulum and even at the plasma membrane (Fig. 1C) (Darlington & Moss, 1968; Schwartz & Roizman, 1969a, b). The paramount question is: how do the nucleocapsids undergoing envelopment in the cytoplasm get there? One school of thought holds that the nucleocapsids become enveloped by the inner lamella of the nuclear membrane, unenveloped by the outer lamella and re-enveloped by cytoplasmic membranes (Stackpole, 1969). Another school of thought holds that envelopment in the cytoplasm occurs only in cells in which the nuclear membrane has ruptured, releasing its contents. Evidence of ruptured membranes coinciding with areas of the cell in which the nuclear membrane folds upon itself in numerous layers has in fact been found (Fig. 1E). Cytoplasmic nucleocapsids in apposition to the outer lamella of the nuclear membrane and undergoing either envelopment or unenvelopment have also been found (Roizman & Spear, 1972). The basic question remains unanswered; the foregoing illustrates not so much the versatility of the herpesviruses as the difficulty in understanding the flow of events from isolated electron photomicrographs.

(iii) A pertinent question concerning envelopment is whether the virus acquires its envelope from a segment of modified pre-existing membrane or only from membrane synthesized *de novo* after infection. Since patches of new membrane could be formed in continuity with existing cellular membranes, it is difficult to distinguish between the alternatives. One puzzling observation that might give credence to the hypothesis that envelopes arise from membranes synthesized *de novo* is that infrequently herpesviruses appear to be enveloped inside the nucleus with membranes that are apparently not continuous with or related to any cellular membrane (Heine, Ablashi & Armstrong, 1971). Another pertinent observation is that purified enveloped nucleocapsids are devoid of host proteins (Spear & Roizman, 1972). This finding indicates that either the envelopes are derived from membranes synthesized *de novo* or the host proteins of pre-existing membranes have been expelled in the course of envelopment.

(iv) The enveloped particles of herpes simplex virus released from infected cells are generally highly infectious; particle to infectious titre ratios of 10:1 have been reported. This is not a general rule for all herpesviruses. In the laboratory, at least some herpesviruses, such as herpes zoster, cytomegaloviruses and the virus associated with Burkitt's lymphoma, are transmitted from culture to culture readily as infected cells and only with great difficulty as cell-free filtrates. On the basis of this finding these viruses have been classified as "cell-associated" (Melnick *et al.*, 1964), in contrast to herpes simplex, pseudorabies and a few others readily obtained as infectious cell-free filtrates. In reality, the term "cell-associated herpesvirus" is tautological and most probably incorrect. In view of the widespread occurrence of varicella and infectious mononucleosis, the production of infectious progeny in nature must be very efficient indeed. It has been suggested that the lack of infectivity of varicella and possibly of other "cell-associated" viruses made in cell culture might be due to the degradation of the virus particle by cellular enzymes (Cook & Stevens, 1968). However, the electron photomicrographs could be just as readily interpreted as indicating that the virus made in these cells is incomplete or defective. Far more likely is the explanation that some herpesviruses require for completion of the biosynthesis of the envelope, enzymes found only in differentiated cells that do not grow in culture. An example in part is Marek's disease herpesvirus, which is "cell-associated" in chick embryo cells (Biggs, 1968; Churchill, 1968), but appears to multiply very efficiently and to produce infectious progeny in feather follicle cells (Nazerian & Witter, 1970). In these cells the envelope of the virus appears to undergo extensive structural modification in cytoplasmic inclusions (Nazerian & Witter, 1970).

Egress

In cells infected with herpes simplex virus subtype 1, enveloped particles accumulate in the space between the inner and outer lamellae of the nuclear membrane, and in the endoplasmic reticulum (Figs. 1A, B, D) (Darlington & Moss, 1968; Schwartz & Roizman, 1969a). The exact process by which the virus emerges from the infected cell is not clear. According to Morgan *et al.* (1959), portions of the endoplasmic reticulum containing the virus particles

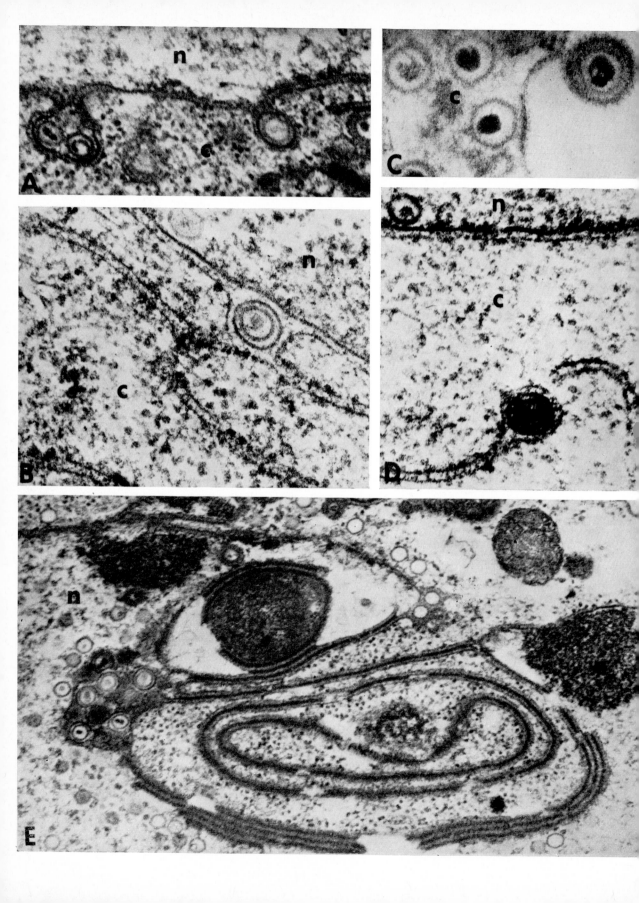

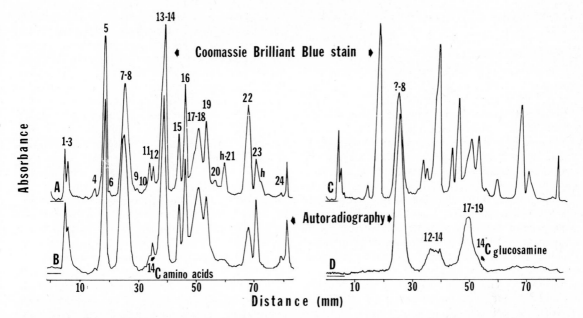

Fig. 2. Electropherogram of the proteins of glycoproteins of enveloped nucleocapsids of HSV-1. HSV-1 enveloped nucleocapsids were purified from the cytoplasm of uninfected cells by banding in a Dextran 10 gradient followed by flotation in a discontinuous sucrose gradient. The infected cells were incubated in medium containing ^{14}C-amino acids or ^{14}C-glucosamine from 5 to 24 hours post infection, i.e., after the complete shut down of all host protein synthesis. The solubilized viral proteins were subjected to electrophoresis in a discontinuous acrylamide gel system containing sodium dodecyl sulphate. After electrophoresis the gels were stained with Coomassie Brilliant Blue stain, scanned with a Gilford spectrophotometer, dried, then placed in contact with X-ray film. The autoradiographic image on the X-ray films was scanned on a Joyce Loebel spectrophotometer. The electrophoresis was from left to right. The numbered protein bands were labelled only after infection. They remain in constant ratio during purification. The protein bands designated with the letter h were identified as cellular by the fact that these bands were labelled only when the cells were incubated in radioactive medium prior to infection or when labelled, uninfected cell cytoplasm was mixed with unlabelled, infected cell cytoplasm prior to purification. The cellular proteins are probably not virus constituents as they are separable from the virus in the final stages of purification. The line under the letters A–D indicates the base line for the absorbance measurements. The unnumbered band at the extreme right consists of material migrating with the buffer front (Spear & Roizman, 1972).

    A. Absorbance of stained bands of purified viral proteins labelled with ^{14}C-amino acids.
    B. Autoradiographic tracing of the same gel.
    C. Absorbance of stained bands of purified viral proteins labelled with ^{14}C-glucosamine.
    D. Autoradiographic tracing of the same gel.

break off, form vesicles, and transport the particles to the extracellular fluid. According to Schwartz & Roizman (1969a), the endoplasmic reticulum connects the perinuclear space with extracellular fluid and serves as a means of egress of the virus from the infected cell. Regardless of the mechanism that actually prevails, it seems clear that enveloped virus is sequestered in an entirely different compartment of the infected cell and does not come in contact with the cytoplasm. One possible explanation of this observation is that exposure of the virus to the cytoplasmic contents is detrimental to it. This is to some extent supported by the observation that substantial amounts of unenveloped particles in the cytoplasm of cells infected with the genital strain (subtype 2) of herpesvirus are found in virus prepa-

---

Fig. 1. Envelopment and egress of herpesviruses. Thin sections of human (HEp-2) cells infected with herpes simplex virus. A, B—Virions presumably enveloped by the inner nuclear membrane and accumulating in the perinuclear space. C—Nucleocapsids of herpes simplex subtype 2 undergoing envelopment at a cytoplasmic membrane. One particle lies in apposition to a modified patch of membrane; another enveloped particle adheres to the outside of the membrane. D—An enveloped particle enclosed within a cytoplasmic tubule. E—Naked nucleocapsids in both the nucleus and cytoplasm coincident with discontinuities in the nuclear membrane. The nuclear membrane appears folded upon itself. Abbreviations: n-nucleus, c-cytoplasm.
A—Spear & Roizman, unpublished studies. B, D—Schwartz & Roizman (1969a). C, E—Roizman (1971a).

rations of relatively low infectivity (Schwartz & Roizman, 1969b).

The fate of the infected cell

Extensive alteration of the host appears to be a universal consequence of productive infection of cells with herpesviruses. Of particular interest to us are the alterations that occur early, before the onset of viral DNA synthesis, and result in the complete cessation of host DNA synthesis (Ben-Porat & Kaplan, 1965; Roizman & Roane, 1964) and protein synthesis (Sydiskis & Roizman, 1967), and in a drastic modification of host RNA metabolism (Roizman et al., 1970). Inhibition of host protein and DNA syntheses begins almost from the time of entrance of the virus into the infected cell and is complete by 3-5 hours post infection. The inhibition of host DNA synthesis coincides with the aggregation and displacement of the chromosomes to the nuclear membrane. Viral DNA is made in a space topologically different from that of host DNA (Roizman, 1969). The inhibition of host protein synthesis coincides with the disaggregation of polyribosomes (Sydiskis & Roizman, 1967). The mechanisms of inhibition of host DNA and protein synthesis are not known. The modification of host RNA synthesis occurs at several levels (Roizman et al., 1970). First, there is a selective reduction in host RNA synthesis. The extent of inhibition varies. Ribosomal precursor RNA synthesis is reduced by as much as 70% as compared to that of uninfected cells. In contrast, $4S$ RNA synthesis is reduced very much less. Second, the host RNA that continues to be made is not processed properly. For example, the $45S$ ribosomal RNA is normally methylated and then cleaved in a series of steps, yielding $18S$ and $28S$ ribosomal RNA. In infected cells a small proportion of the $45S$ RNA continues to be made and is in fact methylated. However, this $45S$ RNA is not cleaved into the $18S$ and $28S$ segments. Third, the small amounts of RNA that are processed are not transported to the cytoplasm in the same way as viral RNA and do not enter the polyribosomal pool to direct host protein synthesis. The biochemical and morphological modifications of the cell appear to be so drastic as to preclude reversal and recovery.

Modification of membranes in productively infected cells

Current information on the alteration of cellular membranes is derived from analyses of infected cells expressing at least some viral productive functions.

As indicated in the preceding section, internal cellular membranes (nuclear membranes, endoplasmic reticulum, etc.) play a decisive part in the envelopment and egress of virus from infected cells. Furthermore, both these membranes and the external cell membrane exhibit structural, functional and immunological changes. The data may be summarized as follows:

(i) The changes in membrane function relate to the retention of intracellular molecules and to the interactions between cells. Thus several laboratories (Kamiya, Ben-Porat & Kaplan, 1965; Wagner & Roizman, 1969a; Nahmias[1]) have observed leakage of large as well as small macromolecules from infected cells. Concurrently, the infected cells fuse, clump, or show no change in their "social behaviour" with adjacent cells (Roizman, 1962; Ejercito, Kieff & Roizman, 1968). The social behaviour of the infected cell is genetically determined by the virus.

(ii) The change in the immunological specificity of the external membranes has been demonstrated in a number of laboratories by cytolytic tests with appropriate antisera and complement (Roizman, 1962; Roane & Roizman, 1964), by haemadsorption (Watkins, 1964), and by immunofluorescence tests on intact cells (Klein et al., 1968). Electron microscope studies with ferritin-conjugated antibody (Nii et al., 1968) demonstrated that all membranes of the infected cells acquire a new immunological specificity. Furthermore, several lines of evidence indicate that the new immunologically determinant groups on the surface of the viral envelope and of the infected cell are similar, if not identical (Roane & Roizman, 1964; Roizman & Spring, 1967; Pearson et al., 1970).

(iii) The change in function and immunological specificity coincides with the appearance of new glycoproteins in the cellular membranes. Several lines of evidence indicate that the proteins are specified by the virus. Thus, glycoproteins made after infection are absent from uninfected cell membranes (Spear, Keller & Roizman, 1970). Perhaps even more significant, the number and structure of the membrane glycoproteins are genetically determined by the virus (Fig. 3). The glycoproteins bind not only to the nuclear membrane, which is involved in the envelopment of nucleocapsids, but also to the smooth cytoplasmic membranes and to the plasma membrane (Fig. 4) (Keller, Spear & Roizman, 1970; Roizman, 1971a). It should be noted that the change in immunological specificity of the membranes is not due to the adhesion of enveloped virus to the surface

---

[1] Personal communication.

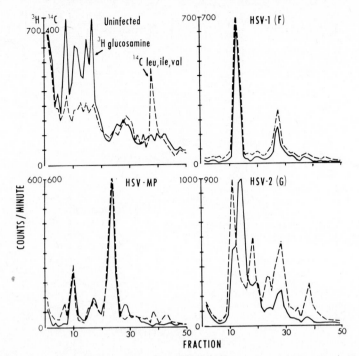

Fig. 3. Electropherograms of proteins and glycoproteins present in smooth membranes of infected and uninfected cells. The cells were simultaneously labelled with D-glucosamine and L-amino acids (leucine, isoleucine, and valine) from 4 to 24 hours post infection. The smooth membranes were prepared as described by Spear & Roizman (1970). Electrophoresis was carried out in 8.5 % acrylamide gels according to the procedure of Laemmli (1970). In this and subsequent electropherograms the direction of migration of the proteins is from left to right.

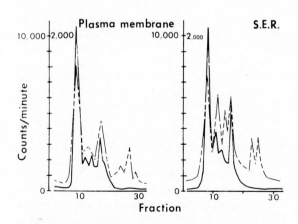

Fig. 4. Electropherograms of proteins and glycoproteins in purified smooth endoplasmic reticulum (SER) and in plasma membranes of HEp-2 cells infected with the F prototype of HSV-1. Membranes were isolated from HEp-2 cells infected for 18 hours with a multiplicity of 10 PFU of herpes simplex virus and labelled with ^{3}H-glucosamine (————) and ^{14}C-amino acids (– – – –) during infection. The cells for plasma membrane isolation were ruptured by microcavitation. Plasma membrane frag-

of the cells. Membranes of infected cells highly purified with respect to virus and soluble proteins also react with antibody (Fig. 5) (Roizman & Spear, 1971c).

(iv) The binding of the viral glycoproteins to membranes coincides with and follows cessation of synthesis and glycosylation of host membrane proteins (Spear, Keller & Roizman, 1970). Pulse-chase experiments indicate that the viral membrane proteins are made in the cytoplasm and become glycosylated after binding to membranes (Spear & Roizman, 1970). The binding of the glycoproteins to

ments were separated from the bulk of other microsomal components by centrifugation on a Dextran 110 cushion. They were purified by floating through a discontinuous sucrose gradient. SER membranes were isolated from cells ruptured by Dounce homogenization. The SER was partially purified by flotation through a discontinuous sucrose gradient and further purified by centrifugation through a Dextran 110 barrier. The fractionation of the membranes was monitored by assays of diaphorase and 5'nucleotidase. Acrylamide gel electrophoresis was done according to the procedure of Laemmli (1970). A detailed description of the above procedure will appear elsewhere (Heine, Spear & Roizman, unpublished data).

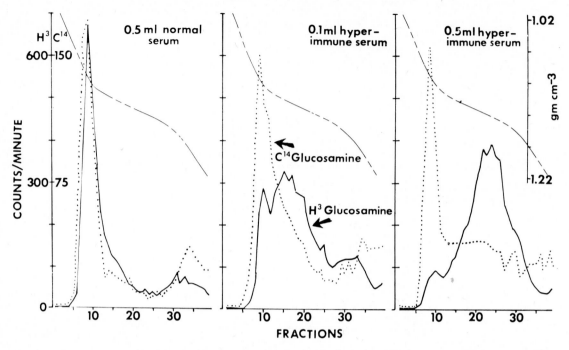

Fig. 5. Flotation of mixtures of infected and uninfected cell membranes in sucrose density gradients after incubation for four hours with normal rabbit serum or varying amounts of hyperimmune serum. The incubation mixtures were made 45 % (by weight) with respect to sucrose, overlaid with linear gradients of 10 to 35 % (by weight) sucrose, and then with 3 ml of saline and centrifuged for 20 hours at 25 000 rpm in a Spinco SW27 rotor. The top of the tube is at the left. Solid line; infected cell membranes labelled with ³H-glucosamine; dotted line: uninfected cell membranes labelled with ¹⁴C-glucosamine. Data from Roizman & Spear (1971a).

membranes is selective, specific and tenacious (Roizman & Spear, 1971b). Specificity is demonstrated by the fact that the cytoplasm of infected cells contains numerous virus-specific structural and nonstructural proteins, yet only the glycoproteins bind to membranes. In fact, the soluble pools are rendered free of these proteins. Selectivity is demonstrated by the finding that, in cells infected with the MP mutant of herpes simplex subtype 1 or with herpes simplex subtype 2, the envelope glycoproteins differ qualitatively and quantitatively from the glycoproteins binding to smooth membranes (Keller, Spear & Roizman, 1970). This indicates that the glycoproteins do not bind to membranes completely at random; rather, the glycoproteins discriminate between the various membranes of the cell. The tenacity of the bond between the membrane and viral glycoproteins is demonstrated by the fact that it is capable of withstanding considerable hydrodynamic stress even when augmented by the binding of antibody to the glycoprotein (Fig. 5) (Roizman & Spear, 1971c).

(v) Several lines of evidence indicate that the changes in the immunological specificity and function of the membranes are determined by the glycoproteins binding to the membranes. Briefly: (a) the glycoproteins have immunological specificity; (b) glycoproteins solubilized from membranes of infected cells react with a high degree of specificity with antisera containing neutralizing and cytolytic antibody (Fig. 6). (vi) Viruses differing in immunological specificity and in their effects on the social behaviour of infected cells specify different membrane glycoproteins (Fig. 3) (Keller, Spear & Roizman, 1970). The correlation between the immunological specificity, social behaviour of infected cells and membrane glycoproteins is particularly significant in the light of experiments with herpes simplex subtype 1 (F) strain and a deletion mutant, MP, derived from it. The F prototype of subtype 1 strain causes cells to round up and clump. It also specifies two major and several minor glycoproteins that bind to all the membranes of the infected cell. The MP mutant contains 4.5×10^{6} Daltons less DNA than the F virus (Bachen-

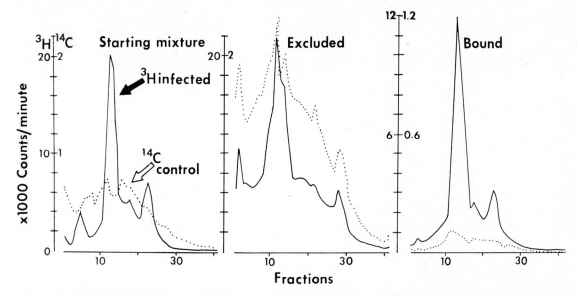

Fig. 6. The affinity of infected and uninfected cell glycoproteins to immunoabsorbent columns of rabbit anti-HSV-1 serum. The rabbits were immunized with HSV-1 as described by Ejercito *et al.* (1968). The immunoabsorbent columns were prepared according to the method of Avrameas & Ternynck (1969). In these experiments, uninfected cells labelled for 20 hours with ^{14}C-glucosamine were mixed with infected cells labelled with ^{3}H-glucosamine from 4 to 24 hours post infection. The mixture was then exposed for 10 minutes at 4°C to 5 % Nonidet P-40 (NP-40) (Shell Oil Co., New York, N.Y.). The nuclei were removed by centrifugation; the cytoplasmic lysate was then centrifuged at 110 000 g for one hour to remove particulate matter, then passed through the column. The immunoabsorbent gel was then washed with 2.5 % NP-40 in isotonic phosphate buffer and eluted with 8M urea and 0.1 % sodium dodecyl sulfate. The panels show from left to right the electropherograms of the glycoproteins in the initial mixture, in the material excluded from the column and in the material bound to the immunoabsorbent gel and eluted from it with 8M urea and 0.1 % sodium dodecyl sulfate. The electrophoresis was carried out according to the procedure described by Spear & Roizman (1968).

heimer [1]). It causes cells to fuse, but the membranes lack one of the major glycoproteins specified by the F virus. As illustrated in Fig. 7, in doubly-infected cells both viruses multiply, but the social behaviour of the infected cells and the composition of the membrane glycoproteins are wholly determined by the dominant F virus (Roizman, 1971a). These data relate the membrane-specific viral glycoproteins to the alteration in immunological specificity and function of the membrane. They also suggest that the various membrane glycoproteins possess unique structural, functional and immunological properties.

Conclusions

Extensive comparative biochemical studies of the putative oncogenic and non-oncogenic herpesviruses are not available. Most of the comparative information is based on electron microscopic and immunological, but relatively few biochemical, studies. How-

ever, the data are sufficient in quantity if not in quality to indicate that the molecular events and their sequence in cells productively infected with the putative oncogenic and non-oncogenic herpesviruses are probably very similar. So far as we can tell, all herpesviruses inhibit host macromolecular metabolism and ultimately destroy the host in which they express viral productive functions. In at least one study (Aurelian & Roizman, 1965), the expression of viral functions and the inhibition of host macromolecular metabolism were found to be linked.

THE EXPRESSION OF VIRAL FUNCTIONS IN
NONPRODUCTIVE INFECTION

The characteristics of the nonproductive infection of cells by herpesviruses have been dealt with in several publications (Roizman, 1965; Terni, 1965; Roizman, 1971a, b, c). The existence of nonproductively infected cells is deduced from the well authenticated fact that cells exhibiting no evidence

[1] Unpublished data.

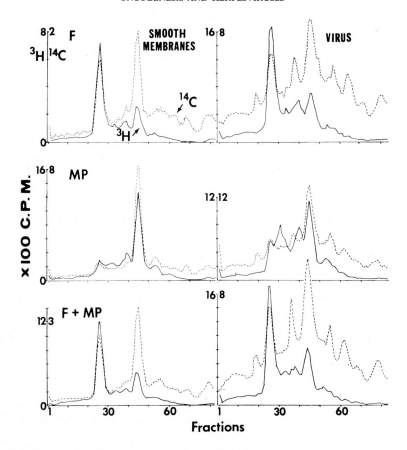

Fig. 7. Electropherograms of membrane and viral proteins prepared from cells singly and doubly infected with virus strains differing with respect to their effect on the social behaviour of cells. The F strain is a prototype of the subtype 1 of herpes simplex virus. The MP strain is a multistep deletion mutant of a subtype 1 strain. The doubly-infected cells were exposed to F and MP strains at identical multiplicities of infection. All infected cells were labelled between 11 and 22 hours post infection in a medium containing ^{14}C-amino acid mixtures (leucine, isoleucine, and valine) and ^{3}H-glucosamine. The acrylamide gel electrophoresis was done according to the procedure of Spear & Roizman (1968). Data from Keller *et al.* (1970).

of infectious virus can be induced to produce virus (Corriell, 1963; Falchi, 1965; Rustigan *et al.*, 1966; Rafferty, 1964; Antonelli & Vignali, 1968; Breidenbach *et al.,* 1971; Stevens & Cook, 1971). This repeated observation indicates that the virus is not harboured in the cell in the form of an infectious virion.

Historically, the best known nonproductive infections are those of herpes simplex and herpes zoster in man. The apparent maintenance of the viral genome in Burkitt's lymphoma cells cultivated *in vitro* (zur Hausen *et al.,* 1970; zur Hausen & Schulte-Holthausen, 1970; Miller, Stitt & Miller, 1970) is another, potentially more amenable to analysis, example of a nonproductive infection. Of particular interest to

us, in regard to the objectives of this paper, are the following two sets of observations.

The first deals with the nature of the viral functions that are expressed in nonproductively infected cells. With the exception of lymphoblasts, these cells generally do not maintain their nonproductive character in culture. Our information concerning somatic nonproductively infected cells is based on limited analyses of observations on multicellular organisms. The fact that the viral genome is perpetuated in the nonproductively infected, multiplying lymphoblasts (Miller *et al.*, 1970; zur Hausen *et al.,* 1970; zur Hausen & Schulte-Holthausen, 1970), indicates that it is also replicated. This raises the question whether any viral functions are expressed in non-

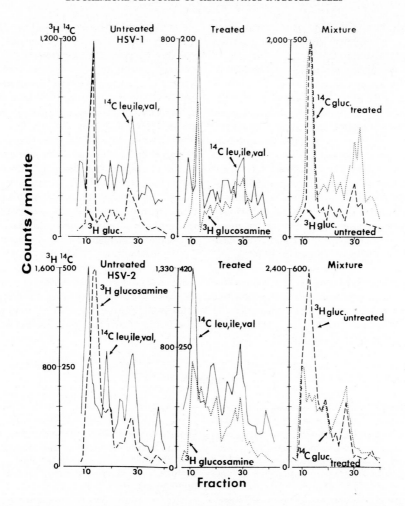

Fig. 8. Electropherograms of proteins and glycoproteins extracted from purified smooth membranes of HEp-2 cells infected with herpes simplex virus subtype 1 and subtype 2 (HSV-1 and HSV-2) and either treated or untreated with cytosine arabinoside (20 μg/ml). The drug was added at the time of exposure of the cells to virus. Radioactive precursors were added at four hours post infection. The left and middle panels of each group show the electropherograms of membrane proteins from cells labelled simultaneously with ¹⁴C-amino acids (leucine, isoleucine, and valine) and ³H-glucosamine. The right panels show the electropherograms of a mixture of the membrane preparations of cells treated and labelled with ¹⁴C-glucosamine and untreated and labelled with ³H-glucosamine. The acrylamide gel electrophoresis was carried out according to the procedure of Laemmli (1970).

Cytosine arabinoside affects two processes. It reduces or inhibits the synthesis of membrane proteins, suggesting that this may be dependent on viral DNA synthesis for maximal expression. At the same time, it appears to enhance the glycosylation of some of the glycoproteins. The results of these experiments are puzzling and not readily amenable to explanation. The hypotheses that we have considered are: (i) that the membrane glycoproteins are "late proteins" and that the reduced amounts made in the presence of the drug are the consequence of incomplete inhibition of DNA synthesis by the drug; and (ii) that each band contains at least one "early" and one "late" protein. However, the first hypothesis does not satisfactorily explain the increased glycosylation of the small glycoproteins whereas the second does not appear tenable on the basis of current evidence regarding the nature of these glycoproteins. A third hypothesis, which is not currently amenable to testing, is that while the viral membrane proteins could in fact be "late" functions of the virus, their glycosylation by hypothetical virus-specific enzymes might be an early function.

In summary, the studies of Gergely, Klein & Einberg (1971a, b, c) suggest that the alteration in cell membrane is an early function dissociated from the cytolytic and viral productive functions of the virus. Our results are compatible with this conclusion. However, in the light of our results, the question arises whether the membrane alterations seen by Gergely *et al.* are brought about by insertion of viral proteins into membranes or by glycosylating enzymes that modify host membrane proteins. The answer is not clear. Preliminary studies in our laboratory indicate that glycopeptides obtained by extensive digestion of glycoproteins with pronase do have immunological specificity and bind to antibody made against the virus.

If the virus-specific membrane proteins were a late viral function associated with the viral productive functions of the virus, could glycosyl transferases and perhaps other enzymes involved in carbohydrate metabolism fit the rôle of the putative oncogenic functions of the virus? The answer must unquestionably be yes. One hypothesis that accounts for the multitude of observations recorded to date is that the virus specifies enzymes involved in lipid and carbohydrate metabolism, that these are made early before DNA synthesis, and that in nonproductively infected cells they could modify host proteins and other constituents of the membranes. The consequence would be new surface-reactive groups, changes in immunological specificity and potentially different cell surface control of cell functions. It is noteworthy that this hypothesis is in accordance with the findings that cells transformed by Papova viruses differ from the nontransformed parent cells largely in the rearrangement of membrane components (Hakomori, Testler & Andrews, 1968; Burger, 1969; Pollack & Burger, 1969) and in the absence of higher gangliosides (Mora *et al.*, 1969; Brady & Mora, 1970), and to a much lesser extent in glycoprotein composition (Hakomori & Murakami, 1968; Meezan *et al.*, 1969). The virtue of this hypothesis is that it can be tested.

SUMMARY

Infection of animal cells with herpesviruses results in either productive infection in which viral progeny are made, or nonproductive infection in which the cell survives and apparently harbours the genome of the virus.

The analysis of productive infection presented here suggests that the molecular nature and sequence of functions expressed by the various herpesviruses studied to date are very similar. Productive infection invariably results in cell death due to irreversible inhibition of the host.

Ability to establish a nonproductive infection appears to be a common characteristic of all herpesviruses. As a rule, nonproductively infected cells can be induced to make virus but perish in the process. If herpesviruses are oncogenic, the putative oncogenic functions may be expressed in both productive and nonproductive infection but can lead to malignant transformation rarely and only in nonproductively infected cells. Full expression of viral productive functions and malignant transformation are mutually exclusive.

Of the known biochemical functions expressed by herpesviruses in the infected cell, the most likely candidate for an oncogenic function is the modification of structure, function and immunological specificity of cellular membranes, as described in the text.

ACKNOWLEDGEMENTS

The studies published in this paper were aided by grants from the United States Public Health Service (CA 08494), the American Cancer Society (E 314F) and the Leukemia Foundation Inc. I should like to express my appreciation to Dr. Patricia Spear for several years of very fruitful collaboration, which formed the basis of this paper, and to both her and Professor George Klein for many most useful discussions.

REFERENCES

Antonelli, A. & Vignali, C. (1968) Ricerche sulla localizzazione del virus dell'herpes simplex dopo guarigione delle recidive cutanee. Coltivazione di cellule della zona colpita da eruzione recidivante. *Riv. ist. sieroter. ital.*, **43**, 43-51

Aurelian, L. & Roizman, B. (1965) Abortive infection of canine cells by herpes simplex virus. II. The alternative supression of synthesis of interferon and viral constituents. *J. molec. Biol.*, **11**, 539-548

Avrameas, S. & Ternynck, T. (1969) The cross-linking of proteins with glutaraldehyde and its use for the preparation of immunoabsorbents. *Immunochemistry*, **6**, 53-66

Ben-Porat, T. & Kaplan, A. S. (1965) Mechanism of inhibition of cellular DNA synthesis by pseudorabies virus. *Virology*, **25**, 22-29

Ben-Porat, T., Shimono, H. & Kaplan, A. S. (1969) Synthesis of proteins in cells infected with herpesvirus. II. Flow of structural viral proteins from cytoplasm to nucleus. *Virology*, **37**, 56-61

Biggs, P. M. (1968) Marek's disease—current state of knowledge. *Curr. Top. Microbiol. Immunol.*, **43**, 93-125

Blank, H. & Brody, M. W. (1950) Recurrent herpes simplex: A psychiatric and laboratory study. *Psychosom. Med.*, **12**, 254-260

Brady, R. O. & Mora, P. T. (1970) Alteration in ganglioside pattern and synthesis in SV-40 and polyoma virus transformed mouse cell lines. *Biochim. biophys. Acta*, **218**, 308-319

Breidenbach, G. P., Skinner, M. S., Wallace, J. H. & Mizell, M. (1971) *In vitro* induction of a herpes-type virus in "Summer-Phase" Lucké-tumor explants. *J. Virol.*, **7**, 679-682

Burger, M. M. (1969) A difference in the architecture of the surface membrane of normal and virally transformed cells. *Proc. nat. Acad. Sci. (Wash.)*, **62**, 994-1001

Carton, C. A. (1953) Effect of previous sensory loss on the appearance of herpes simplex following trigeminal sensory root section. *J. Neurosurg.*, **10**, 463-468

Carton, C. A. & Kilbourne, E. D. (1952) Activation of latent herpes simplex by trigeminal sensory-root section. *New Engl. J. Med.*, **246**, 172-176

Churchill, A. E. (1968) Herpes-type virus isolated in cell culture from tumors of chickens with Marek's disease. I. Studies in cell culture. *J. nat. Cancer Inst.*, **41**, 939-950

Cook, M. L. & Stevens, J. G. (1968) Labile coat: Reason for noninfectious cell-free varicella-zoster virus in culture. *J. Virol.*, **2**, 1458-1464

Coriell, L. L. (1963) Discussion of the paper by B. Roizman. In: *Viruses, nucleic acids and cancer*, Baltimore, Williams & Wilkins, p. 241

Couch, E. F. & Nahmias, A. J. (1969) Filamentous structures of type 2 herpes-virus hominis infection of the chorioallantoic membrane. *J. Virol.*, **3**, 228-232

Cushing, H. (1905) The surgical aspects of major neuralgia of trigeminal nerve; a report of twenty cases of operation on Gasserian ganglion with anatomic and physiologic notes on consequence of its removal. *J. Amer. med. Ass.*, **44**, 773-779, 860-865, 920-929, 1002-1008 & 1088-1093

Darlington, R. W. & Moss, L. H. (1968) Herpesvirus envelopment. *Virology*, **2**, 48-55

Ejercito, P. M., Kieff, E. D. & Roizman, B. (1968) Characterization of herpes simplex virus strains differing in their effect on social behavior of infected cells. *J. gen. Virol.*, **3**, 357-364

Epstein, M. A., Achong, B. G., Barr, Y. M., Zajac, B., Henle, G. & Henle, W. (1966) Morphological and virological investigations on cultured Burkitt tumor lymphoblasts (strain Raji). *J. nat. Cancer Inst.*, **37**, 547-559

Falchi, G. (1925) Herpes sperimentale recidivante nell'uomo. Nota riassuntiva. *Boll. Soc. med.-chir. Pavia*, **37**, 885

Gerber, P. & Hoyer, W. (1971) Induction of cellular DNA synthesis in human leukocytes by Epstein-Barr virus. *Nature (Lond.)*, **231**, 46-47

Gerber, P., Whang-Pang, J. & Monroe, J. H. (1969) Transformation and chromosome changes induced by Epstein-Barr virus in normal human leukocyte cultures. *Proc. nat. Acad. Sci. (Wash.)*, **63**, 740-747

Gergeley, L., Klein, G. & Einberg, I. (1971a) Host cell macromolecular synthesis in cells containing EBV induced early antigens studied by combined immunofluorescence and autoradiography. *Virology*, **45**, 10-21

Gergeley, L., Klein, G. & Einberg, I. (1971b) The action of DNA antagonistic on Epstein-Barr virus (EBV)-associated early antigen (EA) in Burkitt lymphoma lines. *Int. J. Cancer*, **7**, 293-302

Gergeley, L., Klein, G. & Einberg, I. (1971c) Appearance of EBV-associated antigens in infected Raji cells. *Virology*, **45**, 22-29

Hakomori, S. & Murakomi, W. (1968) Glycolipids of hamster fibroblasts and derived malignant-transformed cell lines. *Proc. nat. Acad. Sci. (Wash.)*, **59**, 254-261

Hakomori, S., Testler, C. & Andrews, H. (1968) Organizational difference of cell surface "hematoside" in normal and virally transformed cells. *Biochem. biophys. Res. Commun.*, **33**, 563-568

Heine, U., Ablashi, D. V. & Armstrong, G. R. (1971) Morphological studies on herpes virus saimiri in subhuman and human cell cultures. *Cancer Res.*, (in press)

Henle, W., Diehl, V., Kohn, G., zur-Hausen, H. & Henle, G. (1967) Herpes type virus and chromosome marker in normal leukocytes after growth with irradiated Burkitt cells. *Science*, **157**, 1064

Henle, W. & Henle, G. (1968) Effect of arginine-deficient media on the herpes-type virus associated with cultured Burkitt tumor cells. *J. Virol.*, **2**, 182-191

Kamiya, T., Ben-Porat, T. & Kaplan, A. S. (1965) Control of certain aspects of the infective process by progeny viral DNA. *Virology*, **26**, 577-589

Keir, H. M. (1968) Virus-induced enzymes in mammalian cells infected with DNA viruses. *Mol. Biol. Viruses,* **18**, 67-99

Keller, J. M., Spear, P. G. & Roizman, B. (1970) The proteins specified by herpes simplex virus. III. Viruses differing in their effects on the social behavior of infected cells specify different membrane glycoproteins. *Proc. nat. Acad. Sci. (Wash.),* **65**, 865-871

Kieff, E. D., Bachenheimer, S. L. & Roizman, B. (1971) Size, composition and structure of the DNA of subtypes 1 and 2 herpes simplex virus. *J. Virol.,* **8**, 125-132

Klein, G. (1971) Immunological aspects of Burkitt's lymphoma. *Advanc. Immunol.* (in press)

Klein, G., Pearson, G., Nadkarni, J. S., Nadkarni, J. J., Klein, E., Henle, G., Henle, W. & Clifford, P. (1968) Relation between Epstein-Barr viral and cell membrane immunofluorescence of Burkitt tumor cells. I. Dependence of cell membrane immunofluorescence on presence of EB virus. *J. exp. Med.,* **128**, 1011-1020

Laemmli, U. K. (1970) Cleavage of structural proteins during the assembly of the head of bacteriophage T$_4$. *Nature (Lond.),* **227**, 680-684

Meezan, E., Hu, H. C., Black, P. H. & Robbins, P. H. (1969) Comparative studies on the carbohydrate-containing membrane components of normal and virus-transformed mouse fibroblasts. II. Separation of glycoproteins and glycopeptides by sephadex chromatography. *Biochemistry, N.Y.,* **8**, 2518-2524

Melnick, J. L., Midulla, M., Wimberly, I., Barrera-Oro, J. G. & Levy, B. M. (1964) A new member of the herpesvirus group isolated from South American Marmosets. *J. Immunol.,* **92**, 595-601

Miller, M. H., Stitt, D. & Miller, G. (1970) Epstein-Barr viral antigen in single cell clones of two human lymphocytic lines. *J. Virol.,* **6**, 699-701

Mora, P. T., Brady, R. O., Bradley, R. M. & McFarland, V. W. (1969) Gangliosides in DNA virus-transformed and spontaneously transformed tumorigenic mouse cell lines. *Proc. nat. Acad. Sci. (Wash.),* **63**, 1290-1296

Morgan, J. K. (1968) Herpes gestationis influenced by an oral contraceptive. *Brit. J. Derm.,* **80**, 456-458

Morgan, C., Rose, H. M., Holden, M. & Jones, E. P. (1959) Electron microscopic observations on the development of herpes simplex virus. *J. exp. Med.,* **110**, 643-656

Nadkarni, J. S., Nadkarni, J. J., Klein, G., Henle, W., Henle, G. & Clifford, P. (1970) EB viral antigens in Burkitt tumor biopsies and early cultures. *Int. J. Cancer,* **6**, 10-17

Nazerian, K. & Witter, R. L. (1970) Cell-free transmission and *in vivo* replication of Marek's disease virus. *J. Virol.,* **5**, 388-397

Nii, S., Morgan, C., Rose, H. M. & Hsu, K. C. (1968) Electron microscopy of herpes simplex virus. IV. Studies with ferritin-conjugated antibodies. *J. Virol.,* **2**, 1172-1184

Olshevsky, U., Levitt, J. & Becker, Y. (1967) Studies on the synthesis of herpes simplex virions. *Virology,* **33**, 323-334

Pearson, G., Deney, F., Klein, G., Henle, G. & Henle, W. (1970) Correlation between antibodies to Epstein-Barr virus (EBV)-induced membrane antigens and neutralization of EBV infectivity. *J. nat. Cancer Inst.,* **45**, 989

Pollack, R. E. & Burger, M. M. (1969) Surface-specific characteristics of a contact-inhibited cell line containing the SV40 viral genome. *Proc. nat. Acad. Sci. (Wash.),* **62**, 1074-1076

Pope, J. H., Horne, M. K. & Scott, W. (1968) Transformation of foetal human leukocytes in vitro by filtrates of a human leukaemic cell line containing a herpes-like virus. *Int. J. Cancer,* **3**, 857-866

Pope, J. H., Scott, W., Reedman, B. M. & Walters, M. K. (1971) Herpes-type virus as a biologically active agent. In: Nishioka, K., ed., *Proceedings of the First Princess Takamatsu Cancer Research Fund Symposium on Tumor Virology and Immunology,* Tokyo (in press)

Rafferty, K. A., Jr (1964) Kidney tumors of the leopard frog: a review. *Cancer Res.,* **24**, 169-185

Roane, P. R., Jr & Roizman, B. (1964) Studies of the determinant antigens of viable cells. II. Demonstration of altered antigenic reactivity of HEp-2 cells infected with herpes simplex virus. *Virology,* **22**, 1-8

Roizman, B. (1962) Polykaryocytosis. *Cold Spring Harb. Symp. quant. Biol.,* **27**, 327-342

Roizman, B. (1965) An inquiry into the mechanism of recurrent herpes infections of man. In: Pollard, M., ed., *Perspectives in virology,* New York, Harper-Row, Vol. 4, pp. 283-304

Roizman, B. (1969) The herpes viruses—a biochemical definition of the group. *Curr. Top. Microbiol. Immunol.,* **49**, 1-79

Roizman, B. (1971a) Herpesviruses, membranes and the social behavior of infected cells. In: *Proceedings of the 3rd International Symposium on Applied and Medical Virology,* St. Louis, Warren Green, pp. 37-72

Roizman, B. (1971b) Biochemical features of herpes virus-infected cells. In: Nishioka, K., ed., *Proceedings of the First Princess Takamatsu Cancer Research Fund Symposium on Tumor Virology and Immunology,* Tokyo (in press)

Roizman, B. (1971c) Herpes viruses, man and cancer—or the persistence of the viruses of love. In: Monod, J. & Borek, E. eds, *Of microbes and life,* New York, Columbia University Press, pp. 189-215

Roizman, B., Bachenheimer, S. L., Wagner, E. K. & Savage, T. (1970) Synthesis and transport of RNA in herpesvirus infected mammalian cells. *Cold Spring Harb. Symp. quant. Biol.,* **35**, 753-771

Roizman, B. & Roane, P. R., Jr (1964) The multiplication of herpes simplex virus. II. The relation between protein synthesis and the duplication of viral DNA in infected HEp-2 cells. *Virology,* **22**, 262-269

Roizman, B. & Spear, P. G. (1971a) Herpesviruses: current information on the composition and structure. In: Maramorosch, K. & Kurstak, E, eds, *Comparative virology,* New York, Academic Press, pp. 135-168

Roizman, B. & Spear, P. G. (1971b) The role of herpesvirus glycoproteins in the modification of membranes of infected cells. In: *Proceedings of the First Symposium on Nucleic Acid Synthesis in Viral Infections,* Amsterdam, North-Holland, Vol. 2, pp. 435-480

Roizman, B. & Spear P. G. (1971c) Herpesvirus antigens on cell membranes detected by centrifugation of membrane-antibody complexes. *Science, 171,* 298-300

Roizman, B. & Spear, P. G. (1972) Herpesviruses. In: Dalton & Hagenau, eds, *Atlas of viruses,* New York, Academic Press (in press)

Roizman, B. & Spring, S. B. (1967) Alteration in immunologic specificity of cells infected with cytolytic viruses. In: Trentin, J. J., ed., *Proceedings of the Conference on Cross Reacting Antigens and Neoantigens,* Baltimore, Williams & Wilkins, pp. 85-96

Rustigan, R., Smulow, J. B., Tye, M., Gibson, W. A. & Shindell, E. (1966) Studies on latent infection of skin and oral mucosa in individuals with recurrent herpes simplex. *J. invest. Derm., 47,* 218-221

Schwartz, J. & Roizman, B. (1969a) Concerning the egress of herpes simplex virus from infected cells: electron microscope observations. *Virology, 38,* 42-49

Schwartz, J. & Roizman, B. (1969b) Similarities and differences in the development of laboratory strains and freshly isolated strains of herpes simplex virus in HEp-2 cells: electron microscopy. *J. Virol., 4,* 879-889

Scott, T. F. McN. (1957) Epidemiology of herpetic infections. *Amer. J. Ophthal., 43,* 134-147

Spear, P. G., Keller, J. M. & Roizman, B. (1970) The proteins specified by herpes simplex virus. II. Viral glycoproteins associated with cellular membranes. *J. Virol., 5,* 123-131

Spear, P. G. & Roizman, B. (1968) The proteins specified by herpes simplex virus. I. Time of synthesis, transfer into nuclei, and proteins made in productively infected cells. *Virology, 36,* 545-555

Spear, P. G. & Roizman, B. (1970) The proteins specified by herpes simplex virus. IV. The site of glycosylation and accumulation of viral membrane proteins. *Proc. nat. Acad. Sci. (Wash.), 66,* 730-737

Spear, P. G. & Roizman, B. (1972) Proteins specified by herpes simplex virus. V. Purification and structural proteins of the herpes virion. *J. Virol.* (in press)

Stackpole, C. W. (1969) Herpes-type virus of the frog renal adenocarcinoma. I. Virus development in tumor transplants maintained at low temperature. *J. Virol., 4,* 75-93

Stevens, J. G. & Cook, M. L. (1971) Latent herpes simplex virus in spinal ganglia of mice. *Science, 173,* 843-845

Sydiskis, R. J. & Roizman, B. (1966) Polysomes and protein synthesis in cells infected with a DNA virus. *Science, 153,* 76-78

Sydiskis, R. J. & Roizman, B. (1967) The disaggregation of host polyribosomes in productive and abortive infection with herpes simplex virus. *Virology, 32,* 678-686

Sydiskis, R. J. & Roizman, B. (1968) The sedimentation profiles of cytoplasmic polyribosomes in mammalian cells productively and abortively infected with herpes simplex virus. *Virology, 34,* 562-565

Terni, M. (1965) L'infezione erpetica recidivante: conoscenze e problemi. *Arcisped. S. Anna Ferrara, 18,* 515-532

Toplin, I. & Schidlovsky, G. (1966) Partial purification and electron microscopy of virus in the EB-3 cell line derived from a Burkitt lymphoma. *Science, 152,* 1084-1085

Wagner, E. K. & Roizman, B. (1969a) RNA synthesis in cells infected with herpes simplex virus. I. The patterns of RNA synthesis in productively infected cells. *J. Virol., 4,* 36-46

Wagner, E. K. & Roizman, B. (1969b) RNA synthesis in cells infected with herpes simplex virus. II. Evidence that a class of viral mRNA is derived from a high molecular weight precursor synthesized in the nucleus. *Proc. nat. Acad. Sci. (Wash.), 64,* 626-633

Watkins, J. F. (1964) Adsorption of sensitized sheep erythrocytes to HeLa cells infected with herpes simplex virus. *Nature (Lond.), 202,* 1364-1365

Watkins, J. F. (1965) The relationship of the herpes simplex haemadsorption phenomenon to the virus growth cycle. *Virology, 26,* 746-753

Weinberg, A. & Becker, Y. (1969) Studies on EB virus of Burkitt's lymphoblasts. *Virology, 39,* 312-321

Yata, J., Klein, G., Hewetson, J. & Gergely, L. (1970) Effect of metabolic inhibitors on membrane immunofluorescence reactivity of established Burkitt cell lines. *Int. J. Cancer, 5,* 394-403

zur Hausen, H. & Schulte-Holthausen, H. (1970) Presence of EB virus nucleic acid homology in a "virus-free" line of Burkitt tumour cells. *Nature (Lond.), 227,* 245-248

zur Hausen, H., Schulte-Holthausen, H., Klein, G., Henle, W., Henle, G., Clifford, P. & Santesson, L. (1970) EBV DNA in biopsies of Burkitt tumours and anaplastic carcinomas of the nasopharynx. *Nature (Lond.), 228,* 1056-1058

PATHOLOGY OF MAREK'S DISEASE

Chairman — P. M. Biggs

Rapporteur — J. G. Campbell

Pathogenesis of Marek's Disease—A Review

L. N. PAYNE [1]

Marek's disease (MD) is a contagious lympho-proliferative and neuropathic disease of domestic fowl, caused by a cell-associated herpesvirus (HV). It is the prime cause of loss to the poultry industry in many countries (Biggs, 1971). MD has complex pathological features that are apparently without parallel in mammals, including man. Nevertheless, the disease is of interest in comparative pathology because it has both inflammatory and neoplastic elements that show some similarities to disorders of man, and because of its aetiology.

Study of the pathology of MD is punctuated historically by four major events: (1) first description of the disease in Hungary by Marek (1907), who termed it a polyneuritis. Subsequently the disease was recognized throughout the world and became commonly known as fowl paralysis or neurolymphomatosis. The form of MD in which peripheral neuropathy was the outstanding feature was later referred to by Biggs (1966) as the classical form. It was recognized as long ago as 1926 that lymphoma formation, notably in the ovary, occurred in a small proportion of cases (Pappenheimer, Dunn & Cone, 1926); (2) in 1957 in the USA, Benton & Cover (1957) reported an increased incidence of visceral lymphomatosis in young chickens. This disease, originally termed acute leukosis, was later recognized to be a form of MD in which lymphoma formation was frequent and flock mortality high, and was termed acute MD (Biggs et al., 1965); (3) since 1962 both classical and acute MD have been transmitted experimentally (Sevoian, Chamberlain & Counter, 1962; Biggs & Payne, 1963). Consequently tissues of known provenance have become available for study; (4) the cause of MD was identified as a herpesvirus (HV) (Churchill & Biggs, 1967; Solomon et al., 1968; Nazerian et al., 1968). Fluorescent antibody methods

to identify viral antigen and electron microscopy to identify the HV in tissues have followed and have contributed to knowledge of the pathogenesis of MD.

The pathology of MD is discussed in subsequent sections as follows: changes in the nervous system, lymphoma formation, and changes in other tissues. This approximates to the historical sequence in which the different lesions were recognized and allows the different concepts of the pathology of MD to be discussed in the order in which they evolved. These aspects are followed by discussions on the stimulus for lymphoid proliferation, and on comparative features of MD.

CHANGES IN THE NERVOUS SYSTEM

Peripheral nerves

The majority of fowl with MD have infiltration of their peripheral nerves by lymphoid cells, which often causes nerve enlargement and paretic symptoms. Commonly involved are the brachial, sciatic and vagus nerves and the coeliac plexus, but probably all nerves can be affected. Goodchild (1969) drew attention to the high frequency of affection of the autonomic nerves. The classical strains of MDHV provoke mainly nerve lesions, and occasionally lymphoma formation, especially in the ovary. The more virulent strains of acute MDHV stimulate more extensive lymphoma formation, but nerves are usually affected also.

Two main types of nerve lesions are observed: one is characterized by light to heavy infiltration by proliferating pleomorphic lymphocytes, and the other by interneuritic oedema and usually light infiltration by small lymphocytes and plasma cells. Demyelination and axonal degeneration may accompany both types. From extensive studies, mainly on natural cases of MD, two schools of thought emerged on

[1] Houghton Poultry Research Station, Houghton, Huntingdon, UK.

— 21 —

the nature of the neuropathy in classical MD: (1) that the basic nerve lesion was an inflammatory response to an underlying parenchymatous lesion; (2) that the lymphoid reaction was neoplastic with secondary changes in the neurites.

The inflammation concept was supported by Marek (1907) himself, and by Lerche & Fritzsche (1934), who proposed the name "neurogranulomatosis infectiosa gallinarum". More recent proponents of this view were Campbell (1961) and Wight (1962a). Campbell stated "the impression is that the oedema comes first, followed by a demonstrable demyelination and fatty change, and that cellular infiltration is a late event". This theme was pursued by Wight (1962a), who classified the nerve lesions of natural cases of MD into three histological types: (1) Type I, showing variable infiltration by mainly small lymphocytes and some plasma cells, with increasing numbers of large lymphocytes in more severely affected nerves; (2) Type II, showing interneuritic oedema and a light infiltration by small lymphocytes and plasma cells; and (3) Type III, showing massive neoplasm-like infiltration by lymphoblasts. Wight's studies supported the concept of a progression of nerve lesions Type II → Type I → Type III, being based on the findings that demyelination can occur in the absence of severe lymphocytic infiltration (Wight, 1962a), that loss of axons was less severe in oedematous nerves than in infiltrated nerves (Wight, 1964), and that biochemical changes were less severe in oedematous nerves than in infiltrated nerves (Heald et al., 1964). Wight believed that the nerve lesions originated as an inflammatory response to a primary infection of neurectodermal tissues, which could progress to a neoplastic proliferation of lymphoid tissue.

The second concept, that MD was primarily a neoplastic condition, also had early proponents, notably Pappenheimer, Dunn & Cone (1926), who proposed the name "neurolymphomatosis gallinarum", and Furth (1935), who regarded the inflammatory and degenerative changes as secondary. This view was supported by our experimental study of the pathogenesis of classical MD induced by the HPRS-B14 strain of MDHV (Payne & Biggs, 1967). We modified Wight's classification and described three types of lesions: (1) A-type, characterized by proliferation of lymphoid cells, sometimes with demyelination and proliferation of Schwann cells. The infiltration in the A-type nerve consists of a mixed population of small and medium lymphocytes, blast cells and activated and primitive reticulum cells

(Figs. 1 & 2). Unusual cells termed "Marek's disease cells" are often present. They appear to be degenerating blast cells, possibly a consequence of virus infection. Ubertini & Calnek (1970) found HV in MD cells under the electron microscope in one bird studied; (2) B-type, characterized by diffuse infiltration by plasma cells and small lymphocytes, usually with interneuritic oedema, and sometimes demyelination and Schwann cell proliferation (Figs. 3 & 4); and (3) C-type, characterized by light infiltration by plasma cells and small lymphocytes. We concluded that the nerve lesions followed the sequence: A-type → mixed A+B-type → B-type. The C-type lesion was regarded as a mild form of the B-type lesion. The study indicated that MD was characterized by a primary multifocal proliferation of lymphoid cells in the nerves and other organs, especially the ovary. In some birds the proliferation was progressive and terminal, whereas in others the lymphoid proliferation was arrested and replaced by the more inflammatory change. Demyelination and other nerve changes were believed to be secondary to the lymphoid proliferation.

Although the earliest nerve changes seen in both classical MD (Payne & Biggs, 1967) and acute MD (Purchase & Biggs, 1967) were lymphoproliferative under the light microscope, they did not exclude the possibility of a preceding ultrastructural alteration of nerve parenchyma. The ultrastructure of sciatic nerves from natural cases of MD was examined by Wight (1969). The most striking abnormalities occurred in the Schwann cells, which showed cytoplasmic hypertrophy, ribosomal proliferation and vesicle formation apparently from endoplasmic reticulum. Myelin sheaths showed loss of lamellae and sometimes complete disintegration. The axons were relatively resistant to change. Wight suggested that the Schwann cells could be the primary target cells in the development of nerve lesions, although no virus particles were seen. Demyelination was of a type associated with Schwann cell damage. Ubertini & Calnek (1970) have shown that Schwann cells may be infected with MDHV. In a brachial nerve with much fluorescent and precipitin antigen, morphological changes in Schwann cells similar to those described by Wight (1969) were observed, and HV particles were seen in both normal and abnormal Schwann cells. These changes were exceptional, however, and were not seen in other birds or in another nerve of the same bird having similar lesions, under the light microscope (Calnek et al., 1970a). Apparently Schwann cells are usually either not

Fig. 1. A-type nerve lesion, showing lymphoid infiltration between neurites, in peripheral nerve of 28-days-old chick inoculated with HPRS-B14 strain of MDHV at one-day-old.

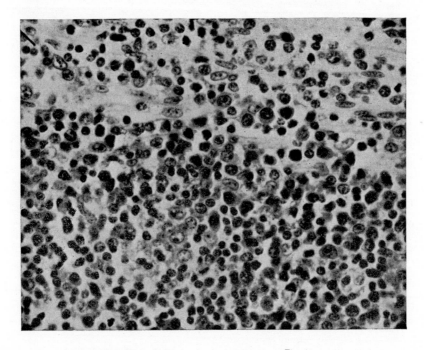

Fig. 2. Mixed population of lymphoid cells in A-type nerve; same case as Fig. 1.

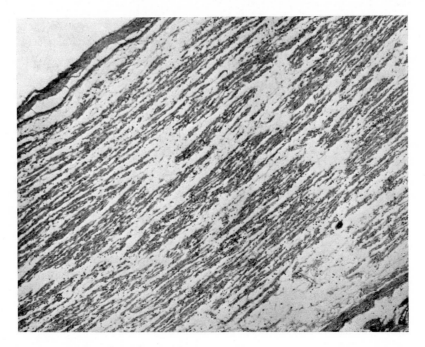

Fig. 3. B-type nerve lesion, showing interneuritic oedema and light cell infiltration, in peripheral nerve of 70-days-old chicken inoculated with HPRS-16 strain of MDHV at one-day-old.

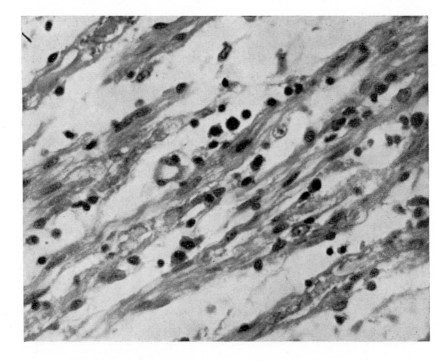

Fig. 4. Plasma cells and lymphocytes in B-type nerve lesion; same case as Fig. 3.

infected, or carry the HV in a hidden form. There is a need for a systematic study of the ultrastructure of nerves in MD, starting before the stage of lymphoid proliferation.

Studies with fluorescent antibodies against MDHV-associated antigens have not clarified the problem of the stimulus to lymphoid proliferation in nerves. In several independent studies there was general agreement that peripheral nerves, whether infiltrated or not, are usually negative for antigen or contain only a few fluorescent cells (Calnek & Hitchner, 1969; Spencer & Calnek, 1970; von Bülow & Payne, 1970; Purchase, 1970). These findings are not surprising in view of the absence of virus particles in nerves, since a good correlation exists between fluorescent antigen and the presence of replicating virus (Calnek, Ubertini & Adldinger, 1970). In view of the possibility of an early infection of Schwann cells, it is noteworthy that Sevoian & Chamberlain (1964) reported an initial proliferation of neurilemmal cells and mesenchymal cells in MD induced by the JM strain virus. These changes have not been recorded with other strains.

Central nervous system

Changes in the brain and spinal cord are variable, consisting of perivascular cuffing by lymphoid cells, endotheliosis and gliosis (Wight, 1962b; Sevoian & Chamberlain, 1964; Purchase & Biggs, 1967; Vickers, Helmboldt & Luginbuhl, 1967). Fluorescent antigen has not been observed in the brain or cord (Calnek & Hitchner, 1969; Purchase, 1970; Payne & Rennie[1]), and no ultrastructural studies have been reported.

LYMPHOMA FORMATION IN MD

The ability of MDHV to induce lymphoma formation depends not only on virus properties but also on host factors, especially genetic constitution. The HPRS-B14 strain of classical MDHV induced lymphomas, mainly in the ovary, in a Rhode Island Red (RIR) strain, but not in a Brown Leghorn strain (Biggs & Payne, 1967). Strains of acute MDHV induced lymphomas in many organs in both strains, but with higher frequency in the RIR (Purchase & Biggs, 1967). The overall distribution of visceral lymphomas in 159 susceptible RIR fowl inoculated with 4 strains of acute MDHV were: gonad – 95.6%, liver – 62.9%, spleen – 30.2%, lung – 30.2%, muscle – 30.2%, heart – 28.9%, kidney – 27.7%, proventriculus and intestine – 12.6%, and mesentery and serosa – 10.7% (Purchase & Biggs, 1967). In highly resistant strains of fowl, acute MDHV produces mainly the neural form of MD (Payne[1]).

The lymphomatous lesions are apparently multifocal in origin and arise at the same time as the first lymphoproliferative changes in peripheral nerves, at about 14 days after infection of day-old chicks (Payne & Biggs, 1967; Purchase & Biggs, 1967). Cytologically, the lymphomas are similar to the proliferating cells in the A-type nerves, consisting of a mixed company of lymphocytes, blast cells, reticulum cells and MD cells (Figs. 5, 6 & 7). Usually the small lymphocyte is the predominating cell, although in some tumours uniformly immature blast-type cells may occur. There is no evidence that the lymphoma arises in either the bursa or thymus; these organs are not invariably involved and MD is not preventable by bursectomy (Payne & Rennie, 1970a; Fernando & Calnek, 1971), or thymectomy (Payne & Rennie[1]). MD thus differs from lymphoid leukosis of the fowl, which arises in the bursa (Cooper *et al.*, 1968) and from lymphoid leukaemia of mice, which originates in the thymus (see Hays, 1968). Macroscopic lymphomas first appear in MD 3–4 weeks after infection. In the study of the pathogenesis of classical MD there was evidence that microscopic lymphomas could regress (Payne & Biggs, 1967).

Nature of lymphomas in MD

There is argument as to whether the lymphomas in MD should be regarded as true neoplasms or as lymphogranulomas. Campbell (1956, 1961) considered ovarian lymphomas to be essentially inflammatory in nature, possibly arising in the adjacent coeliac ganglion. The argument is probably one of semantics rather than pathology. The widespread distribution among different organs of abnormally large masses of proliferating lymphoid cells makes it difficult not to accept that MD fulfills many of the criteria of neoplasia, including: (1) progressive proliferation; (2) qualitative differences from, and excessive increases over, lymphoid hyperplasia produced by many infections of the fowl; (3) multifocal and diffuse origin; and (4) abnormal cells. Biggs (1968) believed that the lymphoproliferative lesions in MD conformed to Marshall's (1956) second group of neoplasms of the reticular tissue, which embraced "multifocal benign neoplasms with progressive extension, but without evidence of true malignancy".

---

[1] Unpublished results.

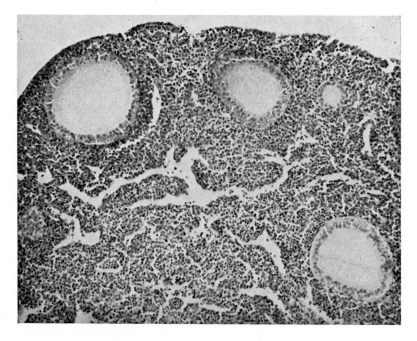

Fig. 5.   Ovarian lymphoma in 32-days-old chick inoculated with HPRS-B14 strain of MDHV at one-day-old.

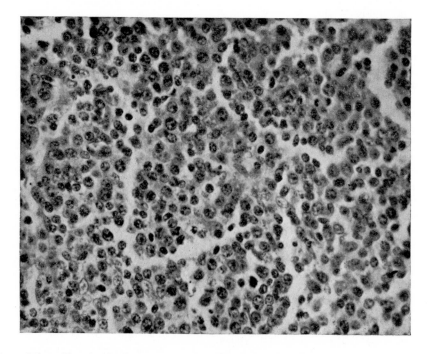

Fig. 6.   Mixed population of lymphoid cells in ovarian lymphoma; same case as Fig. 5.

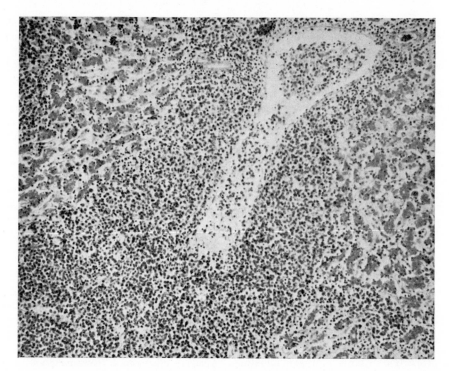

Fig. 7.   Lymphomatous proliferation in liver of 84-days-old chicken inoculated with HPRS-16 strain of MDHV at 3-days-old.

This may be true particularly of classical MD, but acute MD frequently conforms to Group 3, the "malignant neoplasms" showing "local tissue invasion and destruction or evidence of spread by metastasis".

If the origin of the lymphoid cells in MD lesions were known this would help to clarify the question of the nature of the lesions. This approach has been stimulated by the recognition in recent years of a functional and morphological dissociation of the avian lymphoid system into two components: (1) the bursa-dependent lymphoid system, responsible for immunoglobulin production; and (2) the thymus-dependent system, responsible for cell-mediated immune responses (Warner, Szenberg & Burnet, 1962; Cooper *et al.,* 1966). More recent work in fowl and mice suggests that the thymus and bursa may acquire lymphoid stem cells from an outside source, possibly yolk sac or bone marrow (see Miller & Osoba, 1967). There are thus three different sources of cells that might be involved in the formation of MD lymphomas: bursa, thymus and bone marrow.

By removing the bursa and thymus in MD it should be possible to determine their rôles. They may be involved in at least two ways. Firstly, they may be the site or source of target cells, in which case removal should prevent lymphoma formation. For example, for this reason bursectomy of fowl prevents lymphoid leukosis (Peterson *et al.,* 1966; Cooper *et al.,* 1968) and thymectomy in mice prevents lymphoid leukaemia (see Hays, 1968). Secondly, these organs may be involved in immune responses that control tumour growth, and their removal will enhance lymphoma production due to immunosuppression. For this reason bursectomy in fowl increases myeloblastosis (Baluda, 1967), and thymectomy in mammals increases tumour formation by polyoma virus, SV40, adenovirus 12, and methylcholanthrene (see Law, 1967). The target cell or immunologically reactive cell may lie in the peripheral dependent lymphoid tissue, and not in the central lymphoid organ, so that complete removal of both central and peripheral tissues may be required to demonstrate an effect, a requirement often difficult to achieve for technical reasons.

Reports on the effect of bursectomy on MD are discordant. It has been reported as increasing the incidence (Morris, Jerome & Reinhart, 1969), decreasing the incidence (Foster & Moll, 1968) or having no effect (Kenyon *et al.,* 1969). In critical studies in which immune function tests were performed on neonatally bursectomized and X-irradiated fowl from a susceptible strain, we found no influence on MD. The disease was observed in a group of birds which were negative for immunoglobulins IgG and IgM, antibodies to sheep erythrocytes, germinal centres and plasma cells, and we concluded that the bursal system was not essential for the development of MD (Payne & Rennie, 1970a). A similar conclusion was reached by Fernando & Calnek (1971). Bursectomy also had no influence on MD in a resistant strain of fowl (Payne & Rennie[1]).

We have studied the effect of neonatal thymectomy and X-irradiation on MD. In neither genetically susceptible nor genetically resistant strains of fowl was there a significant influence on mortality from gross MD, but several significant effects on the type of disease were observed (Payne & Rennie[1]). Thymectomy of susceptible fowl: (1) decreased the proportion of birds with gross lymphomas; (2) decreased the proportion of birds with lymphoproliferative nerve lesions (A-type and A+B-type); and (3) decreased the proportion of birds with lymphomatous involvement of the bursa. In genetically resistant fowl, thymectomy increased the proportion with gross lymphomas but had no other significant effects. These findings suggest that the thymus-dependent lymphoid system may have two rôles in the pathogenesis of MD: (1) as a source of proliferative target cells; and (2) as mediator of a cellular immune response directed against the target cells. In a susceptible strain of fowl, the cellular immune response may be unable to prevent lymphoma formation, so that reduction of lymphomas by removal of target tissue is the visible effect of thymectomy. In a resistant strain, cell-mediated immunity may normally be effective in preventing proliferation of target cells, but these may be released from restraint by thymectomy, thus increasing lymphoma formation. According to this hypothesis, a MD lymphoma consists of two types of thymus-dependent lymphoid cells: (1) target cells, which may be virus-infected or virus-transformed cells antigenic to the host; and (2) normal or sensitized lymphocytes that are reacting immunologically against the target cells. The transfer of resistance to MD with spleen cells from MD survivors supports a host defensive response (Feld-

bush & Maag, 1969), although the nature of the transferred resistance was not established.

There is evidence for immunologically uncommitted thymus-dependent lymphocytes in MD lymphomas. Lymphoma cells have the ability to evoke a graft-versus-host reaction, as measured by splenomegaly, when inoculated intravenously into chick embryos. This reaction is believed to be caused by thymus-dependent lymphocytes (Cooper *et al.,* 1966). When dose-response curves for lymphoma cells and isologous blood lymphocytes were compared, the relative competency of cells from 5 lymphomas varied from 0.36 to 0.0014 (Payne[1]).

There is also evidence for bursa-dependent lymphoid cells in lymphomas, although these would seem to be nonessential in view of the failure of bursectomy to influence the disease. Seven lymphomas from fowl inoculated with sheep erythrocytes contained on average 16 plaque-forming cells/10^6 tumour cells, detected by the Jerne plaque technique (Roszkowski & Payne[1]). These findings indicate that both thymus-dependent and bursa-dependent lymphoid cells are involved in lymphoma formation. Further studies with antigenically or isotopically labelled cells from different lymphoid sources would be valuable.

CHANGES IN OTHER TISSUES

The ability to produce MD experimentally, and particularly the application of fluorescent antibody (FA) techniques and electron microscopy to the study of MD, have resulted in the recognition of a number of changes, in addition to the neural and lymphomatous lesions, which hitherto had escaped detection. These are discussed below.

Feather follicle

Feather follicle epithelium is a major site of MDHV replication, as shown by detection of virus-associated antigen in FA tests, of cellular and extracellular virions, and of filtrable infectious virus (Calnek & Hitchner, 1969; Nazerian & Witter, 1970; Purchase, 1970). Fluorescent antigen is present as early as 5 days after infection (Purchase, 1970) and virus may be detected at 2 weeks after infection (Nazerian & Witter, 1970). Subsequently, antigen and virus are present in a high proportion of chicks. Calnek & Hitchner (1969) found antigen in feather follicles of 88 % of exposed birds examined 2–6 weeks after infection, and Purchase (1970) found positive fol-

---

[1] Unpublished results.

licles in 65% of MD infected birds. In both studies feather follicle was the tissue most often positive. Infected cellular material derived from feather follicles is probably the major means of spread of MDHV in the environment (Calnek *et al.,* 1970b).

In infected follicles, degenerative changes are seen with the light microscope in cells of the intermediate and transitional layers of the stratum germinativum, and occasional intranuclear inclusions are present in these cells (Nazerian & Witter, 1970; Purchase, 1970; Lapen, Piper & Kenzy, 1970) (Fig. 8). Aggregations of lymphoid tissue in the dermis were specifically associated with fluorescent antigen in feather follicles (Lapen *et al.,* 1970) (Fig. 9). Lymphomas may arise in the dermis, especially around feather follicles (Payne & Biggs, 1967; Sharma, Davis & Kenzy, 1970).

Kidney

Antigen has commonly been found in FA tests in kidney epithelium by some investigators (Calnek & Hitchner, 1969; Spencer & Calnek, 1970) but rarely by others (von Bülow & Payne, 1970; Purchase, 1970); the reason for the discrepancy is not known. Finding of antigen is consistent with the high frequency with which MDHV may be isolated from kidney cells from infected birds grown in culture (Witter, Solomon & Burgoyne, 1969; Nazerian & Witter, 1970; Spencer & Calnek, 1970). Uncoated herpes virions were found in renal epithelium cells by Schidlovsky, Ahmed & Jensen (1969) but not in infiltrating lymphoma cells. In kidneys with few FA positive cells, Calnek *et al.,* (1970a) found no virions under the electron microscope.

No alterations in renal epithelium cells visible with the light microscope have been reported.

Lung

Purchase (1970) found antigen in FA tests in the lungs of 23% of fowl studied, whereas others have not found it (Calnek & Hitchner, 1969).

Other epithelial organs

Small foci of fluorescent cells are seen in FA tests in low frequency in various other epithelial organs, including proventriculus, testis, adrenal and thyroid (Calnek & Hitchner, 1969; Payne & Rennie[1]).

---

[1] Unpublished results.

Bursa of Fabricius

Second to the feather follicle, the bursal medulla is the tissue in which viral antigen is most frequently found in FA tests. Calnek & Hitchner (1969), Spencer & Calnek (1970), and Purchase (1970) found positive bursas in 23%, 94% and 15% of total infected birds, respectively. Von Bülow & Payne (1970) identified antigen in about 50% of infected chickens examined 12–56 days after infection. Fluorescence varies from a few scattered medullary cells to virtually all medullary cells within a follicle. It is not yet clear whether both lymphoid and epithelial elements of the medulla are affected. The presence of antigen is associated with cell necrosis, follicular atrophy and cyst formation (Calnek & Hitchner, 1969; von Bülow & Payne, 1970; Purchase, 1970) and intranuclear inclusion bodies may be found in bursas showing acute cytolytic changes (Payne & Rennie[1]). (Figs. 10 & 11). Viral precipitin antigen may be present, but not filtrable infectious virus (Purchase, 1970). This is consistent with the observation by Calnek *et al.* (1970a) of uncoated virions, but rarely coated virions, in the bursa. The particles were present in immature lymphoid cells. Bursas were positive for antigen in FA tests as early as 5 days after infection (Purchase, 1970).

The early cytolytic effects of MDHV on the bursa may be reflected in a change in bursal size. Jakowski *et al.* (1969) observed a 3-fold reduction in bursal weight 12 days after inoculation of day-old chicks with infected blood. To what extent the lymphoid atrophy is a specific consequence of the cytolytic effects of the HV, or a result of nonspecific changes that commonly occur in the bursa following hormone administration, infections and other stresses (see Glick, 1964), is not known. The cytolytic effects of MDHV are more clearly shown when chicks without passively acquired antibody are infected. Jakowski *et al.* (1970) reported severe destructive effects on the haematopoietic system, causing anaemia and severe degenerative changes in the bursa and thymus, with little tumour formation.

In addition to undergoing atrophy, bursas may also become lymphomatous (Fig. 12). This occurs especially in susceptible chicks infected with virulent virus. The proliferation is interfollicular, and usually consists of pleomorphic lymphocytes, more rarely of reticulum cells or plasmacytoid cells.

Thymus

Scattered single cells or groups of cells containing antigen were found in FA tests in 14% of infected

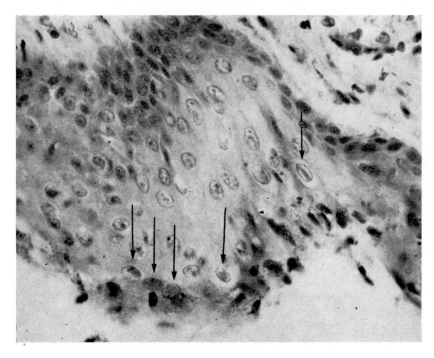

Fig. 8. Intranuclear inclusion bodies (arrows) in epidermis of feather follicle in 22-days-old chick inoculated with HPRS-16 strain of MDHV at one-day-old.

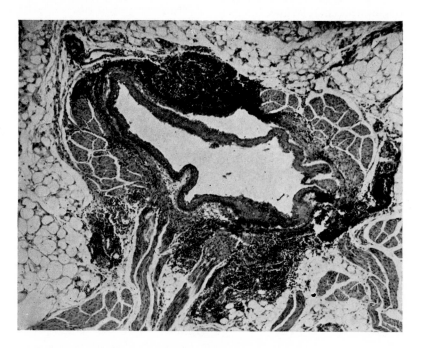

Fig. 9. Accumulations of lymphoid tissue around infected feather follicle in 22-days-old chick inoculated with HPRS-16 strain of MDHV at one-day-old.

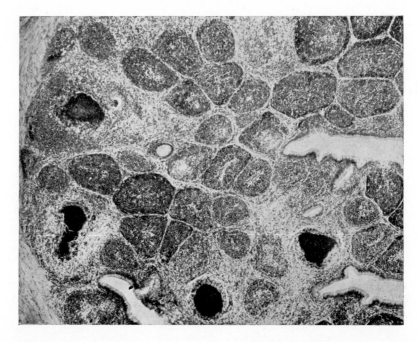

Fig. 10.   Lymphoid atrophy and necrosis in follicles of bursa of Fabricius of 29-days-old chick inoculated with HPRS-16 strain of MDHV at one-day-old.

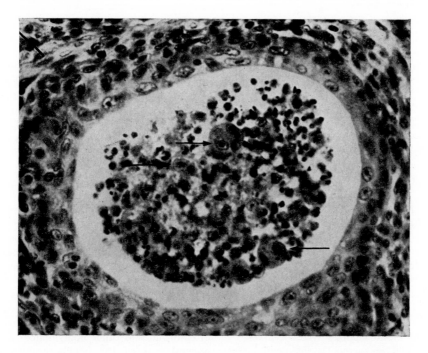

Fig. 11.   Cells containing intranuclear inclusion bodies (arrows) in necrotic tissue in a bursal lymphoid follicle; same case as Fig. 10.

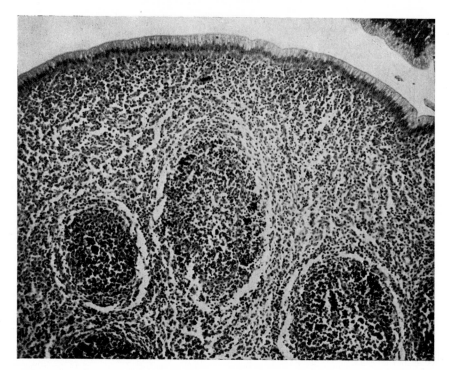

Fig. 12.   Interfollicular lymphomatous proliferation in bursa of Fabricius of 37-days-old chick inoculated with HPRS-16 strain of MDHV at one-day-old.

birds (Purchase, 1970).   Precipitin antigen was detected but not filtrable virus.   Histologically the thymus may be atrophic and more rarely lymphomatous.

Spleen

Calnek & Hitchner (1969) and Purchase (1970) observed antigen in FA tests in splenic cells in a few birds.   In our experience, identification of such cells is made difficult by the occurrence of nonspecifically stained cells.   Purchase (1970) detected precipitin antigen in the spleens of 2 birds.

Splenic changes have not been studied in detail. Lymphomatous involvement is common in acute MD, rare in classical MD.   Occasionally birds with lymphomas elsewhere show excessive macrophage activity in the spleen, suggesting a high death rate of lymphoma cells.

Blood and marrow

No consistent blood changes were observed by early workers, and leukaemia is uncommon (see Biggs, 1968).   Lymphomatous involvement of the marrow occurs in acute MD (Sevoian & Chamberlain, 1964).   The marrow is infected by MDHV, as shown by FA tests (Kottaridis & Luginbuhl, 1968), and this may be responsible for reduction in the packed cell volume (Vickers et al., 1967).   In chicks without protection by passive antibody, infection by MDHV may cause marrow aplasia and marked anaemia (Jakowski et al., 1970).

THE STIMULUS FOR LYMPHOID PROLIFERATION

The nature of the stimulus for lymphoid cells to proliferate in MD is unknown.   There are a number of possibilities; they fall into two groups: (1) those in which the stimulus is intrinsic, within the proliferating lymphoid cells; (2) those where the stimulus is extrinsic, lying outside the cells.   A feature of the first group is that the infective processes in epithelial tissues are not primarily related to the lymphomatous proliferation, whereas in the second group changes in epithelial and other nonlymphoid tissues are an integral part of the lymphoproliferative process, providing the necessary stimulus.

Intrinsic stimulation

In this theory, virus infection of the lymphocyte leads to a change in the properties of the infected cell, endowing it with the properties of a neoplastic cell. Evidence is against HV being present in the majority of lymphoma cells. Fluorescent antigen is either absent from lymphomas or present only in a few scattered cells, which could be stromal cells rather than lymphoid cells (Calnek & Hitchner, 1969; Spencer & Calnek, 1970; von Bülow & Payne, 1970; Purchase, 1970). Attempts to demonstrate herpes virions in lymphoma cells have been almost completely unsuccessful (Schidlovsky *et al.,* 1969; Nazerian & Witter, 1970). Nazerian (quoted by Nazerian & Witter, 1970) found immature herpesvirions in only two gonadal tumours and lymphoid infiltrations in one nerve of more than 100 specimens examined. Furthermore, only a small minority of lymphoma cells are infective in tissue culture, 10^2–10^5 lymphoma cells being required to initiate a microplaque (Churchill & Biggs, 1967; Calnek & Madin, 1969). Additional search for HV in MD cells in tumours would be justified in view of their identification by Ubertini and Calnek (1970) in similar cells in a nerve. The majority of lymphoma cells therefore do not contain replicating HV. Nevertheless, it is possible, as Calnek & Hitchner (1969) have pointed out, that virus genome exists in many or all tumour cells. This possibility is supported by the observation of Campbell and Woode (1970) that virions and fluorescent antigen appear in blood lymphocytes from MD cases after they have undergone blastoid transformation in culture under the influence of phytohaemagglutinin. Possibly the stimulation of cellular DNA is accompanied by derepression and stimulation of HV-DNA. The possibility that MDHV-DNA might be integrated into the host cell genome should be amenable for study by DNA hybridization tests, as described for Burkitt tumour cells by zur Hausen & Schulte-Holthausen (1970). Studies on the state of HV in MD lymphoma cells are handicapped by inability to grow the cells in culture. If cultures could be established it would be possible to determine whether culture at a lower temperature would derepress a virus, as has been observed for a Burkitt's lymphoma line (Hinuma *et al.,* 1967) and for frog kidney tumours (Mizell, Stackpole & Halperen, 1968). Similarly, the effect of arginine deficient medium as a derepressor could be examined (Henle & Henle, 1968a). Only a small proportion of cells in cultures of cloned Burkitt's lymphoma cells show evidence of the presence of virus, indicating viral repression in most cells.

Another possibility is that immunologically competent cells might be modified by virus infection allowing the emergence of pathogenic "forbidden" clones (Burnet, 1969) that react immunologically with host antigens. MD could then be regarded as an autoimmune disease. Such a mechanism has been invoked to explain the numerous abnormal antibodies that appear in infectious mononucleosis in man (Dameshek, 1969). In this connection it is of interest that Zacharia & Sevoian (1970) claimed that MD-infected chickens specifically developed agglutinins to sheep erythrocytes. We have been unable to confirm these findings (Payne & Rennie, 1970b).

Extrinsic stimulation

In this theory it is suggested that the lymphocytes are proliferating in response to an external antigenic stimulus. The reaction would be essentially immunological, its excessive nature, leading to "lymphoma" formation, perhaps being due to some unusual qualitative or quantitative (? persistent) property of the antigen. Possible antigens could be: (1) virus-infected cells; (2) cell-free virus, viral antigen or antigen-antibody complexes; (3) host antigens released by viral damage to cells. Stimulation by virus-infected cells cannot be excluded, because of the possibility of the presence of repressed virus, but available facts do not provide support. Thus, as discussed above, fluorescent antigen and herpes virions are absent or rare in lymphomas. Also, there is no clear association between presence of antigen and virions in a tissue and the predilection of that tissue to lymphoma formation. Feather follicle epithelium and bursa, which are the sites most frequently positive for replicating virus, are not the most frequent sites for lymphoma formation.

Failure to detect antigen and virus in the lymphoma cells could be explained if antigen or virus were produced at some distant site, e.g., feather follicle, and carried via the blood to sites where sensitized lymphocytes had localized. Continued antigenic stimulation, resulting in cellular proliferation and recruitment of additional lymphoid cells, mediated perhaps by soluble "lymphokine" factors (Dumonde *et al.,* 1969), could give rise to lymphoma formation. If this theory is correct, it should be possible to detect specifically sensitized lymphoid cells in lymphomas and to reproduce tumours by continued antigenic stimulation. The possibility of antigen-antibody complexes being specifically involved would appear to be ruled out by the production of MD in agammaglobulinaemic fowl (Payne & Rennie, 1970a).

Release of normally segregated host antigens, which have not been recognised as "self" (Burnet, 1969), as a result of viral damage to cells, is another way (see above) in which autoimmune reactions can occur. Siller (1960) first drew attention to the similarity between the neural lesions of MD and experimental allergic encephalomyelitis (EAE) produced by intramuscular inoculation of homologous spinal cord in Freund's adjuvant. By inoculating guinea pig and chicken sciatic nerves in adjuvant, Petek & Quaglio (1967) produced an experimental allergic neuritis in fowl that was indistinguishable from peripheral nerve lesions of MD. Wight & Siller (1965) found that EAE lesions were suppressed in fowl with spontaneous MD; this may have been related to the known immunosuppressive effect of MDHV (Purchase, Chubb & Biggs, 1968; Payne, 1970). The incidence of spontaneous MD in fowl rendered "tolerant" to EAE was not reduced; on the contrary, there was apparently an increase. These findings raise the possibility that there is an auto-immune component in MD in the pathogenesis of the neural lesions, and possibly in other tissues as well, and it would be worthwhile searching in birds with MD for complement-fixing or precipitin antibodies against normal tissue antigens and for lymphocytes sensitized to normal antigens. Intrinsic and extrinsic stimulation in the formation of lymphomas are not mutually exclusive. I have suggested above, in connection with our thymectomy experiments, that lymphomas consist of: (1) virus-infected target cells, possibly carrying viral antigens, tumour-specific antigens, or transplantation antigens; and (2) reactive immunologically competent cells.

COMPARATIVE PATHOLOGY OF MAREK'S DISEASE, INFECTIOUS MONONUCLEOSIS AND BURKITT'S LYMPHOMA

The role of HV in the aetiology of MD is well established, and strong indirect evidence is available that another herpesvirus, the Epstein-Barr virus (EBV), is the cause of infectious mononucleosis (IM); the same virus may also be the cause of Burkitt's lymphoma (BL) in man (Epstein & Achong, 1970). In view of the fact that HV can cause lymphoproliferative disorders in both man and fowl, it is appropriate in this Symposium to try to draw parallels between these diseases. Dameshek (1969) has suggested that IM is a self-limiting acute leukaemia, and the questions to be answered are: what starts the lymphoid proliferation and what causes it to cease? The same questions may be asked in MD, in which lymphoproliferative lesions in nerves and organs may regress. The possibility has been discussed that lymphoid proliferation in MD depends on continued stimulation of lymphoid tissue by viral or host antigens. Under certain circumstances, e.g., limitation of viral multiplication as immunity develops, stimulation may cease, and the lesions, if not too advanced, may disappear. Possibly extralymphoid sites of virus multiplication, which serve to stimulate the lymphoid tissue, exist in IM and BL, and could be detected by fluorescent antibody techniques or electron microscopy. On the other hand, EBV can be isolated from lymphoid cell cultures derived from IM and BL, suggesting a more direct stimulatory effect (Henle & Henle, 1968b; Epstein & Achong, 1970). Both MDHV and EBV are closely cell-associated viruses; possibly proliferation is the consequence of infection of nonpermissive lymphoid cells by this type of virus. A number of variables determine whether the pathological response in MD is of the limited inflammatory type or of the neoplastic type; these include strain of virus, dose of virus, genetic constitution of the host, age of host, and presence of maternal antibodies at the time of infection. Do these factors influence the outcome of EBV infection?

The neuropathic lesions in MD may have their counterpart in the encephalitis and peripheral neuropathy that occur as complications of IM (Finch, 1969). IM is one cause of acute infective polyneuritis (polyradiculoneuropathy; Guillain-Barré syndrome) in man, which is characterized by neural oedema, demyelination and axonal degeneration, Schwann cell activation, and lymphocytic infiltration (Greenfield et al., 1958). Apparently unlike MD, the oedema precedes the infiltration. It is of interest that the similarities between acute polyneuritis in man and experimental allergic neuritis (EAN) have been remarked upon (Waksman & Adams, 1955; Asbury et al., 1969) as they have between the neural lesions in MD and EAN (Petek & Quaglio, 1967). Borit & Altrocchi (1971), in describing a case of recurrent polyneuropathy in man, specifically drew attention to the similarities between this disorder and MD. Involvement of neural tissue in these diseases may be examples of a predilection of herpesviruses in general to attack nerve cells or supporting cells (Roizman, 1965). Polyneuropathy with invasion of peripheral nerves by tumour cells has also been recorded as a rare complication of lymphoma in man (Moore & Oda, 1962).

The significance of the BL response as opposed to the IM response, assuming both disorders are caused by the same virus, is not known, and the pathological connection between the two has not been determined. This is in contrast to MD where the acute lymphomatous form appears to be a more severe form of a change that is also seen in the more benign classical form. Burkitt (1970) has suggested that BL occurs when the EB virus attacks lymphoid tissue already altered by chronic malaria. It should be possible to devise model systems with MD in chickens to test this idea.

There are both similarities and differences between acute MD and BL. The organs distribution of lymphomas, apart from the absence of skeletal lymphomas in MD, is similar, commonly affected organs being ovary, kidney, liver, lungs and heart (Burkitt & Wright, 1970; Purchase & Biggs, 1968). In both diseases the ovary is more tumour-prone than the testis. In both, the spleen and marrow tend to be spared and leukaemia is uncommon. The lymphoma cells in BL are uniformly immature and strongly pyroninophilic, whereas those in MD are usually mixed and of varying pyroninophilia; sometimes uniformly immature lymphomas occur in MD also. The debris-laden histiocytes, which are responsible for the "starry-sky" appearance of tumours in BL, are not seen in MD tumours, although histiocytic activity is sometimes seen in the spleen and liver. Histiocytic activity in both diseases may reflect a high death rate of lymphoma cells, perhaps due to host defences. Stjernswärd, Clifford & Svedmyr (1970) have observed general depression of immunological reactivity in patients with BL; similar findings have been recorded in MD (Purchase et al., 1968; Payne, 1970). It is possible that these changes may be significant in determining the way in which man or fowl respond to HV infection. A remarkable feature of BL is its marked responsiveness to chemotherapy (Clifford, 1970). This aspect has been scarcely studied in MD.

REFERENCES

Asbury, A. K., Arnason, B. G. & Adams, R. D. (1969) The inflammatory lesion in idiopathic polyneuritis. *Medicine (Baltimore)*, **48**, 173-215

Baluda, M. A. (1967) The role of the bursa-dependent lymphoid tissue on oncogenesis by avian myeloblastosis virus. *Virology*, **32**, 428-437

Benton, W. J. & Cover, M. S. (1957) The increased incidence of visceral lymphomatosis in broiler and replacement birds. *Avian Dis.*, **1**, 320-327

Biggs, P. M. & Payne, L. N. (1963) Transmission experiments with Marek's disease (Fowl paralysis). *Vet. Rec.*, **75**, 177-179

Biggs, P. M., Purchase, H. G., Bee, B. R. & Dalton, P. J. (1965) Preliminary report on acute Marek's disease (fowl paralysis) in Great Britain. *Vet. Rec.*, **77**, 1339-1340

Biggs, P. M. (1966) Avian leukosis and Marek's disease. In: *Thirteenth World's Poultry Congress Symposium. Papers*, Kiev, pp. 91-118

Biggs, P. M. & Payne, L. N. (1967) Studies on Marek's disease 1. Experimental transmission. *J. nat. Cancer Inst.*, **39**, 267-280

Biggs, P. M. (1968) Marek's disease—current state of knowledge. *Curr. Top. Microbiol. Immunol.*, **43**, 91-125

Biggs, P. M. (1971) Marek's disease—recent advances. In: Gordon, R. F. & Freeman, B. M., eds, *Poultry dis-ease and world economy*, Edinburgh, British Poultry Science, pp. 121-133

Borit, A. & Altrocchi, P. H. (1971) Recurrent polyneuropathy and neurolymphomatosis. *Arch. Neurol. (Chic.)*, **24**, 40-49

von Bülow, V. & Payne, L. N. (1970) Direkter Immunofluoreszenztest bei der Marek'schen Krankheit. *Zbl. Vet.-Med.*, **17**, 460-478

Burkitt, D. P. (1970) An alternative hypothesis to a vectored virus. In: Burkitt D. P., & Wright, D. H., eds, *Burkitt's lymphoma*, Edinburgh & London, Livingstone, pp. 210-214

Burkitt, D. P. & Wright, D. H. (1970) *Burkitt's lymphoma*. Edinburgh & London, Livingstone

Burnet, M. (1969) *Self and not-self*. Cambridge, Cambridge University Press

Calnek, B. W. & Hitchner, S. B. (1969) Localization of viral antigen in chickens infected with Marek's disease herpes virus. *J. nat. Cancer Inst.*, **43**, 935-949

Calnek, B. W. & Madin, S. H. (1969) Characteristics of in vitro infection of chicken kidney cell cultures with a herpes virus from Marek's disease. *Amer. J. vet. Res.*, **30**, 1389-1402

Calnek, B. W., Ubertini, T. & Adldinger, H. K. (1970) Viral antigen, virus particles, and infectivity of tissues from chickens with Marek's disease. *J. nat. Cancer Inst.*, **45**, 341-351

Calnek, B. W., Adldinger, H. K. & Kahn, D. E. (1970) Feather follicle epithelium: a source of enveloped and infections cell-free herpes virus from Marek's disease. *Avian Dis.*, **14**, 219-233

Campbell, J. G. (1956) Leucosis and fowl paralysis compared and contrasted. *Vet. Rec.*, **68**, 527-528

Campbell, J. G. (1961) A proposed classification of the leucosis complex and fowl paralysis. *Brit. vet. J.*, **117**, 316-325

Campbell, J. G. & Woode, G. N. (1970) Demonstration of a herpes-type virus in short-term cultured blood lymphocytes associated with Marek's disease. *J. med. Microbiol.*, **3**, 463-473

Churchill, A. E. & Biggs, P. M. (1967) Agent of Marek's disease in tissue culture. *Nature (Lond.)*, **215**, 528-530

Clifford, P. (1970) Treatment. In: Burkitt D. P., & Wright, D. H., eds, *Burkitt's lymphoma*, Edinburgh & London, Livingstone, pp. 52-63

Cooper, M. D., Peterson, R. D. A., South, M. A. & Good, R. A. (1966) The functions of the thymus system and the bursa system in the chicken. *J. exp. Med.*, **123**, 75-102

Cooper, M. D., Payne, L. N., Dent, P. B., Burmester, B. R. & Good, R. A. (1968) Pathogenesis of avian lymphoid leukosis. *J. nat. Cancer Inst.*, **41**, 373-389

Dameshek, W. (1969) Speculations on the nature of infectious mononucleosis. In: Carter, R. L. & Penman, H. G., eds, *Infectious mononucleosis*, Oxford & Edinburgh, Blackwell Scientific Publications, pp. 225-240

Dumonde, D. C., Wolstencroft, R. A., Panayi, G. S., Matthew, M., Morley, J. & Howson, W. T. (1969) "Lymphokines": non-antibody mediators of cellular immunity by lymphocyte activation. *Nature (Lond.)*, **224**, 38-42

Epstein, M. A. & Achong, B. G. (1970) The EB Virus. In: Burkitt, D. P. & Wright, D. H., eds, *Burkitt's lymphoma*, Edinburgh & London, Livingstone, pp. 231-248

Feldbush, T. L. & Maag, T. A. (1969) Passive transfer of resistance to Marek's disease. *Avian Dis.*, **13**, 677-680

Fernando, W. W. D. & Calnek, B. W. (1971) The influence of the bursa of Fabricius on the infection and pathological response of chickens exposed to Marek's disease herpes virus. *Avian Dis.* (in press)

Finch, S. C. (1969) Clinical symptoms and signs of infectious mononucleosis. In: Carter, R. L. & Penman, H. G., eds, *Infectious mononucleosis*, Oxford & Edinburgh, Blackwell Scientific Publications, pp. 19-46

Foster, A. G. & Moll, T. (1968) Effect of immunosuppression on clinical and pathologic manifestations of Marek's disease in chickens. *Amer. J. vet. Res.*, **29**, 1831-1835

Furth, J. (1935) Lymphomatosis in relation to fowl paralysis. *Arch. Path.*, **20**, 329-428

Glick, B. (1964) The bursa of Fabricius and the development of immunologic competence. In: Good, R. A.

& Gabrielson, A. E., eds, *The thymus in immunobiology*, New York, Harper & Row, pp. 343-358

Goodchild, W. M. (1969) Some observations on Marek's disease (Fowl paralysis) *Vet. Rec.*, **84**, 87-89

Greenfield, J. G., Blackwood, W., McMenemey, W. H., Meyer, A. & Norman, R. M. (1958) *Neurology*, London, Edward Arnold (Publishers) Ltd

Hays, E. F. (1968) The role of thymus epithelial reticular cells in viral leukemogenesis. *Cancer Res.*, **28**, 21-26

Heald, P. J., Badman, H. G., Frunival, B. F. & Wight, P. A. L. (1964) Chemical changes in nerves from birds affected by fowl paralysis. *Poult. Sci.*, **43**, 701-710

Henle, W. & Henle, G. (1968a) Effect of arginine-deficient media on the herpes-type virus associated with cultured Burkitt tumor cells. *J. Virol.*, **2**, 182-191

Henle, W. & Henle, G. (1968b) Present status of the herpes-group virus associated with cultures of the hematopoietic system. *Perspect. Virol.*, **6**, 105-117

Hinuma, Y., Konn, M., Yamaguchi, J., Wudarski, D. J., Blakeslee, J. R. & Grace, J. T. (1967) Immunofluorescence and herpes-type virus particles in the P3HR-1 Burkitt lymphoma cell line. *J. Virol.*, **1**, 1045-1051

Jakowski, R. M., Fredrickson, T. N., Luginbuhl, R. E. & Helmboldt, C. F. (1969) Early changes in bursa of Fabricius from Marek's disease. *Avian Dis.*, **13**, 215-222

Jakowski, R. M., Fredrickson, T. N., Chomiak, T. W. & Luginbuhl, R. E. (1970) Hematopoietic destruction in Marek's disease. *Avian Dis.*, **14**, 374-385

Kenyon, A. J., Sevoian, M., Horwitz, M., Jones, N. D. & Helmboldt, C. F. (1969) Lymphoproliferative diseases of fowl—immunologic factors associated with passage of a lymphoblastic leukemia (JM-V). *Avian Dis.*, **13**, 585-595

Kottaridis, S. D. & Luginbuhl, R. E. (1968) Marek's disease III. Immunofluorescent studies. *Avian Dis.*, **12**, 383-393

Lapen, R. F., Piper, R. C., & Kenzy, S. G. (1970) Cutaneous changes associated with Marek's disease of chickens. *J. nat. Cancer Inst.*, **45**, 941-950

Law, L. W. (1967) Function of the thymus in tumor induction by viruses. *Perspect. Virol.*, **5**, 229-250

Lerche, F. & Fritzsche, K. (1934) Histopathologie und Diagnostik der Geflügellähme. *Z. Difekt.-kr. Haustiere*, **45**, 89-109

Marek, J. (1907) Multiple Nervenentzündung (Polyneuritis) bei Hühnern. *Dtsch. tierärztl. Wschr.*, **15**, 417-421

Marshall, A. H. E. (1956) *An outline of the cytology and pathology of the reticular tissue*, Edinburgh, Oliver & Boyd

Miller, J. F. A. P. & Osoba, D. (1967) Current concepts of the immunological function of the thymus. *Physiol. Rev.*, **47**, 437-520

Mizell, M., Stackpole, C. W. & Halperen, S. (1968) Herpes-type virus recovery from "virus-free" frog kidney tumors. *Proc. Soc. exp. Biol. (N.Y.)*, **127**, 808-814

Moore, R. Y. & Oda, Y. (1962) Malignant lymphoma with diffuse involvement of the peripheral nervous system. *Neurology (Minneap.)*, **12**, 186-192

Morris, J. R., Jerome, F. N. & Reinhart, B. S. (1969) Surgical bursectomy and the incidence of Marek's disease (MD) in domestic chickens. *Poult. Sci.*, **48**, 1513-1515

Nazerian, K., Solomon, J. J., Witter, R. L. & Burmester, B. R. (1968) Studies on the etiology of Marek's disease. II. Finding of a herpes virus in cell culture. *Proc. Soc. exp. Biol. (N.Y.)*, **127**, 177-182

Nazerian, K. & Witter, R. L. (1970) Cell-free transmission and *in vivo* replication of Marek's disease virus. *J. Virol.*, **5**, 388-397

Pappenheimer, A. M., Dunn, L. C., & Cone, V. (1926) A study of fowl paralysis (Neurolymphomatosis Gallinarum). *Storrs agric. exp. Stat. Bull.*, No. 143, pp. 186-290

Payne, L. N. & Biggs, P. M. (1967) Studies on Marek's disease. 2. Pathogenesis. *J. nat. Cancer Inst.*, **39**, 281-302

Payne, L. N. (1970) Immunosuppressive effects of avian oncogenic viruses. *Proc. roy. Soc. Med.*, **63**, 16-19

Payne, L. N. & Rennie, M. (1970a) Lack of effect of bursectomy on Marek's disease. *J. nat. Cancer Inst.*, **45**, 387-397

Payne, L. N. & Rennie, M. (1970b) Presence of natural haemagglutinins to sheep erythrocytes in sera from chickens free from Marek's disease. *Vet. Rec.*, **87**, 109-110

Petek, M. & Quaglio, G. L. (1967) Experimental allergic neuritis in the chicken. *Path. vet.*, **4**, 464-476

Peterson, R. D. A., Purchase, H. G., Burmester, B. R., Cooper, M. D. & Good, R. A. (1966) Relationships among visceral lymphomatosis, bursa of Fabricius, and bursa-dependent lymphoid tissue of the chicken. *J. nat. Cancer Inst.*, **36**, 585-598

Purchase, H. G. & Biggs, P. M. (1967) Characterization of five isolates of Marek's disease. *Res. Vet. Sci.*, **8**, 440-449

Purchase, H. G., Chubb, R. C. & Biggs, P. M. (1968) Effect of lymphoid leukosis and Marek's disease on the immunological responsiveness of the chicken. *J. nat. Cancer Inst.*, **40**, 583-592

Purchase, H. G. (1970) Virus-specific immunofluorescent and precipitin antigens and cell-free virus in the tissues of birds infected with Marek's disease. *Cancer Res.*, **30**, 1898-1908

Roizman, B. (1965) An enquiry into the mechanisms of recurrent herpes infections of man. *Perspect. Virol.*, **4**, 283-301

Schidlovsky, G., Ahmed, M., & Jensen, K. E. (1969) Herpes virus in Marek's disease tumors. *Science,* **164**, 959-961

Sevoian, M., Chamberlain, D. M. & Counter, F. (1962) Avian lymphomatosis. Experimental reproduction of neural and visceral forms. *Vet. Med.*, **57**, 500-501

Sevoian, M. & Chamberlain, D. M. (1964) Avian lymphomatosis IV. Pathogenesis. *Avian Dis.*, **8**, 281-310

Sharma, J. M., Davis, W. C. & Kenzy, S. G. (1970) Etiologic relationship of skin tumors (skin leukosis) of chickens to Marek's disease. *J. nat. Cancer Inst.*, **44**, 901-912

Siller, W. G. (1960) Experimental allergic encephalomyelitis in the fowl. *J. Path. Bact.*, **80**, 43-53

Solomon, J. J., Witter, R. L., Nazerian, K. & Burmester, B. R. (1968) Studies on the etiology of Marek's disease. Propagation of the agent in cell culture. *Proc. Soc. exp. Biol. (N.Y.)*, **127**, 173-177

Spencer, J. L. & Calnek, B. W. (1970) Marek's disease: application of immunofluorescence for detection of antigen and antibody. *Amer. J. vet. Res.*, **31**, 345-358

Stjernswärd, J., Clifford, P. & Svedmyr, E. (1970) General and tumour-distinctive cellular immunological reactivity. In: Burkitt, D. P. & Wright, D. H., eds, *Burkitt's lymphoma,* Edinburgh & London, Livingstone, pp. 164-171

Ubertini, T. & Calnek, B. W. (1970) Marek's disease herpes virus in peripheral nerve lesions. *J. nat. Cancer Inst.*, **45**, 507-514

Vickers, J. H., Helmboldt, C. F. & Luginbuhl, R. E. (1967) Pathogenesis of Marek's disease (Connecticut A isolate). *Avian Dis.*, **11**, 531-545

Waksman, B. H. & Adams, R. D. (1955) Allergic neuritis: an experimental disease of rabbits induced by the injection of peripheral nervous tissue and adjuvants. *J. exp. Med.*, **102**, 213-235

Warner, N. L., Szenberg, A. & Burnet, F. M. (1962) The immunological role of different lymphoid organs in the chicken 1. Dissociation of immunological responsiveness. *Aust. J. exp. Biol. med. Sci.*, **40**, 373-388

Wight, P. A. L. (1962a) Variations in peripheral nerve histopathology in fowl paralysis. *J. comp. Path.*, **72**, 40-48

Wight, P. A. L. (1962b) The histopathology of the central nervous system in fowl paralysis. *J. comp. Path.*, **72**, 348-359

Wight, P. A. L. (1964) An analysis of axon number and calibre in sciatic nerves affected by fowl paralysis. *Res. Vet. Sci.*, **5**, 46-55

Wight, P. A. L. & Siller, W. G. (1965) Further studies of experimental allergic encephalomyelitis in the fowl IV. The suppression of the experimental lesions by a naturally-occurring neuritis. *Res. Vet. Sci.*, **6**, 324-329

Wight, P. A. L. (1969) The ultrastructure of sciatic nerves affected by fowl paralysis (Marek's disease). *J. comp. Path.*, **79**, 563-570

Witter, R. L., Solomon, T. T. & Burgoyne, G. H. (1969) Cell culture techniques for primary isolation of Marek's disease-associated herpes virus. *Avian Dis.*, **13**, 101-118

Zacharia, T. P. & Sevoian, M. (1970) Detection of agglutinins in chickens infected with JM leukosis virus. *Appl. Microbiol.*, **19**, 71-72

zur Hausen, H. & Schulte-Holthausen, H. (1970) Presence of EB virus nucleic acid homology in a "virus-free" line of Burkitt tumour cells. *Nature (Lond.)*, **227**, 245-248

Haematological Aspects of Marek's Disease

K. W. WASHBURN [1] & C. S. EIDSON [2]

Although lymphoproliferative lesions are the main criterion of infection in Marek's disease (MD), other pathological effects may also occur. Among these is the haematopoietic destruction in certain cases of MD (Vickers, Helmboldt & Luginbuhl, 1967; Jankowski *et al.*, 1970) resulting in a severe aplastic anaemia with up to 50% reduction in packed cell volume (PCV).

Genetic selection for resistance to MD is usually based on the absence of formation of the lymphoproliferative lesions, but little is known of the physiological mechanism through which the genetic resistance is mediated. Haematopoietic destruction of the extent reported by Vickers *et al.* (1967) and Jankowski *et al.* (1970) would have a profound effect on the physiological system, which could affect the capacity of the individual to respond to the invading organism. Jankowski *et al.* (1970) suggested that chicks without passive protection through maternal antibody are more prone to develop aplasia of the bone marrow. If a haematological response is involved in the development of MD, then genetic variability in the competence of this response could determine whether an individual showed symptoms of the disease. Thus variation in haematological response could in part be the variability on which genetic selection pressure for MD acts.

Coccidiosis, particularly that due to *Eimeria tenella,* causes severe intestinal bleeding sometimes resulting in a 50% reduction in PCV. If a bird that already has an anaemia due either to coccidiosis or MD became infected with the other disease the combined effects on the haematological system might be so severe that death would result from haematological failure.

MATERIALS AND METHODS

Two trials were conducted to determine the effect of the GA isolate of MD on the haematology of Leghorn and broiler chicks, and the interrelationship between coccidiosis caused by *Eimeria tenella* and MD in their effect on haematology. The stock used in Trial 1 were male Babcock Leghorns; in Trial 2, Cobb Broiler males were used in addition to the Leghorns. The birds were placed in Horsfall units at hatching and injected with 0.2 ml GA-MD infective plasma. In Trial 1, the birds in 4 Horsfall units located in one room were injected while the non-injected controls were housed in 4 units in a separate room. A similar procedure was adopted in Trial 2, except that 5 units in each of two houses were used. The following haematological parameters were measured: (i) PCV, determined by the microhaematocrit method at 0, 7, 14, 28, 42 and 49 days of age for Trial 1, and at 0, 7, 14, 21, 29 and 36 days of age for Trial 2; (ii) red blood cell (RBC) counts and mean cell volume (MCV), determined by use of a Model F Coulter Counter; and (iii) haemoglobin content, determined by the acid haematin method at 14 and 28 days for Trial 1, and 29 and 36 days for Trial 2.

At 42 and 29 days for Trials 1 and 2, respectively, the effects of coccidiosis on MD-injected birds were tested. In Trial 1, the MD-injected birds were redistributed on the basis of relatively high or relatively low PCV and all inoculated with 50 000 sporulated oocysts of *E. tenella* at 42 days of age. The non-MD-injected controls were similarly treated except that birds in one unit, all of which had intermediate PCV values, were not inoculated with *E. tenella.* In Trial 2, four units in each house were randomly redistributed and two of them in each house inoculated with 100 000 sporulated oocysts of *E. tenella* at 28 days of age. Lesion scores (Johnson & Reid, 1970) for severity of coccidiosis infection

[1] Poultry Department, University of Georgia, Athens, Georgia, USA.

[2] Poultry Disease Research Center, University of Georgia, Athens, Georgia, USA.

were made at 49 days for Trial 1 and 36 days for Trial 2. Birds were examined for MD lesions at the time the coccidiosis lesion was scored.

RESULTS

The data presented in Table 1 on the effects of injection with the GA-MD isolate on PCV show that MD caused a slight but significant reduction in the PCV of Leghorn chicks, which was first observed at 28 days, but no such reduction was seen in broilers up to 35 days. There was an insignificant reduction in broilers at 28 days, but at 35 days there was no difference. Of equal interest was the effect of MD on the variability of PCV. The coefficients of variation of PCV for the birds in Trial 1 at 42 days were

10.8 for the MD-injected and 6.5 for the controls. The range of PCV values was 18.7–36.1 for the MD-injected and 27.3–35.1 for the control birds. The changes in PCV were due both to changes in RBC numbers (2.67 million cells/mm³ for the controls and 2.60 million/mm³ for the injected) and in cell size (120 μ^3 for the controls and 116 μ^3 for the MD-injected). These data indicate that the injection of the GA-MD isolate caused a slight but significant depression of PCV in some strains of chickens but not in others. In this study, 53% of the birds in Trial 1 were positive for MD by 49 days and in Trial 2, 23% were positive by 36 days. A more severe depression (50% reduction in PCV) in other strains of chickens injected with other isolates of MD virus has been reported (Jankowski *et al.*, 1970).

Table 1. Effect of **GA-MD** injection on packed red blood cell volume of Leghorn and broiler chicks

Chicks	PCV (%) at days after MD injection				
	0	14	28	35	42
Leghorns T1					
Controls	28.9	32.5	31.6	—	33.3
MD-inj.	29.3	31.8	30.7	—	31.3
Broilers T2					
Controls	29.4	31.9	31.0	30.9	—
MD-inj.	29.1	33.0	30.0	31.8	—
Leghorns T2					
Controls	36.1	35.9	32.2	35.0	—
MD-inj.	36.1	36.2	30.3	31.3	—

Although the reduction in the mean PCV was slight, there was a marked reduction in certain individuals. It was therefore of interest to study the effects of a simultaneous outbreak of coccidiosis, caused by species such as *E. tenella*, which causes extensive bleeding. Table 2 shows the effect of coccidiosis on the PCV of MD-injected and control chicks. The PCV of the MD-injected Leghorn chicks used in Trial 1 fell by 24% while that of the controls fell by 40%. It seems clear that the presence of severe MD does not predispose Leghorn chicks to a more severe haematological effect due to coccidiosis. There were also no differences in the coccidiosis lesion scores of the MD-injected and control groups (3.1 for both). The situation is not quite so clear in the broiler stock. In both the MD-injected and control groups that were inoculated with *E. tenella,*

there was a 3% decrease in PCV. However, the values for the MD-injected group which had not been inoculated with *E. tenella,* increased by 14% while those of controls, which did not receive either MD or *E. tenella,* increased only by 2%. If this is taken into account, the fall in the PCV values was more marked in the group inoculated with MD as well as *E. tenella* than in the group inoculated with *E. tenella* alone. In addition, the mean lesion score of the MD-injected (2.5) was higher than for the controls (1.4); this would indicate that the effect was more severe. On examination of the haematological parameters of these birds (Table 3), it is evident that the decrease in PCV is due to decreased cell numbers and that the same trend in the differences between MD-injected and controls can be seen for both RBC and PCV.

Table 2. Effects of coccidiosis on PCV (%)
of MD-injected and control chicks

Chicks	MD-injected	Controls
	Trial 1 (Leghorns)	
Cocc. inoc.	31.5–24.0 (−24%)	32.5–19.5 (−40%)
Controls		33.0–35.0 (+ 6%)
		−46%
	Trial 2 (Broilers)	
Cocc. inoc.	29.3–28.4 (− 3%)	30.6–29.5 (− 3%)
Controls	30.1–34.2 (+14%)	31.8–32.4 (+ 2%)
	−17%	− 5%

Table 3. Changes in haematological parameters of broiler
chicks inoculated with coccidiosis (Trial 2)

Chicks	MD-injected	Control
	RBC counts (× 10 000/mm³)	
Cocc. inoc.	252–220 (−13%)	261–225 (−14%)
Control	258–269 (+ 4%)	269–248 (− 8%)
	−17%	− 6%
	Haemoglobin (gm/100 cm³ blood)	
Cocc. inoc.	7.6–5.8 (−31%)	8.1–5.7 (−30%)
Control	7.9–7.0 (−11%)	8.5–6.2 (−27%)
	−13%	− 3%
	Mean cell volume (μ³)	
Cocc. inoc.	117–129 (+10%)	118–132 (+12%)
Control	117–127 (+ 9%)	118–132 (+12%)
	− 1%	0%

An important question was whether birds that already had a lowered PCV value due to MD would be more severely affected by coccidiosis. The data in Table 4 indicate that this was not the case. For the birds in Trial 1 with a relatively high PCV, the decrease was 21% as compared to 27% for those with a relatively low PCV. For noninjected controls, the decrease was 44% for individuals with a relatively high mean PCV, and 34% for those with low PCV values. If the mean PCV decreases of all high-PCV birds and all low-PCV birds are compared, they are seen to be almost identical.

Table 4. Relationship between severity of anaemia
and coccidiosis effect in Leghorns (Trial 2)

Birds with:	PCV at days after inoculation			Change in PCV
	0	5	7	
	MD-injected			
High PCV	34	35	27	−21%
Low PCV	29	30	21	−27%
	MD control			
High PCV	36	30	20	−44%
Low PCV	29	27	19	−34%
	MD cocc. control			
Med. PCV	33	34	35	+ 6%

SUMMARY

A slight anaemia was observed in Leghorn chicks but not in broiler stock injected with GA-MD infective plasma; severe anaemia was not found in any birds. MD-injected birds were not more severely affected by coccidiosis than controls, and the slight anaemia observed in some birds did not cause a more severe haematological reaction to coccidiosis. The results of this study indicate that genetic differences in resistance to MD do not find expression through haematological response, and that the relationship between outbreaks of MD and coccidiosis is not due to differences in haematological response.

REFERENCES

Jankowski, R. M., Fredrickson, T. N., Chomiak, T. W. & Luginbuhl, R. E. (1970) Haematopoietic destruction in Marek's Disease. *Avian Dis.*, **14**, 374-385

Johnson, J., & Reid, W. M. (1970) Anticoccidial drugs: Lesion scoring techniques in battery and floor-pen experiments with chickens. *Exp. Path.*, **28**, 30-36

Vickers, J. H., Helmboldt, C. F., & Luginbuhl, R. E. (1967) Pathogenesis of Marek's Disease (Connecticut A isolate). *Avian Dis.*, **11**, 531-545

The Effect of Antilymphocytic Globulin on the Clinical Manifestations of Marek's Disease

[D. D. KING,[1] R. W. LOAN [2] & D. P. ANDERSON [3]

Antilymphocytic globulin has been shown to be an immunosuppressant and lymphocyte depleting agent (Levey & Medawar, 1966a, b, 1967; Lawrence, Barnett & Craddock, 1968; Bach, Brashler & Perper, 1970; Lance, 1970). The approach in this study was based on the theory that the manifestations of Marek's disease as a herpesvirus infection, are at least partially dependent on replication of a "target cell" within the host. Even though the work described below does not prove this theory, it is hoped that the information obtained will be of use in further investigations of the immune response of the host to oncogenic virus infections.

MATERIALS AND METHODS

Preparation of antilymphocyte globulin (ALG)

Spleens and thymuses from chicks free from Marek's disease were harvested, and suspensions of single live, intact cells prepared by passing the tissue several times through wire mesh screens. Percentages of live and dead cells were determined by 1% Trypan Blue staining. All cell preparations were shown to contain at least 95% live cells a few minutes prior to inoculation. The recipient animal was injected with antihistamine a few minutes prior to inoculation of the cell preparations. Approximately four billion thymocytes and splenocytes were injected intravenously into a 650-lb mare. This pro-

cedure was repeated at intervals of seven to ten days to give a total of three injections. The mare was bled out, terminally, ten days after the third injection. The serum was harvested and precipitated three times in 33% saturated ammonium sulphate solution. The globulin was dialyzed against physiological saline to remove the ammonium sulphate. The globulin was then heat inactivated (at 56°C for 30 minutes) and adsorbed so as to remove haemolysins by mixing two volumes of globulin with one volume of packed chicken red blood cells at room temperature for two hours.

Skin-graft potency testing of ALG

Allografts from wattles of one strain of chick (SPAFAS) to shanks of another strain (Cobb) were performed by the methods of Purchase (1967). It was found that 1.0 cc ALG inoculated at day –1, +1 and +3 delayed skin-graft rejection for twenty days, as compared with six days for the control skin-grafted group.

Bird inoculation and tissue culture studies

Five groups of twenty birds were used. Group A was injected with JM inoculum at one day of age. Group B was reared free of Marek's disease virus MDV for two weeks, then exposed to birds from Group A. Group C was reared in isolation throughout the entire study with no exposure to MDV. Group D was reared free of MDV for two weeks, then exposed to birds from Group A. Group E was reared in isolation throughout the entire study with no exposure to MDV.

All birds from groups D and E received 1.0 cc ALG subcutaneously, starting at two weeks of age. Thirty injections were made on alternate days. From two weeks of age onwards, and at weekly intervals there-

[1] Animal Health Division, Agricultural Research Service, US Department of Agriculture, Poultry Disease Research Center, College of Veterinary Medicine, University of Georgia, Athens, Georgia, USA.

[2] Department of Microbiology, School of Veterinary Medicine, University of Missouri, Columbia, Missouri, USA.

[3] Department of Avian Medicine, College of Veterinary Medicine, University of Georgia, Athens, Georgia, USA.

after, two birds were removed from each group. Kidney cells (10^6) were collected from each pair of birds and inoculated on to primary chick kidney monolayers grown in minimum essential medium (MEM) by the methods of Calnek & Madin (1969). Virus recovered was expressed as PFU per 10^6 kidney cells inoculated on to the chicken kidney (CK) primary monolayers.

Gross lesions and histopathology

All birds were weighed at time of sacrifice. Gross lesions were recorded and tissues were taken from the brain, heart, liver, kidney, bursa of Fabricius, skin, sciatic nerve and spleen of each bird sacrificed. The histological lesions were scored on a 1 to 4+ scale.

White blood cell counts

After twelve injections on alternate days, five pullets from each group were bled by heart puncture. Total white blood counts were made by the indirect method of Lucas & Jamroz (1961).

RESULTS

Virus recovery, expressed as PFU/10^6 cells inoculated from the infected birds treated with ALG, was somewhat greater early in the study. Later in the study, by the eighth week, virus recovery was reduced from the ALG-treated group and much increased from the untreated-infected group (Fig. 1).

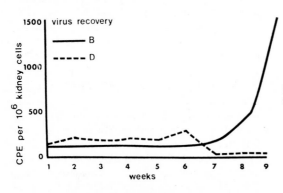

Fig. 1. Virus recovery from JM infected birds (B) and JM infected-ALG treated birds (D).

Gross pathological changes were not observed until the ninth week of age. During the ninth, tenth and eleventh weeks of age, none of the ten birds autopsied in the infected-treated group showed any

gross changes. All the ten birds from the infected-untreated group had gross lesions of Marek's disease.

Histologically, it was noted that the infected-treated group had developed microscopic lesions but that regression of these lesions had occurred (Figs. 2, 3, 4, 5 and 6).

White blood cell (WBC) counts for the five pullets in each of the untreated groups averaged 28 000 per mm³. The ALG-treated groups had an average WBC count of 7500 per mm³.

Gross body weights of the infected-treated groups was 10% greater than that of the infected-untreated group at nine weeks of age.

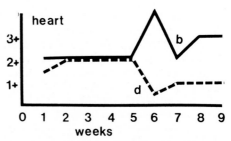

Fig. 2. Microscopic lesion scores of JM infected birds (b) and JM infected-ALG treated birds (d).

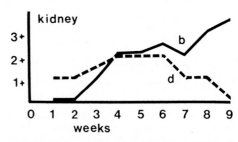

Fig. 3. Microscopic lesion scores of JM infected birds (b) and JM infected-ALG treated birds (d).

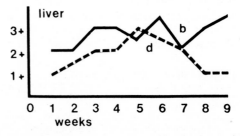

Fig. 4. Microscopic lesion scores of JM infected birds (b) and JM infected-ALG treated birds (d).

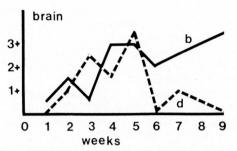

Fig. 5. Microscopic lesion scores of JM infected birds (b) and JM infected-ALG treated birds (d).

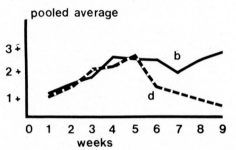

Fig. 6. Microscopic lesion scores of heart, liver, brain and kidney of JM infected birds (b) and JM infected-ALG treated birds (d).

A second trial was performed in which the GA-isolate of MDV was used to infect Athens-Canadian chicks. This trial also incorporated additional bird groups injected with normal horse globulin (NHG). Only fifteen alternate-day injections were performed in all groups in this trial. The results obtained were similar to those in the first trial. No immunosuppression was produced by the NHG, as determined by WBC counts. There was no change, in addition, in Marek's disease lesions in the NHG inoculated-Marek's disease infected group.

DISCUSSION

ALG has been shown to delay skin-graft rejection. This is thought to be the result of inactivation of the circulating small lymphocytes and of most forms of the primary immune response (Strober & Gowans, 1965; Ford & Gowans, 1967). Cellular immunity was shown to be selectively depressed by antilymphocytic-antibody (Lance & Batchelor, 1968). The present study provides evidence that the lympho-proliferative lesions of Marek's disease are dependent upon the immune mechanisms of the host, which may be inactivated by antilymphocytic globulin.

SUMMARY

An investigation of the effect of the prolonged administration of antilymphocytic globulin on the clinical manifestations of the oncogenic virus of Marek's disease was carried out. Antilymphocytic serum was produced in the horse. The globulin fraction was tested for potency by determining its ability to delay skin-graft rejection, and found to possess some activity. Birds from each of the test groups were sacrificed weekly. Blood and kidneys were harvested for use in attempts to isolate the JM virus from tissue cultures. Various organs and tissues were collected for histological study. Virus levels tended to be somewhat greater early in the administration of ALG to the exposed group, as compared with the noninjected exposed group. However, this tendency was reversed later in the study, and virus isolation was significantly higher in the non-ALG inoculated group. During the last five weeks of the study, all the birds sacrificed that had not received ALG showed gross lesions of Marek's disease. None of the ALG-treated birds developed gross lesions.

In a second trial, the GA-isolate of MDV was used to infect Athens-Canadian chicks. Two additional groups were inoculated with normal horse globulin (NHG), one being infected, the other not. The NHG did not effect any change in lesion patterns in the infected group. Other results were similar to those of the first trial.

REFERENCES

Bach, M. K., Brashler, J. R. & Perper, R. J. (1970) An in vitro correlative assay for the immunosuppressive activity of horse antirat lymphocyte sera: Estimation of lymphocytophilic antibody activity using 51 Cr-labeled thymocytes. J. Immunol., 105, 746-754

Calnek, B. W. & Madin, S. H. (1969) Characteristics of in vitro infection of chicken kidney cell cultures with a herpesvirus from Marek's disease. Amer. J. vet. Res., 30, 1389-1402

Ford, W. L. & Gowans, J. L. (1967) The role of lymphocytes in antibody formation. II. The influence of lymphocyte migration on the initiation of antibody formation in the isolated, perfused spleen. Proc. roy. Soc. B, 168, 244

Lance, E. M. (1970) The effect of heterologous anti-lymphocyte serum (ALS) on the humoral antibody response to Salmonella typhi "H" antigen and bovine serum albumin. *J. Immunol.,* **105**, 108-117

Lance, E. M. & Batchelor, J. R. (1968) Selective suppression of cellular immunity by anti-lymphocyte serum. *Transplantation,* **6**, 490

Lawrence, J. H., Barnett, E. V. & Craddock, C. G. (1968) Antisera to neutrophils, lymph node cells and thymus cells of the guinea pig: haematological effects and antibody localization. *Transplantation,* **6**, 70

Levey, R. H. & Medawar, P. B. (1966a) Nature and mode of action of anti-lymphocyte antiserum. *Proc. nat. Acad. Sci. (Wash.),* **56**, 1130

Levey, R. H. & Medawar, P. F. (1966b) Some experiments on the action of anti-lymphoid antisera. *Ann. N.Y. Acad. Sci.,* **129**, 164

Levey, R. H. & Medawar, P. F. (1967) Further experiments on the action of anti-lymphocytic antiserum. *Proc. nat. Acad. Sci. (Wash.),* **58**, 470

Lucas, A. M. & Jamroz, C. (1961) *Atlas of avian hematology.* Agriculture Monograph 25, USDA, pp. 232-234

Purchase, H. G. (1967) A method for multiple skin grafting. *Poult. Sci.,* **46**, 1017-1019

Strober, S. & Gowans, J. L. (1965) The role of lymphocytes in the sensitization of rats to renal homografts. *J. exp. Med.,* **122**, 347

Immunosuppressive Effects of Infection of Chickens with Marek's Disease Herpesvirus

S. H. KLEVEN [1], C. S. EIDSON [1] & D. P. ANDERSON [1]

Several viruses, including the leukaemia viruses of mice and the lymphoid leukosis virus of chickens, have been shown to depress antibody production in the host (Notkins, Mergenhagen & Howard, 1970). Purchase, Chubb & Biggs (1968) demonstrated depression of production of antibody against bovine serum albumin in chickens inoculated with Marek's disease herpesvirus (MDHV) at 1 day of age. The three experiments to be described below were undertaken to determine whether infection with MDHV would cause depression of antibody production against *Mycoplasma synoviae* (MS), thereby making serological diagnosis of MS infection more difficult.

In experiment 1, 10 one-day-old chicks were placed in each of 4 modified Horsfall Bauer units supplied with filtered air under positive pressure. The chicks were progeny of a flock free of MD. The experimental groups consisted of one infected with MS, one infected with MDHV, and one infected with both MD and MS; uninoculated controls formed a fourth group. Chicks were exposed to MDHV by inoculation of 0.2 ml of whole blood from donor chickens infected with the GA isolate of MDHV (Eidson & Schmittle, 1968) at 1 day of age. MS infection was accomplished by aerosol exposure to a 24-hour broth culture of MS in Frey's medium (Frey, Hanson & Anderson, 1968) at 21 days of age. At six weeks of age all survivors were bled and necropsied. Antibodies to MS were measured by the serum plate agglutination test (Olson, Kerr & Campbell, 1963) and the haemagglutination inhibition (HI) test (Vardaman & Yoder, 1969).

In experiment 2, the experimental design was the same as that for experiment 1.

All birds in the groups infected with MDHV showed lesions of MD at the time of death or at the

termination of the experiment. Birds infected with MS in the absence of MDHV showed antibody against MS (15 out of 16 birds in both trials). However, none of the 6 surviving MDHV-infected birds had detectable antibody against MS 3 weeks after aerosol exposure (see Table 1).

Table 1. Serological response [1] to *M. synoviae* [2] after Infection with MDHV [3]. Experiments 1 and 2

Experiment No.	Inoculum			
	None	MD	MS	MD+MS
1	0/9	0/3	7/8	0/3
2	0/8	0/2	8/8	0/3

[1] Number of birds serologically positive for MS up to six weeks of age/number tested.
[2] Chicks infected with MS by aerosol at three weeks of age.
[3] Chicks infected with 0.2 ml of infective whole blood subcutaneously at one day of age.

In experiment 3, 100 day-old chicks were placed in each of 5 isolation rooms. The experimental groups consisted of one group inoculated with MD, one with MS, one with MD+MS, one with turkey herpesvirus (HVT)+MS, and one uninoculated group. Pens receiving MDHV were infected at 1 day of age by inoculating 25 chicks with 0.2 ml of MDHV-infected whole blood, the remaining 75 being left to be infected by contact. Birds receiving HVT were vaccinated at 1 day of age (Eidson & Anderson, 1971). Those infected with MS were exposed by aerosol at 16 days of age. All birds were bled and tested for antibody to MS by the serum plate agglutination and HI tests at 21, 38, and 60 days after exposure to MS.

Table 2 shows the geometric mean HI titres to MS at 21, 38 and 60 days after exposure. As will be

---

[1] Poultry Disease Research Center, University of Georgia, Athens, Georgia, USA.

Table 2. Serological response [1] to *M. synoviae* [2] after infection with MDHV [3] or HVT [4]. Experiment 3

Inoculum	No. of days after exposure to MS		
	21	38	60
MS	190	367	478
HVT+MS	135	274	735
MD+MS	23 [5]	67 [5]	181 [5]
MD	0	0	0
None	0	0	0

[1] Geometric mean HI titre against MS.
[2] Infected by MS by aerosol exposure at 16 days of age.
[3] Twenty-five birds infected with 0.2 ml of whole blood subcutaneously at one day of age, 75 birds exposed by contact.
[4] Vaccinated with HVT at 1 day of age.
[5] Titre significantly less than groups infected with MS only or with HVT+MS (P < 0.001).

seen from the table, birds infected with MDHV had significantly lower titres (P <0.001) than birds not infected with MDHV or birds vaccinated with HVT. Geometric mean MS titres in the MS and the HVT+ MS groups did not differ significantly. However, at 21 and 38 days after MS exposure, HVT appeared slightly to depress antibody response to MS. It is interesting to note that, at 60 days post exposure, the group infected with MS only was found to be accidently infected with MDHV. At that time this group had a lower geometric mean titre than the HVT+MS group. This indicates that MDHV was responsible for depressing MS titres after accidental exposure between 38 and 60 days after infection with MS. It was also noted in experiment 3 that immunodepression occurred in birds exposed by contact to MDHV as well as in those exposed by the parenteral route at 1 day of age. In fact, depression of antibody response was almost as severe in birds exposed by contact as in birds inoculated parenterally. In addition, birds exposed to MDHV between 54 and

76 days of age also appeared to show some inhibition of antibody response.

Although vaccination with HVT did not significantly depress antibody production against MS, titres in HVT-vaccinated birds were somewhat lower than those in unvaccinated birds. Experiments are now in progress to determine whether HVT also has immunodepressive activity or whether HVT vaccination may eliminate the immunodepression due to MDHV. Additional studies are also needed to determine whether MDHV can influence the serological diagnosis of MS infection under field conditions.

The mechanism involved in immunodepression after infection with MDHV has not been determined. MDHV has been reported to cause a slight increase in gamma globulin levels (Washburn & Eidson, 1970). MDHV also may cause atrophy and necrosis in the bursa of Fabricius (Purchase *et al.*, 1968; Fletcher, 1970[1]). However, no bursal lesions have been detected in birds vaccinated with HVT (Fletcher, 1970[1]). It would appear, therefore, as suggested by Purchase *et al.* (1968), that although gamma globulin production is unaffected, some deficiency exists in the ability to commit cells to produce antibody to a particular antigen. Examination of individual birds in these experiments failed to demonstrate any correlation between mortality or lesions due to MD with antibody titre to MS. However, since no histological examination of the bursas was performed, no attempt to find a correlation between antibody titre and bursal lesions could be made.

Although respiratory lesions were observed in several MS-infected birds, neither MDHV or HVT influenced the development of air sac lesions. In addition, the recovery of MS from tracheal cultures was not influenced by the presence of either herpesvirus.

[1] Unpublished data.

SUMMARY

Chickens infected with Marek's disease herpesvirus (MDHV) at 1 day of age failed to synthesize antibody against *Mycoplasma synoviae* (MS) after infection with MS at 1 week of age. At 4 weeks of age, 15 out of 16 MDHV-free birds had antibody against MS, while none of 6 MD-infected birds was positive.

Birds infected with MDHV by contact at 1 day of age produced antibody against MS after aerosol exposure at 16 days of age, but antibody titres were significantly lower than those produced by MDHV-free birds. Birds vaccinated with turkey herpesvirus (HVT) at 1 day of age did not produce antibody to MS as efficiently as unvaccinated birds, but the difference was not statistically significant.

REFERENCES

Eidson, C. S., & Anderson, D. P. (1971) Immunization against Marek's disease. *Avian Dis.,* **15**, 68-81

Eidson, C. S., & Schmittle, S. C. (1968) Studies on acute Marek's disease I. Characteristics of isolate GA in chickens. *Avian Dis.,* **12**, 467-476

Frey, M. L., Hanson, R. P., & Anderson, D. P. (1968) A medium for isolation of avian mycoplasmas. *Amer. J. vet. Res.,* **29**, 2163-2171

Notkins, A. L., Mergenhagen, S. E., & Howard, R. J. (1970) Effect of virus infections on the function of the immune system. *Annu. Rev. Microbiol.,* **24**, 525-538

Olson, N. O., Kerr, K. M., & Campbell, A. (1963) Control of infectious synovitis. 12. Preparation of an agglutination test antigen. *Avian Dis.,* **7**, 310-317

Purchase, H. G., Chubb, R. C., & Biggs, P. M. (1968). Effect of lymphoid leukosis and Marek's disease on the immunological responsiveness of the chicken. *J. nat. Cancer Inst.,* **40**, 583-592

Vardaman, T. H., & Yoder, H. W. (1969) Preparation of mycoplasma synoviae hemagglutinating antigen and its use in the hemagglutination-inhibition test. *Avian Dis.,* **13**, 654-661

Washburn, K. W., & Eidson, C. S. (1970) Changes in concentration of plasma proteins associated with Marek's disease. *Poult. Sci.,* **49**, 784-793

Plasma Membranes of Chicken Lymphoid Cells and Marek's Disease Tumour Cells *

W. L. RAGLAND & J. L. PACE

Although it has been generally conceded that Marek's disease (MD) is a malignant neoplastic disease, conclusive proof is still lacking as it has not been possible to demonstrate clear evidence of metastasis, the hallmark of malignancy. MD could perhaps be an immunoproliferative, non-neoplastic disorder, and it is interesting to note that it was originally considered to be a polyneuritis (Marek, 1907), a view that went unchallenged for several decades (Pappenheimer, Dunn & Cone, 1929). Since recent studies (Sheinin et al., 1971) have revealed striking changes in the chemical composition of the plasma membranes of malignant transformed cells, we have naively assumed that similar changes in the plasma membranes of MD tumour cells would be evidence that MD is indeed a malignant neoplastic disorder.

Plasma membranes were prepared from circulating lymphocytes of 6- to 12-week-old Cornell White Leghorns by means of a procedure similar to that used by Perdue & Sneider (1970) for chick embryo fibroblasts. The chicks were free of lymphoid leukosis and MD and were supplied with filtered air under positive pressure. Blood was obtained by cardiac puncture and mixed with an equal volume of Seligmann's balanced salt solution (Graber, Seligmann & Bernard, 1955) containing EDTA.

Lymphocytes were separated centrifugally from the blood by the Ficoll flotation method, stage I, of Noble, Cutts & Carroll (1968) and collected by means of a Beckman tube slicer. The lymphocytes were washed twice in 0.16 M NaCl at 500 g_{max} for 20 minutes at 4°C. The lymphocytes were resuspended in 0.16 M NaCl and homogenized by 200-400 strokes in a Potter-Elvejhem homogenizer (clearance = 0.002 inch). The cells were examined periodically

during homogenization by phase microscopy, and homogenization discontinued when plasma membranes had been removed from the maximum number of cells (usually 70–80%). The membranes and other cell organelles were sedimented at 200 000 g_{max} for 30 minutes and the soluble fraction discarded. The pellets were resuspended in 0.25 M sucrose and the nuclei, mitochondria and cell debris sedimented at 215 g_{max} for 30 minutes. The supernatant solutions were carefully aspirated and the plasma membranes sedimented at 200 000 g_{max} for 30 minutes. The pellets were resuspended in a small volume of 10% sucrose (all percentages are w/w) and layered on top of discontinuous sucrose gradients composed of 3 ml of 20% sucrose over 4 ml of 35% sucrose over 4 ml of 46% sucrose over 1.5 ml of 55% sucrose. The gradients were spun at 90 000 g_{max} for 12 to 18 hours in a Beckman SW 40 rotor and the bands collected by means of a tube slicer. The fractions were identified by a subscript corresponding to the sucrose solutions on which they floated, e.g., F_{20} was the fraction floating on 20% sucrose. The various fractions were sedimented at 300 000 g_{max} for 90 minutes, resuspended in a small volume of 0.16 M NaCl and subjected to biochemical analysis.

Plasma membranes were prepared from tumour tissue in the following manner. Cornell White Leghorn chicks free of lymphoid leukosis and MD were infected with the GA isolate (Eidson & Schmittle, 1968) of MD herpesvirus by continuous air exposure beginning at one day of age. Moribund birds were killed and tumour tissue collected from large discrete nodules in liver, kidney and breast muscle. Care was taken to eliminate both host organ tissue and haemorrhagic or necrotic tissue from the samples, as confirmed by histological examination. The tissue was homogenized in a Potter-Elvejhem homogenizer and monitored by phase microscopy until the maximum number of cells had been stripped of plasma

* From the University of Georgia, College of Veterinary Medicine and Institute of Comparative Medicine, Athens, Georgia, USA.

membrane. The membranes were prepared as already described.

Biochemical analyses included determination of protein by the method of Lowry et al., (1951) and of sialic acid by the Warren procedure as modified by Kraemer (1966). Phospholipid and cholesterol were extracted with Folch's reagent (Folch, Lees & Sloane-Stanley, 1957). Lipid phosphorus was determined by the method of Baginski, Foá & Zak (1967) and cholesterol by the method of Glick, Fell & Sjølin (1964). Fresh material was used for assay of cytidine triphosphate phosphohydrolase (CTPase) (EC 3.6.1.4) [1] (Perdue & Sneider, 1970) and nicotinamide adenine dinucleotide (NADH)-cytochrome C reductase (EC 1.6.2.1) [1] (Mackler & Green, 1956) activities.

The activity of CTPase and the cholesterol, phospholipid (PLP) and sialic acid content of the four fractions of lymphocyte and MD cell membranes are given in Tables 1 and 2. Fractions F_{20}, F_{35} and F_{46} were relatively rich in CTPase activity, which was indicative of plasma membrane. No activity for two of the microsomal enzyme markers, glucose-6-phosphatase (Harper, 1963) and esterase (Guiditta & Strecker, 1959), was detected in the cell homogenate

and hence these markers could not be used to indicate contamination of the fractions by microsomal membranes. Nicotinamide adenine dinucleotide phosphate (NADH)-cytochrome C reductase activity was present in the cell homogenate but only a small amount of activity was found in the F_{46} fraction. Activities of the marker enzymes of other cell organelles were not determined but it seems unlikely that there would be significant contamination of F_{20} or F_{35}, and probably of F_{46} (Chandrasekhara & Narayan, 1971).

The fractions F_{20}, F_{35} and F_{46} were relatively rich in lipids. The cholesterol to phospholipid molar ratios of F_{20} and F_{35} were about 0.6–0.7, i.e., close to the values found for the light and heavy plasma membrane fractions of chick embryo fibroblasts (Perdue & Sneider, 1970). It was interesting that, even though the tumour F_{20} and F_{35} fractions contained less cholesterol and phospholipid than the normal fractions, the molar ratios remained essentially unchanged. Sialic acid content was less than reported for chick embryo fibroblasts or pig lymphocytes but similar to that for mouse 3T3 fibroblasts (Sheinin et al., 1971).

The F_{20} fraction had a density of 1.09 or less, which was similar to that of the light plasma membrane fraction of chick embryo fibroblasts (Perdue & Snei-

---

[1] Enzyme Commission reference number.

Table 1. Biochemical analysis of chicken lymphocyte membranes

Fraction	CTPase [1]	PLP [2]	Chol [2]	Chol/PLP molar ratio	Sialic acid [3]
F_{20}	8.3	717	215	0.59	13.4
F_{35}	17.2	580	204	0.70	7.6
F_{46}	10.6	274	80	0.57	7.5
F_{55}	3.7	233	63	0.53	7.2

[1] μ moles Pi released/30 min/mg protein.
[2] μg/mg protein.
[3] n moles/mg protein.

Table 2. Analysis of Marek's disease tumour cell membranes

Fraction	CTPase [1]	PLP [2]	Chol [2]	Chol/PLP molar ratio	Sialic acid [3]
F_{20}	7.0	349	100	0.55	10.6
F_{35}	15.4	420	146	0.68	12.9
F_{46}	9.4	328	76	0.45	4.1
F_{55}	4.6	196	43	0.44	3.8

[1] μ moles Pi released/30 min/mg protein.
[2] μg/mg protein.
[3] n moles/mg protein.

der, 1970). The density of the F_{35} fraction was greater than 1.09 but less than 1.16, which was similar to that of the heavy plasma membrane fraction of Perdue & Sneider (1970), and the plasma membranes prepared from pig lymphocytes (Allan & Crumpton, 1970).

Several differences were found between lymphocyte and tumour membranes. The cholesterol and phospholipid contents of F_{20} and F_{35} were lower in tumour membranes than in normal lymphocyte membranes. The sialic acid content of F_{20} was lower in tumour than in lymphocyte, while the sialic acid content of F_{35} was greater in tumour than in lymphocyte.

Additional evidence must be obtained before it can be concluded that MD plasma membranes are transformed membranes. We already have tentative evidence, not presented here, that tumour membranes lack a polypeptide present in normal lymphocyte membranes. Analysis of the gangliosides will be particularly interesting as the ganglioside patterns are considerably simplified in virus-transformed cells as compared with normal cells (Hakamori & Murakami, 1968; Mora et al., 1969; Sheinin et al., 1971).

REFERENCES

Allan, D., & Crumpton, M. J. (1970) Preparation and characterization of the plasma membrane of pig lymphocytes. Biochem. J., 120, 133-143

Baginski, E. S., Foá, P. P., and Zak, B. (1967) Microdetermination of inorganic phosphate, phospholipids and total phosphate in biologic materials. Clin. Chem., 13, 326-332

Chandrasekhara, N. & Narayan, K. A. (1970) A study of the contamination of rat liver plasma membranes with other subcellular components. Life Sci. 9, 1327-1334

Eidson, C. S., & Schmittle, S. C. (1968) Studies on acute Marek's disease. I. Characteristics of isolate GA in chickens. Avian Dis., 12, 467-476

Folch, J., Lees, M., & Sloane-Stanley, G. H. (1957) Simple method for the isolation and purification of total lipids from animal tissues. J. biol. Chem., 226, 497-509

Glick, D., Fell, B. F., & Sjølin, K.-E. (1964) Spectrophotometric determination of nanogram amounts of total cholesterol in microgram quantities of tissue or microliter volumes of serum. Analyt. Chem., 36, 1119-1121

Graber, P., Seligmann, M., & Bernard, J. (1955) Méthodes de préparation d'extraits leucocytaires et de sérums anti-leucocytaires susceptibles d'être utilisés pour des études immunochimiques. Annls Inst. Pasteur, Paris, 88, 548

Guiditta, A., & Strecker, H. S. (1959) Alternate pathways of pyridine nucleotide oxidation in cerebral tissue. J. Neurochem., 5, 50-61

Hakamori, S.-I. & Murakami, W. T. (1968) Glycolipids of hamster fibroblasts and derived malignant-transformed cell lines. Proc. nat. Acad. Sci. (Wash.), 59, 254-261

Harper, A. E. (1963) Glucose-6-phosphatase. In: Bergmeyer, H. V., ed., Methods of enzymatic analysis, Academic Press, New York, pp. 788-792

Kraemer, P. M. (1966) Sialic acid of mammalian cell lines. J. cell. comp. Physiol., 67, 23-34

Lowry, O. H., Rosebrough, N. J., Farr, A. L., & Randall, R. J. (1951) Protein measurement with the Folin phenol reagent. J. biol. Chem., 193, 265-275

Mackler, B. & Green, D. E. (1956) Studies on the electron transport system II. On the "opening" phenomenon. Biochim. biophys. Acta, 21, 1-6

Marek, J. (1907) Multiple Nervenentzündung (Polyneuritis) bei Hühnern. Dtsch. tierärztl. Wschr., 15, 417-421

Mora, P. T., Brady, R. O., Bradley, R. M. & McFarland, V. W. (1969) Gangliosides in DNA virus-transformed and spontaneously transformed tumorigenic mouse cell lines. Proc. nat. Acad. Sci. (Wash.), 63, 1290-1296

Noble, P. B., Cutts, J. H. & Carroll, K. K. (1968) Ficoll flotation for the separation of blood leukocyte types. Blood, 31, 66-73

Pappenheimer, A. M., Dunn, L. C. & Cone, V. (1929) Studies on fowl paralysis (Neurolymphomatosis gallinarum). I. Clinical features and pathology. J. exp. Med., 49, 63-86

Perdue, J. F., & Sneider, J. (1970) The isolation and characterization of the plasma membrane from chick embryo fibroblasts. Biochim. biophys. Acta, 196, 125-140

Sheinin, R., Onodera, K., Yogeeswaran, G., & Murray, R. K. (1971) Studies of components of the surface of normal and virus-transformed mouse cells. Second Le Petits Symposium on Biology of Oncogenic Viruses, Paris (in press)

Pathogenesis of Gross Cutaneous Marek's Disease Chronological Parameters

S. G. KENZY [1] & R. F. LAPEN [2]

Descriptions of gross cutaneous lesions (GCL) of Marek's disease (MD) have been reported by Csermely (1936); Benton & Cover (1957); Helmbolt, Wills & Frazier (1963); McWade & Beasley (1966); Purchase & Biggs (1967); Beasley, Patterson & McWade (1970); and Sharma, Davis & Kenzy (1970).

This paper reports the results of periodic examinations of different avian populations for GCL.

MATERIALS AND METHODS

Experimental chickens

Chicks from strains of two broiler breeders (S and P), commercial broiler (No. 1), commercial White Leghorn (N) and White Leghorn (No. 2) were utilized in this study. All chicks except Strain No. 2 were hatched under commercial conditions; the females were used in the experiments.

Strain No. 1 was selected at random from a 6-week-old female broiler flock with a 12% loss due to GCL.

Strain No. 2 consisted of White Leghorns (WL) obtained from a cross between Cornell-susceptible and Regional Poultry Research Laboratory Line 7 chickens.

Housing and rearing

Trial A was carried out in rooms with doors that opened into a common corridor. Air from each room was mechanically exhausted outside the building.

Trials B and C were carried out in a closed building with a central air ventilation system. Floor space was divided into a series of pens by the use of snow-fencing, which permitted the passage of dust.

Chicks were reared on wood shaving litter stirred regularly to avoid packing, but not changed except for replacement of heavily soiled litter.

Exposure to Marek's disease virus (MDV)

The Id-1 isolate of acute MD, reported by Sharma et al. (1970), was used for the exposure of all chickens. Strain No. 2 birds, which served as a source of MDV for exposure of the chicks in Trials A and B, were exposed by the intra-abdominal inoculation of heparinized blood from fowl with MD.

In Experiment 1 of Trial A, birds were exposed directly to dust [3] clouds for 30 minutes and to the dust that settled into the litter. In Experiment 2 of Trial A, birds were exposed indirectly to dust from Strain No. 2 fowl reared in adjacent rooms.

Exposure to MDV in Trial B, which continued throughout the duration of the experiment, was effected by contact with litter used by Strain No. 2 fowl with clinical signs and lesions of MD, and with dusty air circulated by a central ventilation fan.

In Trial C, broilers were reared for 6 weeks under commercial conditions that included exposure to dust remaining in the broiler house from previous flocks and also to dust carried by air currents from adjacent broiler units.

Diagnosis of GCL

Chickens that died during an experiment and the survivors, electrocuted at the termination, were hard-scalded in water at approximately 62°C to aid in feather removal for examination of the skin.

Presence of enlarged feather follicles with a slightly yellow tinge and/or a definite thickening of the skin,

---

[1] Department of Veterinary Microbiology, Washington State University, Pullman, Washington, USA.

[2] Department of Veterinary Science, Washington State University, Pullman, Washington, USA.

[3] Collected from the pens of Strain No. 2 with MD.

was considered diagnostic of GCL. Diagnosis of MD lesions in viscera and nerves was based on gross enlargements, as reported by Payne & Biggs (1967). Representative numbers of histological preparations from affected tissues were examined for lymphoid changes as reported by Helmbolt *et al.* (1963) and McWade & Beasley (1966) for cutaneous lesions, and by Payne & Biggs (1967) for neural and visceral lesions.

RESULTS

When the observation period was extended from 78 to 110 days, the incidence of GCL increased more than 5-fold in the S strain and 4-fold in the N strain, while visceral and neural lesions increased 2-fold in the S strain and almost 3-fold in the N strain (see Table, Trial A).

In Experiment 1 (Trial B), when approximately equal numbers of S strain were killed at 68, 74 and 81 days after initiation of exposure, the respective levels of GCL were 19.2, 56.7 and 63.1%. The corresponding levels of visceral and neural lesions were 5.0, 13.6 and 13.7% respectively.

The P strain used in Experiment 2 (Trial B) demonstrated an increase in GCL from 7.1 to 24.5% when slaughtered 65 and 86 days respectively after first exposure. In Strain P, the level of gross visceral and neural lesions was greater than that of GCL.

In Trial C, when approximately equal numbers of Strain No. 1 were examined at 44, 58 and 65 days of age, no significant differences were noted in the incidence of GCL or of visceral and neural lesions.

DISCUSSION AND CONCLUSIONS

Some informal field reports indicate that a delay in slaughter for even a few days may result in reduced "skin leukosis" condemnations. The findings of this

Results of periodic examinations of different strains of chickens
for lesions of Marek's disease (MD)

Experiment	Observation period (days)	Exposure level	Number	% with MD	
				GCL[1]	OL[2]
Trial A:					
(S strain)					
Expt. 1	78	Medium[3]	36	58.3	25.0
1	78	} Light[4]	38	10.5	5.3
2	110		38	55.3	10.5
(N strain)					
Expt. 1	78	Medium[3]	40	7.0	42.5
1	78	} Light[4]	91	5.5	11.0
2	110		91	23.1	31.8
Trial B:					
(S strain)					
Expt. 1	68	}	120	19.2	5.0
1	74	Heavy[5]	118	56.7	13.6
1	81		95	63.1	13.7
(P strain)					
Expt. 1	65	} Heavy[5]	56	7.1	1.8
1	86		49	24.5	26.5
Trial C:					
(Strain No. 1)					
Expt. 1	44	}	212	11.3	3.3
1	53	Commercial[6]	204	8.3	8.4
1	65		200	11.0	5.5

[1] Gross cutaneous lesions.

[2] Other gross lesions.

[3] Exposure to an experimentally produced MD-dust aerosol.

[4] Exposure to dust from fowl with MD.

[5] Exposure to dust in a poultry unit in which there were 90% or more deaths due to MD in susceptible WL within 8 weeks.

[6] Exposure to dust in a commercial broiler unit in which there was 12% GCL in 6-week-old chicks.

study show, however, that a delay in slaughter either resulted in a large increase in GCL or had no significant effect.

The incidence of GCL appeared to be affected by genetic factors, since the S strain showed 2 to 8 times more skin lesions than the N strain and more than twice the number observed in the P strain. Such results could be expected in view of the report by Hutt & Cole (1947) that strains of WL with marked differences in visceral-neural tumour response could be selected under conditions of natural exposure.

Skin lesions also increased significantly following heavier exposure to MD-contaminated dust, as shown by S strain in Trial A. The incidence of GCL was not affected in the N strain, but the number of visceral-neural lesions increased markedly.

ACKNOWLEDGEMENTS

This work was supported in part by funds 12-14-100-9889 (44) from Agricultural Research Service, Division of Animal Husbandry, and by funds from Projects 1773 and 5961, Washington Agricultural Research Center, College of Agriculture and Department of Veterinary Microbiology, College of Veterinary Medicine.

REFERENCES

Beasley, J. N., Patterson, L. T. & McWade, D. H. (1970) Transmission of Marek's disease by poultry house dust and chicken dander. *Amer. J. vet. Res., 31*, 339-344

Benton, W. J. & Cover, M. S. (1957) The increased incidence of visceral lymphomatosis in broiler and replacement birds. *Avian Dis., 1*, 320-327

Csermely, H. (1936) A tyukleukosis borelvatosasairol. [Skin changes in fowl leukosis] *Allatorv. lapok., 59*, 279. (Original not seen. Abstract: *Vet. Bull.*, 1938, *8*, 94)

Helmbolt, C. F., Wills, F. K. & Frazier, M. N. (1963) Field observations of the pathology of skin leukosis in Gallus Gallus. *Avian Dis., 7*, 402-411

Hutt, F. B. & Cole, R. K. (1947) Genetic control of lymphomatosis in the fowl. *Science, 106*, 379-384

McWade, D. H. & Beasley, J. N. (1966) Dermal lesions in avian leucosis. *U. S. Livestock San. Assoc., 17*, 607-613

Payne, L. N. & Biggs, P. M. (1967) Studies on Marek's disease. II. Pathogenesis. *J. nat. Cancer Inst., 44*, 901-912

Purchase, H. G. & Biggs, P. M. (1967) Characterization of five isolates of Marek's disease. *Res. Vet. Sci., 8*, 440-449

Sharma, J. M., Davis, W. C. & Kenzy, S. G. (1970) Etiologic relationship of skin tumors (skin leukosis) of chickens to Marek's disease. *J. nat. Cancer Inst., 44*, 901-912

Discussion Summary

J. G. CAMPBELL[1]

The nature of plaque-forming cells and splenomegaly-inducing cells in MD lymphomas. Cellular preparations from MD lymphomas contain plaque-forming cells detected by the Jerne technique and cause splenic enlargement when inoculated into chick embryos. The question was asked whether these were the effects of true lymphoma cells or of non-lymphomatous immunocytes. In reply, it was stated that they were the effects of total populations of cells derived from lymphomas that could contain both types of lymphoid cell. The effects, in some lymphomas at least, were greater than could be accounted for on the basis of peripheral blood lymphocytes circulating in the tumour.

A great deal of discussion took place on the phenomenon of *susceptibility and resistance to Marek's disease.* Although the existence of a genetic factor has been firmly established, the actual basis for resistance remains unknown. All available evidence showed that susceptibility and resistance are not mediated at the cellular level, since no difference in susceptibility to infection with virus had been found *in vitro* in cultures of kidney cells derived from susceptible and resistant strains. Similarly there was no significant difference in the number of foci that developed on the chorioallantoic membrane of eggs from resistant and susceptible strains after inoculation with infective material. However, it was pointed out that kidney cells in tissue culture, for example, were relatively undifferentiated and that their behaviour could not really be used as a basis for any significant conclusion as to the situation *in vivo.* Apparently the susceptibility to infection of lymphoid cells from susceptible and resistant strains of chickens has not been investigated. This might well be worth doing, as also might an investigation of the susceptibility or otherwise of feather-follicle epithelium

or of organ cultures of skin and feather follicles derived from strains with widely divergent susceptibilities to Marek's disease.

Serologically, no detectable difference existed in the immunoglobulin levels of resistant and susceptible strains, and there was ample evidence to show that resistance to Marek's disease did not necessarily indicate resistance to other diseases, and *vice versa.*

Thymectomy plus irradiation in genetically susceptible chickens reduced the number of lymphomas in experimentally induced Marek's disease, whereas in resistant thymectomized birds the number of lymphomas was increased. One explanation of this anomaly was perhaps that there are two kinds of cell in a lymphoma, so that in a susceptible strain, in which cell-mediated immunity is not very effective, the target cells might be more susceptible to thymectomy, whereas in genetically resistant birds the reverse occurred, namely thymectomy had a greater effect on cell-mediated immunity.

Bursectomy of day-old chicks followed by irradiation has been reported from at least two centres to have had no influence on the subsequent incidence of Marek's disease, whether induced by inoculation or exposure. This conflicted with some preliminary results obtained in Edinburgh, where bursectomy of 17-day-old embryos with subsequent irradiation, followed by exposure, resulted in a significant decrease in the incidence of Marek's disease. The odd case of the disease that did develop in birds so treated might be ascribed to incomplete bursectomy, since plasma cells were still detectable in various tissues.

The apparently *anomalous effect of antilymphocytic serum (ALS),* which reduces the incidence of tumours in cases of Marek's disease observed for 6–9 weeks after treatment, instead of increasing it as in other species, was commented upon. ALS enhances tumour growth by immunosuppression, but this enhancement occurs mostly in nonlymphoid tumours. In Marek's disease, in contrast, it may well be that

---

[1] Cancer Campaign Research Unit, Milton Bridge, Midlothian, UK.

the growth of tumour lymphocytes is suppressed by a direct cytocidal effect of ALS.

The *distribution of Marek's antigen in the body* was discussed at some length. It was stated that antigen was detected by immunofluorescence only in lymphoid tissues or tissues where lymphocytes occurred in any number, but apparently not in tumours. The only exceptions to this were the reported detection of formed virus in renal tubule epithelium, and in the feather follicles. With reference to the virus in the kidney, it was not clear whether it was within the epithelial cells primarily, or whether there might be an epithelial-lymphocytic relationship, as seemed possible in view of the demonstration that ALS considerably reduced kidney infection. Lymphocytes might be the means of transport within the body from the site of infection to the site where antigen is produced, or the site where virus could be detected.

Mention was made of the lymphoid aggregates that occur adjacent to feather follicles. In such follicles, light microscopy demonstrated typical Cowdry Type A epithelial cell inclusions, and viral antigen in the form of complete infectious particles could be demonstrated by immunofluorescence within dying and dead superficial skin cells. Shed cells in the form of "dander" or feather dust have proved to be highly infectious via the respiratory tract. The lymphoid aggregates were probably the source of the skin lymphomas in the cutaneous form of Marek's disease.

Virus within blood lymphocytes in Marek's disease. Short-term cultures of lymphocytes from blood, in the presence of phytohaemagglutinin (PHA), become transformed into larger blast-like cells whose nuclei often contain virus particles in various stages of replication. Attention was drawn to the possible significance of "rosettes"—a satellitosis or peripoletic phenomenon due to the clustering of lymphocytes around an immunofluorescent-positive macrophage-like cell, which is presumably of monocytic origin. Immunofluorescence shows a small proportion of cells, thought to be monocytes, in cultures at 48–96 hours, to contain viral antigen. At a later stage the surrounding lymphocytes also exhibit fluorescence and a proportion of them undergo transformation. Electron microscopy shows intranuclear virus particles in these cells. It thus seems possible that the "rosette" phenomenon may be indicative of the transfer of antigen from macrophage to lymphocytes, some of which transform into actively nucleic acid-synthesizing blastoid cells that provide a suitable milieu for replication of complete virus. This would imply an important rôle for tissue macrophages and blood monocytes as cells that phagocytose virus and transport it throughout the body, and that, by transfer of information, stimulate antibody production, sensitize immunocompetent cells, or even induce proliferative changes. In support of this hypothesis it was thought that the widely scattered cells that show positive immunofluorescence in cryotome preparations of various tissues from cases of Marek's disease may in fact be tissue macrophages or histiocytes.

With reference to the *infectivity of Marek's virus in other species,* it was stated that several attempts had been made to transmit the virus to primates, rabbits, hamsters, frog tadpoles and a wide range of avian species, all with negative results. It appeared that, under natural conditions, only the quail was susceptible.

VIROLOGY OF MAREK'S DISEASE

Chairman — B. R. Burmester

Rapporteur — B. W. Calnek

Virology and Immunology of Marek's Disease

A Review

K. NAZERIAN [1]

INTRODUCTION

Although the viral origin of Marek's disease (MD) was suspected for a long time, it was not until 1967 that two groups of investigators, one in England (Churchill & Biggs, 1967) and one in the United States (Solomon et al., 1968; Nazerian et al., 1968), independently and at the same time isolated a herpesvirus in cell cultures derived from diseased chickens and presented good circumstantial evidence implicating the virus in the aetiology of the disease. These findings were further supported by more comprehensive studies by Biggs et al. (1968), Churchill & Biggs (1968), Churchill (1968), Witter, Burgoyne & Solomon (1969), Witter, Solomon & Burgoyne (1969) and many other investigators. In spite of the strong evidence provided by these authors, a definite rôle for this virus and cell-free transmission of the disease could not be demonstrated, mainly due to the strict cell association of the virus in cell culture.

Calnek & Hitchner (1969) detected a viral-specific immunofluorescence (IF) antigen in the feather follicle epithelium of MD-infected chickens, and Calnek, Adldinger & Kahn (1970) subsequently succeeded in isolating infectious cell-free virus from this source and produced the disease in chickens. This work was immediately confirmed by Nazerian & Witter (1970), who produced the disease with cell-free extracts of the feather follicle and also demonstrated the site of replication of the virus in the feather follicle epithelium. These important findings have provided ample proof that this herpesvirus is the aetiological agent of MD.

HOST SUSCEPTIBILITY

The chicken is the natural host for Marek's disease virus (MDV). Although the virus replicates in all genetic lines of chickens examined (Stone & Nazerian [2]) certain lines are resistant to the disease. Chickens in these lines do not develop recognizable clinical signs or gross lesions of the disease, but do develop antibody and the virus replicates in the tissues of such birds. They are also capable of transmitting the disease to more susceptible chickens. This is also true of more susceptible lines of chickens; certain chickens become infected with the virus but do not develop clinical or gross lesions. In the field, most flocks of mature chickens and a great majority of individuals in each flock carry the virus whereas the disease is manifested in only a few.

There is also a strain variability. Purchase & Biggs (1967) referred to a mild or classical (HPRS-17) strain and to more acute (HPRS-16, 18, 19, 20) strains of the virus, which varied in virulence. Certain strains of the virus (GA, RPL-39) (Eidson & Schmittle, 1968; Purchase, Burmester & Cunningham, 1971a) are basically viscerotropic and produce a number of visceral tumours including lymphoid tumours of the gonad whereas other strains, such as JM (Sevoian, Chamberlain & Counter, 1962), are more neurotropic and produce fewer lymphoid tumours in the viscera.

A herpesvirus of turkeys (HVT) (Kawamura, King & Anderson, 1969; Witter et al., 1970) antigenically related to MDV has been isolated in many flocks of turkeys. This virus is apathogenic for turkeys and chickens and is morphologically distinguishable from MDV (Nazerian et al., 1971). Turkeys inoculated

---

[1] Regional Poultry Research Laboratory, 3606 East Mount Hope Road, East Lansing, Michigan, USA.

[2] Unpublished data.

with virulent MDV developed tumours but the virus could not be re-isolated from such birds (Witter [1]).

CELL CULTURE SUSCEPTIBILITY

The virus was first isolated in chick kidney (CK) cultures (Churchill & Biggs, 1967) and duck embryo fibroblast (DEF) cultures (Solomon et al., 1968; Nazerian et al., 1968). Both cultures were found to be susceptible. Although initial isolation of the virus in chicken embryo fibroblast (CEF) cultures was unsuccessful, Nazerian (1968, 1970a) demonstrated the virus in these cultures by co-cultivation with infected DEF. CEF cultures derived from both susceptible and resistant chickens were equally susceptible to the virus (Nazerian, 1968 [1]). CEF cultures were also more susceptible to the virus than DEF (Nazerian, 1970a; Purchase, Burmester & Cunningham, 1971b). Several other cultures derived from other avian species were also found susceptible to the virus (Purchase et al., 1971b). Calnek, Madin & Kniazeff (1969), Nazerian [1], and Sharma (1971) tested the susceptibility of several established mammalian cell lines to MDV and found them all resistant.

Marek's disease virus produces detectable microplaques in all susceptible cultures. These consist of morphologically altered cells grouped together. Many round refractile cells, fusiform cells, and polykaryocytes are seen in the microplaques. The size and number of the polykaryocytes vary with the type of culture and the passage level of the virus. Churchill & Biggs (1967, 1968) found that the HPRS-16 strain of the virus produced microplaques in CK culture, consisting of many round, refractile cells. Some of these cells detached from the monolayer and left a hole behind. Witter, Solomon & Burgoyne (1969) found that the JM strain of MDV produced a somewhat similar microplaque in CK cultures but that only a small number of holes were formed (Fig. 1). In DEF cultures, Solomon et al., (1968) and Nazerian et al. (1968) (Fig. 2) found that the microplaques consisted of a number of round, and fusiform, refractile cells piled on top of each other. Nazerian (1968, 1970a) showed that microplaques in CEF cultures (Fig. 3) consisted of a number of small, rounded, refractile cells. Type A intranuclear inclusions (Fig. 4) were produced in all these cultures.

Churchill, Chubb & Baxendale (1969) and Naze-

rian (1970a) reported a morphological change in the microplaque as a result of continuous cell culture passage. Although these changes were accompanied by a loss of pathogenicity, they were independent events (Purchase et al., 1971). In CK cultures, Churchill, Chubb & Baxendale (1969) noticed a change in the size and shape of the microplaques as a result of cell culture passage, which also resulted in the loss of an antigen referred to as the A antigen and loss of pathogenicity. Microplaque-producing virus (mPA$^+$) was pathogenic and produced the A antigen. Macroplaque-producing virus (MPA$^-$) was apathogenic and lacked the A antigen. The JM strain of the virus also changed its plaque morphology (producing larger plaques and larger and more polykaryocytes in the plaque) and lost its pathogenicity for chickens as a result of continuous passage in DEF cultures (Nazerian, 1970b). Although the virus was highly cell-associated in all these cultures, a small quantity of cell-free virus was detected in the growth fluid of infected CEF cultures (Nazerian, 1970a). Attempts to extract infectious virus from infected cultures were not always successful, but Cook & Sears (1970) reported the extraction of high titre cell-free virus from infected cells in demineralized distilled water, and Calnek, Hitchner & Adldinger (1970) reported the extraction of cell-free virus by sonication of infected cells in SPGA buffer [2] containing 0.2% disodium ethylenediamine tetraacetate (EDTA). The latter method has proved useful in providing amounts of cell-free virus adequate for kinetic studies.

VIRAL ANTIGENS

Several serological techniques have been used for the detection of in vivo and in vitro induced antigens of MDV and antibodies specific to these antigens in infected or hyperimmune chickens. Kottaridis & Luginbuhl (1968) and Kottaridis, Luginbuhl & Chomiak (1968) described immunofluorescent (IF) and agar gel precipitin (AGP) tests for the detection of antigens of MDV. Since at that time little information was available on the nature of the virus, the specificity of these tests remained uncertain. Subsequent to the isolation of the herpesvirus and development of a better in vitro assay system, Chubb & Churchill (1968) described a more specific AGP test using chick kidney cell cultures infected with the

---

[1] Unpublished data.

[2] A phosphate buffer containing sucrose, disodium glutamate, and bovine serum albumin.

herpesvirus of Marek's disease as the antigen against sera from infected chickens. Purchase (1969) also described an indirect IF test using infected tissue culture as antigen and found it more sensitive than the AGP test (Purchase & Burgoyne, 1970). Later, Calnek & Hitchner (1969) and Purchase (1970) detected IF antigens in tissues of infected chickens. An indirect haemagglutination test (Eidson & Schmittle, 1969), a complement fixation test (Cho & Ringen, 1969) and a serum neutralization test (Calnek & Adldinger, 1971) have also been described.

The AGP test (Fig. 5) has been widely used for the detection of specific antibody in infected chickens. Using this test, Churchill, Chubb & Baxendale (1969) detected three groups of antigens referred to as the A, B, and C antigens. A antigen was detected in the supernatant fluid of infected cultures, whereas B and C antigens were detected in cellular material. They noticed that the concentration of the A antigen decreased as the virus was continuously passed in cell culture and that after prolonged passage this antigen was completely absent. Using this test, Chubb & Churchill (1969) found that most flocks of chickens have precipitating antibodies, and that newly hatched chicks from infected dams also had a low level of antibody, which disappeared by three weeks of age. As many as six lines of precipitation were obtained using different sera but there was no indication as to the nature of the antigens thus detected. The A antigen appears to be of low molecular weight and has an approximate sedimentation coefficient of 4S (Velicer [1]).

Similar results (Fig. 6) were observed in the IF test, and three morphologically different antigens were detected in infected cell cultures (Purchase, 1969; Nazerian & Purchase, 1970). These were: (i) diffuse nuclear antigen; (ii) diffuse cytoplasmic antigen; and (iii) a cytoplasmic granular antigen. Chen & Purchase (1970) also found an IF antigen on the surface of infected cells (Fig. 7). Nazerian & Purchase (1970) studied identical infected DEF and CEF cells by means of the IF test and electron microscopy, and found that the IF antigens were quite specific to virus infection and that these antigens were found only in cells containing virus particles. No virus particles were found in cells free from IF antigens. Three structurally different antigens were found in infected cells but these antigens were not necessarily composed of viral particles. In fact, the cytoplasmic granular antigen (Fig. 8) never contained

any recognizable virions. The nuclear antigen (Fig. 9) often contained virus particles but was not composed exclusively of these particles. Diffuse cytoplasmic antigen (Fig. 10) only rarely contained any virions. All three antigens may nevertheless be related to viral structural proteins, but direct and conclusive evidence is lacking. Viral particles in cells treated with antibody were coated with a fine granular substance (Nazerian et al., 1969) indicating the viral specificity of the IF test. Ahmed et al. (1970) also detected this coating on the surface of both naked and enveloped virions.

The surface antigen (Chen & Purchase, 1970) was also specific to the virus. There was no indication that this antigen was different from the internal antigens nor was its nature determined. It may, however, be similar to the viral envelope antigen reported by Roizman & Spear (1971) for herpes simplex virus.

ANTIGENIC DIFFERENCES BETWEEN VIRAL STRAINS

Purchase (1969) studied 8 different virulent isolates of MDV by means of the IF test and failed to detect any antigenic differences. Churchill, Chubb & Baxendale (1969), however, used the AGP test and found a difference between the virulent strains and the cell culture attenuated virus. The latter lacked the A antigen. Using this A antigen as a marker, Purchase et al. (1971a) reported an antigenic difference between a number of virulent and attenuated strains. They found that some of the virulent strains also lacked the A antigen. Churchill, Chubb & Baxendale (1969) and Nazerian (1970a) had already shown that cell culture passage results in a change in the morphology of the microplaque and the pathogenicity of the virus. Purchase et al. (1971a) reported a method of distinguishing between clones of MDV positive or negative for the A antigen on the basis of the IF test. Those positive for the A antigen produced bright staining in flattened cells surrounding the microplaque, whereas those negative for the A antigen did not produce this staining. Witter et al. (1970) found a method of distinguishing between all isolates of MDV and an antigenically related herpesvirus of turkeys. Cells infected with this latter virus stained both in the cytoplasm and the nucleus when reacted with homologous antiserum whereas they stained only in the nucleus when reacted with MDV antiserum.

Using the IF test, Purchase (1969) failed to detect any common antigenicity between MDV and several other herpesviruses. Ono et al. (1970a) recently

---

[1] Personal communication.

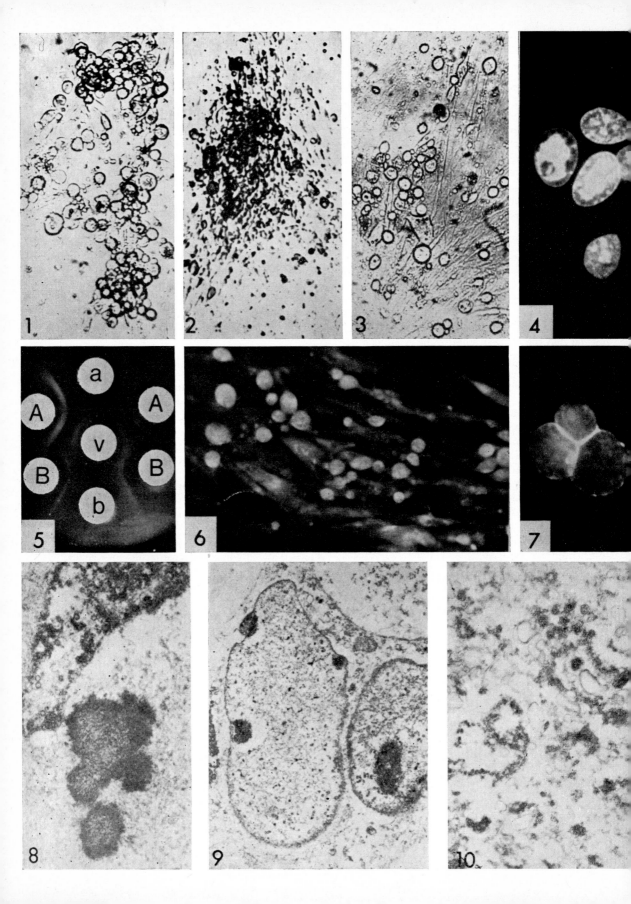

showed some cross-reaction with the Epstein-Barr virus (EBV) of Burkitt's lymphoma by means of this test. Stevens, Kottaridis & Luginbuhl (1971), in contrast, failed to detect any cross serological reaction between MDV and EBV.

STRUCTURE OF THE VIRUS

The fine structure of MDV has been extensively studied with the electron microscope (Nazerian & Burmester, 1968; Epstein *et al.*, 1968; Ahmed & Schidlovsky, 1968; Nazerian & Witter, 1970; Nazerian, 1970b; Nazerian, 1971). As with other known herpesviruses, the nucleocapsid is 90–100 nm in diameter (Fig. 11) and is composed of 162 capsomeres arranged in icosahedral symmetry (Fig. 12). The capsomeres are cylindrical and hollow centred, and measure 6×9 nm (Fig. 13). The centre-to-centre distance between capsomeres is 10 nm. When stain penetrates the nucleocapsid (Fig. 14), another membrane is seen in some virions, resembling the middle capsid reported for the herpes simplex virus (Roizman *et al.*, 1969). This middle capsid surrounds the viral core, which measures 40–50 nm. In negatively stained preparations (Fig. 11), a high proportion of virions are without the outer envelope. Those with the envelope measure 220–250 nm in diameter. The envelope (Fig. 15) resembles a collapsed sack and contains a filamentous structure in addition to the nucleocapsid. Sometimes two to three nucleocapsids are seen within one envelope. The envelope appears to have projections on the surface.

In thin sections (Fig. 16), a similar morphology is seen except that some nucleocapsids lack the core, the morphology of which also varies considerably. Some nucleocapsids have a ring-like core and resemble a double shell, some have a dense central core (Fig. 17), whereas others have structures of bizarre shapes. A group of small (35–40 nm) particles are also seen together with the viral nucleocapsids (Fig. 18). The size of the viral envelope varies. A group of virions measuring 150–180 nm in diameter in which the envelope is tightly attached to the capsid is shown in Fig. 19, whereas other groups (Fig. 20) are considerably larger (220–250 nm) and contain a fine granular matter between the envelope and the capsid. The origin of these two different types of enveloped virions will be discussed when the replication of the virus is considered (see p. 68).

No extensive studies of the biochemical composition of the virus have been made. On the basis of the development of type A intranuclear inclusions, the cytochemical staining and enzymatic digestion of these inclusions with DNase, and the morphological resemblance of the virus to other known herpesviruses, it was concluded that it contains DNA (Churchill & Biggs, 1967; Nazerian & Burmester, 1968). Replication of the virus was also inhibited by both iododeoxyuridine (IUDR) (Churchill & Biggs, 1967; Sharma, Kenzy & Rissberger, 1969) and bromodeoxyuridine (BUDR) (Nazerian, 1970a), indicating the necessity of *de novo* synthesis of DNA. This inhibition was removed by the addition of excess thymidine to the medium (Nazerian, 1970a), and cultures treated with the DNA analogue remained infected and viral replication resumed following its removal.

Different techniques have been used for the purification of the virus. Lee *et al.* (1969) used a combination of sedimentation and flotation in a sucrose gradient for purification of the virus extracted from cells by Dounce homogenization. Later, the same authors (Lee *et al.*, 1971) used extraction of the virus from infected cells with Nonidet P-40 (NP-40) and

---

Figs. 1-3. Light micrographs of unstained MDV-infected cultures of chicken kidney, duck embryo fibroblasts and chicken embryo fibroblasts respectively; ×110 (Fig. 1 kindly provided by R. L. Witter).

Fig. 4. Type A intranuclear inclusion in MDV-infected duck embryo fibroblast culture stained with acridine orange; ×1800.

Fig. 5. Antigen-antibody reaction in MDV agar gel precipitin test. Wells a and b contained the A antigen and B antigen respectively and wells A and B contained serum specific to A and B antigens. Well V contained partially purified and concentrated virus. The virus preparation appears to contain the A and B antigens besides an additional antigen absent or present at a lower concentration in the other two antigen preparations. (Kindly provided by H. G. Purchase.)

Fig. 6. An MDV microplaque in chicken embryo fibroblast culture stained with fluorescine conjugated antibody; ×180.

Fig. 7. Immunofluorescent antigen on the surface of unfixed MDV-infected chicken kidney cells; ×370. (Chen & Purchase, *Virology*, 1970, **40**, 410. Reproduced by permission of the Academic Press.)

Figs. 8, 9 & 10. Composition of the granular (Fig. 8; ×9100) nuclear (Fig. 9; ×3700) and diffuse cytoplasmic (Fig. 10; ×14 000) antigens detected by IF test. Neither granular nor diffuse cytoplasmic antigens contained virions whereas the nuclear antigen contained some virions.

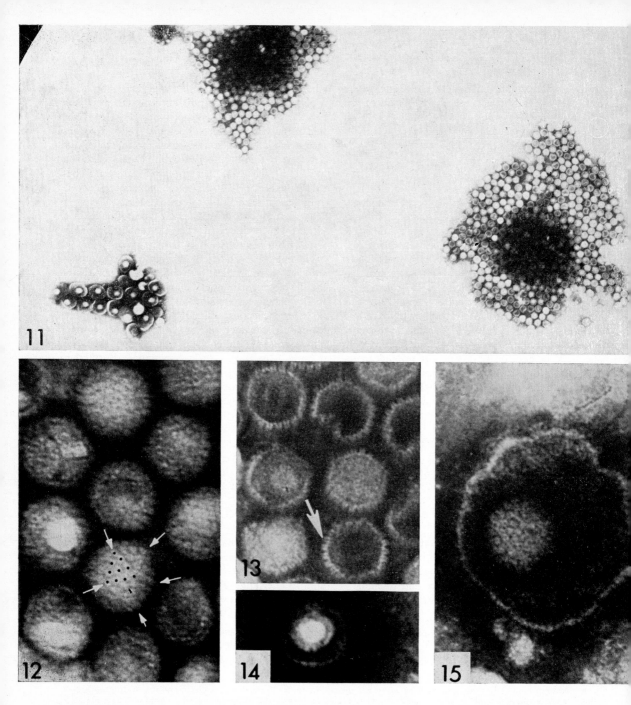

Fig. 11. A negatively stained micrograph of concentrated Marek's disease virions extracted from infected duck embryo fibroblast cultures. Two aggregates of naked virions and a smaller aggregate of enveloped virions are seen in this micrograph; ×15 000.

Figs. 12-15. Higher magnification of naked and enveloped MD virions. Icosahedral symmetry of capsid on a five-folded axis and the number of capsomeres on each facet are demonstrated in Fig. 12. These capsomeres resemble hollow cylinders and measure 6 and 9 nm (Fig. 13). An inner capsid surrounding the core is occasionally seen (Fig. 14). The envelope measures 200-250 nm in diameter and has projections on the surface (Fig. 15).

Fig. 16. Thin section electron micrograph of an MDV-infected duck embryo fibroblast culture. Enveloped virions are seen in vesicles (arrowed) within the nucleus and some naked virions are seen in the nucleoplasm; ×14 000.

Figs. 17-20. Morphology of different forms of MD virions. Naked virions (Fig. 17; ×187 000) measure 90-100 nm in diameter and have cores with varying morphology. Small (35-40 nm) particles (Fig. 18; ×112 000) are occasionally seen along with naked virions in the nucleus. Nuclear enveloped virions (Fig. 19; ×187 000) measure 150-180 nm, accumulate in nuclear vesicles and have an envelope tightly attached to the capsid, whereas cytoplasmic enveloped virions (Fig. 20; ×140 000) measure 200-250 nm in diameter and contain a filamentous material in addition to the nucleocapsid.

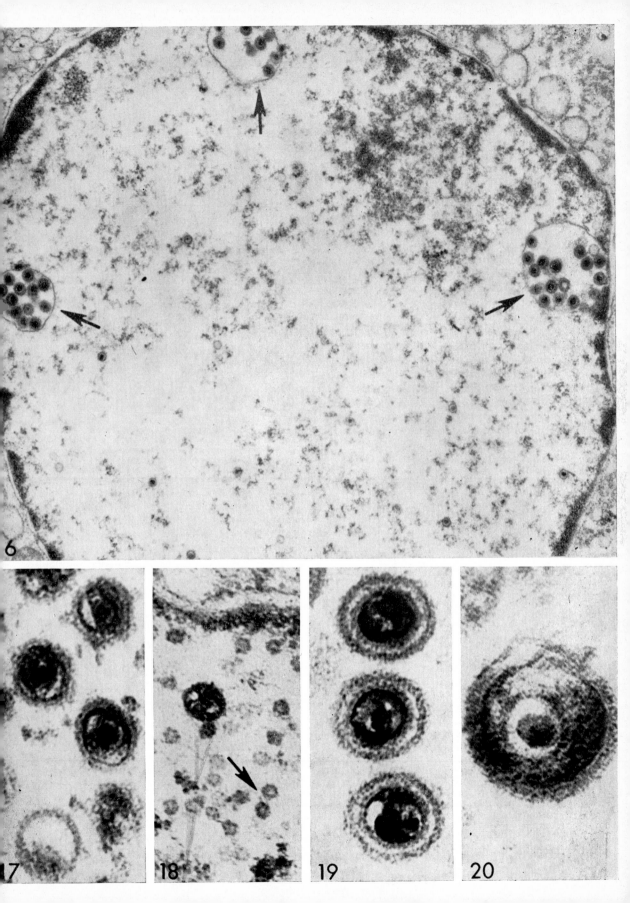

attempted to purify the virus in a linear gradient of 12–52% sucrose. Chen *et al.* (1972) used Freon 113 to extract the virus, and further purified it in 12–52% sucrose gradient. A cleaner viral preparation was obtained by using Freon 113, but there was a significant drop in the number of virus particles observed.

The DNA of the virus was extracted from the partially purified attenuated JM strain of the virus (Lee *et al.*, 1969) with a mixture of sodium dodecyl-sulphate (SDS) and phenol. The buoyant density of the viral DNA was determined by caesium chloride density-gradient centrifugation in a preparative ultra-centrifuge. Based on a buoyant density of 1.716 g/cm³, it was estimated that viral DNA has a guanine +cytosine (G+C) content of 56%. Burg *et al.* (1970), however, used the same procedures and reported a lower buoyant density (1.706 g/cm³) and a lower G+C content (46%) for the GA strain of the virus. This was also shown for the JM strain of the virus by Lee *et al.* (1971), who used purified viral DNA and known DNA markers in a Model E ana-lytical ultracentrifuge (Fig. 21). MDV-DNA was mixed with T4 bacteriophage DNA or herpes simplex virus DNA and centrifuged in a neutral sucrose

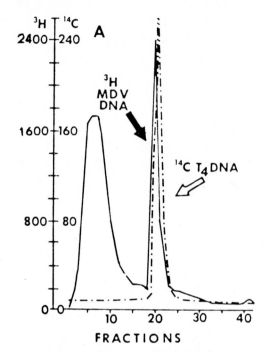

Fig. 22. Sedimentation of native ³H-MDV-DNA with native ¹⁴C-T4-DNA through a linear 10 to 30 % neutral sucrose gradient. (Lee *et al.*, 1971, *J. Virol.*, **7**, 289. Repro-duced with the permission of ASM).

density gradient (Fig. 22). Based on these experi-ments, a sedimentation constant of 56*S* and a molec-ular weight of 1.0×10^8 Daltons was reported for MDV-DNA. This DNA hybridized with the RNA isolated from infected cells, but not with the RNA from uninfected cells, thus showing its viral nature.

PROTEINS OF THE VIRUS

Chen *et al.* (1972) partially purified the virus from the culture fluid of GA-strain-infected cells and analyzed the structural proteins of the virus by poly-acrylamide gel electrophoresis. Nine protein peaks (proteins I–IX) (Fig. 23) were detected in purified preparations containing both enveloped and naked virions. Samples prepared from infected cells by NP-40 extraction were low in enveloped virions and had a low activity corresponding to protein III. This protein, along with proteins I and II, constituted 48% of the total protein, and proteins VI, VII, and IX 33%, whereas the other three proteins were present only in small amounts. The peak for radioactive glucosamine labelled protein corresponded to pro-tein III, thus indicating the glycoprotein nature of

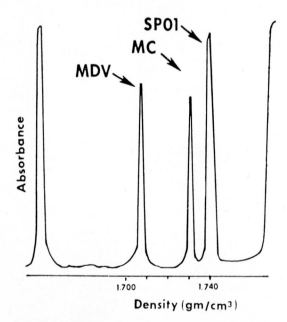

Fig. 21. Ultraviolet absorption profile of MDV-DNA centrifuged with *Micrococcus lysodeikticus* DNA and SPO 1 phage DNA in a Model E analytical ultracentrifuge. (Lee *et al.* 1971, *J. Virol.*, **7**, 289. Reproduced with the per-mission of ASM).

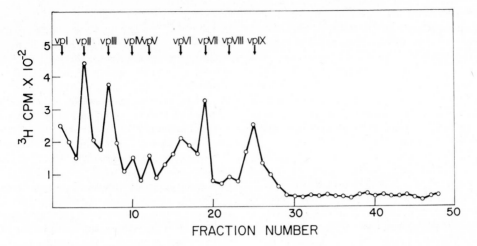

Fig. 23. Electropherogram of solubilized MDV protein analyzed in 7 % polyacrylamide gel, showing 9 species of protein (VPI-VPIX). (Kindly provided by J. H. Chen).

this protein and its association with the viral envelope, since the peak for radioactive glucosamine labelled protein was removed by NP-40 treatment of the virus prior to solubilization of the protein. These nine proteins varied between 126 000 and 33 000 Daltons in molecular weight.

PHYSICOCHEMICAL PROPERTIES OF THE VIRUS

Cell-free virus extracted from the skin of infected chickens (Calnek & Adldinger, 1971) is stable at —65°C, but loses its infectivity at —20°C. It survives several freeze-thaw cycles and is most stable at pH 7.0, but completely loses its infectivity at pH-values below 4 and above 10. The virus was removed from the supernatant fluid by centrifugation at 5000 g for 1 hour. It was completely inactivated after 2 weeks at 4°C, 4 days at 25°C, 18 hours at 37°C, 30 minutes at 56°C and 10 minutes at 60°C.

Cell-free virus obtained from cell culture (Nazerian [1]) was sensitive to ether and formalin. It was moderately sensitive to ultrasonic vibration, but an increase in titre was noticed during the first 45 seconds of sonication. This may be due to the presence of viral aggregates in cell-culture preparations. After incubation for 60 minutes at 25°C, 37°C and 48°C, 40%, 70% and 95% of infectivity was lost respectively. At 56°C, 95% of infectivity was lost within 5 minutes of incubation.

---

[1] Unpublished data.

IN VITRO REPLICATION OF THE VIRUS

Detailed information on the biochemical and antigenic changes induced in infected cell cultures is not available. This is mainly due to the slow growth of the virus, and its cell association.

Initial isolation of the virus from kidney cells, tumour cells or the blood of diseased chickens in CK or DEF cultures usually takes 5–8 days. The growth rate is more rapid, as the virus is adapted to cell culture, but it still takes 3–4 days to establish a highly infected culture. The main criteria for viral replication have been based on morphological changes, but also, to some extent, on antigenic changes. Since the transfer of infection to other cells is usually by contact, the growth cycle of the virus is not known, but is obviously shorter than that required for the detection of morphological changes and the appearance of microplaques.

Sonication of infected cells in SPGA buffer has made it possible to obtain some cell-free virus and to study the initial stages of virus-host interaction, namely adsorption and penetration. Later stages of viral replication leading to the synthesis of progeny virus are not known because of the low titre of input virus and the inability of infected cells to release infectious virus.

Churchill & Biggs (1968) and Sharma et al., (1969) used intact cells as the inoculum and measured the time of contact between infected and noninfected cells necessary for the maximum transfer of infection. A rapid transfer (approximately 50%) in the first half hour of contact was followed by a period of slower transfer of between 72 and 192 hours. With cell-free virus, more than 50% adsorption occurred within 30 minutes at 37°C. The virus was also adsorbed at 25°C and 4°C, but at a much lower rate (Nazerian [1]).

All strains of the virus cause polykaryocytosis in the susceptible cultures tested, but the degree of polykaryocytosis (number of polykaryocytes/culture and number of nuclei/polykaryocyte) depends on the type of culture, strain of virus, and passage level. There is morphological evidence that polykaryocytosis is the result of dissolution of the membrane of two adjacent cells. It is not known when an infected cell can actively engage in polykaryocytosis, when this activity ceases, and what its limits are. Obviously changes in the cell membrane play an important role and these changes are affected by indigenous factors (type of cell), the strain of virus, and the passage level.

All strains of the virus cause intranuclear type A inclusion bodies, which stain bright green with acridine orange (Fig. 4). They are digested by DNase treatment (Churchill & Biggs, 1967). Autoradiographic studies (Ono et al., 1970a) have indicated that in some cells these inclusions are identical with the areas where viral DNA synthesis takes place. They contain naked virions scattered throughout a granular substance, and sometimes also a filamentous structure, the nature of which is not understood. It is not precisely known at what stage of viral replication these inclusions are produced. Some are quite distinct and separated from the nuclear membrane whereas others are diffuse and occupy the entire area of the nucleus.

Cytoplasmic inclusions are rare and are only seen in primary cultures of kidneys from diseased chickens (Ahmed & Schidlovsky, 1968). They are different in composition from nuclear inclusions and contain a large number of enveloped virions.

The viral nucleocapsid is apparently assembled in the nucleus, and is seen in sections of the nucleus as crystals, aggregates or individual particles, which may vary considerably in the shape of the core. Some capsids are apparently without a core whereas others may have a double-shell core, a bar-shaped core, or a round condensed core, or may be completely filled with core material. Besides the nucleocapsids, a group of small (30–40 nm) particles (Fig. 12) are seen in the nucleus. It is not known whether the synthesis

of these particles precedes, occurs at the same time, or follows the assembly of the nucleocapsids. Neither the chemical nature of these particles nor their structural relationship with the viral nucleocapsids is known, but they are found only in infected cells together with viral nucleocapsids. Another rare feature in the nucleus of MDV infected cells is the presence of long microtubules. Both small nuclear particles and microtubules are found associated with other herpesviruses. Their function is unknown.

Viral envelopment appears to take place mainly at the site of the nuclear membrane. The inner nuclear membrane becomes slightly thicker and more electron-dense at the area of envelopment. Naked virions, and often those with a dense central core, bud out from the nucleus into vesicles created by the invagination of the nuclear membrane (Fig. 16). Enveloped in this fashion, the virions measure 150–180 nm in diameter (Fig. 19) and accumulate in these vesicles. The envelope is tightly attached to the capsid and encloses only one nucleocapsid.

There are also indications that envelopment may occur in the cytoplasm, in the Golgi region or within the cytoplasmic inclusions (Nazerian, 1971; Ahmed & Schidlovsky, 1968). These enveloped virions (Fig. 20) are larger than those budding from the nuclear membrane and contain a filamentous structure in addition to one, or rarely two to three, nucleocapsids. These have the same morphology as the extra-cellular enveloped virions. The chemical nature and function of this filamentous structure is not known, but morphologically it resembles the material forming the cytoplasmic inclusions.

IN VIVO REPLICATION

Viral-specific IF and AGP antigens were found as early as five days in chicks experimentally infected at one day of age (Calnek & Hitchner, 1969; Purchase, 1970). Although these antigens were found in epithelial tissues, and only occasionally in the lymphoid tissues of organs such as the bursa of Fabricius, kidney, lung, thymus, and the feather follicle epithelium, it was only in the last-mentioned site that this antigen could be found consistantly. Furthermore, complete viral replication takes place only in the feather follicle epithelium whereas in the other organs viral infection is hidden or incomplete. Different organs and tissues of infected chicken, therefore, may respond differently to infection with MDV, as follows:

[1] Unpublished data.

1. Most organs and tissues of chickens become infected with the virus, but infection remains hidden until *in vitro* conditions are provided. In this case, viral antigen and virus particles are not found.

2. Occasionally (Schidlovsky, Ahmed & Jansen, 1969; Ubertini & Calnek, 1970; Nazerian, 1971) incomplete viral replication proceeds in these tissues and one can detect viral antigens and incomplete virus particles (Fig. 24) in the absence of filterable infectious virus.

3. In the epithelial layer of the feather follicle (Figs. 25, 26, & 27), the virus goes through its complete replication cycle, and cell-free virus is released.

Both incomplete replication of the virus in tumour cells, the bursa of Fabricius, and the kidney, and complete replication in the feather follicle, seem to be degenerative and to cause the eventual death of the cell. The inapparent infection does not seem to be degenerative, but in lymphoid cells it may cause proliferation and eventual formation of tumours. The mechanism by which the virus causes the development of lymphoid tumours and the origin of these tumours is not known. Viral infection, however, is not always followed by tumour development. Factors such as the genetic constitution, age, sex, and immune competence of the chickens and the virulence of the virus, the time of exposure and the dose of virus may influence tumour development.

IMMUNOLOGY

Marek's disease is reported to affect the immune response of chickens. A depression in both cellular and humoral immunity is noticed in MD-infected chickens, as indicated by delayed homograft rejection and poor antibody response (Purchase, Chubb & Biggs, 1968). Evans & Patterson (1970) also observed a poor antibody response in chickens with MD and particularly in those with gross lesions. In spite of this decrease in antibody response, Ringen & Akhtar (1968) and Samadieh, Bankowski & Carroll (1969) found a significant increase in total serum protein and γ globulin. This depression in immunological response is perhaps due to the effect of Marek's disease on the bursa of Fabricius (depression in humoral response) and thymus (depression in cellular immunity) (Payne & Biggs, 1967; Purchase *et al.*, 1968), both related to the immunological competence of chickens (Cooper *et al.*, 1966). However, bursectomy and thymectomy do not affect the development of Marek's disease (Payne & Rennie, 1970), in contrast to lymphoid leukosis, where removal of the bursa prevents the development of the disease.

Viral-specific antibodies are found in the sera of chickens infected with the virus and in the sera of offspring from antibody-positive hens (Chubb & Churchill, 1968). Maternal antibody gradually disappears and is absent by three weeks of age, whereas actively acquired antibody has been detected as early as four weeks of age. The neutralizing ability of this antibody has not been tested *in vitro*, but Chubb & Churchill (1969) reported significant protection against morbidity and mortality in the presence of maternal antibody. A somewhat similar, but less dramatic, effect was obtained when antibody-free chicks were inoculated with specific antibody.

Highly effective vaccines against Marek's disease are now available. Both laboratory experiments (Churchill, Payne & Chubb, 1969; Okazaki, Purchase & Burmester, 1970) and field experiments (Biggs *et al.*, 1970; Purchase, Okazaki & Burmester, 1971) on vaccination with a cell-culture attenuated MDV and a herpesvirus of turkey have shown a significant reduction in the incidence of the disease. Neither vaccine, however, has prevented superinfection with virulent MDV and its subsequent shedding. There is some evidence that vaccination reduces both the level of infection (Purchase [1]) and the rate of shedding of the virulent virus (Purchase & Okazaki, 1971). The mechanism whereby protection is provided by MD vaccine is not known; it does not seem to be due to viral interference, cellular immunity or humoral immunity alone. Perhaps a combination of these factors, and particularly of the last two, is responsible for immunity to Marek's disease.

CONCLUSIONS

The viral origin of MD is now well established and a herpesvirus has been shown to be the causative agent of the disease. Limited information has become available in recent years on both *in vivo* and *in vitro* replication of the virus, but fundamental features of virus-host interactions and the macromolecular changes induced in infected cells remain to be explored. Studies on the molecular pathology of the disease, the biochemical changes related to

---

[1] Unpublished data.

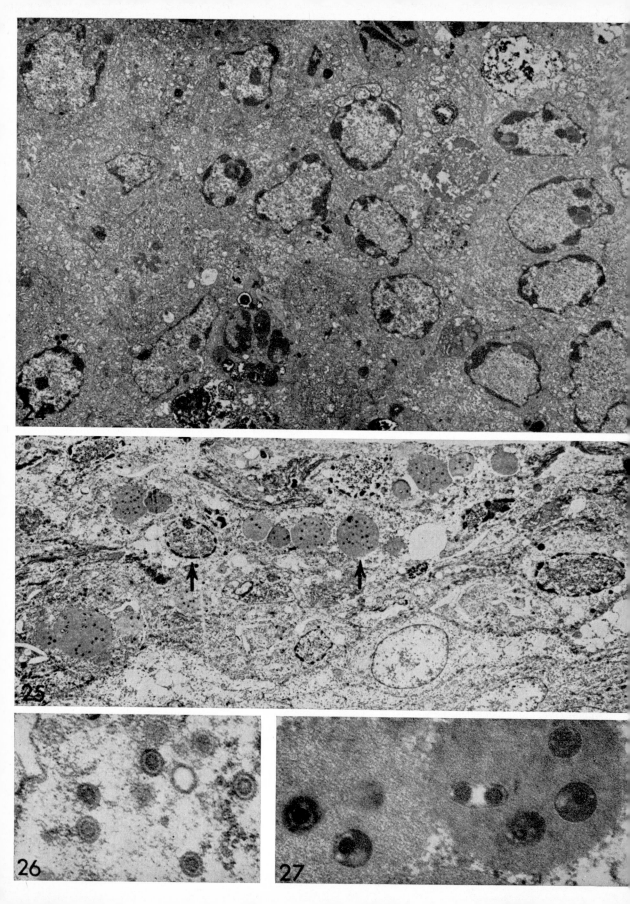

the oncogenic potential of the virus and the possible integration of viral DNA in tumour cell genome are urgently needed in order to gain a clear insight into the process of carcinogenesis by MDV and other herpesviruses. Furthermore, studies on the basic mechanism of natural and acquired immunity to MD could contribute greatly to the field of cancer immunology.

REFERENCES

Ahmed, M. & Schidlovsky, G. (1968) Electron microscopic localization of herpesvirus-type particles in Marek's disease. *J. Virol.*, **2**, 1443-1457

Ahmed, M., Jensen, K. E., Slattery, S. M., Leech, T. B. & Schidlovsky, G. (1970) Detection of Marek's disease herpesvirus antigen by fluorescent and coating antibody. *Avian Dis.*, **14**, 349-363

Biggs, P. M., Churchill, A. E., Rootes, D. G., & Chubb, R. C. (1968) The etiology of Marek's disease—an oncogenic herpes-type virus. *Perspect. Virol.*, **6**, 211-237

Biggs, P. M., Payne, L. N., Milne, B. S., Churchill, A. E. & Chubb, R. C. (1970) Field trials with an attenuated cell associated vaccine for Marek's disease. *Vet. Rec.*, **87**, 704-709

Burg, R. W., Morris, C. A., Adler, S. A., & Maag, T. A. (1970) Characterization of DNA synthesized in duck embryo cells infected with the herpes-like virus of Marek's disease. *Bact. Proc.*, p. 170

Calnek, B. W., Madin, S. H., & Kniazeff, A. J. (1969) Susceptibility of cultured mammalian cells to infection with a herpesvirus from Marek's disease and T-virus from reticuloendotheliosis of chickens. *Amer. J. vet. Res.*, **30**, 1403-1412

Calnek, B. W. & Hitchner, S. B. (1969) Localization of viral antigen in chickens infected with Marek's disease herpesvirus. *J. nat. Cancer Inst.*, **43**, 935-949

Calnek, B. W., Adldinger, H. K., & Kahn, D. E. (1970) Feather follicle epithelium: a source of enveloped and infectious cell-free herpesvirus from Marek's disease. *Avian Dis.*, **14**, 219-233

Calnek, B. W., Hitchner, S. B., & Adldinger, H. K. (1970) Lyophilization of cell-free Marek's disease herpesvirus and a herpesvirus from turkeys. *Appl. Microbiol.*, **20**, 723-726

Calnek, B. W., & Adldinger, H. K. (1971) Some characteristics of cell-free preparations of Marek's disease virus. *Avian Dis.* (in press)

Chen, J. H., & Purchase, H. G. (1970) Surface antigen on chick kidney cells infected with the herpesvirus of Marek's disease. *Virology*, **40**, 410-412

Chen, J. H., Lee, L. F., Nazerian, K. & Burmester, B. R. (1972) Structural proteins of Marek's disease virus. *Virology* (in press)

Cho, H. D., & Ringen, L. M. (1969) Antigenic analysis of soluble tissue antigens from chickens exposed to Marek's disease agents. *Amer. J. vet. Res.*, **30**, 847-852

Chubb, R. C., & Churchill, A. E. (1968) Precipitating antibodies associated with Marek's disease. *Vet. Rec.*, **83**, 4-7

Chubb, R. C. & Churchill, A. E. (1969) The effect of maternal antibody on Marek's disease. *Vet. Rec.*, **85**, 303-305

Churchill, A. E., & Biggs, P. M. (1967) Agent of Marek's disease in tissue culture. *Nature (Lond.)*, **215**, 528-530

Churchill, A. E. (1968) Herpes-type virus isolated in cell culture from tumors of chickens with Marek's disease. I. Studies in cell culture. *J. nat. Cancer Inst.*, **41**, 939-950

Churchill, A. E., & Biggs, P. M. (1968) Herpes-type virus isolated in cell culture from tumors of chickens with Marek's disease. II. Studies *in vivo*. *J. nat. Cancer Inst.*, **41**, 951-956

Churchill, A. D., Chubb, R. C., & Baxendale, W. (1969) The attenuation, with loss of oncogenicity, of the herpes-type virus of Marek's disease (Strain HPRS-16) on passage in cell culture. *J. gen. Virol.*, **4**, 557-563

Churchill, A. E., Payne, L. N., & Chubb, R. C. (1969) Immunization against Marek's disease using a live attenuated virus. *Nature (Lond.)*, **221**, 744-747

Fig. 24. Thin section electron micrograph of a testis tumour. Lymphoid cells of different size are seen in this section. A granulocyte and some degenerating cells are also seen This tumour contained virus particles; these were found only in lymphocytes; ×3800.

Fig. 25. Thin section electron micrograph of the epithelial layer of the feather follicle from an MD-infected chicken. Viral multiplication occurs only in 3-4 layers of epithelial cells. Nuclear inclusions (arrow to the left) and many cytoplasmic inclusions (arrow to the right) are seen in these cells; ×3300.

Figs. 26 & 27. Viral particles in tumour cells were predominantly naked and were found in the nucleus (Fig. 26; ×7600), whereas in the epithelial cells of the feather follicle they were enveloped and were found within cytoplasmic inclusions (Fig. 27; ×48 000).

Cook, M. K., & Sears, J. F. (1970) Preparation of infectious cell-free herpes-type virus associated with Marek's disease. *J. Virol., 5,* 258-261

Cooper, M. D., Peterson, D. A., South, M. A., & Good, R. A. (1966) The functions of the thymus system and the bursa system in chickens. *J. exp. Med., 123,* 75-102

Eidson, C. S. & Schmittle, S. C. (1968) Studies on acute Marek's disease: I. Characteristics of isolate GA in chickens. *Avian Dis., 12,* 467-476

Eidson, C. S., & Schmittle, S. C. (1969) Studies on acute Marek's disease. XII. Detection of antibodies with a tannic acid indirect hemagglutination test. *Avian Dis., 13,* 774-782

Epstein, M. A., Achong, B. G., Churchill, A. E., & Biggs, P. M. (1968) Structure and development of the herpes-type virus of Marek's disease. *J. nat. Cancer Inst., 41,* 805-811

Evans, D. L., & Patterson, L. T. (1970) Correlation of immunological responsiveness with lymphocyte changes in chickens infected with acute leukosis (MD). *Bact. Proc.,* pp. 159-160

Kawamura, M., King, D. J., & Anderson, D. P. (1969) A herpesvirus isolated from kidney cell culture of normal turkeys. *Avian Dis., 13,* 853-863

Kottaridis, S. D., & Luginbuhl, R. E. (1968) Marek's disease. III. Immunofluorescent studies. *Avian Dis., 12,* 383-393

Kottaridis, S. D., Luginbuhl, R. E., & Chomiak, T. W. (1968) Marek's disease. IV. Antigenic components demonstrated by the immuno-diffusion test. *Avian Dis., 12,* 394-401

Lee, L. F., Roizman, B., Spear, P. G., Kieff, E. D., Burmester, B. R. & Nazerian, K. (1969) Marek's disease herpes virus: a cytomegalovirus? *Proc. nat. Acad. Sci. (Wash.), 64,* 952-956

Lee, L. F., Kieff, E. D., Bachenheimer, S. L., Roizman, B., Spear, P. G., Burmester, B. R., & Nazerian, K. (1971) Size and composition of Marek's disease virus deoxyribonucleic acid. *J. Virol., 7,* 289-294

Nazerian, K., Solomon, J. J., Witter, R. L., & Burmester, B. R. (1968) Studies on the etiology of Marek's disease. II. Finding of a herpesvirus in cell culture. *Proc. Soc. exp. Biol. (N.Y.), 127,* 177-182

Nazerian, K. (1968) Electron microscopy of a herpesvirus isolated from Marek's disease in duck and chicken embryo fibroblast cultures. *Proc. electron micr. Soc. Amer., 26,* 222-223

Nazerian, K., & Burmester, B. R. (1968) Electron microscopy of a herpes virus associated with the agent of Marek's disease in cell culture. *Cancer Res., 28,* 2454-2462

Nazerian, K., Sprandel, B. J., & Purchase, H. G. (1969) Localization of Marek's disease herpesvirus and its antigen by electron microscopy and fluorescent antibody techniques. *Proc. electron micr. Soc. Amer., 27,* 230-231

Nazerian, K., & Purchase, H. G. (1970) Combined fluorescent-antibody and electron microscopy study of Marek's disease virus infected cell culture. *J. Virol., 5,* 79-90

Nazerian, K., & Witter, R. L. (1970) Cell-free transmission and *in vivo* replication of Marek's disease virus. *J. Virol., 5,* 388-397

Nazerian, K. (1970a) Attenuation of Marek's disease virus and study of its properties in two different cell cultures. *J. nat. Cancer Inst., 44,* 1257-1267

Nazerian, K. (1970b) *In vivo* and *in vitro* replication of Marek's disease virus ultrastructural studies. In: *Proceedings International Congress of Electron Microscopy, Grenoble, France,* pp. 939-940

Nazerian, K., Lee, L. F., Witter, R. L. & Burmester, B. R. (1971) Ultrastructural studies of a herpesvirus of turkeys antigenically related to Marek's disease virus. *Virology, 43,* 442-452

Nazerian, K. (1971) Further studies on the replication of Marek's disease virus in the chicken and in cell culture. *J. nat. Cancer Inst. 47,* 207-217

Okazaki, W., Purchase, H. G., & Burmester, B. R. (1970) Protection against Marek's disease by vaccination with a herpesvirus of turkeys. *Avian Dis., 14,* 413-429

Ono, K., Kato, S., Iwa, N., & Doi, T. (1970a) Autoradiographic and cytochemical studies on nuclear and cytoplasmic inclusions of duck embryo fibroblasts infected with herpes-type virus isolated from chickens with Marek's disease. *Biken J., 13,* 53-57

Ono, K., Tanabe, S., Naito, M., Doi, T. & Kato, S. (1970b) Antigen common to a herpes type virus from chickens with Marek's disease and EB virus from Burkitt's lymphoma cells. *Biken J., 13,* 213-217

Payne, L. N., & Biggs, P. M. (1967) Studies on Marek's disease. II. Pathogenesis. *J. nat. Cancer Inst., 39,* 281-302

Payne, L. N. & Rennie, M. (1970) Lack of effect of bursectomy on Marek's disease. *J. nat. Cancer Inst., 45,* 387-397

Purchase, H. G. & Biggs, P. M. (1967) Characterization of five isolates of Marek's disease. *Res. Vet. Sci., 8,* 140-149

Purchase, H. G., Chubb, R. E. & Biggs, P. M. (1968). Effect of lymphoid leukosis and Marek's disease on the immunological responsiveness of the chicken. *J. nat. Cancer Inst., 40,* 583-592

Purchase, H. G. (1969) Immunofluorescence in the study of Marek's disease. I. Detection of antigen in cell culture and an antigenic comparison of eight isolates. *J. Virol., 3,* 557-569

Purchase, H. G., & Burgoyne, G. H. (1970) Immunofluorescence in the study of Marek's disease: detection of antibody. *Amer. J. vet. Res., 31,* 117-123

Purchase, H. G. (1970) Virus specific immunofluorescent and precipitin antigens and cell-free virus in the tissues of birds infected with Marek's disease. *Cancer Res., 30,* 1898-1908

Purchase, H. G., Burmester, B. R. & Cunningham, C. H. (1971a) Pathogenicity and antigenicity of clones from strains of Marek's disease virus and the herpesvirus of turkeys. *Inf. and Imm., 3*, 295-303

Purchase, H. G., Burmester, B. R. & Cunningham, C. H. (1971b) Responses of cell culture from various avian species to Marek's disease virus and the herpesvirus of turkeys. *Amer. J. vet. Res.* (in press)

Purchase, H. G., Okazaki, W., & Burmester, B. R. (1971) Field trials with the herpesvirus of turkeys (HVT) strain FC126 as a vaccine against Marek's disease. *Poult. Sci.* **50**, 775-783

Purchase, H. G. & Okazaki, W. (1971) The effect of vaccination with the herpesvirus of turkeys (HVT) on the horizontal spread of Marek's disease herpesvirus. *Avian Dis.* **15**, 391-397

Ringen, L. M., & Akhtar, A. S. (1968) Electrophoretic analysis of serum proteins from paralyzed and unparalyzed chickens exposed to Marek's disease. *Avian Dis., 12*, 4-9

Roizman, B., Spring, S. B., & Schwartz, J. (1969) The herpes virion and its precursors made in productively and abortively infected cells. *Fedn. Proc. Fedn. Am. Socs. exp. Biol., 28*, 1890-1898

Roizman, B. & Spear, P. G. (1971) Herpesvirus antigens on cell membrane detected by centrifugation of membrane-antibody complexes. *Science,* **171**, 298-300

Samadieh, B., Bankowski, R. A., & Carroll, E. J. (1969) Electrophoretic analysis of serum proteins of chickens experimentally infected with Marek's disease agents. *Amer. J. vet. Res., 30*, 837-846

Schidlovsky, G., Ahmed, M., & Jensen, K. E. (1969) Herpesvirus in Marek's disease tumors. *Science,* **164**, 959-961

Sevoian, M., Chamberlain, D. M., & Counter, F. T. (1962) Avian lymphomatosis. I. Experimental reproduction of the neutral and the visceral forms. *Vet. Med.,* **37**, 300-301

Sharma, J. M., Kenzy, S. G., & Rissberger, A. (1969) Propagation and behavior in chicken kidney cultures of the agent associated with classical Marek's disease. *J. nat. Cancer Inst.,* **43**, 907-916

Sharma, J. M. (1971) *In vitro* cell association of Marek's disease herpesvirus. *Amer. J. vet. Res.,* **32**, 291-301

Solomon, J. J., Witter, R. L., Nazerian, K. & Burmester, B. R. (1968) Studies on the etiology of Marek's disease. I. Propagation of the agent in cell culture. *Proc. Soc. exp. Biol. (N.Y.),* **127**, 173-177

Stevens, D. A., Kottaridis, S. D. & Luginbuhl, R. E. (1971) Investigations of antigenic relationship of Marek's disease herpesvirus and EB virus (herpes-type virus) *J. comp. Path.,* **81**, 137-140

Ubertini, T., & Calnek, B. W. (1970) Marek's disease herpesvirus in peripheral nerve lesions. *J. nat. Cancer Inst.,* **45**, 507-514

Witter, R. L., Solomon, J. J., & Burgoyne, G. H. (1969) Cell culture techniques for primary isolation of Marek's disease-associated herpesvirus. *Avian Dis.,* **13**, 101-118

Witter, R. L., Burgoyne, G. H., & Solomon, J. J. (1969) Evidence for a herpesvirus as an etiologic agent of Marek's disease. *Avian Dis.,* **13**, 171-184

Witter, R. L., Nazerian, K., Purchase, H. G., & Burgoyne, G. H. (1970) Isolation from turkeys of a cell-associated herpesvirus antigenically related to Marek's disease virus. *Amer. J. vet. Res.,* **31**, 525-538

Comparative Studies of DNAs of Marek's Disease and Herpes Simplex Viruses

S. L. BACHENHEIMER,[1] E. D. KIEFF [1,2] L. F. LEE [3] & B. ROIZMAN [1]

INTRODUCTION

In the last few years a number of viruses morphologically similar to members of the herpesvirus family have been isolated from normal and neoplastic cells of tumour-bearing men, monkeys, fowl, and frogs. Based on morphological considerations, these viruses have been designated as "herpes-type" or "herpes-like" viruses. This designation raises two questions, namely: first, whether these viruses are in fact herpesviruses; and second, whether they constitute a closely related subgroup. The evidence that the viruses are in fact herpesviruses is based on common biological properties deduced largely from light and electron microscopic and immunological studies, which have already been discussed by Roizman in this Symposium.[4] The objective of the studies described here was to determine whether the viruses associated with neoplasia share common physical and genetic properties differentiating them from other herpesviruses. The problem is not trivial. The classification of viruses into groups is based on their composition and structure. The classification correctly assumes that the information content of the virus can specify a unique structure. It does not preclude or deny the possibility that, if evolution were convergent, the same morphological characteristics could be acquired by two entirely different groups of viruses. Furthermore, the biological properties that we now discern at a superficial level only, could well be dictated by the structure of the virion. In an attempt to deal with this problem, over the last few years our laboratory has examined the DNAs of several herpesviruses (Wagner et al., 1970; Lee et al., 1971; Kieff, Bachenheimer & Roizman, 1971). In this paper we are reporting some preliminary data on the structure of the DNAs of herpes simplex subtype 1 (HSV-1) and subtype 2 (HSV-2) and of Marek's disease herpesvirus (MDV) from the point of view of their relatedness to one another and to other members of the herpesvirus family.

PURIFICATION OF VIRUS AND PREPARATION OF VIRAL DNA

Purification of virus for the extraction of DNA followed one of two procedures. Purified virions were prepared by the velocity centrifugation of cytoplasmic extracts obtained by Dounce homogenization of infected cells on Dextran 10 density gradients, according to a procedure developed by Spear & Roizman.[5] The enveloped virions form a relatively homogeneous band free of naked nucleocapsids. Naked nucleocapsids were obtained by the velocity centrifugation in sucrose density gradients of cytoplasmic lysates obtained by exposure of infected cells to 0.5% Nonidet P-40 (Shell Oil Co., New York, N.Y.) in a hypotonic buffer. The nucleocapsids formed two bands, a top band lacking DNA and a bottom band containing full particles (Roizman & Spear, 1971). In the experiments reported in this paper, the DNA was extracted from: (a) whole cytoplasmic lysate; (b) banded enveloped virions; or (c) banded naked nucleocapsids obtained by the procedure described in detail elsewhere (Lee et al.,

[1] Departments of Microbiology and Biophysics, The University of Chicago, 939 East 57th Street, Chicago, Illinois, USA.
[2] Present address: Department of Medicine, University of Chicago Hospitals and Clinics, 950 East 59th Street, Chicago, Illinois, USA.
[3] Regional Poultry Research laboratory, 3606 East Mount Hope Road, East Lansing, Michigan, USA.
[4] See p. 1 of this publication.

[5] Unpublished data.

1971; Kieff *et al.*, 1971). Briefly, the material containing the virus was suspended in 0.5% sodium dodecyl sulfate (SDS) and 2% sarkosyl, and centrifuged in neutral density gradients. The band containing intact viral DNA was extracted with phenol, followed by chloroform-isoamyl alcohol, and then dialyzed as required.

BASE COMPOSITION OF HSV- AND MDV-DNAs

The base compositions of HSV- and MDV-DNAs were determined from the buoyant density of the DNAs in CsCl solutions, measured in the Spinco model E centrifuge (Fig. 1). As internal standards

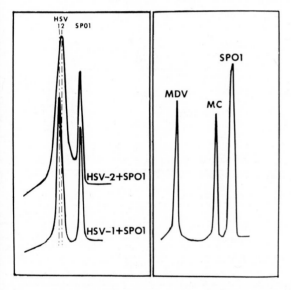

Fig. 1. Buoyant density determinations of HSV-1-, HSV-2- and MDV-DNAs. The buoyant densities of these DNAs were determined by cocentrifugation with SP01-DNA or with both SP01- and microccocal DNA in the Spinco model E centrifuge. The UV absorption photographs were scanned with a Joyce Loebel microdensitometer.

we used SP01-DNA, micrococcal DNA and on some occasions ϕ29-DNA. All of our analytical measurements gave uniform results. The density of the DNAs, computed according to the method of Szybalski (1968), were 1.705 ± 0.0005, 1.726 ± 0.0005 and 1.728 ± 0.0005 gm/cm³ for MDV-, HSV-1- and HSV-2-DNA respectively. The data obtained for HSV-1- and HSV-2-DNA were similar to those obtained by Goodheart, Plummer & Waner (1968).

At this point the question arose whether the difference in buoyant density of the DNA reflected differences in guanine (G)+cytosine (C) content or the presence of unusual bases in HSV-2-DNA. If the difference were due solely to a difference in the G+C content of the DNAs, a 0.002 gm/cm³ difference in buoyant density should correspond to approximately a 1°C difference in the melting temperature. This is precisely what was found (Fig. 2). From these data we conclude that the base composition of the MDV-, HSV-1- and HSV-2-DNAs corresponds to 46, 67 and 69 G+C moles/cent, respectively.

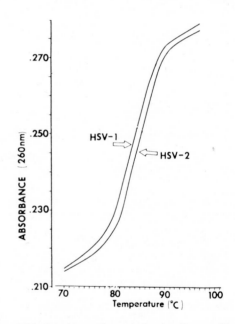

Fig. 2. UV absorbance-thermal denaturation profile of HSV-1- and HSV-2--DNAs in 0.015 M sodium chloride and 0.0015 M sodium citrate, measured in a Gilford Model 2000 spectrophotometer equipped with an automatic temperature and absorbance recorder.

SIZE OF HSV- AND MDV-DNAs

The size of HSV- and MDV-DNAs was determined by the velocity cosedimentation of these DNAs with T4-DNA in neutral and alkaline sucrose density gradients. The choice of the technique was based on two considerations. First, as recently pointed out by Freifelder (1970), all the various procedures for the determination of the size of DNA molecules differ considerably with respect to accuracy and requirements. In fact, absolute measurements of size are available for very few biologically significant DNA molecules. Second, of the various procedures readily applicable for use with HSV- and MDV-DNAs, the one that seemed most appropriate and useful was based on cosedimentation of the DNAs with a DNA of known size. T4-DNA was chosen because the DNA is only slightly larger than HSV-DNA and, moreover, its absolute size is known. In these experiments we measured the sedimentation of the native and alkaline denatured DNAs in neutral and alkaline sucrose density gradients respectively. The results (Figs. 3 & 4) showed the following: (i) In neutral sucrose density gradients, both HSV- and MDV-DNAs sedimented slightly more slowly than T4-DNA (Fig. 3A, B, C and Fig. 4A). MDV-DNA, on cocentrifugation with HSV-1-DNA, sedimented slightly faster (Fig. 4B). On the basis of the relationship of Burgi & Hershey (1963), we calculated the

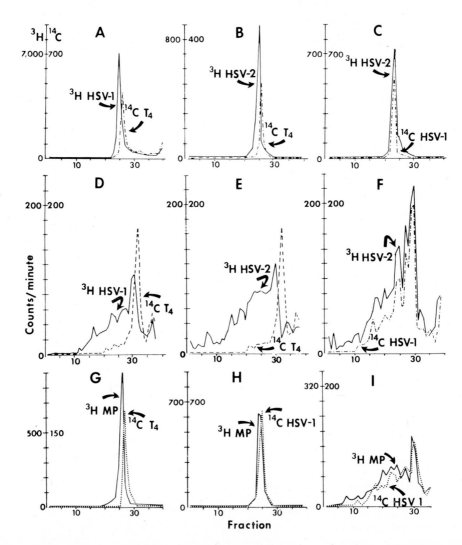

Fig. 3. Zone sedimentation of HSV-DNA in neutral and alkaline sucrose density gradients. HSV-1-DNA, HSV-2-DNA and the DNA of the MP mutant of HSV-1 were labelled with ³H- or ¹⁴C-thymidine and cosedimented with each other (C, F, H, I) and with T4-DNA (A, B, D, G). The DNAs were centrifuged for 3.5 hours in a SW41 rotor at 40 000 rpm and 20°C. ¹⁴C-labelled DNA: dashed line; ³H-labelled DNA: solid line. Direction of sedimentation is to the right. Details of the preparation and denaturation of the DNA, and of the neutral and alkaline sucrose density gradients are given in Kieff et al. (1971).

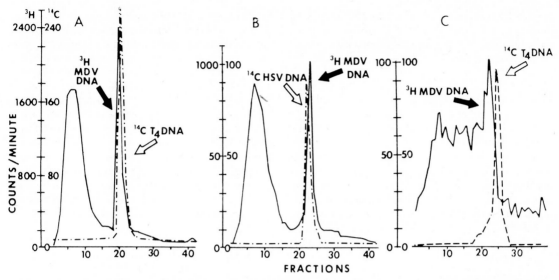

Fig. 4.   Zone sedimentation of MDV-DNA in neutral and alkaline sucrose density gradients.   MDV-DNA labelled with
³H-thymidine was cocentrifuged on neutral sucrose density gradients with ¹⁴C-T4-DNA (A) or with ¹⁴C-HSV-1-DNA
(B) and on alkaline sucrose density gradients with ¹⁴C-T4-DNA (C).   Direction of sedimentation is to the right.   Details
of the preparation of the DNA, etc., are given in Lee *et al.* (1971).

molecular weight of the native HSV-1-, HSV-2- and
MDV-DNAs in neutral sucrose density gradients to
be 99 ± 5, 99 ± 5 and 103 ± 5 million Daltons
respectively.   An interesting finding was that the
DNA of the MP strain, a multistep mutant of HSV-1
(Hoggan & Roizman 1959), sedimented more slowly
than that of HSV-1 (Fig. 3G, H).   The calculated
molecular weight of the MP mutant is 95 ± 5 million
Daltons.   Additional support for the hypothesis that
MP is a deletion mutant is provided by the fact that
this virus is recessive with respect to its effect on the
social behaviour of infected cells and also with
respect to its ability to specify the synthesis of a
virus-specific membrane glycoprotein (Keller, Spear
& Roizman, 1970; Roizman 1971).   (ii) Denatured
intact viral DNA sedimented in alkaline gradients
yielded several discrete bands containing fragments
calculated to range in size from 7 to 48 million
Daltons for HSV and to 50 million Daltons for
MDV (Fig. 3D, E, F and Fig. 4C).   The largest
fragments are of the size expected for intact single
strands.   However, no more than 50% of all HSV-
or MDV-DNA was present as intact single-stranded
DNA.   The alkaline sedimentation data indicated
that the intact double-stranded DNAs of MDV,
HSV-1 and HSV-2 contain single-stranded breaks
and that the single-stranded fragments were not
products of random cleavage of the DNA.

THE SOURCE AND SIGNIFICANCE OF
THE SINGLE-STRANDED BREAKS IN HSV-DNA

HSV-1 served as a prototype for studies of the
source of the single-stranded breaks in the herpesvirus DNA.   The data may be summarized as follows.   (i) The breaks in the strands are not the consequence of manipulation of the virus or the method
of extraction of the DNA (Kieff *et al.,* 1971).   Viral
DNA released from cytoplasmic lysates with SDS
and denatured directly on top of the gradient contained the same single-stranded nicks as DNA extracted from purified virions.   In all these experiments T4-DNA extracted, denatured and cosedimented with HSV-DNA had only trace amounts of
single-stranded breaks.   (ii) The single-stranded
breaks do not appear to be generated by a putative
nuclease present in the virions.

These conclusions are based on the results of
several series of experiments as follows: (i) Incubation of purified virus at 37°C had no effect on the
size of the DNA single strands.   (ii) Exposure of
form 1 SV-40-DNA to intact or sonicated nucleocapsids did not convert the SV-40-DNA to form 2.
(iii) Incubation of intact or sonicated nucleocapsids
with labelled *Escherichia coli* DNA (native or denatured) in solutions under widely varying conditions
of pH and concentration of divalent cations failed

to demonstrate release of labelled nucleotides from the DNA. (iv) Analysis of the DNA present in the nucleus and in the cytoplasm of infected cells showed that both nuclear viral DNA (largely free) and cytoplasmic DNA (all in virions) were complete double-stranded molecules with single-stranded nicks. (v) In the last series of experiments it was found that the DNA made during a 4-hour pulse, between 4 and 8 hours post infection, yielded on denaturation and velocity centrifugation in alkaline sucrose density gradients DNA with an average molecular weight of 4 million Daltons. This highly nicked DNA was present in both the nucleus (free) and in the cytoplasm (in virions). After an 8-hour chase, the DNA made during the pulse appeared to be repaired and/or ligated and yielded fragments of molecular weight 7 to 48 million Daltons (Fig. 5). From these studies

we conclude that HSV-1-DNA and, by extension, HSV-2- and MDV-DNA accumulate in highly nicked form immediately after synthesis and that the nicks are largely but not completely ligated and repaired before encapsidation of the DNA. Combined with the finding that the residual nicks are nonrandom, as indicated earlier in the text, these data suggest that the nicks are not a consequence of some process of random cleavage and repair but rather of an ordered process of DNA cleavage and repair of substantial biological significance. A similar process of single-stranded cleavage and repair appears to control transcription in T4 phages (Riva, Cascino & Geiduschek, 1970). Current studies (Frenkel [1]) indicate that the fragments are preferentially in one strand.

SEQUENCE HOMOLOGY OF HSV-1-, HSV-2- AND MDV-
DNAs

The extent of the nucleotide sequence homology between HSV-1-, HSV-2- and MDV-DNAs was determined by annealing labelled denatured DNA in solution with a 50-fold excess of denatured DNA immobilized on nitrocellulose filters (HAWP 0.45 µ, Millipore Co., Bedford, Mass.).

Two series of experiments were performed. In the first, we chose to estimate the homology between HSV-1- and HSV-2-DNAs by determining the relative rates at which homologous and heterologous labelled DNAs in solution anneal to identical DNA discs. The rationale for this experimental design was as follows: (i) The effect of competitive hybridization of DNA in solution is minimal for short intervals. After 2 hours of incubation at least 80% of the DNA in solution remained as free single strands. (ii) Virtually all ($> 95\%$) of the sequences of HSV-1-DNA re-anneal at the same rate (Frenkel & Roizman, 1971). The initial rate of hybridization therefore reflects the concentration of homologous DNA species in solution. The results summarized in Table 1 indicate that there is 40% sequence homology between HSV-1- and HSV-2-DNAs.

Recent studies on the extent of nucleotide sequence homology between HSV-1- and HSV-2-DNAs by hydroxyapatite chromatography (Kieff et al.[1]) also indicate that 40% of the sequences of the two DNAs are complementary, i.e., capable of forming HSV-1/HSV-2 heteroduplexes. The T_m of HSV-1/HSV-2 heteroduplexes is 9°C less than that of reassociated

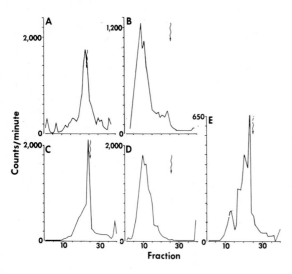

Fig. 5. Sedimentation of HSV-1-DNA extracted from infected cells after a 4-hour pulse with ³H-thymidine and an 8-hour chase.
A—Native nuclear DNA extracted after 4-hour pulse and centrifuged on neutral sucrose density gradients.
B—Alkali-denatured DNA extracted from the nucleus after 4-hour pulse and centrifuged on alkaline sucrose density gradients.
C—Native cytoplasmic DNA extracted after 4-hour pulse and centrifuged on neutral density gradient.
D—Alkali-denatured DNA extracted from the cytoplasm after 4-hour pulse and centrifuged on alkaline sucrose density gradient.
E—Alkali-denatured DNA extracted from cytoplasm of infected cells after 8-hour chase.
The DNAs were centrifuged at 40 000 rpm in a SW41 rotor at 20°C. The arrow indicates the position of ¹⁴C-T4-DNA cosedimented with ³H-HSV-DNA. The direction of centrifugation is to the right.

---

[1] Unpublished data.

HSV-1 or HSV-2, indicating that the HSV-1/HSV-2 heteroduplexes are not capable of complete base pairing. We estimate that 12% of the nucleotides in the related sequences of HSV-1- and HSV-2-DNA are noncomplementary.

Table 1. Sequence homology of DNAs

Labelled DNA in solution	DNA on filters		
	HSV-1	HSV-2	MDV
Experiment 1[1]			
HSV-1	100	33	—
HSV-2	47	100	—
Experiment 2[2]			
HSV-1	100	—	0.5
HSV-2	—	100	3.2
Experiment 3[2]			
HSV-1	100	—	0.6
HSV-2	—	100	1.2

[1] Data based on the relative rate at which heterologous and homologous labelled DNAs hybridized to identical DNA discs. Hybridization was performed at 61°C in a solution containing 5×SSC (Standard saline citrate = 0.15 M sodium chloride, 0.015 M sodium citrate) and 30% formamide. Background level using HEp-2-DNA was less than 1% of homologous hybridization. Homologous hybridization of labelled HSV-1-DNA = 1941 cpm, and for labelled HSV-2-DNA = 1280 cpm.

[2] Data based on the relative amount of DNA bound to heterologous and homologous DNA discs after 24 hours of incubation. Hybridization was performed at 61°C (Experiment 2) or at 53°C (Experiment 3) in solution containing 5×SSC and 30% formamide. Background level using duck embryo fibroblast DNA was 0.5% of homologous hybridization. Homologous hybridization (experiment 2) of labelled HSV-1-DNA = 687 cpm, and for labelled HSV-2-DNA = 486 cpm; homologous hybridization (experiment 3) of labelled HSV-1-DNA = 432 cpm, and for labelled HSV-2-DNA = 840 cpm. Input of labelled HSV-2-DNA in experiments 2 and 3 was 2·4 times that in experiment 1.

In the second series of experiments, labelled HSV-1- and HSV-2-DNAs were incubated with DNA discs containing MDV-DNA or the homologous HSV-DNA. The reaction was extended to 24 hours because of the low level of hybridization of HSV-DNA with MDV-DNA. The results, also shown in Table 1, indicate that the extent of homology between HSV-1- or HSV-2-DNA and MDV-DNA is < 1–2% respectively.

CONCLUSIONS

The studies presented in this paper do not support the hypothesis that the herpesviruses associated with neoplasia have a common derivation or share unique physical properties not shared by other herpesviruses. This conclusion is based on the following arguments:

(i) The difference in the base composition of MDV- and HSV-DNAs is not unexpected. Our laboratory previously reported (Wagner et al., 1970) the DNA base composition of the herpesvirus present in the Lucké adenocarcinoma of winter frogs and confirmed the base composition of the herpesvirus associated with Burkitt's lymphoma reported by Schulte-Holthausen and zur Hausen (1970). The base composition of the DNA of herpesvirus saimiri was reported by Goodheart (1970). The range of base compositions of the DNAs recovered from herpesviruses associated with neoplasia (see Table 2) does not differ significantly from the range of base compositions of all herpesviruses reported to date. (ii) The sizes of the DNAs of the herpesviruses carefully examined to date appear to be very similar. In addition to our data summarized in Table 2, Soehner, Gentry & Randall (1965) calculated the molecular weight of the equine abortion virus DNA to be 84 to 94×10^6 Daltons. HSV-1-DNA ranges in molecular weight from 85 to 115 million Daltons by the Kleinschmidt test (Becker, Dym & Sarov, 1968), while a figure of $99 + 5$ million Daltons is obtained by cosedimentation with T4-DNA (Kieff et al., 1971).

Table 2. Base composition of herpesvirus DNAs

Common name	G+C (moles/cent)
Pseudorabies	74
Infectious bovine rhinotracheitis	71
Herpes saimiri virus	69
Herpes simplex type 2	69
Herpes simplex type 1	67
Equine coital exanthema virus	67
Lapine herpesvirus	66
Bovine mammilitis virus	64
Burkitt's lymphoma herpesvirus	57
Equine abortion virus	57
Cytomegalovirus, human	57
Cytomegalovirus, murine	57
Frog virus No. 4	55
Vervet monkey herpes	51
Marek's disease herpesvirus	47
Turkey herpesviruses	47
Feline herpesvirus	46
Frog herpesvirus (Lucké)	45

It is noteworthy that a similar value (95 million Daltons) was calculated for the genomic size of HSV-1 from the renaturation kinetics of its DNA (Frenkel & Roizman 1971). HSV-1 cannot be differentiated from HSV-2 with respect to the size of the DNA. The significance of the difference in the calculated molecular weights of MDV- and HSV-DNA is not clear. Since the size was calculated from the sedimentation constant and since this is affected by the secondary and tertiary structure of the DNA, it is conceivable that the more rapid sedimentation of MDV-DNA reflects not so much a larger molecule as a less rigid one resulting both from a much lower G+C content and the numerous single-stranded nicks. (iii) Both MDV- and HSV-DNAs are linear, double-stranded molecules with nonrandom breaks. (iv) The DNA-DNA hybridization data presented in this paper confirm the genetic relatedness inferred from the immunological relatedness of HSV-1 and HSV-2, reported by numerous laboratories. Our data concerning the lack of genetic relatedness between MDV and HSV, coupled with the data reported by zur Hausen et al. (1970) and zur Hausen & Schulte-Holthausen (1970) concerning the lack of homology between MDV- or HSV-DNA and the

DNA of the virus associated with Burkitt's lymphoma, indicate that herpesviruses unrelated immunologically, including those associated with tumours, show no evidence of genetic relatedness. It is, perhaps, pertinent to point out here that the failure to find evidence of genetic relatedness by means of DNA-DNA hybridization tests does not justify the conclusion that the various herpesviruses tested to date have arisen either by convergent or divergent evolution. In principle, we observe that viruses that appear to be morphologically similar and to multiply in a similar fashion must in consequence direct the synthesis of fundamentally similar structural and nonstructural proteins. Theoretically, at least, similar or even identical proteins can be specified by entirely dissimilar, nonhomologous deoxynucleotide sequences.

ACKNOWLEDGEMENTS

The studies at The University of Chicago were aided by grants from the American Cancer Society (E314F), the United States Public Health Service (CA 08494) and the Leukemia Research Foundation. S. L. B. is a predoctoral trainee (United States Public Health Service, AI 00238). E. D. K. is a postdoctoral trainee (United States Public Health Service, AI 00238).

REFERENCES

Becker, Y., Dym, H. & Sarov, I. (1968) Herpes simplex virus DNA. *Virology*, **36**, 184-192

Burgi, E. & Hershey, A. D. (1963) Sedimentation rate as a measure of molecular weight of DNA. *Biophys. J.*, **3**, 309-321

Freifelder, D. (1970) Molecular weights of coli phages and coli phage DNA. IV. Molecular weights of DNA from bacteriophage T4, T5, T7 and the general problem of determination of M. *J. molec. Biol.*, **54**, 569-577

Frenkel, N. & Roizman, B. (1971) Herpes simplex virus: studies of the genome size and redundancy by renaturation kinetics. *J. Virol.*, **8**, 591-593

Goodheart, C. R., Plummer, G. & Waner, J. L. (1968) Density differences of DNA of herpes simplex viruses types 1 and 2. *Virology*, **35**, 473-475

Goodheart, C. R. (1970) Herpesviruses and cancer. *J. Amer. med. Ass.*, **211**, 91-96

Hoggan, M. D. & Roizman, B. (1959) The isolation and properties of a variant of herpes simplex producing multinucleated giant cells in monolayer cultures in the presence of antibody. *Amer. J. Hyg.*, **70**, 208-219

Keller, J. M., Spear, P. G. & Roizman, B. (1970) The proteins specified by herpes simplex virus. III. Evidence that membrane glycoproteins made after infection are genetically determined by the virus. *Proc. nat. Acad. Sci. (Wash.)*, **65**, 865-871

Kieff, E. D., Bachenheimer, S. L. & Roizman, B. (1971) Size, composition and structure of the deoxyribonucleic acid of subtypes 1 and 2 herpes simplex viruses. *J. Virol.*, **8**, 125-132

Lee, L., Kieff, E. D., Bachenheimer, S. L., Roizman, B., Spear, P. G., Burmester, B. R. & Nazerian, K. (1971) Size and composition of Marek's disease virus deoxyribonucleic acid. *J. Virol.*, **7**, 289-294

Riva, S., Cascino, A., and Geiduschek, E. P. (1970) Coupling of late transcription to viral replication in bacteriophage T4 development. *J. molec. Biol.*, **54**, 85-102

Roizman, B. (1971) Herpesviruses, membranes and the social behavior of infected cells. In: *Proceedings of the 3rd International Symposium on applied and medical virology, Fort Lauderdale, Florida*, pp. 37-72

Roizman, B. & Spear, P. G. (1971) Herpesviruses: current information on the composition and structure. In: Maramorosch, K. & Kurstak, E., eds, *Comparative virology*, New York, Academic Press, pp. 135-168

Schulte-Holthausen, H. and zur Hausen, H. (1970) Partial purification of the Epstein-Barr virus and some properties of HS DNA. *Virology*, **40**, 776-779

Soehner, R. L., Gentry, C. A. & Randall, C. C. (1965) Some physicochemical characteristics of equine abortion virus nucleuc acid. *Virology*, **26**, 394-405

Szybalski, W. (1968) Use of cesium sulfate for equilibrium dersity centrifugation. *Methods in enzymology*, **12**, 330-360

Wagner, E. K., Roizman, B., Savage, T., Spear, P. G., Mizell, M., Durr, F. E. & Sypowicz, D. (1970) Characterization of the DNA of herpes viruses associated with Lucké adenocarcinoma of the frog, and Burkitt lymphoma of man. *Virology*, **42**, 257-261

zur Hausen, H. & Schulte-Holthausen, H. (1970) Presence of EB virus nucleic acid homology in a "virus-free" line of Burkitt tumour cells. *Nature (Lond.)*, **227**, 245-248

zur Hausen, H., Schulte-Holthausen, H., Klein, G., Henle, W., Henle, G., Clifford, P. & Santesson, L. (1970) EBV DNA in biopsies of Burkitt tumors and anaplastic carcinomas of the nasopharynx. *Nature (Lond.)*, **228**, 1056

Some Characteristics of a Herpesvirus of Turkeys and its DNA*

O. KAADEN & B. DIETZSCHOLD

The isolation of a herpesvirus of turkeys (HVT), designated as strain FC-126, was reported by Witter *et al.* (1970). The biological properties of this virus in various tissue culture systems were similar to those of group B herpesviruses. Because of its serological relationship to Marek's disease herpesvirus (MDHV), it was of interest to study further the biological and biochemical features of HVT.

Our experiments were performed with the FC-126 strain of the Regional Poultry Research Laboratory, East Lansing. The HVT FC-126 used in these experiments was propagated by serial passaging in cell cultures from 10-day-old chicken embryo fibroblasts (CEF) grown in roller bottles (40 × 7 cm). The growth medium consisted of TCM 199 enriched with 10% (v/v) calf serum. After infection with HVT, the TCM 199 was replaced by a maintenance medium consisting of TCM 199 and Hank's solution in equal parts; 0.25% (w/v) lactalbumine hydrolysate and 0.25% (v/v) calf serum were added. Each roller flask containing approximately $1.0–1.2 \times 10^8$ cells was infected with 0.5 to 0.6 PFU per cell. The specificity of the HVT stock material used was confirmed by the agar gel precipitin test using known positive HVT and MDHV antisera.

In order to obtain labelled virus particles or viral DNA, respectively, infected CEF cells in roller bottles were incubated at 37° with fresh maintenance medium lacking thymine. Eight to ten hours post infection (p.i.) 20 μCi/ml ^{3}H-thymidine (Buchler & Co., Brunswick; specific activity 1000 μCi/mM) were added. The cultures were allowed to incorporate for 24 to 32 hours. In some experiments radioactive labelling was effected by addition of 30 μCi/ml ^{32}P-orthophosphate (Buchler & Co., Brunswick) using a tissue culture medium with a reduced phosphate content. The acid-insoluble radioactivity of the viral particles or DNA isolated was determined after precipitation with ice-cold 5% (w/v) trichloroacetic acid (TCA) containing 0.1 M pyrophosphate. The precipitates were collected on membrane filters (SM 11305, Sartorius, Membranfilter GmbH, Göttingen) and washed five times with TCA. Dried filters were transferred to glass vials containing Insta-Gel^R (Packard, Frankfurt) or to a toluene scintillator for the determination of radioactivity in a liquid scintillation spectrometer (Tri-Carb, Packard model 3375). For isolation of cell-free HVT, heavily infected CEF cells were scraped from the glass surface; after low-speed centrifugation, they were suspended in a small volume of tris buffer, pH 7.4, with addition of 0.5 M sucrose, 0.01 M $MgCl_2$, and 0.001 M EDTA, and homogenized in a tight-fitting Dounce homogenizer. The clarified suspensions were immediately diluted for virus titration. The effect of this treatment on the virus titre and on the recovery of infectious HVT is shown in the Table. The homogenization of HVT-infected cells in a sucrose-containing tris buffer was shown to be an effective method for the isolation of biologically active subcellular structures, yielding a mean recovery of infectivity of 4–6%. The use of phenylmethylsulphonylfluoride (PMSF) (Mann Research Laboratories, USA) as a potent protease inhibitor only slightly increased the virus yield after homogenization. The suitability of PMSF for the isolation of extracellular infectious HVT is being studied further. The isolated extracellular HVT, supplemented with 10% (v/v) calf serum and 5% (v/v) dimethylsulphoxide (DMSO), remained infectious even after storage at —80°.

For electron microscopic examination, cell-free virus suspensions were layered on top of a linear potassium citrate gradient (density range 1.04–1.32 g/cm^3) and centrifuged in a SW 39 rotor of the Spinco

* From the Federal Research Institute for Animal Virus Diseases, Tübingen, Federal Republic of Germany.

Effects of isolation procedure [1] on the titre
of cell-associated and cell-free turkey herpesvirus (HVT)
grown in chicken embryo fibroblasts

Origin of HVT	Infectivity (PFU/ml)	Recovery of infectivity (%)
Cell-associated HVT	5.04×10^5	100
Cell-free tissue culture supernatant	0.7×10^2	0.01
Cell-free virus (homogenized and clarified cell suspension)	2.7×10^4	5.4
Cell-free HVT as above, but filtered through Millipore filters (HA, 0.45 μ)	1.4×10^2	0.03

[1] Cultured cells (1.8×10^6/ml) were suspended in 0.01 M tris-HCl, 0.5 M sucrose, 0.01 M $MgCl_2$, and 0.001 M EDTA, pH 7.6, and homogenized by means of a tight-fitting Dounce homogenizer.

model L ultracentrifuge for 2 hours at 100 000 g. Generally, two broad zones could be seen after such density gradient centrifugation of homogenized HVT-infected cells. Aliquots of the gradient were negatively stained with phosphotungstic acid (PTA) at pH 6.5, and examined in a Siemens Elmiskop 101. Fig. 1 shows four different morphological types of virus particles appearing in the aliquots of the gradient. The terms "full" or "naked" are descriptive only and based on the observation that the core of the viral particles appeared either to be penetrated or not penetrated by PTA (Watson, Russell & Wildy, 1963). Noninfectious "naked" nucleocapsids of 100 to 110 nm diameter were found predominantly in the upper zone. From the point of view of the distribution of PTA, the zone material contained "full" as well as "empty" particles. Many of the particles showed a cross-shaped structure within the core, as reported by Nazerian et al. (1971), and the typical icosahedral structure of herpesviruses with hollow elongated capsomeres on their surface (Fig. 1E). The lower band consisted of enveloped particles ("full" or "empty"), capable of infecting cultured cells and inducing typical microplaques in CEF in the same way as HVT and MDHV. In some preparations, distinct projections on the surface of the envelope appeared as a small border.

Fig. 1. These micrographs show different morphological types of HVT particles negatively stained with PTA, pH 6.5.

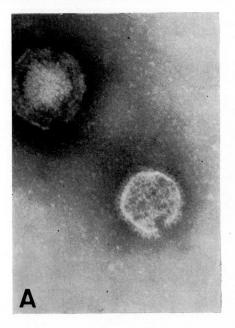
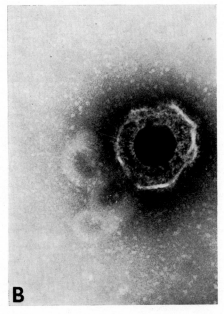

A—HVT particle isolated from infected CEF cells. Note envelope surrounding capsid. × 120 000.
B—An enveloped "empty" HVT particle. There are projections on the surface of the envelope. × 120 000.

Fig. 1 *(cont.)*

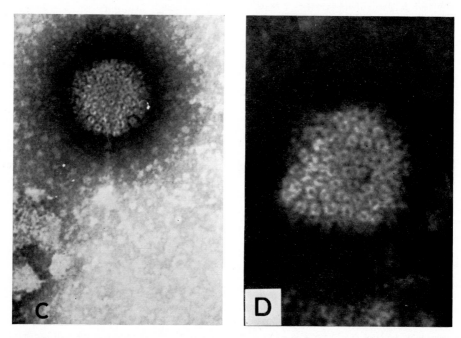

C—HVT nucleocapsid.   ×240 000.
D—HVT nucleocapsid at higher magnification.   ×320 000.

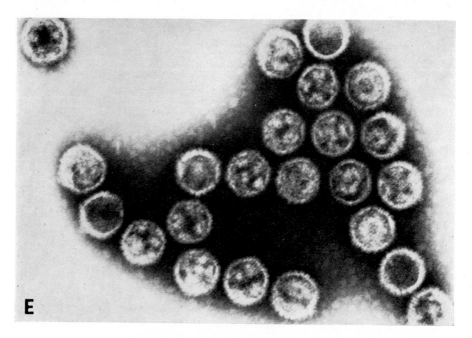

E—A group of HVT nucleocapsids isolated from cytoplasm of infected CEF cells.   Note particles showing a cross-shaped structure in their cores.   ×120 000.

Ultrahistological methods (Nazerian *et al.,* 1971) and our isolation experiments showed that mainly nucleocapsids but also "empty" enveloped particles or particles with membrane alterations are present in HVT-infected CEF cells. The envelope of HVT appeared to be extremely labile and often disrupted (Figs. 1, A & B). These observations led us to test the adsorption behaviour of isolated nucleocapsids. The preparations labelled with ^{32}P-orthophosphate and isolated by sucrose gradient centrifugation were almost free of viral envelopes, as shown by electron microscopic examination and sedimentation analysis. After several washings, the CEF grown in Petri dishes were inoculated with HVT-nucleocapsid preparations containing 10 000 cpm ^{32}P. A second group of tissue cultures was pretreated with DEAE-dextran dissolved in phosphate-buffered saline (PBS) (1 mg/ml). The infected cells were then incubated at 37°C. Samples of washed cells obtained at various times p.i. were removed, centrifuged, lysed with Hyamin^R (Packard, Frankfurt), and the radioactivity counted. From 5 to 120 minutes p.i., less than 0.5% of the inoculum radioactivity was adsorbed by the cells. Most of the activity could be removed by three consecutive washings. We obtained the same results after treatment of viral preparations with 0.1% (w/v) Nonidet P-40, known to be useful as a nonionic detergent for stripping off viral envelopes and cell membranes. Our results seem to confirm that the lack of infectiousness of HVT nucleocapsids is caused by the loss of the viral envelopes rather than by a lack of infectious DNA.

Viral DNA was isolated from ^{3}H-thymidine-labelled nucleocapsids. The material used in the isolation procedure was a suspension of HVT-infected CEF cells in 0.1 M tris buffer, pH 7.4, supplemented with 0.1% (w/v) NP-40. The nuclei of the cells were immediately sedimented by centrifuging for 3 minutes at 500 g, resuspended in a hypotonic tris buffer and disrupted in a Dounce homogenizer. After restoration of isotonicity, cytoplasmic and nuclear fractions were mixed and used for the isolation of viral nucleocapsids. The suspension was clarified at 7000 g for 10 minutes, centrifuged through a 25% (w/v) sucrose cushion and the resuspended pellet layered on to a sucrose gradient (15–30% w/v) made in reticulocyte standard buffer, pH 7.4. The sedimentation profile of the material is shown in Fig. 2. By electron microscopic examination, the ^{3}H-thymidine activity could be located in the viral nucleocapsids. In control preparations of uninfected cells no radioactivity was found at this site. The buoyant density of the peak material in CsCl density gradients was estimated to be 1.27 g/cm³.

Viral DNA was isolated by the method of Becker, Dym & Sarov (1968). Sodium dodecyl sulfate (final concentration 1% w/v) and pronase heated for 10 minutes at 80°C (final concentration 1 mg/ml) were added to the concentrated virus preparations and incubated in a water bath at 37°C overnight. The digested preparations were extracted with a chloroform—isoamyl alcohol mixture, centrifuged, and the aqueous phase was then removed for further studies. Using the method of Burgi & Hershey (1963), the

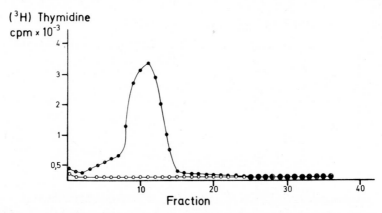

Fig. 2. Sucrose gradient centrifugation of HVT nucleocapsids from infected CEF cells. The cells were incubated in the presence of ^{3}H-thymidine for 24 hours, dissolved in 0.1% (w/v) NP-40 and analysed in a 15-30% (w/v) sucrose gradient prepared in reticulocyte standard buffer, pH 7.4. The gradient was centrifuged for 70 minutes at 40 000 g. Uninfected CEF cultures were treated in the same way.

●——————● HVT-infected CEF cells.
○——————○ Uninfected CEF cultures.

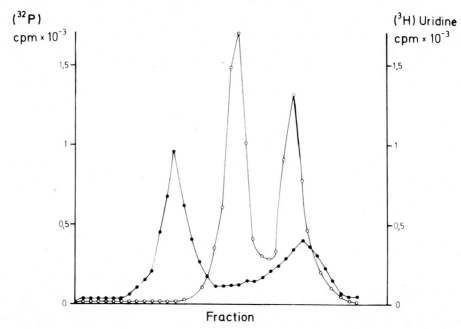

Fig. 3. Zonal centrifugation of native HVT-DNA. DNA was prepared from ³H-thymidine-labelled nucleocapsids by extraction with SDS/pronase, layered on a 15-30 % (w/v) sucrose gradient containing 0.1 M NaCl, and centrifuged for 13 hours at 60 000 g in a SW 25 rotor; 28S and 16S ribosomal RNA from BHK cells were used as sedimentation markers.
●———● HVT-DNA.
○———○ 28S and 16S ribosomal BHK-RNA.

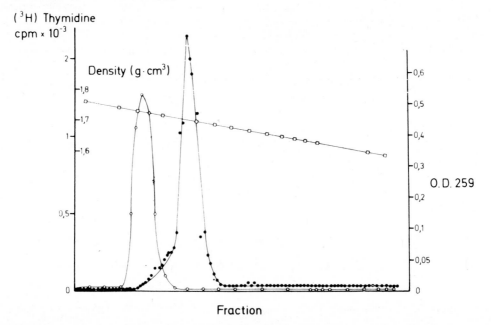

Fig. 4. Equilibrium centrifugation of HVT-DNA extracted from purified nucleocapsids. CsCl was added to give a final density of 1.700 g/cm³ and the tubes were centrifuged in the SW 39 rotor of the Spinco model L ultracentrifuge for 60 hours at 100 000 g. DNA from *Micrococcus lysodeicticus* was cosedimented as a density marker.
●———● HVT-DNA.   ○———○ *M. lysodeicticus*-DNA.   □———□ CsCl density (g/cm³).

sedimentation analysis of the HVT-DNA, as com-pared with 28S and 16S ribosomal BHK cell RNA, showed a sedimentation coefficient of about 40–45S. A slower component sedimented about 10S behind the 16S RNA marker (Fig. 3). After CsCl-equilibri-um centrifugation in the SW 39 rotor of the pre-parative ultracentrifuge for 60 hours at 100 000 g and 20°C, a homogeneous peak of radioactivity was found, banding at a buoyant density of 1.716 g/cm³ (Fig. 4). DNase treatment (50 μg/ml) of the peak fractions after dialysis resulted in acid-soluble radio-activity. In order to confirm the sedimentation coefficient and the buoyant density of the HVT-DNA, additional experiments in the analytical ultracentrif-uge are to be carried out.

Finally, we also determined the melting profile obtained by the method of Marmur & Doty (1962). HVT-DNA was dissolved at 35 μg/ml in standard saline citrate (SSC) solution, pH 7.4, and heated by circulating water through a jacketed cell. A therm-istor immersed in the DNA solution measured its temperature. The melting experiments yielded T_m values for HVT-DNA varying between 92 and 93°C (Fig. 5). According to the formula of Marmur & Doty (1962) and Schildkraut et al. (1962), the HVT-DNA has a buoyant density and a melting temperature corresponding to 56–57 mole percent guanine + cyto-sine. This value is in agreement with the data for MDHV-DNA, as reported by Lee et al. (1969). The data from the determination of buoyant density, melting temperature, and G+C content suggest that

HVT virions contain a linear double-stranded mole-cule of DNA. We were recently able to demonstrate the viral DNA molecules by electron microscopy and to confirm the linear conformation. Because of the strong cell-associated infectivity and the character-istics of the viral DNA, we propose to include the HVT FC-126 strain in the B group of herpesviruses.

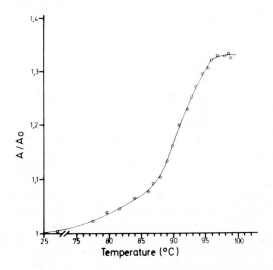

Fig. 5. Melting profile of HVT-DNA. Viral DNA was dissolved at 35 μg/ml in 1 × SSC (pH 7.4) and the tempera-ture of the solution was raised by 1° every 5 min. A/A₀ is the ratio of the adsorbances at 259 nm at the given temper-ature (A) and at 25°C (A₀).

ACKNOWLEDGEMENTS

We should like to express our gratitude to Dr B. R. Burmester and Dr V. von Bülow for providing HVT isolate FC-126 and MDHV-antiserum, respectively.

REFERENCES

Becker, Y., Dym, H. & Sarov, I. (1968) Herpes simplex virus DNA. *Virology*, 36, 184-192

Burgi, E. & Hershey, A. D. (1963) Sedimentation rate as a measure of molecular weight of DNA. *Biophys. J.*, 3, 309-321

Lee, L. F., Roizman, B., Spear, P. G., Kieff, E. D., Bur-mester, B. R., & Nazerian, K. (1969) Marek's disease herpes virus: A cytomegalovirus? *Proc. nat. Acad. Sci. (Wash.)*, 64, 852-856

Marmur, J., & Doty, P. (1962) Determination of the base composition of deoxyribonucleic acid from its thermal denaturation temperature. *J. molec. Biol.*, 5, 109-118

Nazerian, K., Lee, L. F., Witter, R. L. & Burmester, B. R. (1971) Ultrastructural studies of a herpesvirus of turkeys

antigenically related to Marek's disease virus. *Virology*, 43, 442-452

Schildkraut, C. L., Marmur, J., & Doty, P. (1962) Deter-mination of the base composition of deoxyribonucleic acid from its buoyant density in CsCl. *J. molec. Biol.*, 4, 430-443

Watson, D. H., Russell, W. C., & Wildy, P. (1963) Electron microscopic particle counts on herpes virus using the phosphotungstate negative staining technique. *Virology*, 19, 250-260

Witter, R. L., Nazerian, K., Purchase, H. G., & Burgoyne, G. H. (1970) Isolation from turkeys of a cell-associated herpesvirus antigenically related to Marek's disease virus. *Amer. J. vet. Res.*, 31, 525-538

Biological Properties of a Number of Marek's Disease Virus Isolates

P. M. BIGGS [1] & B. S. MILNE [1]

INTRODUCTION

For a number of years it has been recognized that isolates of Marek's disease (MD) differ in their pathogenicity (Biggs *et al., 1965*; Biggs, 1966; Purchase & Biggs, 1967). For example, the HPRS-B14 and WSU-GF strains produce classical Marek's disease (Biggs & Payne, 1967; Sharma, Kenzy & Rissberger, 1969) and the HPRS-16, JM and GA strains produce acute Marek's disease (Purchase & Biggs, 1967; Sevoian, Chamberlain & Counter, 1962; Eidson & Schmittle, 1968). The former is characterized primarily by lymphoid involvement of the peripheral nerves, and by lower mortality than is seen in acute Marek's disease, in which lymphoid proliferation in visceral organs is a prominent feature.

The object of this study was to examine a number of field isolates for their pathogenicity using standardized test conditions, and to determine whether isolates of different pathogenicity behave differently in cell culture. Because Marek's disease virus (MDV) is avidly cell-associated in cell culture these experiments were undertaken with cell-associated virus.

MATERIALS AND METHODS

Field isolates

Blood samples were collected from chickens in flocks at a number of farms where the incidence of MD varied greatly. Buffy coat cells prepared by two cycles of low-speed centrifugation were used to inoculate primary cultures of chick kidney cells. Cultures were passaged until adequate numbers of infected cells were obtained. Infected cells were frozen and stored in liquid nitrogen as already described for MD vaccine (Biggs *et al., 1970*). Most isolates were passaged between 3 and 6 times. Some of the slower growing isolates had to be passaged 9 times before adequate numbers of infected cells were obtained.

Virus strains

Stocks of cell-associated HPRS-16 strain of acute MD and its attenuated derivative, HPRS-16/att (45th chick kidney cell culture passage), and a turkey herpesvirus recently isolated in Britain (HPRS-26; Biggs & Milne [2]) were used in some experiments.

Cultured chick kidney cells

Primary cultured chick kidney cells (CKC) were derived from isolator reared Sykes line B Rhode Island Red chicks (Biggs, Thorpe & Payne, 1968) between one and four weeks of age. The techniques of culture were similar to those described by Churchill (1968) with minor modifications.

Virus assay

Cell-associated virus was assayed in CKC, as described by Churchill (1968), and titres expressed in plaque forming units (PFU).

Antibody

Antibody to the A antigen found in the supernatant fluid of infected CKC (Churchill, Chubb & Baxendale, 1969) was detected using the agar-gel double diffusion precipitin test (Chubb & Churchill, 1968).

Pathogenicity tests

Pathogenicity tests on virus isolates were undertaken using the following technique. In the early tests, groups of not less than 10 one-day-old Hough-

---

[1] Houghton Poultry Research Station, Houghton, Huntingdon, UK.

[2] Unpublished data.

ton Poultry Research Station Rhode Island Red chicks (Biggs & Payne, 1967) were inoculated intra-abdominally with an estimated 500 PFU. The inoculum used in each group was assayed in CKC to determine the exact dose used. Each group of chicks were kept in separate isolators for a period of 50 days. Chicks dying during the experiment and those killed at the termination of the experiment were examined *post mortem*. A score was given for lesions in the peripheral nerves and in visceral organs and tissues as follows: Lesions apparent to the naked eye were given a score of five, and those chicks with no lesions apparent to the naked eye were examined histologically. The left and right brachial and sciatic plexuses, the coeliac plexus and left gonad were examined and the nerves and gonad were given a score from one to four according to the severity of the lymphoid lesion. The scores for each chick in a group were added together and a mean calculated.

A group of controls inoculated with phosphate buffered saline were similarly treated. The final score given to an isolate was derived by subtracting the mean score for the control group from the mean score derived from the chicks inoculated with the isolate.

A group of chicks inoculated with a standard dose of HPRS-16 was included to provide an estimate of the sensitivity of the chicks used in each test. If the score was below that expected the results of that test were discarded. At the termination of the test, blood samples were collected and the serum examined for the presence of antibody.

Some modifications were made to later tests in the light of experience, and because chicks free of maternally derived antibody became available. In these tests 15 one-day-old antibody-free RIR were inoculated intra-abdominally with an estimated 1000 PFU and the test was terminated 42 days later. The

Table 1. Results of pathogenicity tests and plaque typing of 25 field isolates of Marek's disease virus

Test no.	Isolate	Length of test (days)	Antibody status of day-old chicks	Dose (PFU/chick)	Gross lesions [1]	Score [2] N	Score [2] V	Plaque type [3]	Category of patho-genicity
1	HAP 6	51	+	500	5/11	0.4	2.4	MP	Acute
	HAP 19	51	+	220	3/12	1.7	0.2	MP	Classical
	HAP 9	51	+	755	2/13	1.2	0.4	MP	Classical
	HAP 15	51	+	152	0/12	0.0	0.0	SP	Apathogenic
	HAP 3	51	+	185	0/12	0.3	0.0	SP	Apathogenic
2	C 5	50	+	184	4/10	2.1	2.3	MP	Acute
	HF 6/5	50	+	178	5/10	3.0	2.7	MP	Acute
	HF 6/20	50	+	85	0/10	0.0	0.0	SP	Apathogenic
	HF 6/10	50	+	388	2/10	1.3	0.6	MP	Classical
	HF 6/17	50	+	365	5/10	0.4	2.7	MP	Acute
3	HG 16	55	+	398	2/10	1.5	0.6	MP	Classical
	A 5	55	+	207	0/10	0.0	0.0	SP	Apathogenic
	A 1	55	+	264	0/8	0.0	0.3	SP	Apathogenic
	A 3	55	+	151	0/10	0.1	0.0	SP	Apathogenic
	A 10	55	+	222	0/10	0.0	0.2	SP	Apathogenic
4	C 6	50	+	453	9/10	4.4	4.3	MP	Acute
9	HG 17	42	−	5850	9/10	4.6	2.6	MP	Acute
	CW	42	−	1750	8/10	4.1	3.3	MP	Acute
	CZ	42	−	1770	10/10	4.4	4.3	MP	Acute
	HG 10	42	−	935	6/9	3.2	3.3	MP	Acute
	F 11	42	−	1075	10/10	4.9	3.1	MP	Acute
	F 16	42	−	1010	10/10	4.9	3.4	MP	Acute
	F 12	42	−	1425	10/10	4.9	5.0	MP	Acute
10	BBB 5	42	−	855	0/10	0.2	0.0	SP	Apathogenic
	BBB 13	42	−	895	5/10	2.4	0.0	SP	Classical

[1] Number of chicks with gross lesions of Marek's disease over number in group.
[2] N = Peripheral nerves; V = Visceral organs.
[3] MP = Medium plaque; SP = Small plaque.

number of chicks surviving at 14 days of age were reduced to 10 at random and these 10 formed the experimental number in the group. The use of anti-body-free chicks was considered to increase the accuracy of the test.

RESULTS

Pathogenicity of isolates

The results of pathogenicity tests on 25 isolates of MDV are shown in Table 1. The virus isolates have been classified into three categories of pathogenicity on the basis of their score in the tests (Table 2): (1) those that produced acute MD in which the incidence of gross lesions was high and lymphoid involvement of visceral organs as well as peripheral nerves was a prominent feature; (2) those that produced classical MD in which the incidence of gross lesions was lower and lymphoid involvement was predominantly in the peripheral nerves; and (3) apathogenic isolates that produced no or only very minor histological lesions.

A number of replicate experiments, both within and between tests, on a series of virus isolates con-

Table 2. Peripheral nerve and visceral organ scores for the three categories of pathogenicity

Category of pathogenicity	Peripheral nerves	Visceral organs
Apathogenic	<1.0	<1.0
Classical	>1.0	<1.5
Acute	0–5	>1.5

firmed the accuracy of these categories (Table 3). There was only one discrepancy in categorization in the series of replicates.

All virus isolates, regardless of their pathogenicity, induced antibodies to the A antigen in infected chickens. Isolates of all three types were derived from single flocks of chickens, and more than one type was isolated from individual pens of about 300 chickens. There was no readily apparent relationship between the type, or types, of isolate made from a group of chickens and the incidence of Marek's disease. Acute MDV isolates were made from pens of chickens with an incidence of MD varying from 0 to 24%.

Table 3. Between and within test replicate pathogenicity studies of Marek's disease virus

Test no.	Isolate	Length of test (days)	Antibody status of day-old chicks	Dose (PFU/chick)	Score [1] N	Score [1] V	Category of pathogenicity
3	A1	55	+	264	0.0	0.3	Apathogenic
6	A1	50	+	445	0.6	0.7	Apathogenic
9	A1	42	−	1565	0.8	0.0	Apathogenic
11	A1	42	−	1065	0.3	0.0	Apathogenic
3	A3	55	+	151	0.1	0.0	Apathogenic
6	A3	50	+	685	0.0	0.0	Apathogenic
7	A3	50	+	660	0.0	0.0	Apathogenic
7	A3	50	+	660	0.0	0.0	Apathogenic
10	A3	42	−	735	0.4	0.0	Apathogenic
1	HAP6	51	+	500	0.4	2.4	Acute
7	HAP6	50	+	385	1.3	1.1	Classical
7	HAP6	50	+	385	2.2	2.3	Acute
1	HAP19	51	+	220	1.7	0.2	Classical
7	HAP19	50	+	460	2.4	0.2	Classical
7	HAP19	50	+	460	1.6	1.2	Classical
7	HPRS-16 (Stock 8) [2]	50	+	925	3.8	3.2	Acute
7	HPRS-16 (Stock 8)	50	+	925	4.1	3.6	Acute
5	HPRS-16 (Stock 9) [3]	50	+	1000	4.8	4.5	Acute
5	HPRS-16 (Stock 9)	50	−	1000	5.0	4.7	Acute
9	HPRS-16 (Stock 9)	42	−	1265	4.9	4.4	Acute
10	HPRS-16 (Stock 9)	42	−	980	4.2	4.8	Acute
11	HPRS-16 (Stock 9)	42	−	1240	4.3	4.5	Acute

[1] N = Peripheral nerves; V = Visceral organs.
[2] 45th chicken passage of HPRS-16 prepared in cultured chick kidney cells.
[3] First chicken passage of HPRS-16 prepared in cultured chick kidney cells.

Growth in cell culture: plaque type

Although growth in CKC revealed a variation in plaque size for each isolate, isolates could be divided into two categories on the basis of plaque characteristics. Some isolates gave rise to small and slowly developing plaques characterized by a focal grouping of a small number of large and small rounded refractile cells (Fig. 1). Occasionally these cells retracted around a small hole in the monolayer. Large refractile cells were frequently a prominent feature of these plaques although their incidence sometimes declined on passage. Other isolates gave rise to larger plaques characterized initially by focal grouping of mainly small rounded refractile cells (Fig. 2). Retraction of these cells around a hole in the monolayer occurred more frequently than with the small plaques. These will be referred to as small and medium plaques, because the attenuated MDV and the herpesvirus isolated from turkeys produce larger plaques (Figs. 3 & 4). Of the 25 isolates examined, 9 were of the small plaque category and 16 of the medium plaque category.

Growth rate in CKC

Preparations of infected cells of the small plaque isolates had titres ranging from 10^3 to 1.2×10^4 PFU per $2-4 \times 10^6$ CKC, with a geometric mean of 4.4×10^3. The titres of similar preparations of medium plaque isolates varied between 5×10^3 and 10^5 PFU, with a geometric mean of 2.1×10^4. The small number of affected cells in each plaque of the small plaque isolates compared with the number in plaques of medium plaque isolates, and the lower titres of preparations of infected cells of small plaque isolates compared with those of the medium plaque isolates, suggest that the small plaque isolates spread from cell to cell more slowly than the medium plaque isolates. Experiments were undertaken to examine the rate of spreading in CKC of a number of selected small and medium plaque isolates, together with the HPRS-16 attenuated MDV and turkey herpesvirus as representatives of strains that produce large plaques.

A number of Petri dishes of CKC cultures for each isolate were inoculated with approximately 100 PFU per Petri dish. Chick kidney cells were harvested from two randomly selected Petri dishes for each isolate at intervals after infection. The cells were used to inoculate primary CKC and the plaques subsequently formed were counted. The number of infectious cells in the culture per infectious cell in the inoculum was calculated by dividing the total number of PFU in the harvest of CKC at each interval of time by the number of PFU in the inoculum.

After a period of slow increase in the number of infectious cells, a phase of rapid increase was seen for each isolate. This was followed by a phase in which the number of infectious cells remained constant and finally declined with the deterioration of the culture. The results of a representative experiment are shown in Fig. 5. The length of time between inoculation of the cultures and the phase of rapid increase in infectious cells was indirectly related to plaque size. This period was shortest for the large plaque strains, longest for the small plaque isolates and intermediate for the medium plaque isolates.

Relationship between plaque type and pathogenicity

Eight of the 9 small plaque isolates were apathogenic, the other produced classical Marek's disease. Of the 16 medium plaque isolates, 4 produced classical and 12 acute Marek's disease (Table 1).

DISCUSSION AND CONCLUSIONS

All isolates examined in this study possessed the A antigen but they could be separated into at least three categories on the basis of their pathogenicity, and into two categories on the basis of plaque and growth characteristics in cultured CKC. The small plaque isolates were apathogenic or mildly pathogenic, whereas all the medium plaque isolates were pathogenic.

It has been well established that chickens vaccinated with an attenuated strain of MDV can be superinfected with a field strain (Churchill, Payne & Chubb, 1969; Biggs et al., 1970). Marek's disease virus is ubiquitous in poultry populations and this study has shown that viruses of more than one type can be present in single flocks of chickens and in small groups of chickens within a flock. It therefore seems probable that individual chickens may be infected by more than one strain of MDV. For this reason, many of the isolates used in this study could be mixtures of viruses. It is possible that this is the explanation for the variability in plaque size seen with each virus isolate. It would be profitable to examine the properties of viruses derived from clones of these isolates.

Pre-infection with an apathogenic small plaque isolate offered a degree of protection to later challenge with an acute MDV strain (Biggs & Milne[1]). Similar

---

[1] Unpublished data.

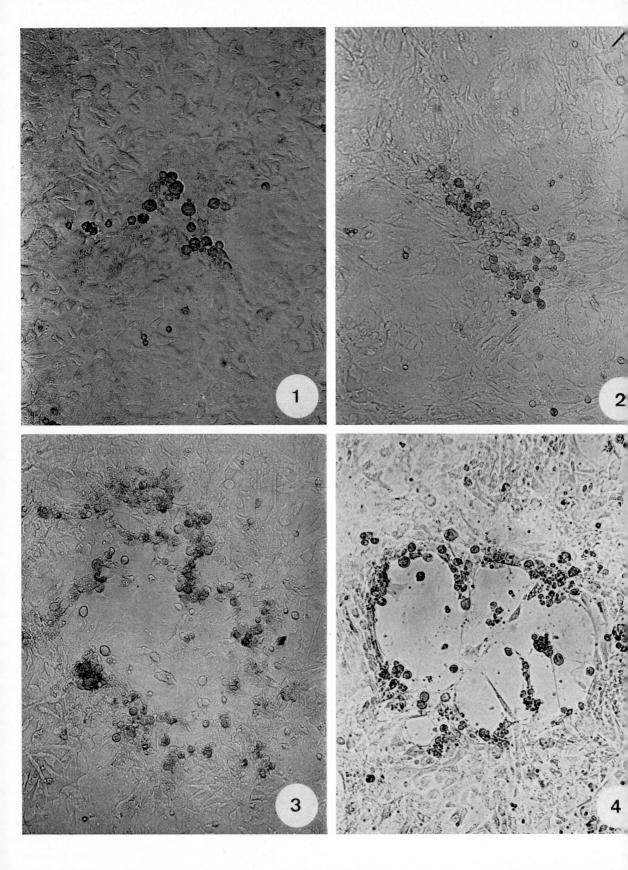

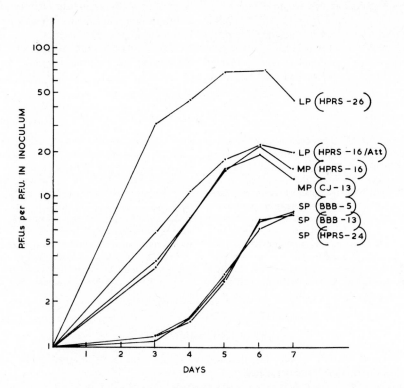

Fig. 5. Numbers of infectious cells in the culture per infectious cell in the inoculum at intervals after inoculation of cultured primary chick kidney cells with a number of strains and isolates of Marek's disease virus (expressed as PFU in the culture per PFU in the inoculum).

HPRS-26 (HTV)	– turkey herpesvirus (large plaque)
HPRS-16/att	– Attenuated HPRS-16 strain of acute Marek's disease (large plaque)
HPRS-16	– Strain of acute Marek's disease (medium plaque)
CJ 13	– Isolate of acute Marek's disease (medium plaque)
HPRS-24	– Isolate of apathogenic Marek's disease virus (small plaque)
BBB13	– Isolate of classical Marek's disease virus (small plaque)
BBB5	– Isolate of apathogenic Marek's disease virus (small plaque)
LP	– Large plaque
MP	– Medium plaque
SP	– Small plaque

Fig. 1. A small plaque 7 days after the inoculation of cultured primary chick kidney cells with isolate BBB5 infected cells. ×160.

Fig. 2. A medium plaque 7 days after the inoculation of cultured primary chick kidney cells with HPRS-16 infected cells. ×160.

Fig. 3. A large plaque 7 days after the inoculation of cultured primary chick kidney cells with HPRS-16/att infected cells. ×160.

Fig. 4. A large plaque 4 days after the inoculation of cultured primary chick kidney cells with HPRS-26 strain of turkey herpesvirus infected cells. ×160.

observations have been described for an apathogenic field virus by Rispens, van Vloten & Maas (1969). Because flocks of chickens may be infected with more than one type of MDV and because acute MDV can be found in flocks with a low as well as high incidence of MD, it is possible that the incidence of MD may be related to the sequence of infections. If a virus of low or no pathogenicity infects a group of chickens before a highly pathogenic virus, the subsequent incidence of MD could be less than if infection in the reverse order had taken place.

SUMMARY

The pathogenicity of 25 isolates of Marek's disease virus made from flocks at a number of farms was examined under standard conditions. The isolates could be divided into three categories: (1) those that produced acute Marek's disease in which the incidence of lesions was high and lymphoid involvement of visceral organs was a prominent feature; (2) those that produced classical Marek's disease in which the incidence of lesions was lower than in (1) and lymphoid involvement was predominantly in the peripheral nerves; and (3) apathogenic isolates that produced no or only very minor histological lesions. Isolates of all three types were made from single flocks of chickens and more than one type was isolated from individual pens of about 300 chickens within a flock. All virus isolates, regardless of their pathogenicity, induced antibodies to the A antigen in infected chickens.

On the basis of growth rate and plaque characteristics in cultured chick kidney cells, isolates could be divided into two categories: (1) small plaque isolates; and (2) larger plaque isolates that have been termed normal plaque isolates. Small plaque isolates spread from cell to cell more slowly than normal plaque isolates.

Eight of the 9 small plaque isolates were apathogenic, the other produced classical Marek's disease. Of the 16 normal plaque isolates, 4 produced classical and 12 acute Marek's disease.

ACKNOWLEDGMENTS

We wish to thank K. Howes for his able technical assistance and D. G. Powell, Dr. A. E. Churchill and G. Carrington for providing the primary isolates.

REFERENCES

Biggs, P. M., Purchase, H. G., Bee, B. R. & Dalton, P. J. (1965) Preliminary report on acute Marek's disease (fowl paralysis) in Great Britain. *Vet. Rec., 77*, 1339-1340

Biggs, P. M. (1966) Avian leukosis and Marek's disease. In: *Thirteenth World's Poultry Congress Symposium Papers,* Kiev, pp. 91-118

Biggs, P. M. & Payne, L. N. (1967) Studies on Marek's disease. I. Experimental transmission. *J. nat. Cancer Inst., 39*, 267-280

Biggs, P. M., Thorpe, R. J. & Payne, L. N. (1968) Studies on genetic resistance to Marek's disease in the domestic chicken. *Brit. poult. Sci., 9*, 37-52

Biggs, P. M., Payne, L. N., Milne, B. S., Churchill, A. E., Chubb, R. C., Powell, D. G. & Harris, A. H. (1970) Field trials with an attenuated cell associated vaccine for Marek's disease. *Vet. Rec., 87*, 704-709

Chubb, R. C. & Churchill, A. E. (1968) Precipitating antibodies associated with Marek's disease. *Vet. Rec., 83*, 4-7

Churchill, A. E. (1968) Herpes-type virus isolated in cell culture from tumors of chickens with Marek's disease. I. Studies in cell culture. *J. nat. Cancer Inst., 41*, 939-950

Churchill, A. E., Chubb, R. C. & Baxendale, W. (1969) The attenuation with loss of oncogenicity of the herpes-type virus of Marek's disease (Strain HPRS-16) in passage in cell culture. *J. gen. Virol., 4*, 557-564

Churchill, A. E., Payne, L. N. & Chubb, R. C. (1969) Immunization against Marek's disease using a live attenuated virus. *Nature (Lond.), 221*, 744-747

Eidson, C. S. & Schmittle, S. C. (1968) Studies on acute Marek's disease. I. Characteristics of isolate GA in chickens. *Avian Dis., 12*, 467-476

Purchase, H. G. & Biggs, P. M. (1967) Characterization of five isolates of Marek's disease. *Res. Vet. Sci., 8*, 440-449

Rispens, B. H., van Vloten, J. & Maas, H. J. L. (1969) Some virological and serological observations on Marek's disease: a preliminary report. *Brit. vet. J., 125*, 445-453

Sevoian, M., Chamberlain, D. M. & Counter, F. (1962) Avian lymphomatosis. Experimental reproduction of the neural and visceral forms. *Vet. Med., 57*, 500-501

Sharma, J. M., Kenzy, S. G. & Rissberger, A. (1969) Propagation and behaviour in chicken kidney cultures of the agent associated with classical Marek's disease. *J. nat. Cancer Inst., 43*, 907-916

Biological Markers for Purified Strains of Marek's Disease Virus and the Herpesvirus of Turkeys

H. G. PURCHASE [1]

INTRODUCTION

Closely related viruses can often be distinguished from one another by the effects they produce in biological systems. Such biological markers are useful in identifying viruses, distinguishing between closely related viruses, and in detecting changes such as mutations and adaptations that occur during the serial propagation of viruses. Some changes in properties, e.g., increase in growth rate, give a selective advantage to the altered virus as compared with the original parent type. During serial passage the altered virus may therefore predominate and eventually exclude the original parent type. Other changes, e.g., the loss of the ability to produce a certain antigen or the loss of pathogenicity, may be obscured by the effect of the parent virus still present in the mixed population. In order to detect such markers it is extremely important to purify the altered virus and to eliminate all parent types.

Biological markers have been identified capable of distinguishing between strains of Marek's disease herpesvirus (MDV). An MDV attenuated by serial passage in cell culture has been shown to protect chickens against Marek's disease (MD) (Churchill, Payne & Chubb, 1969). Similarly a herpesvirus isolated from turkeys (HVT), which is antigenically related to MDV, will also offer protection against MD (Okazaki, Purchase & Burmester, 1970). This paper will discuss some of the biological markers used to distinguish between virulent MDV, attenuated MDV and HVT.

PATHOGENICITY

Isolates of MDV from the field vary greatly in their pathogenicity to susceptible chickens (Purchase & Biggs, 1967). Some produce mainly visceral lesions and others predominantly nerve lesions or skin lesions (Sharma, Davis & Kenzy, 1970). The incubation period before symptoms develop and the median latent period to death may also vary. Since the pathogenicity of individual isolates is often highly reproducible, the distribution of lesions and time to death may be used as markers. More complex estimates of the pathogenicity involving the grading of the severity and distribution of lesions have been developed (Biggs & Milne [2]). Another useful measure is the number of infectious doses required to produce lesions (Chubb, 1969). In the case of properties such as haematopoietic destruction and changes in the bursa of Fabricius, it is not clear whether the virus or the strain of chicken is responsible (Jakowski et al., 1969, 1970).

On passage in cell culture, MDV may lose its pathogenicity (Churchill, Chubb & Baxendale, 1969; Nazerian, 1970). Attempts to demonstrate naturally nonpathogenic viruses in the field have failed (Purchase, Okazaki & Burmester [3]). However, populations of virus reduced in pathogenicity can be isolated in cell culture from a virulent stock of MDV (Purchase, Burmester & Cunningham, 1971a). There was no indication whether this virus was present in the original stock or whether it became apathogenic on passage in cell culture.

---

[1] Regional Poultry Research Laboratory, 3606 East Mount Hope Road, East Lansing, Michigan, USA.

[2] See p. 88 of this publication.
[3] Unpublished observations.

It has now been well established that HVT is non-pathogenic in chickens and turkeys (Purchase, Okazaki & Burmester [1]).

PLAQUE MORPHOLOGY AND REPLICATION CYCLE

There is no evidence that MDVs producing plaques of different morphology occur naturally. On initial isolation, all strains of MDV produce microplaques but on passage in cell culture a macroplaque variant appears (Churchill, Chubb & Baxendale, 1969; Nazerian, 1970; Mikami & Bankowski, 1970, 1971). The growth cycle of the macroplaque variant is shortened, the cytopathic effect is more severe and the social behaviour of the infected cells is altered. Thus, compared to the original microplaque producing agent, the macroplaque variant produces larger cytopathic areas in a shorter time. Affected cells are more highly vacuolated and frequently lyse, giving a hole in the monolayer. The virus produces larger syncytia and there are many more intranuclear inclusions. In many respects, HVT is similar to the macroplaque variant of MDV in cell culture but with experience the cytopathic changes produced by HVT can be distinguished from those produced by the microplaque and macroplaque MDVs.

It is probable that many stages in the replication cycle of these viruses may differ. However, they have not been studied in sufficient detail for them to be considered as markers.

VIRUS PRODUCTION

Macroplaque variants differ in the amount of virus they produce. Some produce no detectable virus (Churchill, Chubb & Baxendale, 1969), others only produce extractable virus (Nazerian, 1970) whereas the type 2 plaque-producing agent produces considerable amounts of virus (Mikami & Bankowski, 1970). Enveloped virions would presumably be observed in cells infected with this type of virus whereas they are rare in cells infected with the microplaque agent. Isolates of HVT also vary in the amount of virus produced. Thus the WHG isolate produces considerable amounts (Eidson & Anderson, 1971) whereas no cell-free virus can be detected in the supernatants of cells infected with FC126 (Witter et al., 1970).

Recently, improved techniques have been developed for the extraction of cell-free MDV and HVT (Calnek, Hitchner & Adldinger, 1970). The presence of EDTA (disodium ethylenediamine tetraacetate) in the medium used for extraction increased the yield of MDV but not that of HVT.

ANTIGENIC DIFFERENCES

Antigens produced by MDV and antibody to these antigens have been detected by the agar gel precipitation test (Chubb & Churchill, 1968; Okazaki, Purchase & Noll, 1970) and by both the direct and indirect fluorescent antibody tests (Purchase, 1969; Purchase & Burgoyne, 1970). An antigenic change has been described in MDV passaged repeatedly in chicken kidney cell cultures (Churchill, Chubb & Baxendale, 1969; Purchase et al., 1971a). During passage the virus lost an antigen (referred to as the A antigen) that was usually found in the supernatant fluids of cultures infected with the original strain (Figs. 1, 2 & 3). Purified strains of virus lacking the ability to produce this antigen have also been isolated from virulent stocks of MDV (Purchase et al., 1971a). In contrast, the type 1 or microplaque and type 2 or macroplaque producing agents derived from the Cal-1 strain of MDV have at least one antigen in common and one each that differs (Mikami & Bankowski, 1971). HVT has two A antigens and also other antigens in common with MDV (Purchase et al., 1971a).

Because of the highly cell-associated nature of MDV in cell culture, neutralization tests have been difficult to perform; nevertheless, neutralization of the type 2 or macroplaque producing agent has been possible (Mikami & Bankowski, 1971). Serum from chickens infected with MDV did not neutralize this virus, but neutralization did occur with homologous antiserum. Cell-free HVT is neutralized by antiserum to MDV and HVT, but to a greater extent by homologous antiserum (Witter et al., 1970).

Using the fluorescent antibody test, it was possible to distinguish between antibody to HVT and antibody to MDV (Witter et al., 1970). Thus antigen in both the nucleus and the cytoplasm of HVT-infected cells stained with homologous antiserum whereas only the nuclear antigen stained with MDV antiserum. There was also some indication of antigenic differences between purified strains of MDV when examined by this test (Purchase et al., 1971a). Although a surface antigen has been detected on chick

---

[1] Unpublished observations.

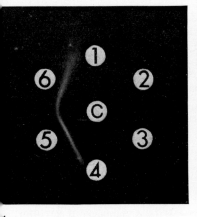

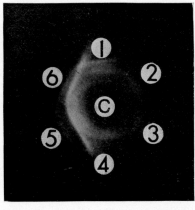

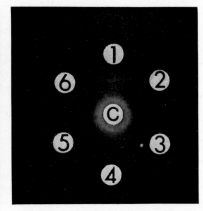

1   2   3

Figs. 1 to 3. Agar gel precipitation test using cloned preparations of Marek's disease virus (MDV) and selected sera. In all figures wells 1 to 4 contain serum from birds inoculated with pathogenic clones of MDV that do not produce A antigen (clones JM30 and JM31) and wells 5 and 6 contain serum from birds inoculated with a clone of low pathogenicity that does produce A antigen (clone JM32). The centre wells contain antigen as follows: Figs. 1 and 2: supernatant and cellular preparations respectively from cell cultures infected with a pathogenic clone of MDV producing A antigen (clone JM 19); and Fig. 3: a cellular preparation from cell cultures infected with an apathogenic clone of MDV passaged repeatedly in cell culture (clone JMHP). The supernatant preparation contains predominantly A antigen whereas the cellular preparations contain B and other antigens in addition to the A antigen, except in clone JMHP, which lacks the A antigen. For the origin of the clones, see Purchase et al., 1971a.

kidney cells infected with MDV, there is no indication as to whether this antigen is the same for all isolates of MDV and HVT (Chen & Purchase, 1970).

HOST RANGE IN VIVO

Even though both turkeys and chickens are susceptible to tumour development when inoculated with MDV, virus cannot be re-isolated from turkeys (Witter et al., 1970). It is not known whether the macroplaque producing variants of MDV will infect turkeys. In contrast, HVT appears to infect both chickens and turkeys equally well. However, this virus and the macroplaque variant of MDV do not spread from inoculated chickens to uninoculated ones in contact (Nazerian & Witter, 1970; Okazaki, Purchase & Burmester, 1970).

HOST RANGE IN VITRO

Although MDV does not produce plaques in chicken embryo fibroblasts (CEF) on initial isolation it will do so after a few passages in susceptible cell cultures, and its ability to produce plaques in CEF increases with passage (Purchase et al., 1971b). There are no great differences between purified strains of MDV in their ability to produce cytopathic effects

in various avian cell cultures. However, HVT produces plaques much more readily in CEF than in any other cell type tested. None of these viruses will grow in mammalian cells.

In contrast to variants of other herpesviruses, pocks produced on the chorioallantoic membrane of embryonated eggs by purified strains of MDV and by HVT are indistinguishable.

OTHER MARKERS

Although other markers have been described for viruses, e.g., virion density, heat stability, requirements for nutrients, etc., none of these have been examined in MDV or HVT.

CONCLUSIONS

Although numerous markers have been described for MDV and HVT, it is often not clear whether viruses found in the field possess them or whether they are a result of changes during in vitro cultivation. Apathogenic MDVs and viruses lacking the A antigen have not been found in the field. However, viruses of differing pathogenicity and viruses lacking the A antigen have been obtained from stocks of MDV. Similarly, macroplaque variants of MDV should be easily observable even on initial isolation and none have been recovered directly in the field. It would therefore appear that most of these changes in MDV are the result of passage in cell culture. However, since purified strains of virus have been obtained both with single markers and with a number of markers in combination, it appears that each change is independent of the others. However, with continued passage and selection, apathogenic macro-plaque producing agents lacking the ability to produce the A antigen eventually predominate, to the exclusion of the parent type.

REFERENCES

Calnek, B. W., Hitchner, S. B. & Adldinger, H. K. (1970) Lyophilization of cell-free Marek's disease herpesvirus and a herpesvirus from turkeys. *Appl. Microbiol., 20,* 723-726

Chen, J. H. & Purchase, H. G. (1970) Surface antigen on chick kidney cells infected with the herpesvirus of Marek's disease. *Virology, 40,* 410-412

Chubb, R. C. (1969) A comparison of two assay methods for Marek's disease. *Brit. vet. J., 125,* 16-17

Chubb, R. C. & Churchill, A. W. (1968) Precipitating antibodies associated with Marek's disease. *Vet. Rec., 83,* 4-7

Churchill, A. E., Chubb, R. C. & Baxendale, W. (1969) The attenuation, with loss of oncogenicity, of the herpes-type virus of Marek's disease (Strain HPRS-16) on passage in cell culture. *J. gen. Virol., 4,* 557-564

Churchill, A. E., Payne, L. N. & Chubb, R. C. (1969) Immunization against Marek's disease using a live attenuated virus. *Nature (Lond.), 211,* 744-747

Eidson, C. S. & Anderson, D. P. (1971) Immunization against Marek's disease. *Avian Dis., 15,* 68-81

Jakowski, R. M., Fredrickson, T. N., Luginbuhl, R. E. & Helmboldt, C. F. (1969) Early changes in the bursa of Fabricius from Marek's disease. *Avian Dis., 13,* 215-222

Jakowski, R. M., Fredrickson, T. N., Chomiak, T. W. & Luginbuhl, R. E. (1970) Hematopoietic destruction in Marek's disease. *Avian Dis., 14,* 374-385

Mikami, T. & Bankowski, R. A. (1970) Plaque types and cell-free virus from tissue cultures infected with Cal-1 strain of herpesvirus associated with Marek's disease. *J. nat. Cancer Inst., 45,* 319-333

Mikami, T. & Bankowski, R. A. (1971) Pathogenic and serologic studies of Type I and Type 2 plaque-producing agents derived from Cal-1 strains of Marek's disease virus. *Amer. J. vet. Res., 32,* 303-317

Nazerian, K. (1970) Attenuation of Marek's disease virus and study of its properties in two different cell cultures. *J. nat. Cancer Inst., 44,* 1257-1267

Nazerian, K. & Witter, R. L. (1970) Cell-free transmission and in vivo replication of Marek's disease virus. *J. Virol., 5,* 388-397

Okazaki, W., Purchase, H. G. & Burmester, B. R. (1970) Protection against Marek's disease by vaccination with a herpesvirus of turkeys. *Avian Dis., 14,* 413-429

Okazaki, W., Purchase, H. G. & Noll, L. (1970) Effect of different conditions on precipitation in agar between Marek's disease antigen and antibody. *Avian Dis., 14,* 532-537

Purchase, H. G. (1969) Immunofluorescence in the study of Marek's disease. I. Detection of antigen in cell culture and an antigenic comparison of eight isolates. *J. Virol., 3,* 557-563

Purchase, H. G. & Biggs, P. M. (1967) Characterization of five isolates of Marek's disease. *Res. Vet. Sci., 8,* 440-449

Purchase, H. G. & Burgoyne, G. H. (1970) Immunofluorescence in the study of Marek's disease: Detection of antibody. *Amer. J. vet. Res., 31,* 117-123

Purchase, H. G., Burmester, B. R. & Cunningham, C. H. (1971a) Pathogenicity and antigenicity of clones from different strains of Marek's disease virus and the herpesvirus of turkeys. *Inf. and Imm., 3,* 295-303

Purchase, H. G., Burmester, B. R. & Cunningham, C. H. (1971b) Responses of cultured cells of different avian species to Marek's disease virus and the herpesvirus of turkeys. *Amer. J. vet. Res.* (in press)

Sharma, J. M., Davis, W. C. & Kenzy, S. G. (1970) Etiologic relationship of skin tumors (skin leukosis) of chickens to Marek's disease. *J. nat. Cancer Inst., 44,* 901-912

Witter, R. L., Nazerian, K., Purchase, H. G. & Burgoyne, G. H. (1970) Isolation from turkeys of a cell-associated herpesvirus antigenically related to Marek's disease virus. *Amer. J. vet. Res., 31,* 525-538

Effect of Chelators on *in vitro* Infection with Marek's Disease Virus

H. K. ADLDINGER [1] & B. W. CALNEK [2]

The phenomenon of cell-association, as described for herpesviruses of the B-subgroup (Melnick *et al.,* 1964), is a prominent feature in infection with Marek's disease virus (MDV) as expressed both *in vivo* and *in vitro* (Biggs & Payne, 1967; Churchill & Biggs, 1967, 1968; Solomon *et al.,* 1968). Virus-specific antigens, and numerous intranuclear virus particles without envelopes are present in cells of several different types in MDV-infected chickens, but only a small number of enveloped cytoplasmic virions can be found in most infected cells (Calnek, Ubertini & Adldinger, 1970). As an exception, the cells of the superficial, keratinizing layers of the feather follicle epithelium (FFE) contain a larger number of mature, cytoplasmic virions, and so far this has been the only tissue from which infectious cell-free MDV has been obtained (Calnek, Adldinger & Kahn, 1970; Calnek, Ubertini & Adldinger, 1970; Nazerian & Witter, 1970). However, the quantities of virus that can be recovered from the FFE are still small, with *in vitro* infectivity titres of 10^3 focus forming units (FFU) per ml, and the extraction of cell-free MDV from infected cell cultures is even less efficient (Nazerian 1970; Mikami & Bankowski, 1970; Adldinger, 1971).

In experiments originally aimed at improving the recovery rate of cell-free MDV, it was observed that the continuing presence of additional Mg^{2+} in the fluid of cell cultures during the virus assay caused a sharp decrease in the expected number of cytopathic foci. In contrast, the introduction of the chelating agent EDTA (disodium ethylenediamine tetraacetate)

into infectious cell extracts to be assayed resulted in higher focus counts than were otherwise obtained. This effect of the chelator could be consistently reproduced both with skin extracts from infected birds and with extracts of infected chicken kidney (CK) cell cultures. It was also found that the increased number of foci appearing in the assay cultures after inoculation with EDTA-containing extracts were actually induced by MDV in the inoculum. An *in vitro* assay procedure for cell-free MDV, in which a concentration of 5 mM EDTA in the inoculum produces at least a 10-fold titre increase, has since been developed and described (Adldinger & Calnek, 1971). The precise mode of action of the chelator, however, remained obscure.

This report describes further studies on the EDTA effect and gives a tentative interpretation of the phenomenon.

MATERIALS AND METHODS

The JM-isolate of MDV (Sevoian, Chamberlain & Counter, 1962) was used either as skin extract or after serial passage in CK cell cultures. Stock virus suspensions from these sources were prepared by homogenization, sonication, and clarification by centrifugation as described previously (Calnek, Hitchner & Adldinger, 1970), and stored at $-60°C$. For assay of cell-free virus, drained cultures were normally inoculated with 0.25 ml of test suspension. Virus diluents were phosphate-buffered saline (PBS), or a phosphate buffer containing sucrose, disodium glutamate, and bovine serum albumin (SPGA; Bovarnick, Miller & Snyder, 1950), at pH-values between 6 and 7. When the inoculum contained EDTA, the cells responded immediately with rounding, and after a few minutes the culture had the appearance of freshly seeded cells. After a primary virus adsorption

[1] Department of Avian Diseases, New York State Veterinary College, Cornell University, Ithaca, New York, USA. Present address: Department of Veterinary Microbiology, School of Veterinary Medicine, University of Missouri — Columbia, Columbia, Mo., USA.

[2] Department of Avian Diseases, New York State Veterinary College, Cornell University, Ithaca, New York, USA.

period of 30 minutes at 38°C, 5 ml of maintenance medium were added, care being taken not to detach the cells. With EDTA concentrations up to 10 mM, the cells began to stretch out a few minutes after the addition of maintenance medium, which diluted the chelator approximately 20-fold and replenished divalent cations. The inoculum was not removed, and this allowed for a secondary adsorption period of 20 to 24 hours before the fluid was replaced by an overlay containing 0.8% agar. Cell-associated infectivity was assayed by placing the inoculum in the maintenance medium for 20 to 24 hours. The fluid was then removed and the agar overlay applied. Cytopathic foci were counted microscopically on post-inoculation days 6, 8, and 10. After 10 days no significant increase of foci occurred. The mean count of replicate cultures was used to estimate focus forming units per ml (FFU/ml) of test suspension. Direct fluorescent antibody (FA) tests for the detection of viral antigens were performed on CK cells grown on coverslips in 60-mm plastic Petri

dishes. The conjugate and the procedure were as described earlier (Spencer & Calnek, 1970). Anti-MDV γ-globulins (Calnek & Adldinger, 1971) were used in a final concentration of 1:50.

EXPERIMENTAL DESIGN AND RESULTS

It was important to know whether the reduction in the number of cytopathic foci in infectivity assays caused by excess Mg^{2+}, hereafter referred to as the Mg effect, and the contrary action of EDTA, the EDTA effect, were directly related or independent events. To resolve this question, the dose-response curves obtained with Mg^{2+} and EDTA were compared with those obtained with Ca^{2+} and ethylene-glycol tetraacetate (EGTA), respectively. EGTA is a chelator selectively complexing Ca^{2+} in the presence of Mg^{2+} (Nakas, Higashino & Loewenstein, 1966), whereas EDTA has affinity for both ions. Fig. 1 shows that increasing concentrations of EDTA

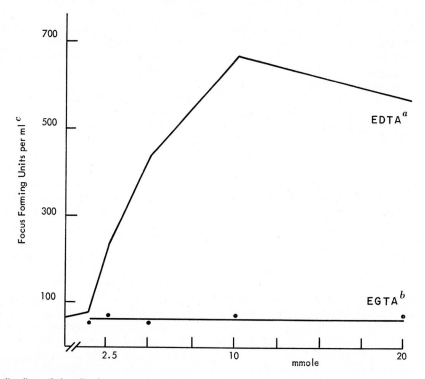

[a] EDTA, disodium ethylenediamine tetraacetate.
[b] EGTA, ethyleneglycol tetraacetate.
[c] 10% (w/v) skin extract was sonicated for 15 seconds, diluted 1:2 in SPGA, centrifuged at 650 g for 10 minutes, and the supernatant diluted 1:2 in SPGA containing the respective chelator to give the final concentrations indicated. Points of the curves represent mean focus counts from 3 or 4 replicate cultures.

Fig. 1. Effect of chelators on the infectivity of a skin extract prepared from MDV-infected chickens.

in the inoculum, in this case MDV-containing skin extract, produced increasing numbers of foci in CK cell cultures. EGTA had no effect in concentrations up to 20 mM. The fall in the EDTA curve at the 20 mM level was probably due to cell damage by the chelator, since the monolayers did not fully recover after the primary adsorption period. Fig. 2 shows that, unlike Mg^{2+}, Ca^{2+} had no effect on the number of foci. This, and the ineffectiveness of EGTA, indicated that Ca^{2+} played no role in the EDTA effect. Because of their low concentration in cell cultures, other divalent cations were not tested, and it was assumed, as a working hypothesis, that the increase in number of foci was a consequence of the chelation of Mg^{2+}-ions.

FA tests with coverslip cultures revealed that neither Mg^{2+} nor EDTA influenced the time needed for focus development, and the number of positive cells per focus as well as the distribution of antigens were unchanged. Secondary focus formation was not involved, since equal numbers of foci developed

in EDTA-treated cultures held under liquid or agar medium. Obviously, the EDTA effect did not enhance the release of virus from the infected cells, but was rather an earlier event in the infectious cycle or an action on the cell-free virus.

MDV infectivity in cell extracts shows strongly the characteristics of particle aggregation or association with subcellular material. Typical results, shown in Fig. 3, indicate that a large fraction of the total demonstrable infectivity was lost on low-speed centrifugation and/or filtration. Sonication, in contrast, increased the titre. However, the presence of up to 10 mM of EDTA in the extracts did not prevent the sedimentation at low-speed centrifugation and/or retention by Millipore filtration (650 nm pore diameter) of a large part of the infectivity, and essentially the same pattern as in Fig. 3 was obtained. This suggested that the EDTA effect did not consist simply in the dispersion of clumps entrapping the virus. Evidence to support this finding came from trials in which cell-free virus preparations were pretreated

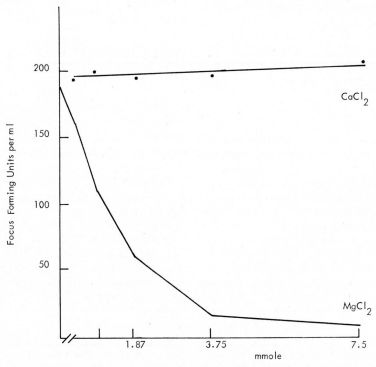

Fig. 2. Effect of excess Mg^{2+} and Ca^{2+} ion concentration on the infectivity of a skin extract[a] prepared from MDV-infected chickens.
[a] Similar preparation of skin extract. Cultures were inoculated by placing 0.5 ml of virus suspension into the maintenance medium containing the indicated concentrations of $CaCl_2$ or $MgCl_2$ in addition to the physiological levels. Cultures were maintained under liquid media throughout the observation period of 10 days.

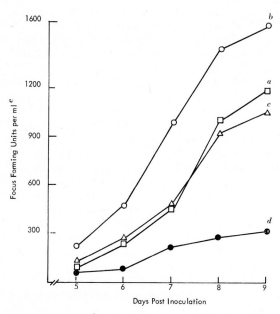

Fig. 3. Influence of sequential physical treatments on the infectivity of a skin extract prepared from MDV-infected chickens.

[a] Starting material: 10 % (w/v) skin extract in PBS.
[b] Starting material sonicated for 15 seconds at 4°C.
[c] Sonicated suspension clarified by centrifugation at 650 g for 10 minutes at 4°C.
[d] Clarified suspension filtered through a 650 nm average pore diameter Millipore membrane presoaked in bovine foetal serum.
[e] Samples were taken after each step, diluted 1:2 in SPGA containing 10 mM EDTA and assayed immediately in 4 replicate cultures.

Table 1. Pretreatment of cell-free Marek's disease virus with EDTA[1]

Virus preparation [2]	Pretreatment [3] (EDTA concentration in mM)	EDTA added for assay (in mM)	Focus forming units per ml
Clarified suspension	0	0	168
	0	5	1792
	5	0	1664
Filtered suspension (0.45 μ)	0	0	56
	0	5	688
	5	0	944

[1] Disodium ethylenediaminetetraacetate,
[2] 10 % (w/v) skin extract in PBS diluted 1:2 in PBS, sonicated 15 seconds, centrifuged at 650 g for 10 minutes (clarified); supernatant filtered through 450 nm average pore diameter Millipore filter presoaked in bovine foetal serum.
[3] Test suspension diluted 1:2 in SPGA with or without EDTA and held at 4°C for 6 hours.

Table 2. Pretreatment of cell-free Marek's disease virus with EDTA[1] or divalent cations

Pretreatment	Compound added for assay	Free EDTA in inoculum (mM) [2]	Focus forming units per ml
EDTA	EDTA	4.1	736
EDTA	Mg	1.2	268
EDTA	Ca	0	192
Ca	EDTA	3.7	604
Mg	EDTA	2.9	320

[1] Disodium ethylenediaminetetraacetate.
[2] Concentrations of unbound EDTA are estimated, ion contents of cells not considered.

with EDTA or divalent cations. The data in Table 1 indicate that exposure of the virus to the chelator for 6 hours produced no greater increase in infectivity titres than did the addition of the EDTA immediately preceding the inoculation of the assay cultures. This could be demonstrated with the same virus preparation before and after filtration through a filter membrane with an average pore diameter of 450 nm. In another experiment, the results of which are shown in Table 2, various amounts of Mg^{2+} or Ca^{2+} ions were added to a virus suspension that had been pretreated for 1 hour with EDTA. Replicate samples of the virus were treated in reverse order. Assay of the samples produced titres positively correlated with the concentration of uncomplexed EDTA in the inoculum.

A chelator treatment delayed until after the usual virus adsorption period also resulted in a titre increase. Drained cultures were inoculated with infectious cell extract with or without EDTA, and with primary and secondary adsorption periods, as already described. After 20 hours, the cultures were washed twice with PBS and treated as outlined in Table 3. It was found that an EDTA effect could be produced in which the increase in the number of cytopathic foci was only half that obtained with the immediate chelator treatment. In other experiments, not described here, the incorporation of anti-MDV γ-globulins into the maintenance medium resulted in titre reductions indicating that 20 hours after inoculation of the cultures 40 to 50% of the virus could still be neutralized. The simultaneous application of antibodies and EDTA to CK cells 20 hours post inoculation did not seem to influence the EDTA effect. However, when antibody treatment after 20 hours was followed by addition of the chelator after 30 minutes, as shown in Table 3 (Experiment II), the EDTA effect was almost completely eliminated. Slightly decreasing infectivity titres, or no effect at all was produced in the presence of EDTA when infectious suspensions of intact tumour or CK cells were used as inoculum. In contrast, the Mg effect was demonstrable with both cell-free virus and infectious cell suspensions, and a delayed inhibitory action was shown with cell-free virus.

Table 3. Delayed EDTA[1] treatment of chicken kidney cell cultures inoculated with Marek's disease virus

Experiment	Inoculation technique [2]	Treatment after 20 hours [3]	Focus forming units per 0.25 ml
I	5 mM EDTA	None	383
	No EDTA	Drained, none	48
	No EDTA	Drained, 5 mM EDTA	196
II	5 mM EDTA	γ glob., none	268
	No EDTA	γ glob., drained, none	32
	No EDTA	γ glob., drained, 5 mM EDTA	59

[1] Disodium ethylenediamine tetraacetate.
[2] Drained cultures were inoculated with 0.25 ml of virus suspension with or without the addition of EDTA. The primary adsorption period was 30 minutes, the secondary adsorption period under liquid medium was 20 hours.
[3] The treatments resembled the inoculation procedure, except that no virus was introduced. Adsorption periods were the same as under 2. Addition of γ globulins preceded the treatment by 30 minutes.

No EDTA effect was observed with a canine herpesvirus assayed in dog kidney cell monolayers, although the cells responded to the EDTA action by rounding. The canine herpesvirus is not regarded as a member of the B-subgroup, as cell-free virus can be readily obtained from cell cultures (Carmichael, Strandberg & Barnes, 1965). Assay results with a herpesvirus of turkeys (Witter *et al.*, 1970), which is not as strongly cell-associated as MDV, were indicative of a much weaker EDTA effect, or no effect at all. Earlier trials (Calnek, Hitchner & Adldinger, 1970) have shown that a six-fold increase in titre can be produced when the virus is extracted and diluted in PBS with EDTA. EDTA added to virus in SPGA diluent had no appreciable effect. SPGA does not contain Mg^{2+} ions.

In other assays by the chelator technique, extracts of lymphoid and tumour cell suspensions did not yield cell-free MDV. Thymus, spleen and bursa tissue from MDV-infected S-strain chickens were collected over a period ranging from 5 to 16 days after infection. Most of the tissues contained the virus in cell-associated form, and were positive in FA tests.

DISCUSSION

Low multiplicities of cell-free virus, slow virus adsorption by susceptible cells, and unavailability of purified infectious virus are major obstacles to the study of virus-host cell interactions with MDV. Because of these difficulties, the data presented here permit only preliminary interpretations. Nevertheless, it appears that the EDTA effect, as defined, is a consequence of the chelation of Mg^{2+}-ions. As far as the site of these critical ions is concerned, several results indicate that they are associated with the host cell rather than with the virus. EDTA-pretreatment of virus suspensions resulted in no greater titre increase than did the addition of the chelator immediately preceding the assay. No appreciable liberation of virus from aggregates or cellular debris by EDTA could be demonstrated in centrifugation and filtration experiments. It was consistently observed that the presence of relatively high concentrations of free EDTA in the inoculum produced the greatest titre increases. Obviously, unbound chelator was needed to act on the cell, most probably at the surface membrane. This interpretation was supported by the results of antibody and/or delayed chelator treatment. The finding that 20 hours after inoculation 40 to 50% of adsorbed virus was susceptible to neutralizing antibodies revealed that this virus had not penetrated into the cell. In agreement with this, a chelator treatment at this time produced only 50% of the titre increase obtained by a similar treatment at the time of inoculation. It was conceivable, and has been confirmed recently (Calnek & Adldinger, 1971), that thermal inactivation considerably reduces the infectivity of cell-free MDV during prolonged exposure at 37°C. Moreover, the increase in titre caused by chelator treatment at 20 hours post inoculation could be largely prevented by applying the antibodies 30 minutes before the EDTA. It was therefore concluded that a large fraction of the virus contained in cell extracts is not capable of penetrating into susceptible cells although it can be adsorbed. Insufficient disaggregation or separation from subcellular material, even after sonic vibration, as reported for herpes simplex virus (Smith, 1963), was observed with MDV preparations in these studies and might have been the cause of the supposed lack of penetration. The exact rôle of the EDTA in the promotion of penetration is not clear. Perhaps the "surface bubbling" induced in cells exposed to EDTA, as described by Dornfeld & Owczarzak (1958), more gently disaggregates the virus, while at the same time exposing further cellular adsorption sites in close proximity. The absence of an EDTA effect on virus transfer from infected to uninfected cells in culture may also be explained by the phenomenon of "surface bubbling". It is possible that it may prevent close apposition and temporary bridge formation between cells, a mechanism recently considered effective in cell-to-cell spread of MDV (Hlozanek, 1970).

As long as the exact timing of both the EDTA action and the Mg effect are unknown, however, the possibility cannot be ruled out completely that stages later than penetration may be involved. One pos-sibility is the breakdown of host-specific polysomes by the temporary chelation of intracellular Mg^{2+}-ions. Reassembled ribosomes then may become available more easily for virus-specific messenger.

The EDTA effect seemed to be directly related to the phenomenon of cell-association. It was most pronounced with MDV, possibly intermediate with a herpesvirus of turkeys, and absent with the canine herpesvirus. The application of the chelator technique to other cell-associated herpesviruses, especially those recently associated with neoplastic diseases, would be of interest.

SUMMARY

Circumstantial evidence is presented for the theory that a major cause of the cell-association of MDV is the inability of a large proportion of extracted virions to initiate infection.

After a 20-hour period of adsorption of extracted MDV by CK cells *in vitro*, 40 to 50% of the virus was still susceptible to neutralizing antibodies. A brief treatment of the cultures with EDTA at this time increased the numbers of cytopathic foci significantly as compared with those in untreated cultures. A similar treatment given at the time of inoculation increased the infectivity titres 2-fold as compared with those obtained with the delayed treatment, and at least 10-fold as compared with those of untreated controls. The phenomenon involved the chelation of Mg^{2+}-ions in the cultures. The transfer of MDV from infected to uninfected CK cells could not be enhanced by similar chelator treatments.

The data suggest that the EDTA somehow aids in the penetration of the virus into the cell, although an action on a later stage of the infectious cycle could not be ruled out. Similar conditions may be present with other cell-associated herpesviruses, and the chelator technique may facilitate the demonstration of cell-free virus.

ACKNOWLEDGEMENTS

This investigation was supported in part by Public Health Service Training Grant No. 5T01 A 100355 from the National Institutes of Health and by Public Health Service Research Grant No. CA 06709-08 from the National Cancer Institute, National Institutes of Health.

REFERENCES

Adldinger, H. K. (1971) Thesis, Cornell

Adldinger, H. K. & Calnek, B. W. (1971) An improved *in vitro* assay for cell-free Marek's disease virus. *Arch. ges. Virusforsch.*, **34**, 391-395

Biggs, P. M. & Payne, L. N. (1967) Studies on Marek's disease. I. Experimental transmission. *J. nat. Cancer Inst.*, **39**, 267-280

Bovarnick, M. R., Miller, J. C. & Snyder, J. C. (1950) The influence of certain salts, amino acids, sugars, and proteins on the stability of Rickettsiae. *J. Bact.*, **59**, 509-522

Calnek, B. W. & Adldinger, H. K. (1971) Some characteristics of cell-free preparations of Marek's disease virus. *Avian Dis.* (in press)

Calnek, B. W., Adldinger, H. K. & Kahn, D. E. (1970) Feather follicle epithelium: a source of enveloped and infectious herpesvirus from Marek's disease. *Avian Dis.,* **14**, 219-233

Calnek, B. W., Ubertini, T. & Adldinger, H. K. (1970) Viral antigen, virus particles, and infectivity of tissues from chickens with Marek's disease. *J. nat. Cancer Inst.,* **45**, 341-351

Calnek, B. W., Hitchner, S. B. & Adldinger, H. K. (1970) Lyophilization of cell-free Marek's disease herpesvirus and a herpesvirus from turkeys. *Appl. Microbiol., 20*, 723-726

Carmichael, L. E., Strandberg, J. D. & Barnes, F. D. (1965) Identification of a cytopathogenic agent infectious for puppies as a canine herpesvirus. *Proc. Soc. exp. Biol. (N.Y.), 120*, 644-647

Churchill, A. E. & Biggs, P. M. (1967) Agent of Marek's disease in tissue culture. *Nature (Lond.), 215*, 528-530

Churchill, A. E. & Biggs, P. M. (1968) Herpes-type virus isolated in cell culture from tumors of chickens with Marek's disease: II. Studies *in vivo. J. nat. Cancer Inst., 41*, 951-956

Dornfeld, E. J. & Owczarzak, A. (1958) Surface response in cultured fibroblasts elicited by EDTA. *J. biophys. biochem. Cytol., 4*, 243-253

Hlozanek, I. (1970) The influence of ultraviolet-inactivated Sendai virus on Marek's disease virus infection in tissue culture. *J. gen. Virol., 9*, 45-50

Melnick, J. L., Midulla, M., Wimberly, J., Barrera-Oro, J. G. & Levy, B. M. (1964) A new member of the herpesvirus group isolated from South American marmosets. *J. Immunol., 92*, 596-601

Mikami, T. & Bankowski, R. A. (1970) Plaque types and cell-free virus from tissue cultures infected with Cal-1 strain of herpesvirus associated with Marek's disease. *J. nat. Cancer Inst., 45*, 319-333

Nakas, M., Higashino, S. & Loewenstein, W. R. (1966) Uncoupling of an epithelial cell membrane junction by calcium-ion removal. *Science, 151*, 89-91

Nazerian, K. (1970) Attenuation of Marek's disease virus and study of its properties in two different cell cultures. *J. nat. Cancer Inst., 44*, 1257-1267

Nazerian, K. & Witter, R. L. (1970) Cell-free transmission and *in vivo* replication of Marek's disease virus. *J. Virol., 5*, 388-397

Sevoian, M., Chamberlain, D. M. & Counter, F. (1962) Avian lymphomatosis. Experimental reproduction of the neural and visceral forms. *Vet. Med., 57*, 500-501

Smith, K. O. (1963) Physical and biological observations on herpesvirus. *J. Bact., 86*, 999-1009

Solomon, J. J., Witter, R. L., Nazerian, K. & Burmester B. R. (1968) Studies on the etiology of Marek's disease. I. Propagation of the agent in cell culture. *Proc. Soc. exp. Biol. (N.Y.), 127*, 173-177

Spencer, J. L. & Calnek, B. W. (1970) Marek's disease: Application of immunofluorescence for detection of antigen and antibody. *Amer. J. vet. Res., 31*, 345-358

Witter, R. L., Nazerian, K., Purchase, H. G. & Burgoyne, G. H. (1970) Isolation from turkeys of a cell-associated herpesvirus antigenically related to Marek's disease virus. *Amer. J. vet. Res., 31*, 525-538

Discussion Summary

B. W. CALNEK [1]

Several important points were made in connection with biological markers. Kato presented new data describing a temperature-sensitive mutant of Marek's disease virus (MDV) grown in quail embryo fibroblasts. Parental virus grew equally well at 36°C or 41°C, but while the mutant replicated as well as parental virus at 36°C, it failed to grow at 41°C. Replication ceased or proceeded when cultures were changed from the permissive to the nonpermissive temperature or *vice versa*. There were no differences in plaque morphology or virus pathogenicity as between the parental and mutant strains.

Because repeated cell culture passage can result in the loss of the A antigen with MDV, the number of passages and the cell culture type employed by Purchase, in the cloning experiments, were thought significant. This is especially true in the case of the pathogenic MDV, which was described as A-minus, and was contrasted with Biggs' failure to find any A-minus field strains, whether pathogenic or nonpathogenic. It was pointed out that the 13 passages conducted during the cloning experiment may have been of considerable significance since chicken embryo fibroblasts (CEF) were employed, a cell type that has been found especially good at "removing" the A antigen. Equally important is the possibility of an A-negative virulent strain being masked by the presence of A-positive virus in noncloned preparations.

The apparent relationship between plaque morphology and pathogenicity was discussed in connection with another possible misinterpretation that may arise when noncloned virus is studied. Contaminating plaques might mask the true character of the more obvious majority type. It was pointed out that plaque morphology with herpes simplex changes,

even after cloning. Small plaques isolated from early type 1 herpes simplex virus (HSV) strains produced only small plaques while type 2 HSV produced both small and large plaques. After cloning experiments, the large plaques gave rise to both large and small plaques, while the progeny of small plaques produced only small plaques. The absolute size of HSV plaques was not suitable as a marker since that characteristic changed according to the type of cell culture and the culture conditions. Nevertheless, the *relative* size of plaques (large and small) remained constant regardless of the system.

In reply to a question whether maternal antibody might protect young chicks against infection, it was pointed out that passive antibody was protective against brief contact exposure. Further, this effect appears to be most pronounced in the case of the less virulent MDV isolates. Features characteristic of maternal antibody protection, other than lowered incidence of disease, include a lowered rate of virus isolation from kidneys and decreased amounts of, and less widespread, viral antigen in tissues.

The question of the relationship between MDV and turkey herpesvirus (HVT) was raised. Three relationships were considered possible: (1) virus originated in turkeys and at some time spread to chickens and became pathogenic in that species; (2) the reverse occurred with a concomitant loss of pathogenicity; or (3) both HVT and MDV evolved separately but happened to contain some common antigens. The information now available does not permit a choice between these alternatives, so that the two strains are linked only by the observable antigenic relationship. There is no evidence as to whether early or late antigens are involved in the difference between the two viruses. Cross-neutralization occurs, but a complete two-way cross has not been observed.

Evidence pointing to nicking in one strand of viral DNA from HSV raised the question of whether or not

[1] Department of Avian Diseases, Cornell University, Ithaca, New York, USA.

an endonuclease might be present at some location within the virion. Data were reported that failed to indicate the presence of the enzyme in nucleo-capsids or enveloped virions, but it is always danger-ous to draw conclusions from negative data. Prob-ably more pertinent was the observation that only one strand of the DNA was nicked. If this was not artefactual and an endonuclease was responsible for nicking, then both strands should be nicked. Data from self-annealing experiments would tend to rule out any nicking of both strands.

Newton presented data from some preliminary experiments on the buoyant density of the DNA of attenuated HPRS-16 MDV. Using a caesium chlo-ride gradient in a preparative centrifuge, she deter-mined the density to be 1.718. This contrasted with results of studies by others in which the density of the DNA from various MDV isolates ranged from 1.705 to 1.707. The use of an analytical centrifuge rather than a preparative centrifuge was suggested as a means of avoiding discrepancies, and a warning against overloading of gradients was given.

EPIDEMIOLOGY OF MAREK'S DISEASE

Chairman — B. W. Calnek

Rapporteur — B. R. Burmester

Epidemiology of Marek's Disease—A Review

R. L. WITTER [1]

INTRODUCTION

Marek's disease (MD) is a widespread, readily transmissible, lymphoproliferative disease of chickens. Because the lesions frequently have neoplastic characteristics (Payne & Biggs, 1967) and recent studies have revealed a cell-associated herpesvirus to be the aetiological agent (Churchill & Biggs, 1967; Nazerian et al., 1968), the disease can justifiably be termed a virus-induced malignancy. The present nomenclature (Biggs, 1961) honours Joseph Marek, who first described the disease in 1907. The disease is of interest not only because of its rôle as a model for naturally occurring neoplasia, but also because of its immense economic importance to the commercial poultry industry. In recent years MD has been acknowledged as by far the most important disease of chickens, causing annual losses of more than $150 million in the United States alone.

The epidemiology of MD has been the subject of intensive study, largely because knowledge of the natural spread of infection seemed necessary to provide a basis for practical control of the disease. Recent progress in this area has been facilitated by sensitive in vitro assays for virus and antibody (Chubb & Churchill, 1968; Witter, Solomon & Burgoyne, 1969b; AAAP Report, 1970; Purchase & Burgoyne, 1970), developed subsequently to the identification of the causative agent.

This paper deals principally with the current state of knowledge of the epidemiology of MD and thus omits much of the historical background, which is comprehensively dealt with elsewhere (Biggs, 1968; Calnek & Witter, 1972).

Some epidemiological characteristics of a newly discovered herpesvirus of turkeys (HVT) will also be considered since HVT is antigenically related to MD virus, is widespread in turkeys and, moreover, is in current use as a vaccine for the control of MD in chickens.

INCIDENCE AND DISTRIBUTION

During the last half century, the character of field outbreaks of MD has undergone a marked change. Early reports referred to MD as fowl or range paralysis (Pappenheimer, Dunn & Cone, 1929; Patterson et al., 1932), names indicative both of its predilection for chickens reared on range and its most characteristic clinical sign. Typically, losses were sporadic, occurring after the 15th week and involving less than 10% of the flock. Lesions were often confined to the peripheral nerves although visceral tumours were occasionally seen; such tumours tended particularly to involve the gonad. This syndrome, which has been called the "classical" form of MD (Purchase & Biggs, 1967), was widespread and, indeed, is still observed to some extent in virtually all flocks.

A more virulent form of the disease was probably present prior to 1950 in the United States, but the inability to differentiate between MD and lymphoid leukosis during this period undoubtedly obscured the identity and thus the significance of the new syndrome. Ultimately termed "acute leukosis" (Dunlop et al., 1965) or "acute" MD (Biggs et al., 1965), this new syndrome was characterized by a mortality of 20 to 30% (occasionally up to 60%) in yound birds. Losses started as early as the 4th week, usually reached a peak between the 10th and 25th weeks, and often continued at low levels for long periods. Nerve lesions could be seen in some birds but the most prevalent lesions were lymphoid tumours involving the skin, skeletal muscle and a variety of visceral organs (Biggs et al., 1965; Purchase & Biggs, 1967). Although the virulent form of MD is the most conspicuous, the disease occurs

[1] Regional Poultry Research Laboratory, 3606 East Mount Hope Road, East Lansing, Michigan, USA.

as a variety of syndromes, ranging from acute to subclinical. Since viral isolates vary in pathogenicity (Purchase & Biggs, 1967), the clinical expression of the disease is probably determined, at least in part, by the virulence of the infecting virus.

The magnitude of the economic problem increased sharply with the spread of virulent MD in the United States. Statistics for chicken flocks reared for egg production are not available, but the 15-fold increase in condemnations of 7- to 9-week-old broiler chickens for lymphoid tumours (assumed to be MD lesions) from 1960 to 1970 illustrates the rapidity with which one aspect of economic loss from this disease has grown (Fig. 1).

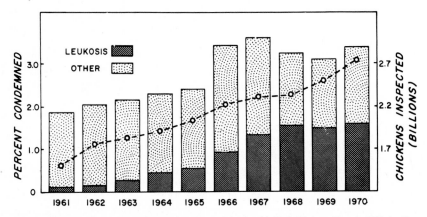

Fig. 1. Condemnation rates of young chickens slaughtered in the United States, March 1961 to December 1970, as reported by the United States Department of Agriculture. Note the increase in leukosis condemnations, both as a percent age of chickens inspected and in relation to condemnations for all causes. O— — —O Number of chickens inspected (billions).

Within the United States, virulent outbreaks of MD were first noted in those states on the Eastern seaboard near the Chesapeake Bay (Benton & Cover, 1957). Subsequently, high MD losses were observed in the Southern states, and nearly all important poultry producing areas of both the United States and the world as a whole eventually experienced at least some increased losses from the disease. Fig. 2 illustrates this relationship and also brings out other regional differences in MD losses. Condemnation losses in Georgia have been consistently higher than in Arkansas although the rearing of broiler chickens is intensive in both states. Washington, a geographically remote state with a low density of poultry, suffered no appreciable increased losses from leukosis condemnations until 1967. Recent trends in leukosis condemnations have been up for Georgia and Washington but down for Delaware and Arkansas. Such differences cannot so far be explained.

Recent surveys have shown MD virus to be virtually universally distributed among commercial chicken flocks (Churchill, 1968; Chubb & Churchill, 1968; Sevoian, 1969; Witter, Burgoyne & Solomon, 1969; Ianconescu & Samberg, 1971) but the incidence of MD losses can vary greatly between regions or between individual farms within regions (AAAP Report, 1967; Calnek & Witter, 1972), as is also illustrated in Fig. 2.

There is no satisfactory explanation for the origin of the virulent disease or the manner in which it became widespread. It seems plausible that virulent strains of MD virus were derived by mutation from less virulent strains, but the increased prevalence of acute forms of the disease might also be related in some way to the genetic constitution of commercial chickens or to techniques of poultry husbandry, both of which have changed dramatically in recent years.

SOURCES OF INFECTION

Potential sources of infectious MD virus may be classed as exogenous, i.e., from the environment, or endogenous, i.e., via the embryo. Recent work has shown that infection is invariably derived from exogenous sources.

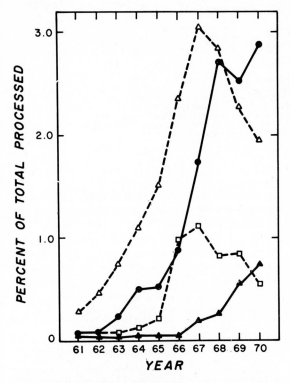

Fig. 2. Leukosis condemnation rates for young chickens in selected states, March 1961 to December 1970, as reported by the United States Department of Agriculture. Note differences in time of onset of increased losses, magnitude of losses and their recent trend. △— — —△ Delaware ●————● Georgia □— — —□ Arkansas ▲————▲ Washington.

Animate reservoirs

Except for chickens, natural infection with MD virus has been demonstrated only in quail (Kenzy & Cho, 1969; Witter [1]) but there is no information on the distribution of infection in this species or whether it is an important natural reservoir for the virus. In fact, Calnek found newly-hatched *Coturnix coturnix japonica* to be refractory to virulent MD virus (Calnek & Witter, 1972). Lesions resembling those of MD have been noted in pheasants (Harris, 1939; Jungherr, 1939), ducks (Cottral & Winton, 1953), owls (Halliwell, 1971), quail (Wight, 1963), swans (Blomberg, 1949) and partridges (Jennings, 1954) but the aetiological specificity of these lesions has not been determined. Turkeys seem to be a special case. Lymphoproliferative lesions have been ob-

served in turkeys (Andrews & Glover, 1939; Harnden & Smith, 1943; Simpson, Anthony & Young, 1957; Busch & Williams, 1970) and the species is susceptible to tumour induction when inoculated with virulent MD virus (Sevoian, Chamberlain & Larose, 1963b; Witter *et al.*, 1970b). However, Witter *et al.*, (1970b) found no virological or serological evidence of infection in turkeys inoculated with MD virus, even in those developing neoplasms. Virulent MD virus was isolated on one occasion from uninoculated turkeys (Witter & Solomon, 1971), but is probably not widespread. However, surveys for MD virus in turkeys are complicated by the extreme prevalence of HVT, a nonpathogenic, antigenically related herpesvirus.

Attempts to demonstrate the presence of MD virus or antibody in several species of free-flying birds, pheasants, waterfowl, and pigeons have been unsuccessful (Baxendale, 1969; Ianconescu & Samberg, 1971; Morgan, 1971; Witter [2]). Species apparently refractory to laboratory exposure to MD virus include sparrows (Kenzy & Cho, 1969), pigeons (Schettler & Witter [2]), and various mammals (Churchill & Biggs, 1968; Calnek [3]; Witter, Deinhardt and Landon [2]); however, Baxendale (1969) demonstrated an antibody response in ducks. It is clear that chickens themselves constitute the principal, if not the sole, animate reservoir of infection.

Inanimate reservoirs

Infectious material passing from chicken to chicken can be considered to remain for variable periods outside the host. Various components of the environment have been identified as sources of infection (Table 1).

Air from the vicinity of infected chickens usually contains MD virus, as shown by the high efficiency of airborne transmission (Sevoian *et al.*, 1963a; Colwell & Schmittle, 1968). This infectivity is removed by passage through filters with efficiency ratings of over 95% for 2–10μ particles (Solomon *et al.*, 1970) but passes 30% efficient filters (Burmester & Witter [2]). Experimentally, less than 30 minutes exposure to infectious air can produce infection in newly hatched chicks (Chen & Witter [2]).

Infectivity has also been associated with litter and droppings from infected chicken environments (Witter, Burgoyne & Burmester, 1968). The observed inverse relationship between moisture content and infectivity of litter (Witter *et al.*, 1968) could

---

[1] Unpublished data.

---

[2] Unpublished data.
[3] Personal communication.

Table 1. Infectivity of various components from the environment of chickens infected with MD virus

Material	Infectivity	Duration	Reference
Air (from infected birds)	+	Not measured	Sevoian et. al. (1963a)
Air (from used litter)	+	Not measured	Colwell & Schmittle (1968)
Litter and droppings	+	≥ 16 weeks	Witter et al. (1968)
Dust and dander	+	4 weeks	Beasley et al. (1970)
Dust and dander	+	≥ 7 weeks	Jurajda & Klimes (1970)
Chopped dry feathers	+	3 weeks	Calnek, Adldinger & Kahn (1970)
Chopped dry feathers	+	≥ 1 year	Calnek [1]

[1] Personal communication.

indicate either that this material must become aerosolized to effect infection or that the virus is more stable in the dry state. Both hypotheses are favoured by the finding that poultry house dust, which consists of small, easily aerosolized particles and has very little moisture, is also a potent source of infection (Beasley, Patterson & McWade, 1970; Jurajda & Klimes, 1970).

The viability of MD virus under environmental conditions has not yet been studied in depth but it is clear that infectivity can survive for long periods (Witter et al., 1968; Beasley et al., 1970; Jurajda & Klimes, 1970). Calnek, Adldinger & Kahn (1970) reported that desiccated cell-free MD virus was viable for up to three weeks, and it was found subsequently (Calnek [1]) that such material was infectious for over one year at 4°C.

It is likely that a common physical unit of infectivity exists in all of these environmental reservoirs of infection. It is easily airborne, relatively stable (and thus independent of living cells) but sufficiently particulate to be removed by high efficiency filters. This infectious unit is presumably enveloped, cell-free infectious MD virus derived from the feather follicle epithelium (Calnek, Adldinger & Kahn, 1970; Calnek, Ubertini & Adldinger, 1970) and probably exists in aggregates or in association with larger particles rather than as individual virions.

Transport hosts

Eidson et al. (1966) reported that the darkling beetle *(Alphitobius diaperinus)*, a common resident of poultry house litter, was a carrier of MD virus. However, attempts to associate transmission of MD with mosquitoes (Brewer et al., 1969), litter mites (Witter et al., 1968) or other insects common to the environment of chickens including beetles (Lancaster

& Beasley [2]) were unsuccessful. Likewise, the coccidial oocyst proved unimportant as a reservoir of MD virus (Long, Kenzy & Biggs, 1968; Brewer et al., 1968). Although the existence of transport hosts for MD virus cannot be ruled out, the importance of such a phenomenon has not yet been convincingly demonstrated. The physical transport of infectious material by man and other fomites is probably a more important mechanism for disseminating infection.

Embryo

Early evidence against transmission of MD via the embryo (Cole & Hutt, 1951) was recently challenged by Sevoian (1968), who reported the isolation of MD virus from tissue of Cornell S-line and other embryos by bioassay in Cornell S-line chickens. However, other workers using chicks presumably derived from infected breeders, obtained flocks free of MD infection with relative ease by strict isolation rearing (Witter, 1970; Rispens, van Vloten & Maas, 1969; Vielitz & Landgraf, 1970). Furthermore, Solomon et al. (1970) failed to detect MD virus in over 2500 chicks and embryos derived from 9 different infected breeder flocks (Table 2). Thus, it seems that embryo transmission, if it occurs at all, is an exceedingly rare event and is of little importance in the natural transmission of the disease.

VIRUS-HOST RELATIONSHIPS

Portions of the virus-host interaction cycle, and particularly the mechanisms of initial infection and virus excretion, are critical to the epidemiology of this disease.

Infection

Although chickens become infected readily when placed in contact with infected chickens or materials

---

[1] Personal communication.   [2] Unpublished data.

Table 2. Absence of infection in the progeny of nine MD-infected breeder flocks [1]

Flock	No. of progeny	Test	Age at test	Result
1	596	⎫	10 weeks	0/535
2	222	⎪	⩾ 10 weeks	0/180
3–6	587	⎬ Isolation rearing [2]	26 weeks	0/363
7–8	48	⎪	9–13 weeks	0/48
9	41	⎭	22 weeks	0/35
1	117	Chick bioassay [3]	emb (18)	0/27
2	863	⎫ Cell culture [4]	⩾ 4 days	0/119
7–8	200	⎭	⩾ 1 day	0/200
Total	2664			0/1507

[1] Solomon *et al.*, 1970.
[2] Chicks were reared in strict isolation and evaluated for MD precipitating antibodies.
[3] Pooled suspensions of embryo tissue were inoculated into susceptible chicks, which were subsequently examined for MD lesions and antibodies.
[4] Individual or pooled kidney sample were cultured and examined for cytopathic changes typical of MD virus.

from their environment, the precise route by which the infectious material enters the host and initiates infection is not known.

Studies with MD inocula containing viable cells, although not strictly relevant to the mechanism of natural infection, have shown that exposure via the external nares or the trachea induced infection whereas the ocular route was less effective and intracrop instillation was unsuccessful (Calnek & Hitchner, 1969). However, the oral route was routinely successful for infecting chicks with plasma from donors infected with the GA isolate (Eidson & Schmittle, 1968).

Administration of chopped dry feathers or dust by inhalation or intratracheal insufflation have also been successful in transmitting MD (Calnek, Adldinger & Kahn, 1970; Fredrickson [1]). Thus the respiratory epithelium may be an important portal of entry for the virus. The localization of material containing virus-specific immunofluorescent antigen adjacent to the pulmonary epithelium of contact-exposed chickens (Purchase, 1970), while not proving infection by this route, gives some support to this possibility.

Shedding

The ease with which MD is transmitted by contact indicates that virus is released or excreted by infected chickens in substantial quantities and in an infectious form. Early work demonstrated infectivity in oral washings (Witter & Burmester, 1967; Kenzy & Biggs,

1967) and faeces from infected chickens (Witter & Burmester, 1967) but the importance of these materials in natural transmission could not be determined.

Virtually all tissues of infected birds are sources of cell-associated virus (Biggs & Payne, 1967; Witter, Solomon & Burgoyne, 1969), but the tissue distribution of viral antigens is more restricted (Calnek & Hitchner, 1969). Virus-specific immunofluorescent antigens were found in many tissues with access to the exterior of the chicken, e.g., the feather follicle, bursa of Fabricius, kidney, uropygial gland, Harder's gland and portions of the alimentary tract (Calnek & Hitchner, 1969). However, the most likely route of egress seemed to involve the epithelium of the feather follicle since this tissue was unique in its permissiveness for replication of complete, enveloped virus that was infectious in the absence of cells (Calnek, Adldinger & Kahn, 1970; Calnek, Ubertini & Adldinger, 1970; Nazerian & Witter, 1970). Details of the shedding process are lacking, but Calnek, Adldinger & Kahn (1970) considered moulted feathers, sloughed epithelial cells and direct release of virions all to be possible mechanisms. The connexion between virus replication in the integument and contact transmision is supported by circumstantial evidence (Nazerian & Witter, 1970; Witter, *et al.*, 1971).

Temporal relations

In chicks inoculated at hatching with virulent MD virus, shedding can be detected by contact-exposure experiments within two weeks (Kenzy & Biggs, 1967), but maximum levels of shedding occur between the

[1] Personal communication.

third and fifth weeks (Witter [1]). These data agree well with the temporal development of virus-specific immunofluorescent antigens in the feather follicle epithelium (Calnek & Hitchner, 1969).

Persistence of MD virus in naturally infected chickens, as indicated by the high rates of virus isolation from older flocks (Churchill & Biggs, 1968; Witter et al., 1969), also suggests that virus shedding may be prolonged. Contact transmission was effected by donor chickens at ages of 20 weeks (Kenzy & Biggs, 1967) and up to 24 months (Kenzy & Cho, 1969). Witter et al. (1971) demonstrated shedding in 5 out of 8 naturally infected chickens at 76 weeks of age, and found this to be directly related both to the presence of integumentary virus and to levels of viraemia. Although virus titres were often very low, there was evidence that infection persisted indefinitely in virtually all birds (Witter et al., 1971). Thus, infected chickens of any age are probable sources of infection.

FLOCK INFECTION

Although laboratory studies of the experimentally induced disease have contributed greatly to our understanding of MD epidemiology, valuable information has also been obtained from studies of natural infection in chicken flocks raised under commercial conditions.

The early events consequent to MD infection in four commercial broiler flocks were studied by Witter et al. (1970a). Infection occurred early (by the 9th day in one flock) and spread rapidly so that by the 8th week virus could be isolated from virtually all individuals and the incidence of precipitating antibody ranged from 20 to 90%. The time interval between 50% incidence of virus and 50% incidence of antibody was about two weeks (Fig. 3). Similar data were reported by Bankowski, Mikami & Reynolds (1970).

Later characteristics of natural infection were studied in one flock of White Leghorn chickens (Witter et al., 1971). Virus and antibody responses were already maximal at the 8th week and, despite high mortality from MD, remained at fairly constant levels through a 76-week observation period (Fig. 4). Although titres of precipitating antibody decreased in some chickens, the level of antibody detected by the indirect fluorescent antibody test remained stable.

A summary of the epidemiological events relevant to natural infection with MD virus is presented

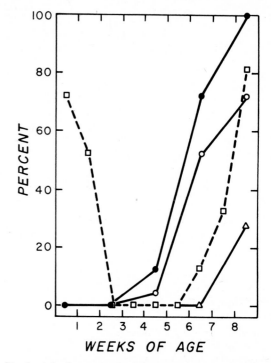

Fig. 3. Infection response in a commercial broiler flock determined to have been exposed to Marek's disease prior to the 9th day (Witter et al., 1970b). □— — —□ Antibody ●————● Virus isolation O————O Microscopic lesions △————△ Gross lesions.

schematically in Fig. 5. Newly-hatched chicks remain free of infection for variable periods of time depending on their proximity to the MD-contaminated environment. Some commercial flocks apparently escape exposure for substantial periods (Witter, 1970), but this is exceedingly rare. Once infected, the chicken exists henceforth in a contaminated environment, continually replenishing the level of environmental infectivity through virus shedding from the integument. Death may occur at any age, but mortality rates are highly variable and most birds survive indefinitely.

Infection has been prevented by rearing chickens in special houses supplied with filtered air under positive pressure (Drury, Patterson & Beard, 1969; Vielitz & Landgraf, 1970) and, indeed, this technique may have promise for the control and possible eradication of the disease. The success of isolation rearing is dependent on complete elimination of pre-existing virus from the housing facilities and prevention of subsequent contamination from environmental sources (Witter, 1970); commercially feasible techniques are not yet available.

---

[1] Unpublished data.

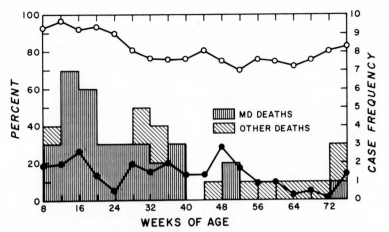

WEEKS OF AGE

Fig. 4. Long-term infection response of 100 naturally exposed White Leghorn chickens (Witter *et al.*, 1971). Note the persistence of infection despite 33 % mortality from Marek's disease. O——————O Precipitating antibody. ●—————● Virus isolation from whole blood in duck embryo fibroblast cultures.

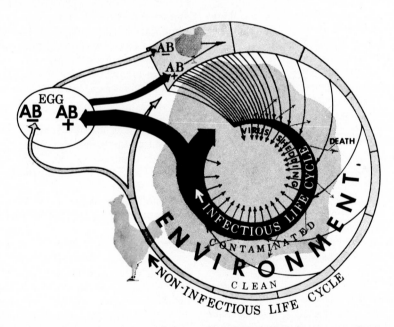

Fig. 5. Schematic representation of epidemiological relationships in Marek's disease showing the source and frequency of natural infection, the time and frequency of mortality, the shedding of virus by infected chickens and the discontinuous nature of the infection cycle (chicks are always free of infection at hatching).

DETERMINANTS OF NATURAL TRANSMISSION AND DISEASE

Interactions between host, agent and environment form the essence of epidemiology; some of the factors known to influence this dynamic relationship in the case of MD therefore warrant brief consideration.

Genetics

Genetic resistance to MD has been well documented (Hutt & Cole, 1947) and breeding programmes for the selection of resistant stock have been proposed for the control of the disease (Cole, 1968; Biggs, Thorpe & Payne, 1968). The nature of this resistance, however, has not been defined. Cells

from genetically resistant chickens are fully susceptible to infection *in vitro* (Spencer, 1969), thus indicating that the resistance mechanism is operative
at a higher level. However, *in vivo* virus replication
is less productive in genetically resistant stock than
in susceptible stock (Spencer, 1969; Stone [1]). Furthermore, Calnek & Hitchner (1969) found a higher
incidence and wider tissue distribution of virus-
specific immunofluorescent antigen in genetically
susceptible S-line chickens than in more resistant
stock. Both observations suggest a possible effect
of genetic constitution on the natural transmission
of infection.

Age

The MD susceptibility of chickens has been reported to decrease with increasing age (Sevoian &
Chamberlain, 1963; Biggs & Payne, 1967), but in
these trials the possibility of natural infection with
MD prior to challenge could not be excluded. However, more recent studies in which isolated infection-
free stock was challenged at various ages supported
the existence of an "age" resistance to lesion development (Anderson, Eidson & Richey, 1971; Vielitz &
Landgraf, 1970). The extent of this resistance seems
variable, as does its time of development. We have
found substantial resistance against lesion development but only slight resistance against infection in
chickens at 20 weeks of age (Witter, Solomon &
Burmester [2]).

Antibody

Under natural conditions virtually all chicks possess levels of maternal antibody at hatching unless
obtained from special specific pathogen-free sources.
Since such chicks are more resistant to the disease
than antibody-free chicks (Chubb & Churchill, 1969),
immunization of parent stock, presumably in order
to increase levels of maternal antibody in chicks, has
been recommended as a control measure (Sevoian,
1969) and used with some success. There is some
evidence that degenerative lesions in the bursa of
Fabricius (Jakowski & Fredrickson, 1969) and feather
follicle epithelium (Calnek, Adldinger & Kahn,
1970a), are more common in exposed chicks lacking
maternal antibody than in chicks with antibody
(Calnek [1]). To the extent that such degenerative
lesions parallel virus replication, the lack of maternal
antibody might increase the spread of infection.

The role of humoral antibody actively synthesized
by infected chickens as a determinant of disease is
less clear (Bankowski *et al.*, 1970). Although a correlation has been noted between antibody titre and
survival (Eidson & Schmittle, 1969; Witter *et al.*,
1971), it is uncertain whether this is due to an inability of severely affected birds to synthesize antibody or to a protective effect of the antibody against
the disease (Witter *et al.*, 1971).

Viral pathogenicity

Differences in the ability of certain MD virus isolates to induce lesions in chickens have been noted
(Purchase & Biggs, 1967) and it is obvious that the
pathological response to natural MD infection is
greatly influenced by the pathogenicity of the infecting virus. However, the finding of wide variations in the pathogenicity of MD viruses naturally
present in a single environment (Biggs & Milne [3];
Witter *et al.*, 1971) or cloned from established laboratory strains (Purchase, Burmester & Cunningham,
1971) indicates that host-virus interrelationships may
be far more complex in nature than formerly believed.
Clearly, monotypic infection rarely occurs under
either natural or experimental conditions.

Prior infection

A virulent or attenuated viruses induce resistance
against lesion induction by virulent viruses (Churchill,
Payne & Chubb, 1969; Okazaki, Purchase & Burmester, 1970; Rispens *et al.*, 1969). This phenomenon, which is widely utilized at present in the application of vaccines against MD, undoubtedly also
occurs naturally, since mild or avirulent isolates of
MD virus are commonly found in field chickens.
Such natural vaccination could account for the
beneficial effects of so-called "controlled exposure"
and, furthermore, might provide the solutions to a
variety of epidemiological enigmas, such as the
variations in MD mortality among pens or houses
on a single premises. The high incidence of disease
commonly observed in new poultry houses could
also result from a lack of the avirulent virus flora
usually present in established chicken environments.
Even simultaneous exposure with avirulent and
virulent viruses might result in partial protection, as
has been shown in vaccination trials (Okazaki and
Purchase [4]). Furthermore, chickens infected with
avirulent virus prior to exposure to virulent virus

---

[1] Personal communication.
[2] Unpublished data.

[3] See p. 88 of this publication.
[4] Personal communication.

are apparently less effective transmitters of the disease than birds infected only with virulent virus (Purchase & Okazaki, 1971), thereby possibly influencing the rate of horizontal transmission within a flock. Admittedly, the existence of most of these relationships remains to be confirmed but at least the question is now amenable to study. Progress may be hampered by the lack of a precise assay method for viral pathogenicity, but certain techniques have been proposed (AAAP Report, 1970; Witter *et al.*, 1971) and improvements can be expected.

Stress

The influence of stress on MD has not been extensively studied, but there is some evidence that both social and physiological stress may play a part in the natural epidemiology of the disease. Gross [1] has found that exposed birds moved continuously among unfamiliar cagemates develop a higher incidence of MD than birds under conditions of low social stress. He also noted depression of MD mortality following the administration of adrenal-blocking compounds. Proudfoot & Aitken (1969) reported significantly reduced MD losses in chickens fed on low protein diets. Physical stresses such as heat, cold, water deprivation, etc., may also be of importance but require further study.

Other diseases have been studied as possible influencing or triggering factors for MD but as yet no such relationship has been proved (AAAP Report, 1967). Coccidiosis has no direct effect on MD but is, conversely, more severe in dually-infected chickens because MD interferes with the development of coccidial immunity (Biggs *et al.*, 1968; Kenzy, Long & Biggs, 1970). Vaccination against avian encephalomyelitis, in order to prevent the possible stress of later infection with this virus, did not influence MD condemnations in broiler flocks (Davis, Dawe & Brown, 1970). Infectious bursal agent was reported to enhance gross nerve lesions but had no effect on mortality or tumour frequency in dually-infected birds (Cho, 1970).

TURKEY HERPESVIRUS

The epidemiological characteristics of HVT infection are only currently being elucidated. The virus occurs commonly in turkeys (Kawamura, King & Anderson, 1969; Witter *et al.*, 1970b) and limited observations suggest that the virus is transmitted horizontally but not vertically in this species (Witter *et al.*, 1970b). The virus probably does not occur naturally in chickens but is now reasonably prevalent in commercial flocks due to its widespread use as a vaccine for MD, particularly in the United States.

Our recent studies on HVT in turkeys (Witter & Solomon, 1971) confirmed that the virus spreads horizontally with ease. In two commercial flocks of market turkeys, virus was first isolated at the 5th and 6th weeks, and by the 7th to 10th weeks both virus and antibody were present in 100% of the birds. The interval between the 50% response times for virus and antibody was about one week, a considerably shorter interval than that previously found for MD virus in chickens (Witter *et al.*, 1970a). Like MD, HVT seemed to persist indefinitely in infected turkeys, but the virus titres of blood did not exceed 1000 cell culture infectious doses per ml. Turkeys infected at hatching began to shed virus between the 12th and 16th day. Cell-free infectious virus was recovered from the integument of infected turkeys (Witter, Nazerian & Solomon [2]) in a manner similar to that described for MD virus in chickens and was probably responsible for its contagiousness.

Embryo transmission of HVT in turkeys seemed unlikely since turkeys from infected breeders could usually be maintained free of infection when reared in isolation cages. In contrast, infectivity could be demonstrated in the environment of infected turkeys (Witter & Solomon, 1971), indicating that, like MD, exogenous sources of infection were more important than endogenous sources.

In chickens, HVT is clearly less contagious than in turkeys (Witter *et al.*, 1970b; Okazaki *et al.*, 1970). Although Nazerian & Witter (1970) failed to demonstrate virions in the feather follicle of HVT-inoculated chickens, more comprehensive studies (Nazerian [3]) revealed integument-associated virus in an occasional bird. A somewhat greater degree of contact transmission has been reported from donor chickens inoculated with high doses of HVT at 8 weeks of age (Cho [3]). Also, attempts to demonstrate virus in the progeny of HVT-inoculated breeder chickens were unsuccessful (Witter [2]). Thus infection with HVT in chickens seems to be self-limiting.

HVT appears to lack oncogenicity for both turkeys and chickens (Witter *et al.*, 1970b; Okazaki *et al.*, 1970). No lesions have been seen in HVT-inoculated chickens after one year (Okazaki & Purchase [2]). Also,

---

[1] Personal communication.

[2] Unpublished data.

[3] Personal communication.

turkeys of three strains, which were inoculated with high doses of HVT at hatching, have not developed neoplasms after 12 months of observation (Witter[1]).

CONCLUSIONS

The potential value of epidemiological approaches in the study of neoplastic disease is well exemplified by MD, where such studies have lead not only to the definition of its aetiology but also to the development of methods for its control. Epidemiological investigations of some other neoplasms, notably Burkitt's lymphoma (MacMahon, 1968; Burkitt,

[1] Personal communication.

1969) and bovine leukaemia (Conner *et al.*, 1966), have provided leads to their fundamental nature, but the lack of tools and experimental hosts has restricted progress. Thus MD may have unique value as a model for herpesvirus-induced neoplastic disease, particularly because of its widespread occurrence, transmissibility, defined aetiology, convenient *in vitro* assay techniques for virus and antibody, and infection-free experimental hosts. Perhaps the "difficult years" of MD research (like those described by Weller (1970) for cytomegaloviruses) have nearly reached their conclusion, so that workers, by using MD as a model, may be able to open up new areas of knowledge in herpesvirus-induced oncogenesis.

REFERENCES

American Association of Avian Pathologists (1967) Leukosis Committee Workshop Report. *Avian Dis.,* **11**, 694-702

American Association of Avian Pathologists (1970) Leukosis Committee Workshop Report: Methods in Marek's disease research. *Avian Dis.,* **14**, 820-828

Anderson, D. P., Eidson, C. S. & Richey, D. J. (1971) Age susceptibility of chickens to Marek's disease. *Amer. J. vet. Res.,* **32**, 935-938

Andrews, C. N. & Glover, R. E. (1939) A case of neuro-lymphomatosis in a turkey. *Vet. Rec.,* **51**, 934-935

Bankowski, R. A., Mikami, T. & Reynolds, B. (1970) The relation between infection of chickens with Marek's disease and the presence of precipitin antibodies. *Avian Dis.,* **14**, 723-737

Baxendale, W. (1969) Preliminary observations on Marek's disease in ducks and other avian species. *Vet. Rec.,* **85**, 341-342

Beasley, J. N., Patterson, L. T. & McWade, D. H. (1970) Transmission of Marek's disease by poultry house dust and chicken dander. *Amer. J. vet. Res.* **31**, 339-344

Benton, W. J. & Cover, M. S. (1957) The increased incidence of visceral lymphomatosis in broiler and replacement birds. *Avian Dis.,* **1**, 320-327

Biggs, P. M. (1961) A discussion on the classification of the avian leucosis complex and fowl paralysis. *Brit. vet. J.,* **117**, 326-334

Biggs, P. M., Purchase, H. G., Bee, B. R. & Dalton, P. J. (1965) Preliminary report on acute Marek's disease (fowl paralysis). *Vet. Rec.,* **77**, 1339-1340

Biggs, P. M. & Payne, L. N. (1967). Studies on Marek's disease. I. Experimental transmission. *J. nat. Cancer Inst.,* **39**, 267-280

Biggs, P. M. (1968) Marek's disease—current state of knowledge. *Curr. Top. Microbiol. Immunol.* **43**, 93-125

Biggs, P. M., Long, P. L., Kenzy, S. G. & Rootes, D. G. (1968) Relationship between Marek's disease and coccidiosis. I. The effect of Marek's disease on the susceptibility of chickens to coccidial infection. *Vet. Rec.,* **83**, 284-289

Biggs, P. M., Thorpe, R. J. & Payne, L. N. (1968) Studies on genetic resistance to Marek's disease in the domestic chicken. *Brit. poult. Sci.,* **9**, 37-52

Blomberg, H. (1949) Ett fall av neurolympfomatos hos svan. *Nord. vet. Med.,* **1**, 719-725

Brewer, R. N., Reid, W. M., Botero, H. & Schmittle, S. C. (1968) Studies on acute Marek's disease. 2. The role of coccidia in transmission and induction. *Poult. Sci.,* **47**, 2003-2012

Brewer, R. N., Reid, W. M., Johnson, J. & Schmittle, S. C. (1969) Studies on acute Marek's disease. VIII. The role of mosquitoes in transmission under experimental conditions. *Avian Dis.* **13**, 83-88

Burkitt, D. P. (1969) Etiology of Burkitt's lymphoma—an alternative hypothesis to a vectored virus. *J. nat. Cancer Inst.,* **42**, 19-28

Busch, R. H. & Williams, L. E., Jr. (1970) Case report—A Marek's disease-like condition in Florida turkeys. *Avian Dis.,* **14**, 550-554

Calnek, B. W. & Hitchner, S. B. (1969) Localization of viral antigen in chickens infected with Marek's disease herpesvirus. *J. nat. Cancer Inst.,* **43**, 935-950

Calnek, B. W., Adldinger, H. K. & Kahn, D. E. (1970) Feather follicle epithelium: a source of enveloped and infectious cell-free herpesvirus from Marek's disease. *Avian Dis.,* **14**, 219-233

Calnek, B. W., Ubertini, T. & Adldinger, H. K. (1970) Viral antigen, virus particles, and infectivity of tissues from chickens with Marek's disease. *J. nat. Cancer Inst.,* **45**, 341-351

Calnek, B. W. & Witter, R. L. (1972) Neoplastic diseases. In: Hofstad, M. S., ed., *Diseases of poultry*, Iowa State University Press (in press)

Cho, B. R. (1970) Experimental dual infections of chickens with infectious bursal Marek's disease and agents. I. Preliminary observation on the effect of infectious bursal agent on Marek's disease. *Avian Dis.*, **14**, 665-675

Chubb, R. C. & Churchill, A. E. (1968) Precipitating antibodies associated with Marek's disease. *Vet. Rec.*, **83**, 4-7

Chubb, R. C. & Churchill, A. E. (1969) The effect of maternal antibody on Marek's disease. *Vet. Rec.*, **85**, 303-305

Churchill, A. E. & Biggs, P. M. (1967) Agent of Marek's disease in tissue culture. *Nature (Lond.)*, **215**, 528-530

Churchill, A. E. (1968) Herpes-type virus isolated in cell culture from tumors of chickens with Marek's disease. I. Studies in cell culture. *J. nat. Cancer Inst.*, **41**, 939-950

Churchill, A. E. & Biggs, P. M. (1968) Herpes-type virus isolated in cell culture from tumors of chickens with Marek's disease. II. Studies *in vivo. J. nat. Cancer Inst.*, **41**, 951-956

Churchill, A. E., Payne, L. N. & Chubb, R. C. (1969) Immunization against Marek's disease using a live attenuated virus. *Nature (Lond.)*, **221**, 744-747

Cole, R. K. & Hutt, F. B. (1951) Evidence that eggs do not transmit leucosis. *Poult. Sci.*, **30**, 205-212

Cole, R. K. (1968) Studies on genetic resistance to Marek's disease. *Avian Dis.*, **12**, 9-28

Colwell, W. M. & Schmittle, S. C. (1968) Studies on acute Marek's disease. VII. Airborne transmission of the GA isolate. *Avian Dis.*, **12**, 724-729

Conner, G. H., LaBelle, J. A., Langham, R. F. & Crittenden, M. (1966) Studies on the epidemiology of bovine leukemia. *J. nat. Cancer Inst.*, **36**, 383-388

Cottral, G. E. & Winton, B. (1953) Paralysis in ducks stimulating neural lymphomatosis in chickens. *Poult. Sci.*, **32**, 585-588

Davis, R. B., Dawe, D. L. & Brown, J. (1969) Effect of avian encephalomyelitis vaccination on condemnation rates of broilers for Marek's disease. *Avian Dis.*, **13**, 872-875

Drury, L. N., Patterson, W. C. & Beard, C. W. (1969) Ventilating poultry houses with filtered air under positive pressure to prevent airborne diseases. *Poult. Sci.*, **48**, 1640-1646

Dunlop, W. R., Kottaridis, S. D., Gallagher, S. C., Smith, S. C. & Strout, R. G. (1965) The detection of acute avian leucosis as a contagious disease. *Poult. Sci.*, **44**, 1537-1540

Eidson, C. S., Schmittle. S. C., Goode, R. B. & Lal, J. B. (1966) Induction of leukosis tumors with the beetle *Alphitobius diaperinus. Amer. J. vet. Res.*, **27**, 1053-1057

Eidson, C. S. & Schmittle, S. C. (1968) Studies on acute Marek's disease: I. Characteristics of isolate GA in chickens. *Avian Dis.*, **12**, 467-476

Eidson, C. S. & Schmittle, S. C. (1969) Studies on acute Marek's disease. XII. Detection of antibodies with a tannic acid indirect hemagglutination test. *Avian Dis.*, **13**, 774-782

Halliwell, W. H. (1971). Lesions of Marek's disease in a great horned owl. *Avian Dis.*, **15**, 49-55

Harnden, E. E. & Smith, H. C. (1943) A case of turkey lymphomatosis. *Poult. Sci.*, **22**, 331-333

Harris, S. T. (1939) Lymphomatosis (Fowl paralysis) in the pheasant. *Vet. J.*, **95**, 104-106

Hutt, F. B. & Cole, R. K. (1947) Genetic control of lymphomatosis in the fowl. *Science*, **106**, 379-384

Ianconescu, M. & Samberg, Y. (1971) Etiological and immunological studies in Marek's disease. II. Incidence of Marek's disease precipitating antibodies in commercial flocks and in eggs. *Avian Dis.*, **15**, 177-186

Jakowski, R. M. & Fredrickson, T. N. (1969) Early changes in bursa of Fabricius from Marek's disease. *Avian Dis.*, **13**, 215-222

Jennings, A. R. (1954) Disease in wild birds. *J. comp. Path.*, **64**, 356-359

Jungherr, E. (1939) Neurolymphomatosis phasianorum. *J. Amer. vet. med. Ass.*, **94**, 49-52

Jurajda, V. & Klimes, B. (1970) Presence and survival of Marek's disease agent in dust. *Avian Dis.*, **14**, 188-190

Kawamura, H., King, D. J., Jr & Anderson, D. P. (1969) A herpesvirus isolated from kidney cell culture of normal turkeys. *Avian Dis.*, **13**, 853-863

Kenzy, S. G. & Biggs, P. M. (1967) Excretion of the Marek's disease agent by infected chickens. *Vet. Rec.*, **80**, 565-569

Kenzy, S. G. & Cho, B. R. (1969) Transmission of classical Marek's disease by affected and carrier birds. *Avian Dis.*, **13**, 211-214

Kenzy, S. G., Long, P. L. & Biggs, P. M. (1970) Relationship between Marek's disease and coccidiosis. III. Effects of infection with *Eimeria mivati* on Marek's disease in the chicken. *Vet. Rec.*, **86**, 100-104

Long, P. L., Kenzy, S. G. & Biggs, P. M. (1968) Relationship between Marek's disease and coccidiosis. I. Attempted transmission of Marek's disease by avian coccidia. *Vet. Rec.*, **83**, 260-262

MacMahon, B. (1968) Epidemiologic aspects of acute leukemia and Burkitt's tumor. *Cancer*, **21**, 558-562

Marek, J. (1907) Multiple Nervenentzündung (Polyneuritis) bei Hühnern. *Dtsch. tierärztl. Wschr.*, **15**, 417-421

Morgan, H. R. (1971) Research note—Antibodies for Marek's disease in sera from domestic chickens and wild fowl in Kenya. *Avian Dis.*, **15**, 611-613

Nazerian, K., Solomon, J. J., Witter, R. L. & Burmester, B. R. (1968) Studies on etiology of Marek's disease. II. Finding of a herpesvirus in cell culture. *Proc. Soc. exp. Biol. (N.Y.)*, **127**, 177-182

Nazerian, K. & Witter, R. L. (1970) Cell-free transmission and *in vivo* replication of Marek's disease virus (MDV). *J. Virol.*, **5**, 388-397

Okazaki, W., Purchase, H. G. & Burmester, B. R. (1970) Protection against Marek's disease by vaccination with a herpesvirus of turkeys. *Avian Dis.,* **14,** 413-429

Pappenheimer, A. M., Dunn, L. C. & Cone, V. (1929) Studies on fowl paralysis (neurolymphomatosis gallinarum). I. Clinical features and pathology. *J. exp. Med.,* **49,** 63-86

Patterson, F. D., Wilcke, H. L., Murray, C. & Henderson, E. W. (1932) So-called range paralysis of the chicken. *J. Amer. vet. Ass.,* **81,** 746-767

Payne, L. N. & Biggs, P. M. (1967) Studies on Marek's disease. II. Pathogenesis. *J. nat. Cancer Inst.,* **39,** 281-302

Proudfoot, F. G. & Aitken, J. R. (1969) The effect of diet on mortality attributed to Marek's disease among leghorn genotypes. *Poult. Sci.,* **48,** 1457-1459

Purchase, H. G. & Biggs, P. M. (1967) Characterization of five isolates of Marek's disease. *Res. vet. Sci,* **8,** 440-449

Purchase, H. G. (1970) Virus-specific immunofluorescent and precipitin antigens and cell-free virus in the tissues of birds infected with Marek's disease. *Cancer Res.,* **30,** 1898-1908

Purchase, H. G. & Burgoyne, G. H. (1970) Immunofluorescence in the study of Marek's disease: detection of antibody. *Amer. J. vet. Res.,* **31,** 117-123

Purchase, H. G., Burmester, B. R. & Cunningham, C. H. (1971) Pathogenicity and antigenicity of clones from strains of Marek's disease virus and the herpesvirus of turkeys. *Inf. and Imm.,* **3,** 295-303

Purchase, H. G. & Okazaki, W. (1971) The effect of vaccination with the herpesvirus of turkeys (HVT) on the horizontal spread of Marek's disease herpesvirus. *Avian Dis.,* **15,** 391-397

Rispens, B. H., van Vloten, J. & Maas, H. F. L. (1969) Some virological and serological observations on Marek's disease: A preliminary report. *Brit. vet. J.,* **9,** 445-453

Sevoian, M., Chamberlain, D. M. & Larose, R. N. (1963a) Avian lymphomatosis. V. Air-borne transmission. *Avian Dis.,* **7,** 102-105

Sevoian, M., Chamberlain, D. M. & Larose, R. N. (1963b) Avian lymphomatosis. VII. New support for etiologic unity. *Proceedings 17th World's Veterinary Congress,* Vol. 2, pp. 1475-1476

Sevoian, M. & Chamberlain, D. M. (1963) Avian lymphomatosis. III. Incidence and manifestations in experimentally infected chickens of various ages. *Avian Dis.,* **7,** 97-102

Sevoian, M. (1968) Egg transmission studies of type II leukosis infection (Marek's disease). *Poult. Sci.,* **47,** 1644-1646

Sevoian, M. (1969) Studies for the control of type II (Marek's) leukosis. *Proceedings 4th Congress World's Veterinary Poultry Association,* pp. 259-266

Simpson, C. E., Anthony, D. W. & Young, F. (1957) Visceral lymphomatosis in a flock of turkeys. *J. Amer. vet. med. Ass.,* **130,** 93-96

Solomon, J. J., Witter, R. L., Stone, H. A. & Champion, L. R. (1970) Evidence against embryo transmission of Marek's disease. *Avian Dis.,* **14,** 752-762

Spencer, J. L. (1969) Marek's disease herpesvirus: *In vivo* and *in vitro* infection of kidney cells of different genetic strains of chickens. *Avian Dis.,* **13,** 753-761

Vielitz, E. & Landgraf, H. (1970) Contribution to the epidemiology and control of Marek's disease. *Dtsch. tierärztl. Wschr.,* **77,** 357-362

Weller, T. H. (1970) Cytomegaloviruses: The difficult years. *J. infect. Dis.,* **122,** 532-539

Wight, P. A. L. (1963) Lymphoid leucosis and fowl paralysis in the quail. *Vet. Rec.,* **75,** 685-687

Witter, R. L. & Burmester, B. R. (1967) Transmission of Marek's disease with oral washings and feces from infected chickens. *Proc. Soc. exp. Biol. (N.Y.),* **124,** 59-62

Witter, R. L., Burgoyne, G. H. & Burmester, B. R. (1968) Survival of Marek's disease agent in litter and droppings. *Avian Dis.,* **12,** 522-530

Witter, R. L., Burgoyne, G. H. & Solomon, J. J. (1969) Evidence for a herpesvirus as an etiologic agent of Marek's disease. *Avian Dis.,* **13,** 171-184

Witter, R. L., Solomon, J. J. & Burgoyne, G. H. (1969) Cell culture techniques for primary isolation of Marek's disease associated-herpesvirus. *Avian Dis.,* **13,** 101-118

Witter, R. L. (1970) Epidemiological studies relating to the control of Marek's disease. *Wld's Poult. Sci. J.,* **26,** 755-762

Witter, R. L., Moulthrop, J. I., Jr., Burgoyne, G. H. & Connell, H. C. (1970a) Studies on the epidemiology of Marek's disease herpesvirus in broiler flocks. *Avian Dis.,* **14,** 255-267

Witter, R. L., Nazerian, K., Purchase, H. G. & Burgoyne, G. H. (1970b) Isolation from turkeys of a cell-associated herpesvirus antigenically related to Marek's disease virus. *Amer. J. vet. Res.,* **31,** 525-538

Witter, R. L., Solomon, J. J., Champion, L. R. & Nazerian, K. (1971) Long-term studies of Marek's disease infection in individual chickens. *Avian Dis.,* **15,** 346-365

Witter, R. L. & Solomon, J. J. (1971) Epidemiology of a herpesvirus of turkeys: possible sources and spread of infection in turkey flocks. *Inf. and Imm.,* **4,** 356-361

The Genetics of Resistance to Marek's Disease

R. K. COLE [1]

Genetic differences in response to infection with Marek's disease virus (MDV) have usually been found whenever sought. The chicken, as a species, is well adapted to breeding procedures required to make the most of such differences, and it should also be possible to take advantage of them in the commercial poultry industry. They have been accentuated by the use of appropriate selection procedures (Cole, 1968; Morris, Ferguson & Jerome, 1970). To do so, at the present time, requires the deliberate exposure of fully-pedigreed chicks to MDV and the selection of breeding stock on the basis of a sib-test or a combination of sib- and progeny-tests.

Genetic resistance to MD, as expressed by freedom from mortality or lesions, is not dependent upon resistance to MDV, but resides in the ability of the bird to prevent transformation of the infected cells into neoplastic ones. While several factors may influence the final outcome of an infection with MDV, only those dependent upon gene action within the host can be utilized to produce permanent changes. Such genetic gains in resistance as may be obtained by a few generations of selection are cumulative and become a permanent part of the inheritance of the particular stock. There are reasons to believe that the genes responsible for resistance accumulated as a consequence of a selection programme are not dissipated after selection has ceased. Such a method of controlling MD would be preferable to one requiring annual vaccination or modification of the environment in an attempt to keep it free of MDV.

Most commercial-type chickens are produced by crossing two or more breeds, strains, or inbred lines. It is important to know what happens to the resistance of the hybrid progeny when stocks that vary in levels of genetic resistance to MD are used as parents or grandparents.

Where crosses have been made, the level of resistance in the F_1 population is usually intermediate between that of the two parent strains. For example, in a study of 10 broiler strains and 21 crosses between them, Hartman (1969) observed the hybrids to have a mortality, presumably from MD, about 6% below the mid-parent level, depending on the method of exposure. In his series, mortality by 12 weeks of age exceeded 40% when test chicks were inoculated, and ranged from 35 to 40% by 18 weeks of age following exposure by contact.

There are few reports in the literature on crosses between strains with known levels of susceptibility to MD. Where parent stocks did differ in susceptibility, they had not been bred specifically with this end in view.

In crosses of inbred lines of White Leghorns from East Lansing, Schmittle & Eidson (1968), using the GA isolate of MDV, and Stone (1969), using the JM isolate, observed similar results. Line 6 was quite resistant and Line 7 very susceptible to both viruses, following inoculation. Hybrid progeny from the resistant Line-6 dams were more resistant than those from Line-7 dams, indicating an effect associated with dam, possibly due to egg-borne antibodies. The results of further studies of these lines (Stone *et al.,* 1970) were interpreted as indicating that resistance was dominant and resulted from the action of a small number of genes. Higher levels of precipitating antibody were present in the resistant line. When exposure was by contact, hence delayed, Schmittle & Eidson (1968) observed that progeny from Line-7 dams were actually a little more resistant than those from Line-6 dams. This would suggest that the effect of antibody is greater when dosage of virus is high and given early in life. Cole (1968) reported certain results that were explained on this basis.

[1] Department of Poultry Science, Cornell University, Ithaca, New York, USA.

In each of three reciprocal crosses between breeds, which varied in susceptibility to JM-induced MD only from 64.5 to 74.7%, the hybrids were more susceptible than the purebreds (Han *et al.*, 1969). When the dams came from the more susceptible of the two breeds crossed, the hybrid progeny were somewhat more resistant. For Lines 6 and 7, it was the resistant-line dam that produced the more resistant progeny.

The susceptible Cornell S strain and the relatively resistant Cornell K strain, and a cross between them, were tested for susceptibility to the JM strain of MDV by Sevoian (1968). The reciprocal logarithm of the dilution of an inoculum required to induce a 50% incidence of lesions by 3 weeks, was 4.6 for the S strain and only 1.5 for the K strain. The cross of K ♂♂ × S ♀♀ gave the intermediate value of 2.8, and was therefore much more resistant than the average for the two parents, even though they came from susceptible-strain dams. In this case eggs came from parents that had received identical exposure to naturally-occurring MDV.

In a reciprocal cross between the K and S strains, Grunder *et al.* (1969) observed a variable response. They had no pure-strain controls. There were no differences between the crosses following exposure by contact of chicks from 7-month-old breeders, although those from the S-strain dams were slightly more resistant. When progeny from the same dams, but then 13 months of age, were inoculated with MDV, those from the S-strain dams were significantly more susceptible.

That some of the differences observed by various investigators could be due to egg-borne antibody seems probable. Chubb & Churchill (1969) demonstrated that maternal antibodies did markedly affect the incidence of MD and that when the level of exposure was extremely high or the virus more virulent, the antibody influenced the survival time rather than the level of mortality. When an effect associated with the dam is present, it could result from different levels of exposure or from differences between dams in ability to produce and transmit antibody via the egg.

Tests to evaluate the transmission of genetic resistance to MD have been carried out, using the JM-N line of White Leghorns. This line (Cole, 1968), now selected for 4 generations, exhibits a very high level of resistance to MD, as compared to its companion line selected for susceptibility (JM-P) or the S strain. It has been used in reciprocal crosses with strains having known levels of response to MDV.

MATERIALS AND METHODS

Mating procedures

Matings were so organized that both pure-line and strain-cross progeny of each sire and each dam were hatched on the same day. The parents had been raised under similar conditions of management, often as mixed populations, so that any effect of level of exposure of the dam to MDV on egg-borne antibodies would be similar for the two crosses compared. The crosses tested have been produced over a period of 4 years and hence represent samples of the JM-N and JM-P lines taken at different stages during their development.

Inoculum

Fresh tumour tissue (gonadal, from 15 to 20 S-strain chicks previously inoculated with or exposed to JM-MDV) as a 2.5% tissue emulsion in sterile saline was inoculated intra-abdominally at the rate of 0.25 ml per chick for all tests prior to 1969 and in 1969 for selection within the N and P lines. The crosses tested in 1969 and 1970 were inoculated with a frozen (since January 1968) tumour-tissue preparation and at a dose of only one-tenth that used previously.

Test procedure

All chicks in a given test were randomized, then inoculated on day 2, and raised in batteries to an age of 8 weeks. Necropsies were performed on all dead or obviously sick birds and at the end of the test period on all survivors. In most of the tests, the location of the MD lesions was also recorded. In each test, both S-strain and JM-P chicks were included as controls.

RESULTS

The results (Table 1) show that when the JM-N line was used in a cross the susceptibility of the F_1 progeny was much closer to the level for the N-line controls than to the mean of the susceptibilities of the two parental stocks.

The number of chicks tested per pure strain or cross averaged 129 and ranged from 57 to 178. The susceptibility (21.1%) of the sample of N-line chicks tested in 1970 was not typical for this stock. In other tests of 4th generation N-line chicks in 1969, the susceptibility was 3.6% (mortality and lesions to 8 weeks) among 527 chicks and 4.5% (mortality only,

Table 1. Susceptibility to Marek's disease of chicks from various pure strains and their crosses with the JM-N line

| Year | Generation of N and P lines | Pure strains | | | | | Crosses | | |
| | | | | | | | Parental mean | N line as: | |
		JM-N	JM-P	K	C	S		sire	dam
1967	2	14.7%	95.1 %[1]	38.1%	—	100 %	54.9%	23.9%	30.2%
1968	3	6.2%	100 %[1]	39.4%	—	100 %	53.1%	16.7%	26.7%
1969	4	4.0%	91.5%	15.6%[1]	—	97.2%	9.8%	4.5%	3.4%
1970	4	21.1%[2]	96.0%	26.4%	44.2%[1]	97.5%	32.6%	6.1%	11.3%
1970	4	6.5%	94.4%	—	—	—	—	—	—
1970		Commercial hybrid No. I (40.7% in 1964) [2]						6.4%	
1970		Commercial hybrid No. II (73.6% in 1964) [2]						12.2%	

[1] Strain used with JM-N for production of crosses.
[2] See text for explanations.

to 16 weeks) among 529 other chicks. Second samples from the same parents in each of the N and P lines were therefore tested. The results (Table 1) obtained with these samples (6.5% for N and 94.4% for P) were in complete agreement with all other previous tests and suggest that there had been some error in the identity of the eggs initially set as JM-N line.

Hybrid females from two commercial sources were available for use in crosses with N-line sires. (Some indication of their probable level of genetic susceptibility to MD was available from data of 1964, when male chicks from the same sources had been tested). The progeny from both crosses were quite resistant and the results indicate that N-line sires are able to transmit high levels of genetic resistance when mated to dams that are hybrids.

If the data are arranged in a different form (Table 2) by listing, in descending order of susceptibility,

the pure strains crossed with the N line, it will be seen that they also affect the resistance of the hybrid progeny.

Contribution of each parent

For three of the four reciprocal crosses compared (Table 1), those from resistant JM-N-line sires were less susceptible than those from JM-N-line dams. None of these differences is statistically significant, but the evidence indicates that the male has as much influence upon the response of the progeny as does the female which, in addition to genes, may also transmit protective antibodies.

In 7 of the 8 possible comparisons, the female progeny had higher levels of MD than did their male siblings (Table 3). This is in agreement with most reports in the literature.

If sex-linked genes are a factor in reciprocal crosses, the hybrid daughters from the N-line sires should

Table 2. Contribution of parents to resistance of hybrid progeny from N-line breeders

| Strain | Susceptibility of the: | | |
	Pure strain	Hybrids from N-line sires	Hybrids from N-line dams
P[1]	97.6%	20.8%	28.4%
Commercial hybrid No. II	(73.6%) [2]	12.2%	—
C	44.2%	6.1%	11.3%
Commercial hybrid No.I	(40.7%) [2]	6.4%	—
K	15.6%	4.5%	3.4%

[1] Data for 1967 and 1968 combined.
[2] Minimum, because data are based on male chicks only.

Table 3. Contribution of resistant-line parent (JM-N line) to response of hybrid progeny to MDV

Year	Other stock	Susceptibility of:							
		♂♂ progeny, from:				♀♀ progeny, from:			
		N-sires		N-dams		N-sires		N-dams	
		No.	MD	No.	MD	No.	MD	No.	MD
1967	JM-P	45	15.6 %	40	32.5 %	43	32.6 %	46	28.2 %
1968	JM-P	52	11.5 %	47	17.0 %	62	21.0 %	54	35.2 %
1969	K	92	4.3 %	94	2.1 %	87	4.6 %	84	4.8 %
1970	C	88	5.7 %	76	5.3 %	59	6.8 %	74	17.6 %

have a lower level of MD than those from the N-line dams and there should be no difference between the two sources of male offspring. The data, as summarized in Table 3, indicate that neither sex-linked genes nor protective antibodies transmitted by N-line dams were of any importance. The resistance observed in these reciprocal crosses, therefore, appears to depend primarily upon genes borne on the autosomes.

Types and distribution of lesions

The distribution of gross lesions of MD seen at necropsy (Table 4) varied for the different strains and crosses. For those with high levels of susceptibility, the frequencies of gross lesions of the nerves and gonads were high and many chicks had lesions at both sites, but none elsewhere. In contrast, lesions in other visceral organs varied from stock to stock. Fewer S-strain chicks had lesions of visceral organs,

other than the gonads, than did JM-P-line chicks. This resulted, presumably, from the fact that earlier death in the S strain, associated with nerve lesions and paralysis, eliminated many birds before lesions in other organs developed far enough to be recognized.

Within each pure strain, there was considerable uniformity in the distribution of lesions from one test to another. The one glaring exception was for the JM-N line in 1970 in which, among those susceptible, the frequency of lesions of the nerves and gonads differed markedly from that for the other two tests; this fact, as well as the abnormally high level of susceptibility, suggests some error in the identity of the JM-N chicks used in that test.

In addition to the excellent level of resistance shown by the crosses in 1970, involvement of nerves and gonads was at a very low level among the few birds affected. This may have been due, in part, to

Table 4. Distribution of gross lesions of MD in susceptible chicks from various sources

Test	Stock	Chicks		Lesions in susceptible chicks			
		Tested (No.)	Susceptible (No.)	Small only [1] (%)	Nerves (%)	Gonads (%)	Other organs (%)
1970a	JM-N	147	31	0	90	90	77
1970b	JM-N	154	10	30	40	40	80
1967–68 [2]	JM-N	172	17	59	12	24	65
1970a	JM-P	151	145	0.7	91	85	84
1970b	JM-P	125	118	0.8	92	82	81
1967–68	JM-P	192	188	1.1	93	70	54
1970a	S	158	154	0.6	98	79	39
1967–68	S	164	164	0.6	97	84	34
1970a	K	159	42	12	60	48	50
1967–68	K	188	73	8	79	36	44
1970a	C	104	46	22	43	61	80
1970a	N × C	147	9	67	11	22	77
1970a	C × N	150	17	65	12	12	94
1970a	N × Com. I	157	10	70	0	0	100
1970a	N × Com. II	156	19	68	11	5	95
1967–68	N × P	202	40	12	50	60	55
1967–68	P × N	187	53	15	66	49	53

[1] See text for explanation.
[2] Data for 1967 and 1968 combined.

the inclusion as "susceptible" of those chicks that, on necropsy, showed only very small lesions on the heart, and/or occasionally on the liver, testis, or kidney. These, for the most part, resembled smooth colonies of *Salmonella sp.* growing on plain agar plates. Such lesions were seen at necropsy of the symptom-free chicks at the end of the test and were often limited to 2 or 3 such foci on the heart. They were more common among the survivors of the more resistant stocks. Of the strain-cross chicks of 1970 that were considered as susceptible, about two-thirds in each cross showed lesions of this type alone. In a few instances, such lesions were recorded for chicks that also had more typical MD lesions in other organs.

It is probable that these small lesions do *not* indicate MD but rather lymphoid leukosis or some other condition that causes localized accumulations of lymphoid cells. However, one can arrange a series of hearts to show a complete range from normal, to hearts with 2 or 3 foci, those with several or a large number of foci, and finally to those so severely affected that the identity of the lesion as that of MD is unquestionable. The problem is where to draw the line between those cases that should and those that should not be called MD. For the more susceptible stocks, birds having only lesions of this type are not common (Table 4). To classify then as MD would not materially change the level of susceptibility. For the more resistant stocks, especially the crosses involving the JM-N line, the level of susceptibility would be markedly influenced. This would lead to the conclusion that these crosses are more susceptible than, in fact, they are. It seems better to err in this direction than to call them resistant.

DISCUSSION AND CONCLUSIONS

If genetic differences in resistance to MD are to be used for the purpose of improving the viability of commercial stocks of chickens, the parent and/or grandparent strains used must be so selected and combined that hybrid offspring possess a high level of resistance. The mode of inheritance becomes, therefore, of some significance. To take advantage of the known mechanism of resistance to the leukosis-sarcoma viruses (Crittenden *et al.,* 1967) requires that all parental strains used to produce the hybrid must be homozygous and recessive for the two genes (*tva* and *tvb*) that control susceptibility at the cellular level to the two major types of leukosis virus. At the present time there is no evidence to show that resistance at the cellular level exists with respect to

MDV. All birds can apparently be infected, hence any resistance to MD observed must result from another type of mechanism.

The fact that selection for resistance and susceptibility to MD was very effective (Cole, 1968) and that the inbred lines from East Lansing were either very resistant (Line 6) or very susceptible (Line 7) suggests that a small number of major genes are probably involved. In fact, Stone *et al.* (1970) suggest that only 2 to 4 genes may be of much significance (in crosses studied by them) and that they induce resistance as a dominant trait.

In a majority of the crosses reported so far there is evidence that the hybrids are more resistant than would be expected if resistance resulted solely from additive genes. They are more resistant than the average for the parent strains used, a characteristic that one might associate with hybrid vigour as much as with dominance. However, the progeny of the crosses reported by Han *et al.* (1969) should have been characterized by high levels of hybrid vigour because they came from breed crosses, yet they were more susceptible than the pure breeds. Thus it appears that heterosis *per se* does not confer resistance to MD.

For the crosses that have been made between strains that vary considerably in resistance to MD (Lines 6 and 7, strains K and S, as well as those reported in this paper) there is evidence that the resistant stock transmits the trait in question effectively to the hybrid offspring. This is fortunate, for it suggests dominance of resistance so that all parental strains or lines need not be selected for high levels of resistance to MD. However, the data indicate that both parents contribute to the resistance of the hybrid progeny and that higher levels of resistance will probably occur when both or all parental stocks possess some degree of genetic resistance to MD.

That the sire is as effective as the dam in influencing the resistance of the hybrid progeny, especially when maternal antibody is not a major factor, seems clear. In inoculation trials, where exposure is severe, the significance of maternal antibody is greater, although the genes for resistance within the host do express themselves effectively. There is no evidence that sex-linked genes are particularly concerned with resistance to MD. The resistant pure-line sire (JM-N) can transmit high levels of resistance when mated to hybrid dams. These findings, taken together, indicate that effective control of MD can be achieved by developing a strain with a high level of sesistance and using it in two-way crosses or as a rire in 3-way crosses to produce commercial chicks.

ACKNOWLEDGEMENTS

This work was supported in part by the Northeastern Regional Poultry Breeding Project (NE-60).

REFERENCES

Chubb, R. C. & Churchill, A. E. (1969) Effect of maternal antibody on Marek's disease. *Vet. Rec.,* 85, 303-305

Cole, R. K. (1968) Studies on genetic resistance to Marek's disease. *Avian Dis.,* **12**, 9-28

Crittenden, L. B., Stone, H. A., Reamer, R. H. & Okazaki, W. (1967) Two loci controlling genetic cellular resistance to avian leukosis-sarcoma viruses. *J. Virol.,* 1, 898-904

Grunder, A. A., Dickerson, G. E., Robertson, A. & Morin, E. (1969) Incidence of Marek's disease as related to phenotypes of serum alkaline phosphatase. *Poult. Sci.,* **48**, 1608-1611

Han, F. S., Smyth, J. R. Jr., Sevoian, M. & Dickinson, F. N. (1969) Genetic resistance to leukosis caused by the JM virus in the fowl. *Poult. Sci.,* **48**, 76-87

Hartman, W. (1969) Studies on genetic resistance to Marek's disease in broiler strains. In: *Report of British Poultry Breeders Roundtable Conference,* Harrogate

Morris, J. R., Ferguson, A. E. & Jerome, F. N. (1970) Genetic resistance and susceptibility to Marek's disease. *Can. J. Anim. Sci.,* **50**, 69-81

Schmittle, S. C. & Eidson, C. S. (1968) Studies on acute Marek's disease. IV Relative susceptibility of different lines and crosses of chickens to GA isolate. *Avian Dis.,* **12**, 571-576

Sevoian, M. (1968) Variations in susceptibility of three selected strains of chickens to cell suspensions of JM virus. *Poult. Sci.,* **47**, 688-689

Stone, H. A. (1969) Investigations of the genetic control of Marek's disease. *Poult. Sci.,* **48**, 1879

Stone, H. A., Holly, E. A., Burmester, B. R. & Coleman, T. H. (1970) Genetic control of Marek's disease. *Poult. Sci.,* **49**, 1441-1442

Antibody Development in Chickens Exposed to Marek's Disease Virus

B. W. CALNEK [1]

Methodology in Marek's disease (MD) research has progressed rapidly since the very significant breakthrough in identifying a group B herpesvirus as the aetiological agent (Churchill & Biggs, 1967; Nazerian et al., 1968). In addition to permitting virus assay in vitro, the cell culture propagation of virus has provided antigens for antibody detected by agar gel immunodiffusion (Chubb & Churchill, 1968), indirect immunofluorescence (Spencer & Calnek, 1970; Purchase & Burgoyne, 1970), and indirect haemagglutination tests (Eidson & Schmittle, 1969). The recent discovery of the feather follicle epithelium of infected birds as a source of enveloped, infectious, cell-free virus preparations (Calnek, Adldinger & Kahn, 1970) has made possible the use of a virus-neutralization test for antibody detection (Calnek & Adldinger, 1971).

While it was thought earlier that Marek's disease virus (MDV) was not particularly antigenic, the results following the application of the above tests provided strong evidence that an immune response was indeed induced in exposed chickens. Furthermore, studies by Sevoian [2] and Chubb & Churchill (1969) indicated that maternal antibody passively acquired by chicks hatched from infected dams was protective. That observation warranted consideration of an immune response as significant in the pathogenesis of the disease. An earlier study (Calnek & Hitchner, 1969) had suggested that the precipitating-antibody response of two genetically different strains of chickens differed and that there might be a correlation between immunological competence and survival after MDV infection. However, the data were meagre and thus equivocal. The studies reported here were undertaken to determine whether, in fact, genetic strains of chickens known to respond differently to MD infection could be shown to differ with regard to their immunological responsiveness as well.

MATERIALS AND METHODS

Experimental chickens and holding facilities

All chickens were Single Comb White Leghorns derived from flocks known to be infected with MDV; thus, the chicks carried maternal antibody at hatching. Experiments were conducted when the birds were 4 or 8 weeks old, so that maternal antibody would not interfere with infection; the immunological incompetence sometimes associated with newly hatched chicks would also be reduced. Prior to the beginning of the trials, all birds intended for use in an experiment were kept together as a single group. For exposure to MDV, the birds were divided into appropriate groups and each group placed in a wire-floored battery in a separate isolation unit. However, birds of different genetic strains that were to be compared were kept in the same isolation pen.

Four genetic strains were studied. The relatively resistant departmental PDRC flock had previously been compared with the genetically susceptible S-strain (Hutt & Cole, 1947) in studies of virus localization (Calnek & Hitchner, 1969); these two strains were employed in the first experiment. Chickens for the second experiment were from two lines selected from the Regional Cornell Random bred stock, one for susceptibility, the JM-P line (P-line), and the second for resistance, the JM-N line (N-line), to Marek's disease virus (Cole, 1968). They represented progeny of the fourth generation of selected lines, the MD mortality for the P-line being generally

[1] Department of Avian Diseases, New York State Veterinary College, Cornell University, Ithaca, New York, USA.
[2] Personal communication.

greater than 90% and that of the N-line less than 10% following inoculation with MDV at one day of age (Cole [1]).

Virus exposure

The JM isolate (Sevoian, Chamberlain & Counter, 1962) of MDV was employed throughout. For Experiment 1, birds were inoculated intra-abdominally with 0.2 ml of a suspension of JM-infected gonadal tumour cells (Batch 236/239) or intratracheally with 0.1 ml of a cell-free suspension of JM virus (Batch X-300) prepared from the skin of infected S-strain chickens, as described elsewhere (Calnek, Hitchner & Adldinger, 1970). Each dose of the two preparations, as calculated by assay on chicken kidney (CK) cell cultures, contained about 480 and 280 focus forming units (FFU) per bird, respectively. Cellular inoculum for Experiment 2 consisted of JM-infected blood and tumour cells (Batch 269-2). Birds were inoculated intra-abdominally with 0.1 ml containing an estimated 6 FFU. Intratracheal inoculations were performed with the same cell-free virus preparation and dose as in the first experiment.

Serological tests

The agar gel precipitin (AGP) test described by Chubb & Churchill (1968) was performed as described elsewhere (Calnek, Ubertini, & Adldinger, 1970). Antigen in a centre well consisted of an extract of skin from JM-infected, S-strain chicks. Sera or dilutions of sera in pH-6.8 phosphate-buffered saline (PBS) were placed in each of six surrounding wells. Each test serum was placed in a well adjacent to one containing a known positive serum so as to increase the sensitivity of the test and to verify the specificity of lines of precipitation. The precipitating antibody titre was expressed as the reciprocal of the highest dilution causing a line of precipitation within a 72-hour observation period.

For the virus neutralization (VN) test, a cell-free extract from the skin of JM-infected S-strain chickens (Batch X-364) was diluted 1:10 in a stabilizer, SPGA[2] (Calnek, Hitchner & Adldinger, 1970), containing 0.2% disodium ethylenediamine tetraacetate (EDTA). This diluent has been shown to increase virus titres by as much as 10-fold compared to virus diluted in PBS (Adldinger & Calnek, 1971) and not to interfere with virus neutralization by antibody

(Calnek [3]). Two-fold serial dilutions of serum were prepared from an initial 1:5 dilution in PBS. One part of serum dilution was mixed with 4 parts of the virus dilution. PBS was substituted for serum for a virus control, and known positive and negative sera were included in each test as controls. After a 30-minute reaction time at 37°C, 0.25 ml of the mixture was inoculated on to each of two drained, 24-hour, CK monolayers in 50-mm Petri dishes. Liquid medium (5 ml) was added after 30 minutes and an agar overlay was applied the next day. Focal lesions were enumerated after a total of 7 or 8 days of incubation at 38–39°C. Virus control cultures developed about 100 foci each. The neutralizing antibody titre was expressed as the reciprocal of the highest serum dilution causing at least a 50% reduction in the virus titre. Survival of virus after reaction with a positive serum dilution was usually less than 20%.

EXPERIMENTAL DESIGN AND RESULTS

Experiment 1

Eight-week-old PDRC and S-strain birds were divided into three groups. Six PDRC and four S-strain birds served as unexposed controls. Thirteen PDRC and 11 S-strain birds were inoculated intra-abdominally with cell-associated virus. Cell-free virus was administered intratracheally to 15 and 10 birds of the respective groups. Sera were obtained at 0, 7, 10, 14, 21, 28, 54, and 97 days post exposure. AGP tests were conducted with undiluted serum in all instances and titrations were conducted with some sera. Neutralizing antibody titres were determined for all serum samples collected at 28 and 54 days post exposure. Samples collected earlier were tested only if antibody was detected in the 28-day sample. If the 54-day sample was negative, additional samples taken later were tested.

Birds that died were examined for gross lesions indicative of MD. At 97 days post exposure, all surviving birds were examined for clinical signs of MD and then killed and examined for gross lesions. The few birds that died during the course of the experiment due to causes other than MD were not included in the data.

The results obtained in Experiment 1 are recorded in Tables 1 and 2. As expected, the PDRC birds experienced a high rate of infection but a low incidence of MD. A single PDRC bird, inoculated

---

[1] Personal communication.
[2] Phosphate buffer containing sucrose, disodium glutamate and bovine serum albumin.

[3] Unpublished data.

Table 1. Experiment 1: Serological response and incidence of Marek's disease in PDRC and S-strain chickens exposed to Marek's disease virus at 8 weeks of age [1]

Day post exposure	Cell-associated virus inoculated intra-abdominally						Cell-free virus inoculated intratracheally					
	PDRC			S-strain			PDRC			S-strain		
	Cum. MD [2]	Serology [3]		Cum. MD [2]	Serology [3]		Cum. MD [2]	Serology [3]		Cum. MD [2]	Serology [3]	
		AGP	VN		AGP	VN		AGP	VN		AGP	VN
0	—	0/13	...	—	0/11	...	—	0/15	...	—	0/10	...
7	—	0/13	...	—	0/11	...	—	...	...	—	...	...
10	—	5/13	1/13	—	0/11	...	—	2/15	...	—	2/10	...
14	—	8/13	...	—	1/11	...	—	2/15	0/15	—	2/10	...
21	—	9/13	11/13	—	4/11	0/11	—	2/15	1/14	—	3/10	...
28	0/13	11/13	12/13	0/11	4/11	0/11	0/15	2/15	2/14	1/10	2/9	0/9
54	1/13	12/12	12/12	10/11	1/1	0/1	0/15	14/15	5/15	4/10	4/6	0/6
97	1/13	12/12	...	11/11	...	...	0/15	15/15	15/15	9/10	1/1	1/1

[1] Six PDRC and 4 S-strain birds served as unexposed controls. AGP and VN tests on sera obtained at 0, 21, 28, 54, and 97 days were all negative and all birds remained free of MD throughout the experiment. All results are expressed as the ratio: number positive / number examined; where ... appears, none were examined.
[2] Cumulative incidence of Marek's disease; includes MD mortality and survivors with gross lesions of MD at 97 days post exposure.
[3] Serological tests—AGP: agar gel precipitin; VN: virus neutralization.

Table 2. Experiment 1: Precipitating and neutralizing antibody titres in sera from PDRC and S-strain chickens exposed to Marek's disease virus at 8 weeks of age [1]

| Strain | Type of inoculum | Day post exposure | AGP test [2] No. of birds yielding sera with precipitating antibody titre of: | | | | | VN test [3] No. of birds yielding sera with neutralizing antibody titre of: | | | | |
			1	2	4	8	≥ 16	5	10	20	40	≥ 80
PDRC	Cell-associated	10	2	—	1	—	—	—	1	—	—	—
		21	1	2	3	1	1	5	3	2	1	—
		28	—	—	—	—	—	7	1	1	1	2
		54	—	—	2	3	4	—	—	2	3	7
	Cell-free	10	2	—	—	—	—	—	—	—	—	—
		21	1	—	—	—	1	—	—	1	—	—
		28	—	—	—	—	—	1	1	—	—	—
		54	—	1	1	5	3	—	1	2	—	2
		97	—	—	—	—	—	—	2	2	4	7
S-strain	Cell-associated	21	1	1	2	—	—	—	—	—	—	—
		28	1	—	—	—	—	—	—	—	—	—
		54	—	—	—	1	—	—	—	—	—	—
	Cell-free	10	2	—	—	—	—	—	—	—	—	—
		14	2	—	—	—	—	—	—	—	—	—
		21	2	1	—	—	—	—	—	—	—	—
		54	1	—	—	1	2	—	—	—	—	—
		97	—	—	—	—	—	—	—	—	1	—

[1] Data include only positive sera for which titrations were done. Negative sera and positive but non-titrated sera are not included.
[2] Reciprocal of highest serum dilution giving a line of precipitation within 72 hours after reaction with MDV antigen.
[3] Reciprocal of highest serum dilution capable of reducing the virus titre by 50% or more.

with cell-associated virus, succumbed at 49 days; all others survived and were free of MD at 97 days.

In contrast, 20 of 21 inoculated S-strain birds died of MD during the 97-day period. A single survivor was paralyzed at the termination of the experiment. The mean time to death (MTD) for those inoculated with cell-associated virus was 41.7 days; for those exposed to cell-free virus, it was 66.6 days. Both the serological response and the mortality pattern suggested that the intratracheal inoculations resulted in the infection of only a few birds, which later spread the infection to the remainder of the group. In the case of the PDRC strain inoculated with cell-free virus, two birds were serologically positive by 10 days but no additional positive birds were detected until 54 days, when 14 of 15 were positive.

Serological responses, as detected by both the AGP and VN tests, were more frequent and occurred earlier in the PDRC strain than in the S-strain birds. Titres of positive sera, however, covered about the same range. It was interesting that the PDRC bird that did not develop neutralizing antibody was the only one that died of MD and, conversely, the S-strain bird that survived to 97 days was the only one that did develop neutralizing antibody. Several S-strain birds developed precipitin, but no association could be made between that response and the MTD. When birds with and without precipitin before death were compared, the MTD was 43.8 and 39.2 days respectively for those receiving cell-associated inoculum, and 62.6 and 65.5 days respectively following infection with cell-free inocula.

All serological tests on controls were negative at 0, 28, 54 and 97 days and no evidence of Marek's disease developed in that group.

Experiment 2

For the second experiment, 4-week-old N-line and P-line birds were divided into three groups and either exposed by inoculation with cell-associated or cell-free virus or left uninoculated. Sera were obtained from all birds at 0, 11, 14, 21, 29 and 56 days post exposure. AGP tests were conducted on all samples; VN tests were performed only on sera collected from the third week onwards. Titrations were not performed in this experiment; samples were tested as undiluted serum for the AGP test and as 1:5 serum dilution for the VN test, and were characterized only as positive or negative. This experiment was terminated at 60 days. All dead birds and those live birds with clinical signs of MD at the termination of the experiment were examined for gross lesions.

The results obtained are shown in Table 3. All pre-inoculation tests were negative. However, accidental infection must have occurred with at least one group, since MD mortality and serological responses were observed in the uninoculated controls. It is not certain when this accidental exposure took place. It may have occurred before the birds were separated for various experimental exposures at 4 weeks of age; those birds inoculated with cellular or cell-free virus preparations may thus also have experienced prior additional virus exposure. This possibility is supported by the fact that one of the P-line birds (in the intra-abdominally inoculated group) died from MD at only 13 days after inoculation. Nevertheless, the exposure of the two genetic lines in each group would have been the same since all birds had been kept together since hatching. While the data cannot

Table 3. Experiment 2: Serological response and incidence of Marek's disease in N-line and P-line chickens exposed to MDV at 4 weeks of age [1]

Exposure group	Genetic strain	Precipitating antibody (AGP test)						Neutralizing antibody (VN test)			Marek's disease incidence (cumulative) [2]				
		Days post exposure						Days post exposure			Days post exposure				
		0	11	14	21	29	53	21	29	53	14	21	29	53	60
Control	N-line	0/15	...	0/15	0/15	3/15	13/15	9/15	15/15	...	0/15	0/15	0/15	0/15	0/15
	P-line	0/16	...	1/16	8/16	12/16	3/3	2/16	4/16	2/3	0/16	0/16	1/16	13/16	14/16
Cell-associated virus, intra-abdominal	N-line	0/13	0/13	0/13	1/13	2/13	9/13	11/13	13/13	...	0/13	0/13	0/13	0/13	0/13
	P-line	0/15	2/15	1/14	7/14	8/13	1/1	1/13	1/13	0/1	1/15	1/15	2/15	14/15	15/15
Cell-free virus, intratracheal	N-line	0/13	0/13	0/13	0/13	0/13	5/13	8/13	13/13	...	0/13	0/13	0/13	0/13	0/13
	P-line	0/14	1/14	3/14	7/14	8/14	5/5	1/14	1/14	2/5	0/14	0/14	0/14	9/14	10/14

[1] All results are expressed as the ratio: number positive / number examined; where ... appears, none were examined.
[2] Cumulative incidence of Marek's disease includes MD mortality and survivors at 60 days with clinical and pathological evidence of MD.

be relied upon with regard to the time required for antibody production in any single strain, the comparisons between N- and P-line birds are considered entirely valid.

During the 60-day "post-exposure" period, a total of 39 out of 45 (87%) P-line birds succumbed to MD while none of the 41 N-line birds died or showed clinical evidence of infection. In contrast to the situation with S-strain and PDRC birds, the more resistant N-line birds were much less efficient than the susceptible P-line birds in the development of precipitin; at 29 days post exposure, positive birds in the two groups amounted to 12.2 and 66.7% respectively. Even after 53 days, 14 out of 41 N-line birds still had no detectable precipitating antibody while all 9 P-line birds surviving at that time were positive. In marked contrast was the situation with neutralizing antibody. At 29 days, all of the 41 N-line sera were positive compared with only 5 out of 42 (11.9%)

of the P-line sera. Only 4 out of the 9 P-line birds that survived 53 days gave positive VN tests. Those 4 birds remained free of MD; however, 3 out of the 5 birds without neutralizing antibody succumbed to MD during the next week.

Again, the development of precipitin did not appear to affect the mortality pattern. The MTD for birds with or without precipitin before death was 43.0 and 40.9 days respectively. In this experiment, 4 P-line birds did develop VN antibody prior to death from MD. The MTD for those birds was 43.3 days.

Table 4 illustrates the various patterns of antibody development that were observed in the two experiments and serves to bring out the apparently complete independence of the two antibody responses. Development of one kind of antibody was certainly not an indication that the other kind might also be found. Further, even if both types of antibody were

Table 4. Examples of antibody-response patterns

Strain	Bird no.	Test [1]	Weeks post exposure [2]						Marek's disease [3]
			1½	2	3	4	8	14	
PDRC	807	AGP	−	−	+	+	+	+	—
		VN	+	+	+	+	+	+	
	823	AGP	−	−	−	−	+	+	—
		VN	−	nt	+	+	+	nt	
	806	AGP	−	+	+	+	+	+	—
		VN	−	−	+	+	+	+	
	808	AGP	−	+	+	+			Dead 49 days post exposure
		VN	−	−	−	−			
	884	AGP	+	+	+	+	+	+	—
		VN	nt	−	nt	+	+	+	
S-strain	916	AGP	−	−	−	−	−		Dead 82 days post exposure
		VN	−	−	−	−	−		
	918	AGP	+	+	+	+			Dead 49 days post exposure
		VN	nt	nt	nt	−			
	924	AGP	−	−	−	−	−	+	Paralyzed at 97 days
		VN	−	−	−	−	−	+	
N-line	75	AGP	nt	−	−	−	+		—
		VN	nt	nt	+	+	+		
P-line	136	AGP	+	+	+	+	+		Dead 54 days post exposure
		VN	nt	nt	nt	−	−		
	134	AGP	−	−	−	+			Dead 43 days post exposure
		VN	nt	nt	+	+			

[1] AGP—agar gel precipitin test; VN—virus neutralization test.
[2] nt—not tested.
[3] Marek's disease mortality or clinical disease up to 97 days for PDRC and S-strain and up to 60 days for N-line and P-line birds.

detected in the same bird, the first appearance of each was quite variable and no consistent pattern was observed; in some birds, precipitating antibody preceded neutralizing antibody while in others the reverse was true. Some birds were consistently negative in both tests and yet were infected, since they died of MD. MD could develop in the presence of either kind of antibody. However, the data in Table 5 suggest that there was perhaps a relationship between neutralizing antibody present at 4 weeks after exposure and survival. Generally, it can be seen (Table 5) that those birds that developed neutralizing antibody by 4 weeks post exposure usually survived. Of the 16 birds that had neither kind of antibody but survived, 13 were PDRC birds exposed by intratracheal inoculation with cell-free virus. As pointed out earlier, that exposure may not have resulted in infection of many of the birds.

DISCUSSION

The serological data recorded here are new in many respects. Thus, while other studies have followed the development of precipitin in young chicks, these experiments were concerned with older birds free from the possibly restrictive influences of maternal antibody or immunological immaturity. In the first of the two experiments, where exposure prior to experimental inoculation did not take place, some PDRC birds were shown to have antibody as early as 10 days, and a majority of the intra-abdominally inoculated birds had both precipitating and neutralizing antibody by 21 days post exposure. Also, these data are the first to characterize the production of neutralizing antibody following infection with MDV.

More significant were the findings regarding the variability of responses among the four genetic strains studied and the observed association between the development of neutralizing antibody and resistance to Marek's disease. The N- and P-line birds were considered to be especially valuable for studies such as these since susceptibility and resistance to MD were the only traits for which selection was made (Cole, 1968). Otherwise, they should be genetically similar.

The observations constitute support for, but not proof of, an immunological basis for genetic resistance to Marek's disease. This thesis is particularly attractive. If efficient immunological responsiveness following MDV infection is directly and causally related to survival, the same phenomenon might also explain resistance associated with age and the efficacy of vaccination with attenuated (Churchill, Payne & Chubb, 1969) or naturally avirulent but antigenically related (Okazaki, Purchase & Burmester, 1970) herpesviruses.

There are at least three possible explanations for the observations reported here:

(1) Certain birds may have genetically controlled defective immunogenic systems. Thus, they may be incompetent to produce either precipitating or neutralizing antibody (or both) against MDV antigens. Failure to respond rapidly and efficiently in the production of neutralizing antibody could permit MDV infection to become unusually intense and widespread, and thereby increase the likelihood of an overt pathological response.

(2) Alternatively, a genetically controlled defect in immunological competence might occur but not be at all related to susceptibility to MD; in other words, the correlation observed in these studies could have

Table 5. Relationship between precipitating and/or neutralizing antibody developed by 4 weeks post exposure and the subsequent incidence of Marek's disease

Antibody at 4 weeks post exposure	No. of birds with subsequent Marek's disease/total no. of birds [1]				
	Experiment 1		Experiment 2		Total
	PDRC	S-strain	N-line	S-strain	
None	0/13	10/12	...	9/10	19/35
Precipitating alone	1/1	8/8	...	22/25	31/34
Neutralizing alone	0/1	...	0/35	1/3	1/39
Precipitating and neutralizing	0/13	...	0/5	2/3	2/18

[1] Where ... appears, there were no birds in the category concerned.

been coincidental. This appears to have been the case regarding precipitating antibody and it might also be true of neutralizing antibody.

(3) The failure to make antibody may have been the result of the more intense infection known to occur in birds susceptible to MD because of genetic make-up, early age, or absence of maternal antibody. Organs associated with the immune respone (bursa of Fabricius, thymus, and spleen) are sites of early infection, and severe infections may destroy certain populations of immunocompetent cells. If this is the reason for the observations, cells responsible for making the two kinds of antibody must have different susceptibilities, since many of the susceptible

P-line and S-strain birds had precipitating but not neutralizing antibody. It is important to find out whether the genetically susceptible strains fail to produce neutralizing antibody when challenged with inactivated MDV.

The first of the three possible explanations seems quite plausible in view of the known protective effect of maternal antibody (Chubb & Churchill, 1969) and the fact that infection does appear to be restricted in birds known to be unlikely to succumb to MD because of age at exposure or genetic resistance (Calnek & Hitchner, 1969). However, additional studies will be required before any of the above, or any other, explanations can be considered correct.

SUMMARY

In two experiments, four strains of chickens, of which two were susceptible and two resistant to Marek's disease (MD), were exposed to Marek's disease virus (MDV) at 4 or 8 weeks of age. The subsequent development of precipitin and neutralizing antibody proceeded independently. The various strains of chickens differed appreciably in their immunocompetence, as judged by the rapidity of the response and the percentage of positive birds. Precipitin appeared frequently and occured early in the relatively MD-resistant PDRC birds and in the susceptible P-line chickens. It was less frequently seen and occurred later in the MD-resistant N-line and susceptible S-strain birds. No association existed between precipitin and survival. However, neutralizing antibody occurred in nearly 100 % of the resistant strains compared to its rare occurrence in the susceptible strains. Whether the relationship between competence to make neutralizing antibody and survival was one of cause and effect was not determined.

ACKNOWLEDGMENTS

The author is indebted to Dr R. K. Cole, Department of Poultry Science, Cornell University, for making available the S-strain and N- and P-line eggs from which the experimental chicks were derived, and to Dr P. P. Levine for his valuable suggestions during the preparation of this manuscript.

This investigation was supported in part by Public Health Service research grant 5 R01 CA 6709-09 from the National Cancer Institute.

REFERENCES

Adldinger, H. K. & Calnek, B. W. (1971) An improved *in vitro* assay for cell-free Marek's disease virus. *Arch. ges. Virusforsch.* (in press)

Calnek, B. W. & Adldinger, H. K. (1971) Some characteristics of cell-free preparations of Marek's disease virus. *Avian Dis.,* **15,** 508-517

Calnek, B. W., Adldinger, H. K. & Kahn, D. E. (1970) Feather follicle epithelium : a source of enveloped and infectious cell-free herpesvirus from Marek's disease. *Avian Dis.,* **14,** 219-233

Calnek, B. W. & Hitchner, S. B. (1969) Localization of viral antigen in chickens infected with Marek's disease herpesvirus. *J. nat. Cancer Inst.,* **43,** 935-949

Calnek, B. W., Hitchner, S. B. & Adldinger, H. K. (1970) Lyophilization of cell-free Marek's disease herpesvirus and a herpesvirus from turkeys. *Appl. Microbiol.,* **20,** 723-726

Calnek, B. W., Ubertini, T. & Adldinger, H. K. (1970) Viral antigen, virus particles, and infectivity of tissues from chickens with Marek's disease. *J. nat. Cancer Inst.,* **45,** 341-351

Chubb, R. C. & Churchill, A. E. (1968) Precipitating antibodies associated with Marek's disease. *Vet. Rec.*, **83**, 4-7

Chubb, R. C. & Churchill, A. E. (1969) The effect of
maternal antibody on Marek's disease. *Vet. Rec.*, **85**,
303-305

Churchill, A. E. & Biggs, P. M. (1967) Agent of Marek's
disease in tissue culture. *Nature (Lond.)*, **215**, 528-530

Churchill, A. E., Payne, L. N. & Chubb, R. C. (1969)
Immunization against Marek's disease using a live
attenuated virus. *Nature (Lond.)*, **221**, 744-747

Cole, R. K. (1968). Studies on genetic resistance to
Marek's disease. *Avian Dis.*, **12**, 9-28

Eidson, C. S. & Schmittle, S. C. (1969) Studies on acute
Marek's disease. XII. Detection of antibodies with a
tannic acid indirect hemagglutination test. *Avian Dis.*,
13, 774-782

Hutt, F. B. & Cole, R. K. (1947) Genetic control of
lymphomatosis in the fowl. *Science*, **106**, 379-384

Nazerian, K., Solomon, J. J., Witter, R. L. & Burmester,
B. R. (1968) Studies on the etiology of Marek's disease.
II. Finding of a herpesvirus in cell culture. *Proc. Soc.
exp. Biol. (N.Y.)*, **127**, 177-182.

Okazaki, W., Purchase, H. G. & Burmester, B. R. (1970)
Protection against Marek's disease by vaccination with
a herpesvirus of turkeys. *Avian Dis.*, **14**, 413-429

Purchase, H. G. & Burgoyne, G. H. (1970) Immunofluorescence in the study of Marek's disease: detection of
antibody. *Amer. J. vet. Res.*, **31**, 117-123

Sevoian, M., Chamberlain, D. M. & Counter, F. T. (1962)
Avian lymphomatosis. I. Experimental reproduction
of the neural and visceral forms. *Vet. Med.*, **57**, 500-
501

Spencer, J. L. & Calnek, B. W. (1970) Marek's disease:
application of immunofluorescence for detection of
antigen and antibody. *Amer. J. vet. Res.*, **31**, 345-
358

Observations on the Epidemiology of Classical Marek's Disease *

G. N. WOODE & J. G. CAMPBELL

We reported recently on the demonstration of a herpesvirus in short-term cultured blood lymphocytes, taken from birds suffering from both classical and acute Marek's disease (Campbell & Woode, 1970). Most of the cases were taken from the flock of the Poultry Research Centre, Edinburgh (Flock A), in which classical Marek's disease was endemic. A few clinically normal birds from the flock were also shown to be infected with the virus. A flock in which Marek's disease had not been recorded was used as a source of controls (Flock B).

It was proposed, in this study, to examine the incidence of virus infection and antibody to Marek's disease in these two flocks and to compare this with the incidence of antibody in commercial flocks in Scotland and commercial and local village flocks in Nigeria.

In order to determine the distribution of the virus in Flocks A and B, lymphocyte cultures were examined for herpesvirus using the technique described by Campbell & Woode (1970), and birds of different ages were killed and cultures of their kidney cells (CK) prepared. The latter technique has been shown to be the most sensitive for isolation of Marek's disease virus from an infected bird (Witter, Solomon & Burgoyne, 1969). For 17-day-old embryos and 3-day-old chicks, kidney cells were cultured for 14 days, after which they were removed with trypsinversene, and inoculated on to virus-free CK cells. The inoculation of primary cultures of kidney cells with blood was also used on occasion for isolation of Marek's disease virus (Witter et al., 1969). Sera from birds of all flocks were tested for antibody to

Marek's disease by the gel precipitation technique (Chubb & Churchill, 1968) and the immunofluorescent technique (Campbell & Woode, 1970).

RESULTS

The proportion of virus-positive lymphocyte cultures from clinically ill and clinically normal birds from Flock A was 21/26. One bird was shown to be infected persistently over a period of 18 months, by repeated lymphocyte culture and inoculation of CK cells with blood cells. During this period precipitating antibodies were also present. The distribution of antibody-positive birds in Flock A was similar to that reported by Witter (1970). No virus was demonstrated in birds from Flock B by kidney culture, but one bird out of a group of four at 20 weeks of age was shown to be infected, by lymphocyte culture. The bird subsequently died of Marek's disease. Serum samples taken from Flock B were negative for Marek's disease antibody by gel diffusion and immunofluorescent studies with infected CK cells, for the first two years of the flock's existence. Thereafter, an increasing incidence of positive sera from birds of 16 weeks of age or older was observed.

From the results (see Table), it was concluded that all the birds in Flock A were infected with the virus within 7 days of age, but that 17-day-old embryos and 3-day-old chicks were apparently uninfected. Lymphocyte culture and electron microscope examination proved to be a useful technique for demonstrating infection in an adult bird in Flock B, at a time when the flock appeared to be free of infection.

The presence of antibody was always correlated with infection with virus, once the birds were five weeks of age, as in Flock A; in Flock B, however, all those birds shown to be virus-free were also free

* From the Department of Veterinary Pathology, Royal (Dick) School of Veterinary Studies, Edinburgh, UK, and the Cancer Research Campaign Unit, Department of Veterinary Pathology, Royal (Dick) School of Veterinary Studies, Edinburgh, UK.

of precipitating antibodies. From these results we assumed that birds free of antibody at 5 weeks of age or older were either free of virus or only recently infected.

By gel diffusion and immunofluorescence, commercial flocks of chickens in Scotland and Nigeria were compared with Flock A. In four flocks in Scotland, at nineteen weeks of age, the proportion of positive sera varied between 75 and 95%. One flock was serologically negative at 6 weeks of age. In adult Nigerian chickens, the incidence of infection varied between 17 and 22%, and there was no significant difference between imported commercial birds of North American origin, and the local village birds, although the latter were isolated by many miles of bush. One flock of birds held at Ibadan University had been imported from the USA, and designated Marek's free. Virus was not isolated by kidney culture of 3 six-week-old birds, and sera from all the adults were serologically negative. Marek's disease is known to occur in imported commercial birds in Nigeria but there is no record of the incidence of disease in the indigenous village bird.

From the serological studies, the commercial flocks would appear to have had less infection present shortly after hatching, possibly due to strict hatchery hygiene. However, there was no apparent correlation between age of infection and percentage mortality, for the latter was similar in Flock A and in the commercial flocks.

We have no explanation for the low incidence of serologically positive birds from the Nigerian sources. It was particularly interesting to note that the results for village birds were similar to those for the imported, intensively husbanded commercial flocks.

Incidence of Marek's disease virus in cultured kidney cells

Age of bird	Flock A		Flock B	
	Virus isolation [1]	Serological status [2]	Virus isolation [1]	Serological status [2]
17-day embryos	0/3	N.T.	N.T.	N.T.
3 days	0/6	+	N.T.	N.T.
7 days	4/4	+	N.T.	N.T.
2-3 weeks	9/9	—	N.T.	N.T.
4 weeks	11/11	—	0/1	—
5 weeks	10/10	+	0/3	—
6 weeks	3/3	+	0/2	—
9 weeks	1/1	+	0/5	—
13 weeks	3/3	+	0/5	—
19 weeks	N.T.	+	0/1	—

[1] Number infected/number examined.
[2] By gel precipitation. N.T. = not tested.

REFERENCES

Campbell, J. G. & Woode, G. N. (1970) Demonstration of a herpes-type virus in short term cultured blood lymphocytes associated with Marek's disease. *J. med. Microbiol.*, **3**, 463-473

Chubb, R. C. & Churchill, A. E. (1968) Precipitating antibodies associated with Marek's disease. *Vet. Rec.*, **83**, 4-7

Witter, R. L., Solomon, J. J. & Burgoyne, G. H. (1969) Cell culture techniques for primary isolations of Marek's disease-associated herpesvirus. *Avian Dis.*, **12**, 101-118

Witter, R. L. (1970) Epidemiological studies relating to the control of Marek's disease. *Wld's Poult. Sci. J.*, **26**, 755-762

A Vaccination Study with an Attenuated Marek's Disease Virus

P. M. BIGGS,[1] C. A. W. JACKSON,[2] R. A. BELL,[3] F. M. LANCASTER[4]
& B. S. MILNE[1]

INTRODUCTION

Experimental laboratory studies have shown that attenuated strains of Marek's disease virus (MDV) have been able to offer a considerable degree of protection against later infection with a virulent strain of MDV (Churchill, Payne & Chubb, 1969; Vielitz & Landgraf, 1970; Eidson & Anderson, 1971; von Bülow, 1971). These observations were recently confirmed under field conditions (Biggs *et al.*, 1970; Rispens & Maas, 1971).

The object of this study was to examine in greater detail the response to vaccination with an attenuated MDV under field conditions. The effect of vaccination on mortality, body weight, food consumption and egg production was examined. The incidence of precipitating antibodies and viraemia was also examined at intervals throughout the duration of the trial in an attempt to determine those factors of importance in the protection offered by vaccines of this kind.

MATERIALS AND METHODS

General design

Approximately 2000 day-old White Leghorn hybrid chickens were obtained from one supply flock. Chicks were individually numbered and distributed randomly between 10 pens in a deep litter brooder house. One of each consecutive pair of pens was selected at random and vaccinated with the attenuated HPRS-16 strain of MDV (HPRS-16/att). At 8 weeks of age, half of the surviving chickens in each pen were selected at random and transferred to 10 pens in a deep litter rearing house. At 16 weeks of age, the surviving chickens in each house were culled at random to reduce to 72 the numbers of chickens in each of the 20 pens. The selected chickens were distributed at random in laying cages in a single house.

Blood samples were collected from 79 day-old chicks. From one week of age, blood samples were collected from 16 randomly selected chickens from each pen, making a total of 80 vaccinated and 80 unvaccinated chickens. All samples from each pen were examined for precipitating antibodies and 6 from each pen for MDV. Where selected chickens died they were replaced by random selection from the remaining chickens in the same pen. At 8 weeks of age, half the chickens used for sampling in each pen were selected at random and moved to the appropriate pen in the rearing house.

Blood samples were collected and examined for viraemia and precipitating antibodies at weekly intervals until the chickens were 11 weeks of age, and then at 7-week intervals to 32 weeks of age and finally at 52 weeks of age. Additional samples were collected and examined for precipitating antibodies only at 12, 13, 15, 42 and 62 weeks of age.

Vaccine

HPRS-16/att vaccine was used, derived from the 45th passage of the HPRS-16 strain of MDV in cultured chick kidney cells (CKC). The preparation and use of the vaccine were as previously described

---

[1] Houghton Poultry Research Station, Houghton, Huntingdon, UK.
[2] Formerly Houghton Poultry Research Station, Houghton, Huntingdon, UK. Present address: Veterinary Research Station, Glenfield, NSW, Australia.
[3] Veterinary Inestigation Centre, "Woodthorne", Wolverhampton, UK.
[4] Harper Adams Agricultural College, National Institute of Poultry Husbandry, Newport, Shropshire, UK.

(Biggs *et al.*, 1970). Each chick was vaccinated at day old by the intra-abdominal inoculation of an estimated 800 plaque forming units (PFU) in 0.25 ml, using an A.R.H. pipetting unit.[1] Diluted vaccine was assayed in CKC on return to the laboratory 6 hours after the vaccine was prepared. The average PFU/0.25 ml was 308.

Isolation of MDV

Two or 4 ml of blood were collected in sodium citrate and stored overnight at 4°C. The following day, buffy coat cells were prepared from each blood sample by two cycles of centrifugation, and inoculated on to monolayers of CKC. Plaques derived from at least half the buffy coat cells were counted and the number per ml of blood originally collected was determined.

Precipitation test

All sera were tested for the presence of antibody to the A antigen of MDV (Churchill, Chubb & Baxendale, 1969) by the agar gel precipitation test described by Chubb & Churchill (1968) without the use of a buffer. Antibodies to BC antigens were similarly tested for, but an agar concentration of 0.5% was used. Antigens from MDV isolates were prepared by precipitating supernatants from cultures showing at least 25% cytopathic effect, followed by concentrating 10–20 times. Concentrated antigens were tested for the presence of A antigen by the agar gel precipitation test.

Mortality

All chickens that died were examined *post mortem* for lesions of MD. The examination procedure was as outlined by Biggs *et al.* (1970). Depending on whether lesions of MD were present or not, each chicken was classifield as MD or nonspecific disease (NSD). Incidence of MD and NSD was calculated up to 8 and between 8 and 16 weeks of age on the basis of the number of chicks present in each group at day old and 8 weeks old respectively and after 16 weeks on the basis of the number placed in the laying house in each group.

Body weight

A random sample of 162 chickens from the vaccinated and 166 chickens from the unvaccinated groups were weighed at 4 weeks of age. At 8 and

---

[1] Arnold R. Howell Ltd., 2 Grangeway, Kilburn High Road, London, NW6.

16 weeks of age, all vaccinated and unvaccinated chickens were weighed.

Food consumption

Food consumed was recorded throughout the entire experiment, which was divided into the following periods: 0 to 8 weeks, 9 to 16 weeks and four weekly periods thereafter. Food consumption was expressed on a bird day basis and was calculated as follows:

$$\text{Bird day food consumption for period in question} = \frac{\text{Total food consumed for period}}{\text{Total bird days}} \times \text{number of days in period}$$

where total bird days equals the sum of the number of birds present on each day of the period in question.

Egg production

Daily egg production of vaccinated and unvaccinated chickens was recorded. The hen housed production and hen day production were calculated for the period 16 to 24 weeks of age and for four weekly periods thereafter. Hen housed production was calculated by dividing the number of eggs produced in the period by the number of chickens placed in the laying house at 16 weeks of age. Hen day production was calculated by dividing the number of eggs produced on each day by the number of chickens alive on that day; these figures were then summed for the period in question.

Statistical analysis

The χ square test was used to examine the difference between the incidence of MD in vaccinated and unvaccinated chickens after correction for NSD.

Analysis of variance was carried out on the average figures for each pen at each age for body weight, and for each period for egg production.

Regression of period on average food consumption for each age group for vaccinated and unvaccinated chickens was shown to be not significant. The differences between food consumption for the vaccinated and unvaccinated chickens were therefore examined with the paired t-test.

RESULTS

Mortality

The accumulated percentage mortality from MD and NSD for the unvaccinated and vaccinated groups

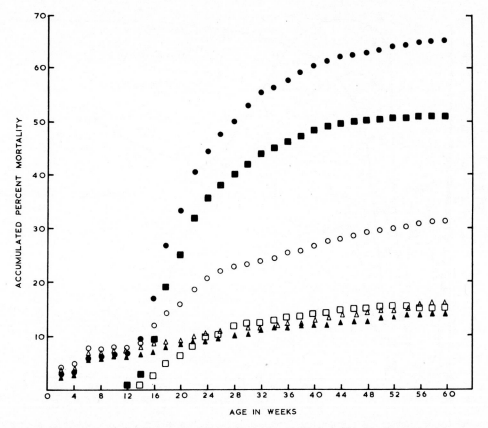

Fig. 1. Accumulated percentage incidence of total mortality and mortality from Marek's disease and from nonspecific disease in unvaccinated chickens and chickens vaccinated with an attenuated Marek's disease virus.

Unvaccinated chickens
● Total mortality
■ Mortality from Marek's disease
▲ Mortality from nonspecific disease

Vaccinated chickens
○ Total mortality
□ Mortality from Marek's disease
△ Mortality from nonspecific disease

is shown in Fig. 1. Mortality up to 8 weeks of age in the unvaccinated group was 5.8% and in the vaccinated group 6.6%. Marek's disease was first diagnosed when the chickens were 77 days old. There was little difference between the unvaccinated and vaccinated groups in mortality from NSD throughout the period of the experiment. However, the incidence of MD was always less in the vaccinated group, and at 60 weeks of age it was 51.1% in the unvaccinated and 15.2% in the vaccinated chickens (P < 0.01).

Viraemia

The percentage of blood samples from which MDV was isolated at various ages is shown in Fig. 2. MDV was first isolated from unvaccinated chickens

when they were five weeks of age, and by 10 weeks viraemia was present in all 30 sampled chickens. All sampled unvaccinated chickens were still viraemic at 18 weeks of age but the incidence of detectable viraemia fell gradually to 50% at 52 weeks of age. In contrast, there were two peaks in incidence of viraemia in vaccinated chickens. Viraemia was first noted at two weeks of age, reached a peak incidence of 53% at 5 weeks and declined to 18% by 8 weeks of age. The incidence then rose to 84% by 10 weeks and then slowly declined to 57% at 32 weeks of age. It was 70 % at 52 weeks of age. The primary rise in incidence of viraemia in the vaccinated chickens appeared, from plaque size, to be due to vaccine virus. Field virus with A antigen was first noted in both unvaccinated and vaccinated chickens at

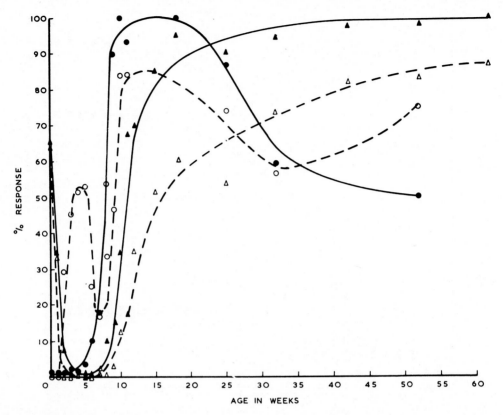

Fig. 2. Incidence of viraemia and precipitating antibodies in unvaccinated chickens and chickens vaccinated with an attenuated Marek's disease virus.

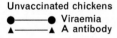

Unvaccinated chickens	Vaccinated chickens
●————● Viraemia	○— — —○ Viraemia
▲————▲ A antibody	△— — —△ A antibody

8 weeks of age. The secondary rise in incidence of viraemia seen in the vaccinated chickens was considered to be due to field virus infection.

The levels of viraemia in PFU/ml of whole blood in unvaccinated and vaccinated chickens are shown in Fig. 3. The levels of viraemia during the primary rise in incidence of vaccine virus were low with a mean of 3.6 PFU/ml. The secondary rise in incidence of viraemia in vaccinated chickens, coincident with the isolation of field virus and with the rise in incidence of viraemia in unvaccinated chickens, was associated with a much greater rise in mean titres of viraemia. In both groups the mean titres rose to a maximum at 25 weeks after which they rapidly declined to low levels at 32 weeks of age. The mean titres of viraemia were always lower in vaccinated chickens than in unvaccinated chickens. The mean viraemic titre in vaccinated chickens was at least five

times smaller in those that had had a detectable vaccine viraemia compared with those that had not.

Antibody

The incidence of antibodies to the A antigen in unvaccinated and vaccinated chickens is shown in Fig. 2. The incidence of maternal antibody was 65% in the day-old chicks. This incidence declined to 34 and 35% respectively in unvaccinated and vaccinated chickens by one week of age. By the second week of age, maternal antibody could not be identified in vaccinated chicks and was found only in 7% of unvaccinated chicks. At three weeks of age, maternal antibody was not present in chicks of either group.

Antibody to A antigen reappeared in the vaccinated group at 7 weeks of age and in the unvaccinated group at 8 weeks of age. The incidence of A antibodies rose more slowly and reached a lower peak in vac-

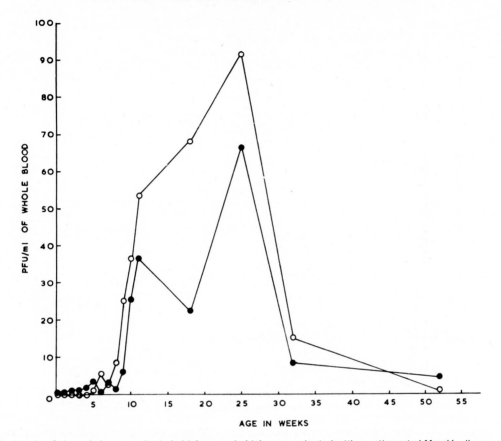

Fig. 3. Levels of viraemia in unvaccinated chickens and chickens vaccinated with an attenuated Marek's disease virus.

○————○ Unvaccinated chickens
●————● Vaccinated chickens

cinated than in unvaccinated chickens. Of the sampled unvaccinated chickens, 50 % had A antibody at 72 days of age whereas this incidence of A antibodies was not reached in vaccinated chickens until they were 102 days of age. Of unvaccinated chickens, 95 % had antibodies at 18 weeks of age and the

incidence gradually rose to 100 % at 62 weeks of age. In the vaccinated chickens, A antibodies were present only in 60 % at 18 weeks, but reached 87 % by 62 weeks of age.

The incidence of antibodies rose more slowly and reached a lower level in vaccinated chickens that had

Table 1. Mean body weight at 4, 8 and 16 weeks of age of chickens vaccinated against Marek's disease with HPRS-16/att and of unvaccinated chickens

Age in weeks	Unvaccinated		Vaccinated		Significance level
	No. of chickens	Weight (g)	No. of chickens	Weight (g)	
4	162	292.2	166	292.5	NS[1]
8	942	733.6	956	774.9	0.05
16	877	1291.5	776	1337.7	NS[1]

[1] Not significant.

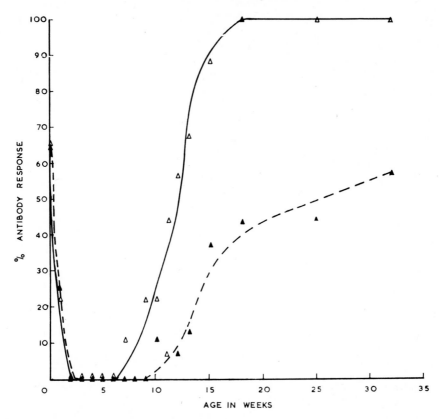

Fig. 4. Incidence of precipitating antibodies to the A antigen in chickens with and without a viraemia after vaccination with an attenuated Marek's disease virus.

△————△ Chickens without vaccine viraemia
▲— — —▲ Chickens with vaccine viraemia

had a detectable vaccine viraemia than in those vaccinated chickens that had not (Fig. 4). Of chickens in the latter group, 50% had antibodies by 81 days whereas it was 182 days before 50% of vaccinated chickens that had had a detectable vaccine viraemia had antibodies. The figure for the chickens that had not experienced a detectable vaccine viraemia was similar to that seen in the unvaccinated chickens.

Antibodies to BC antigen were detected in 6% of vaccinated chickens at 7 weeks of age.

Body weight

The mean body weights of unvaccinated and vaccinated chickens at 4, 8 and 16 weeks of age are shown in Table 1. There was no significant difference between body weights of unvaccinated and vaccinated chickens at 4 and 16 weeks of age, but at 8 weeks of age body weights were 5.6% greater in vaccinated chickens ($P < 0.05$).

Food consumption

The bird day food consumption figures are given in Table 2. The total food consumed per bird from day old to 60 weeks of age was 3% greater in the vaccinated than in the unvaccinated chickens ($P < 0.01$).

Egg production

The hen housed average and hen day average egg production for the unvaccinated and vaccinated chickens are shown in Table 2. The hen housed average egg production was 59.2% greater ($P < 0.001$) and the hen day average 8.2% greater ($P < 0.01$) in vaccinated than unvaccinated chickens.

Table 2. Food consumption and egg production in chickens vaccinated against Marek's disease with HPRS-16/att and in unvaccinated chickens

| Age in weeks | Bird day food consumption (kg) | | Egg production | | | |
| | | | HHA[1] | | HDA[2] | |
	Unvacci-nated	Vacci-nated	Unvacci-nated	Vacci-nated	Unvacci-nated	Vacci-nated
0–8	2.25	2.27				
8–16	4.10	4.37				
16–20	2.22	2.40	3.8	7.7	5.2	8.4
20–24	2.21	2.48				
24–28	2.93	3.09	9.6	15.6	14.4	17.6
28–32	2.93	3.10	12.3	18.9	19.8	21.4
32–36	3.05	2.90	12.3	18.9	20.6	21.9
36–40	3.41	3.49	10.8	17.1	18.9	20.1
40–44	3.57	3.70	11.5	18.6	20.7	22.2
44–48	3.46	3.52	11.5	18.2	21.0	22.0
48–52	3.40	3.46	11.1	17.7	20.7	21.5
52–56	3.44	3.47	10.7	16.8	20.2	20.7
56–60	3.53	3.44	10.1	15.8	19.2	19.7
0–60	40.50[3]	41.69[3]				
16–60			103.7[4]	165.3[4]	180.7[3]	195.5[3]

[1] HHA: Hen housed average.
[2] HDA: Hen day average.
[3] $P < 0.01$.
[4] $P < 0.001$.

DISCUSSION AND CONCLUSIONS

The very high incidence of MD in the unvaccinated chickens and the isolation of field virus from the blood at 5 weeks of age indicates that this flock was heavily exposed to MDV early in its life.

The reduction in incidence in MD in the vaccinated chickens was about 3.5-fold, or 70%, a figure similar in order to that seen in previous trials where a heavy challenge was experienced and a similar dose of vaccine used (Biggs et al., 1970).

The vaccine apparently had no detrimental effect on body weight or egg production. At four and 16 weeks of age there was no significant difference in mean body weight of the unvaccinated and vaccinated chickens, but at 8 weeks of age the vaccinated chickens were significantly heavier than those not vaccinated.

The much greater egg production of the vaccinated chickens on a hen housed basis is undoubtedly due to the greater mortality in the unvaccinated chickens. It is interesting that vaccinated chickens produced more eggs than unvaccinated chickens even when calculated on a hen day basis. Because the differences in egg production between the two groups was greater early in the laying period when mortality from MD was highest, the lower egg production in unvaccinated chickens could have been due to a cessation of lay prior to death from MD.

The greater food consumption of the vaccinated chickens, as compared with the unvaccinated chickens, could have been due to the greater productivity and better general level of health of the vaccinated chickens.

Lower titres of viraemia due to field virus and a delay in appearance and lower incidence of A antibodies were seen in vaccinated as compared with unvaccinated chickens. Vaccine virus viraemia was detected in only half the vaccinated chickens and these had lower titres of viraemia due to field virus and a delay in appearance and lower incidence of A antibodies, as compared with vaccinated chickens for which a vaccine viraemia was not detected. These observations indicate that the level of superinfection with field virus is inversely proportional to the level of vaccine viraemia. It is interesting to note that by 32 weeks of age, two out of 11 vaccinated chickens that had not had a vaccine viraemia died from MD whereas none died from MD out of 21 that had had a vaccine viraemia. These observations suggest that a higher dose would increase the efficacy of the vaccine and that the protection offered by vaccination with an attenuated virus is mediated through an interference with the multiplication of field virus in the host.

SUMMARY

Approximately 2000 day-old White Leghorn chicks were distributed at random into 10 pens in a deep litter house. Five pens selected at random were vaccinated intra-abdominally with an estimated 800 PFU per chick of the attenuated strain of HPRS-16. At eight weeks of age, half the chicks in each pen were moved to another deep litter house divided into 10 pens. At 16 weeks of age, the chickens were randomly placed in battery cages in a laying house.

At 60 weeks of age, mortality from all causes and from Marek's disease was 31.2% and 15.2% respectively in vaccinated chickens compared with 65.3% and 51.1% in unvaccinated chickens. There was no difference in body weights of vaccinated and unvaccinated chickens at 4 and 16 weeks of age, but at 8 weeks body weights were 5.6%

greater in the vaccinated chickens. Vaccinated chickens consumed more food than unvaccinated chickens. Hen house and hen day egg production were respectively 59.2% and 8.2% greater in vaccinated than unvaccinated chickens.

Active antibody production to the A antigen occurred later and in a smaller proportion of vaccinated than unvaccinated chickens. In vaccinated chickens vaccine viraemia was first noted at 2 weeks of age. Viraemia due to field virus occurred at 5 weeks in unvaccinated chickens and soon after in vaccinated chickens. The incidence and titres of viraemia due to field virus were higher in unvaccinated compared with vaccinated chickens. The data indicate that the level of superinfection with field virus is inversely proportional to the level of vaccine viraemia.

ACKNOWLEDGEMENTS

We wish to thank Dr H. Temperton for his advice and for making available the facilities for this trial. We also wish to thank Mr Dudley for advice and Dr P. K. Pani for undertaking the statistical analyses. We are also grateful to K. Howes, D. Lyall, G. Baldock and J. White for their invaluable assistance.

REFERENCES

Biggs, P. M., Payne, L. N., Milne, B. S., Churchill, A. E., Chubb, R. C., Powell, D. G. & Harris, A. H. (1970) Field trials with an attenuated cell associated vaccine for Marek's disease. *Vet. Rec.*, **87**, 704-709

Chubb, R. C. & Churchill, A. E. (1968) Precipitating antibodies associated with Marek's disease. *Vet. Rec.*, **83**, 4-7

Churchill, A. E., Chubb, R. C. & Baxendale, W. (1969) The attenuation, with loss of oncogenicity, of the herpes-type virus of Marek's disease (strain HPRS-16) on passage in cell culture. *J. gen. Virol.*, **4**, 557-564

Churchill, A. E., Payne, L. N. & Chubb, R. C. (1969)

Immunization against Marek's disease using a live attenuated virus. *Nature (Lond.)*, **221**, 744-747

Eidson, C. S. & Anderson, D. P. (1971) Immunization against Marek's disease. *Avian Dis.*, **15**, 49-55

Rispens, B. H. & Maas, H. J. L. (1971) Control of Marek's disease by means of vaccination. In: *Proceedings of Symposium on Avian Leukosis, Sofia, Bulgaria, 1970* (in press)

Vielitz, E. & Landgraf, H. (1970) Beitrag zur Epidemiologie und Kontrolle der Marek'schen Krankheit. *Dtsch. tierärztl. Wschr.*, **77**, 357-392

von Bülow, V. (1971) Studies on the diagnosis and biology of Marek's disease virus. *Amer. J. vet. Res.*, **32**, 1275-1288

Vaccination Against Marek's Disease

C. S. EIDSON [1], S. H. KLEVEN [1] & D. P. ANDERSON [1]

INTRODUCTION

Marek's disease (MD) is a lymphoproliferative disease of chickens characterized by the presence of lymphoid tumours in various organs (Eidson & Schmittle, 1968; Purchase & Biggs, 1967; Sevoian & Chamberlain, 1964). MD is of great economic importance to the poultry industry in the United States; in spite of the many attempts that have been made to control the disease, it has been estimated that it is still costing the poultry industry approximately 200 million dollars annually.

Initial attempts to propagate the MD agent *in vitro* were unsuccessful, but Churchill & Biggs (1967) and Solomon *et al.* (1968) were successful in propagating it in cell culture. Churchill & Biggs (1967) and Nazerian *et al.* (1968) identified the aetiological agent as a cell-associated group B herpesvirus. Churchill, Payne & Chubb (1969) found that a highly pathogenic strain of this virus became attenuated after passage in cell culture; when administered to chicks in laboratory trials, it protected against the development of MD tumours. Biggs *et al.* (1970) found that the attenuated cell-associated strain of virus was nononcogenic and that it conferred a degree of protection against exposure to virulent Marek's disease virus (MDV) under field conditions.

Kawamura, King & Anderson (1969) and Witter *et al.* (1970) isolated a herpesvirus from turkeys not having lymphoid tumours. The virus produced a syncytial cytopathology and type A intranuclear inclusion bodies. Okazaki, Purchase & Burmester (1970) and Eidson & Anderson (1971) found that the turkey herpesvirus (HVT), when administered intraabdominally or subcutaneously, could protect chickens when immunity was challenged with virulent MDV.

This paper presents the results of field trials undertaken to examine the safety and efficacy of HVT and attenuated MDV isolates when used as vaccines against MD. It was also found that MD antigen develops in the feather follicle epithelium of chickens vaccinated at one day of age with HVT and subsequently exposed to virulent MDV.

MATERIALS AND METHODS

Source of experimental chickens

A flock of White Leghorn chickens derived from Poultry Disease Research Center (PDRC) stock was maintained in a filtered-air, positive-pressure poultry house at the PDRC. These birds were free of most recognized poultry pathogens, including resistant-inducing factor (RIF) viruses and MDV.

Origin of herpesviruses

The MDV isolate designated as GA has been maintained at PDRC since 1964 (Eidson & Schmittle, 1968). The HVT isolate, designated as WHG, was isolated by Kawamura *et al.* (1969) and was recently described by Eidson & Anderson (1971). The HVT isolate FC126 was supplied by W. Okazaki, East Lansing, Michigan.

Preparation of vaccines

HVT (WHG and FC126) and MDV vaccines were prepared in cultured chick embryo fibroblasts as described by Eidson & Anderson (1971).

Fluorescent antibody (FA) procedures

The conjugate against WHG was prepared from the serum of a chicken inoculated subcutaneously with WHG at 1 day and 6 weeks of age. Serum was collected 9 days after the second inoculation. A line of precipitation was produced when WHG cell culture antigen was tested against this serum in the agar gel precipitin test.

[1] Poultry Disease Research Center, College of Veterinary Medicine, University of Georgia, Athens, Georgia, USA.

The serum was conjugated with fluorescein iso-thiocyanate by the method of Braune & Gentry (1965). A 1:4 dilution of this conjugate in phosphate-buffered saline gave a fluorescence specific for the WHG herpesvirus-associated antigen, whereas no fluorescence was obtained on MDV-infected or uninfected chick embryo fibroblasts. A conjugate against MDV was similarly prepared and gave a fluorescence specific for the MDV-associated antigen.

Experimental design

In the first trial, four houses 30 yards apart were utilized. Two of the houses were enclosed and had forced-air ventilation, while the other two were curtain-type houses. One enclosed and one curtain-type house were cleaned and had new litter; the other two houses had old litter. Each house was divided into 7 pens by wire partitions. Each pen was capable of holding up to 2000 Leghorn chickens for 20 weeks. Pens 1, 3, 5 and 7 contained unvaccinated controls; chickens in pens 2, 4 and 6 were vaccinated with 0.2 ml of WHG, cell-free WHG (this will not be discussed in this study) or MDV. After birds were placed in the first house, the remaining three houses were filled at 10-day intervals. The three pens of vaccinated birds were marked for identification by toe-clipping. Mortality was the criterion used in determining the effectiveness of the three vaccines. However, at 2-week intervals, birds were removed from the four houses and examined at *post mortem* for gross lesions of MD.

At the conclusion of trial 1, the four houses were cleaned and new litter added to each house. In contrast to the experimental design of trial 1, all the Leghorn birds in trial 2 were vaccinated with either the WHG- or FC126-HVT isolate in three of the houses, while the remaining house served as an unvaccinated control. Mortality was the criterion used in determining the effectiveness of the vaccines. Approximately 14 700 chicks were placed in each house and were moved to laying houses at 20 weeks of age.

Three thoroughly cleaned curtain-type houses were used in trial 3. Each house held 10 000 Leghorn chicks. These chicks were not of the same genetic background as those in trials 1 or 2. Of the birds in each house, 80% were vaccinated with either MDV or WHG. The control birds were separated from the vaccinated birds by a plywood and wire partition. All the birds dying after 4 weeks of age were necropsied and examined for macroscopic lesions of MD.

The birds in trial 4 were broiler breeders started in a curtain-type house that had been thoroughly cleaned and had fresh shavings. Five thousand birds were vaccinated with WHG and 2500 birds served as unvaccinated controls. Although the two groups were separated by a wire partition, the vaccinated birds were toe-clipped to ensure identification. At the end of 9 weeks, 300 of the vaccinated and 100 of the control birds were necropsied. At this time the remaining vaccinated and control birds were moved to a second farm. These birds were placed in two houses containing other birds, but were separated from these birds by a feed room located in the middle of the house.

In trial 5, half the broiler chickens in four flocks were vaccinated at one day of age with HVT (2000 PFU/bird; the vaccine was prepared by PDRC) and half of one flock was vaccinated with 200 PFU/bird (vaccine not prepared at University of Georgia). The vaccinated and control birds were housed in separate buildings. At the end of an 8–9 week grow-out period, the birds were examined at a poultry processing plant.

In trial 6, 9 groups of one-day-old chicks (10 per group) were housed in modified Horsfall-Bauer units under filtered air at positive pressure. Three groups were vaccinated subcutaneously with 0.2 ml (2.5×10^3 PFU) of WHG and three groups with FC126. At 3 weeks of age, one group from each of the vaccinated groups was challenged with 0.2 ml of MD-infective plasma. A second group of vaccinated birds was challenged by contact exposure. The third group of vaccinated birds was not exposed to MDV. There were three control groups: (i) positive controls that received 0.2 ml of MD-infective plasma when one day of age; (ii) a second group that received 0.2 ml of MD-infective plasma when 3 weeks old; (iii) a third untreated group.

Beginning with the fourth week post vaccination (1 week post challenge), two birds from each unit were removed, killed, and examined for gross lesions. Skin samples were collected and stained with a HVT or MDV FA conjugate. Minced skin samples were also placed on normal chick kidney monolayers. Samples were collected from the fourth to the eighth week.

At the seventh week, blood was collected from each group and injected subcutaneously into one-day-old birds. This was done to determine whether vaccinated birds challenged with MDV had an MDV viraemia.

In trial 7, 9 groups of one-day-old chicks were housed in modified Horsfall-Bauer units. Six out of 12 birds in units 4, 5, 6, 7, 8 and 9 were vaccinated subcutaneously with 0.2 ml (2.5 × 10³ PFU) of WHG or FC126 and the remainder of each group served as contact controls. At 3 weeks of age, the vaccinated birds from 5, 6, 8 and 9 were challenged subcutaneously with 0.2 ml of MD-infective plasma; the contact controls in the same units were not challenged. None of the birds in units 4 and 7 were exposed to MDV. There were two control groups: (1) an untreated group (unit 1); (2) a group, half of which was challenged with MDV while the remainder served as contact controls (units 2 and 3). At the end of 9 weeks, all the birds were killed and examined for gross lesions.

RESULTS

Trial 1

The Leghorn chicks in houses 1 and 2 were placed on old litter in an uncleaned house and considerable mortality was observed in the controls as well as the vaccinated birds (Table 1). Between the 8th and 12th weeks, the birds in these two houses suffered severely from enteritis. Although enteritis was observed in some of the vaccinated birds, it was much more severe in the controls. Medication was added to the feed, and the enteritis disappeared temporarily, but reappeared after the medication was withdrawn. The enteritis did not return after a second addition of medication. Marek's disease was first diagnosed in the control birds in the 6th week. By the 13th week, MD was observed in all the groups. The protection afforded by vaccination with WHG was most

marked during the peak of mortality between the 12th and 20th week. Houses 3 and 4 were cleaned out and the birds reared on new litter. The mortality in these houses was much lower overall and particularly in the groups vaccinated with WHG and MDV. Marek's disease was first diagnosed in the control birds in both of these houses during the 8th week. It was present in all groups by the 14th week.

Trial 2

Mortality figures (as shown in Table 1) indicate that birds from the three houses vaccinated with WHG or FC126 had significantly lower mortality than the control birds. Peak mortality occurred in the control and vaccinated birds between 14 and 20 weeks.

Trial 3

These Leghorn birds were reared on new litter in three "super"-sanitized curtain-type houses. Total mortality in these three houses, as presented in Table 2, was low when compared to mortality in the first two trials; however, in this trial all birds dying after 4 weeks were examined for gross lesions. Marek's disease was first diagnosed in these birds at 9 weeks. Mortality due to MD in the WHG vaccinated birds was 0.23% (2.78% total mortality) compared to 3.60% (5.75% total mortality) in the controls. The two houses of birds vaccinated with MDV experienced 2.92% (5.77% total mortality) and 1.91% (4.93% total mortality) MD mortality.

Table 1. Field trials of vaccines for Marek's disease (Trials 1 and 2)

Trial	Vaccine	Mortality from 4–21 weeks							
		House 1[1]		House 2[2]		House 3[3]		House 4[4]	
		No. died/No. started	Percent died	No. died/No. started	Percent died	No. died/No. started	Percent died	No. died/No. started	Percent died
1	WHG	290/1823	15.9	270/1777	15.2	64/1723	3.7	163/1683	9.7
	MDHV	549/1771	31.0	359/1751	20.5	238/1723	13.8	243/1633	14.9
	Control	2705/7233	37.4	1921/7114	27.0	1308/7227	18.1	1714/6696	25.6
2	WHG	—	—	—	—	—	—	546/14 818	3.7
	FC 126	—	—	490/14 762	3.3	340/14 600	2.3	—	—
	Control	2753/14 746	18.7	—	—	—	—	—	—

[1] Curtain-type house, old litter.
[2] Enclosed house, old litter.
[3] Enclosed house, new litter.
[4] Curtain-type house, new litter.

Table 2.   Total mortality and mortality due to MD in vaccinated and unvaccinated chickens (Trial 3) [1]

Vaccine	No. of chicks	Mortality (4–21 weeks)						Significance [2]
		Total	Percent	MD	Percent	Other	Percent	
MDHV	8316	480	5.8	243	2.9	237	2.8	$P < 0.05$
MDHV	8441	416	4.9	161	1.9	255	3.0	$P < 0.05$
WHG	8455	235	2.8	19	0.2	216	2.6	$P < 0.01$
Control [3]	5760	331	5.7	207	3.6	124	2.2	

[1] Curtain-type house; new litter.
[2] Tested by analysis of variance of MD mortality.
[3] Mortality in the controls from the 3 houses was combined because it was approximately the same in each house.

Trial 4

Broiler breeders were started on new litter in a curtain-type house isolated in the mountains of North Georgia. At the end of 9 weeks there was very little difference in mortality between the controls (3.3%) and the vaccinated (WHG) group (2.3%). At 9 weeks of age, 100 controls and 300 vaccinated birds were necropsied and examined for gross lesions of MD. None of the birds was found to have such lesions. At this time the remaining birds were moved to another farm and placed in two poultry houses containing birds of the same age that were already experiencing MD mortality. These vaccinated and control birds were separated from MD-infected birds by a feed room in the middle of the house. The vaccinated birds experienced significantly lower mortality ($P < 0.001$) than the unvaccinated controls up to 29 weeks of age (Table 3).

Trial 5

Broiler chicks vaccinated with 2000 PFU were protected against MD; this is shown by MD condemnations of such chicks ranging from 0.2 to 0.42% as compared with 2.01% to 12.0% MD condemnations of unvaccinated controls at the poultry pro-cessing plant. The chicks vaccinated with 200 PFU had 2.0% condemnations due to MD whereas the controls had 4.92% condemnations (Table 4).

Trial 6

At 4 weeks post inoculation, the MD antigen was detected by the direct FA test in the feather follicle epithelium of unvaccinated positive controls (group 1) exposed at 1 day of age to MDV, and at 6 weeks in positive controls (group 2) exposed when 3 weeks old (Table 5). The MD antigen could not be detected by the direct FA test in the feather follicle epithelium of uninoculated controls (group 3). The antigen was also detected in the follicle epithelium at 6 weeks in vaccinated birds (groups 4 and 7) challenged with plasma at 3 weeks of age; in vaccinated birds challenged at 3 weeks by contact exposure (groups 5 and 8) it was not detected until the birds were 7 or 8 weeks old. MDV was isolated from sonicated skin samples from vaccinated birds challenged with MDV. Birds were given injections of 0.2 ml plasma from the controls and vaccinated birds in groups 2–9. Plasma from the positive control group (group 1) could not be injected since there were no survivors by 7 weeks post inoculation.

Table 3.   Field trials of vaccines for Marek's disease (Trial 4)

Group	No. started	Vaccine	Birds after 3–9 weeks			Birds after 9–29 weeks		Statistical analysis [2]
			Mor-tality	Per-cent	No. +/No. necropsied	No. died/ No. started [1]	Per-cent	
1	2500	Control	83	3.3	0/100	310/1180	26.3	—
2	5000	WHG	116	2.3	0/300	407/4177	9.7	$P < 0.001$

[1] Only 1180 birds were selected from the control group.
[2] Comparison of vaccinated group with control group.

Table 4. Condemnations of broiler chicks vaccinated at one day of age with HVT (Trial 5)

Flock	No. of birds	No. of PFU	Condemnations		
			Total (%)	MD (%)	Other conditions (%)
1	10 000	2000	1.20	0.20	1.00
	10 000	—	3.24	2.42	0.82
2	6500	2000	0.49	0.30	0.19
	5500	—	12.88	12.00	0.88
3	10 200	2000	0.53	0.42	0.11
	10 200	—	2.35	2.06	0.29
4	2600	2000	0.38	0.31	0.07
	10 000	—	2.25	2.01	0.24
5	6500	200	2.40	2.00	0.40
	6500	—	5.65	4.92	0.73

Table 5. Birds vaccinated at one day of age with HVT and challenged at 3 weeks of age (Trial 6)

Group	No. of birds	Vaccine virus	Challenge	Gross MD lesions at various weeks					FA reaction in feather follicle epithelium at various weeks					No. of chicks with gross MD lesions after injection of plasma from groups 2–9
				4	5	6	7	8	4	5	6	7	8	
1 [1]	10	None	Plasma	2/2	2/2	2/2	ND [2]	ND [2]	2/2	2/2	2/2	ND [2]	ND [2]	ND [2]
2 [3]	10	None	Plasma	0/2	0/2	0/2	1/2	2/2	0/2	0/2	1/2	2/2	2/2	11/15
3 [4]	10	None	None	0/2	0/2	0/2	0/2	0/2	0/2	0/2	0/2	0/2	0/2	0/15
4	10	FC 126	Plasma	0/2	0/2	0/2	0/2	0/2	0/2	0/2	2/2	2/2	2/2	5/15
5	10	FC 126	Contact	0/2	0/2	0/2	0/2	0/2	0/2	0/2	0/2	1/2	2/2	3/14
6	10	FC 126	None	0/2	0/2	0/2	0/2	0/2	0/2	0/2	0/2	0/2	0/2	0/15
7	10	WHG	Plasma	0/2	0/2	0/2	0/2	0/2	0/2	0/2	2/2	2/2	2/2	8/15
8	10	WHG	Contact	0/2	0/2	0/2	0/2	0/2	0/2	0/2	0/2	2/2	2/2	3/13
9	10	WHG	None	0/2	0/2	0/2	0/2	0/2	0/2	0/2	0/2	0/2	0/2	0/14

[1] These birds were injected with virulent MDV at one day of age and served as positive controls.
[2] Not done, since the remaining four birds had died of MD by 7 weeks.
[3] These birds were exposed at three weeks and served as positive controls.
[4] Uninoculated controls.

Table 6. Horizontal transmission of MD to contact birds from birds vaccinated at 1 day of age with HVT and challenged at 3 weeks with virulent MDV (Trial 7)

Horsfall unit	Vaccine virus	Challenge [1]	Virus isolation [2]	No. of gross MD lesions
1	None	None	0/11	0/11
2	None	Plasma	6/6	6/6
	CC [3]	None	5/6	2/6
3	None	Plasma	6/6	6/6
	CC [3]	None	6/6	4/6
4	FC 126	None	6/6	0/6
	CC [3]	None	0/6	0/6
5	FC 126	Plasma	6/6	0/6
	CC [3]	None	4/6	2/6
6	FC 126	Plasma	6/6	0/6
	CC [3]	None	5/6	3/6
7	WHG	None	6/6	0/6
	CC [3]	None	0/6	0/6
8	WHG	Plasma	6/6	0/6
	CC [3]	None	5/6	3/6
9	WHG	Plasma	6/6	0/6
	CC [3]	None	6/6	3/6

[1] All birds except the contact controls from groups 2–9 were challenged at 3 weeks of age.
[2] Kidneys and monolayers were examined for areas of syncytial cytopathology.
[3] Contact control.

Birds that received plasma from groups 4, 5, 7 and 8 developed MD lesions, indicating that although the vaccine protected the birds against tumour development, it did not prevent them from becoming viraemic with MDV.

Trial 7

The data in Table 6 indicate that both vaccinated and control birds challenged with virulent MDV transmit MDV to contact birds. Birds in contact with unchallenged vaccinated birds did not develop gross tumours nor was virus isolation possible, but HVT was demonstrated in the kidneys of the vaccinated birds.

DISCUSSION

The results of all field trials reported in this study indicate that birds vaccinated with either the WHG or FC126 strains of HVT had a statistically significant lower mortality than birds vaccinated with MDV (P <0.01), while birds vaccinated with WHG,

FC126 or MDV had a statistically significant lower mortality than unvaccinated controls.

Although both HVT vaccines prevented tumour development and clinical disease, they did not prevent infection or replication of virulent MDV. MDV was isolated from sonicated skin samples of vaccinated birds challenged with MDV. Vaccinated birds had both HVT and virulent MDV and antibodies against each. The MD, but not the HVT, antigen could be detected by the direct fluorescent antibody test in the feather follicle epithelium of birds vaccinated at one day of age with either the FC126- or WHG-HVT isolates and subsequently challenged with MDV when 3 weeks old. This indicates that vaccinated birds shed virulent MDV in the feather follicle epithelium in the same way as unvaccinated, infected birds. One-day-old chicks given injections of plasma from vaccinated, infected donor birds, containing both HVT and MDV, developed MD lesions. Also, birds reared in direct contact with vaccinated birds developed MD when the vaccinated birds were challenged with MDV.

REFERENCES

Biggs, P. M., Payne, L. N., Milne, B. S., Churchill, A. E., Powell, D. G. & Harris, A. H. (1970) Field trials with an attenuated cell associated vaccine for Marek's disease. *Vet. Rec.*, **87**, 704-709

Braune, M. D. & Gentry, R. F. (1965) Standardization of the fluorescent antibody technique for the detection of avian respiratory viruses. *Avian Dis.*, **9**, 535-545

Churchill, A. E. & Biggs, P. M. (1967) Agent of Marek's disease in tissue culture. *Nature (Lond.)*, **215**, 528-530

Churchill, A. E., Payne, L. N. & Chubb, R. C. (1969) Immunization against Marek's disease using a live attenuated virus. *Nature (Lond.)*, **221**, 744-747

Eidson, C. S. & Anderson, D. P. (1971) Immunization against Marek's disease. *Avian Dis.*, **15**, 68-81

Eidson, C. S. & Schmittle, S. C. (1968) Studies on acute Marek's disease. I. Characteristics of isolate GA in chickens. *Avian Dis.*, **12**, 467-476

Kawamura, H., King, Jr., D. J. & Anderson, D. P. (1969) A herpesvirus isolated from kidney cell culture of normal turkeys. *Avian Dis.*, **13**, 853-863

Nazerian, K., Solomon, J. J., Witter, R. L. & Burmester, B. R. (1968) Studies on the etiology of Marek's disease. II. Finding of a herpesvirus in cell culture. *Proc. Soc. exp. Biol. (N.Y.)*, **127**, 177-182

Okazaki, W., Purchase, H. G. & Burmester, B. R. (1970) Protection against Marek's disease by vaccination with a herpesvirus of turkey. *Avian Dis.*, **14**, 413-429

Purchase, H. G. & Biggs, P. M. (1967) Characterization of five isolates of Marek's disease. *Vet. Sci.*, **8**, 440-449

Sevoian, M. & Chamberlain, D. M. (1964) Avian lymphomatosis. IV. Pathogenesis. *Avian Dis.*, **8**, 281-310

Solomon, J. J., Witter, R. L., Nazerian, K. & Burmester, B. R. (1968) Studies on the etiology of Marek's disease. I. Propagation of the agent in cell culture. *Proc. Soc. exp. Biol. (N.Y.)*, **127**, 173-177

Witter, R. L., Nazerian, K., Purchase, H. G. & Burgoyne, G. H. (1970) Isolation from turkeys of a cell-associated herpesvirus antigenically related to Marek's disease virus. *Amer. J. vet. Res.*, **31**, 525-538

Discussion Summary

B. R. BURMESTER [1]

1. *Infection cycle and antibody response*

Early questions were directed towards comparison between Burkitt's lymphoma and Marek's disease (MD) with regard to the respective antibody levels in relation to tumour development and regression. Long-term data on chicken flocks were cited, but such data on individual chickens are not available so that a direct comparison could not be made.

In Marek's disease, the disease process itself induces a degree of immune incompetence, and one must therefore question which is cause and which is effect. The lowered antibody responses in diseased birds compared to nondiseased birds could be due to the disease process, or birds without antibody may be more susceptible to disease than those that develop antibody.

The patterns of occurrence of precipitating and immunofluorescent antibody in natural infections are much the same as those induced experimentally. Variations that occur are related to the time of infection and the degree of exposure. Thus, when birds are exposed after maternal antibody has disappeared, a more dramatic serological response ensues. Long-term studies of a natural outbreak revealed that all chickens became infected within three months of age. During an 18-month period, about one third developed clinical signs of disease, but all developed antibodies; the level of precipitating antibody tended to fall with time, but the immunofluorescent antibody persisted at high levels through the entire observation period of 68 weeks. Although the proportion of viraemic birds decreased with time, many had a persistent viraemia. Generally speaking, birds that developed tumours had lower precipitating antibodies and higher levels of viraemia than did birds of the same flock that never developed clinical signs. This difference was quantitative, but there

were exceptions. In fact, some birds with high antibody titres developed tumours but, in most instances, the titre fell just before death.

The indirect immunofluorescence test detected antibodies earlier, for a longer duration, and at higher titres than did the agar gel precipitin test. It is probable that different antibodies were detected by the two tests.

Experiences with two flocks in Connecticut were described. One, consisting of 2000 birds, was reared in isolation under conditions of continuous positive pressure with filtered air and was negative for precipitating antibodies for the first 50 weeks. Thereafter, the flock converted serologically, reaching 30% positive, but the level then fell to about 17% positive by the 70th week. None of the birds developed tumours. The second flock consisted of chicks without maternal antibodies which were injected with a high dose of virus. Those birds that died of Marek's disease in a short time did not develop antibodies, whereas those that survived developed precipitating antibodies.

Data obtained from five strains of chickens deliberately exposed to Marek's disease virus (MDV) showed that at 55 days all chickens were free of precipitating antibody whereas at 97 days approximately 95% of all chickens tested were reactors. Within all strains there were poor producers and good producers of antibody. However, the titres were markedly higher in the genetically resistant Cornell K strain than in the highly susceptible Cornell S strain. In two other resistant strains and one susceptible strain, titres at 293 days could not be related to resistance to MD.

It is quite clear that the majority of MDV infections do not lead to clinical Marek's disease. It is probable that many minor and some major lesions completely regress, and that there are many factors that influence the outcome of MDV infection. Some of these factors are known but some, perhaps the

[1] Regional Poultry Research Laboratory, East Lansing, Michigan, USA.

most important ones, have as yet not been defined. Certainly the immune response has a direct effect on the outcome of MDV infection; the time and strength of the response appear to be most important. The genetic differences are also important and these may be mediated through the immune response. Also of great importance is the pathogenicity of the virus. Probably the first virus that becomes established in a flock determines the disease outcome.

2. *Immunity to Marek's disease*

There was some discussion on the mechanism by which protection is conferred by vaccines against MD. The possibility was mentioned that the vaccine may prevent infection of the target cells with virulent virus, since it obviously did not prevent infection of epithelial cells. Against this was the fact that the virulent virus was present in the leukocytes (buffy coat), and presumably in the lymphocytes, of vaccinated birds, and lymphocytes were the most likely target cells.

Since the mode of infection with MDV is probably via the respiratory tract, it was suggested that immunity might be mediated by IgA-type antibodies in the respiratory tract and that these might develop more rapidly if birds were immunized by this route. However, the observation that the genetically resistant chickens become infected but fail to develop disease does not support this hypothesis.

Maternally transmitted antibody reduced the intensity of infection with Marek's disease. The number of birds with disease, the number of tissues with antigen demonstrated by the fluorescent antibody test, the number of focal areas of antigen in the positive organs and the number of cells in each focus were usually markedly reduced when maternal antibody was present at the time of infection. How the effect was mediated was unknown. With extracellular virus in the plasma, the mechanism is easy to understand but with cell-transported virus it is not so clear. It is possible that by reducing the level of extracellular virus the spread of virus within the body is reduced. In addition, maternal antibody to MDV did not prevent the "take" of either cell-associated or cell-free virus vaccines.

It was reported that passive transfer of antibody by inoculation of serum with high precipitating antibody titre did reduce the incidence of MD after challenge by a natural route but it was not as effective as maternal antibody. Chickens that received

antibody had a lower level of virus in their kidneys than those that did not.

Since birds infected with either MDV or the herpesvirus of turkeys (HVT) remained infected and the blood leukocytes carried the virus indefinitely, it was not possible to perform passive transfer experiments with lymphocytes from immune birds since the virus would be transferred also. An examination of the response of birds vaccinated with inactivated vaccines would be interesting.

In comparisons with Epstein-Barr virus (EBV) it was emphasized that more than one antibody was directed against various types of antigens. Antibodies to membrane antigen present on the surface of cells and on the viral envelope were probably the same as the MD neutralizing antibodies described, provided that viral envelope components are inserted into the membrane of productively and nonproductively infected cells, by analogy with the EBV system. Such an antibody would have an effect on the cells whether they make virus or not. Antibodies against the corresponding EBV-determined membrane antigen tend to rise with tumour regression, and to remain at a steady level, but decline with recurrence. The immunoprecipitating antibodies, directed against the soluble antigen and the early antigen show a different behaviour. The level of these antibodies is usually low in regression and rises with recurrence of the tumour. Cell-mediated immunity would be directed against surface antigens and would be expected to have the same specificity as antibody to cell membranes and viral envelope antigens.

Humoral and cellular immunity directed against membrane antigens and reflected as neutralizing antibody may be responsible for suppression or rejection of tumours in birds not susceptible to MD. Regression of large visceral tumours has been reported and the disappearance of clinical signs of nerve impairment has been observed by many investigators, but formal reports are lacking.

3. *Population density*

A question concerning the effect of population density was posed. No specific studies could be cited, but undoubtedly the rate of spread of MDV was related to such factors as the distance between birds and the frequency of bird-to-bird contact. It was thought that only a small percentage of chickens became infected from contact with a contaminated environment and that the remainder were infected by their pen mates. Therefore, the higher the population density the more rapid was the spread of virus to noninfected pen mates. Airborne transmission was very efficient, so that direct contact was not required for the transfer of infection. However, population density may affect the rate of transmission by virtue of its effect on the concentration of infectious particles in the air.

4. *Pathogenesis of Marek's disease*

The means of spread of MDV from one tissue to another is not known. The plasma of chickens infected with certain strains of virus contain cell-free virus, but this is unusual. The virus is present in the leukocytes of the blood, probably in either the macrophages or lymphocytes, and so could co-exist with antibody. These cells are probably responsible for the spread of infection from one tissue to another. Virus has never been detected in the plasma of birds infected with the Houghton Poultry Research Station strains.

5. *Transformation*

The use of the word "transformation" was then clarified. Firstly, lymphocytes undergo transformation *in vitro* when stimulated in a number of ways. This is a morphological change in which cells become lymphoblastoid and has nothing to do with malignancy. Secondly, there is a malignant transformation in which normal cells become malignant tumour cells on infection with tumour viruses. Thirdly, there is "lymphoblastoid transformation", which is sometimes seen in long-term cultures. This is a poor term and may refer only to a change in morphology of the cells. Sometimes the predominant cell type is overgrown by a cell previously in the minority, so that the overall appearance of the culture may change from, for example, fibroblastic to lymphoblastoid. This does not imply that any change in the cells themselves has occurred.

Avian lymphocyte cultures are usually difficult to maintain for any length of time. Lymphocytes attached to glass usually undergo blast transformation similar to that seen in stimulated immunologically competent cells. When chicks were inoculated with such cells derived from MD-infected birds, MD resulted but without tumours.

6. *RNA virus and Marek's disease*

The question of the possibility of an RNA virus acting as a cocarcinogen or even as the primary inducing virus in Marek's disease was posed. Flocks of chickens have been maintained for years free of lymphoid leukosis virus, but tissues of embryos from such flocks still have the group-specific (gs) antigen. Such flocks have remained free of lymphoid leukosis tumours but developed Marek's disease on natural or experimental exposure to the herpesvirus. It was further explained that the gs antigen naturally occurs commonly in chickens and may represent a leukosis virus present in a repressed, possibly genetically integrated, form; thus the proponents of the theory that Marek's disease is caused by an RNA virus might say that all that the herpesvirus does is to derepress the integrated RNA virus and this is the actual cause of the disease. Observations on one inbred line of birds are particularly pertinent. The Reaseheath C line lacks the gs antigen, and it has, in fact, been shown that this antigen is inherited as a gene. Birds which lack the gs antigen, and hence lack a possibly integrated virus, are still able to develop MD tumours after inoculation with MDV. However, it should be pointed out that in the process of infecting them, cells with the MDV were injected, and quite possibly these cells also carried the integrated virus (gs antigen). However, this argument is quite tenuous, and the consensus was that there can be little doubt that the herpesvirus is the primary aetiological agent of MD.

LUCKÉ FROG RENAL CARCINOMA

Chairman — A. Granoff

Pathology of Amphibian Renal Carcinoma—A Review

K. A. RAFFERTY, Jr[1]

INTRODUCTION

The classical pathology of the renal adenocarcinoma of the North American leopard frog, *Rana pipiens,* has been described in a series of publications by Lucké and co-workers (especially Lucké, 1934, 1952; Lucké & Schlumberger, 1949). In the paper first cited, Lucké brought the tumour to scientific attention, and for a 20-year period was almost the only person seriously at work with it. Thus, the tumour bears his name.

As is the case with most or all tumour entities, there is considerable variation in the histological and biological forms of the frog kidney tumour, but in this instance it has become increasingly probable that all forms are caused by a virus, as Lucké originally believed. It therefore seems reasonable to retain the term "Lucké tumour" as a collective name for a series of malignant forms, in the same sense that corresponding human kidney carcinomas, including variants such as cystadenoma and clear cell carcinoma, are commonly lumped together under the synonymous terms, "Grawitz's tumour", "renal cell carcinoma", "hypernephroma", etc. In the present paper it is proposed to review briefly the salient features of the pathology of the Lucké tumour, to compare it with human renal cell carcinoma, and to describe some new observations.

BASIC PATHOLOGY OF THE LUCKÉ TUMOUR

The tumours are frank adenocarcinomas, consisting of rather large cells and tending toward well-differentiated forms, although anaplasia has been reported as a rarity (Duryee *et al.,* 1960; Rafferty, 1964).

Employing standard criteria of malignancy, Duryee (1956) and Duryee *et al.* (1960) have identified all grades of malignancy in the series examined, but in the present author's experience, grade III histology is rare. In spite of this, however, the tumours are unencapsulated, and, in most instances, highly malignant on biological grounds. Metastasis is almost entirely temperature-dependent and occurs in at least 80% of cases in which temporal, nutritional and thermal factors are favorable (Lucké & Schlumberger, 1949). The tumours may become very large, reaching weights of 14 grams and more, and comprising as much as half the body weight of the animal. There seems to be little toxicity associated with the tumour, since afflicted frogs are often quite active until shortly before death. Frogs with advanced tumours often become moribund quite suddenly, and after death it may be difficult to find remnants of normal-appearing kidney. In addition, many tumours do not become cystic in morphology, suggesting that malignant tubules may maintain a connection between glomerulus and collecting duct, or perhaps even retain some function. The skin in these animals also functions as an excretory organ, however, so the point is obscure. About half the tumours seen in field frogs are unilateral, and even in tumour-bearing laboratory frogs that are kept for long periods, one kidney occasionally remains free of tumour even after metastasis has developed elsewhere. Central necrosis tends to develop in the larger tumours, and stromal proliferation (see below) is variable. There are no reports of large, palpable, tumours having regressed, but there is now some rather convincing histological and epidemiological evidence, to be noted below, that indicates that a substantial proportion of small tumours may undergo spontaneous remission in the field.

An example of a typical tumour *in situ* is shown in Fig. 1.

[1] Department of Anatomy, University of Illinois Medical Center, Chicago, Illinois, USA.

A remarkable feature of the tumours is the frequent occurrence of intranuclear inclusions of the Cowdry type A form, which are generally associated with acute viral disease (Cowdry, 1934). For a number of years considerable confusion existed with respect to the occurrence of these inclusions, since they had been observed only by Lucké (1952) and by Fawcett (1956), while other workers were unable to find them. It was finally appreciated that inclusions (and virus particles, as visualized in the electron microscope and isolated in tissue culture) occur only in tumour-bearing frogs brought fresh from the field during the winter months, but not in those captured in the summer months, or kept at room temperature in the laboratory for a month or more (Rafferty, 1964; McKinnell[1]).

An example of the most commonly occurring form of the Lucké tumour is shown in Fig. 2; a similar form, but occurring in a frog captured during the winter and containing characteristic intranuclear inclusions, is illustrated in Fig. 3. All evidence to date indicates that a herpes-like virus is always associated with inclusion-bearing tumour cells. A point of further interest is that mitosis is seldom or never seen in inclusion-cell regions of tumours, and that inclusion-containing cells appear moribund on cytological grounds. Roizman notes[2] that productive herpesvirus infections *in vitro* always result in cell death. It is not impossible, therefore, that a

high degree of spontaneous remission may occur through destruction of tumours during the period of inclusion formation and virus production. This hypothesis is made stronger by study of many tumours of over-wintering frogs, in which it is hardly possible to find tumour cells that appear viable.

The only tumours that have been reported in the frog kidney, save for certain rare cases (Balls, 1962), are carcinomas of undisputed parenchymal cell origin. Autochthonous tumours seem to be strictly limited to the geographical races of the northern subspecies of the North American leopard frog, *Rana pipiens pipiens* (Rafferty, 1968; McKinnell[1],) although under unusual circumstances the tumours may be induced in a few closely related forms following intraocular transplantation (Tweedell, 1955). Field incidences vary between about 3% and 9% in different populations (Lucké, 1952; McKinnell, 1965), figures that are astonishingly high in view of the fact that they represent portions of the population seen at a particular moment, and since the tumours are highly malignant and usually rapidly progressive (in some summertime populations of susceptible frogs in Wisconsin and Minnesota, virtually no tumours are seen, but this was shown by McKinnell (1969) to be a seasonal phenomenon). Summer tumours do occur in Vermont populations (Rafferty & Rafferty, 1961).

High as field incidences are, laboratory incidences are even more impressive, rising to an apparent maximum of some 25% when large adults are kept in the laboratory for eight months at 25°C with no

---

[1] See p. 183 of this publication.
[2] See p. 1 of this publication.

PLATE I

Fig. 1. Tumour-bearing frog RT 11-65. Female frog with large spontaneous tumour partly obscured by ovaries and oviducts. Portions of the tumour, which is bilateral, are indicated by arrows. Metastastic nodules in liver and lung are also indicated.

Fig. 2. Typical architecture of Lucké adenocarcinoma, with a small amount of connective tissue present. The tumour was relatively slow-growing; the frog was moribund two months after initial palpation, with metastasis in orbit only. Occasional vacuolated cells are present in this specimen and may be similar to individual epithelial cells of human "clear cell" carcinoma. RT 1075. Giemsa stain; final magnification: ×120.

Fig. 3. Lucké renal adenocarcinoma consisting of inclusion-containing and inclusion-free cells, the latter most readily identified by the presence of prominent nucleoli. Inclusion-containing cells tend to occur in groups (I), and are not seen in adjacent normal and transitional renal tubules. The frog arrived from Vermont on 23 March and was killed two days later. RT 1043. Giemsa stain; final magnification: ×120.

Fig. 4. Human renal cell carcinoma (Grawitz's tumour). A relatively well-differentiated example of the papillary form of adenocarcinoma. Comparison of this example with the frog tumour of Fig. 2 will show the basic similarity. Slide provided by Dr. Roger Smith, University of Illinois College of Medicine. Slide No. 10659. Hematoxylin and eosin; final magnification: ×120.

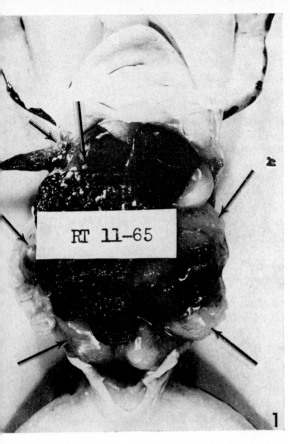

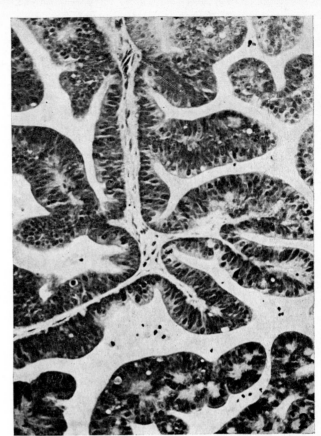

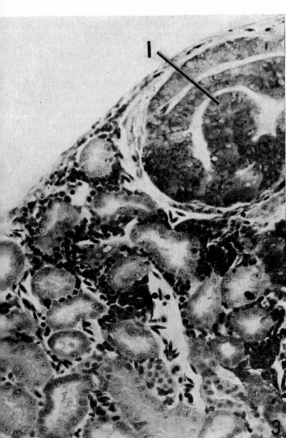

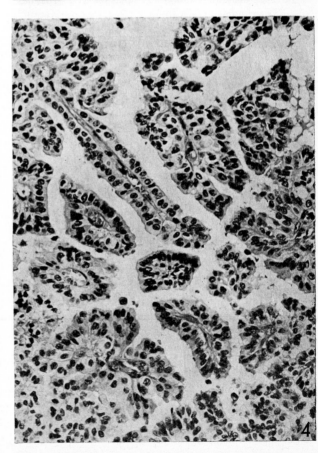

treatment other than adequate feeding and care (Rafferty & Rafferty, 1961; Rafferty, 1963a, b). The remaining 75% appear to be essentially immune, and tumours are seen only rarely in juvenile frogs (DiBerardino & King, 1965). Some of Lucké's experiments (1952) seemed to suggest that contact transmission of the tumour occurred, but others failed to confirm this suspicion, and a matched pair experiment by Rafferty (1963b), using only large adult frogs, gave no evidence of horizontal, or contact, transmission. The experiments conducted by DiBerardino & King (noted above) did, however, indicate that transmission may occur between adults and tadpoles, or horizontally among tadpoles, even at room temperature. This observation is remarkable in view of the fact that all viruses of the herpes type (which are the only ones observed with certainty in association with tumour cells *per se*) are seen only in cells that have undergone prolonged cold incubation and contain intranuclear inclusions. The frogs used by DiBerardino and King lacked inclusions and presumably the herpes-virus as well, at least in otherwise detectable form. The point must remain obscure at the moment but is of interest in that at least three kinds of virus (two kinds of herpes-type virus and Granoff's cytoplasmic frog virus) have been found in Lucké tumours (see review by Granoff[1]).

AETIOLOGY AND HISTOLOGICAL TYPE

It is perhaps easier to accept the observation that various agents can cause similar tumours than that similar agents cause different tumours of the same cell type. The fact that an array of cytological and histological tumour types occurs in the frog kidney was first emphasized by Duryee and co-workers (1960), who, however, were largely concerned with assessing malignancy in terms of experience with human tumours and whose attention was chiefly focussed upon cytological features. It is also clear, however, from the present author's examination of some 500 Lucké tumours, that at least three clearly differentiated histological types occur. Further comparisons with human tumours are even more instructive from the viewpoint of histological type, just as were those of Duryee *et al.* (1960) from the cytological viewpoint. An illustration of the most common type of human renal carcinoma (i.e., renal cell carcinoma, or Grawitz's tumour) is given in

---

[1] See p. 171 of this publication.

Fig. 4. It should be compared with the Lucké tumour of Fig. 2.

Fig. 2 illustrates a "typical" Lucké tumour, characterized by complex epithelial infolding, obviously the result of overgrowth of transformed tubular components of the kidney. The cells involved are generally healthy-looking, and mitotic figures are frequent in tumours of frogs that are well-nourished and acclimatized to room temperature. Some vacuolated cells are seen in this case, and may be analogous to the cells of human "clear cell" variants.

An inclusion tumour is illustrated in Fig. 3. It should be noted that inclusions are seen only in tumour cells, and never in those of normal tubules.

In terms of histological type, frog tumours also include cystic forms of adenocarcinoma, sometimes associated with rather extreme fibrosis, or overdevelopment of the stromal component of the tumour; this condition is illustrated in Figs. 5 and 6. Cystadenoma is also shown in Fig. 6.

The impression gained from a study of fibrotic or cystic tumours is that mitotic figures are relatively infrequent, even in otherwise actively metabolizing frogs. Histologically, tumours of this type resemble several cases from human pathology in which clinical regression or remission was in progress. In the human series, features related to remission are cystic form, fibrosis, and calcification (see review by Goodwin *et al.,* 1968). As is made clear by McKinnell's evidence (1969), regression must occur rather often in the frog, since high field incidences of small asymptomatic tumours of Wisconsin frogs are not matched by similar incidences in those captured only a few months later in the summer, while such tumours are found in Vermont frogs at the same season.

MALIGNANCY IN THE HUMAN KIDNEY

The prevailing evidence that the Lucké tumour and at least three different types of human malignancy (Burkitt's lymphoma, nasopharyngeal carcinoma and cervical carcinoma) may be caused by herpes-type viruses emphasizes the potential value of comparisons between frog and human kidney tumours.

In man, the incidence of kidney tumours is, of course, far lower than in the populations of susceptible frogs. About 3% of all human tumours are malignant tumours of the kidney, and the great majority of these (83%) are parenchymal cell malignancies of cellular origin similar to those seen in the frog (Hewitt, 1967).

Mostofi (1967) notes that this largest class of human malignancies is variously termed renal cell carcinoma (the preferred designation), Grawitz's tumor, and hypernephroma. Within this group he recognizes five histological types. These consist of variously differentiated types (papillary; differentiated glandular; differentiated; clear cell, or vacuolated; and mixed, with acinar and undifferentiated regions). They vary characteristically in development of the fibrous stromal component and cystic configuration. Hence, all of these types have fairly exact counterparts in the frog. The remaining 17% of human renal tumours consist of lymphosarcoma, nephroblastoma (Wilms's tumour), and squamous cell tumour, the last of which may be secondary.

Renal carcinoma of parenchymal origin is not commonly seen in other animals. Wessing (1959) reported such tumours in fish. Spontaneous cases have been reported in the mouse (Claude, 1962) and various other rodents. Nephroblastomas have been induced in chickens by avian myeloblastosis virus (Ishiguro et al., 1962; Diamond, 1968). While these contain epithelioid cell nodules, cartilage, keratinization and sarcoma are also notable features of these tumours. Hass et al. (1968) and other workers have induced tumours of the renal cell carcinoma type in rats and rabbits by feeding lead compounds and supplementary additives, in situations where there is no indication of viral involvement.

THE TRANSITIONAL PROCESS IN MALIGNANCY

The kidney is an exceptionally interesting organ in which to study spontaneous malignant transformation, since the process occurs often in the frog (and hence is easy to observe and even predictable in onset). It is also of interest because the kidney contains many cell types organized into a structure of great complexity. Thus, this system provides an exceptional opportunity to study the rapidity of cell transformation; the type of cells that do and do not transform, and the stages through which cells pass as they transform. The special value of the frog kidney in malignant transformation was first recognized by Duryee (1956), who provided convincing histological and cytological evidence of the stages through which the transformation of nephrons from normal to malignant appeared to pass; this evidence was further reinforced by the fact that such intermediate stages were usually associated with frank carcinoma elsewhere in the kidney, or were frequently found adjacent to carcinomatous nodules. Transitional stages were later also observed by Rafferty (1962) and by Barch et al. (1965).

Examples of intermediate tubular forms that are interpreted as transitional are shown in Figs. 7 and 8. Presumed transformation is marked, in addition to the criteria cited above, by greatly increased basophilia, nucleolar enlargement, a tendency towards reduction of chromatin clumping within the nuclei, and, perhaps most significantly, by the occasional appearance of mitoses, which are virtually never seen in normal tubules (mitosis may, however, occur within haematopoietic nodules, since in amphibia this process often takes place within the kidney).

An additional point of interest in terms of the pathology of the frog tumours is the tendency towards localization of the process of malignant transformation. There are wide differences in the rapidity with which normal renal tissue is recruited into expanding malignant centres as well as great variation in the number of malignant nodules or transforming centres. Although these processes were clearly shown by Lucké & Schlumberger (1949) and by Rafferty (1962) to be temperature-regulated to a very large extent, pronounced individual "biological" or "pathological" differences also exist in individuals. This is shown, for example, by the occurrence of metastases in animals with unilateral tumours. In one case, a unilateral tumour weighed more than 8 grams, or 25% of the total body weight of the host. Small metastases were present in the liver, but careful gross inspection of the contralateral kidney revealed no sign of a tumour nodule. Other such cases have been recorded. Individual tumour-bearing frogs also vary greatly in the degree to which apparently transitional tubules are seen adjacent to tumours, and in rapidity of growth of known tumours. The unconfirmed impression is that these features are not closely correlated with the occurrence of metastasis.

It is possible, therefore, that some form of local immunity may be operative in restricting the spread of tumorigenesis. It should be remembered that the three-chambered amphibian heart permits some mixing of arterial and venous blood, and would therefore presumably provide opportunities for circulating tumour cells to reach the renal artery (of the contralateral kidney) without passing through the pulmonary, or any other, capillary bed. Lucké tumour metastases are seen most frequently in liver and lung, but tumour cells that reach the venous circulation should have reasonably high chances of

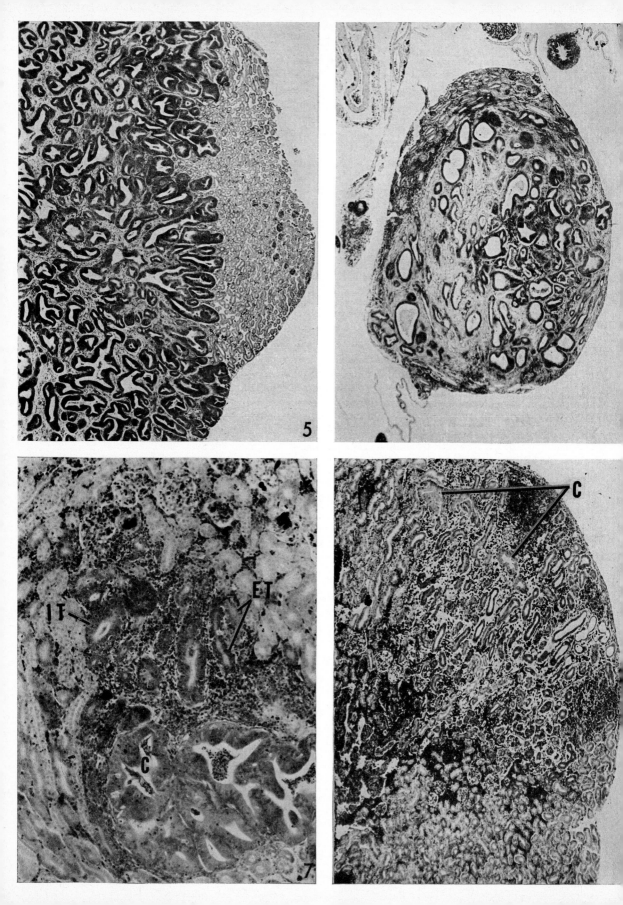

returning to the kidneys again. In the relatively infrequent instances of tumour-free contralateral kidneys associated with metastases in other organs, it seems possible that some form of partial local immunity may occur.

There is also wide variation, as just noted, in frequency of foci within individual kidneys. The extreme case is that in which an entire kidney is apparently transformed simultaneously, a suggestion made by Duryee (1956) following his observation of several tumours that had retained the shape and conformation of the normal kidney while being considerably enlarged and retaining few, if any, normal renal tubules. Duryee called these "replica" tumours. Gross specimens have subsequently been observed by the author, and Fig. 8 may represent histologically an early stage in the simultaneous or near-simultaneous transformation of a sizable volume of kidney.

The case at the opposite extreme is, of course, that in which nephrons are transformed singly. That this also occurs is shown in Fig. 9, in which the transformed nephron was not grossly visible, but seen in sections of a block of apparently normal kidney. In the interpretation of Fig. 9, it should be recalled that the frog kidney is dorsoventrally flattened, and that the nephron, although contorted somewhat, runs back and forth between the two surfaces three times, occupying about the same area as that of the malignant tubular component illustrated.

Some cases are also seen in which less than a whole nephron is transformed, and in these and other instances one may, with patience and diligence, locate segments in which normal joins neoplastic. Figs. 10 and 11 illustrate cases in which only a part of a single nephron is transformed or transitional, and a case in which a transformed nephron joins an apparently normal collecting duct.

An additional point of interest is the fact that, contrary to Lucké's belief, malignant transformation occurs in cells other than those of the proximal tubular portion. This is shown in part by the absence of recognizable distal and collecting duct segments in massive growths or in replica tumours. More to the point, however, is the direct observation of transformed ureters (Wolffian ducts) and longitudinal collecting ducts. Because of the architecture of the kidney, the original character of the neoplastic ducts may be deduced, in these cases, from their course and position. In addition, however, normal-neoplastic junctions have been observed in some instances, and in the case of small tumours (which presumably have been recently transformed), the "histological fingerprints" of normal collecting duct or Wolffian duct morphology may be retained, at least for a time. An example of this type is shown in Fig. 12, which shows a section of a Wolffian duct. One half of the duct epithelium is transformed (and includes a mitotic figure), while the other appears normal. This region was sectioned serially, and, as shown in Fig. 13, the duct merged with a nodule of "typical" Lucké adenocarcinoma. From this specimen, therefore, it is unclear whether the transition site arose within the Wolffian duct or within a nephron, or at some

---

PLATE II

Fig. 5. Lucké adenocarcinoma showing some tendency towards increase in the amount of connective tissue present, and also towards the cystic development often seen in such cases. Normal kidney tissue appears at right, with little evidence of transitional cells or nephrons. RT 1148. Giemsa stain; final magnification: ×20.

Fig. 6. Lucké adenocarcinoma (papillary adenoma or cystadenoma) showing extreme fibrotic development accompanied by cystic enlargement of carcinomatous portions. A small area of rather infiltrated normal kidney appears at the top (especially upper left), but also seems to lack transitional forms. RT 1146. Giemsa stain; final magnification: ×20.

Fig. 7. Lucké adenocarcinoma showing numerous apparent transitional forms. Frank adenocarcinomatous cysts and tubules (C) connect with intermediate forms (IT) in an adjacent, heavily infiltrated, region. Farther removed, apparently early transitional stages (ET), are also seen. RT 1051. Giemsa stain; final magnification: ×45.

Fig. 8. Section of frog kidney in which extensive areas seem to be undergoing simultaneous malignant transformation. Normal-appearing regions of tubules are seen at bottom and at upper left. The apparently transforming area is heavily infiltrated and contains some tubular components (C) that, in isolation, would probably be diagnosed as malignant. Unequivocal carcinoma occurs elsewhere in the same kidney, which had been injected earlier with extract of a non-inclusion tumour. Giemsa stain; final magnification: ×20.

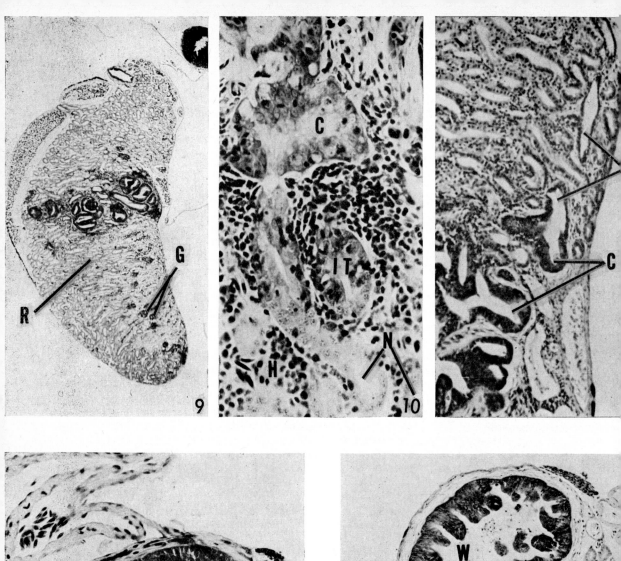

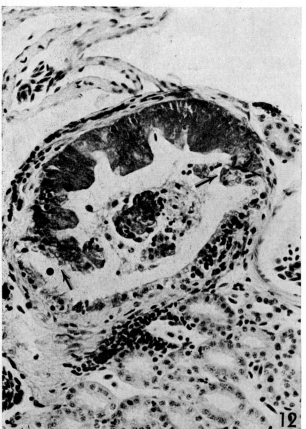

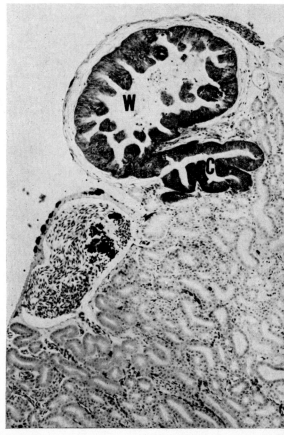

other site. Other specimens strongly suggest, however, that primary transformation may occur both in the Wolffian duct and in collecting ducts. Because of the tendency for whole nephrons with their collecting ducts and the Wolffian duct eventually to form large, completely transformed nodules, it is very probable that transformation occurring within a single duct or tubule cell may spread with greater or less rapidity (doubtless varying from frog to frog) to adjacent cells. The suggestion is that this is a relatively slow process, however, and it would seem most likely that replica transformation is a consequence of the existence of multiple transformation centres. This hypothesis is supported by observations of tumours in which, as an estimate, only every second or third nephron, on the average, seems to be transformed or in process of transformation.

On the basis of cytological evidence (specifically the retention of a subdued brush border), Lucké postulated that the tumours consist of transformed proximal segment cells of renal nephrons. Lunger, Darlington & Granoff (1965) and Zambernard & Mizell (1965) had confirmed that Lucké tumour cells generally possess a poorly organized microvillous surface, which adds weight to the contention that many or most of the tumours arise as transformed proximal segment cells. There is now no doubt, however, that distal segment and collecting (including Wolffian) duct cells are recruited in advanced tumours, as noted, and selected specimens strongly suggest that the Wolffian duct may be a primary site of origin as well. In any case, a minimum of four distinctive cell types participate in malignant transformation, namely distal and proximal tubule segment cells, collecting duct cells, and ureter or Wolffian duct cells.

ANATOMICAL FEATURES POSSIBLY INVOLVED
IN TUMOUR FORMATION

Based upon his observation of more than 11 000 tumour-bearing frogs, Lucké (1952) determined that in field frogs the tumour occurs twice as often in males as in females. In males, modified nephrons (vasa efferentia) connect the testis with some of the transverse collecting ducts, permitting access of spermatozoa to the Wolffian duct and hence to the cloaca and externum. Since spermatozoa are produced throughout the year (although appearing in the Wolffian duct only during the spawning season), the ducts involved are larger and may permit more refluxing of contents, perhaps including virus. Much evidence (Rafferty, 1964; McKinnell [1]) suggests that a virus associated with experimental tumorigenic action is liberated in the spring at the time of spawning, when the duct system of the male is dilated and maximally active. Factors of this sort may obviously

---

[1] See p. 183 of this publication.

PLATE III

Fig. 9. Transverse section through a frog kidney showing an adenocarcinomatous nodule apparently derived from one or a few transformed nephrons. This tumour was encountered in a random section and was not detected grossly. Glomeruli (G) of normal nephrons are located in the ventral portion of the kidney, and tubules of associated nephrons traverse the kidney three times before ending in dorsal collecting ducts. The system of rays thus formed is indicated (R) and may explain the configuration of this type of early carcinoma. RT 1154. Giemsa stain; final magnification: ×10.

Fig. 10. Junction between normal proximal convoluted tubule (N) and malignant (C) in a single nephric tubule. Immediate area is heavily infiltrated, and haematopoietic centres (H) are present to a varying extent. Intermediate tubule sections (IT) are also seen. RT 1150. Giemsa stain; final magnification: ×200.

Fig. 11. Junction between malignant renal tubule (C) and normal transverse collecting duct (T). A transversely oriented small elongated tumour occurred elsewhere in the same kidney, implying extensive transformation of a transverse collecting duct. RT 1144. Hematoxylin and eosin; final magnification: ×70.

Fig. 12. Ureter (Wolffian duct) showing junctions (indicated by arrows) between normal epithelium in the lower portion of the duct and apparently malignant epithelium in the upper portion. In a few sections through the same preparation, as shown in Fig. 13, the epithelium is seen to merge with unequivocal carcinoma. RT 1116. Giemsa stain; final magnification: ×120.

Fig. 13. Section through the preparation illustrated in Fig. 12, showing continuity between typical carcinoma of tubular origin (C) and neoplastic Wolffian duct component (W). Final magnification: ×45.

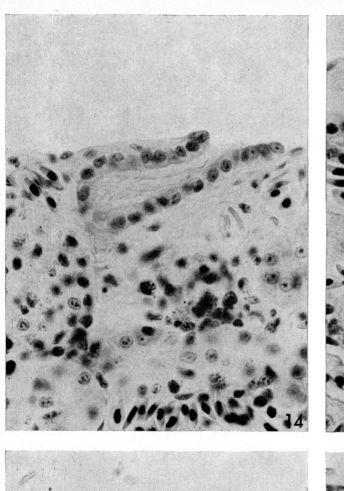

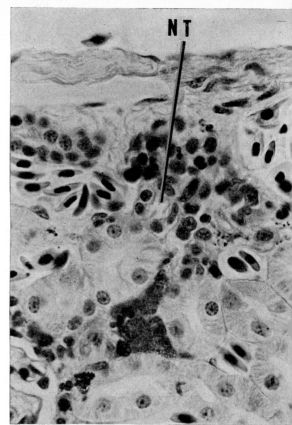

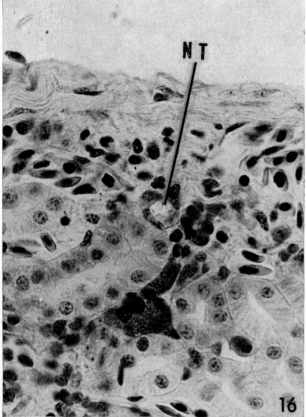

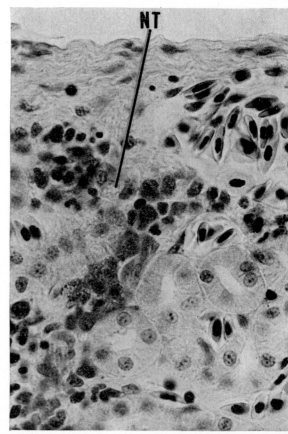

serve to explain differential susceptibility, although other factors (hormonal ones, for example) may, of course, also be involved.

Other possible routes of viral communication within the body may also exist. Many amphibians, particularly primitive forms such as urodeles, possess nephrostomes retaining the primitive connection between renal tubule and coelom; this might permit virus appearing in the lumen of a malignant tubule to make its way to the body cavity. Clearly, the reverse route might also be followed since it is the natural direction of movement, the nephrostomes serving to scavenge fluid from the coelom (see Noble, 1954, for review). The more specialized amphibians (*Salientia* or *Anura*) possess nephrostomes, but at least some of these open into renal veins instead of into nephric tubules (Sweet, 1908). The point remains a controversial one in that there is a lack of agreement

as to whether some nephric connections are retained in these forms, which include *Rana pipiens*. The present author, however, has spent a number of tedious hours following the course of nephrostomes in serial sections, and in every case they have opened only into veins. Figs. 14–17 illustrate selected serial sections of such a connexion within a nephrostome. The point is of marginal interest in view of the finding by Granoff [1] of virus in the ascitic fluid of tumour-bearing frogs. Simpler explanations for the appearance of virus in ascites—such as direct emission in ulceration—seem equally probable. Nevertheless, there may be extensive opportunities for movement of virus or other small particles in body spaces, and there seems little doubt that some viruses found in tumour cells also appear in the urine (Rafferty, 1965).

[1] Personal communication.

REFERENCES

Balls, M. (1962) Spontaneous neoplasms in amphibia: A review and description of six new cases. *Cancer Res.,* **22**, 1142-1155

Barch, S. H., Shaver, J. R. & Wilson, G. B. (1965) Some aspects of the ultrastructure of cells of the Lucké renal adenocarcinoma. *Ann. N.Y. Acad. Sci.,* **126**, 188-203

Claude, A. (1962) Spontaneous, transplantable renal carcinoma of the mouse: Electron microscope study of the cells and an associated virus-like particle. *J. Ultrastruct.,* **6**, 1-18

Cowdry, E. V. (1934) The problem of intranuclear inclusions in virus diseases. *Arch. Path.,* **18**, 527-542

Diamond, L. (1968) Virus-induced kidney tumors in laboratory animals. In: King, J. S. Jr, ed., *Renal neoplasia*, Boston, Little, Brown & Co., pp. 317-343

DiBerardino, M.A. & King, T. J. (1965) Renal carcinomas induced by crowded conditions in laboratory frogs. *Cancer Res.,* **25**, 1910-1912

Duryee, W. R. (1956) Precancer cells in amphibian adenocarcinoma. *Ann. N.Y. Acad. Sci.,* **63**, 1280-1302

Duryee, W. R., Long, M. E., Taylor, H. C., Jr., McKelway, W. P. & Ehrmann, R. L. (1960) Human and amphibian neoplasms compared. *Science,* **131**, 276-280

Fawcett, D. E. (1956) Electron microscope observations on intracellular virus-like particles associated with the cells of the Lucké renal adenocarcinoma. *J. biophys. biochem. Cytol.,* **2**, 725-742

Goodwin, W. E., Mims, M. M., Kaufman, J. J., Cockett, A. T. H. & Martin, D. C. (1968) Under what circumstances does "regression" of hypernephroma occur?

PLATE IV

Fig. 14. Nephrostome opening on peritoneal surface of normal frog kidney. For nearby sections of the same preparation, photographed under identical conditions, see Figs. 15–17. Giemsa stain; final magnification: ×430.

Fig. 15. Continuation of ciliated nephrostome tubule (NT) a few sections removed, seen in association with venous sinusoid.

Fig. 16. Further continuation of the same nephrostome tubule (NT), showing reduction in calibre but retention of cilia and closer association with venous sinusoid.

Fig. 17. Terminus of nephrostome tubule (NT), illustrating thinning and discontinuity of tubule wall in intimate association with venous cul-de-sac, the wall of which is sectioned tangentially just to the left of the thinned tubular portion. Cilia are retained; tubular contents are thought to enter venous sinusoids at this point.

In: King, J. S. Jr, ed., *Renal neoplasia,* Boston, Little, Brown & Co., pp. 13-40

Hass, G. M., McDonald, J. H., Oyasu, R., Battifora, H.A. & Paloucek, J. T. (1968) Renal neoplasia induced by combinations of dietary lead subacetate and N-2 Fluorenylacetamide. In: King, J. S. Jr, ed., *Renal neoplasia,* Boston, Little Brown & Co., pp. 377-412

Hewitt, C. B. (1967) Renal carcinoma: A clinical challenge. In: King, J. S. Jr, ed., *Renal carcinoma,* Boston, Little, Brown & Co., pp. 3-12

Ishiguro, H., Beard, D., Sommer, J. R., Heine, U., de-Thé, G. & Beard, J. W. (1962) Multiplicity of cell response to the BAI strain A (myeloblastosis) avian tumor virus: I. Nephroblastoma (Wilms' tumor); gross and microscopic pathology. *J. nat. Cancer Inst., 29,* 1-40

Lucké, B. (1934) A neoplastic disease of the kidney of the frog, *Rana pipiens. Amer. J. Cancer, 20,* 352-379

Lucké, B. (1952) Kidney carcinoma of the leopard frog: a virus tumor. *Ann. N.Y. Acad. Sci., 54,* 1093-1109

Lucké, B. & Schlumberger, H. G. (1949) Induction of metastasis of frog carcinoma by increase of environmental temperature. *J. exp. Med., 89,* 269-279

Lunger, P. D. (1969) Fine structure studies of cytoplasmic viruses associated with frog tumors. In: Mizell, M., ed., *Biology of amphibian tumors,* New York, Heidelberg, Berlin, Springer-Verlag, pp. 206-309

Lunger, P., Darlington, R. W. & Granoff, A. (1965) Cell-virus relationships in the Lucké renal adenocarcinoma: An ultrastructural study. *Ann. N.Y. Acad. Sci., 126,* 289-314

McKinnell, R. G. (1965) Incidence and histology of renal tumors of leopard frogs from the North Central States. *Ann. N.Y. Acad. Sci., 128,* 85-98

McKinnell, R. G. (1969) Lucké renal carcinoma: Epidemiological aspects. In: Mizell, M., ed., *Biology of amphibian tumors,* New York, Heidelberg & Berlin, Springer-Verlag, pp. 254-260

Mostofi, F. K. (1968) Pathology and spread of renal cell carcinoma. In: King, J. S. Jr, ed., *Renal neoplasia,* Boston, Little, Brown & Co., pp. 41-86

Noble, G. K. (1954) *The biology of the amphibia.* New York, Dover Publications

Rafferty, K. A., Jr (1962) Age and environmental temperature as factors influencing development of kidney tumors in uninoculated frogs. *J. nat. Cancer Inst., 29,* 253-265

Rafferty, K. A., Jr (1963a) Effect of injected frog kidney tumor extracts on development of tumors under promoting conditions. *J. nat. Cancer Inst., 30,* 1103-1113

Rafferty, K. A., Jr (1963b) Spontaneous kidney tumors in the frog: rate of occurrence in isolated adults. *Science, 141,* 720-721

Rafferty, K. A., Jr (1964) Kidney tumors of the leopard frog: A review. *Cancer Res., 24,* 169-185

Rafferty, K. A., Jr (1965) The cultivation of inclusion-associated viruses from Lucké tumor frogs. *Ann. N.Y. Acad. Sci., 126,* 3-21

Rafferty, K. A., Jr (1968) The biology of spontaneous renal carcinoma of the frog. In: King, J. S. Jr., ed., *Renal neoplasia,* Boston, Little, Brown & Co., pp. 301-316

Rafferty, K. A., Jr & Rafferty, N. S. (1961) High incidences of transmissible kidney tumors in uninoculated frogs maintained in a laboratory. *Science, 133,* 702

Sweet, G. (1908) The anatomy of some Australian amphibia; Part I. A. The openings of the nephrostomes from the coelom; B. The connection of the vasa efferentia with the kidney. *Proc. roy. Soc. Vict., 20,* 222-249

Tweedell, K. S. (1955) Adaptation of an amphibian renal carcinoma in kindred races. *Cancer Res., 15,* 410-418

Wessing, A. von B. & Bargen, G. von (1959) Untersuchungen über einen virusbedingten Tumor bei Fischen. *Arch. ges. Virusforsch., 9,* 521-536

Zambernard, J. & Mizell, M. (1965) Viral particles of the frog renal adenocarcinoma: causative agent or passenger virus? I. Fine structure of primary tumors and subsequent intraocular transplants. *Ann. N.Y. Acad. Sci., 126,* 127-145

Lucké Tumour-associated Viruses—A Review

A. GRANOFF [1]

INTRODUCTION

It is only in recent years that members of the herpesvirus group have been recognized as candidate oncogenic agents of man and lower animals. However, the first indication that a herpesvirus might be oncogenic dates back to 1956 when Fawcett reported the presence of "virus particles" morphologically similar to herpes simplex virus in nuclear-inclusion-bearing cells of the Lucké tumour, a renal adeno-carcinoma of *Rana pipiens* occurring frequently in a wild, heterogeneous animal population (Lucké, 1934). The presence of acidophilic nuclear inclusions in certain of these tumours, and cell-free transmission experiments, had suggested a viral aetiology (Lucké, 1934, 1938) but the nature of the putative virus was unknown. Although a number of laboratories actively engaged in a variety of studies on this tumour prior to and after Fawcett's observation, it was not until the early sixties that several laboratories, primarily Rafferty's at Johns Hopkins University and mine at St. Jude Children's Research Hospital, attempted to isolate and propagate virus from Lucké tumours. In the ensuing years, several viruses were isolated and propagated *in vitro* (see review, Granoff, 1969) and additional information on the herpesvirus seen in the tumour cells was also obtained. In this review, I shall discuss these results and their relationships to the question of the viral aetiology of the Lucké tumour.

PROPERTIES OF THE HERPESVIRUS CUSTOMARILY OBSERVED IN LUCKÉ TUMOURS

The Lucké tumour and its pathology, as well as its distribution in *Rana pipiens* in nature, have been described by Rafferty [2] and McKinnell [3], and will not be discussed in depth here. Lucké (1952) first recognized that intranuclear inclusions in renal tumour cells were more frequent in over-wintering frogs than in frogs obtained during the warmer months. Fawcett (1956) was able to demonstrate the presence of herpes-type virus in tumour cells only in those tumours containing intranuclear inclusions. This observation was subsequently corroborated by a number of workers (Lunger, Darlington & Granoff, 1965; Zambernard & Mizell, 1965; Zambernard, Vatter & McKinnell, 1966) and a developmental cycle has been proposed for the virus on the basis of electron microscopic pictures (Lunger *et al.*, 1965; Stackpole, 1969). All the above reports contained detailed information on the ultrastructural aspects of the herpesvirus seen in tumour cells, hereafter referred to as Lucké herpesvirus (LHV), and this subject will not be dealt with here. A positive correlation between the presence of intranuclear inclusions, as observed by light microscopy, and the presence of herpesvirus, demonstrated by electron microscopy, has invariably been found.

Lunger (1964) was the first to extract virus from the tumour and to establish its morphological relationship to the herpesvirus group (162 capsomeres). Zambernard & Vatter (1966) used combined enzyme treatment and electron microscopy of virus-containing tumours, and presented evidence for the presence of DNA in the virus. More recently, the DNA of LHV extracted from tumours has been shown to be double-stranded with a base composition of 45–47% guanine plus cytosine (G+C) (Wagner *et al.*, 1970; Gravell, 1971).

Fig. 1a shows an electron micrograph of a thin section of an inclusion-bearing tumour cell having large numbers of typical herpesvirus particles in various stages of development in the nucleus. Fig. 1b illustrates enveloped extracellular virus, and the ultrastructure of a negatively stained nonenveloped particle is shown in Fig. 1c.

[1] Laboratories of Virology and Immunology, St. Jude Children's Research Hospital, and The University of Tennessee Medical Units, Memphis, Tennessee, USA.
[2] See p. 159 of this publication.
[3] See p. 183 of this publication.

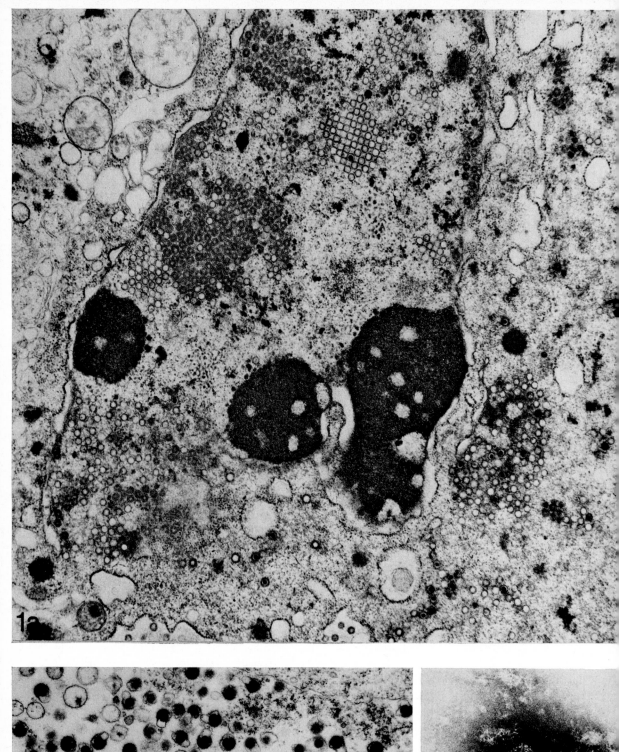

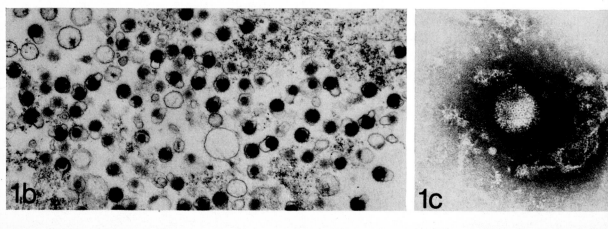

RELATIONSHIP OF INTRANUCLEAR INCLUSIONS AND VIRUS PRODUCTION TO TEMPERATURE

The relationship between temperature and the presence or absence of herpesvirus in these tumours is of particular interest and deserves further comment. The first well documented study on the temperature dependence of inclusion-body formation was made by Roberts (1963), who found that the renal tumours of frogs held at 20–25°C never had typical intranuclear inclusions whereas those of frogs held at 5°C did. The first experimental approach to this problem was by Rafferty (1965), who used biopsy techniques to demonstrate the development of intranuclear inclusions in the tumours of frogs after low-temperature treatment. Mizell, Stackpole & Halperen (1968) and Mizell, Stackpole & Isaacs (1969), by use of the anterior eye chamber of *R. pipiens* as well as of other species, followed the development of virus by electron microscopy in tumours transplanted to the eye chambers. The animals were subsequently held at low temperature and the transplanted tissue was examined for the presence of virus at various intervals thereafter. Rafferty's and Mizell's results supported the hypothesis that inclusion formation and virus production are temperature dependent. Granoff and Darlington (1969) have found that urine from some tumour-bearing frogs maintained at 4°C contains large amounts of herpesvirus while urine from tumour-bearing frogs held at 25°C does not. Moreover, ascitic fluid from tumour-bearing frogs held at low temperature also contains herpesvirus (Naegele, Granoff & Darlington [1].)

By the use of organ cultures of noninclusion tumours with subsequent treatment at low temperature (9°C), Morek & Tweedell [2] have observed the development of intranuclear inclusions in tumour cells. Recently, Breidenbach *et al.* (1971) reported the induction of herpesvirus as viewed by electron microscopy in tissue culture explants of noninclusion Lucké tumours that had been maintained *in vitro* at 7.5°C. These last two experiments indicate that the intact host is not required for induction of virus production at low temperature.

[1] Unpublished data.
[2] Personal communication.

It is clear from the evidence thus far accumulated that replication of LHV is dependent on temperature both *in vivo* and *in vitro*. Although the mechanism of this dependence is unknown, its similarity to lysogeny has been pointed out (Fawcett, 1956; Mizell & Zambernard, 1965; Mizell, Stackpole & Isaacs, 1969). However, the full significance of the rôle of temperature in virus replication and Lucké tumour formation remains to be discovered.

TUMOUR INDUCTION BY CELL-FREE TUMOUR EXTRACTS

When Rafferty (1964) reviewed the status of transmission experiments with cell-free preparations of Lucké tumours, the experiments had been carried out in adult frogs and the results were generally equivocal. No definitive conclusions could be reached regarding the cell-free transmission of the tumour, particularly as related to the presence or absence of virus in the inocula.

The first significant advance in cell-free transmission experiments since Lucké's original experiments was made by Tweedell (1967) who demonstrated that a high percentage of *R. pipiens* embryos or larvae receiving injections of cytoplasmic fractions of inclusion-containing Lucké tumours developed typical renal tumours as they reached metamorphosis. The material injected contained large numbers of herpesvirus as determined by electron microscopy, and filtrates of this material retained their activity, although this was somewhat diminished. Fractions of normal adult *R. pipiens* kidneys or of tumours not containing intranuclear inclusions failed to induce tumours.

The experiments of Tweedell conclusively demonstrated the induction of tumours by some subcellular factor, presumably a virus. A cytoplasmic fraction from virus-containing Lucké tumours, purified by zonal centrifugation and containing predominantly enveloped herpesvirus, has also been shown to have oncogenic activity (Mizell, Toplin & Isaacs, 1969; Mizell, 1969). The authors suggested that the viral envelope is required for activity, as is believed to be the case with other members of the herpesvirus group (Darlington & Moss, 1969).

Fig. 1a. A thin section of an inclusion-bearing Lucké tumour cell with typical herpesvirus particles in various stages of development in the nucleus. ×19 400.
b. Enveloped virions found extracellularly. ×24 000.
c. A negatively stained nonenveloped particle showing typical herpesvirus morphology. ×110 000.

A noteworthy observation that conflicts with our present ideas of the relationship of temperature to virus production was made by Tweedell (1969). Oocytes were exposed to a virus-containing cytoplasmic fraction by intraperitoneal injection of frogs during ovulation. The eggs were then fertilized, and the young adults developed renal tumours 5 to 12 months later. These tumours were frequently metastatic and intranuclear inclusions were "sometimes" present. Intranuclear inclusions occurred in the tumour cells of these animals, which were bred at 19–20°C, without low-temperature incubation.

The oncogenic activity of herpesvirus-containing tumour fractions certainly suggests a rôle for this virus in tumour formation. However, since Lucké tumours have been found to contain other viruses, as described below, and since herpesvirus has been detected only by electron microscopy, the presence of other viruses in amounts too low for detection by electron microscopic examination cannot be ruled out.

ANTIGENIC RELATIONSHIP OF LHV TO OTHER HERPESVIRUSES

Some information is available on the antigenic relationship of LHV to other herpesviruses. Microimmunodiffusion reactions suggest that LHV and the Epstein-Barr herpesvirus associated with Burkitt tumours share at least one antigen (Fink, King & Mizell, 1968, 1969; Kirkwood et al., 1969). Interestingly, rabbit anti-LHV was reported also to precipitate nucleocapsid preparations of herpes simplex virus and several strains of cytomegalovirus (Kirkwood et al., 1969). Apparently the antiserum used detected a group-specific antigen of herpesviruses, presumably located on the virus nucleocapsid. Naturally occurring antibody to LHV was found also in both normal (9/115) and tumour-bearing (4/5) frogs. Additional information on antigenic relationships among various herpesviruses was presented by Kirkwood et al. at this Symposium.[1]

Unfortunately, the hope that candidate oncogenic herpesviruses would have distinctive and shared antigenic components appears not to be realized, unless

---

[1] See p. 479 of this publication.

it is suggested that the herpesviruses tested (herpes simplex and cytomegalovirus) are also associated with malignancies hitherto unrecognized as virus-induced. Herpes simplex type 2 has been implicated on a sero-epidemiological basis as a causative factor in cervical carcinoma (Rawls et al., 1968).

ISOLATION, PROPAGATION, AND CHARACTERIZATION OF VIRUSES FROM LUCKÉ TUMOURS

Polyhedral cytoplasmic deoxyribovirus (PCDV)

The first isolation of an amphibian virus was made from plaques appearing spontaneously in monolayer cultures of normal frog kidney cells overlaid with agar (Granoff, Came & Rafferty, 1965). Subsequently, additional isolations of virus were made from normal cultured adult frog kidney cells, liver homogenates of normal or tumour-bearing frogs, and homogenates of Lucké tumours (Granoff et al., 1965; Granoff, Came & Breeze, 1966). One of these isolates from a Lucké tumour, frog virus 3 (FV 3), has been studied in some detail, and I have recently reviewed the properties of this agent and of isolates made from other amphibia (Granoff, 1969). In marked contrast to the herpesvirus found in tumour cells, this virus is a polyhedral cytoplasmic deoxyribovirus not fitting into any current animal virus classification scheme. In one experiment the incidence of virus isolation from normal liver was 10% (3/30) and from Lucké tumours 26% (5/19) (Granoff, Gravell & Darlington, 1969). Careful electron microscopic examination of tumours has revealed that this virus can be visualized in about 10% of tumours examined (Lunger, 1969). However, the virus appears limited to stromal cells and is not seen in tumour cells. An electron micrograph of a thin section of a fathead minnow cell infected with FV 3 is shown in Fig. 2. A cytoplasmic virus synthesis site has displaced the nucleus, and several loosely packed virus crystals are present in the cytoplasm. Virus particles can be seen budding from cytoplasmic membranes, where they acquire a cell-derived envelope. Virions in the cytoplasm measure about 120×130 nm, and Fig. 2 (insert) shows a negative stain of an unenveloped virus particle. Attempts to visualize ultrastructural details of the capsid of the

---

Fig. 2. A thin section of a fathead minnow cell infected with the **FV 3** isolate of **PCDV**. A cytoplasmic synthesis site has displaced the nucleus and several virus crystals are present in the cytoplasm.   ×14 000.
Insert. A negatively stained unenveloped FV 3 virion.   ×100 000.

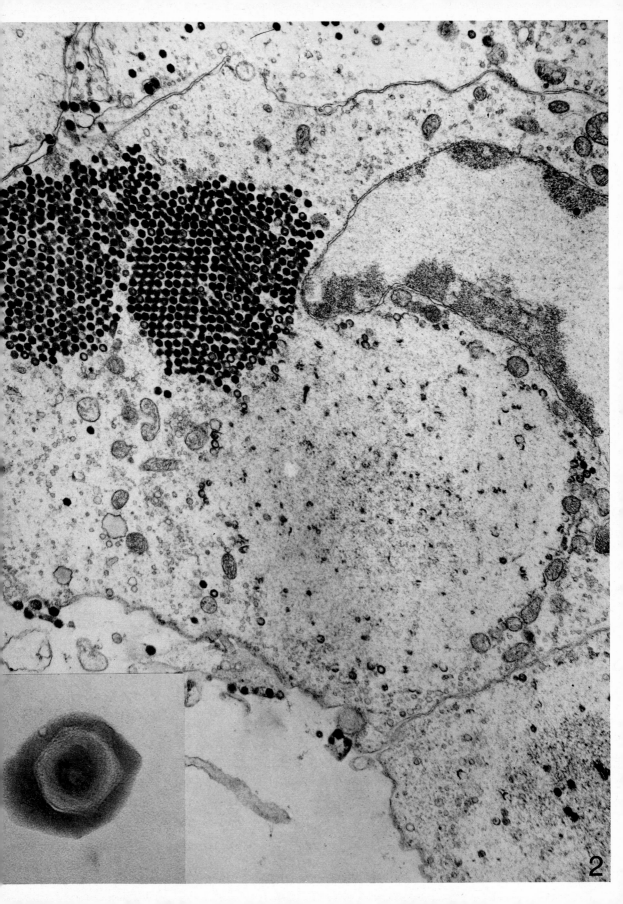

2

virus have been unsuccessful, although it appears to have cubic symmetry. The viral DNA has a G+C content of 53–54% and a molecular weight of 130×10^6 Daltons (Maes & Granoff, 1967; Houts, Gravell & Darlington, 1970). The distinction between this agent and LHV is quite clear.

Since this virus does occur with a relatively high frequency in Lucké tumours, attempts were made to induce tumours in frogs by the embryo inoculation method of Tweedell. All embryos died within 15 days, the time of death being dose-dependent (Tweedell & Granoff, 1968). Larvae (17–26 mm) were killed by high concentrations ($2.7–4.5 \times 10^6$ PFU) of virus but survived lower ones (10^5 PFU), and survivors carried through metamorphosis did not develop renal tumours. Adult R. pipiens infected with 10^8 PFU also failed to develop tumours by eight months after inoculation. All indications to date are that PCDV plays no aetiological role in Lucké tumour formation, but the possibility that it may act as a helper virus cannot as yet be ruled out. For example, an extract of Lucké tumour that contains LHV, as determined by electron microscopy, may also contain PCDV although in a concentration too low to be detected with the electron microscope. The presence of PCDV can sometimes be established only by isolation and growth in tissue culture (Granoff[1]).

Herpesvirus

The isolation of a second distinct virus from homogenates of kidney tumours and from the pooled urines of tumour-bearing frogs was reported in 1965 by Rafferty. These isolates were designated frog viruses 4 to 7. In contrast to PCDV, which has a broad host range in cultured cells, Rafferty's isolates multiplied only in a frog embryo cell line derived from R. sylvatica. Virus multiplication was slow at 25°C and was accompanied by a cytopathic effect (CPE) characterized by elongation of cells and eventual detachment from the glass. Lightly eosinophilic intranuclear inclusions containing DNA developed as a result of virus infection. The isolates were obviously different from those we had been

working with, and the cytopathic effect was certainly suggestive of a herpesvirus. Unfortunately, work on the agent was not possible for several years because of loss of the embryo cell line that supported its replication. Several years later, using two R. pipiens embryo cell lines (RPE) (Freed, Mezger-Freed & Shatz, 1969), we were able to grow the virus and to characterize it (Gravell, Granoff & Darlington, 1968). The isolate used was FV 4 derived from the pooled urine of tumour-bearing frogs.

A cytopathic effect in R. pipiens embryo cells was observed 10–21 days after infection at 25°C; it was characterized by elongation and contraction of infected cells followed by rounding, vacuolization, enlargement of nuclei, and polykaryon formation. Typical intranuclear inclusions, characteristic of herpesvirus infection, were formed and contained DNA. The virus can be quantitatively assayed in RPE cells and, more recently, it has been shown that an adult frog kidney cell line can be used for this purpose (Gravell, 1971). Fourteen to fifteen days are required for plaque formation. Virus yields in liquid media are generally low, ranging from 1×10^5 to 2×10^6 PFU/ml; this represents, on the average, 1–10 PFU/cell. Cell-free infectious virus can be obtained by brief exposure of infected cells to sonic vibration. Exposure to ethyl ether or to pH 3 rapidly inactivates the virus. Although it has not been possible to determine single-cycle growth curves, because of low virus titres, the virus does appear to have a rather long growth cycle compared to other herpesviruses. We have also observed virus multiplication at 10°C, but at a much slower rate than at 25°C.

Electron microscopy of infected RPE cells and negative staining of cell-free virus have established the site of synthesis and the morphology of this agent. The results are consistent with a herpesvirus. Fig. 3a shows an infected RPE cell containing a nuclear crystalline array of typical herpesvirus in various stages of development. Fig. 3b illustrates extracellular enveloped virus and Fig. 3c a negative stain of an unenveloped virus, sedimented from infected culture fluid. The particle contains 162 capsomeres. Fig. 3 may be compared with Fig. 1, which shows similar virus particles in a Lucké

---

[1] Unpublished data.

Fig. 3a. A thin section of an FV 4-infected RPE cell containing a nuclear crystalline array of typical herpesvirus in various stages of development. ×39 500.
b. Extracellular enveloped FV 4. ×20 000.
c. A negative stain of an unenveloped FV 4 particle sedimented from infected fluid. ×180 000.

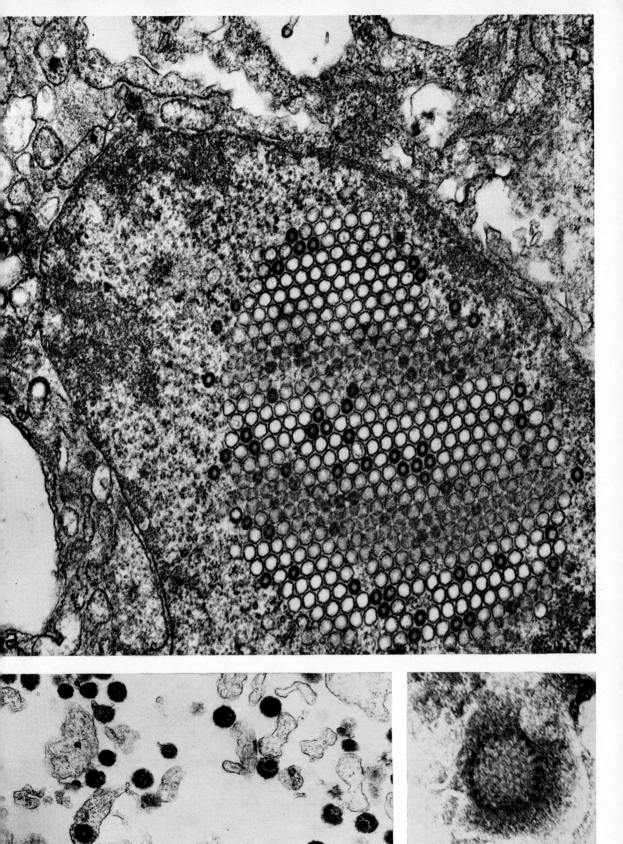

tumour cell. It is quite clear that FV 4 and LHV, the herpesvirus seen in Lucké tumour cells, are indistinguishable morphologically and in their nuclear site of synthesis.

In early experiments, Rafferty (1965) was unable to induce tumours in *R. pipiens* tadpoles exposed to virus in the water in which the animals were kept. In collaboration with Tweedell,[1] we have tested FV 4 for tumorigenic activity in developing *R. pipiens* embryos and larvae; animals inoculated with 400 PFU of FV 4 failed to develop tumours (Granoff *et al.*, 1969).

Studies were recently initiated in my laboratory to determine whether FV 4 and the herpesvirus extracted from naturally occurring Lucké tumours were identical (Gravell, 1971). FV 4 was grown in an adult frog kidney cell line and purified by differential and isopyknic centrifugation. LHV was extracted from Lucké tumours and similarly purified. The base composition of the viral DNA, extracted by standard procedures, was estimated by two independent methods: thermal denaturation profiles and equilibrium buoyant density in caesium chloride. The DNA from LHV contained 45–47% G+C and that from FV 4 54–56% G+C. The value for LHV is in excellent agreement with the value reported by Wagner *et al.* (1970). In spite of the differences in G+C content, the DNA from FV 4 and that from LHV could still possess considerable base sequence homology, but no such homology between FV 4 and LHV could be detected by DNA-DNA hybridization.

FV 4 and LHV were also compared immunologically. Antiserum prepared in rabbits against LHV did not suppress plaque production by FV 4, whereas homologous FV 4 antiserum completely suppressed it. Although the corresponding experiment with LHV was not possible (namely, determination of the capacity of LHV antiserum to neutralize homologous virus) since this virus has not been propagated *in vitro,* the serum used reacted positively with extracted herpesvirus in a micro-Ouchterlony immunoprecipitation test (Kirkwood *et al.,* 1969). It is

quite clear from these experiments, the results of which are summarized in the Table, that FV 4 and the herpesvirus physically extracted from Lucké tumours are different viruses. Whether FV 5 to 7 are similar

Comparison of two herpesviruses isolated from Lucké tumour-bearing frogs

Virus	Base composition (% G+C)	Base sequence homology [1]		Immunological relatedness [2]	
		FV 4	LHV	FV 4	LHV
FV 4	54–56	+	—	+	—
LHV	45–47	—	NT	NT	NT

[1] By DNA-DNA hybridization.
[2] By neutralization by specific antiserum.

to FV 4 or to LHV remains to be determined, since no additional information, other than on the cytopathic effect originally reported by Rafferty (1965), is available on these particular isolates.

Papova-like virus

Because of the susceptibility of RPE cells to infection by FV 4 we attempted to make a fresh isolation of herpesvirus from several tumour homogenates. Of the six tumours chosen, five contained LHV, as determined by the presence of intranuclear inclusions or of virus by electron microscopy; the sixth tumour was free of detectable herpesvirus by these criteria. Homogenates of each tumour were also tested for the presence of PCDV by isolation procedures (Granoff *et al.,* 1966) and were free of this agent. The results of several of these experiments (carried out by Gravell) have been reported previously (Granoff *et al.,* 1969; Granoff, 1969) and will be reviewed only briefly here.

One of the RPE cell lines, when inoculated with a 10% homogenate of tumour, showed a moderate CPE at 23–25°C; the effect was characterized by stranding of cells and clumping, and could be serially transmitted by cell-free extracts. Somewhat different cell alterations were obtained with another RPE cell line, characterized by little discernible cellular change

---

[1] Department of Biology, University of Notre Dame, College of Science, Notre Dame, Indiana, USA.

Fig. 4. A thin section of a Lucké tumour cell showing small particles about 46 nm in diameter in the nucleus, which also contains herpesvirus. ×68 000.
Insert. A negative stain of virus particles about 45 nm in diameter obtained from the growth medium of RPE cells infected with a Lucké tumour homogenate. ×95 000.

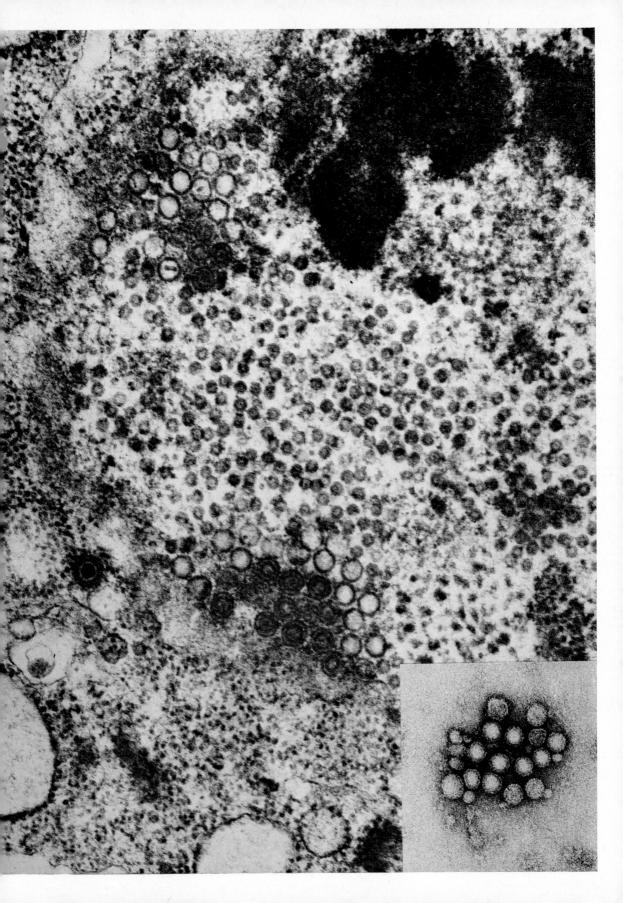

but a striking difference in cell morphology. The cells, originally epithelioid, assumed a more fibroblastic nature and grew in whorls. Although these cells could be serially passaged several times, they multiplied less vigorously than uninfected cells and did not multiply noticeably when placed in the anterior eye chamber of *R. pipiens* (Gravell and Granoff [1]). Electron microscopic examination of thin sections of both infected RPE cell lines failed to reveal evidence of virus infection, but the presence of an apparently icosahedral virus about 45 nm in diameter was observed by negative staining of infected culture medium (Fig. 4; insert). Although there is some indication of viral capsomeres, it has not so far been possible to determine their number. Preliminary chemical analysis (Gravell [1]) indicates that the virus contains DNA. The limited amount of evidence so far available (size, morphology, nucleic acid) suggests that this agent, which has been isolated from both inclusion and noninclusion tumours, resembles members of the papova group of viruses.

High concentrations of the virus when inoculated into developing frog embryos failed to elicit tumour formation (Gravell, Tweedell & Granoff [1]). In addition, inoculation of frog embryos with mixtures of FV 4 and the papova-like virus also failed to induce tumours.

It is worth pointing out that 45-55-nm particles have been observed in thin sections of Lucké tumour cells that also contain herpesvirus (Lunger *et al.,* 1965; Lunger, 1966; Stackpole & Mizell, 1968). These particles have generally been considered to be in some way related to herpesvirus replication; the interpretation that they represent another virus has been less widely accepted. Similar 46-nm particles have been observed by us in herpesvirus-containing Lucké tumour cells (Granoff *et al.,* 1969). Fig. 4 is an electron micrograph of a thin section of a Lucké tumour cell showing small particles about 46 nm in diameter in the nucleus, which also contains herpesvirus. We believe these small particles are identical with the papova-like virus isolated from RPE cells inoculated with tumour extracts. In several instances we have isolated PCDV from a tumour that, by electron microscopy, also contained herpesvirus and the 46-nm particles. Thus, certain tumours in some instances may contain three distinct viruses and possibly four, since FV 4 and LHV cannot be distinguished morphologically.

---

[1] Unpublished data.

CONCLUSIONS

It is important to recognize the fact that none of our attempts to isolate LHV have been successful. The reasons for this are still obscure but the possibility that LHV is defective or that susceptible host cells have not been tested may be considered. Although the production of tumours by inoculation of semipurified herpesvirus obtained from Lucké tumours should be rather convincing evidence for its role in tumour formation, proof that the subcellular inducer is solely the herpesvirus is lacking. As mentioned earlier, electron microscopy has been the sole method for determining virus content (Tweedell, 1967; Mizell, 1969; Mizell, Toplin & Isaacs, 1969). In view of the fact that other viruses are present in Lucké tumours, sometimes in numbers too few to be detected by electron microscopy, the conclusions to be drawn from the transmission experiments so far performed are subject to certain reservations.

The association of at least three other viruses with Lucké tumours (PCDV, FV 4, and the papova-like virus) suggests that the relationship of virus to tumour may not be simple but may involve a defective virus-helper virus-cell interaction. Our attempts to solve this problem have been directed towards the isolation and propagation of the aetiological agent(s) in tissue culture, followed by their characterization, particularly from the point of view of oncogenic potential. Without this, the aetiology of the Lucké tumour will remain in question. The fact that two herpesviruses may be associated with tumours or tumour-bearing frogs provides a further challenge to us to achieve this goal. Our inability to isolate another herpesvirus of the FV 4 type from tumours is puzzling, especially since the cells used were susceptible to infection with Rafferty's FV 4. Perhaps FV 4 is not as widely distributed in frog populations as is LHV or PCDV. It is also clear that the herpesviruses of *R. pipiens* will have a restricted host range, so that the search for additional susceptible host cells must continue.

The Lucké tumour provides one of the few experimental systems for the study of the oncogenicity of a herpesvirus—if a herpesvirus proves unequivocally to be the inducer of the tumour. Unlike the tumours induced by other DNA viruses (except the benign virus-induced papilloma), the Lucké tumour occurs spontaneously in nature and is not a laboratory phenomenon. Additionally, most virus-induced tumours, whether caused by DNA or RNA viruses, are usually lymphoproliferative or sarcomas (with the exception of the mouse mammary tumour). Thus the Lucké tumour offers a unique opportunity to study the natural history of what appears to be a virus-induced tumour, with the additional advantage of permitting a broad range and variety of experimental designs not possible with other systems (Habel, 1969). With the development of tissue culture systems that support the replication of viruses from *R. pipiens*, with increased knowledge of the nature of the different viruses isolated from frogs, and with the application of the *in vivo* test for tumorigenicity, the precise role of virus or viruses in the aetiology of the Lucké tumour should be clarified in the near future.

ACKNOWLEDGEMENTS

I am indebted to Dr R. W. Darlington for the electron micrographs and to Mr G. Edwin Houts for the negatively stained PCDV particle.

This work was supported by Public Health Research Grant CA 07055, Childhood Cancer Research Center Grant CA 08480, and Multidisciplinary Cancer Research Training Grant CA 05176 from the National Cancer Institute, Grant DRG 1073 from the Damon Runyon Memorial Fund for Cancer Research, and by ALSAC.

REFERENCES

Breidenbach, G. P., Skinner, M. S., Wallace, J. H. & Mizell, M. (1971) *In vitro* induction of a herpes-type virus in "summer-phase" Lucké tumor explants. *J. Virol.*, **7**, 679-682

Darlington, R. W. & Moss, L. H. (1969) The envelope of herpesvirus. *Progr. med. Virol.*, **11**, 16-45

Fawcett, D. W. (1956) Electron microscope observations of intracellular virus-like particles associated with the cells of the Lucké renal adenocarcinoma. *J. biophys. biochem. Cytol.*, **2**, 725-742

Fink, M. A., King, G. S. & Mizell, M. (1968) Preliminary note: Identity of a herpesvirus antigen from Burkitt lymphoma of man and the Lucké adenocarcinoma of frogs. *J. nat. Cancer Inst.*, **41**, 1477-1478

Fink, M. A., King, G. & Mizell, M. (1969) Reactivity of serum from frogs and other species with a herpesvirus antigen extracted from a Burkitt lymphoma cultured cell line. In: *Recent results in cancer research*, New York, Springer-Verlag, pp. 358-364

Freed, J. J., Mezger-Freed, L. & Shatz, S. A. (1969) Characteristics of cell lines from haploid and diploid anuran embryos. In: *Recent results in cancer research*, New York, Springer-Verlag, pp. 101-111

Granoff, A. (1969) Viruses of amphibia. *Curr. Top. Microbiol. Immunol.*, **50**, 107-137

Granoff, A., Came, P. E. & Breeze, D. C. (1966) Viruses and renal carcinoma of *Rana pipiens*. I. The isolation and properties of virus from normal and tumor tissue. *Virology*, **29**, 133-148

Granoff, A., Came, P. E. & Rafferty, K. A., Jr (1965) The isolation and properties of viruses from *Rana pipiens*: Their possible relationship to the renal adenocarcinoma of the leopard frog. *Ann. N.Y. Acad. Sci.*, **126**, 237-255

Granoff, A. & Darlington, R. W. (1969) Viruses and renal carcinoma of *Rana pipiens*. VIII. Electron microscopic evidence for the presence of herpesvirus in the urine of a Lucké tumor-bearing frog. *Virology*, **38**, 197-200

Granoff, A., Gravell, M. & Darlington, R. W. (1969) Studies on the viral etiology of the renal adenocarcinoma of *Rana pipiens* (Lucké tumor). In: *Recent results in cancer research*, New York, Springer-Verlag, pp. 279-295

Gravell, M. (1971) Viruses and renal carcinoma of *Rana pipiens*. X. Comparison of herpes-type viruses associated with Lucké tumor-bearing frogs. *Virology*, **43**, 730-733

Gravell, M., Granoff, A. & Darlington, R. W. (1968) Viruses and renal carcinoma of *Rana pipiens*. VII. Propagation of a herpes-type frog virus. *Virology*, **36**, 467-475

Habel, K. (1969) Summation and perspectives. In: *Recent results in cancer research*, New York, Springer-Verlag, pp. 482-484.

Houts, G. E., Gravell, M. & Darlington, R. W. (1970) Base composition and molecular weight of DNA from a frog polyhedral cytoplasmic deoxyribovirus. *Proc. Soc. exp. Biol. (N.Y.)*, **135**, 232-236

Kirkwood, J. M., Geering, G., Old, L. J., Mizell, M. & Wallace, J. (1969) A preliminary report on the serology of Lucké and Burkitt herpes-type viruses: A shared antigen. In: *Recent results in cancer research*, New York, Springer-Verlag, pp. 365-367

Lucké, B. (1934) A neoplastic disease of the kidney of the frog, *Rana pipiens*. *Amer. J. Cancer*, **20**, 352-379

Lucké, B. (1938) Carcinoma in the leopard frog: Its probable causation by a virus. *J. exp. Med.*, **68**, 457-468

Lucké, B. (1952) Kidney carcinoma in the leopard frog: A virus tumor. *Ann. N.Y. Acad. Sci.,* **54**, 1093-1109

Lunger, P. D. (1964) The isolation and morphology of the Lucké frog kidney tumor virus. *Virology,* **24**, 138-145

Lunger, P. D. (1966) A new intranuclear inclusion body in the frog renal adenocarcinoma. *J. Morph.,* **118**, 581-588

Lunger, P. D. (1969) Fine structure studies of cytoplasmic viruses associated with frog tumors. In: *Recent results in cancer research,* New York, Springer-Verlag, pp. 296-309

Lunger, P. D., Darlington, R. W. & Granoff, A. (1965) Cell-virus relationships in the Lucké renal adenocarcinoma: An ultrastructure study. *Ann. N.Y. Acad. Sci.,* **126**, 289-314

Maes, R. & Granoff, A. (1967) Viruses and renal carcinoma of *Rana pipiens.* IV. Nucleic acid synthesis in frog virus 3-infected BHK 21/13 cells. *Virology,* **33**, 491-502

Mizell, M. (1969) State of the art: Lucké renal adenocarcinoma. In: *Recent results in cancer research,* New York, Springer-Verlag, pp. 1-25

Mizell, M., Stackpole, C. W. & Halperen, S. (1968) Herpes-type virus recovery from "virus-free" frog kidney tumors. *Proc. Soc. exp. Biol. (N.Y.),* **127**, 808-814

Mizell, M., Stackpole, C. W. & Isaacs, J. J. (1969) Herpes-type virus latency in the Lucké tumor. In: *Recent results in cancer research,* New York, Springer-Verlag, pp. 337-347

Mizell, M., Toplin, I. & Isaacs, J. J. (1969) Tumor induction in developing frog kidneys by a zonal centrifuge purified fraction of the frog herpes-type virus. *Science,* **165**, 1134-1137

Mizell, M. & Zambernard, J. (1965) Viral particles of the frog renal adenocarcinoma: Causative agent or passenger virus? II. A promising model system for the demonstration of a "lysogenic" state in a metazoan tumor. *Ann. N.Y. Acad. Sci.,* **126**, 146-169

Rafferty, K. A., Jr (1964) Kidney tumors of the leopard frog: A review. *Cancer Res.,* **24**, 169-185

Rafferty, K. A., Jr (1965) The cultivation of inclusion-associated viruses from Lucké tumor frogs. *Ann. N.Y. Acad. Sci.,* **126**, 3-21

Rawls, W. E., Tompkins, W. A. F., Figueroa, M. E. & Melnick, J. L. (1968) Herpesvirus type 2: Association with carcinoma of the cervix. *Science,* **161**, 1255-1256

Roberts, M. E. (1963) Studies on the transmissibility and cytology of the renal carcinoma of *Rana pipiens. Cancer Res.,* **23**, 1709-1714

Stackpole, C. W. (1969) Herpes-type virus of the frog renal adenocarcinoma. I. Virus development in tumor transplants maintained at low temperature. *J. Virol.,* **4**, 75-93

Stackpole, C. W. & Mizell, M. (1968) Electron microscopic observations on herpes-type virus-related structures in the frog renal adenocarcinoma. *Virology,* **36**, 63-72

Tweedell, K. S. (1967) Induced oncogenesis in developing frog kidney cells. *Cancer Res.,* **27**, 2042-2052

Tweedell, K. S. (1969) Simulated transmission of renal tumors in oocytes and embryos of *Rana pipiens.* In: *Recent results in cancer research,* New York, Springer-Verlag, pp. 229-239

Tweedell, K. S. & Granoff, A. (1968) Viruses and renal carcinoma of *Rana pipiens.* V. Effect of frog virus 3 on developing frog embryos and larvae. *J. nat. Cancer Inst.,* **40**, 407-410

Wagner, E. K., Roizman, B., Savage, T., Spear, P. G., Mizell, M., Durr, F. E. & Sypowicz, D. (1970) Characterization of the DNA of herpesviruses associated with Lucké adenocarcinoma of the frog and Burkitt lymphoma of man. *Virology,* **42**, 257-261

Zambernard, J. & Mizell, M. (1965) Virus particles of the frog renal adenocarcinoma: Causative agent or passenger virus? I. Fine structure of primary tumors and subsequent intraocular transplants. *Ann. N.Y. Acad. Sci.,* **126**, 127-145

Zambernard, J. & Vatter, A. E. (1966) The fine structural cytochemistry of virus particles found in renal tumors of leopard frogs. I. An enzymatic study of the viral nucleoid. *Virology,* **28**, 318-324

Zambernard, J., Vatter, A. E. & McKinnell, R. G. (1966) The fine structure of nuclear and cytoplasmic inclusions in primary renal tumors of mutant leopard frogs. *Cancer Res.,* **26**, 1688-1700

Epidemiology of the Frog Renal Tumour and the Significance of Tumour Nuclear Transplantation Studies to a Viral Aetiology of the Tumour—A Review

R. G. McKINNELL[1] & V. L. ELLIS [1]

INTRODUCTION

Much is known about the symmetry and fine structure of the herpesvirus associated with the renal tumours of the North American leopard frog, *Rana pipiens*. We can isolate and make relatively pure virus preparations. We know something about the DNA of the virus and can speak knowledgeably concerning the effect of chilling in evoking virus replication from latency. However, viruses do not have cancer; frogs do. Frogs for one thousand miles across the northern United States and southern Canada are afflicted with the tumour initially described by Lucké (1934). If the frog tumour is to be related to other viral tumours, the prevalence, distribution, mode of transmission, and other epidemiological aspects of the tumour must be understood. It will be the purpose of this paper to review what is known concerning the epidemiology of the frog renal adenocarcinoma, with particular reference to virus replication in the tumours of frogs from natural populations.

Viruses have been thought to be associated with the renal adenocarcinoma of *R. pipiens* since the pioneering observations of Lucké (1938). These early suspicions were strengthened by an electron microscope study by Fawcett (1956). Fawcett studied four tumours by electron microscopy, and described cells containing "virus-like particles". Fawcett and others have reported that the frog tumour has a pleomorphic cytology, characterized chiefly by the presence or absence of virus particles (Lunger, Darlington & Granoff, 1965). The pleomorphism revealed by electron microscopy corresponds well with the variation in tumour cell morphology reported in light microscope studies (Roberts, 1963).

Laboratory studies indicated that a tumour of one cytological type was easily converted into the other. Thus, virus-containing tumour cells lose the viruses and become "virus-free" with increase in temperature (Zambernard & Vatter, 1966). Conversely, "virus-free" tumours readily replicate virus particles, detectable with the electron microscope, after chilling (Tweedell, 1967; Mizell, Stackpole & Halperen, 1968; Stackpole, 1969; Collins & Nace, 1970).

The environment of frogs in the northern United States, and particularly in Minnesota, is as changeable as the tumour fine structure. *R. pipiens* starts life in a breeding pond. Breeding ponds are shallow, evanescent bodies of water that are usually dry by midsummer. The ponds warm quickly in the spring sun but chill rapidly at night. Frogs in Minnesota move to the breeding ponds from lakes in mid-April. Amplexus occurs within a few days. The eggs hatch in the breeding ponds, and metamorphosis is complete by the time the ponds dry up. Juvenile and adult frogs forage in field and meadow throughout the remaining warm period that lasts until approximately mid-October. Frogs then congregate at the margins of certain lakes, and when the temperature is appropriate, enter the lakes and remain there until the following April. They emerge after their cold weather residence in the lakes to go again to breeding ponds and repeat the cycle. The behaviour of *R. pipiens* is thus beautifully adapted for survival in boreal Minnesota. Although the air temperature falls to many degrees below freezing during the cold season, the lakes never freeze solid,

[1] Department of Zoology, University of Minnesota, Minneapolis, Minnesota, USA.

thus making survival possible under an ice sheet. We suggest that the replication and dispersal of the presumed aetiological agent of the renal adenocarcinoma is attuned to the varied environment of its host.

REASONS FOR MAKING AN ENVIRONMENTAL STUDY OF FROG TUMOURS

Since frogs are commercially available and cheap, and since virus growth can be controlled in the laboratory by a simple temperature change, is an environmental study of any value ? We are interested in knowing whether laboratory phenomena have equivalent expressions in natural populations of frogs. In addition, we seek information that is not easily available from laboratory experiments. No study has been made of frog tumours obtained during the cold period. Viral synthesis has been studied in laboratory-chilled frogs, but a number of questions remain unanswered, including the following: When does virus replication begin in the tumours of a natural population ? Do all tumours replicate viruses during the cold season ? Are mature virions released throughout the cold season or are virions released only after emergence of the host frog from lakes in the spring ? Is virion release pre- or post-amplexus ? If virion release is pre-amplexus, contagion may be either vertical (by gamete infection) or horizontal (spread of infection to other adults and young). Transovarial transmission lacks plausibility if virion release occurs after amplexus. Is virion production over before blood-sucking insects become abundant in the spring; i.e., is an insect vector possible ? How are these questions to be answered without scrutiny of natural populations ?

Although it was thought earlier that renal tumours were found in frogs primarily from the northeastern United States (Dmochowski, 1959; Medes & Reimann, 1963), it was subsequently shown that such tumours were abundant in frogs from the north-central parts of the United States (McKinnell, 1965; Tweedell, 1965). The early studies were of frogs obtained from commercial dealers in amphibians. We wished to find tumours in natural populations of frogs because information on the pedigree and the region of origin is not available when frogs are bought from dealers. American frog dealers buy from commercial collectors in the United States, Canada, Mexico, and Japan. Although the Japanese frog can be distinguished from North American frogs, it is unlikely that a frog systematist would be able to distinguish a Canadian frog from a South Dakota

frog. In addition, tumour cell fine structure may change while the batrachian host is in the hands of a dealer as rapidly as in a tumour laboratory. Accordingly, we started gathering leopard frogs in rural Minnesota with the aim of answering some of the questions we had asked.

Another reason for our interest in the environmental aspects of the frog renal adenocarcinoma is that we are studying the capacity of nuclei from these tumours to programme for normal embryonic development. Since the tumour is pleomorphic with respect to its nucleus (among other things), and since we are interested in cell products from the tumour genome, it is of great interest to us to know what factors control tumour nuclear physiology.

MINNESOTA GEOGRAPHY

We are seeking answers through a continuing epidemiological study of spontaneous tumours

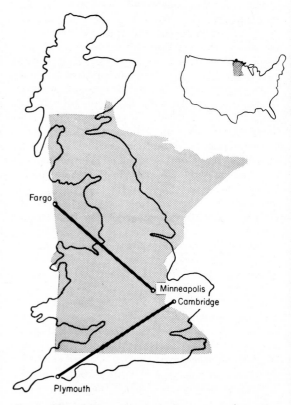

Fig. 1. Map of Minnesota drawn to same scale as map of England, Scotland and Wales, to illustrate relative sizes. Insert in upper right-hand corner shows location of Minnesota in the United States.

among Minnesota leopard frogs. Minnesota is a particularly appropriate geographical area to study frog tumour epidemiology. While it is but one of the 50 United States, it exceeds England plus Scotland in area and is only slightly smaller than the combined areas of England, Wales, and Scotland (Fig. 1). It has 15 291 lakes ten acres in size or larger. One county, Otter Tail, has 1048 lakes with a combined total of 173 851 acres (Minnesota Department of Conservation, 1968). The lakes are surrounded for

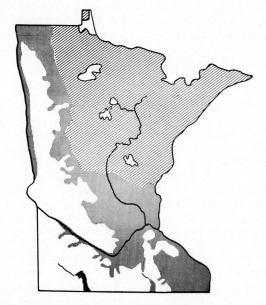

Fig. 2. Outline map of Minnesota showing generalized vegetation at time of earliest white settlement.

▨ Northern coniferous forest

▦ Deciduous forest

▢ Grassland

the most part by breeding ponds and marshy areas such that the state supports an enormous population of leopard frogs. Natural vegetation varies throughout the state. North-eastern Minnesota was originally northern coniferous forest. South-eastern Minnesota was primitively prairie. A mixed deciduous forest separated the north-eastern forests from the tall grass prairie (Fig. 2). Climate varies also. The mean daily temperature for the summer months is as low as 14°C in the north-east and 22°C in the south-east. The average date of the first fall freeze is 6 October for Minneapolis and Saint Paul but a month

earlier in northern Minnesota. Average annual rainfall varies from 30 inches in the south-eastern part of the state to only 20 inches at the opposite corner of the state. Lakes are not uniformly distributed but are found primarily in the glacial "Big Moraine" belt, which extends from south of Minneapolis and Saint Paul to Detroit Lakes in the north-west, then back eastward to Grand Rapids and Brainerd (Borchert & Yaeger, 1968). With such diversity of the environment, one would not expect to find uniformity in the prevalence, distribution and pathology of frog renal tumours.

PRELIMINARY STUDIES OF FIELD-COLLECTED FROGS

Our early studies were concerned simply with finding areas containing tumour-prone populations of frogs (McKinnell, 1966). The studies were complicated because of an apparent seasonal fluctuation of tumour prevalence. Tumour prevalence seemed to be high during the cold months when frogs were difficult to obtain and very low during the warm times of the year when frogs could be taken with ease in the field (McKinnell, 1967, 1969; McKinnell & McKinnell, 1968). After we had identified tumour-prone populations and the period of the year during which they were most readily available, we were then in a position to answer certain questions.

The first question had to do with the effect of temperature on tumours obtained from field collections. We already knew of the effect of warmth and cold on tumours in laboratory frogs. Frogs in natural populations are rarely exposed to a constant temperature, but frogs of a single population are presumably exposed to similar temperatures (i.e., although the temperature may vary, it probably affects all the frogs of a given population similarly).

Eleven renal tumours were obtained in the spring of 1967 by autopsy of frogs collected in three Minnesota counties, either from lakes, just prior to emergence, or from breeding ponds. Lake water temperature varied from 2 to 7°C. Nuclear, cytoplasmic and extracytoplasmic virus particles were observed in each of the eleven tumours (McKinnell & Zambernard, 1968).

Eleven renal tumours were obtained from frogs in late summer and early autumn of the same year. The hosts, which were collected prior to entering the lakes for the cold period, had experienced a mean minimum temperature of 7°C and a mean maximum temperature of 18°C for about two weeks prior to the time of collection. None of these tumours collected in Minnesota had virus particles detectable with the electron microscope (Zambernard & McKinnell, 1969).

CONCERNING TERMINOLOGY

Before proceeding further, perhaps a comment concerning terminology would be in order. We shall use the adjective "algid" to describe tumours that are engaged in viral replication and have inclusion bodies. Such tumours have been described elsewhere as "winter" tumours (Mizell, 1969; Collins & Nace, 1970). Winter alludes to an environmental situation but the virus-replicating phase can be induced by a temperature drop in the laboratory at *any* season. "Algid" seems to be a more appropriate term because, as we shall see below, tumours containing viruses can be taken from their natural habitat in the autumn, winter and spring. The term "algid" is derived from the Latin *algidus,* and both the Latin and its English equivalent simply mean cold. We could use "cold" but that term would perhaps lack the specific meaning that we wish to imply with the adjective "algid".

Tumours that are "virus-free" will be described by the adjective "calid". We believe that "calid" is preferable to "summer" because such tumours are found in natural populations during summer and autumn and, of course, the characteristic structure can be produced by warming a virus-containing tumour. Perhaps even more relevant to the appropriateness of the term "calid" is that "virus-free" suggests that there is a tumour phase in which the virus genome is lacking in the tumour cell. We know of no study that demonstrates the absence of the viral genome from the tumour cells. Finally, we prefer the use of "calid" to that of "summer" because naturally occurring frog renal adenocarcinoma is exceedingly rare during summer (McKinnell & McKinnell, 1968). Only two frog tumours have ever been reported in the scientific literature that were collected by a scientist during the summer (Zambernard & McKinnell, 1969), although commercially obtained Vermont frogs are reported to have a tumour frequency of about 5% during July and August (K. A. Rafferty, Jr [1]). "Calid" is derived from a Latin word (*calidus*) meaning "warm", as does its English equivalent.

A third kind of tumour will be described below. It will be designated "transitional" because, although it may contain viruses, most of the virus particles

---

[1] Personal communication.

are found in the cellular debris of tumour tubules. "Transitional" tumours are found in frogs collected during the spring.

Sometimes frogs that inhabit the water under ice during winter are referred to as "hibernating". Although the etymology of "hibernation" suggests winter, the term has a rather specific meaning to physiologists and the appropriateness of the term with respect to amphibians has been questioned (see discussion following Tester & Breckenridge, 1964). Whether one uses the term hibernation or alternative terms such as torpidity or dormancy, one should be aware that frogs in cold water are capable of quick and vigorous responses. They are not asleep during the cold season, and will swim energetically while eluding a collector.

FIELD STUDIES DURING 1970–1971

During this past season, we wanted to circumscribe more closely the time of virus replication onset, the time or times of virus release, and the time of the return of tumours to the calid state. None of this information was previously available. The following is an account of 26 spontaneous tumours collected in various localities in Minnesota from 18 October 1970 to 3 June 1971 (see Table).

Methods for electron microscopic examination of tumours

Tumours were detected by autopsy within 24 hours after collection in the field. Small fragments of tumour, about 1 mm³ in size, were fixed for one hour in 4% cold glutaraldehyde buffered with S-collidine (pH 7.3–7.6). The fixed tissue was rinsed and stored in 0.2 M S-collidine buffer. The tissue was postfixed in cold 2% osmium tetroxide buffered with S-collidine (pH 7.3 to 7.6) for one hour. The tumour fragments were then rapidly dehydrated through a graded ethanol series, cleared in propylene oxide, and embedded in Epon. Polymerization was at 60°C for 36 hours. Thin sections were cut with a Reichert OM U2 Ultramicrotome and placed on Parlodion-coated grids. The sections were stained with 2% aqueous uranyl acetate and 1% lead citrate and examined with an AEI EM 801 electron microscope.

Calid tumours

Two tumours were collected on 18 October 1970 (see Table). The fine structure of these tumours

Spontaneous renal tumours collected in 1970–1971 season

Date of collection	Collection site	Number of tumours		
		Calid	Algid	Transi-tional
18 October 1970	Diamond Lake, Kandiyohi County	2	—	—
1 December 1970	Calhoun Lake, Kandiyohi County	—	1	—
19 February 1971	Calhoun Lake, Kandiyohi County	—	1	—
18 March 1971	Glacial Lake, Pope County	—	1	—
18 March 1971	Westport Lake, Pope County	—	1	—
26 March 1971	Westport Lake, Pope County	—	1	—
26 March 1971	Terrace River, Pope County	—	7	—
13 April 1971	Diamond Lake, Kandiyohi County	—	1	—
16 April 1971	Diamond Lake, Kandiyohi County	—	2	—
23 April 1971	Diamond Lake, Kandiyohi County	—	2	—
4 May 1971	Diamond Lake, Kandiyohi County	—	1	—
12 May 1971	Block Lake, Otter Tail County	—	—	3
3 June 1971	Diamond Lake, Kandiyohi County	—	—	3
Totals		2	18	6

agreed with that described earlier by Zambernard & McKinnell (1969). We were unable to detect any virus particles in nuclei, cytoplasm, or the lumena of the renal tumours (Fig. 3), and no cytolysis was observed. Frogs at that time were clustered in a weedy area adjacent to the east side of Diamond Lake in Kandiyohi County, Minnesota. The temperature varied between 15°C and 18°C on the day of collection.

Algid tumours

The first renal tumour obtained from a hibernating frog was taken on 1 December 1970 (see Table). The tumour contained virus particles (Fig. 4). This tumour was obtained from a frog taken from Calhoun Lake in Kandiyohi County, which is 5 miles north of Diamond Lake. A comparison of the Diamond Lake collection of 18 October 1970 and the Calhoun Lake collection of 1 December 1970 would suggest that virus replication in nature can occur within 44 days. It thus occurs in natural populations in about half the time that it is reported to take in the anterior eye chamber (Stackpole, 1969) or on agar slants (Breidenbach *et al.*, 1971). Frogs will replicate viruses extremely rapidly in the laboratory under certain circumstances. Thus, Collins & Nace (1970) reported that one tumour contained "intranuclear inclusion bodies" after only 11 days at 4°C. Whether or not virus replication, as ascertained with electron microscopy, occurs as soon as 11 days among frogs that have entered lakes for the cold season awaits our further study.

Eighteen tumours were collected between December 1970 and May 1971 (see Table). The spontaneous, naturally occurring tumours contained viruses in various stages of maturation from the time that they were first detected (December) until the last tumour was obtained prior to emergence from hibernation. It should be noted that *mature* viruses were detected in December and virions were noted in the lumena of tumour tubules (Fig. 4). Crystals of incomplete viruses, thought by some to be viruses that will never mature (Roizman, 1969), were observed for the first time in the tumours collected on 18 March 1971. The tumour cells that contained the crystals also contained mature viruses in the cytoplasm, suggesting that crystal formation is not incompatible with the capacity to form infectious particles.

All cold weather frogs contained nuclear and cytoplasmic virus particles that were both structured and of a size to allow for presumptive identification as herpes-type viruses. In one of the cold weather tumours, *in addition* to the omnipresent herpesviruses was an unidentified nuclear virus about one-half the diameter of the former (Fig. 5). We anticipated that, in an examination of a large series of tumours, other viruses would be present. Indeed, we would have been surprised had we not found other viruses. Just as we are not dismayed that humans are subject to a variety of viral infections, so we are not unhappy to find other viruses in frogs. We do not consider the small virus to be a "fly in the ointment" (Roizman, 1969) but rather as an expected phenomenon. Filaments of an unknown nature were observed in some nuclei (Fig. 6).

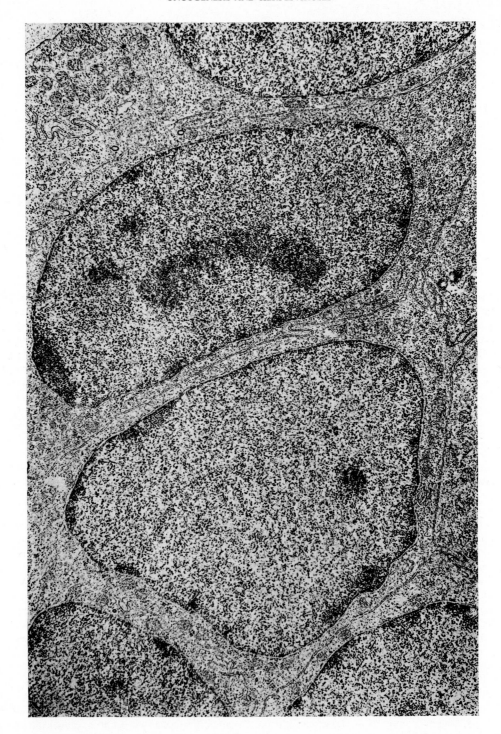

Fig. 3. Calid tumour obtained from a frog collected in Kandiyohi County, Minnesota, 18 October 1970. Note the absence of formed virus particles.   ×11 200.

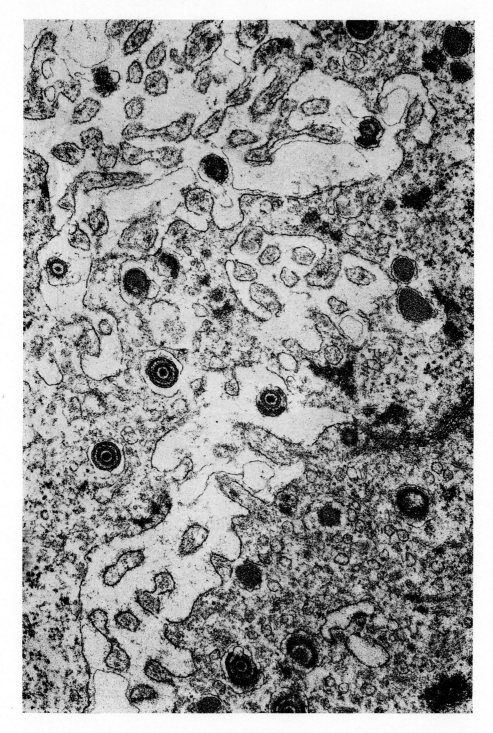

Fig. 4.   Mature virus particles in lumen of algid tumour collected in Kandiyohi County, Minnesota, 1 December 1970.
× 38 400.

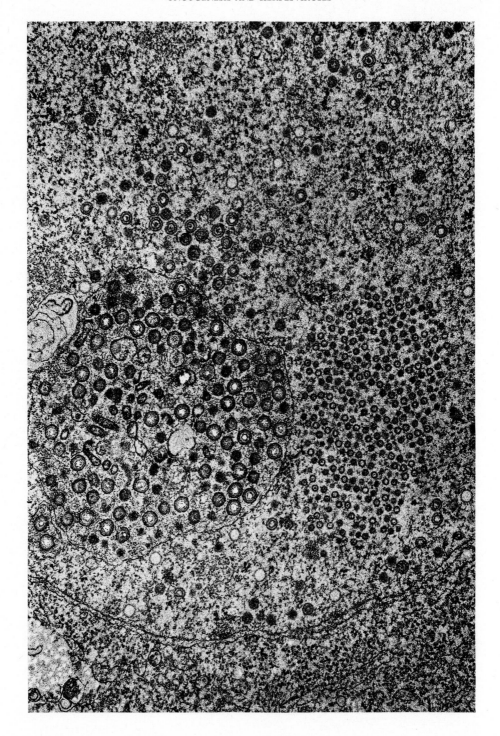

Fig. 5. An algid tumour with concurrent infections of a herpesvirus and a smaller, unidentified virus. Tumour collected in Kandiyohi County, Minnesota, 13 April 1971. ×31 000.

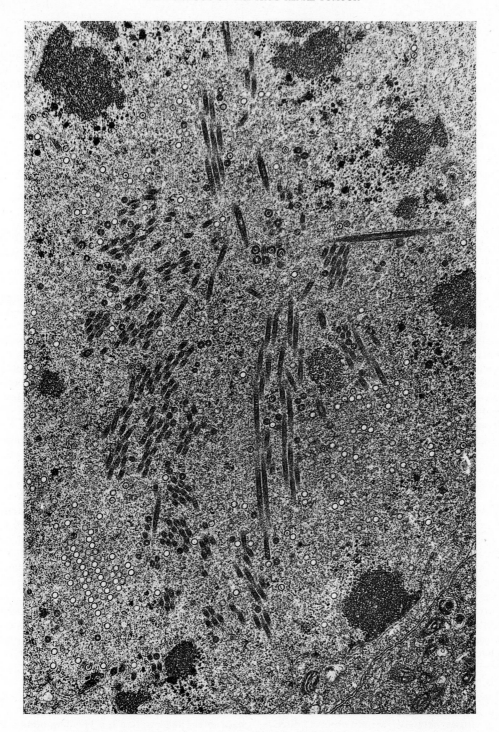

Fig. 6. Nucleus of calid tumour containing filaments of unknown significance. Tumour collected in Kandiyohi County, Minnesota, 4 May 1971. ×16 000.

Transitional tumours

Six tumours were detected from 12 May 1971 to 3 June 1971 (see Table). We have designated these tumours as "transitional". The June frogs were collected 41 days after egg masses were first observed in that particular area. Tadpoles approximately one inch in total length were seen on the day of collection. The air temperature was 28°C at the time of collection of these frogs. The transitional tumours were characterized by reduced epithelium that was largely virus-free, necrotic cells in the epithelium and free in the lumen, extensive cellular debris in the lumen, and viruses in an occasional epithelial cell and in the lumen. Are these tumours regressing? Our study of transitional tumours will be reported in greater detail elsewhere. We should like to emphasize here, however, that viruses are present in tumours well after the period of sexual reproduction.

CONCERNING THE OMNIPRESENCE OF AN ACTIVE VIRAL GENOME IN ALGID TUMOURS

Goodheart (1970) has suggested that one of the problems associated with the herpesvirus of the frog tumour is that infectious virus is produced *only* in frogs maintained at low temperature. We should like to comment that *all* frogs in Minnesota survive a boreal climate for at least 6 months of the year. Every one of the 24 tumours taken after the onset of the cold season contained herpesviruses. The presence of the virus during the cold season does not surprise us, nor do we consider it to be a difficulty, because it is correlated with the social behaviour of frogs and the stages in the life cycle of the frog susceptible to virus infection (see discussion below). What would surprise us would be the presence of virus when frogs are dispersed, with little opportunity for contagion to occur.

It would seem unlikely that a virus would be invariably present in spontaneous tumours if the virus were not related to the tumour in a causal way. One would not expect a capricious relationship between a virus and a tumour if the virus was indeed the aetiological agent of the tumour. Although the detection of a virus in all phases in the growth of a tumour is not a prerequisite for establishing that a virus is the aetiological agent of the tumour, the *invariable* presence of the virus in any phase of the tumour is strongly suggestive of a causal relationship. If viruses themselves are not observed directly, it has been suggested that detection of the viral genome by nucleic acid hybridization would be highly

relevant to the aetiology of the tumour (Green, 1969). The visualization of a finger is more concrete evidence of a viable finger than is the detection of a fingerprint. Accordingly, we feel that the invariable presence of electron microscope-detectable virus particles in 24 consecutive tumours cannot be considered a chance relationship of the virus with the tumour but most likely reflects a causal relationship.

VIRION RELEASE AND POSSIBLE ROUTES OF CONTAGION

Where do mature enveloped viruses go from the kidney? They can be detected in the lumena of kidney tubules and it would seem probable that the bulk of them move to the bladder with the urine. Indeed, viruses have been detected in urine (Granoff & Darlington, 1969). A frog with virions in its urine presumably voids both the urine and the virions to the environment. Mature, enveloped virions are believed to be the infective particle of the Lucké tumour (Mizell, Toplin & Isaacs, 1969).

Mature, enveloped virions have been detected in the lumena of tumours prior to the host's emergence from the lake (i.e., mature viruses have been observed in the tumours of hibernating frogs). Accordingly, it would seem possible that horizontal transmission could occur in lakes. We are unaware of the extent of crowding that frogs are subjected to at the bottom of lakes but it seems probable that congestion frequently occurs. If the mature virions that are known to be present in the frog tumour are voided while frogs are congregated at lake bottoms, transmission could occur between adult frogs. Are adult frogs susceptible to viral infection? Although data from tumour brei injection experiments suggest that immature frogs are very susceptible to the virus (Tweedell, 1967; McKinnell & Tweedell, 1970), there is also evidence that suggests that crowding may elicit tumour formation in adults (DiBerardino & King, 1965a). Whether or not free virions occur in lake water is unknown but techniques are now available for the recovery of viruses from water (see below). Lake water during times of "hibernation" should be examined for the presence of infectious virus particles. We plan to do this.

Tumour cells lose their viruses within a week at room temperature in the laboratory (Zambernard & Vatter, 1966). The first warm environment a frog encounters in the spring is a breeding pond. Lakes may be 10 to 20°C colder than adjacent breeding ponds in mid-April. Breeding ponds are where frogs are found in amplexus, where eggs are deposited, and where tadpoles develop. Frogs have tumours containing viruses while they are in breeding ponds. The lumena of the tumours are rich in mature virions. These virions presumably may pass to the environment via the urine. Accordingly, it seems reasonable that contagion may occur in breeding ponds. If contagion is via the water route, then the isolation of the infective virus should be possible.

CONCERNING THE POSSIBILITY OF VIRUS RECOVERY FROM WATER

If both infectious particle release and a susceptible stage in the life cycle are synchronized and occur in the same place, a reasonable hypothesis is that contagion is horizontal and occurs in breeding ponds. Therefore, it would be desirable to recover and identify the Lucké tumour herpesvirus from breeding pond water. A number of procedures are currently available that may be of use in recovering the frog tumour virus from pond water. Continuous-flow isopyknic-zonal centrifugation has been used for the concentration of viruses present in water (Anderson et al., 1967). An effective and economical method of virus concentration is the aqueous polymer two-phase separation method that involves liquid-liquid partitioning (Shuval et al., 1969). Soluble alginate filters are now commercially available and provide probably the simplest and most effective method for concentrating viruses from environmental waters (Gärtner, 1967). These and other methods for the detection and concentration of viruses present in environmental water have been reviewed and criticized by Hill, Akin & Benton (1971).

The isolation of viruses from water, suggesting horizontal transmission in breeding ponds, would be in harmony with the life cycle of the Lucké tumour virus proposed by Rafferty (1964).

Roizman (1969) has commented on the instability of herpesviruses. How long will enveloped viruses survive in pond water? We are unwilling to prognosticate but should like to suggest that pond water, characterized by abundant organic matter, is not very similar to distilled water or tissue culture media. Perhaps herpesviruses survive for longer periods in ponds than they do in the laboratory. Several enteric viruses survive longer in heavily contaminated water than in river water (e.g., there is a reduction in titre of 99.9 at 4°C after 26 days in water of low contamination while 130 days are required for an equivalent reduction of titre in heavily contaminated water

(Clarke et al., 1964). The survival of enteric viruses in ground and surface waters has been reviewed by Akin, Benton & Hill (1971).

IS THERE AN INSECT VECTOR?

Calid tumours are devoid of infective particles (Tweedell, 1967). Accordingly, there seems little reason to speculate about transmission in the summer and early autumn. Infectious agents derived from amphibian neoplasms have been reported to have retained infectivity after passage through an insect (Nayar, Arthur & Balls, 1970). Reptiles and amphibia are known to be bitten by virus-carrying mosquitoes (Downes, 1971). Culex territans has been observed feeding on amphibians, and this mosquito occurs throughout the entire state of Minnesota (Barr, 1958). It would seem that the possibility of vector transmission of the virus agent would be a fruitful area for exploration. Several critical unknowns exist. Do mature virions occur in the blood of algid or transitional tumour-bearing frogs? Does the frog-biting insect emerge from overwintering prior to the loss of virions from the frog in the spring?

DOES CONTAGION OCCUR BY INFECTION OF GAMETES WITH VIRUSES?

Transovarial transmission would require the ovary of a tumorous animal to be challenged with infective particles. The most likely time for this to occur is when the tumour is actively synthesizing viruses. Virus replication occurs during the autumn, winter and much of spring. Virions could move via the circulatory system or the body cavity fluid to the gonads. Ova may be infected while in the body cavity (Tweedell, 1969) and herpesvirus has been observed in ascitic fluid (Naegele, Granoff & Darlington[1]). A demonstration of contagion via one route does not eliminate the possibility that alternative routes are also used.

DISTRIBUTION AND PREVALENCE OF RENAL TUMOURS

Few field studies have been made. Auclair (1961) collected frogs in northern Vermont and Canada in April of that year as frogs emerged from hibernation. He found tumours in frogs from a tributary of the Pike River in Quebec Province about 15 miles north

---

[1] Unpublished observations.

of Alburg, Vermont. He also reported finding frogs with tumours from the Swanton River east of Alburg and in a swamp near St. Albans, Vermont. Frogs collected in several localities in northern United States and Canada and maintained in the laboratory for several months developed tumours (Rafferty, 1967). "Wisconsin" frogs (commercially obtained) have been reported to have renal neoplasms (Rose & Rose, 1952; Rafferty, 1964; Tweedell, 1965).

Our studies of tumour distribution have been primarily of Minnesota. We were unable to find tumours in either North Dakota or Louisiana leopard frogs despite the fact that such frogs are susceptible to tumours in the laboratory (McKinnell & Duplantier, 1970). We have now identified 12 Minnesota counties with populations of frogs with renal adenocarcinoma, namely: Otter Tail, Douglas, Todd, Pope, Kandiyohi, Aitkin, Kanabec, Sherburne, Hennepin, Scott, LeSueur, and Cottonwood. These counties, with the exception of Cottonwood, are all characterized by varying amounts of glacial moraine. Pothole lakes are common throughout the moraine. All of the counties, with the exception of Cottonwood and Kanabec, have regions with a minimum of 10 lake basins per township (township = 36 square miles) and most have areas where the lake density exceeds 20 per township.

Our observations in the field suggest that the frog population density is directly related to the frequency of lakes. We have the impression that tumours are found where frogs are most abundant. An example which supports this impression is our experience with Otter Tail and Kandiyohi counties. These two counties have served for many years as sources for commercial frogs because commercial frog collectors obtain their frogs where frogs are abundant. We have regularly taken tumour-bearing frogs from Otter Tail and Kandiyohi counties. If tumour frequency is shown to be related to frog abundance, then the situation would be not dissimilar to that found with leukaemia in cattle, where herd size is directly correlated with leukaemia frequency (Anderson et al., 1971).

NUCLEAR TRANSPLANTATION STUDIES AND VIRAL AETIOLOGY

Nuclear transplantation in multicellular animals was first accomplished in R. pipiens (Briggs & King, 1952). The capacity of renal tumour nuclei to programme for embryonic development was tested within the decade (King & McKinnell, 1960). Tad-

poles with considerable embryonic differentiation ensue when tumour nuclei are inserted into eggs bereft of maternal genetic material (McKinnell, 1962; King & DiBerardino, 1965; DiBerardino & King, 1965b). Polyploidy has been used as a nuclear marker in the tumour nuclear transfer experiments (McKinnell, Deggins & Labat, 1969). Triploid tumour nuclei, when transplanted to enucleated eggs, result in tadpoles that are also triploid, thus eliminating virtually all possibility that the tadpoles develop by gynogenesis (Fig. 7). What relevance do the nuclear transplantation experiments have to a viral aetiology of the frog tumour?

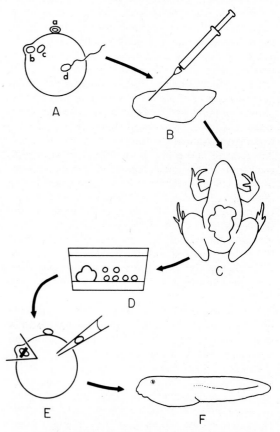

Fig. 7. Triploid tumour nuclear transplantation. Triploidy was induced by hydrostatic repression of the second polar body (A). Embryos that were injected with a herpesvirus-containing preparation (B) developed into tumour-bearing triploid frogs (C). Triploid tumour cells were dissociated (D), and transplanted into previously enucleated recipient eggs (E). The triploidy of the ensuing tumour nuclear transplant tadpoles (F) witnessed to the origin of the nucleus, i.e., it is exceedingly unlikely that triploid gynogenetic tadpoles developed from these experiments.

We predict that it may be possible to detect the virus in the tumour nuclear transplant tadpoles. If the herpesvirus DNA is latent in the frog tumour nucleus that is transplanted to the enucleated egg, then it may well be that the tumour nuclear transplant tadpoles carry viral DNA in every cell of their highly differentiated body. How could it be detected?

Mesonephric tissue from tumour nuclear transplant tadpoles could be implanted in the anterior eye chamber of an anuran of another species. A nonhomologous species as a host would eliminate the possibility of a *R. pipiens* having viruses that would infect the implant. The host, with its grafted embryonic tissue, would then be chilled (at 7.5°C). After 8 to 10 weeks, the graft would be examined for virus particles by electron microscopy. Presence of herpesvirus in the mesonephros of the tumour nuclear transplant tadpole tissue would provide substantial evidence in addition to the already impressive array of arguments in favour of a viral aetiology of the Lucké renal adenocarcinoma. Why would this kind of evidence be important. "It is difficult to conceive of a passenger virus which would be present in all Lucké carcinomas" (Green, 1969). If a passenger virus is difficult to conceive, as Green suggests, in all Lucké tumours, it is even *more* difficult to imagine passenger viruses finding their way to the mesonephros of a tadpole produced by tumour nuclear transfer.

If virus particles cannot be evoked from latency in the tumour nuclear transplant tadpoles, alternative procedures are available to detect the presence of viral activity in such tadpoles. Molecular hybridization of labelled tadpole RNA with viral DNA in order to detect viral messenger RNA is a means of obtaining evidence of expressed viral nucleotide sequences in the tadpole cells. The synthesis of RNA complementary to viral DNA and the hybridization of the cRNA with tumour nuclear transplant DNA would attest to the presence of viral genome in the tadpole. The latter procedure would be useful if the viral DNA were not metabolically active. Localization of viral antigens in the tumour nuclear transplant embryo by means of labelled antibody cytochemical tests is an alternative and feasible procedure.

It should be noted here that, although swimming tadpoles result from the tumour nuclear transplantation procedure, the tadpoles are *not* normal. Normal tadpoles develop into frogs. Tumour nuclear transplant tadpoles do not metamorphose. Thus, it may be premature to suggest that the studies have shown that the tumour nucleus is not altered by presence of the virus (Braun, 1970). A recent article in the New York *Times* magazine (March 7, 1971, page 69), when describing tumour nuclear transplantation, stated: "Evidently the cancerous state had failed to take anything away or add anything to the nucleus." The *Times* statement is premature. We hope, however, that future experiments will provide information on whether or not viral DNA is integrated into the genome of tadpoles produced by tumour nuclear transplantation.

ACKNOWLEDGEMENTS

The authors wish to thank William Boernke and David Dapkus, graduate students at the University of Minnesota, who assisted with the collection of frogs during the 1970-1971 season. We express our appreciation to Acting Director F. W. Johnson and Officer Miles Pooler of the State of Minnesota Department of Natural Resources, Division of Enforcement and Field Services, for information and assistance in the collection of frogs. We acknowledge Mrs. Marilyn Steere's preparation of the art work. We thank Professor Kenyon S. Tweedell, University of Notre Dame, for his helpful comments concerning the manuscript. This work was aided by Grant DRG 990-B from the Damon Runyon Memorial Fund for Cancer Research, Inc., New York.

REFERENCES

Akin, E. W., Benton, W. H. & Hill, Jr, W. F. (1971) Enteric viruses in ground and surface waters: A review of their occurrence and survival. In: Snoeyink, V. L., ed., *Virus and water quality: occurrence and control,* Urbana, University of Illinois, pp. 59-74 *(University of Illinois Bulletin,* Vol. 69, No. 1)

Anderson, N. G., Cline, G. B., Harris, W. W. & Green, J. G. (1967) Isolation of viral particles from large fluid volumes. In: Berg, G., ed., *Transmission of viruses by the water route,* New York, Interscience Publishers, pp. 75-88

Anderson, R. K., Sorensen, D. K., Perman, V., Dirks, V. A., Snyder, M. M. & Bearman, J. E. (1971) Selected epizootiologic aspects of bovine leukemia in Minnesota (1961-1965). *Amer. J. vet. Res., 32,* 563-577

Auclair, W. (1961) Monolayer culture of *Rana pipiens* kidney and ecological factors. In: Duryee, W. R. & Warner, L., eds, *Proceedings of the frog kidney adenocarcinoma conference,* Bethesda, National Institutes of Health, pp. 107-113

Barr, A. R. (1958) *The mosquitoes of Minnesota.* Saint Paul, University of Minnesota Agricultural Experiment Station (Technical Bulletin 228)

Borchert, J. R. & Yaeger, D. P. (1968) *Atlas of Minnesota resources and settlement.* Saint Paul, Documents Section, State of Minnesota

Braun, A. C. (1970) On the origin of the cancer cells. *Amer. Scient.,* **58,** 307-320

Breidenbach, G. P., Skinner, M. S., Wallace, J. H. & Mizell, M. (1971) In vitro induction of a herpes-type virus in "summerphase" Lucké tumor explants. *J. Virol.,* **7,** 679-682

Briggs, R. & King, T. J. (1952) Transplantation of living nuclei from blastula cells into enucleated frogs' eggs. *Proc. nat. Acad. Sci. (Wash.),* **38,** 455-463

Clarke, N. A., Berg, G., Kabler, P. W. & Chang, S. L. (1964) Human enteric viruses in water: Source, survival, and removability. In: Eckenfelder, W. W., ed., *International conference on water pollution research, London, 1962,* New York, Pergamon, 523-541

Collins, S. L. & Nace, G. W. (1970) The rapid induction of viral inclusions in frog renal adenocarcinoma. *Amer. Zool.,* **10,** 531-532

DiBerardino, M. A. & King, T. J. (1965a) Renal adenocarcinoma promoted by crowded conditions in laboratory frogs. *Cancer Res.,* **25,** 1910-1912

DiBerardino, M. A. & King, T. J. (1965b) Transplantation of nuclei from the frog renal adenocarcinoma. II. Chromosomal and histologic analysis of tumor nuclear transplant embryos. *Devl. Biol.,* **11,** 217-242

Dmochowski, L. (1959) Viruses and tumors. An old problem in the light of recent advances. *Bact. Rev.,* **23,** 18-40

Downes, J. A. (1971) The ecology of blood-sucking diptera: An evolutionary perspective. In: Fallis, A. M., ed., *Ecology and physiology of parasites,* Toronto, University of Toronto Press, pp. 232-258

Fawcett, D. W. (1956) Electron microscope observations on intracellular virus-like particles associated with cells of the Lucké renal adenocarcinoma. *J. biophys. biochem. Cytol.,* **2,** 725-742

Gärtner, H. (1967) Retention and recovery of polioviruses on a soluble ultrafilter. In: Berg, G., ed., *Transmission of viruses by the water route,* New York, Interscience Publishers, pp. 121-127

Goodheart, C. R. (1970) Herpes viruses and cancer. *J. Amer. med. Ass.,* **211,** 91-96

Granoff, A. & Darlington, R. W. (1969) Viruses and renal carcinoma of *Rana pipiens.* VIII. Electron microscopic evidence for the presence of herpes virus in the urine of a Lucké tumor-bearing frog. *J. Virol.,* **38,** 197-200

Green, M. (1969) Nucleic acid homology as applied to investigations on the relationship of viruses to neoplastic diseases. In: Mizell, M., ed., *Biology of amphibian tumors,* New York, Springer-Verlag, pp. 445-454

Hill, W. F., Akin, E. W., & Benton, W. H. (1971) Detection of viruses in water: A review of methods and application. In: Snoeyink, V. L., ed., *Virus and water*

quality: occurrence and control, Urbana, University of Illinois, pp. 17-46 (*University of Illinois Bulletin,* Vol. 69, No. 1)

King, T. J. & McKinnell, R. G. (1960) An attempt to determine the developmental potentialities of the cancer cell nucleus by means of transplantation. In: *Cell physiology of neoplasia,* Austin, University of Texas Press, pp. 591-617

King, T. J. & DiBerardino, M. A. (1965) Transplantation of nuclei from the frog renal adenocarcinoma. I. Development of tumor nuclear-transplant embryos. *Ann. N.Y. Acad. Sci.,* **126,** 115-126

Lucké, B. (1934) A neoplastic disease of the kidney of the frog, *Rana pipiens. Amer. J. Cancer,* **20,** 352-379

Lucké, B. (1938) Carcinoma in the leopard frog: Its probable causation by a virus. *J. exp. Med.,* **68,** 457-468

Lunger, P. D., Darlington, R. W. & Granoff, A. (1965) Cell-virus relationships in the Lucké renal adenocarcinoma: an ultrastructure study. *Ann. N.Y. Acad. Sci.,* **126,** 289-314

McKinnell, R. G. (1962) Development of *Rana pipiens* eggs transplanted with Lucké tumor cells. *Amer. Zool.,* **2,** 430-431

McKinnell, R. G. (1965) Incidence and histology of renal tumors of leopard frogs from the north central states. *Ann. N.Y. Acad. Sci.,* **126,** 85-98

McKinnell, R. G. (1966) Renal tumors obtained from pre-breeding Minnesota lake frogs. *Amer. Zool.,* **6,** 558

McKinnell, R. G. (1967) Evidence for seasonal variation in incidence of renal adenocarcinoma in *Rana pipiens. J. Minnesota Acad. Sci.,* **34,** 173-175

McKinnell, R. G. (1969) Lucké tumor: Epidemiological aspects. In: Mizell, M., ed., *Biology of amphibian tumors,* New York, Springer-Verlag, pp. 254-260

McKinnell, R. G. & McKinnell, B. K. (1968) Seasonal fluctuation of frog renal adenocarcinoma prevalence in natural populations. *Cancer Res.,* **28,** 440-444

McKinnell, R. G. & Zambernard, J. (1968) Virus particles in renal tumors from spring *Rana pipiens* of known geographic origin. *Cancer Res.,* **28,** 684-688

McKinnell, R. G., Deggins, B. A. & Labat, D. D. (1969) Transplantation of pluripotential nuclei from triploid frog tumors. *Science,* **165,** 394-396

McKinnell, R. G. & Tweedell, K. S. (1970) Induction of renal tumors in triploid frogs. *J. nat. Cancer Inst.,* **44,** 1161-1166

McKinnell, R. G. & Duplantier, D. P. (1970) Are there renal adenocarcinoma-free populations of leopard frogs? *Cancer Res.,* **30,** 2730-2735

Medes, G. & Reimann, S. P. (1963) *Normal growth and cancer,* Philadelphia, J. P. Lippincott

Minnesota Department of Conservation (1968) *An inventory of Minnesota lakes,* Saint Paul (Bulletin No. 25)

Mizell, M. (1969) State of the art: Lucké renal adenocarcinoma. In: Mizell, M., ed., *Biology of amphibian tumors,* New York, Springer-Verlag, pp. 1-25

Mizell, M., Stackpole, C. W., & Halperen, S. (1968) Herpes-type virus recovery from "virus-free" frog kidney tumors. *Proc. Soc. exp. Biol. (N.Y.)*, **127**, 808-814

Mizell, M., Toplin, I. & Isaacs, J. J. (1969) Tumor induction in developing frog kidneys by a zonal centrifuge purified fraction of the frog herpes-type virus. *Science*, **165**, 1134-1137

Nayar, K. K., Arthur, E. & Balls, M. (1970) Transmission of an amphibian lymphosarcoma to and through insects. *Oncology*, **24**, 370-377

Rafferty, Jr, K. A. (1964) Kidney tumors of the leopard frog: A review. *Cancer Res.*, **24**, 169-185

Rafferty, Jr, K. A. (1967) The biology of spontaneous renal carcinoma of the frog. In: King, J. S., ed., *Renal neoplasia*, Boston, Little Brown & Co., pp. 301-315

Roberts, M. E. (1963) Studies on the transmissibility and cytology of the renal carcinoma of *Rana pipiens*. *Cancer Res.*, **23**, 1709-1714

Roizman, B. (1969) The herpesviruses—a biochemical definition of the group. *Curr. Top. Microbiol. Immunol.* **49**, 1-79

Rose, S. M. & Rose, F. C. (1952) Tumor agent transformation in amphibia. *Cancer Res.*, **12**, 1-12

Shuval, H. I., Fattal, B., Cymbalista, S. & Goldblum, N. (1969) The phase-separation method for the concentration and detection of viruses in water. *Water Res.*, **3**, 225-240

Stackpole, C. W. (1969) Herpes-type virus of the frog renal adenocarcinoma. I. Virus development in tumor transplants maintained at low temperature. *J. Virol.*, **4**, 75-93

Tester, J. R. & Breckenridge, W. J. (1964) Winter behavior patterns of the Manitoba toad, Bufo hemiophrys, in northwestern Minnesota. *Ann. Acad. Scient. Fennicae, Ser. A, IV*, **71**, 421-431

Tweedell, K. S. (1965) Renal tumors in a western population of *Rana pipiens*. *Amer. Midl. Nat.*, **73**, 285-292

Tweedell, K. S. (1967) Induced oncogenesis in developing frog kidney cells. *Cancer Res.*, **27**, 2042-2052

Tweedell, K. S. (1969) Simulated transmission of renal tumors in oocytes and embryos of *Rana pipiens*. In: Mizell, M., ed., *Biology of amphibian tumors*, New York, Springer-Verlag, pp. 229-239

Zambernard, J. & Vatter, A. E. (1966) The effect of temperature change upon inclusion-containing renal tumors of leopard frogs. *Cancer Res.*, **26**, 2148-2153

Zambernard, J. & McKinnell, R. G. (1969) "Virus-free" renal tumors obtained from pre-hibernating leopard frogs of known geographic origin. *Cancer Res.*, **29**, 653-657

Bioassay of Frog Renal Tumour Viruses

K. S. TWEEDELL,[1] F. J. MICHALSKI [1] & D. M. MOREK [1]

INTRODUCTION

Since the time that a virus was first implicated (Lucké, 1938; Fawcett, 1956) in the transmission of the frog renal adenocarcinoma, a series of frog cell viruses have been isolated as potential causative agents. Essentially three viral types have been identified (Granoff, 1969): a herpes-type virus (HTV), a polyhedral cytoplasmic, DNA virus (PCDV), and more recently an adeno-type virus. Two of these viruses have been isolated and tested as tumour-inducing agents *in vivo*, namely the cytoplasmic virus FV 3 (Tweedell & Granoff, 1968) and the HTV (FV 4) (Granoff, Gravell & Darlington, 1969), but so far without success. However, recent data indicate that FV 4 and HTV from Lucké tumours are not the same (Gravell, 1971). Cell extracts from the renal tumour have been oncogenic in adults (Duryee, 1956), and HTV derived from tumour cell fractions have been tumorigenic when inoculated into the embryo (Tweedell, 1967; Mizell, Toplin & Isaacs, 1969) or when contacted with the developing oocyte (Tweedell, 1969).

The Lucké herpes-type virus (L-HTV) always seen in inclusion body tumours (Zambernard, Vatter & McKinnell, 1966) is strongly suspect since nuclear inclusion body tumours are alone oncogenically active when experimentally transmitted through embryos (Tweedell, 1967, 1969). Since the tumour cell fractions have proved oncogenicity, it seemed plausible that identification of the rôles of the viruses found in them might help define those that are oncogenic. A frog embryo cell line has proved to be suitable for testing the infectivity of viruses isolated from subcellular fractions of frog renal inclusion body tumours (Michalski, 1971). Such infections have produced both cytopathic effects (CPE) characteristic of known frog viruses and cell transfor-

mations. The effects produced by cell fractions or filtrates of the infected cell line after injection into frog embryos and incubation of larval kidney in organ culture were therefore studied.

MATERIALS AND METHODS

Preparation of virus-containing filtrates of tumour cells

The nuclear inclusion renal carcinomas were obtained either from *Rana pipiens burnsi* (from the north-central United States) or *Rana pipiens pipiens* (from the north-eastern United States). Prior to use, the frogs were kept at 9°C for 8 or more weeks to promote viral replication (Rafferty, 1965), then tumours collected and cryostat sections examined for the presence of intranuclear (Cowdry type A) inclusions, a sign of Lucké herpes-type virus (L-HTV). The tumours were removed aseptically, weighed, cut into 3-mm pieces and washed in successive changes of sterile, cold amphibian Ringer's solution containing penicillin and streptomycin. Aseptic techniques were used in all remaining procedures, which were carried out under a transfer hood.

Tumours were homogenized in sterile 0.02 M phosphate buffer or 0.005 M tris buffer at 0°C. Subsequent differential centrifugation finally yielded a mitochondrial virus fraction (P_2) (Tweedell, 1967) containing both nuclear Lucké herpes-type (L-HTV) and cytoplasmic viruses (PCDV). Dilutions of these fractions (10^{-1} to 10^{-5}) were made with 0.005 M tris buffer, pH 7.2, and passed through a 0.45-μ Millipore filter.

As a check for tumour oncogenicity, 10^{-2} dilutions of the filtrates were injected into the pronephric region of the nephrogenic ridge of stage-17 embryos by the method of Tweedell (1967).

Tissue culture procedures

An established frog embryo cell line (E-191), primarily epithelial in character, was derived from

---

[1] Department of Biology, University of Notre Dame, Indiana, USA.

S18 *R. pipiens burnsi* embryos by Michalski (1971). These cells were grown and maintained in Leibovitz (L-15) culture medium containing 10% foetal calf serum, penicillin, 100 U/ml, streptomycin, 100 μg/ml and Fungizone 2 μg/ml. Cultivation was carried out in 6-oz glass flasks or Leighton tubes with cover slips at either 9°C or 25°C.

Inoculation of tissue cultures. Nearly confluent monolayers were exposed to 0.2 to 1.0 ml of 10^{-2} filtrates of the P_2 fractions from inclusion tumours. Adsorption lasted 2–4 hours at 25°C. The inoculum was removed and replaced by L-15 medium. Cultures were then maintained at 25°C, although some were subsequently kept at 9°C.

Bioassay of infected tissue cultures. Subcellular fractions from both noninfected and infected cultures were obtained by triturating the cells from the glass. Cells plus the original medium were sonicated in ice with a Willems Polytron (PT) sonicator at 22 000 cps for 30–60 seconds, and then submitted to the same centrifugation regimen as that used to produce a comparable mitochondrial (P_2) pellet. Dilutions for various inoculations were made with 0.02 M phosphate buffer, and filtered through a 0.45-μ filter.

Fractions of noninfected and infected tissue cultures were injected into stage-16 or stage-17 (tailbud) embryos. Dilutions from 10^{-1} to 10^{-5} were delivered in 0.2 μl amounts from a microlitre syringe. Inoculation was made along the left nephrogenic ridge into the pronephric region. Embryos were reared at 12°C for 10 days, then transferred to 24°C.

Organ culture assay

Infected tissue culture fractions were applied to normal kidney *in vitro*, after which the latter were placed in organ culture. An *in vitro* vertical Millipore organ culture technique developed by Morek was utilized (Morek & Tweedell, 1969). One ml of the total tissue culture homogenate was diluted in 10 ml of L-15 culture fluid and filtered through a 0.45-μ Millipore filter. A1 :10 dilution of the filtrate was then used to bathe the kidney pieces for 19 hours at 12°C. Control cultures were bathed in an equal amount of plain growth medium. After incubation the cultures were washed and fresh medium applied. Cultures were maintained at 12° or 25°C.

Intraocular assay

Large buds of cell sheets scraped from cultures were implanted into the anterior eyechamber of adult frogs anaesthetized with ethyl-*m*-amino benzoate methane sulphonic acid (Sigma). Small slits were made in the cornea with an iris knife and cell masses wedged between the cornea and iris. Operated eyes were sprinkled with sulfadiazine surgical powder and the animals maintained at 24°C.

RESULTS

Cultures of the embryonic cell line E-191 (Fig. 1) that supported viral infection were harvested for further bioassay. The first successful virus infection was obtained from a fresh subcellular fraction of an inclusion body renal tumour (RT 192) derived from an adult *R. pipiens,* and the infected line designated E-191/V-192. It was a mixed infection and caused a characteristic cytopathology that included plaque formation, cell sheet shedding and bud-like proliferations. Other cytopathic effects included cytoplasmic inclusions that were predominately DNA containing and smaller RNA inclusions (Fig. 2) (Michalski, 1971). These inclusions appear similar to the cytoplasmic inclusions produced by the PCDV virus. Evidence for a second virus of lower incidence that may be a HTV comes from the presence of intranuclear inclusions in about 5% of the cells. These inclusions are surrounded by a clear halo and almost fill the nucleus, but engulf and do not displace the nucleoli (Figs. 3A & B). They are also DNA-positive. Electron microscopic studies of infected E-191 cells have shown that, morphologically, one virus may be identical with the cytoplasmic PCDV (Kajima & Michalski[1]).

Inoculation of virus-infected tissue cultures into embryos

Virus-infected cultures of E-191 embryo cells were used to inoculate embryos of *R. pipiens* in the tailbud stage (stage 16–17). Viruses from the 7th to the 17th passage were used in five different experiments. Initially, a centrifuged subcellular pellet of the sonicated infected cells was injected in concentrations ranging from undiluted to a 10^{-5} dilution in phosphate buffer. Each embryo received 0.2 μl of the virus-subcellular fraction. In the first three experiments, the embryos were incubated at 12°C for 4, 6 or 8 days, then gradually brought to 24°C.

In early experiments, inoculation of the embryos was rapidly lethal, and the response was dose-dependent (Table 1). At the higher concentrations, i.e., undiluted to a 10^{-3} dilution, the infected embryos

---

[1] Unpublished data.

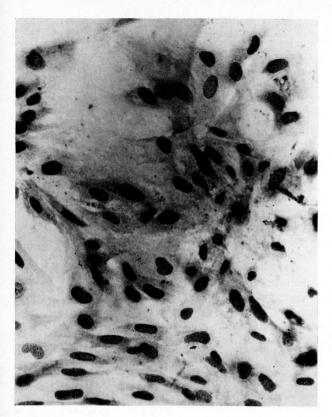

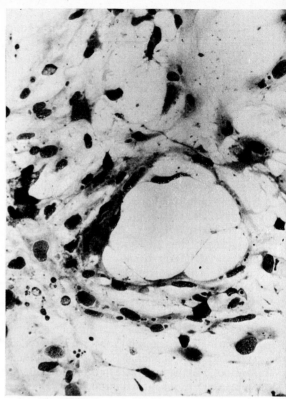

1

2

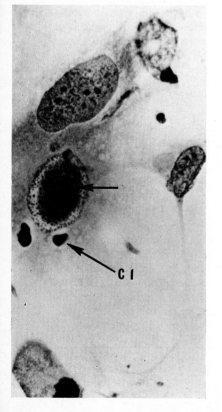

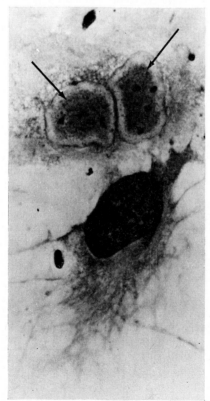

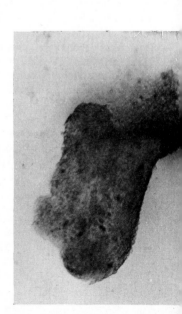

3A

3B

4

Table 1. Effect of sonicated cell-viral fractions inoculated into *R. pipiens* embryos

Source of inoculum	No. of exper-iments	Inoculum	No. In-oculated	Survivors		
				No./ days	No./pro-meta.	No./meta-morpho-sis
Infected tissue culture E-191/V-192	3	P$_2$ concen.	97	0/6-9	0	0
	2	P$_2$ 10^{-1}	48	0/13-17	0	0
	1	P$_2$ 10^{-2}	22	0/17	0	0
	3	P$_2$ 10^{-3}	60	23/16	18	16
	1	P$_2$ 10^{-5}	24	24/16	20	17
Noninfected tissue culture E-191/—	2	P$_2$ concen.	74	57/16-19	50	31
None	1	—	24	24/19	23	15
Re-isolated virus (Embryo)	1	P$_2$ concen.	13	0/6-9	0	0
Infected tissue culture + em-bryo E-191/V-192/embryo	1	P$_2$ concen.	32	0/17	0	0
E-191/V-192 filtrate/embryo	1	P$_2$ concen.	32	2/17	0	0
Embryo	1	P$_2$ concen.	24	12/17	12	8
None	1	—	19	19/17	12	—

exhibited a typical syndrome characterized by lor-dosis, stunted gills, epithelial papillae and sloughing, and a disproportionate overgrowth of the trunk region. Dilutions up to 10^{-3} generally resulted in the death of all embryos after 6 to 17 days, depending upon the dose and length of the post-inoculation cold period.

An attempt to improve the HTV activity was made in one experiment in which the infected tissue culture was incubated at 9°C for 1 month prior to harvest. At concentrations of 10^{-0} to 10^{-2}, injection of each infected fraction into 24 embryos still resulted in death in all cases.

Similar undiluted P$_2$ fractions obtained from non-infected E-191 cells were also injected into embryos. No features of the syndrome characteristic of inocu-lation with the infected fraction were seen in these embryos. Of 57 survivors at 3 weeks, 31 embryos completed metamorphosis with no visible abnormal effects.

An attempt was then made to dilute out the lethla concentrations of virus. Embryos receiving undi-luted and 10^{-1} dilutions died as before, but some of those injected with 10^{-3} or 10^{-5} dilutions were more successful. One quarter of those inoculated with a 10^{-3} dilution survived through metamorphosis, and 70% of those inoculated with a 10^{-5} dilution survived without ill effects. Presumably the oncogenic virus was also diluted, and to date no tumours have appeared in the survivors.

Analysis of tissue culture-virus interaction

The lethal effect of the infective inoculum might conceivably be reduced by prior exposure to frog embryonic cells. Consequently, untreated *R. pipiens* embryos (stage 19) were sonicated in L-15 culture fluid (5 embryos/ml). One portion of the sonicated embryos was incubated with an equal volume of sonicated E-191/V-192 infected cells grown at 9°C for one month, since L-HTV is known to replicate

Fig. 1. Frog embryo cell line (E-191) derived from *Rana pipiens burnsi* embryos. Two-day culture at 25°C. (Jenner-Giemsa) × 90.

Fig. 2. An infected frog embryo culture (E-191/V-192) inoculated with fractions from an inclusion body tumour. Two-day culture. Note cytoplasmic inclusions (arrows). (Jenner-Giemsa) × 90.

Fig. 3A & B. The infected cell line E-191/V-192 showing a cell containing both cytoplasmic inclusions (C I) and nuclear inclusions (arrows) adjacent to noninclusion nuclei. (Jenner-Giemsa) × 300.

Fig. 4. An isolated proliferating bud of tissue in an infected culture E-191/V-201 kept at 9°C for 30 days after infection. The attached buds continue to proliferate slowly above the monolayer surface. Living, unstained, in tissue flask. × 10.

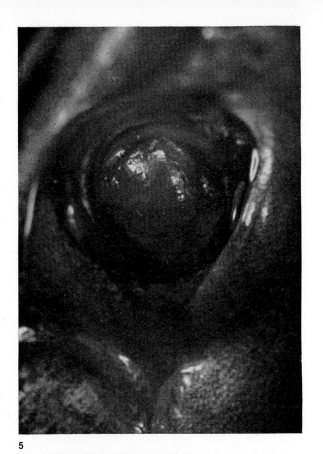

5

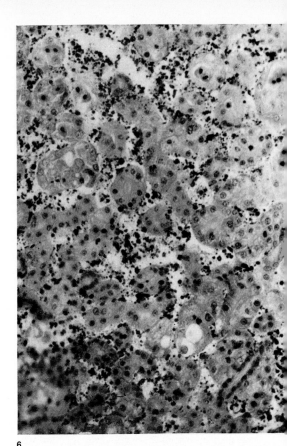

6

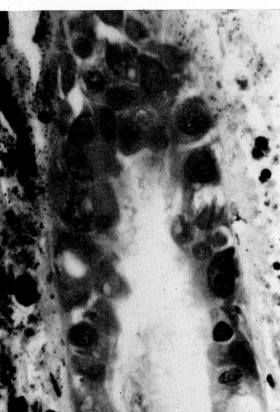

7

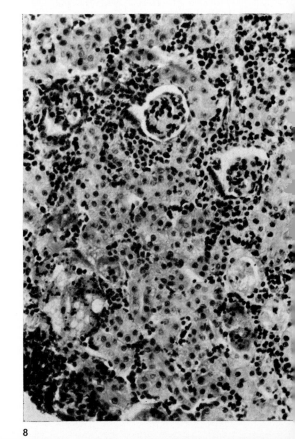

8

at this temperature. A separate embryo sonicate was incubated in 5 ml of a 0.45-μ filtrate of the culture fluid from the infected culture and the third uninfected embryo sonicate was incubated in L-15 alone. Incubation was at $4°C$ for 1 hour. Subsequently, each of the incubated groups was used to produce a P_2 pellet. After resuspension, the P_2 fractions were injected separately into embryos.

The injected embryos maintained at $12°C$ began to react to the combined infected tissue culture-embryo fractions by the 4th day after injection. Simultaneously, the embryos inoculated with the viral filtrate-embryo sonicate displayed papillary epidermal growths on the abdomen and heart region, oedema, and epidermal shedding (Table 1).

The embryos injected with noninfected embryo sonicate or with noninfected tissue culture sonicate were unaffected and half of them completed metamorphosis with no sign of abnormalities.

Re-isolation of viruses

In order to ascertain whether the embryos were dying from the viruses present in the infected cultures, inoculated embryos were reared for 6–9 days until they developed the usual pathological symptoms. Prior to death, the embryos were homogenized, sonicated, and 0.45-μ filtrates used to inoculate noninfected E-191 cells. After one cell passage, the cell culture was used to reinfect new embryos. The inoculated embryos died after 6–9 days with symptoms identical to those seen following the original inoculation of E-191/V-192. Embryos that were inoculated with a subcellular fraction from noninfected embryos were unaffected after rearing through metamorphosis (Table 1).

Cell transformation

Local changes in the cell density and growth habits of the infected embryo cell monolayers suggested that some cells had transformed. Infection of a confluent sheet of E-191/V-192 cells caused the sheet to fall off and roll up into large pieces. Subsequent tumour fraction isolations (E-191/V-200 and E-191/

V-201) of virus have produced additional transformations of the E-191 embryo cell line at $9°C$ resulting in partial detachment and withdrawal of the cell sheet. Isolated foci of massive buds of proliferating cells develop along the ridges (Fig. 4).

Subsequent tests have shown that the E-191/V-192 cells proliferate in the eye of adults. When intraocular implants are made into adult frogs (Michalski, 1971), the infected culture proliferates along the cornea and often bulges or erodes through it (Fig. 5). Eyes inoculated with noninfected E-191 cell sheets remain normal and the implanted cells gradually regress.

Organ culture analysis of viral infection

To determine whether the lethal response of embryos to the infected tissue cultures was a general systemic or organ-specific reaction, filtrates (0.45 μ). from the infected tissue cultures fractions used in the previous embryo interactions were used to infect larval mesonephroi prior to organ culture.

As an *in vivo* control, 12 larvae, 65–70 mm in length, were injected with 5–10 μl of the same P_2 fraction from the incubated tissue culture-embryo cell fractions. Soon after injection, all larvae developed enlarged abdomens, oedema and general subdermal hemorrhage in the hind limbs and vent areas. The pleural-peritoneal fluid was haemorrhagic; oedema and haemorrhagic lesions on the mesonephroi were also seen. All larvae were dead at the end of two weeks.

Previously infected mesonephroi from 50–70-mm larvae were simultaneously placed in vertical organ culture.

Organ cultures were harvested just after incubation and from 5 days up to 9 weeks post incubation (Table 2). Out of 22 infected pieces of larval kidney, 20 showed total stromal destruction and 2 moderate destruction (Fig. 6). Degenerating stroma was first observed after 5 days of incubation at either temperature, although cultures fixed immediately after incubation had pyknotic stromal cell nuclei. In 12 controls, only 1 exhibited any stromal destruction.

Fig. 5. Eye-chamber growth of an intraocular implant of E-191/V-192 cells that had detached and rounded up. Taken $2\frac{1}{2}$ months after eye implantation. ×10.

Fig. 6. Larval kidney infected with cell-viral filtrate, then placed in organ culture for 5 days at 25°C. Note the normal tubules and degenerating stromal tissue. (Jenner-Giemsa) ×128.

Fig. 7. An infected kidney in organ culture with rapidly proliferating parenchymal cells showing cell separation, large nucleoli, increased basophilia and changes in nuclear polarity. Necrotic stroma adjacent. After 3 weeks at 25°C. (Jenner-Giemsa) ×385.

Fig. 8. Normal kidney control in organ culture for 5 days at 25°C. (Jenner-Giemsa) ×128.

In distinct contrast, the tubule cells maintained their normal cytology or showed proliferative changes in at least 18 of the cultures, as seen in Fig. 7. The

changes in the parenchymal cells of the kidney tubule were slow to develop, becoming evident between the 3rd and 9th week. A total of 7 cases of "proliferative

Table 2. Effect of organ culture infection with tissue culture filtrates

Culture at:	Type of culture	No. of cultures	Stroma		Tubules		
			Normal	Degen-erating	Normal	Prolifer-ative changes	Degen-erating
12°C	Infected	12	0	12	8	4	0
	Controls	6	5	1	6	0	0
24°C	Infected	10	0	10	3	3	4
	Controls	6	6	0	6	0	0

change" in the kidney tubule epithelium was noted. The tubular changes indicating transformation between 8–20 days in the 24°C cultures were: nucleolar enlargement, increased basophilia, mitotic figures, clumping of chromatin, increased variation in nuclear size and shape, and the presence of multinucleated cells. Similar changes took 6–9 weeks in the 12°C cultures. A very limited number of intranuclear inclusions were seen in some tubules after 8 weeks at 12°C. These inclusions were solid, amorphous and eosinophilic; they were usually small, but sometimes almost filled the nucleus, marginating the chromatin and displacing or engulfing the nucleoli. Control cultures showed no modification other than that normal for kidney in organ culture (Fig. 8).

In summary, the viruses derived from the frog tumour fractions consist of a PCDV that causes early death in embryos and larvae and specifically attacks kidney stromal tissue in organ culture. A second virus with the characteristics of a herpesvirus appears to cause cell transformation in tissue culture, proliferates in the eye-chamber and produces latent proliferation of the kidney tubules in organ culture. Attempts to select for the latter agent *in vitro* so as to test its oncogenicity are under way.

DISCUSSION

It seems likely that one of the viruses present in the infected cell line is identical with the polyhedral cytoplasmic virus isolated by Granoff (1969). Thus, characteristic DNA cytoplasmic inclusions, like those produced by PCDV, appear in the cytoplasm of the infected embryo cells. Furthermore, when embryos are inoculated with the more concentrated subcellular

fractions of filtrates from the infected cell line, the syndrome leading to their death is typical of that seen in earlier experiments on the injection of PCDV (FV 3) into similar embryos (Tweedell & Granoff, 1968). The mass destruction of the stroma of larval kidney in organ culture after exposure to filtrates from the infected culture fluid would further suggest that this virus affects the stroma preferentially. It should be noted that inoculated embryos begin to die at about the 5th day, which corresponds to the time of stromal destruction *in vitro*. The cytoplasmic virus has been isolated from normal kidney cells (Granoff, Came & Rafferty, 1965) as well as other frog tissues, including tumour tissue. The PCDV is found only in the cytoplasm of infected cells (Came & Lunger, 1966), and electron microscopic examination indicates that the particles in tumours are restricted to renal stromal cells (Lunger, 1969).

The growth characteristics of some of the infected cell cultures, leading to the formation of localized buds of proliferation within the monolayer, suggested cell transformation. There were unusual changes in the bioassays used that also suggested transformation. The latent proliferation of parenchymal cells in some of the organ cultures after exposure to the mixed virus infection is evidence of this. Such tubule proliferations were not obtained in the controls. We attribute this proliferation to the second virus present in the inoculum, since the PCDV is specific for stromal tissue.

Cell transformation of the tissue culture line has also been demonstrated *in vivo*. Following inoculation of the eye-chambers of the adult frog with the proliferating buds described above, the cells soon filled the eye-chamber and often eroded through the cornea to the outside (Michalski, 1971). Control implants of noninfected tissue merely faded away. This proliferative reaction is very similar to that obtained when intraocular tumour implants are made (Tweedell, 1955).

REFERENCES

Came, P. E. & Lunger, P. D. (1966) Viruses isolated from frogs and their relationship to the Lucké tumor. *Arch. ges. Virusforsch.,* **4**, 464-468

Duryee, W. R. (1956) Precancer cells in amphibian adenocarcinoma. *Ann. N.Y. Acad. Sci.,* **63**, 1280-1302

Fawcett, D. W. (1956) Electron microscope observations on intracellular virus-like particles associated with the cells of the Lucké renal adenocarcinoma. *J. biophys. biochem. Cytol.,* **2**, 725-742

Granoff, A. (1969) Viruses of amphibia. *Curr. Top. Microbiol. Immund.,* **50**, 107-137

Granoff, A., Came, P. E. & Rafferty, K. A. (1965) The isolation and properties of viruses from *Rana pipiens:* Their possible relationship to the renal adenocarcinoma of the leopard frog. *Ann. N.Y. Acad. Sci.,* **126**, 237-255

Granoff, A., Gravell, M. & Darlington, R. W. (1969) Studies on the viral etiology of the frog renal carcinoma (Lucké tumor). In: Mizell, M., ed., *Biology of amphibian tumors,* New York, Springer-Verlag, pp. 279-295

Gravell, M. (1971) Viruses and renal carcinoma of *Rana pipiens.* X. Comparison of herpes-type viruses associated with Lucké tumor-bearing frogs. *Virology,* **43**, 730-733

Lucké, B. (1938) Carcinoma in the leopard frog: its probable causation by a virus. *J. exp. Med.,* **68**, 457-468

Lunger, P. D. (1969) Fine structure studies of cytoplasmic viruses associated with frog tumors. In: Mizell, M., ed., *Biology of amphibian tumors,* New York, Springer-Verlag, pp. 269-309

Michalski, F. J. (1971) *The effect of viruses from frog renal tumor fractions on amphibian cells in vitro.* (Thesis, Notre Dame)

Mizell, M., Stackpole, C. W. & Halperen, S. (1968) Herpes-type virus recovery from "virus-free" frog kidney tumors. *Proc. Soc. exp. Biol. (N.Y.),* **127**, 808-814

Mizell, M., Toplin, I. & Isaacs, J. J. (1969) Tumor induction in developing frog kidneys by a zonal centrifuge purified fraction of the frog Herpes-type virus. *Science,* **165**, 1134-1137

Morek, D. M. & Tweedell, K. S. (1969) Interaction of normal and malignant tissues in organ culture. *Amer. Zool.,* **9**, 1125

Rafferty, K. A. (1965) Cultivation of an inclusion-associated virus from Lucké tumor frogs. *Ann. N.Y. Acad. Sci.,* **126**, 3-21

Tweedell, K. S. (1955) Adaptation of an amphibian renal carcinoma in kindred races. *Cancer Res.,* **15**, 410-418

Tweedell, K. S. (1967) Induced oncogenesis in developing frog kidney cells. *Cancer Res.,* **27**, 2042-2052

Tweedell, K. S. (1969) Simulated transmission of renal tumors in oocytes and embryos of *Rana pipiens.* In: Mizell, M., ed., *Biology of amphibian tumors,* New York, Springer-Verlag, pp. 229-239

Tweedell, K. S. & Granoff, A. (1968) Viruses and renal carcinoma of *Rana pipiens.* V. Effect of frog virus 3 on developing frog embryos and larvae. *J. nat. Cancer Inst.,* **40**, 407-410

Zambernard, J., Vatter, A. E. & McKinnell, R. G. (1966) The fine structure of nuclear and cytoplasmic inclusions in primary renal tumors of mutant leopard frogs. *Cancer Res.,* **26**, 1688-1700

The Lucké Tumour Herpesvirus—Its Presence and Expression in Tumour Cells

M. MIZELL[1]

The very characteristic that distinguishes the frog renal adenocarcinoma (Lucké tumour) also has promise of providing a model system for gaining insight into the other vertebrate tumours associated with herpestype viruses (HTV). The Lucké tumour exists in two naturally occurring seasonal forms and is especially well suited for oncogenic virological studies because large quantities of its associated herpesvirus are produced in the *in situ* "winter" tumour phase. Active herpesvirus synthesis occurs in the parenchymal cells of this neoplasm when the tumour or tumour-bearing frog is subjected to low temperature. Whether this is by natural winter hibernation or by placing in a low-temperature laboratory incubator, prolonged periods at low temperature suppress cellular DNA synthesis and stimulate virus production (Mizell, Stackpole & Halpern, 1968). In contrast, at normal or elevated temperatures, features typical of the "summer" phase of this

[1] Laboratory of Tumor Cell Biology, Tulane University, New Orleans, Louisiana, USA.

tumour prevail; cellular DNA synthesis proceeds with resultant cellular division and absence of virus production.

LUCKÉ TUMOUR HERPESVIRUS—LATENCY AND DETECTION

Our laboratory's earlier *in vivo* studies (including alloplastic and xenoplastic tumour explantation, as well as *in situ* experiments) have shown that the carcinomas of "summer" and "winter" frogs are temperature-mediated phases of the same tumour (Mizell *et al.,* 1968). These early studies, plus our subsequent investigations, have also provided substantial evidence for the existence of the complete herpesvirus genome in "virus-free" summer-tumour cells (Mizell *et al.,* 1968; Mizell, Stackpole & Isaacs, 1969). The recent successful culture, accompanied by low-temperature induction of virus *in vitro,* has now provided additional evidence unequivocally demonstrating that factors present *only* in the frog

---

PLATE I

Fig. 1. (a) Lucké renal adenocarcinoma that developed and was grown to a weight of 9 grams at room temperature ("summer tumour"); the frog was then subjected to prolonged (four months) low temperature (7.5°C) to induce virus production ("winter tumour"). Scale in millimeters. (b) Electron micrograph of frog herpesvirus particles in the cytoplasm of a cell from this tumour. Approx. ×26 900. (c) Electron micrograph of negatively stained frog herpesvirus, showing capsomere detail. Approx. ×102 000. (From Mizell, 1969.)

Fig. 2. Electron micrograph of "winter"-tumour cell. A large intranuclear sac containing a cluster of enveloped nucleated virus particles can be seen in the centre. Chromatin is marginated and the nuclear membrane continuity is interrupted in several places in this virus-infected cell. Approx. ×7200. (From Mizell, 1969.)

Fig. 3. Histological cross-section through a premetamorphic tadpole (Taylor-Kollros Stage IV) 107 days after injection with a zonal fraction enriched with the tightly enveloped form of virus illustrated in Fig. 2 (prepared from the nuclear pellet of a winter tumour). The injection was made into the region of the developing right pronephric kidney; the resulting huge right pronephric adenocarcinoma is readily apparent and fills the entire right half of the photograph. (The digestive tract was removed before fixation). n = notochord; sc = spinal cord; g = anlage of left and right gonads. Approx. ×10.

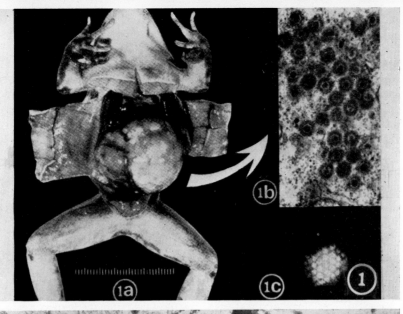

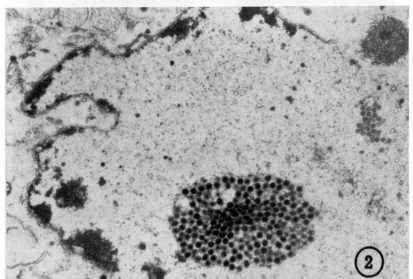

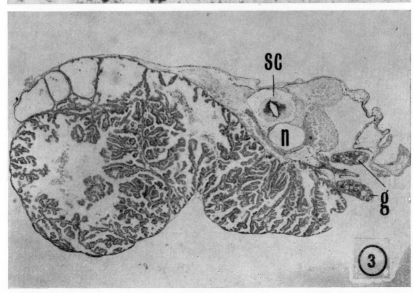

host are *not* required for the induction of virus in these "virus-free" tumours (Breidenbach *et al.*, 1971).

Furthermore, direct evidence for the existence of virus-specific messenger RNA in summer-tumour cells has recently been obtained by DNA-RNA hybridization. These elegant hybridization techniques, perfected for DNA oncogenic viruses by Green and his colleagues (see Fujinaga & Green, 1968), have revealed the covert existence of herpesvirus genetic information in the genome of summer-tumour cells. Herpesvirus-specific DNA was transcribed in relatively large quantities (4% for a 7-hour ^{3}H-uridine label) (Collard *et al.* [1]). As previously postulated, transcription of viral genes in the virus-free tumour phase would "provide quite strong evidence for the herpesvirus aetiology of the Lucké tumour" (Green, 1969).

OVERT EXPRESSION—VIRUS PRODUCTION *IN VIVO*

The characteristic that makes the Lucké tumour especially well suited for herpesvirus oncological studies and also distinguishes it from other tumour-virus systems, is the high viral yield that can be harvested from *in situ* winter tumours. An average tumour (Fig. 1) produced 10^{11} virus particles per

---

[1] Unpublished data.

gram of tumour (Toplin, Brandt & Sottong, 1969). Not only are large quantities of virus produced, but an appreciable percentage of the virus consists of complete nucleated virions. As a result, sufficient quantities of purified Lucké tumour herpesvirus DNA could be prepared from winter tumours for characterization and, in addition, comparison with the HTVs associated with other neoplasms was made possible (Wagner *et al.*, 1970).

Fig. 2 illustrates a typical example of a winter tumour parenchymal cell with an intranuclear sac of virus. It should be noted that each virus particle within this membrane-bound sac possesses a thick outer envelope, which is closely applied to its nucleocapsid. Moreover, these tightly enveloped herpes virions are prime candidates for the oncogenic form of the Lucké tumour herpesvirus: zonal centrifuge purified fractions containing high concentrations of this tightly enveloped form were found to be oncogenic when injected into frog embryos, whereas adjacent fractions of the frog herpesvirus lacking this type of particle failed to produce tumours (Mizell, Toplin & Isaacs, 1969; Mizell, 1969).

A histological section through a frog tadpole with a massive unilateral pronephric tumour can be seen in Fig. 3. This animal was injected, as a 5-day-old embryo, in the region of the then developing right pronephric kidney with 0.2 μl of a dilution (10^{-7} in DMSO) of the zonal fraction in which the tightly enveloped particle was concentrated. Further iso-

---

PLATE II

Fig. 4. Ventral view of another young tadpole 100 days after similar injection with a zonal fraction prepared from the mitochondrial pellet from a different winter tumour. The digestive tract was removed before the photograph was taken to show the full extent of the right pronephric tumour (arrow). Compare with histological cross-section in Fig. 3. Scale in millimeters.

Fig. 5. Histological sections through Lucké renal adenocarcinoma—before and after virus induction. (a) "Summer"-tumour biopsy. Note evenly dispersed chromatin; also note mitotic figure bordering the lumen in upper right-hand portion of figure. (b) Same tumour after 4 months of low-temperature treatment. Several cells now display typical "winter"-tumour characteristics: enlarged nuclei containing Cowdry type A intranuclear inclusions (arrows) with chromatin margination along the nuclear membrane. (H & E) Approx. ×560. (From Mizell, 1969.)

Fig. 6. Results of immunodiffusion tests with rabbit anti-Lucké herpesvirus serum (anti-LV) and: (1) Normal frog kidney homogenate (diluted 1:10); (2) Summer-tumour homogenate (diluted 1:10); (3) Herpesvirus from zonal fraction of winter tumour. (Virus recharged at 24 hours); (4) Herpesvirus from zonal fraction of another winter tumour (same tumour as shown in Fig. 1). Note the lack of precipitation between 1; the single faint line of precipitation (arrow) between 2; and the two precipitation lines between both 3 and 4. (The vertical tear in the agar occurred during drying of agar before staining.)

Fig. 7. Indirect immunofluorescent staining of winter-tumour cells. In this preparation the fluorescence is especially prominent in the cells along the lumenal surface. Approx. ×930.

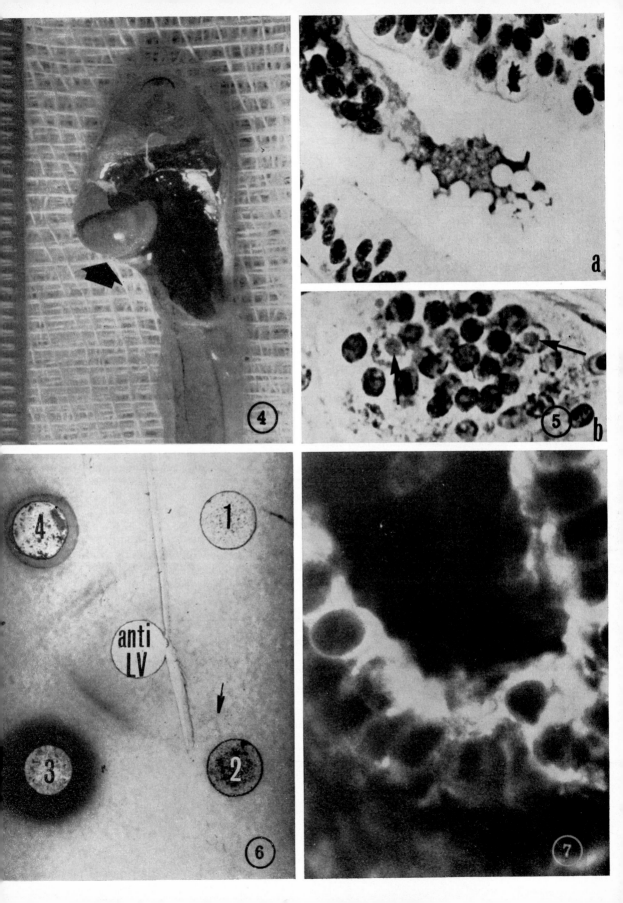

pyknic separation of this fraction was successful, and this particle banded at a density of 1.20 in both sucrose and potassium tartrate (Toplin *et al.,* 1971). Fig. 4 illustrates the external appearance of the adenocarcinomas produced in tadpoles that received injections of purified fractions of this enveloped form of herpes virion.

Summer-phase tumours characteristically display a uniformly vigorous histological pattern (Fig. 5a); however, when subjected to prolonged low temperature, histological features heralding the presence of virus become evident (Fig. 5b). Controls injected with zonal fractions prepared from summer-phase tumours, obtained from an equivalent region of the zonal sucrose gradient, failed to induce tumours. Moreover, the high degree of separation and concentration of virus attained by the use of the zonal centrifuge upon winter tumours failed to reveal *any* virus in summer tumours (Toplin *et al.,* 1971).

Although demonstration of the covert existence of Lucké herpesvirus information in summer-phase tumour supports the hypothesis of a herpesvirus aetiology, many fundamental questions of virus-tumour cell relationship remain unanswered.

IMMUNOLOGICAL EXPRESSION AND DETECTION

Once these virus-containing tumours had been subjected to zonal centrifugation, purified virus preparations became available for use as antigen in immunological experiments. Heterologous antiserum was produced, and early collaborative immunodiffusion studies revealed the unexpected sharing of a common antigen with the EB virus of Burkitt's lymphoma (Fink, King & Mizell, 1968;

Fink *et al.,* 1969; Kirkwood *et al.,* 1969). Our laboratory has since been engaged in developing sensitive serological methods capable of distinguishing between summer ("virus-free") adenocarcinoma cells and winter or virus-containing adenocarcinoma cells. These tests should also differentiate between "normal" kidney parenchymal cells from the two tumour phases.

We have recently been able to perfect a specific immunofluorescence test for detecting herpesvirus antigen in cells of winter-phase tumour (Paul *et al.,* 1972). Before this test was devised, *in situ* virus detection could be achieved only by electron microscopy. Zonal fractions, enriched with frog HTV, were used to prepare immune rabbit serum (IRS). The heterologous antiserum was repeatedly absorbed with normal frog tissue homogenates of kidney, liver, and spleen, and then tested in standard agar gel diffusion plates. After four of five absorptions the antiserum ceased to react with normal kidney antigen (Fig. 6; well no. 1). However, this absorbed IRS elicited a faint precipitation band with summer-tumour antigen(s) (well no. 2). When this was removed by an additional absorption with nonviral summer-tumour homogenate, the resulting IRS still retained the ability to react strongly with frog herpesvirus preparations.

Fluorescein-conjugated goat antirabbit serum was absorbed twice with rehydrated mouse liver powder and once with normal frog kidney homogenate before being used in the examination of summer- and winter-phase tumours. The strong reaction of the antisera with virus prepared from different neoplasms in gel immunodiffusion tests (see Fig. 6) was also reflected in the immunofluorescence testing of histo-

PLATE III

Fig. 8. Histological section of summer-tumour explant maintained on agar slant in screw cap tube *in vitro* for 4 months at 7.5°C. Although long-term culture has resulted in connective tissue (C.T.) encircling the tumour tubules, the morphology of the tubules and tumour cells appears similar to that of an *in vivo* induced "winter" tumour. (H & E) Approx. ×95.

Fig. 9. Higher magnification of explant in Fig. 8, showing the swollen nuclei (arrows) characteristic of virus-containing "winter"-tumour cells. (H & E) Approx. ×420.

Fig. 10. Unstained phase-microscope photograph of new tubular outgrowth from the periphery of a "summer"-phase tumour explant after 16 days in culture (plastic flask) at 15.5°C. Approx. ×45.

Fig. 11. Low-power phase-microscope photograph of the same explant one month later. After 46 days in culture at 15.5°C, a large tongue of new growth (arrow) has almost doubled the mass of the explant. (Unstained.) Approx. ×10.

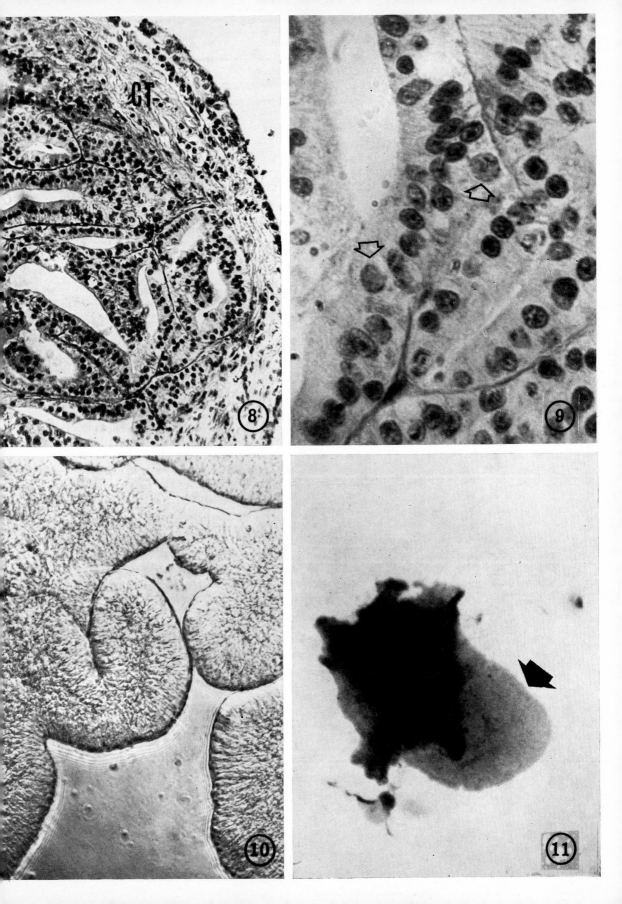

logical sections of winter tumour. Fig. 7 demonstrates the bright apple-green fluorescence occurring in winter tumours when stained with IRS in the indirect immunofluorescence test. The number of cells staining, and their location with respect to tumour histology, varied with the individual winter tumours tested. The example shown in Fig. 7 demonstrated bright fluorescence in the epithelial cells lining a tumour tubule; other winter tumours characteristically showed fluorescence in cells and fragments of cells within the tubular lumen; still others demonstrated bright fluorescence in groups of epithelial cells scattered throughout the section. In all cases, subsequent electron microscopy of these different winter tumours confirmed that the virus particles were localized in the areas concerned. Both intranuclear and cytoplasmic fluorescence was observed, but freezing and thawing artefacts made more precise localization impossible. Normal kidney and summer ("virus-free") carcinomas consistently failed to fluoresce.

The exact nature of the antigen being detected is still unknown, but it is obviously an antigen of the herpesvirus characteristic of the winter state of the Lucké tumour. Continued investigation and especially comparative studies with some of the other HTV antigens reported at this meeting will undoubtedly prove instructive.

MONITORING VIRUS PRODUCTION *IN VITRO*

The Lucké HTV immunofluorescence test now provides the means to monitor the induction of vegetative virus production in summer-tumour cells. Screw cap tubes with agar slants have been used to demonstrate the *in vitro* induction of virus in summer-tumour explants (see Figs. 8 & 9; electron microscopic examination of comparable explant samples in this experiment confirmed the presence of herpesvirus). Nevertheless, connective tissue growth often impairs visualization of the tumour parenchymal cells (see Fig. 8) and, furthermore, the curved surface of the tube makes microscopic visualization impossible. Close monitoring is obviously impossible in the adult tumour-bearing frog. We have therefore devoted considerable effort to developing an *in vitro* culture system that would overcome the above difficulties and permit close monitoring of vegetative virus production. The techniques previously employed in our laboratory for *in vitro* virus induction (Breidenbach *et al.*, 1971) were adapted for culturing

explants in plastic Falcon flasks (Blazek *et al.* [1]). In the quest for explant strains that would manifest vigorous epithelioid growth in culture, several hundred explants were prepared from the primary tumours of adult frogs, primary "embryonic" tumours of tadpoles (see Figs. 3 & 4) and serially transplanted eye-chamber-adapted tumour growths. Several incubation temperatures were employed. If an explant exhibited vigorous epithelioid growth *in vitro* it was split, and an explant strain was developed and maintained. An example of one of these vigorous growths is shown in Figs. 10 & 11. While these explants are being monitored during herpesvirus induction, the factors encouraging tongues of epithelial growth, similar to the one illustrated in Fig. 11, are also being studied. The knowledge gained from this model system of closely monitored virus rescue should be of immeasurable value for other systems of known or suspected HTV aetiology.

CONCLUSIONS AND DISCUSSION

In summary, the summer or "virus-free" tumour phase consists of tumour cells lacking demonstrable virus. Virus is undetectable by electron microscopy, and zonal centrifugation, with its high degree of separation and concentration, does not reveal any virus. Furthermore, virus-specific antigen is lacking or is undiscernible in a sensitive indirect immunofluorescence test. Nevertheless, Lucké-virus-specific messenger RNA is transcribed in these neoplastic cells, and *in vitro* culture plus low-temperature treatment results in virus induction. These findings thus demonstrate the absence of this viral antigen or virus, *per se,* but at the same time indicate the existence of the herpesvirus genome in these virus-free summer-tumour cells.

How does one explain this phenomenon? It was pointed out in 1965 that "several hypothetical schemes can be put forward, which in themselves are exciting as 'mental gymnastic exercises'." (Mizell & Zambernard, 1965); among the suggested possibilities were a parallel to bacterial lysogeny or even a helper-virus phenomenon, similar to Rous sarcoma virus (Mizell & Zambernard, 1965). A choice between the numerous possibilities should soon be feasible. Whatever the mechanism, we now have conclusive proof that the herpesvirus genome is present in the summer tumour, and the presence and

---

[1] Unpublished data.

expression of the virus are undoubtedly linked to a temperature-sensitive step in the development of the tumour cells (Mizell, Stackpole & Isaacs, 1969). Furthermore, the sensitive monitoring system mentioned above should soon yield detailed knowledge of viral replication and lead to a firm understanding of this fascinating system.

Finally, my background as a developmental physiologist leads me to conjecture that the reasons why the viral genome is not fully expressed in the summer-tumour cells may be similar to the reasons why

normal differentiated kidney cells do not transcribe muscle information.

ACKNOWLEDGEMENTS

These studies were supported by US Public Health Service grant CA-011901 from the National Cancer Institute, and Damon Runyon Memorial Fund grant DRG-1055. The library facilities of the Woods Hole Biological Laboratories aided immensely in the preparation of this paper and are gratefully acknowledged.

REFERENCES

Breidenbach, G. P., Skinner, M. S., Wallace, J. H. & Mizell, M. (1971) In vitro induction of a herpes-type virus in "summer-phase" Lucké tumor explants. *J. Virol.,* 7, 679-682

Fink, M. A., King, G. S. & Mizell, M. (1968) Preliminary note: identity of a herpesvirus antigen from Burkitt lymphoma of man and the Lucké adenocarcinoma of frogs. *J. nat. Cancer Inst.,* 41, 1477-1478

Fink, M. A., King, G. S. & Mizell, M. (1969) Reactivity of serum from frogs and other species with a herpesvirus antigen extracted from a Burkitt lymphoma cultured cell line. In: Mizell, M., ed., *Biology of amphibian tumors,* New York, Springer-Verlag, pp. 358-364

Fujinaga, K. & Green, M. (1968) The mechanism of viral carcinogenesis by DNA mammalian viruses. V. Properties of purified viral-specific RNA from human adenovirus induced tumor cells. *J. molec. Biol.,* 31, 63-73

Green, M. (1969) Nucleic acid homology as applied to investigations on the relationship of viruses to neoplastic diseases. In: Mizell, M., ed., *Biology of amphibian tumors,* New York, Springer-Verlag, pp. 445-454

Kirkwood, J. M., Geering, G., Old, L. J., Mizell, M. & Wallace, J. (1969) A preliminary report on the serology of Lucké and Burkitt herpes-type viruses: a shared antigen. In: Mizell, M., ed., *Biology of amphibian tumors,* New York, Springer-Verlag, pp. 365-367

Mizell, M. (1969) State of the art: Lucké renal adenocarcinoma. In: Mizell, M., ed., *Biology of amphibian tumors,* New York, Springer-Verlag, pp. 1-25

Mizell, M., Stackpole, C. W. & Halpern, S. (1968) Herpes-type virus recovery from "virus-free" frog kidney tumors. *Proc. Soc. exp. Biol. (N.Y.),* 127, 808-814

Mizell, M., Stackpole, C. W. & Isaacs, J. J. (1969) Herpes-type virus latency in the Lucké tumor. In: Mizell, M., ed., *Biology of amphibian tumors,* New York, Springer-Verlag, pp. 337-347

Mizell, M., Toplin, I., & Isaacs, J. J. (1969) Tumor induction in developing frog kidneys by a zonal centrifuge purified fraction of the frog herpes-type virus. *Science,* 165, 1134-1137

Mizell, M. & Zambernard, J. (1965) Viral particles of the frog renal adenocarcinoma: causative agent or passenger virus? II. A promising model system for the demonstration of a "lysogenic" state in a metazoan tumor. *Ann. N.Y. Acad. Sci.,* 126, 146-169

Paul, S. M., Mizell, M., Craige, B., Blazek, J. & Skinner, M. (1972) Specific immunofluorescence test for detection of herpesvirus antigen in cells of the Lucké renal adenocarcinoma. *Proc. Soc. exp. Biol. (N. Y.)* (in press)

Toplin, I., Brandt, P., & Sottong, P. (1969) Density gradient centrifugation studies on the herpes-type virus of the Lucké tumor. In: Mizell, M., ed., *Biology of amphibian tumors,* New York, Springer-Verlag, pp. 348-357

Toplin, I., Mizell, M., Sottong, P. & Monroe, J. (1971) Zonal centrifuge applied to the purification of herpesvirus in the Lucké frog kidney tumor. *Appl. Microbiol.,* 21, 132-139

Wagner, E. K., Roizman, B., Savage, T., Spear, T., Mizell, M., Durr, F. E. & Sypowicz, D. (1970) Characterization of the DNA of herpesviruses associated with Lucké adenocarcinoma of the frog and Burkitt lymphoma of man. *Virology,* 42, 257-261

PATHOLOGY OF BURKITT'S LYMPHOMA, INFECTIOUS MONONUCLEOSIS AND NASOPHARYNGEAL CARCINOMA

Chairman — J. B. Moloney

Rapporteur — C. S. Muir

The Pathology of Burkitt's Lymphoma—A Review

D. H. WRIGHT [1]

INTRODUCTION

The gross and microscopic features of Burkitt's lymphoma have been described in detail previously (Wright, 1970a, b). It is my intention in this paper to discuss the problems involved in the diagnosis of this tumour, and its relationship to other neoplasms of the lymphoreticular system.

Burkitt described a clinical entity seen in African children and characterized by jaw tumours and multifocal visceral tumours (Burkitt, 1958). On the basis of these characteristic and unusual features, he mapped out the geographical distribution of the tumour in Africa; this in turn gave rise to the suspicion that the tumour might have an infectious aetiology and to the search for aetiological agents of the tumour. While it is true that the clinical and gross anatomical features of Burkitt's lymphoma are distinctive and form a satisfactory basis for broad epidemiological studies, they are not in any way specific. The most characteristic feature is the presence of multiple jaw tumours. However, a number of other neoplastic and non-neoplastic lesions in the jaw may mimic Burkitt's lymphoma to varying degrees, and some of these can be distinguished with certainty only by their microscopic features (Ziegler, Wright & Kyalwazi, 1971). Perhaps of greater importance is the fact that jaw tumours in Burkitt's lymphoma are age-dependent and that, when patients of all ages are considered, only half the total have jaw or facial involvement. The remainder present mainly with abdominal and neurological symptoms that may be quite nonspecific in character, although bilateral ovarian tumours, multiple renal tumours and thyroid tumours are almost as characteristic as the jaw tumours, particularly if they occur in conjunction with one another.

It was because of these problems that attempts were made to define Burkitt's lymphoma on the basis of microscopical features alone. By means of diagnostic criteria based on such features, it was possible to identify adult cases of Burkitt's lymphoma, as well as cases with unusual clinical features, such as massive bilateral breast swellings associated with pregnancy and lactation in young adults (Shepherd & Wright, 1967). It was also possible to identify cases of Burkitt's lymphoma outside Africa, both in tropical Papua-New Guinea and in nontropical regions, such as Europe and North America. Non-African cases identified on the basis of microscopical features had in general an anatomical tumour distribution similar to that of the African patients, but different from that of "conventional" lymphosarcoma (Wright, 1966).

The questions that remain to be answered are: What are the limits of variation of the microscopical features of Burkitt's lymphoma; what is the relationship of Burkitt's lymphoma to other malignant lymphomas; and what are the essential criteria for the diagnosis of Burkitt's lymphoma? Before attempting to answer these questions, I shall outline the microscopical features of Burkitt's lymphoma, as seen in the majority of African patients with characteristic clinical features.

CYTOLOGY

Burkitt's lymphoma cells do not form or excite the production of reticulin and the tumours are very soft, except when infiltrating pre-existing tissues. As a consequence, the tumour cells are readily detached when touched on to glass slides, and make very satisfactory imprint preparations. The imprint technique is perhaps the simplest and most satisfactory method for the identification of Burkitt's lymphoma cells and is less liable to yield artefacts

[1] Reader in Pathology, The University of Birmingham, The Medical School, Birmingham, UK. Formerly Reader in Pathology, Makerere University College Medical School, Kampala, Uganda.

than the histological technique. Smears of body fluids are equally satisfactory for making cytological preparations of Burkitt's lymphoma cells. Nevertheless, artefacts do occur with this method. Thick "wet" preparations give rounded, darkly staining cells (Bluming & Templeton, 1971), and "dry" preparations result in the rupture of cells and artefactual nuclear pleomorphism.

In cytological preparations (Figs. 1 & 2) Burkitt's lymphoma cells vary in size from 10μ to 25μ. This variation in size does not correspond to any apparent maturation of the tumour cells. The cytoplasm is well defined, deeply basophilic, apart from a pale staining area opposite the nuclear indentation, and nongranular. It contains a variable number of clear rounded vacuoles 1–2 μ in diameter.

The nucleus is usually rounded or indented, but in some cases shows deep divisions or clefts. Two to five nucleoli are visible in most cells. These are surrounded by condensed parachromatin, the remainder of the chromatin being evenly distributed and having a stippled or granular character.

HISTOLOGY

The examination of sections of paraffin-embedded tissue is the standard method of the diagnostic histopathologist. While this is satisfactory for most purposes, it has limitations when applied to the diagnosis of lymphomas. Paraffin embedding may cause considerable cell shrinkage and loss of the cell detail on which the identification of lymphoma cells rests. Variations in cell morphology may also be produced by fixation artefacts. These limitations of the histological technique are responsible in part for the difficulties experienced in the diagnosis of Burkitt's

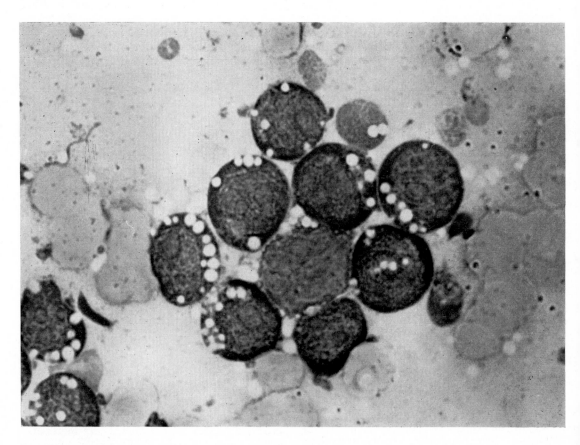

Fig. 1. Imprint of Burkitt's lymphoma cells showing well defined rim of intensely basophilic cytoplasm containing prominent vacuoles. (May-Grunwald-Giemsa stain) ×1700.

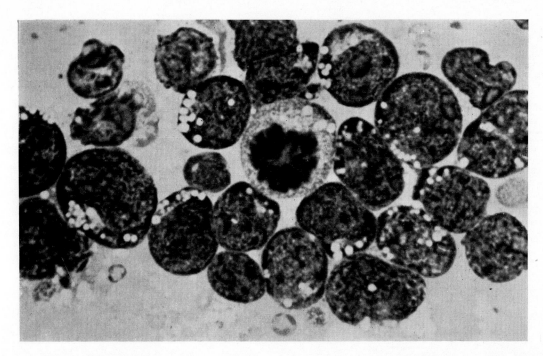

Fig. 2. Imprint of Burkitt's lymphoma cells. The cells show variation in size but all contain nucleoli, have deeply basophilic cytoplasm and show no apparent maturation. (May-Grunwald-Giemsa stain) ×1560.

lymphoma and its separation from other lympho-reticular neoplasms.

One of the most striking features of histological sections of Burkitt's lymphoma is the presence of large clear or foamy histiocytes interspersed between the lymphoid cells (Fig. 3). These histiocytes are usually laden with cell debris, pyknotic nuclei and even whole tumour cells, and give the so-called "starry-sky" appearance to the tumour. This appearance is characteristic of most cases of Burkitt's lymphoma, but is not diagnostic and may occasionally be seen in other lymphomas and even carcinomas. It may be related to the high rate of proliferation of Burkitt's lymphoma cells and the consequent high rate of cell death. The foamy appearance of the histiocytes is due, at least in part, to the presence of coarse globules of neutral fat (Figs. 4 & 5) derived, presumably, from ingested lymphoid cells.

In histological sections, the lymphoid cells have a uniform appearance with rounded or indented nuclei (Fig. 3). The appearance of the nuclear chromatin varies with the fixation, being granular in well fixed areas but tending to become vesicular, often towards the centre of the section, where fixation has been delayed. Two or more nucleoli are usually visible in the cells with vesicular nuclei, but are inconspicuous in well fixed cells. In well fixed tissues, the tumour cells have a clearly defined narrow eccentric rim of cytoplasm. This has an amphophilic staining quality in haematoxylin and eosin stained sections, due to the high cytoplasmic content of RNA. Inspection of the cytoplasm with an oil immersion lens usually reveals occasional vacuoles corresponding to the vacuoles seen in imprint preparations. These features are much better seen in plastic-embedded tissues (Figs. 6 & 7). Plastic embedding eliminates much of the shrinkage artefact caused by hot paraffin wax, and the thin sections that may be cut using this technique are much more amenable to the critical study of cell cytology. The disadvantage of the technique is that some special stains are difficult to use with plastic-embedded tissue; it also requires greater technical skill than routine processing in paraffin wax.

A number of special stains may aid the diagnosis of Burkitt's lymphoma. Methyl green-pyronin staining shows a high and uniform cytoplasmic pyroninophilia throughout the tumour, provided

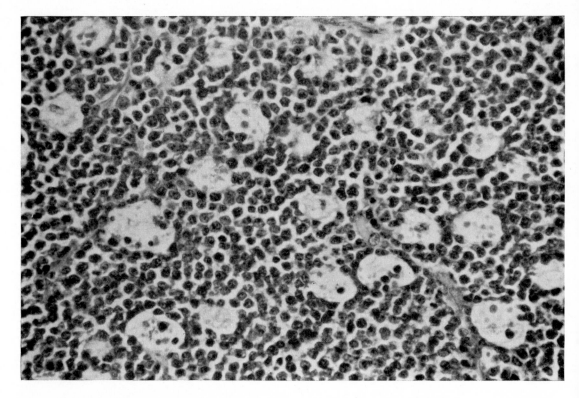

Fig. 3. Section of Burkitt's lymphoma showing rounded uniform immature lymphoid cells with scattered large clear histiocytes containing pyknotic cell nuclei and cell debris. (Haematoxylin and eosin) ×530.

that the tissue has been well fixed (Wright & McAlpine, 1966). Pyronin staining more clearly delineates the cytoplasm than eosin; the amount of cytoplasm and the cytoplasmic vacuoles are thus more readily seen with this stain. Other undifferentiated lymphomas and carcinomas may show a uniform intense cytoplasmic pyroninophilia. Malignant lymphoma, lymphocytic type, poorly differentiated, is, however, composed of cells with a varying cytoplasmic content of RNA, and a range of cytoplasmic pyroninophilia can be seen throughout this tumour.

The vacuoles seen in imprint preparations of Burkitt's lymphoma cells are caused by neutral fat droplets that are dissolved out of the cells by methyl alcohol fixation. Frozen sections of formalin-fixed Burkitt's lymphoma tissue usually reveal coarse droplets of neutral fat in the cytoplasm of the lymphoid cells, and the histiocytes responsible for the "starry-sky" pattern are usually laden with coarse lipid droplets (Figs. 4 & 5) (Wright, 1968). Occasional Burkitt's lymphomas contain little lipid, and

lipid may be present in large quantities in histiocytic lymphomas and carcinomas, so this feature is not in itself diagnostic. PAS staining [1] of histological sections is of little diagnostic value, except that the histiocytes in Burkitt's lymphoma usually contain diastase-resistant coarse PAS-positive granules. This stain may therefore bring out the histiocytes in situations in which they are not otherwise prominent, e.g. in *post mortem* tissue, in which they often appear collapsed.

Burkitt's lymphoma cells do not produce reticulin or excite its formation. The reticulin content of large tumours is therefore usually scanty.

FINE STRUCTURE

At low magnification, the lymphoid cells exhibit a striking monomorphism. The characteristic ultra-

---

[1] Staining with periodic acid-Schiff-haematoxylin stain.

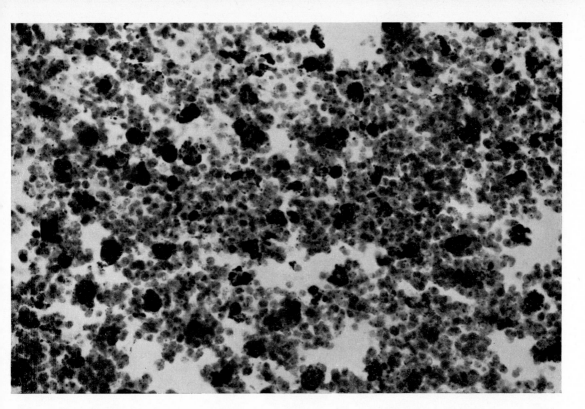

Fig. 4. Frozen section of formalin fixed Burkitt's lymphoma tissue showing small lipid droplets in lymphoid cells. The histiocytes scattered throughout the tumour are filled with lipid. (Oil red 0 stain) ×270.

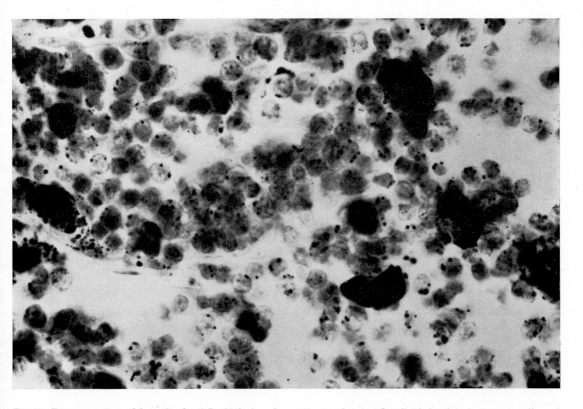

Fig. 5. Frozen section of formalin fixed Burkitt's lymphoma tissue showing fine lipid droplets in tumour cells and coarse aggregates of lipid in histiocytes. (Oil red 0 stain) ×690.

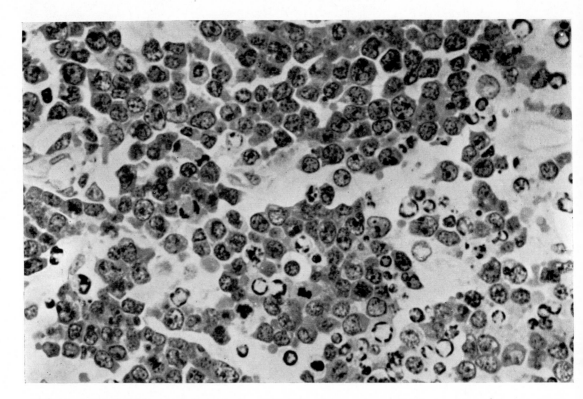

Fig. 6. Section of plastic embedded Burkitt's lymphoma tissue. Note regular appearance of lymphoid cells and the well defined darkly staining cytoplasm of these cells. Several degenerate cells show the characteristic ring of condensed nuclear chromatin. (Haematoxylin and eosin) ×510.

structural features of these cells are as follows (Bernhard, 1970):

1. The *nucleus* is rounded oval or indented and rarely deeply invaginated. Characteristic projections of the nuclear envelope, often enclosing islands of cytoplasm, have been described (Achong & Epstein, 1966); these are not, however, specific for Burkitt's lymphoma cells. Similarly, a characteristic but non-specific type of nuclear degeneration has been described (Epstein & Herdson, 1962; Bernhard & Lambert, 1963). This may also be seen at the light microscope level (Fig. 6).

Chromatin is clumped at the nuclear membrane and around the nucleoli and the interchromatinic substance is relatively clear. Nucleoli are quite large and the nucleolonemas are usually visible.

2. The *cytoplasm* is characterized by a large number of polyribosomes with a paucity of endoplasmic reticulum. Mitochondria are few and tend to occur at one pole of the cell.

In a study of 46 lymphoid neoplasms from patients with the clinical features of Burkitt's lymphoma, Bernhard (1970) classified 38 as undifferentiated lymphoid cell sarcomas, three as differentiated lymphoid cell sarcomas and five as reticulum cell sarcomas. The distinction between these three categories was not clear cut and was to some extent arbitrary, although two of the reticulum cell sarcomas were said to be highly undifferentiated and to have no resemblance to the usual Burkitt's lymphoma cells. These two cases were described as "typical Burkitt jaw tumours", although one of them was a man 38 years of age.

DEFINITION OF BURKITT'S LYMPHOMA

On the basis of the morphological features described above, a panel of haematopathologists and cytologists defined Burkitt's lymphoma in the following terms (Berard *et al.*, 1969):

"Burkitt's tumour is a malignant neoplasm of

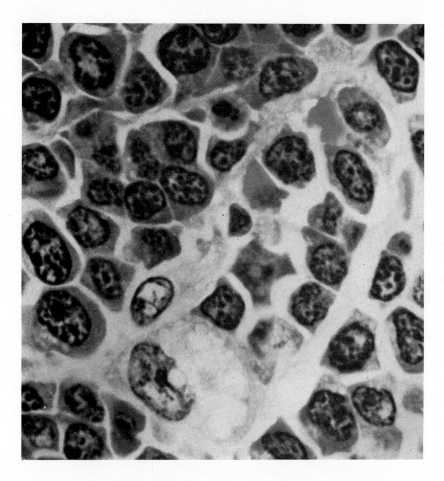

Fig. 7. Section of plastic-embedded Burkitt's lymphoma tissue. Characteristic vacuoles can be seen in the darkly staining cytoplasm of some of the cells. There is a non-neoplastic histiocyte with vacuolated cytoplasm to the bottom of the photomicrograph. (Haematoxylin and eosin) ×1300.

the haematopoietic system and is more specifically designated: malignant lymphoma, undifferentiated, Burkitt's type. The predominant and characteristic cells are undifferentiated lymphoreticular, or primitive stem cells, showing moderate nuclear and cytoplasmic variations interpretable either as biological variations within the same cell type or as limited differentiation to histiocytic or lymphocytic cell types."

RELATIONSHIP OF BURKITT'S LYMPHOMA TO OTHER TUMOURS OF THE LYMPHORETICULAR SYSTEM

Burkitt's lymphoma is a neoplasm of stem cells, or undifferentiated lymphoreticular cells. With adequate histological or cytological preparations, therefore, it can be readily separated from differentiated neoplasms of the lymphoreticular system. Hodgkin's disease and well differentiated lymphocytic lymphoma can be clearly distinguished from Burkitt's lymphoma on cytological and histological criteria, and have an entirely different clinical and anatomical presentation. Histiocytic lymphomas, even when poorly differentiated, can be separated from Burkitt's lymphoma in histological preparations by their large, folded, vesicular nuclei. In imprint preparations, these cells have indented or pleomorphic nuclei, fine nuclear chromatin, a single nucleolus and abundant cytoplasm that may contain fine vacuoles (Fig. 8). Ziegler et al. (1970) classified six

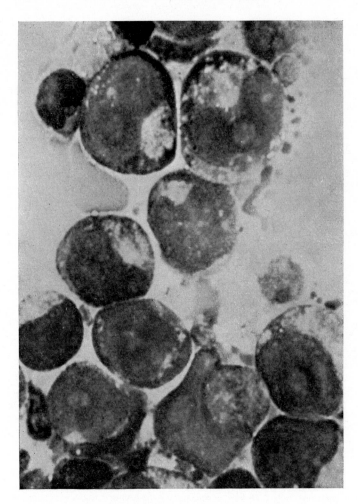

Fig. 8. Imprint of poorly differentiated histiocytic lymphoma (reticulum cell sarcoma) showing indented nucleus with single nucleolus and abundant cytoplasm containing very fine vacuoles. (May-Grunwald-Giemsa stain) ×1560.

out of 66 children admitted to the Lymphoma Treatment Centre, Kampala, as having histiocytic lymphoma. These children differed from those with Burkitt's lymphoma in age, clinical features and response to cytotoxic therapy.

Confusion between Burkitt's lymphoma and malignant lymphoma, lymphocytic type, poorly differentiated (lymphosarcoma) has existed in the past. This is in part due to semantics. Lymphocytic lymphoma, poorly differentiated, is composed of cells that range in appearance from lymphoblasts to small lymphocytes (Fig. 9).

Clinically, these tumours present mainly as lymphadenopathy, superior vena caval obstruction and intestinal obstruction (Wright, 1967). With adequate histological and cytological preparations, the distinction between them and Burkitt's lymphoma is relatively well defined. There are, however, other undifferentiated lymphoid tumours that are composed of uniformly immature cells (Fig. 10); these are often categorised as lymphoblastomas. In some instances they may represent a tissue phase of acute lymphoblastic leukaemia. Burchenal (1966) has described gonadal tumours in patients with lymphoblastic leukaemia in remission. Such cases can be separated from Burkitt's lymphoma on the basis of cytological, histological and clinical features.

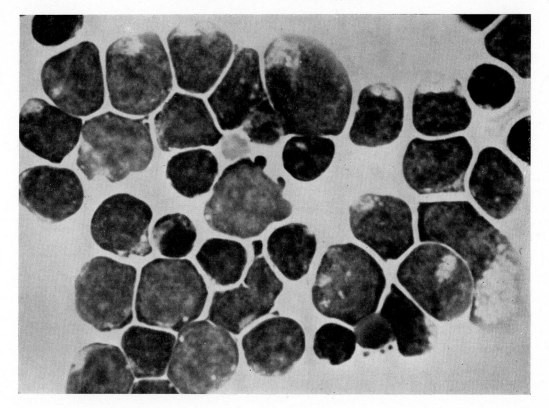

Fig. 9. Imprint of malignant lymphoma, poorly differentiated, lymphocytic type, showing cells that range in maturity from lymphoblasts to small lymphocytes. (May-Grunwald-Giemsa stain) ×1560.

ATYPICAL BURKITT'S LYMPHOMAS

Marked cytological atypia is seen occasionally in Burkitt's lymphomas that have recurred following chemotherapy, and Wright (1970c) has described Reed-Sternberg-like cells in such cases (Fig. 11). Cytogenetic studies of Burkitt's lymphoma cells from pretreatment biopsies have shown that the majority of cells have a chromosome number around the diploid mode (Jacobs, Tough & Wright, 1963; Gripenberg, Levan & Clifford, 1969). However, in the few recurrent Burkitt's lymphomas studied there has been a high proportion of tetraploid cells (Clifford et al., 1968). Sinkovics et al. (1970), studying the emergence of polyploid immunoresistant cells from a rodent lymphoma, suggest that, when selective factors are not active, diploid cells have a proliferative advantage over tetraploid cells, but that in the presence of host immunity the emergence of tetraploid immunoresistant cells may be favoured. This proliferative advantage of tetraploid cells could

account for the pleomorphism of some Burkitt's lymphomas following chemotherapy. Similarly, polyploid cells might selectively proliferate in patients with high levels of tumour immunity before chemotherapy has been given. Since most histopathologists equate pleomorphism in lymphomas with histiocytic differentiation, such cases are likely to be diagnosed as histiocytic lymphomas or as Burkitt's lymphomas showing histiocytic differentiation.

In their study of 66 children with malignant lymphoma admitted to the Lymphoma Treatment Centre, Kampala, Ziegler et al. (1970) diagnosed six cases as histiocytic lymphoma and six cases as stem cell or lymphoblastic lymphoma. Five of the cases diagnosed as lymphoblastic lymphoma had facial or jaw tumours, of which three were clinically typical of Burkitt's lymphoma but two showed atypical features. The response of the cases of lymphoblastic lymphoma to chemotherapy was not as good as that of the 54 cases of Burkitt's lymphoma, but was better than that of the patients with histiocytic lympho-

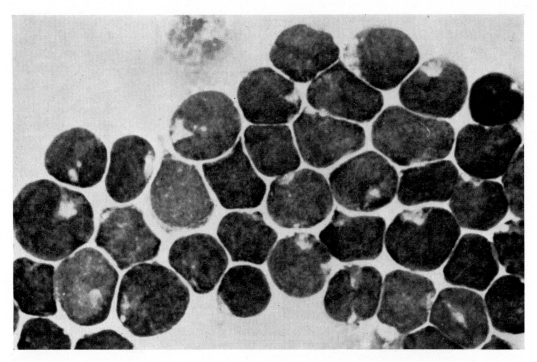

Fig. 10. Imprint of malignant lymphoma composed of uniformly immature lymphoid cells (lymphoblastoma). (May-Grunwald-Giemsa stain) ×1560.

mas. The authors raise the question of whether these cases of lymphoblastic lymphoma should be classified as Burkitt's lymphoma or atypical Burkitt's lymphoma.

The cytology of a case similar to those classified by Ziegler *et al.* as lymphoblastic lymphoma is illustrated in Fig. 12. This shows an imprint of a jaw tumour from a 5-year-old boy who also had paraplegia. I would interpret this as a stem cell lymphoma, since the evidence for lymphocytic differentiation is debatable, although many pathologists would accept the roundness of the cells and the small size of some of them as evidence of differentiation. In my opinion, the classification of these cases hinges on the tumour cytology. If there is unequivocal evidence of differentiation in the tumour it should not be classified as a Burkitt's lymphoma, and in my experience it is unlikely that such cases will show the characteristic clinical features of this tumour. Conversely, if the tumour is undifferentiated but shows the clinical features of Burkitt's lymphoma, such as the case illustrated in Fig. 12, it should probably be classified as an atypical Burkitt's lymphoma.

DIAGNOSTIC CRITERIA FOR BURKITT'S LYMPHOMA

The categorization of African patients showing the characteristic clinical and cytological features of Burkitt's lymphoma presents no problem provided that adequate histological and cytological preparations are made from the tumour. Similar cases are seen in non-African patients and these are also usually diagnosed as Burkitt's lymphoma or, by the more cautious, as "Burkitt-like" lymphomas. The recognition of these cases is important from the point of view of treatment and prognosis (*Lancet, 1968*). It is also important from the point of view both of epidemiological studies and the search for aetiological agents of the tumour that such cases should be accurately identified, although morphological identity does not necessarily imply a common aetiology.

Tumours showing the cytological features of Burkitt's lymphoma but not showing the characteristic clinical features should be classified as Burkitt's lymphoma, or at least as "clinically atypical Burkitt's lymphoma". The clinical presentations of the tumour are very varied and by themselves cannot be used to make or exclude the diagnosis.

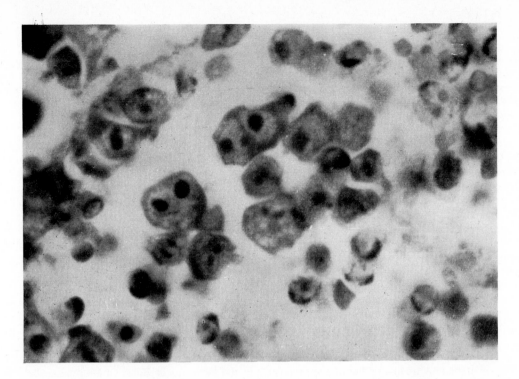

Fig. 11. Section of recurrent Burkitt's lymphoma showing cellular pleomorphism and binucleate Reed-Sternberg-like cells. (Haematoxylin and eosin) ×1500.

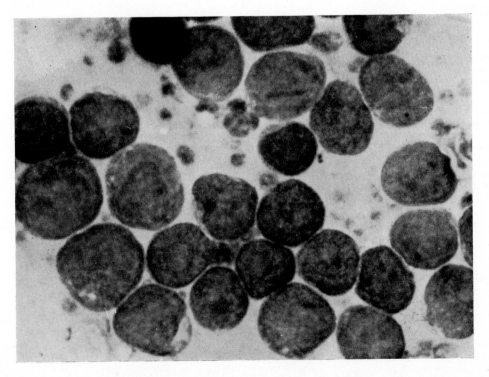

Fig. 12. Imprint of an atypical Burkitt's lymphoma from a 5-year-old boy with a maxillary tumour and paraplegia. The roundness of the cells suggests lymphocytic differentiation but most cells, including the small darkly staining one in the centre of the field, contain a single large nucleolus. (May-Grunwald-Giemsa stain) ×1560.

Patients with the clinical features of Burkitt's lymphoma but with atypical cytological features represent the major diagnostic problem. It has been shown that a number of tumours, such as embryonal rhabdomyosarcoma, retinoblastoma and chloroma may be indistinguishable from Burkitt's lymphoma clinically (Ziegler, Wright & Kyalwazi, 1970). These cases can be identified by critical histological and cytological examination. Patients with differentiated lymphomas can similarly be identified by critical tumour cytology. Such cases rarely mimic Burkitt's lymphoma clinically. On the other hand undifferentiated, or stem cell, lymphomas showing the clinical features of Burkitt's lymphoma should probably be classified as atypical Burkitt's lymphoma, as discussed above.

It must be recognised, therefore, that the diagnosis of Burkitt's lymphoma on histological criteria, although more precise than the use of clinical features alone, is not entirely satisfactory. If one adheres rigidly to the characteristic cytological features for the recognition of the tumour, cases that show cytological atypia but are in all other respects typical examples of Burkitt's lymphoma, such as that illustrated in Fig. 12, will be excluded. Conversely,

if such cases are accepted as Burkitt's lymphoma one should logically accept all cases with a similar cytology irrespective of the clinical features. For example, if the case illustrated in Fig. 12 is accepted as Burkitt's lymphoma, a tumour showing the same cytology but presenting as a retroperitoneal mass would also have to be accepted. In practice, such cases are likely to be diagnosed as stem cell lymphomas, particularly if they occur in non-African patients. It is obvious that this problem will not be solved by the use of both clinical and cytological features for the diagnosis of the tumour.

Such problems arise because we do not know whether Burkitt's lymphoma is an aetiological entity. If it is, we may one day be able to recognise it on the basis of aetiological or antigenic features and to categorize more precisely the cytologically atypical cases. Until such time as these other criteria are established, if ever, it will be necessary for atypical Burkitt's lymphomas to be included in a separate category. It will also remain essential that skilled and critical judgement should be used in the diagnosis of Burkitt's lymphoma, both from the point of view of the patient's welfare and of the epidemiological and laboratory studies of this tumour.

SUMMARY

Burkitt's lymphoma was first described as a clinical entity yet it cannot be precisely defined or diagnosed on the basis of clinical features alone. The characteristic cytology of the tumour provides a more satisfactory criterion for identifying the tumour. On this basis it can be clearly separated from nonlymphomatous tumours that may closely mimic its clinical features. It can also be separated from lymphomas that show differentiation towards recognizable differentiated cell types.

Occasional tumours composed of undifferentiated lymphoreticular cells without the characteristic features of Burkitt's lymphoma cells, but presenting the clinical features of this tumour have been seen. These should probably be diagnosed as atypical Burkitt's lymphomas. If this categorization is accepted, tumours with a similar cytology but not necessarily showing the characteristic clinical features of Burkitt's lymphoma should logically be placed in the same grouping. Ultimately Burkitt's lymphoma may be defined more precisely on the basis of aetiological or antigenic features. Until such time as these features are recognized, it is essential that skilled and critical judgement of good quality cytological and histological preparations should be used in the diagnosis of this tumour.

REFERENCES

Achong, B. G. & Epstein, M. A. (1966) Fine structure of the Burkitt tumour. *J. nat. Cancer Inst.,* **36**, 877-897

Berard, C., O'Conor, G. T., Thomas, L. B., & Torloni, H. (1969) Histopathological definition of Burkitt's tumour. *Bull. Wld Hlth Org.,* **40**, 601-607

Bernhard, W. & Lambert, D. (1964) Ultrastructure des tumeurs de Burkitt de l'enfant africain. In: Roulet, F. C., ed., *Symposium on lymphoreticular tumours in Africa,* Basel, Karger, pp. 270-284

Bernhard, W. (1970) Fine structure of Burkitt's lymphoma. In: Burkitt, D. P. & Wright, D. H., eds, *Burkitt's lymphoma,* Edinburgh & London, Livingstone, pp. 103-117

Bluming, A. Z. & Templeton, A. C. (1971) Histological diagnosis of Burkitt's lymphoma: A cautionary tale. *Brit. med. J.,* **1**, 89-90

Burchenal, J. H. (1966) Geographic chemotherapy— Burkitt's tumor as a stalking horse for leukemia. *Cancer Res.,* **26**, 2393-2405

Burkitt, D. (1958) Sarcoma involving jaws in African children. *Brit. J. Surg.* **46**, 218-223

Clifford, P., Gripenberg, U., Klein, E., Fenyo, E. M. & Manalov, G. (1968) Treatment of Burkitt's lymphoma. *Lancet,* **ii**, 517-518

Epstein, M. A. & Herdson, P. B. (1963) Cellular degeneration associated with characteristic nuclear fine structural changes in the cells from two cases of Burkitt's malignant lymphoma syndrome. *Brit. J. Cancer,* **17**, 56-58

Gripenberg, U., Levan, A. & Clifford, P. (1969) Chromosomes in Burkitt lymphomas. 1. Serial studies in a case with bilateral tumours showing different chromosomal stemlines. *Int. J. Cancer,* **4**, 334-349

Jacobs, P. A., Tough, I. M. & Wright, D. H. (1963) Cytogenetic studies in Burkitt's lymphoma. *Lancet* **ii**, 1144-1146.

Lancet (1968) Survival in Burkitt's lymphoma, **ii**, 200

Shepherd, J. J. & Wright, D. H. (1967) Burkitt's tumour presenting as bilateral swelling of the breast in women of child-bearing age. *Brit. J. Surg.,* **54**, 776-780

Sinkovics, J. G., Drewinko, B. & Thornell, E. (1970) Immunoresistant tetraploid lymphoma cells. *Lancet,* **i**, 139-140

Wright, D. H. (1966) Burkitt's tumour in England, a comparison with childhood lymphosarcoma. *Int. J. Cancer,* **1**, 503-514

Wright, D. H. & McAlpine, J. C. (1966) Ribonucleic acid content of Burkitt tumour cells. *J. clin. Path.,* **19**, 257-259

Wright, D. H. (1967) Burkitt's tumor and childhood lymphosarcoma. *Clin. Pediat.,* **6**, 116-123

Wright, D. H. (1968) Lipid content of malignant lymphomas. *J. clin. Path.,* **21**, 643-649

Wright, D. H. (1970a) Gross distribution and haematology. In: Burkitt, D. P. & Wright, D. H., eds, *Burkitt's lymphoma,* Edinburgh & London, Livingstone, pp. 64-81

Wright, D. H. (1970b) Microscopic features, histochemistry, histogenesis and diagnosis. In: Burkitt, D. P. & Wright, D. H., eds, *Burkitt's lymphoma,* Edinburgh & London, Livingstone. pp. 82-102

Wright, D. H. (1970c) Reed-Sternberg-like cells in recurrent Burkitt lymphomas. *Lancet,* **i**, 1052-1053

Ziegler, J. L., Morrow, R. H., Templeton, A. C., Templeton, C., Bluming, A. Z., Fass, L. & Kyalwazi, S. K. (1970) Clinical features and treatment of childhood malignant lymphoma in Uganda. *Int. J. Cancer,* **5**, 415-425

Ziegler, J. L., Wright, D. H. & Kyalwazi, S. K. (1971) Differential diagnosis of Burkitt's lymphoma of the face and jaws. *Cancer,* **27**, 503-514

The Pathology of Infectious Mononucleosis—A Review

R. L. CARTER [1]

This account of the pathology of infectious mono-nucleosis (IM) concentrates on two main aspects—haematology and histopathology; immunology and virology are discussed elsewhere.

Perhaps the best known laboratory feature of IM is the atypical mononuclear cell found in the blood. The nature of these cells is difficult to determine on the basis of their appearance in stained smears, but their ultrastructural features indicate that they are similar to lymphocytes transformed *in vitro* by contact with antigens or plant mitogens such as phyto-haemagglutinin (Inman & Cooper, 1965). Several of the ultrastructural features shown in Fig. 1 suggest that the atypical lymphocytes are metabolically active in synthesizing nucleic acids. This has now been confirmed and it is clear that the atypical lymphocytes in IM are characterized by remarkably intense DNA synthesis and cell proliferation. Some of the main findings are summarized below (the information has been taken from Gavosto, Pileri & Maraini, 1959; Bertino, Simmons & Donohue, 1962; Carter, 1965; MacKinney, 1967; Cooper, 1969):

1. Isolated dividing cells are sometimes seen in routine blood smears; they occur regularly in leukocyte concentrates.

2. Of the circulating atypical lymphocytes, 5–15% take up ^{3}H-thymidine *in vitro;* labelling is greatest during the acute phase of the disease.

3. Increased ^{3}H-thymidine uptake coincides with raised levels of folic acid enzyme systems (formate-activating enzyme, N^5–N^{10}-methylenetetrahydro-folate dehydrogenase, dihydrofolate reductase); there is also increased incorporation of ^{14}C-for-mate.

4. Colchicine-blocked cultures of ^{3}H-thymidine-labelled cells show that the G_2 period is short and that many labelled cells divide.

5. Karyotypes of freshly isolated leukocytes are apparently normal.

These results, derived solely from observations on peripheral blood, give only a partial picture of cell proliferation in IM. In particular, there is a discrepancy between the large numbers of cells that take up ^{3}H-thymidine and the small numbers that divide in the circulation. It was formerly thought that the increased ^{3}H-thymidine uptake might indicate impending *viral* multiplication (Gavosto et al., 1959; Bertino et al., 1962), but the absence of virus particles in freshly isolated leukocytes makes this unlikely. It is more probable that many postsynthetic cells return to the lymphoid tissues to divide (Epstein & Brecher, 1965) but this phase is not open to investigation as the procedures involved—administration of radio-isotopes *in vivo* and multiple biopsies—are unjustifiable in young adults with a benign disease. It is a tantalizing situation as the few tissues from patients with IM that have become available for histopathological examination show very striking abnormalities (Carter & Penman, 1969; Penman, 1970).

Morphological changes are most marked in the lymph nodes, spleen and tonsils (Figs. 2–10). These tissues show intense reactive hyperplasia and contain

---

[1] Chester Beatty Research Institute, Institute of Cancer Research: Royal Cancer Hospital, Fulham Road, London, S.W.3, UK.

Fig. 1. Atypical lymphocyte from peripheral blood. The nucleus has been sectioned in three places due to its convoluted structure. Note the rather prominent mitochondria and ribosomes; there is no organized endoplasmic reticulum. ×16 800.

Fig. 2. Lymph node. There is infiltration of the capsule and surrounding connective tissues, compression of peripheral sinuses, and hyperplasia of the pulp. Haematoxylin and eosin (H & E) ×200.

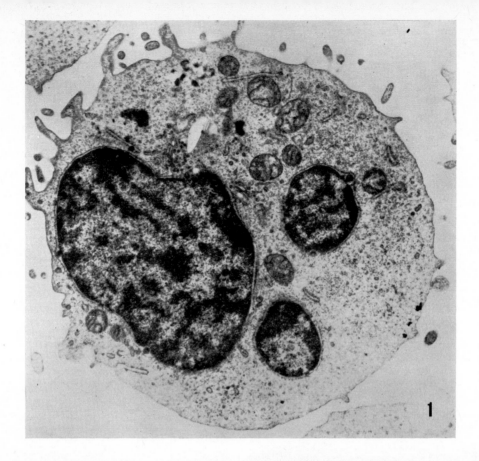

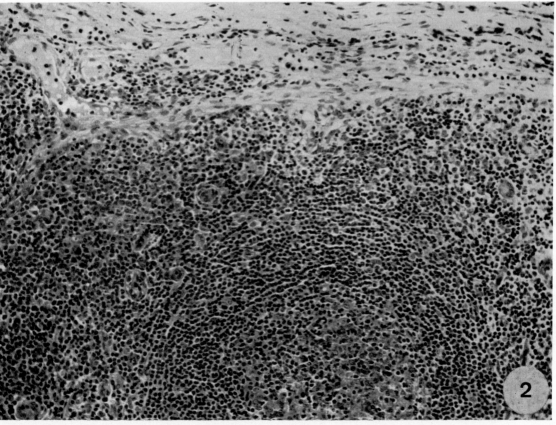

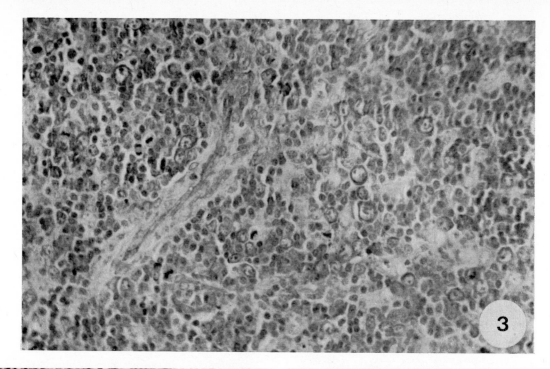

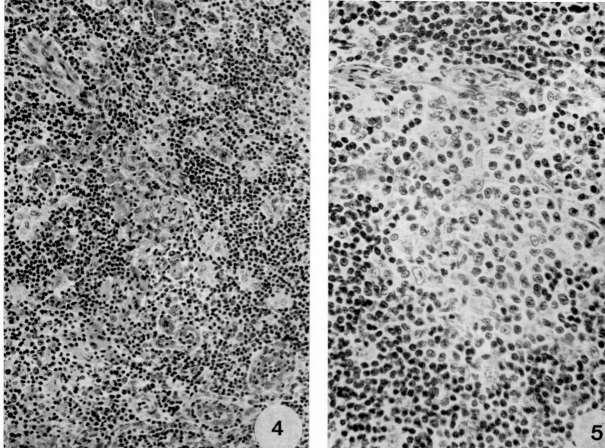

Fig. 3.   Lymph node.   Hyperplastic cortex with many proliferating blast cells.   (Giemsa) ×400.
Fig. 4.   Lymph node.   Cortex with numerous collections of histiocytes.   (H & E) ×200.
Fig. 5.   Lymph node.   Collection of histiocytes from Fig. 4, shown at a higher magnification.   (H & E) ×400.

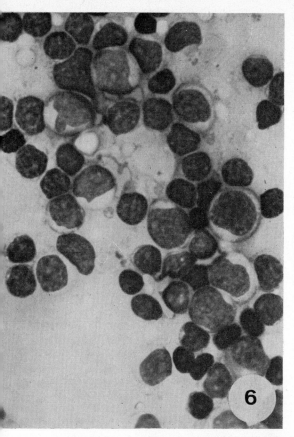

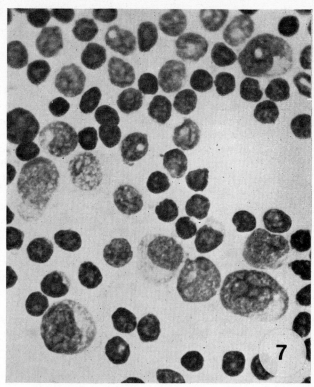

7

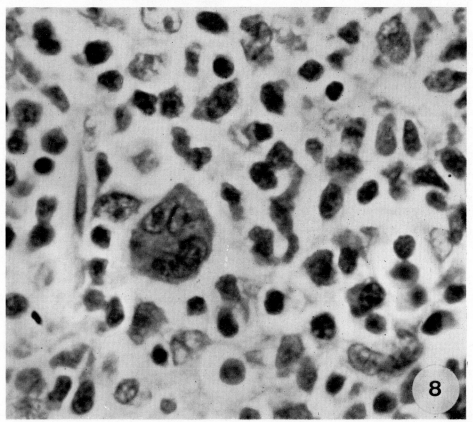

6

8

(See captions on page 234)

numerous immature-looking cells, many of them pyroninophilic and in mitosis. Imprint preparations show numerous large lymphocytes similar to the circulating atypical cells. Elements resembling Reed-Sternberg cells may also be encountered. Their significance is obscure, as similar cells may occur in a number of conditions and their relationship to Reed-Sternberg cells from Hodgkin's disease is unknown (Lukes, Tindle & Parker, 1969; Wright, 1969; Strum, Park & Rappaport, 1970). The proliferating lymphocytes extend widely within the lymphoid tissues and normal components may be almost effaced. The capsules of lymph nodes are often infiltrated, and there are striking lymphoid infiltrates within the walls of the smaller blood vessels in the spleen. Histiocytes are also increased in number and tend to congregate in small congeries reminiscent of those found in toxoplasmosis.

Nonlymphoid structures are also widely infiltrated in IM (see Figs. 11–15). In organs such as the liver, the histological changes are accompanied by definite functional disturbances; in other organs (e.g., heart, kidney) there is usually little or no clinical evidence that the structure in question is involved in the disease process.

The principal point that emerges is that IM is a generalized lymphoproliferative disease, affecting virtually all the tissues and organs of the body. This conclusion has been strongly argued before (Custer & Smith, 1948; Goldberg, 1962; Dameshek, 1965) but it is worth repeating as it is essential for any basic understanding of IM. Comparison with other generalized lymphoproliferative diseases, notably the leukaemias, is inescapable. Some of the changes found in IM are not unlike those of acute lymphoblastic leukaemia, but two fundamental differences exist: (i) the bone marrow is not involved in IM. There is no infiltration by atypical lymphocytes and haematopoiesis is almost always normal or slightly increased;[1]

---

[1] It is worth noting that transient thrombocytopenia and granulocytopenia commonly occur in IM (Carter, 1969), but the formation and release of platelets and granulocytes from the marrow are both normal and it is likely that—in contrast to acute lymphoblastic leukaemia—the platelet and leukocyte changes in IM are due to mechanisms acting peripherally rather than centrally.

(ii) in contrast to acute lymphoblastic leukaemia, the whole proliferative process in IM is self-limiting. The crucial question here is still not so much: What starts such activity? as: What stops it?

The histopathological changes in IM raise many questions. Large numbers of atypical lymphocytes are, for instance, present in the lymphoid tissues but it is impossible to determine whether they are being formed de novo or are derived from mature lymphocytes transformed by viral or some other antigen; both processes may occur. The function of the atypical lymphocytes is equally obscure but the intense antibody response in IM has often raised the question of the rôle of these cells in antibody formation. The antibodies formed in IM are predominantly γM globulins (Carter, 1969) and there is a striking increase in overall serum γM globulin levels during the acute phase of the disease (Wollheim, 1968): are these γM globulins synthesized by the atypical lymphocytes? Certain features of the cells are compatible with such a suggestion—cytoplasmic basophilia, the abundance of ribosomes, and their ability to take up RNA precursors such as ^{3}H-cytidine and ^{3}H-uridine. However, they lack organized endoplasmic reticulum, and most experiments with fresh cells from the blood have failed to show antibody formation, as judged by immunofluorescence, mixed cell agglutination and Jerne plaque techniques (MacKinney, 1968; Glade & Chessin, 1968). In contrast, there is some evidence that increased numbers of lymphocytes aspirated from bone marrow and lymph nodes are stained specifically with fluorescent γM globulin sera (Carter, 1966).

These observations emphasize the need to distinguish between findings based on atypical lymphocytes in the blood and those based on atypical lymphocytes obtained from tissues. Recent work indicates that it is even more important to distinguish between the properties of freshly isolated IM leukocytes and those of IM leukocytes maintained in vitro in continuous culture for months or years (Chessin et al., 1968). The reasons for the comparative ease with which IM leukocytes can be established in vitro are still obscure (Glade et al., 1969) but it is obvious

---

Figs. 6 & 7. Aspirates from lymph nodes showing various types of atypical lymphocytes. (May-Grünwald-Giemsa) ×880.

Fig. 8.   Tonsil. Intense hyperplasia with a large multinucleate giant cell. (H & E) ×800.

Fig. 9.   Spleen. High-power view of sinus filled with pleomorphic mononuclear cells. (H & E) ×580.

Fig. 10.   Spleen. Trabecular venule showing subintimal lymphatics distended with lymphoid cells. (H & E) ×190.

Fig. 11.   Liver. Low-power view of cellular infiltrate, extending along portal tracts. (H & E) ×190.

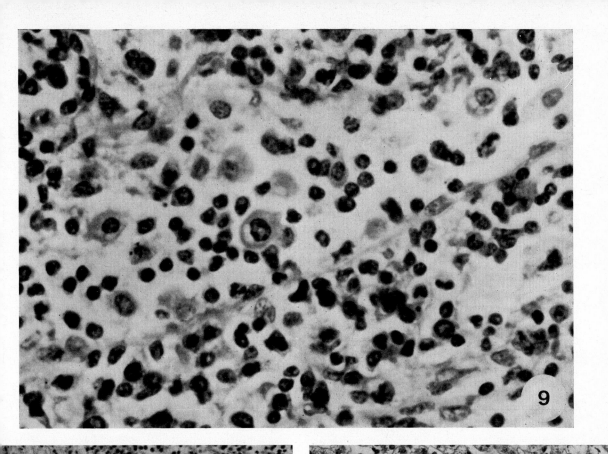

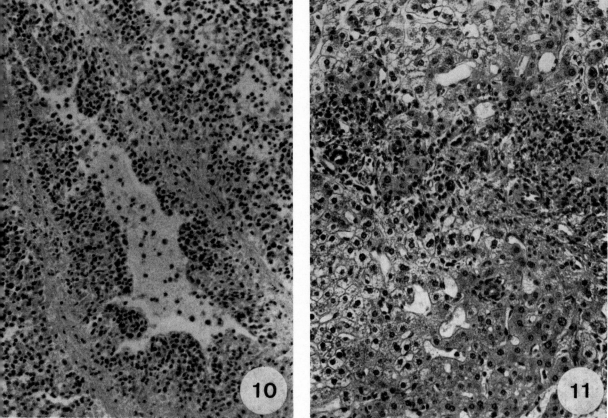

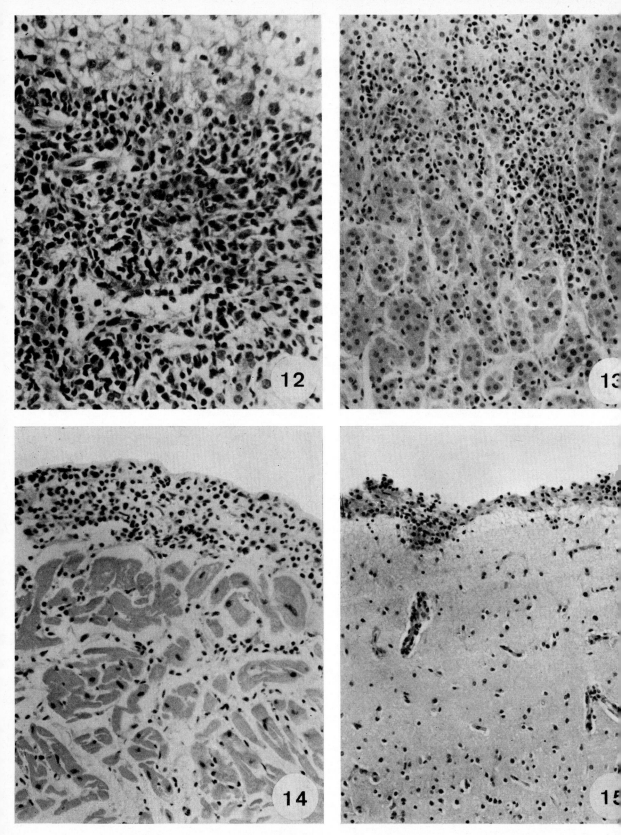

Fig. 12. Liver. High-power view of cellular infiltrate in portal tracts. (H & E) ×380.
Fig. 13. Adrenal gland. Lymphoid infiltrates in adrenal cortex. (H & E) ×190.
Fig. 14. Heart. Lymphoid infiltrates in pericardium. (H & E) ×190.
Fig. 15. Brain. Meningeal infiltration by lymphoid cells. (H & E) ×190.

that cells from such lines differ from freshly isolated cells in many ways—e.g., in ultrastructure, the presence of virus particles and viral antigens, karyotypes, heterotransplantability, and the ability to synthesize immunoglobulins and other products (Chessin *et al.*, 1968; Klein *et al.*, 1968; Glade & Hirschhorn, 1970; Steel & Hardy, 1970). At the same time that the properties of cultured IM cells diverge from those of freshly isolated leukocytes they come increasingly to resemble those of long-cultured leukocytes from other sources, notably Burkitt's lymphoma, acute leukaemia, and even normal subjects. It seems increasingly likely that the common factor here is the presence of Epstein-Barr virus (EBV); the evidence for this view is discussed elsewhere. In the case of IM and Burkitt's lymphoma, there are strong grounds for regarding EBV as a common aetiological agent, so that these two seemingly disparate diseases may be closely related. No immunological differences have yet been found between EBV obtained from cell lines in the two diseases (Stevens *et al.*, 1970) and it appears to be *host* rather than viral factors which determine the clinical course of EBV infections in man (Burkitt, 1969).

Much of this new work raises the long-debated question of the relationship between infectious mononucleosis and lymphoid neoplasia (Dameshek, 1969); the question is still open but there is little doubt that further investigations into the basic pathology of IM should throw valuable light on both reactive and neoplastic processes in the lymphoid system.

ACKNOWLEDGEMENTS

Most of the illustrations in this paper have been reproduced from *Infectious mononucleosis,* edited by R. L. Carter & H. G. Penman, by kind permission of the publishers, Blackwell Scientific Publications, Oxford and Edinburgh.

REFERENCES

Bertino, J. R., Simmons, B. M. & Donohue, D. M. (1962) Increased activity of some folic acid enzyme systems in infectious mononucleosis. *Blood,* 19, 587-592

Burkitt, D. P. (1969) Etiology of Burkitt's lymphoma—an alternative hypothesis to a vectored virus. *J. nat. Cancer Inst.,* 42, 19-28

Carter, R. L. (1965) The mitotic activity of circulating atypical mononuclear cells in infectious mononucleosis. *Blood,* 25, 279-586

Carter, R. L. (1966) Infectious mononucleosis: some observations on the cellular localisation of immune globulin synthesis. *Amer. J. clin. Path.,* 45, 574-580

Carter, R. L. (1969) Some recent advances in the clinical pathology of infectious mononucleosis. *Proc. roy. Soc. Med.,* 62, 1282-1285

Carter, R. L. & Penman, H. G., eds (1969) *Infectious mononucleosis.* Oxford & Edinburgh, Blackwell Scientific Publications

Chessin, L. N., Glade, P. R., Kasel, J. A., Moses, H. L., Herberman, R. B., & Hirshaut, Y. (1968) The circulating lymphocyte—its role in infectious mononucleosis. *Ann. intern. Med.,* 69, 333-359

Cooper, E. H. (1969) Experimental studies of the atypical mononuclear cells in infectious mononucleosis. In: Carter, R. L. & Penman, H. G., eds, *Infectious mononucleosis,* Oxford & Edinburgh, Blackwell Scientific Publications, pp. 121-146

Custer, R. P. & Smith, E. B. (1948) Pathology of infectious mononucleosis. *Blood,* 3, 830-857

Dameshek, W. (1965) Immunologic proliferation and its relationship to certain forms of leukaemia and related disorders. *Israel J. med. Sci.,* 1, 1304-1315

Dameshek, W. (1969) Speculations on the nature of infectious mononucleosis. In: Carter, R. L. & Penman, H. G., eds, *Infectious mononucleosis,* Oxford & Edinburgh, Blackwell Scientific Publications, pp. 225-241

Epstein, L. B. & Brecher, G. (1965) DNA and RNA synthesis of circulating atypical lymphocytes in infectious mononucleosis. *Blood,* 25, 197-203

Gavosto, F., Pileri, A., & Maraini, G. (1959) Incorporation of thymidine labelled with tritium by circulating cells of infectious mononucleosis. *Nature (Lond.),* 183, 1691-1692

Glade, P. R. & Chessin, L. N. (1968) Infectious mononucleosis: immunoglobulin synthesis by cell lines. *J. clin. Invest.,* 47, 2391-2401

Glade, P. R., Hirshaut, Y., Stites, D. P., & Chessin, L. N. (1969) Infectious mononucleosis: *in vitro* evidence for limited lymphoproliferation. *Blood,* 33, 292-299

Glade, P. R., & Hirschhorn, K. (1970) Products of lymphoid cells in continuous culture. *Amer. J. Path.* 60, 483-492

Goldberg, G. M. (1962) A study of malignant lymphomas and leukemias. V. Lymphogenous leukemia and infectious mononucleosis: the lymphatics in benign and malignant lymphoproliferative diseases. *Cancer (Philad.),* 15, 869-881

Inman, D. R., & Cooper, E. H. (1965) The relation of ultrastructure to DNA synthesis in human leucocytes. *Acta Haemat. (Basel)*, **33**, 257-278

Klein, G., Pearson, G., Henle, G., Henle, W., Diehl, V. & Niederman, J. C. (1968) Relation between Epstein-Barr viral and cell-membrane immunofluorescence in Burkitt tumor cells. II. Comparison of cells and sera from patients with Burkitt's lymphoma and infectious mononucleosis. *J. exp. med.*, **128**, 1021-1030

Lukes, R. J., Tindle, B. H., & Parker, J. W. (1969) Reed-Sternberg-like cells in infectious mononucleosis. *Lancet*, **ii**, 1003-1004

MacKinney, A. A. Jr (1967) Division of leucocytes already in DNA synthesis from patients with acute leukemia and infectious mononucleosis. *Acta Haemat. (Basel)*, **38**, 163-169

MacKinney, A. A. Jr (1968) Studies of plasma protein synthesis by peripheral cells from normal persons and patients with infectious mononucleosis. *Blood*, **32**, 217-224

Penman, H. G. (1970) Fatal infectious mononucleosis: a critical review. *J. clin. Path.*, **23**, 765-771

Steel, C. M. & Hardy, D. A. (1970) Evidence of altered antigenicity in cultured lymphoid cells from patients with infectious mononucleosis. *Lancet*, **i**, 1322-1323

Stevens, D. A., Pry, T. W., Blackham, E. A., & Manaker, R. A. (1970) Immunodiffusion studies of EB virus (Herpes-type virus)-infected and -uninfected hemic cell lines. *Int. J. Cancer*, **5**, 229-237

Strum, S. B., Park, J. K., & Rappaport, H. (1970) Observations of cells resembling Sternberg-Reed cells in conditions other than Hodgkin's Disease. *Cancer (N.Y.)*, **26**, 176-190

Wollheim, F. A. (1968) Immunoglobulin changes in the course of infectious mononucleosis. *Scand. J. Haemat.*, **5**, 97-106

Wright, D. H. (1969) Reed-Sternberg-like cells in recurrent Burkitt lymphomas. *Lancet*, **ii**, 1052-1053

The Pathology of Nasopharyngeal Carcinoma
A Review

K. SHANMUGARATNAM [1]

Nasopharyngeal carcinoma differs from both Burkitt's lymphoma and infectious mononucleosis, which are the other human diseases under consideration at this Symposium, in that it is an epithelial neoplasm. The similarities in the virological and immunological features of these diseases, which have indicated their association with the Epstein-Barr virus, are not reflected in their pathological or epidemiological characteristics. This paper will be restricted to the pathology of nasopharyngeal carcinoma.

GROSS PATHOLOGY

The nasopharynx is situated behind the nasal cavity, below the sphenoid bone, in front of the basilar part of the occipital bone and the first and second cervical vertebrae and above the soft palate. The nasopharyngeal mucosa is thrown into several folds and has a surface area of approximately 50 cm² (Ali, 1965). The nasopharyngeal cavity communicates anteriorly with the nasal cavity, laterally with the middle ears and inferiorly with the oropharynx. Anatomically and developmentally, the nasopharynx is part of the pharynx but, as an annex to the nasal cavity, it is functionally related to the respiratory system.

Macroscopically, the neoplasms may appear as polypoid or exophytic lesions, ulcerating lesions, infiltrating lesions or lesions with mixed features. The great majority of neoplasms (more than 80%) are unilateral, and the right and left sides are affected with approximately equal frequency. Approximately 40% of tumours arise from the lateral walls, especially from the pharyngeal recesses or fossae of Rosenmüller and the Eustachian prominences. About 30% of neoplasms take their origin from the

[1] Department of Pathology, University of Singapore, Outram Road, Singapore.

superoposterior wall, which arches downwards from the superior margin of the choanae to the level of the free border of the soft palate. Less than 10% of neoplasms arise from the anterior and inferior walls, which consist of the choanae and the superior surface of the soft palate. The remaining 20% of neoplasms involve more than one wall of the nasopharyngeal cavity.

MODES OF SPREAD

The neoplasms have marked invasive and metastatic properties, and spread frequently by the direct, lymphatic and haematogenous routes.

Local tumour infiltrates and extensions may be found in the nasal fossae, the paranasal air sinuses, the orbital cavities, the Eustachian tubes, the lateral parapharyngeal spaces (where it comes into close relationship with cranial nerves IX, X, XI, XII and the cervical sympathetic trunk), the oropharynx and buccal cavity, the parotid gland, the neighbouring soft tissues and bone, and the cranial cavity. The neoplasm may extend into the cranial cavity and intracranial venous sinuses by invading bone or, more frequently, by traversing the cranial foramina; these intracranial extensions are usually extradural. The foramen lacerum, which lies only 1 cm above the fossa of Rosenmüller, provides easy access to the cavernous sinus, which contains cranial nerves III, IV, V & VI. The jugular foramen (containing cranial nerves IX, X and XI) is more frequently involved than the hypoglossal canal (XII) or the stylomastoid foramen (VII). The olfactory foramen (I) and the auditory canal (VIII) are only rarely involved.

Lymphatic spread is exceedingly common. The lymph nodes most frequently involved are the retropharyngeal nodes and the upper and posteriorly placed deep cervical nodes (Figs. 1 & 2); metastases

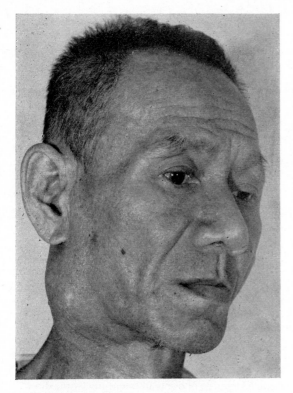

Fig. 1. Patient with nasopharyngeal carcinoma with metastatic enlargement of right upper deep cervical lymph nodes.

Fig. 2. Patient with nasopharyngeal carcinoma with metastases in left upper deep cervical lymph nodes, and left oculomotor paralysis.

in these nodes are found early in the course of the disease. In the later stages of the disease, lymphnode metastases may be found in other cervical nodes and, less commonly, in the axillary and thoracoabdominal nodes.

Blood-borne metastases may occur in any organ but are most frequently found in the bones, liver and lungs.

The gross pathology and modes of spread of nasopharyngeal carcinoma are responsible for the symptomatology of this disease. The main symptoms in order of frequency are:

(1) Cervical lymphadenopathy. This is the presenting symptom in more than 50% of cases; in the late stages of the disease it is present in almost all cases. The lymph nodes most frequently affected are those lying behind the angle of the jaw.

(2) Naso-respiratory symptoms, such as bleeding and nasal obstruction. These symptoms are referable to the primary growth.

(3) Neurological symptoms, such as headache and nerve palsies. Cranial nerve palsies are found in approximately 30–40% of all cases. The nerves

Fig. 3. Undifferentiated nasopharyngeal carcinoma, showing tumour cells with pale vesicular nuclei, prominent nucleoli and indistinct cell margins. Haematoxylin and eosin (H & E) ×580.

Fig. 4. Undifferentiated nasopharyngeal carcinoma showing alveolar tumour masses infiltrated by lymphocytes—lymphoepithelioma of the Regaud type. (H & E) ×350.

Fig. 5. Undifferentiated nasopharyngeal carcinoma showing pale tumour cells loosely scattered in a lymphoid stroma—lymphoepithelioma of the Schmincke type. (H & E) ×580.

Fig. 6. Undifferentiated nasopharyngeal carcinoma of the "lymphoepithelioma" type, showing alveolar tumour masses with marked lymphoid infiltration. (H & E) ×180.

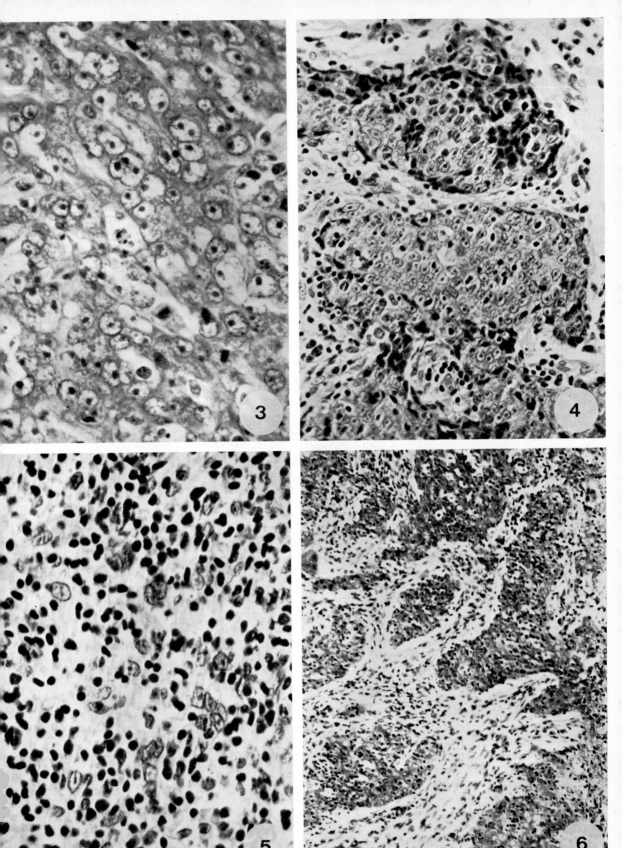

most commonly involved are VI and V; next in frequency are III, IV, IX, X, XI, XII and the cervical sympathetic nerve. Involvement of cranial nerves II and VII is less frequent; I and VIII are only rarely involved.

(4) Auditory symptoms, such as earache, tinnitus, conduction deafness and otitis media. These symptoms are found in 5–10% of cases and are due to involvement of the Eustachian tubes.

(5) Symptoms referable to other forms of local invasion. These are present in approximately 5% of cases and include exophthalmos, ophthalmoplegia and enlargements of the parotid gland, muscles and soft tissues of the head and neck.

(6) Symptoms referable to distant metastases. These include bone pains, hepatomegaly, cough, haemoptysis and pulmonary osteoarthropathy.

HISTOPATHOLOGY

The nasopharynx is lined by columnar ciliated, stratified squamous and transitional epithelium and its walls contain various glandular, lymphoid and connective tissue elements. It is not surprising, therefore, that a wide variety of malignant neoplasms has been identified in this area. Several histological classifications of nasopharyngeal cancer have been suggested (Frank *et al.*, 1941; Yeh, 1962; Liang *et al.*, 1962; Shanmugaratnam, 1967). In some of these, nasopharyngeal carcinomas have been subdivided into a large variety of histological subtypes. A simpler classification, separating malignant neoplasms of the nasopharynx into four broad groups (Shanmugaratnam, 1971), is given below:

MALIGNANT NEOPLASMS OF THE NASOPHARYNX

1. CARCINOMA ("NASOPHARYNGEAL CARCINOMA")
    Undifferentiated carcinoma (including lymphoepithelioma, transitional carcinoma and spindle-cell carcinoma)
    Squamous-cell carcinoma
    Other carcinomas (including clear-cell carcinoma, basaloid carcinoma, pleomorrphic cacinoma and indeterminate carcinoma).

2. ADENOCARCINOMAS
    (*a*) Adenoid cystic carcinoma
    (*b*) Mucoepidermoid carcinoma
    (*c*) Other adenocarcinomas

3. SARCOMAS
    (*a*) Malignant lymphomas
    (*b*) Other sarcomas, viz. fibrosarcoma, myosarcoma, neurosarcoma, etc.

4. OTHER CANCERS
    (*a*) Chordoma
    (*b*) Malignant teratoma
    (*c*) Malignant melanoma.

Nasopharyngeal carcinoma, the tumour under consideration at this Symposium, is the commonest form of nasopharyngeal cancer in man. In China and in South East Asia, where the incidence of nasopharyngeal cancer is high, the ratio of nasopharyngeal carcinoma to all other forms of nasopharyngeal cancer is approximately 99:1. In contrast, in Europe and America, where the incidence is low, the reported ratios have ranged between 2:1 and 9:1. This ratio, therefore, gives a good indication of the incidence of nasopharyngeal carcinoma in any country (Shanmugaratnam, 1967, 1971).

The majority (more than 60%) of nasopharyngeal carcinomas are undifferentiated carcinomas. These tumours exhibit histological appearances that are sufficiently characteristic to enable a presumptive diagnosis of nasopharyngeal carcinoma to be made, even when they are identified in lymph-nodal or other metastatic deposits. The tumour cells are of moderate size and have indistinct cell membranes, which give the tumour masses a syncytial appearance (Fig. 3). The tumour cells are arranged in irregular but fairly well defined alveolar masses (Fig. 4), the so-called Regaud type, or in thin strands of loosely connected cells with syncytial or pseudo-sarcomatous appearances (Fig. 5), the so-called Schmincke type. In spite of its undifferentiated cytological features, the structure of this group of neoplasms leaves no doubt of its epithelial origin. The nuclei are round or oval, pale and vesicular and exhibit prominent nucleoli (Fig. 3). Neoplasms with these features are sometimes described, for reasons that are not obvious,

Fig. 7. Undifferentiated nasopharyngeal carcinoma of the "lymphoepithelioma" type, showing numerous lymphocytes among tumour cells. (H & E) ×580.

Fig. 8. Undifferentiated nasopharyngeal carcinoma with spindle-cell features. (H & E) ×580.

Fig. 9. Nasopharyngeal carcinoma showing moderate squamous-cell differentiation. (H & E) ×180.

Fig. 10. Nasopharyngeal carcinoma with well developed squamous-cell features. (H & E) ×580.

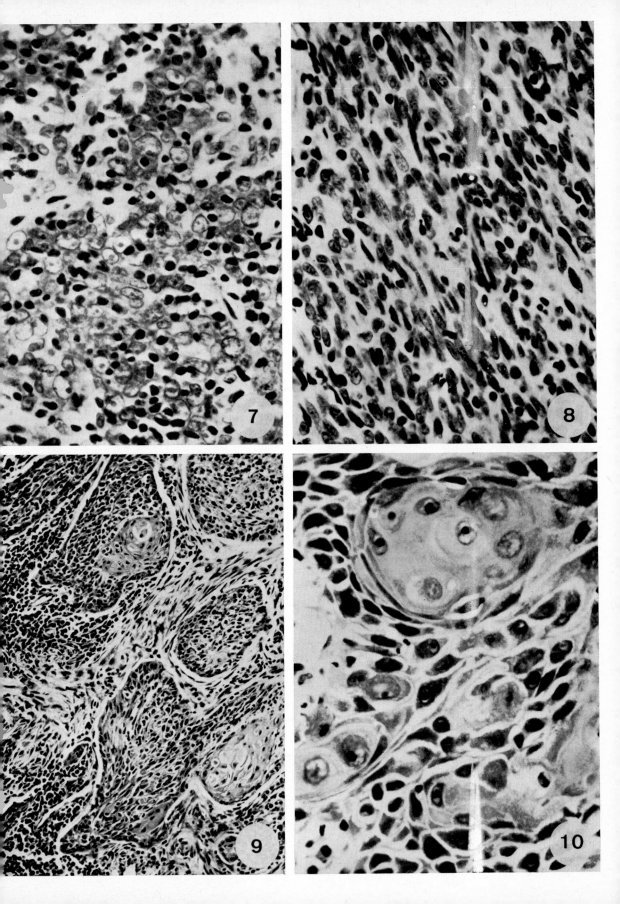

as transitional-cell carcinomas. Many of these neoplasms exhibit an intimate admixture with lymphocytes (Figs. 6 & 7); these have been termed lymphoepitheliomas (Regaud: see discussion on Reverchon & Coutard, 1921; Schmincke, 1921). A few undifferentiated carcinomas may present spindle-celled appearances (Fig. 8). The view that undifferentiated nasopharyngeal carcinomas are variants of squamous-cell carcinoma (New & Kirch, 1928) has been supported by many investigators (Hauser & Brownell, 1938; Teoh, 1957; Yeh, 1962; Shanmugaratnam & Muir, 1967) and confirmed by electron microscopy. However, as nasopharyngeal carcinomas have certain cytological and structural features that are in some ways different from those of squamous-cell carcinomas in other sites, it would appear preferable to designate them simply as "nasopharyngeal carcinoma". The neoplasms are PAS-negative. The stroma is usually delicate but may be fibrous. The histological picture may be complicated by necrosis, ulceration and superimposed inflammatory reaction.

Approximately 30% of nasopharyngeal carcinomas are classical squamous-cell carcinomas, i.e., they may be recognized as such by light microscopy (Figs. 9 & 10). However, the histological evidence of squamous differentiation, such as keratinization and intercellular bridges, is usually quite subtle and it is not often that the neoplasm exhibits the keratinizing nests seen in epidermoid carcinomas of other sites. Less than 10% of nasopharyngeal carcinomas are neoplasms with clear-celled (Fig. 11), basaloid (Fig. 12), pleomorphic (Fig. 13) and indeterminate characteristics; these, like the undifferentiated carcinomas, are probably variants of squamous-cell carcinoma.

All the histological types of nasopharyngeal carcinoma described previously are variants of a homogeneous group of tumours. They frequently co-exist in the same case and their subdivision often rests on such tenuous grounds that the value of their separation may be questioned. The various histological types of nasopharyngeal carcinoma that have been documented (Yeh, 1962; Liang et al., 1962; Shanmugaratnam, 1967) are of descriptive value but there is as yet no good evidence that they represent differ-

ences in aetiology, histogenesis or biological behaviour. While some authors have claimed relationship between histological type and biological behaviour (Quick & Cutler, 1927; Frank, Lev & Blahd, 1941; Liang et al., 1962), this has been denied by others (Hauser & Brownell, 1938; Yeh, 1962; Martin, 1939). In view of this disagreement, and the fact that many nasopharyngeal carcinomas are histological hybrids, it is preferable, in aetiological and epidemiological studies, to group all the histological variants of this neoplasm together as nasopharyngeal carcinoma.

The lymphocytes that are found frequently in nasopharyngeal carcinoma are not neoplastic but represent an immunological reaction; their presence may also be due to neoplastic invasion of the nasopharyngeal lymphoid tissues. Marked lymphocytic infiltration is evident in more than 50% of nasopharyngeal carcinomas. It may be found in the undifferentiated, squamous or other varieties of nasopharyngeal carcinoma and may be identified both in the primary neoplasms and in the metastatic deposits.

The histogenesis of this neoplasm has been variously attributed to transitional epithelium (Ewing, 1929), ciliated columnar epithelium (Digby, 1951) or stratified squamous epithelium (Teoh, 1957). Beck & Guttman (1932) and Kramer (1950) expressed the view that squamous-cell carcinoma arises from squamous epithelium, transitional-cell carcinoma from the transitional or columnar ciliated epithelium, and lymphoepithelioma from a so-called lymphoepithelium. However, as malignant neoplasms often exhibit variations in structure, such a strict correlation between structure and histogenesis is not acceptable. Studies of in situ malignant change (Fig. 14) and of continuity of the neoplasm with the surface epithelium have led to the view that all histological types of nasopharyngeal carcinoma may arise from the squamous, columnar ciliated or transitional-cell epithelia lining the surface and crypts of the nasopharynx (Shanmugaratnam & Muir, 1967).

ULTRASTRUCTURE

Ultrastructural evidence of squamous differentiation, such as the presence of cytoplasmic aggregates

Fig. 11. Nasopharyngeal carcinoma with clear-cell characteristics. (H & E) ×580.

Fig. 12. Nasopharyngeal carcinoma showing a "basaloid" structure. (H & E) ×180.

Fig. 13. Nasopharyngeal carcinoma with pleomorphic appearance. (H & E) ×580.

Fig. 14. Carcinoma-in-situ of nasopharyngeal stratified squamous epithelium. This field was close to a focus of invasive undifferentiated carcinoma of the nasopharynx. (H & E) ×180.

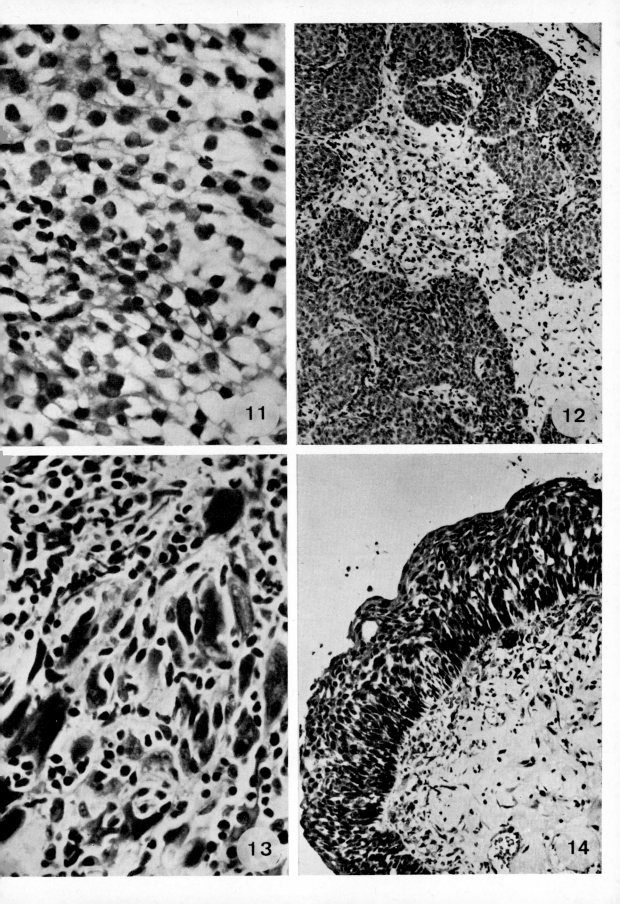

of tonofibrils, bodies resembling membrane coating vesicles and desmosomes between adjacent cells, are consistently found in all histological varieties of nasopharyngeal carcinoma, including undifferentiated carcinomas (Svoboda, Kirchner & Shanmugaratnam, 1965; Papadimitriou & Shanmugaratnam [1]). Papadimitriou & Shanmugaratnam [1] found cytoplasmic aggregates of fibrillar material measuring 2–2.5 μ in diameter; these were composed of a tangle of fibrils measuring 8–11 nm in thickness, and were distinct from tonofibrils.

Lymphocytes, many of which possess abundant cytoplasm with increased numbers of ribosomal aggregates, are frequently found closely apposed to the neoplastic cells. Reticular osmiophilic cells,

which resemble the tumour cells in many respects but differ by lacking membrane coating vesicles and by possessing more mitochondria and endoplasmic reticulum, are found scattered among the paler cells of undifferentiated carcinoma.

The cells of undifferentiated nasopharyngeal carcinoma occasionally exhibit thin nuclear projections similar to those that have been described in the cells of Burkitt's lymphomas (Achong & Epstein, 1966; Epstein & Achong, 1967) and other human lymphomata (McDuffie, 1967; Dorfman, 1967). While the significance of these projections is not known, the possibility remains that their presence may be nonspecific and that they may represent bridges between neighbouring lobes of irregularly shaped nuclei (Papadimitriou & Shanmugaratnam [1]).

Nuclear bodies or inclusions varying in size from

---

[1] Unpublished data.

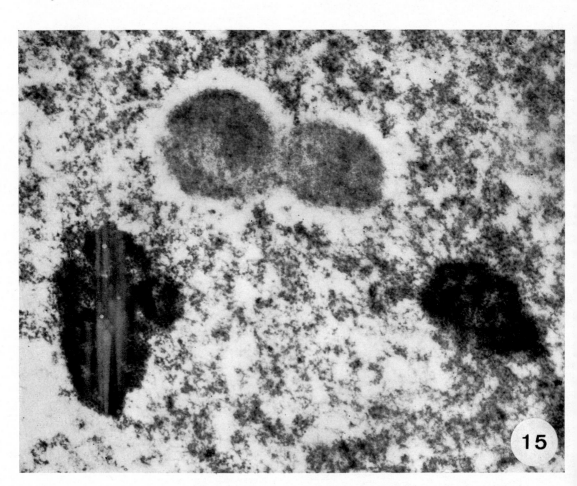

Fig. 15. Electron micrograph showing fibrillar nuclear bodies or inclusions between two nucleoli. (×32 000).

0.6–1.8 μ are found in many neoplastic cells (Fig. 15). Most of these are composed of fine fibrillar material measuring from 4.5–5.5 nm, but some also exhibited more electron-dense and granular components measuring 15–25 nm. In a few of these inclusions, the granular components are centrally placed and

surrounded by a fibrillar cortex (Papadimitriou & Shanmugaratnam [1]). The significance of these bodies is not clear. Similar inclusions were found by Lin et al. (1969), who drew attention to their resemblance to bodies which have been reported to co-exist with herpes simplex virus particles (Morgan et al., 1959; Swanson et al., 1966). No virus particles have yet been found in blocks prepared directly from tumour biopsies.

---

[1] Unpublished data.

SUMMARY

Nasopharyngeal carcinoma is an epithelial neoplasm that arises from the epithelium lining the surface and crypts of the nasopharynx.

Macroscopically, the neoplasm may present polypoid, exophytic, ulcerative or diffusely infiltrative appearances. The neoplasm exhibits marked invasive powers and spreads by the direct, lymphatic and haematogenous routes.

By light microscopy, the majority of nasopharyngeal carcinomas (more than 60%) are undifferentiated neoplasms with vesicular nuclei, prominent nucleoli and indistinct cell margins. Many neoplasms exhibit intimate admixture with lymphocytes. These lymphocytes are not neoplastic but probably represent an immunological reaction to the tumour. Approximately 30% of nasopharyngeal carcinomas are classical squamous-cell carcinomas. The squamous differentiation, however, is often quite subtle and it is only rarely that the neoplasm exhibits the keratinizing nests of epidermoid carcinoma. Other varieties of nasopharyngeal carcinoma are clear-cell carcinoma, pleomorphic carcinoma, basaloid carcinoma and indeterminate carcinoma. All the above-mentioned histological types of nasopharyngeal carcinoma are variants of a homogeneous group of neoplasms.

All histological variants of nasopharyngeal carcinoma have consistently shown ultrastructural evidence of squamous differentiation. One of the striking ultrastructural features is the presence in many neoplastic cells of nuclear bodies or inclusions varying from 0.6–1.8 μ; the significance of these bodies is not known. No virus particles have yet been found in blocks prepared directly from tumour biopsies.

REFERENCES

Achong, B. G. & Epstein, M. A. (1966) Fine structure of the Burkitt tumour. *J. nat. Cancer Inst.*, **36**, 877-897

Ali, M. Y. (1965) Histology of the human nasopharyngeal mucosa. *J. Anat. (Lond.)*, **99**, 657-672

Beck, J. C. & Guttman, M. R. (1932) Relation of histopathology of nasopharyngeal neoplasms to their radiosensitivity. *Ann. Otol. (St. Louis)*, **41**, 349-358

Digby, K. H. (1951) Nasopharyngeal carcinoma. *Ann. roy. Coll. Surg. Engl.*, **9**, 253-265

Dorfman, R. F. (1967) The fine structure of a malignant lymphoma in a child from St.Louis, Missouri. *J. nat. Cancer Inst.*, **38**, 491-504

Epstein, M. A. & Achong, B. G. (1967) Formal discussions: immunologic relationships of the herpes-like EB virus of cultured Burkitt lymphoblasts. *Cancer Res.*, **27**, 2489-2493

Ewing, J. (1929) Radiosensitive epidermoid carcinoma. *Amer. J. Roentgenol.*, **21**, 313-321

Frank, I., Lev, M. & Blahd, M. (1941) Transitional cell carcinoma of upper respiratory tract. *Ann. Otol. Rhinol. Laryngol.*, **50**, 393-420

Hauser, I. J. & Brownell, D. H. (1938) Malignant neoplasms of the nasopharynx. *J. Amer. med. Ass.*, **111**, 2467-2473

Kramer, S. (1950) Treatment of malignant tumours of nasopharynx. *Proc. roy. Soc. Med.*, **43**, 867-874

Liang, P. C., Ch'en, C. C., Chu, C. C., Hu, Y. F., Chu, H. M. & Tsung, Y. S. (1962) The histopathologic classification, biologic characteristics and histogenesis of nasopharyngeal carcinomas. *Chin. med. J.*, **81**, 629-658

Lin, H. S., Lin, C. S., Yeh, S. & Tu, S. M. (1969) Fine structure of nasopharyngeal carcinoma with special reference to the anaplastic type. *Cancer*, **23**, 390-405

Martin, C. L. (1939) Complications produced by malignant tumors of nasopharynx. *Amer. J. Roentgenol.*, **41**, 377-390

McDuffie, N. G. (1967) Nuclear blebs in human leukaemia. *Nature (Lond.)*, **214**, 1341-1342

Morgan, C., Rose, H., Holden, M. & Jones, E. P. (1959) Electron microscopic observations on the development of herpes simplex virus. *J. exp. Med.*, **110**, 643-656

New, G. B. & Kirch, W. (1928) Tumors of nose and throat. A review of literature. *Arch. Otolaryngol.*, **8**, 600-607

Quick, D. & Cutler, M. (1927) Transitional cell epidermoid carcinoma, radiosensitive type of intra-oval tumour. *Surg. Gynec. Obstet.*, **45**, 320-331

Reverchon, L. & Coutard, H. (1921) Lymphoépithéliome de l'hypopharynx traité par la roentgenthérapie. *Bull. Soc. fr. Oto-Rhino-Laryngol.*, **34**, 209-214

Schminke, A. (1921) Über lympho-epitheliale Geschwulste. *Zieglers Beitr. path. Anat.*, **58**, 161-170

Shanmugaratnam, K. (1967) *Racial and geographical factors in tumour incidence*, Edinburgh Press Monograph, p. 169

Shanmugaratnam, K. (1971) Studies on the aetiology of nasopharyngeal carcinoma. *Int. Rev. exp. Path.* (in press)

Shanmugaratnam, K. & Muir, C. S. (1967) *Cancer of the nasopharynx*. Munksgaard, Copenhagen, pp. 153-162 (*UICC Monograph Series*, Vol. 1)

Svoboda, D., Kirchner, F. & Shanmugaratnam, K. (1965) Ultrastructure of nasopharyngeal carcinomas in American and Chinese patients; an application of electron microscopy to geographic pathology. *Exp. molec. Path.*, **4**, 189-204

Swanson, J. L., Craighead, J. E. & Reynolds, E. S. (1966) Electron microscopic observation on herpes virus hominis (herpes simplex virus) encephalitis in man. *Lab. Invest.*, **15**, 1966-1981

Teoh, T. B. (1957) Epidermoid carcinoma of the nasopharynx among Chinese, a study of 31 necropsies. *J. path. Bact.*, **73**, 451-465

Yeh, S. (1962) A histological classification of carcinomas of the nasopharynx with a critical review as to the existence of lymphoepitheliomas. *Cancer*, **15**, 895-920

Polyclonal Origin of Lymphoblastoid Cell Lines Established from Normal and Neoplastic Human Lymphoid Tissues

J. M. BÉCHET [1], K. NILSSON [1] & G. KLEIN [2]

In recent years, continuously growing cell lines have been established from the peripheral blood or solid lymphatic tissue of patients with leukaemia, lymphoma or infectious mononucleosis, and from healthy individuals. An important question is whether the lines originating from tumour material are derived from the neoplastic cells or from other cells present in the tumour. This question can be answered by comparing various functional and morphological characteristics of the cells in the established lines with those of the tumour, and also by looking for differences between lines of different origins. Owing to the importance of established cell lines from Burkitt's lymphomas as a tool in the study of this disease, it is of particular interest to compare these lines with lymphoid lines derived from other tumours and from non-neoplastic lymphoid tissue.

Immunoglobulin production *in vitro* is a useful functional marker. The number of different species of immunoglobulins produced by a cell line also gives information on its mono- or polyclonal origin and thereby on its relation to the tumour, since Burkitt's lymphoma (BL) is a monoclonal tumour (Fialkow *et al.*, 1970). Immunoglobulin production in permanent lymphoblastoid cell lines has been studied in several laboratories (Tanigaki *et al.*, 1966; Finegold, Fahey & Granger, 1967; Wakefield *et al.*, 1967; Glade & Chessin, 1968; Nilsson, Pontén & Philipson, 1968; Osunkoya *et al.*, 1968; Takahashi *et al.*, 1969a; Nilsson, 1971). Differences were found between the BL-derived lines and those from

other sources; the latter frequently produced several species of immunoglobulins, while the BL lines produced either a single species, usually IgM, or none. However, the donors of these lines were not of comparable age and geographical origin, the cultures had been established and propagated under different conditions, and the BL lines were usually much older than the lines of other origins.

The purpose of this work was therefore to compare the pattern of immunoglobulin secretion in BL lines with that of suitable control lines. Cultures were initiated from BL biopsies and from biopsies of solid tissues (four tonsil biopsies from patients with tonsillitis and a lymph node from a patient with Hodgkin's disease) from young African patients living in the area of Central Africa in which BL is endemic. The biopsies were fragmented, layered on gelatin foam grids, and cultured as described elsewhere (Nilsson, 1971). When the amount of tissue was sufficient, a number (up to fifteen) of independent cultures were set up from each biopsy. As soon as the lines had become established, their spent culture media were concentrated and tested for the presence of immunoglobulins by double immunodiffusion against rabbit antisera specific for the γ, μ, α and δ heavy chains and $\varkappa$ and λ light chains. All antisera were absorbed with calf serum. The cell lines were maintained in culture and tested for immunoglobulin production at intervals of a few months.

The results are shown in the Table. Out of 29 tonsil lines, only six produced initially a single species of immunoglobulin (i.e., one heavy and one light chain), while 14 synthesized all three heavy-chain classes and both types of light chains. The lines from the Hodgkin's disease lymph node showed a similar pattern: out of six lines, only one made a single species of

[1] The Wallenberg Laboratory, University of Uppsala, Sweden.
[2] Department of Tumor Biology, Karolinska Institutet, Stockholm, Sweden.

Immunoglobulin production by established lymphoid lines

Tissue	Patient	Cell line designation	Time of establish-ment (days)	First test Heavy α	γ	μ	Light κ	λ	Time be-tween first and last tests (days)	Last test Heavy α	γ	μ	Light κ	λ
Tonsil	Ambala	40 568	66	−	+	+	+	−	336	−	+	−	+	−
		40 566	83	+	−	+	−	+	275	−	+	−	+	−
		40 567	83	+	+	+	+	+	275	−	+	−	+	−
		40 565	90	+	+	+	+	+	275	−	+	−	+	−
		40 574	142	−	+	−	−	+						
		40 571	154	+	+	+	+	+	275	+	+	−	+	+
		40 569	197	+	+	+	+	+	217	−	+	−	−	+
	Okumu	42 593	59	−	+	+	+	+						
		42 592	69	−	+	−	−	+	167	−	+	−	+	−
		42 597	69	−	+	+	+	+	217	−	+	−	+	+
		42 594	80	+	+	+	+	+	167	+	+	+	+	+
		42 596	80	−	+	+	−	+	217	−	+	+	−	+
		42 598	92	−	+	+	+	−						
	Tado	43 913	30	+	+	+	+	+	243	−	−	+	−	+
		43 914	30	+	+	+	+	+	243	−	+	−	+	+
		43 907	66	+	+	+	+	+	243	−	+	−	+	+
		43 912	66	+	+	+	+	+	243	−	+	+	+	+
		43 915	66	+	+	+	+	+	134	−	+	+	+	+
		43 903	73	+	+	+	+	+	243	+	−	−	−	+
		43 905	73	+	+	+	+	+	243	−	+	−	+	+
		43 911	73	+	+	+	+	+						
		43 909	79	+	+	+	+	+	243	−	+	−	+	+
		43 902	114	−	+	+	+	−	90	−	−	+	−	+
		43 906	114	−	+	+	+	+	199	−	+	−	−	+
		43 904	128	−	+	−	−	+	109	−	+	−	−	+
		43 910	128	−	−	+	−	+	109	−	−	+	+	+
	Mwikali	46 206	168	−	+	+	+	−	55	−	+	+	+	−
		46 202	223	−	+	−	+	−						
		46 203	223	−	+	−	+	−						
Lymph node (Hodgkin's disease)	Cypriano	42 936	34	+	−	+	+	+	224	+	+	+	+	+
		42 937	34	+	+	+	+	+	301	−	+	+	+	+
		42 938	34	−	−	+	+	−	224	−	+	+	+	+
		42 939	34	+	+	+	+	+	224	−	+	+	+	+
		42 943	41	−	+	+	+	+	224	−	+	+	+	+
		42 941	78	−	+	+	+	−						
Burkitt's lymphoma	Muriungi (16. 6. 70)	447	25	−	−	−	−	−	64					
	Muriungi (25. 8. 70)	47 703	10	−	−	−	−	−						
		47 704	10	−	−	−	−	−	14	−	−	−	−	−
		47 701	20	−	−	−	−	−	153	−	−	−	−	−
		47 702	20	−	−	−	−	−	153	−	−	−	−	−
	Muriungi (13. 10. 70)	499	84	−	−	−	−	−						
	Akinyı	49 303	34	−	−	+	+	−	77	−	+	+	+	−
		49 301	55	−	+	+	+	−	48	−	+	+	+	−
		49 304	81	−	+	+	+	+	48	−	+	+	+	−
	Sulubu	52 001	69	−	−	−	+	−	42	−	−	−	−	−
	Chaka	52 201	54	−	−	−	−	+						
		52 203	54	−	−	+	−	+						
	Radiro	4884	45	−	+	+	+	−						
		4886	58	−	−	+	+	−						

immunoglobulin. Two lines made three heavy and two light chains.

After some months in culture, there was a reduction in the number of immunoglobulins made by individual lines from tonsils. The seven lines from the oldest tonsil biopsy in this study (Ambala) may be taken as an example. In the first test, only one line made a single immunoglobulin, while two made two heavy chains and one light chain, and four made three heavy chains and two light chains. After 9 months in culture, all lines except one made a single immunoglobulin. In contrast, little change in the pattern of immunoglobulin production was observed during 7 months in culture in the lines from the Hodgkin's disease lymph node.

The BL lines gave a different picture. From the point of view of the pattern of synthesis, these lines could be divided into three groups: (1) lines of the control type, making one or several heavy and light chains; (2) lines making only a single light chain; and (3) nonproductive lines. It is noteworthy that all lines derived from three consecutive biopsies from the same patient (Muriungi) were of the third type, and that all lines derived from patients Akinyi and Radiro produced complete immunoglobulins. The line from patient Sulubu, which initially produced light chains of a single type, lost this marker in less than two months.

Immunofluorescence studies (Takahashi et al., 1968; Finegold, Fahey & Dutcher, 1968) and cloning experiments (Hinuma & Grace, 1967; Takahashi et al., 1969b; Bloom, Choi & Lamb, 1971) have shown that individual cells of lymphoid lines may produce two heavy chains. However, production of three heavy chains or two light chains by a single cell has never been observed (Takahashi et al., 1969a). A line producing three heavy chains or two light chains can reasonably be assumed, therefore, to be heterogene-ous. As will be seen from the Table, this was initially the case for the majority of the tonsil and Hodgkin lines. The reduction with time of the number of immunoglobulin species produced by each line is best explained by a process of cell selection, which is expected to occur in a mixed population. Our results therefore suggest that the lines originating from normal lymphoid tissue, and from lymphoid tumours other than BL, were derived from a non-neoplastic heterogeneous cell population.

The BL lines that do not produce immunoglobulins or produce only light chains, must be different in nature from the control lines, which suggests that they are derived from the tumour cells. The immuno-globulin marker, when present, is of a single type, in agreement with the known clonal origin of Burkitt tumours (Fialkow et al., 1970). The BL lines (Radiro and Akinyi) that produce two classes of heavy chains could either be a homogeneous population of double-producing cells, possibly of tumorous origin, or heterogeneous lines of the control type, derived from non-neoplastic cells present in the tumour. Of interest in this con-nection is the fact that these lines are morpholog-ically closer to the control lines than to the typical BL lines.

ACKNOWLEDGEMENTS

This work was supported by grant No. 70:23 from the King Gustav V Jubilee Fund, and was undertaken during the tenure of a Research Training Fellowship of the International Agency for Research on Cancer. We thank Dr Pontén for valuable help and advice. The skilful technical assistance of Miss A. Jordell is gratefully acknowledged.

REFERENCES

Bloom, A. D., Choi, K. W. & Lamb, B. J. (1971) Im-munoglobulin production by human lymphocytoid lines and clones: Absence of genic exclusion. *Science,* **172**, 382-384

Fialkow, P. J., Klein, G., Gartler, S. M. & Clifford, P. (1970) Clonal origin for individual Burkitt tumours. *Lancet, i*, 384-386

Finegold, I., Fahey, J. L. & Granger, H. (1967) Synthesis of immunoglobulins by human cell lines in tissue culture. *J. Immunol.,* **99**, 839-848

Finegold, I., Fahey, J. L. & Dutcher, T. F. (1968) Im-munofluorescent studies of immunoglobulins in human lymphoid cells in continuous culture. *J. Immunol.,* **101**, 366-373

Glade, P. R. & Chessin, L. N. (1968) Infectious mono-nucleosis: Immunoglobulin synthesis by cell lines. *J. clin. Invest.,* **47**, 2391-2401

Hinuma, Y. & Grace, J. T., Jr (1967) Cloning of im-munoglobulin-producing human leukemic and lym-phoma cells in long-term cultures. *Proc. Soc. exp. Biol. (N.Y.),* **124**, 107-111

Nilsson, K. (1971) High-frequency establishment of human immunoglobulin-producing lymphoblastoid lines from normal and malignant lymphoid tissue and peripheral blood. *Int. J. Cancer,* **8**, 432-442

Nilsson, K., Pontén, J., & Philipson, L. (1968) Devel-opment of immunocytes and immunoglobulin produc-tion in long-term cultures from normal and malignant human lymph nodes. *Int. J. Cancer,* **3**, 183-190

Osunkoya, B. O., McFarlane, H., Luzzatto, L., Udeozo, I. O. K., Mottram, F. C., Williams, A. I. O. & Ngu, V. A. (1968) Immunoglobulin synthesis by fresh biopsy cells and established cell lines from Burkitt's lymphoma. *Immunology,* **14**, 851-860

Takahashi, M., Tanigaki, N., Yagi, Y., Moore, G. E. &
  Pressman, D. (1968) Presence of two different immunoglobulin heavy chains in individual cells of established
  human hematopoietic cell lines. *J. Immunol.*, **100**,
  1176-1183

Takahashi, M., Yagi, Y., Moore, G. E. & Pressman, D.
  (1969a) Pattern of immunoglobulin production in
  individual cells of human hematopoietic origin in
  established culture. *J. Immunol.*, **102**, 1274-1283

Takahashi, M., Takagi, N., Yagi, Y., Moore, G. E. &
  Pressman, D. (1969b) Immunoglobulin production in
  cloned sublines of a human lymphocytoid cell line.
  J. Immunol., **102**, 1388-1393

Tanikagi, N., Yagi, Y., Moore, G. E. & Pressman, D.
  (1966) Immunoglobulin production in human leukemia
  cell lines. *J. Immunol.*, **97**, 634-646

Wakefield, J. D., Thorbecke, G. J., Old, L. J. & Boyse,
  E. A. (1967) Production of immunoglobulins and their
  subunits by human tissue culture cell lines. *J. Immunol.*,
  99, 308-319

Immunoglobulin Synthesis as Cellular Marker of Malignant Lymphoid Cells *

E. KLEIN, R. van FURTH, B. JOHANSSON, I. ERNBERG & P. CLIFFORD

Many lymphoblast culture lines established from biopsies of Burkitt's lymphoma and from the buffy coat from peripheral blood of patients with leukaemia, lymphoma, infectious mononucleosis and normal donors have been shown to synthesize one or more classes of immunoglobulins or only heavy or light chains (Fahey *et al.*, 1966; Wakefield *et al.*, 1967; Osunkoya *et al.*, 1968; Takahashi *et al.*, 1969; Yagi, 1970).

Of 35 Burkitt's lymphoma biopsies, 21 were found to secrete immunoglobulin (see Table 1) (van Furth

Table 1. Immunoglobulin production by Burkitt's lymphoma cells detected by autoradiography

No. of biopsies [1]	Chains				
	γ	α	μ	$\varkappa$	λ
14	–	–	–	–	–
4 (287, 959, 788, 911)	+	–	–	–	–
3 (1044, 1041, 829)	+	–	–	–	+
4 (875, 827, 834, 819)	+	–	–	+	+
5 (929, 805, 865, 1046, 759)	+	–	–	+	–
3 (1035, 300, 952)	+	–	+	+	+
1 (967)	+	–	+	+	–
1 (812)	+	–	+	–	+

[1] Patients identified by Kenya Cancer Council number.

* From the Department of Tumor Biology, Karolinska Institutet, and Radiumhemmet, Karolinska Sjukhuset, Sweden; Department of Microbial Diseases, University Hospital, Leiden, The Netherlands; and Department of Head and Neck Surgery, Kenyatta National Hospital, Nairobi, Kenya.

et al., 1971). The study was performed by autoradiography of the immunoelectrophoretic patterns from cell suspension cultures in media containing radioactive amino acids. Most commonly the tumour cells synthesized IgG with type $\varkappa$ and/or type λ light chains. IgM synthesis was rarely seen while none of the biopsy cells or cell culture lines synthesized IgA.

In two cases immunoglobulin synthesis was studied and found to be identical in the primary tumour and recurrences after 5 and 12 months. One case was a nonproducer, while the other synthesized $\varkappa$ and λ chains. This finding may serve as proof that the recurrence was most probably derived from the cell population of the primary tumour and not the result of a fresh induction. In one case the tumours were located at different anatomical sites.

In 12 cases we were able to compare the biopsy with the culture line developed from it (Table 2). The age of the culture lines varied between 1 and 6 months. Identical patterns were obtained in 9. In two cases the culture lines produced additional light chains, which might be due to a superior capacity for protein synthesis of the established lines during the 24-hour incubation period in the labelled media as compared to the biopsy. In one line the synthesis of one light chain was lost. It is also possible, however, that these differences are real and indicate that the cell lines are not derived from the tumour cells in all cases.

The good agreement between the patterns of immunoglobulin secretion indicates that generally the cells of established lines can be considered to represent the *in vivo* tumour cell population.

The need to compare such established lines with cells of the tumours from which they originated is indicated by the studies of Fialkow *et al.* (1971), who found that the isoenzyme patterns of 6 out of 20

Table 2. Immunoglobulin production by Burkitt's lymphoma biopsies and by established cell lines detected by autoradiography

Patients [1]	Source of material	Chains				
		γ	α	μ	ϰ	λ
Ekesa (816; 1) Opasa (766; 6)	Biopsy and culture line	−	−	−	−	−
Dalmas (829; 6) Abwao (812; 1)		(+)	−	−	−	(+)
Omolo (929; 1) Nelson (805; 2)						
Margret (759; 2)		+	−	−	+	−
Kiliopa (834; 2)		+	−	−	+	+
Agnes (967; 3)		+	−	+	+	−
Isaac (788; 1)	Biopsy	+	−	−	−	−
	Culture line	+	−	(+)	+ +	−
Shegani (841; 6)	Biopsy	−	−	−	−	−
	Culture line	−	−	−	+	−
Juma (827; 5)	Biopsy	+	−	−	+	+
	Culture line	+ +	−	−	+	−

[1] KCC number and age of culture line at time of test (in months).

cultures differed from those of the original tumours. While laboratory contamination could not be entirely excluded it was considered to be more likely that the cells giving rise to the culture lines were present in the tumour and were introduced by blood transfusion. These studies were unfortunately not carried out on the material which was the subject of the present paper, but it may be pointed out the frequency of the differences was similar to that for immunoglobulin synthesis (3/12).

The clonal origin of Burkitt tumours is suggested by the G-6PD enzyme pattern in all seven patients studied who were heterozygous at this locus, since only one type was found in each of their tumours (Fialkow *et al.*, 1970). If one accepts the hypothesis that one cell synthesizes only one class of immunoglobulin, not all Burkitt biopsies studied can be regarded as monoclonal. Among the 21 biopsies positive for immunoglobulin production, 4 synthesized both γ and μ heavy chains and 8 synthesized both ϰ and λ light chains. The difference in these results may be caused by the use of different markers, whose methods of detection differ in sensitivity. It is known that the isoenzyme method does not detect the enzyme if it is present in a proportion lower than 10% (Fialkow *et al.*, 1971), whereas immunoglobulin synthesis may be detected with a smaller

proportion of cells. The other possibility is that the hypothesis that only single classes of immunoglobulin are produced by single cells or clones, which has emerged from studies on myeloma cells, does not hold true for lymphocytes. In fact, there is evidence indicating that this may be the case. Takahashi *et al.* (1968) demonstrated by means of immunofluorescence that single cells of established lymphoblastoid lines produce both γ and α heavy chains. Bloom, Choi & Lamb (1971) cloned cells of three lines and found that 23 out of 25 immunoglobulin-producing clones produced both γ and μ chains. It is thus possible that the clone which gives rise to the *in vitro* stem line is still pluripotential.

Another indication of heterogeneity is the fact that, in the majority of radioautographs in our studies, the entire γ line was labelled, indicating that different subclasses had been synthesized.

In addition to the immunoglobulins detected in studies on cell cultures, several biopsies from Burkitt's lymphoma cases were found to have cell-membrane-associated immunoglobulin moieties revealed by the reactivity of live cells for anti-immunoglobulin reagents (Table 3) (Klein, E. *et al.*, 1968; Clifford *et al.*, 1968). In all cases, the reactions indicated the presence of a μ chain with or without a ϰ light chain. This contrasts with the preponderance of biopsies which synthesized γ chains. Definite proof that the membrane-bound immunoglobulin is a cell product comes from its maintenance in some of the derived culture lines, which shows that it cannot represent an antibody coat attached to the cells *in vivo*.

The biopsy cells of different patients contained different amounts of immunoglobulin as judged by the intensity of fluorescence when the cells were exposed to conjugated antisera. Among the four (KCC Nos. 829, 805, 827 and 300) biopsies which had surface-localized IgM and which were studied for immunoglobulin synthesis, only one (KCC No. 300) synthesized detectable amounts of μ chains, but all synthesized γ chains. The intensity of the fluorescence suggested that these cells were richest in IgM. Perhaps the synthesis of the membrane-bound IgM proceeds at a slow rate and is therefore not detectable by the labelling and autoradiography procedure used.

The question arises whether or not the same cell synthesizes both γ and μ chains—the latter in this case remaining on the cell membrane. This is probably the case since fluorescence staining shows that nearly 100% of the cells carry the membrane-

Table 3. Reactivity of lymphoid malignancies with fluorescein-conjugated anti-IgM serum

Diagnosis and source of cells	Reactivity with anti-IgM serum [1]			Total no. of patients
	Absent or weak	Intermediate	Strong	
Burkitt's lymphoma [2]				
a. Cells from tumours	10 (28 %)	17 (47 %)	9 (25 %) [3]	36
Chronic lymphocytic leukaemia and lymphocyte-lymphoblast lymphoma				41
a. Leukaemia cells from peripheral blood	10 (37%)	12 (39 %)	12 (24 %)	
b. Leukaemia cells from bone marrow or lymph nodes	5	4	1	
Chronic granulocytic leukaemia				2
a. Leukaemia cells from peripheral blood	1	—	—	
b. Leukaemia cells from bone marrow	1	—	—	
Undifferentiated leukaemia				6
a. Leukaemia cells from peripheral blood	2	—	—	
b. Leukaemia cells from bone marrow or lymph nodes	4	—	—	
Polycythaemia vera				1
a. Normal granulocytes from peripheral blood	1	—	—	
Plasma cell tumours				2
a. Leukaemia cells from peripheral blood (multiple myeloma)	1	—	—	
b. Plasma cells from cutaneous lesion	1	—	—	
Hodgkin's disease				1
a. Cells from lymph-node lesion	1	—	—	
Essential monoclonal IgG proteinaemia				1
a. Normal lymphocytes from peripheral blood	1	—	—	
Tuberculosis				1
a. Cells from lymph-node lesion	1	—	—	

[1] The evaluation is based on the brilliance and extent of the reactivity.
[2] Culture lines were established from all these biopsies.
[3] Of the culture lines, five had cell-surface-bound IgM.

bound immunoglobulin. However, this question must be resolved by further studies. In this study the synthesis of γ chains was determined for the entire cell population, which may contain only a minority of IgG-secreting cells.

The cell-membrane-bound IgM is an excellent cell marker since it is present on all cells in a particular biopsy. The maintenance of this marker in the cell line is thus a proof of the tumour derivation of the cultured cells. Its loss, however, obviously does not necessarily indicate the converse.

Search for cell-membrane-bound IgM in other lymphoid malignancies revealed that it can be found in chronic lymphocytic leukaemia and lymphocyte-lymphoblast lymphoma (Table 3) (Johansson & Klein, E., 1970). In these diseases, judged by the intensity of the fluorescence reaction with anti-IgM serum, the amount of IgM on cells varies from patient to patient but is similar for different samples taken during the course of the disease in any given patient. It was present on almost all cells in the cases with strong reactivity.

This finding is important from the point of view of the relationship between chronic lymphocytic leukaemia and Burkitt's lymphoma. In spite of the dissimilarity in clinical course and age of onset, it indicates that a common cell type is the target of neoplastic transformation.

Lymphoblastoid cells with strong membrane reactivity with anti-μ and anti-κ serum were found to occur in normal cell populations derived from the bone marrow, thymus and liver of human foetuses (Klein E. *et al.*, 1970), indicating that the membrane-bound immunoglobulin is not connected with the malignant transformation but may represent a state of differentiation which is maintained in the tumour. Evidence that normal lymphoid cells synthesize and accumulate immunoglobulin on the cell membrane is provided by other experiments in which anti-immunoglobulin antibodies were also observed to bind to the surface of lymphoid cells or to influence the immunological performance of such cells. From our studies, the antigen determinants of immunoglobulin molecules have alone been detected on the surface of the lymphoid cells and there is no information on their antibody specificity.

Since it can be postulated that the proliferation of these cells was triggered by an antigen reacting with a cell-membrane-localized immunoglobulin receptor, its specificity for the Epstein-Barr virus (EBV) was investigated in one case of Burkitt's lymphoma.

It has been demonstrated that certain lymphoblast lines can be infected with EBV (Henle *et al.*, 1970). The infection results in an abortive cycle. The attachment of virus particles to the cell surface and the appearance of virus-specific antigens—early antigen and membrane antigen—can be demonstrated by reactions with fluorescein-conjugated antisera (Gergely, Klein & Ernberg, 1971). The infectivity of the virus can be neutralized by sera containing antibodies against EBV (Pearson *et al.*, 1970). When Daudi cells, which carry surface-bound IgM, were infected with the virus the results were similar to those obtained with the Raji line (see Fig.), the cells of

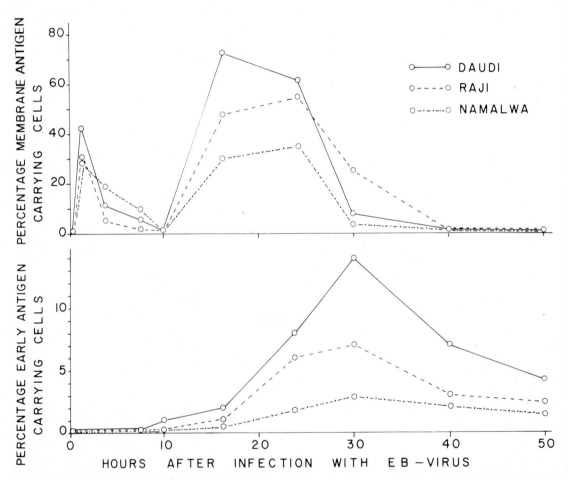

Infection of Daudi, Raji and Namalwa Burkitt cell lines with EBV. The virus preparation was obtained from Chas. Pfizer & Co., Inc., and was estimated to contain 2.4 × 10⁹ virus particles/ml. It was diluted to 1:10 with Eagle's minimum essential medium (MEM) without foetal calf serum (FCS) before use. One ml of cell suspension containing 2 × 10⁷ cells was incubated with 1 ml of virus suspension for 1 hour at 37°C. Thereafter the cells were washed three times with MEM. The suspension was diluted to contain 2.4 × 10⁵/ml with MEM + 20 % FCS. After 5 hours' incubation, the cells were washed again. Samples were stained at intervals for the presence of membrane antigen by exposing living cells to FITC-conjugated "Mutua" Burkitt serum (dilution 1:4). The membrane reactivity soon after infection reveals the adsorbed virus, the second wave of membrane reactivity the appearance of EBV-determined cell-membrane antigen. Early antigen (EA) was detected on fixed preparations by staining with FITC-conjugated "Ciromberia" nasopharyngeal carcinoma (CaPNS) serum (dilution 1:10).

which are commonly used for infection studies and which do not carry surface-bound immunoglobulin. Similar results were obtained with Namalwa cells which carry μ chains on their surface. Thus the surface-bound IgM on the Daudi cells did not neutralize EBV.

The feasibility of showing anti-EBV reactivity of surface-attached immunoglobulin by virus adsorption was investigated in an experiment in which two cell lines (Onesmus and LY28) were reacted with Burkitt serum "Agnes" (known to react with the EBV-determined cell-membrane antigen and the virus itself) conjugated with fluorescein isothyocyanate (FITC). To cover all antigen sites, the cells were then reacted with nonconjugated undiluted "Agnes" serum. After washing, the cells were incubated with EBV as in the infection studies. Thereafter the cells were stained again with "Agnes" serum conjugated with rhodamine. Observation of the cells showed that all green fluorescing areas, i.e., membrane-antigen-reactive antibody, were also stained with the red conjugate, i.e., antibody reacting with adsorbed virus. This indicates that virus was bound by the antibodies that, by their reactivity with the cell-membrane antigen, were attached to the cell surface. On a number of cells of the Onesmus line, the viral receptor sites could be seen as red fluorescent spots that were not stained by the green conjugate. In the Daudi line, all cells carry IgM moieties, but only 40% of the cells adsorbed virus; this finding also suggests that the IgM is not directed against EBV.

ACKNOWLEDGEMENTS

This work was supported by grants from contract No. 69-2005 within the Special Virus Cancer Program of the National Cancer Institute, the Swedish Cancer Society, the Åke Wiberg Foundation, and the Lotten Bohman Fund.

REFERENCES

Bloom, A. D., Choi, K. W. & Lamb, B. (1971) Immunoglobulin production by human lymphocytoid lines and clones: Absence of genic exclusion. *Science,* **172,** 382-383

Clifford, P., Gripenberg, U., Klein, E., Fenyö, E. M. & Manolow, G. (1968) Treatment of Burkitt's lymphoma. *Lancet,* **ii,** 517

Fahey, J. L., Finegold, I., Rabson, A. S. & Manaker, R.A. (1966) Immunoglobulin synthesis *in vitro* by established human cell lines. *Science,* **152,** 1259-1261

Fialkow, P. J., Klein, G., Gartler, S. M. & Clifford, P. (1970) Clonal origin for individual Burkitt tumours. *Lancet,* **i,** 384-386

Fialkow, P. J., Giblett, E., Klein, G., Gothoskar, B. & Clifford, P., (1971) Foreign-cell contamination in Burkitt tumours. *Lancet,* **i,** 883-886

Gergely, L., Klein, G. & Ernberg, I. (1971) Appearance of EBV-associated antigens in infected Raji-cells. *Virology,* **45,** 10-21

Henle, W., Henle, G., Zajac, B. A., Pearson, G., Waubke, R. & Scriba, M. (1970) Differential reactivity of human serums with early antigens induced by Epstein-Barr virus. *Science,* **169,** 188-190

Johansson, B. & Klein, E. (1970) Cell surface localized IgM-kappa immunoglobulin reactivity in a case of chronic lymphocytic leukaemia. *Clin. exp. Immunol.,* **6,** 421-428

Klein, E., Klein, G., Nadkarni, J. S., Nadkarni, J. J., Wigzell, H. & Clifford, P. (1968) Surface IgM-kappa specificity on a Burkitt lymphoma cell *in vivo* and in derived culture lines. *Cancer Res.,* **28,** 1300-1310

Klein, E., Eskeland, T., Inoue, M., Strom, R. & Johansson, B., (1970) Surface immunoglobulin-moieties on lymphoid cells. *Expl Cell Res.,* **62,** 133-148

Osunkoya, B. O., McFarlane, H., Luzzatto, L., Udeozo, I. O. K., Mottram, F. C., Williams, A. I. O. & Ngu, V. A. (1968) Immunoglobulin synthesis by fresh biopsy cells and established cell lines from Burkitt's lymphoma. *Immunology,* **14,** 851-860

Pearson, G., Dewey, F., Klein, G., Henle, G., and Henle, W. (1970) Relation between neutralization of Epstein-Barr virus and antibodies to cell-membrane antigens induced by the virus. *J. nat. Cancer Inst.,* **45,** 989-997

Takahashi, M., Tanigaki, N., Yagi, Y., Moore, G. E. & Pressman, D. (1968) Presence of two different immunoglobulin heavy chains in individual cells of established human hematopoietic cell lines. *J. Immunol.,* **100,** 1176-1183

Takahashi, M., Yagi, Y., Moore, G. E., & Pressman, D. (1969) Pattern of immunoglobulin production in individual cells of human hematopoietic origin in established culture. *J. Immunol.,* **102,** 1274-1283

van Furth, R., Gorter, H., Nadkarni, J. S., Nadkarni, J. J., Klein, E. & Clifford, P. (1971) Synthesis of immunoglobulins by tissue biopsies and cell lines from Burkitt's lymphoma. *Immunology* (in press)

Wakefield, J. D., Thorbecke, G. J., Old, L. J. & Boyse, E. A. (1967) Production of immunoglobulins and their subunits by human tissue culture cell lines. *J. Immunol.,* **91,** 308-319

Yagi, Y. (1970) Production of immunoglobulin by cells of established human lymphocytoid cell lines. *Symp. int. soc. Cell Biol.,* **9,** 121-133

Discussion Summary

C. S. MUIR [1]

Burkitt's lymphoma (BL)

The anatomical location of the lesions was discussed. It was noted that localization in the jaws was common in younger children and was thus frequent in those regions where BL was common. The organ most affected was the kidney; by contrast, the spleen was never involved. The peripheral lymph nodes were affected in perhaps 1 % of cases, whereas lesions in the abdominal lymph nodes were common. Breast lesions had been seen in lactating teen-age African mothers.

That a malignant neoplasm of lymphoid tissue should arise in precisely those organs having little lymphoid tissue seemed paradoxical. It was suggested that if the neoplasm started to grow in lymphoid cell aggregates, it might be suppressed whereas if it originated in an organ poor in lymphoid tissue growth would proceed.

It was observed that, in treated patients who relapsed early, the recurrence often arose in the same location as the original neoplasm, whereas in later relapses, say after one year, recurrence usually occurred elsewhere.

Regeneration, after treatment, of the normal structures, such as the jaw, which had been destroyed by the tumour, remained a mystery.

While indubitable cases of BL occur throughout the world, there seemed to be little information on possible differences in the pathology of BL in high- and low-incidence areas. In the latter, however, early bone marrow involvement had been noted.

Nasopharyngeal carcinoma (NPC)

Unlike the cervix uteri, there had been no studies of the temporal relationship between dysplasia and in situ and invasive NPC. All the in situ cancers demonstrated had been present close to an invasive neoplasm.

The bone marrow and the lymph nodes of NPC patients had been studied only in relation to the presence of metastatic deposits.

Infectious mononucleosis (IM)

There was some discussion on whether the atypical mononuclear cells (which, because they are found in other diseases caused by viruses, might be called "virocytes") were due to lymphocytotoxins. It was noted that in Hodgkin's disease, apart from the early stages of the disease, there were no typical abnormal peripheral mononuclear cells, although lymphocytotoxins were present. With regard to the absence of bone marrow changes in IM, it was true that small granulomata were found, but these were non-specific and could be found in infectious hepatitis and brucellosis.

It was noted that IM rarely progressed to leukaemia or lymphosarcoma, although mention was made of IM cases developing BL, Hodgkin's disease and acute leukaemia shortly after IM was diagnosed. The appearance of IM in leukaemic patients has been recorded, but this was possibly due to transfusion of infected blood. Mention was made of a patient with concurrent IM and lymphoblastic leukaemia who exhibited unusually rapid relapses associated with lymph-node enlargement.

Polyclonal origin of lymphoblastoid cell lines

It was noted that, in addition to the immunoglobulins, the isozyme system could be used to indicate whether a given cell line was monoclonal or polyclonal.

[1] International Agency for Research on Cancer, Lyon, France.

VIROLOGY OF BURKITT'S LYMPHOMA, INFECTIOUS MONONUCLEOSIS AND NASOPHARYNGEAL CARCINOMA

Chairman — R. J. C. Harris

Rapporteur — P. Gerber

Virology and Immunology of Epstein-Barr Virus (EBV) in Burkitt's Lymphoma—A Review

M. A. EPSTEIN [1]

HISTORICAL BACKGROUND

Burkitt's lymphoma was first recognized as a specific syndrome during the middle years of the 1950s in Uganda (Burkitt, 1958). This first report excited little interest at the time, but studies nevertheless continued in East Africa and the geographical distribution of the tumour was found to be dependent in some way on temperature and rainfall (Burkitt, 1962a, b).

In 1961 in London, Burkitt gave the first account outside Africa of the dramatic tumour that now bears his name and included in this talk details of his geographical distribution studies. It was immediately clear that, if the distribution of the tumour was indeed determined by temperature and rainfall, some biological factor must be playing an aetiological role, and an arthropod-vectored causative viral agent seemed the most likely. Because of this intriguing possibility, extensive investigations of Burkitt's lymphoma were immediately undertaken, a particular effort being made to find and identify any viruses associated with the tumour.

Biopsy samples from confirmed cases of Burkitt's lymphoma were removed in Uganda and flown overnight to London. Repeated testing of this material by conventional virological techniques using tissue-culture systems, eggs, and various laboratory animals, failed to give positive results over a period of many months. Following this, thin sections of tumour samples were searched directly in the electron microscope in an effort to find virus that might not be demonstrable by standard biological isolation techniques, but this too gave uniformly negative results (Epstein & Herdson, 1963).

It was then considered (Epstein, Barr & Achong, 1964) that progress might be made if tumour cells could be grown in vitro away from host defences, so that an inapparent oncogenic virus might then be able to replicate, as happens with cultured cells from certain virus-induced animal tumours (Bonar et al., 1960). This concept provided the basis for attempts to propagate Burkitt's lymphoma cells in serial long-term culture and led therefore to the discovery of the Epstein-Barr virus (EBV).

DISCOVERY OF EBV

At the time, the prospects for establishing in vitro strains of malignant lymphoid cells from Burkitt tumours were not good, for no member of the human lymphocytic series had ever been grown in continuous culture despite innumerable efforts to do this (Woodliff, 1964).

However, success was achieved in 1963 when two strains of Burkitt cells were established simultaneously from tumour material flown from Uganda to London (Epstein & Barr, 1964) and from a patient with a Burkitt's lymphoma in Nigeria (Pulvertaft, 1964).

As soon as sufficient cells were available from the first strain of cultured lymphoblasts (EB1), they were examined in the electron microscope for the presence of virus. By a curious twist of fortune, a cell containing unmistakable virus particles was found in the first grid square examined, and the virus was immediately recognized as having the typical morphology of a member of the herpes family (Epstein, Achong & Barr, 1964).

Tests were naturally undertaken to determine which herpesvirus was involved, and preparations from virus-bearing cultures were inoculated into

[1] Department of Pathology, The Medical School, University of Bristol, UK.

various tissue-culture systems, into eggs, and intra-cerebrally into suckling mice. When these tests proved uniformly negative it became clear that the agent present in the EB1 cells was highly unusual in that it showed a biological behaviour unlike that of any known member of the herpes group of viruses; confirmation of this biological inertness was rapidly obtained in further extensive experiments (Epstein *et al.*, 1965).

CONFIRMATION OF EBV IN CULTURED BURKITT CELLS

Since this early work, the presence of a morphologically similar and biologically inert herpesvirus has been confirmed in numerous cell strains from Burkitt tumours in further patients from Uganda, in patients from Nigeria, and in patients from New Guinea; in addition to this work with material from areas where Burkitt's lymphoma is endemic, the virus has also been found in cells grown from sporadic cases in other zones (for review, see Epstein, 1970).

MORPHOLOGICAL STUDIES OF EBV

The particles in all strains of cultured Burkitt cells are structurally indistinguishable and show the typical morphology of herpesviruses both in thin sections and negative-contrast preparations (Figs. 1 & 2). Characteristically, the hexagonal immature particles bud at cellular membranes, both those of the nuclear envelope and in the cytoplasm, so

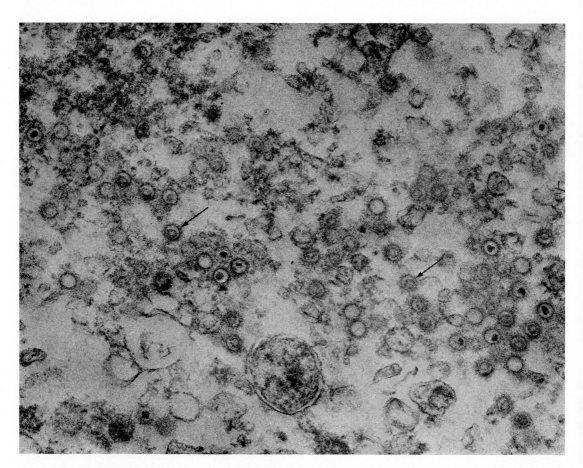

Fig. 1. Electron micrograph of a thin section cut through a disintegrating cultured Burkitt lymphoblast. Numerous hexagonal immature EBV particles are scattered throughout the field; some particles are empty, some have developing ring-shaped nucleoids (arrows) and some have dense central nucleoids. Glutaraldehyde followed by osmium fixation and epon embedding. ×64 000.

that the mature particles with an additional outer envelope of cell-membrane origin come to lie in the perinuclear space or in membrane-bounded cytoplasmic spaces. Mature particles have also been observed lying at the cell surface, just outside the plasmalemma, so that budding at this site, too, is likely. The morphogenesis of EBV maturation has been described in detail elsewhere (Epstein *et al.*, 1965).

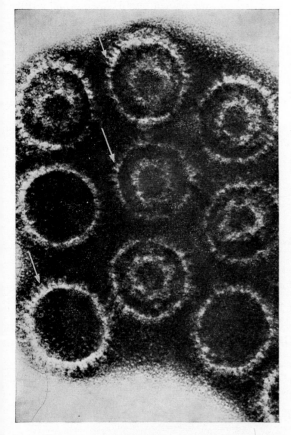

Fig. 2. Electron micrograph of whole-mount negative-contrast preparation of a group of immature EBV particles, either empty or with central ring-shaped nucleoids. The hollow tubular surface capsomeres are well seen in profile (arrows). Potassium phosphotungstate at pH 4.5. ×206 000.

Negative-contrast preparations have confirmed the herpes nature of EBV by revealing the typical hollow tubular capsomeres surrounding the immature particle (Toplin & Schidlovsky, 1966; Hummeler, Henle & Henle, 1966) (Fig. 2).

IMMUNOLOGICAL UNIQUENESS OF EBV

Although EBV has all the morphological attributes of a member of the herpes family, its biological inertness suggested the possibility that it might be a new member of this group. Immunological studies were therefore undertaken and it was found that antisera to known herpesviruses failed to react in immunofluorescence tests with cultured cells carrying EBV (Henle & Henle, 1966a, b). At the same time it was found that some human sera that did give a positive response to the virus would not react with cells infected with known herpesviruses (Henle & Henle, 1966a, b). In addition, an antiserum to purified EBV prepared in rabbits and found to be specific for the virus (Epstein & Achong, 1968a) would likewise not react with cells infected with known human herpesviruses (Epstein & Achong, 1968b). These original studies have been extended and confirmed in many subsequent publications (for review, see Epstein, 1970) and it is now quite clear that EBV is a new and distinct member of the herpes group.

POSSIBLE SIGNIFICANCE OF EBV IN BURKITT'S LYMPHOMA

Several important findings indicate that EBV is intimately associated with Burkitt's lymphoma, even if an aetiological relationship still remains to be established:

1. Sero-epidemiological studies have indicated that all patients with Burkitt's lymphoma have high titres of antibodies to EBV, in comparison with lower titres in only about 50% of appropriate African controls (Levy & Henle, 1966).

2. There is evidence that EBV is always associated with cell lines derived from Burkitt's lymphomas; virus production takes place in a small percentage of cells in most lines (Epstein & Achong, 1970); viral antigens detectable by immunofluorescence (Henle & Henle, 1966b) or by complement fixation (Pope, Horne & Wetters, 1969) are present in the cultured cells; even in the absence of virus production, the cells carry virus-determined neo-antigens (Klein, Klein & Clifford, 1967); and the viral genome can be detected in the cells (zur Hausen & Schulte-Holthausen, 1970).

3. EBV can infect normal blood cells *in vitro* (Henle *et al.*, 1967; Pope, Horne & Scott, 1968) and when it does so the following changes are brought about:

(*a*) cell morphology is altered to a blastoid form;
(*b*) the cultures acquire the power of unlimited proliferation (Henle *et al.*, 1967; Pope *et al.*, 1968);
(*c*) contact inhibition appears to be lost, since the cells tend to grow in clumps;
(*d*) the cells acquire what has been described as a specific chromosomal marker (Henle *et al.*, 1967; Pope *et al.*, 1968);
(*e*) virus-determined surface neo-antigens appear on the cells (Klein *et al.*, 1966; Klein *et al.*, 1967).

If the changes described above were to occur when a known animal tumour virus infects normal cells, they would indicate that the latter had undergone malignant transformation. Ultimate proof for such transformation in animal systems can be obtained when the transformed cells grow to form tumours on transplantation into isologous hosts, but for obvious reasons this step cannot be undertaken in relation to human material.

In addition to the foregoing effects of EBV in *in vitro* systems, it is now known that virus-determined surface neo-antigens are present on the actual tumour cells from Burkitt's lymphomas (Klein *et al.*, 1968b; Klein *et al.*, 1968a; Klein *et al.*, 1969) and that the viral genome is present in such cells taken directly from the patient (zur Hausen *et al.*, 1970). It is now also known that EBV is the aetiological agent of infectious mononucleosis (Henle, Henle & Diehl, 1968; Niederman *et al.*, 1968); although, of course, this disease is self-limiting, it is in many other respects very similar to the early stages of leukaemia or malignant lymphoma.

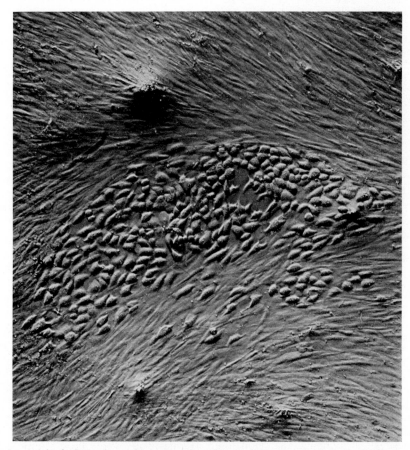

Fig. 3. Photomicrograph of a live culture of human embryo fibroblasts 9 days after infection in suspension with EBV. Within the confluent sheet of normal fibroblasts (above and below), a colony of morphologically transformed, polygonal cells can be seen. Oblique illumination. ×76.

ONCOGENIC POTENTIAL OF EBV

With the foregoing attributes and its particularly intimate association with Burkitt's lymphoma, it appears very probable that EBV is the aetiological agent of this peculiar tumour. Whether it acts alone or together with some cofactor is not known, but there is much circumstantial evidence suggesting that, in areas where the tumour is endemic, EBV may act in conjunction with holo- or hyperendemic malaria to cause the lymphoma (Burkitt, 1969); if this is the case, the curious geographical distribution of Burkitt's lymphoma becomes comprehensible, although it would not be the virus whose spread depends on temperature and rainfall but the co-factor.

With a suspected human tumour virus there are enormous difficulties in devising experiments to show that the suspect in fact plays an aetiological rôle. Various approaches have been suggested ranging from experimental vaccination programmes (Epstein, 1970) to elaborate large-scale serological and epidemiological surveys on populations at risk (Geser & de-Thé, 1971[1]).

At the experimental level it was considered that new information on the oncogenic potential of EBV might be obtained if some more conventional demonstration of *in vitro* transformation could be achieved. It has long been known that EBV cannot be made to infect any of a wide variety of monolayer test tissue cultures when standard techniques are used (Epstein *et al.*, 1965). It was therefore thought that some special manipulation might allow infection to take place and that this might lead either to malignant transformation, thus demonstrating viral oncogenicity, or to fully productive viral replication, thus providing methods for the preparation of virus in quantity. Experiments were accordingly initiated

---

[1] See p. 372 of this publication.

Fig. 4. Plastic culture bottle 19 days after seeding with human fibroblasts infected in suspension with EBV. The confluent cell sheet contains macroscopically visible colonies of morphologically transformed cells. ×1½.

along these lines and the following preliminary results seem to justify the hope that the first of the two alternatives mentioned has been achieved.

Partially purified preparations of EBV were mixed in suspension with human embryo skin/muscle fibroblasts and the virus was allowed to adsorb on to the cells for 1½ hours at 37°C in a slowly rotating container. Thereafter some of the cells were seeded directly in plastic bottles and others were exposed to UV-light-inactivated Sendai virus for 90 minutes before seeding in the same way. In addition, a sample of the fibroblasts not exposed to EBV was also treated with the inactivated Sendai virus as a control. After 9 days growth, the fibroblasts had formed a monolayer in all bottles, but in the cultures where cells had been exposed in suspension to EBV, small colonies of polygonal cells could be seen scattered amongst the fibroblasts (Fig. 3). In the bottles of cells exposed in suspension to EBV and subsequently treated with inactivated Sendai virus, about 7 times as many groups of polygonal cells were observed. No colonies were present in the control bottles. After a further 10 days incubation, the colonies of polygonal cells were macroscopically visible (Fig. 4) and were seen to be heaped up in the shape of a cone with a central crater (Fig. 5).

Passage of picked colonies of polygonal cells gave rise to cultures in which such cells outgrew the fibroblastic elements and appeared to be present as a uniform population.

It can thus be said that EBV appears to have brought about morphological transformation of human embryo fibroblasts *in vitro* when infection was carried out in an appropriate manner, and that the transformation rate was considerably increased where cell fusion was induced after the cells were

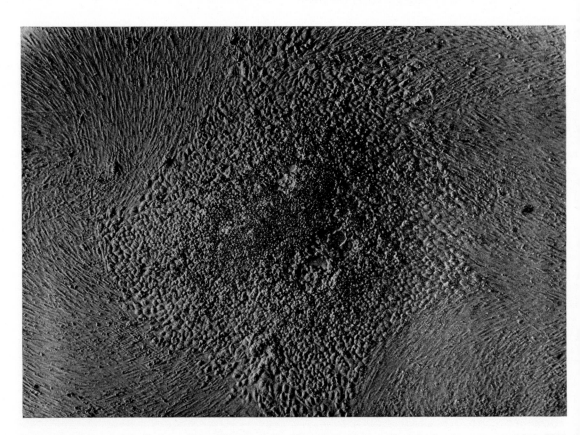

Fig. 5. Photomicrograph of a live culture of human embryo fibroblasts 19 days after infection in suspension with EBV. The morphologically transformed polygonal cells have grown up within the monolayer of normal fibroblasts into a cone-shaped colony with a central crater. Oblique illumination. ×45.

exposed to the virus. The foci of morphological transformation gave rise to recognizable plaques from which apparently pure cultures of rapidly growing abnormal cells have been obtained. A preliminary account of these findings has been presented elsewhere (Probert & Epstein, 1971).

Whether this morphological transformation indeed corresponds to malignant transformation remains to be determined, but in any event it is considered of sufficient significance to merit further extensive investigation since it may lead to the finding of oncogenic properties for EBV *in vitro*.

REFERENCES

Bonar, R. A., Weinstein, D., Sommer, J. R., Beard, D. & Beard, J. W. (1960) Virus of avian myeloblastosis. XVII. Morphology of progressive virus—myeloblast interactions *in vitro*. *Nat. Cancer Inst. Monogr.*, **4**, 251-290

Burkitt, D. (1958) A sarcoma involving the jaws in African children. *Brit. J. Surg.*, **46**, 218-223

Burkitt, D. (1962a) A children's cancer dependent on climatic factors. *Nature (Lond.)*, **194**, 232-234

Burkitt, D. (1962b) Determining the climatic limitations of a children's cancer common in Africa. *Brit. med. J.*, **2**, 1019-1023

Burkitt, D. P. (1969) Etiology of Burkitt's lymphoma—an alternative hypothesis to a vectored virus. *J. nat. Cancer Inst.*, **42**, 19-28

Epstein, M. A. (1970) Aspects of the EB virus. *Advanc. Cancer Res.*, **13**, 383-411

Epstein, M. A. & Achong, B. G. (1968a) Specific immunofluorescence test for the herpes-type EB virus of Burkitt lymphoblasts, authenticated by electron microscopy. *J. nat. Cancer Inst.*, **40**, 593-607

Epstein, M. A. & Achong, B. G. (1968b) Observations on the nature of the herpes-type EB virus in cultured Burkitt lymphoblasts, using a specific immunofluorescence test. *J. nat. Cancer Inst.*, **40**, 609-621

Epstein, M. A. & Achong, B. G. (1970) The EB virus. In: Burkitt, D. P. & Wright, D. H., eds, *Burkitt's lymphoma*, Edinburgh & London, Livingstone, p. 231

Epstein, M. A., Achong, B. G. & Barr, Y. M. (1964) Virus particles in cultured lymphoblasts from Burkitt's lymphoma. *Lancet*, **i**, 702-703

Epstein, M. A. & Barr, Y. M. (1964) Cultivation *in vitro* of human lymphoblasts from Burkitt's malignant lymphoma. *Lancet*, **i**, 252-253

Epstein, M. A., Barr, Y. M. & Achong, B. G. (1964) Avian tumor virus behavior as a guide in the investigation of a human neoplasm. *Nat. Cancer Inst. Monogr.*, **17**, 637-650

Epstein, M. A., Henle, G., Achong, B. G. & Barr, Y. M. (1965) Morphological and biological studies on a virus in cultured lymphoblasts from Burkitt's lymphoma. *J. exp. Med.*, **121**, 761-770

Epstein, M. A. & Herdson, P. B. (1963) Cellular degeneration associated with characteristic nuclear fine structural changes in the cells from two cases of Burkitt's malignant lymphoma syndrome. *Brit. J. Cancer*, **17**, 56-58

Henle, G. & Henle, W. (1966a) Immunofluorescence in cells derived from Burkitt's lymphoma. *J. Bact.*, **91**, 1248-1256

Henle, G. & Henle, W. (1966b) Studies on cell lines derived from Burkitt's lymphoma. *Trans. N.Y. Acad. Sci.*, **29**, 71-74

Henle, G., Henle, W. & Diehl, V. (1968) Relation of Burkitt's tumor-associated herpes-type virus to infectious mononucleosis. *Proc. nat. Acad. Sci. (Wash.)*, **59**, 94-101

Henle, W., Diehl, V., Kohn, G., zur Hausen, H. & Henle, G. (1967) Herpes-type virus and chromosome marker in normal leukocytes after growth with irradiated Burkitt cells. *Science*, **157**, 1064-1065

Hummeler, K., Henle, G., & Henle, W. (1966) Fine structure of a virus in cultured lymphoblasts from Burkitt lymphoma. *J. Bact.*, **91**, 1366-1368

Klein, G., Clifford, P., Klein, E. & Stjernswärd, J. (1966) Search for tumor-specific immune reactions in Burkitt lymphoma patients by the membrane immunofluorescence reaction. *Proc. nat. Acad. Sci. (Wash.)*, **55**, 1628-1635

Klein, G., Klein, E. & Clifford, P. (1967) Search for host defenses in Burkitt lymphoma: membrane immunofluorescence tests on biopsies and tissue culture lines. *Cancer Res.*, **27**, 2510-2520

Klein, G., Pearson, G., Henle, G., Henle, W., Diehl, V. & Niederman, J. C. (1968a) Relation between Epstein-Barr viral and cell membrane immunofluorescence in Burkitt tumor cells II. Comparison of cells and sera from patients with Burkitt's lymphoma and infectious mononucleosis. *J. exp. Med.*, **128**, 1021-1030

Klein, G., Pearson, G., Nadkarni, J. S., Nadkarni, J. J., Klein, E., Henle, G., Henle, W. & Clifford, P. (1968b) Relation between Epstein-Barr viral and cell membrane immunofluorescence of Burkitt tumor cells I. Dependence of cell membrane immunofluorescence on presence of EB virus. *J. exp. Med.*, **128**, 1011-1020

Klein, G., Pearson, G., Henle, G., Henle, W., Goldstein, G. & Clifford, P. (1969) Relation between Epstein-Barr viral and cell membrane immunofluorescence in Burkitt tumor cells. III. Comparison of blocking of direct membrane immunofluorescence and anti-EBV reactivities of different sera. *J. exp. Med.*, **129**, 697-705

Levy, J. A. & Henle, G. (1966) Indirect immunofluorescence tests with sera from African children and cultured Burkitt lymphoma cells. *J. Bact.*, **92**, 275-276

Niederman, J. C., McCollum, R. W., Henle, G. & Henle, W. (1968) Infectious mononucleosis. Clinical manifestations in relation to EB virus antibodies. *J. Amer. med. Assoc.,* **203**, 205-209

Pope, J. H., Horne, M. K. & Scott, W. (1968) Transformation of foetal human leukocytes *in vitro* by filtrates of a human leukaemic cell line containing herpes-like virus. *Int. J. Cancer,* **3**, 857-866

Pope, J. H., Horne, M. K. & Wetters, E. J. (1969) Significance of a complement-fixing antigen associated with herpes-like virus and detected in the Raji cell line. *Nature (Lond.),* **222**, 186-187

Probert, M. & Epstein, M. A. (1971) Morphological transformation *in vitro* of human fibroblasts by Epstein-Barr virus. *Science,* **175**, 202–203

Pulvertaft, R. J. V. (1964) Cytology of Burkitt's tumour (African lymphoma). *Lancet,* **i**, 238-240

Toplin, I. & Schidlovsky, G. (1966) Partial purification and electron microscopy of the virus in the EB-3 cell line derived from a Burkitt lymphoma. *Science,* **152**, 1084-1085

Woodliff, H. J. (1964) *Blood and bone marrow cell culture,* London, Eyre & Spottiswoode

zur Hausen, H. & Schulte-Holthausen, H. (1970) Presence of EB virus nucleic acid homology in a "virus-free" line of Burkitt tumour cells. *Nature (Lond.),* **227**, 245-248

zur Hausen, H., Schulte-Holthausen, H., Klein, G., Henle, W., Henle, G., Clifford, P. & Santesson, L. (1970) EBV DNA in biopsies of Burkitt tumours and anaplastic carcinomas of the nasopharynx. *Nature (Lond.),* **228**, 1056-1058

Epstein-Barr Virus: The Cause of Infectious Mononucleosis—A Review *

W. HENLE[1] & G. HENLE

The discovery that the Epstein-Barr virus (EBV) is the cause of infectious mononucleosis (IM) (Henle, G., Henle, W. & Diehl, 1968; Niederman et al., 1968) should in retrospect not have been as surprising as it appears to have been. IM was known to be a lymphoproliferative illness and had been termed a self-terminating malignant disease (Dameshek, 1966). It had been shown that lymphoid cells in the blood of patients in the acute stage of IM were temporarily stimulated to develop on culture into continuous lines of lymphoblastoid cells (Pope, 1967; Glade et al., 1968). EBV was known to depend for replication on cells of the lymphoid series and attempts to transmit it to other types of cells had failed (Epstein et al., 1965), observations which are still valid. Certain experiments had suggested that EBV exerts a growth-stimulating effect on cultured normal peripheral lymphocytes (Henle, W. et al., 1967). Moreover, another herpes group virus, cytomegalovirus, had been shown to cause IM-like disease (Klemola & Kääriäinen, 1965). Yet, evidence for a causal link between EBV and IM was obtained by serendipity rather than by design based on the above considerations. To be sure, a search for diseases caused by EBV was in progress when the initial clue was obtained but it was limited to paediatric patients who were thought, mistakenly, to offer a better prospect of success than adult patients. Failure to link EBV with childhood illnesses in itself provides support for the ultimate conclusion, since IM in children is mostly mild, often not recognized and usually unaccompanied by heterophil antibody responses, a key diagnostic criterion.

The clue was provided by a laboratory technician who had no antibodies to EBV but seroconverted as a result of IM (Henle, G. et al., 1968). Her leukocytes during the next two months, but not prior to or four months after the illness, yielded lymphoblastoid cell lines harbouring EBV in some of the cells. Similar results were obtained when the only remaining antibody-negative technician developed IM several months later. These observations sparked off intensive studies with results that leave little doubt that EBV is the cause of IM.

The present evidence for the causal relation of EBV to IM will be reviewed together with new information on antibody responses to various EBV-related antigens.

DEMONSTRATION OF EBV ACTIVITY IN PATIENTS WITH IM

Leukocyte cultures

Cultures of leukocytes from 10 to 20 ml of blood from 24 patients in the acute phase of IM regularly developed into lines (Diehl et al., 1968). Growth commenced within two to four weeks and all lines revealed EBV in a small proportion of cells. In contrast, cultures from by now over 60 children or adults not previously exposed to EBV, i.e., without antibodies to EBV, failed to grow under these conditions with two exceptions. In both of these, the donors were incubating primary EBV infections; one, the second technician mentioned above, presented signs of IM 11 days later and the other, a child with recurrent respiratory illnesses, had antibodies when re-bled after two months. Leukocytes from healthy donors or patients with illnesses other than IM, who had histories of this disease and/or antibodies to EBV, yielded intermediate results,

* From the Virus Laboratories, Children's Hospital, Philadelphia and School of Medicine, University of Pennsylvania.
[1] Recipient of Career Award 5-K6-AI-22,683 from the National Institutes of Health, US Public Health Service.

i.e., about 40% of the cultures yielded continuous lines that began to grow only after long delays (>5 weeks) but then revealed EBV, at least temporarily, in a fraction of the cells.

These data supported the suggestion that EBV stimulates the growth of lymphoid cells since, in its absence, no lines developed; this does not, however, exclude the possibility that other viruses may occasionally have the same effect. In an acute EBV infection, the target cells are highly stimulated and regularly yield continuous cultures. After primary infections, EBV evidently persists, as do other herpesviruses, in this instance in the lymphoreticular system. Depending upon the extent of the persistent infection, the number of virus-bearing lymphoid cells entering the circulation might vary and with it the success of leukocyte cultures. This interpretation is supported by the fact that leukocytes from anti-EBV negative individuals could be stimulated to grow into lines by addition of EBV-containing materials, whether lethally X-irradiated blastoid cells from virus-positive lines (Henle, W. et al., 1967; Diehl et al., 1969), cell-free culture media from such lines (Pope, Horne & Scott, 1968) or virus separated therefrom (Gerber, Whang-Peng & Monroe, 1969; Henle & Henle, 1970). No growth was obtained when X-irradiated cells from EBV-negative lines were used, when the virus was removed by filtration, inactivated by heat, or neutralized by antibody.

These observations were extended by Nilsson,[1] who established continuous cell lines from lymph nodes of antibody-positive individuals at a higher frequency (91%) than from their peripheral leukocytes (45%), suggesting that more EBV-containing cells are present in lymph nodes than in the blood. Furthermore, Nilsson failed to establish lines from foetal lymphoid tissues. This finding, as well as the unsuccessful attempts to establish cultures of lymphocytes from anti-EBV negative infants and children, denote that EBV is not vertically transmitted but spread postnatally by horizontal transmission.

Antibody responses

Initially, serological tests were restricted to indirect immunofluorescence with acetone-fixed smears of cells from EBV-positive lines of Burkitt's tumour cells (Henle, G. et al., 1968; Niederman et al., 1968; Evans, Niederman & McCollum, 1968; Pereira, Blake & Macrae, 1969; Banatvala & Gryllis, 1969;

Hirshaut et al., 1969). This test determines essentially antibody titres to viral capsid antigens (VCA) (see below). Every patient in the acute stage of IM was found to have anti-VCA (formerly anti-EBV) in titres ranging from 1:40 to 1:640. Individuals with well-documented histories of IM as long as 37 years ago also regularly showed anti-VCA, but usually at lower titres (1:5 to 1:80). Most important, pre-illness sera, whether available by chance (Henle, G. et al., 1968; Pereira et al., 1969) or by design from prospective studies of IM (Niederman et al., 1968, 1970; Evans et al., 1968; Wahren et al., 1970) were always devoid of antibodies (<1:2). These results were confirmed by complement-fixation tests with concentrated EBV suspensions as antigen (Gerber et al., 1968) and by demonstration of *de novo* formation of antibodies to EBV-determined membrane antigens (MA) on cultured lymphoblasts of Burkitt's tumour or of IM leukocyte origin (Klein et al., 1968). The MA complex seems also to be present in the viral envelope since anti-MA was found to parallel the virus-neutralizing activity of sera (Pearson et al., 1970).

The incubation period of IM is long and patients often seek medical aid late, when maximal anti-VCA titres may have been reached. It is uncommon, therefore, to observe diagnostically significant increases in titre (≥4-fold) in subsequent sera. This serodiagnostic handicap is compounded by the fact that the titres attained might on occasion not exceed those persisting after past infections. Thus, only high anti-VCA titres permit a presumptive diagnosis. The anti-VCA titres decline, after recovery, to lower levels, which is also of serodiagnostic significance, but persistent titres may be less than 4-fold below the maximum recorded, especially when the first serum was not taken at the peak. The heterophil antibody test has therefore remained the primary serodiagnostic tool, but some patients, mainly in the paediatric age range, fail to develop this antibody. Clearly, additional, specific tests are needed and the one described below has provided a partial answer.

ANTIBODIES TO EBV-INDUCED EARLY ANTIGENS (EA) IN IM

Exposure of lymphoblastoid cell lines free of detectable EBV or EBV-related antigens, but not necessarily of viral genomes (zur Hausen & Schulte-Holthausen, 1970), to EBV derived from carrier

---

[1] See p. 285 of this publication.

cultures (HR1-K) was found to cause mainly abortive cellular infections (Henle, W. *et al.*, 1970a). Most, if not all, invaded cells synthesized antigens that were detectable by indirect immunofluorescence with sera from many *patients* with IM, Burkitt's lymphoma (BL) or nasopharyngeal carcinoma (NPC), but rarely with sera from *healthy donors*, even if they showed high antibody titres when tested on cell smears from EBV carrier lines (EB3 or HR1-K). The infectious process was usually arrested at this stage, and few, if any, cells went on to produce antigens detectable by healthy donor sera. The antigens detected solely by patients' sera were synthesized well in advance of antigens stainable by donor sera, and therefore called early antigens (EA). Indeed, EA production was not prevented when DNA synthesis was inhibited by cytosine arabinoside (Gergely, Klein & Ernberg, 1971). Since the donor sera contained antibodies which coated viral nucleocapsids and caused their agglutination (Henle, W., Hummeler & Henle, G., 1966; Mayyasi *et al.*, 1967), the antigens detected by them were called viral capsid antigens (VCA). Positive cells in EBV carrier cultures undoubtedly contain EA as well as VCA, but the titres of the patients' sera in this system were taken to reflect anti-VCA since they usually exceeded by a factor of $\geqslant 2$ the anti-EA titres determined with abortively infected cells.

Anti-EA was found to arise in about 75% of over 200 IM patients in titres ranging from 1:5 to 1:320 (Henle, W. *et al.*, 1971b). This response depended to a large degree upon the severity of illness. Although some patients failed to produce anti-EA, limiting the usefulness of the test, this handicap was partly offset by the fact that anti-EA tended to arise later than anti-VCA and disappeared again within a few months, in contrast to the persistance of anti-VCA for years, if not for life. Thus, diagnostically significant increases or decreases in anti-EA were more frequent ($>50\%$ of patients) than corresponding changes in anti-VCA (about 15%) and would undoubtedly have been more frequent if additional sera had been available from some patients during the acute or late convalescent stages. Furthermore, mere presence of anti-EA in a patient's serum is in itself of diagnostic import since it is rarely found in healthy individuals. Its rare presence in sarcoidosis or systemic lupus erythematosis, two diseases with increased incidences of high anti-VCA titres (Hirshaut *et al.*, 1970; Evans, Rothfield & Niederman, 1971), should not cause confusion.

In examining numerous sera from IM, BL and NPC patients, it was noted that the pattern of immunofluorescent staining elicited by some differed from that shown by the majority. In the most common pattern, positive cells were *diffusely* stained involving the nucleus as well as the cytoplasm (this will be referred to as D). In the other pattern, staining was *restricted* to aggregates in the cytoplasm (this will be referred to as R). With some dominantly anti-D reactive sera, R aggregates were discernible through the diffuse staining. The vast majority of IM sera yielded D staining without evidence of anti-R reactivity, which was confirmed when means were found to differentially denature the two types of early antigens.

It was observed, in addition, that the R antigen was destroyed when abortively infected cells were fixed in 95% ethanol or methanol instead of acetone. All but one of 80 anti-EA positive IM sera reacted as well with ethanol-fixed as with acetone-fixed cells, confirming their anti-D reactivity. The one anti-R reactive serum came from a patient who presented with extensive urticaria five days before classical signs of IM developed (Niederman[1]). Whether anti-D sera might contain low anti-R levels could not reliably be ascertained until it was found that D antigen in acetone-fixed smears was denatured by brief exposure to proteolytic enzymes without seriously reducing R antigen activity. Of 20 anti-D reactive sera from IM patients, none reacted with pronase-treated smears, suggesting that anti-R responses in IM are rare, and, if observed, involve unusual cases.

These results show that anti-D represents a transitory response to primary EBV infections, depending upon the severity of the illness. Results obtained with sera from BL patients differed in several respects. Anti-EA, if found, usually persisted at nearly constant levels as long as the patients lived (Henle, G. *et al.*, 1971). A decline in titre, observed occasionally, was indicative of a decrease in likelihood of tumour recurrence whereas a rise in anti-EA had apparently the opposite prognosis. In contrast to IM, many of the anti-EA-positive BL sera revealed only or predominantly anti-R. In a few BL sera, anti-D exceeded anti-R titres or was the only antibody detected. Anti-EA-positive NPC sera usually contained anti-D, often accompanied by lower levels of anti-R and, rarely, only anti-R. The relation of anti-D and anti-R and of their titres

---

[1] Personal communication.

to the clinical status and prognosis of BL and NPC remain to be ascertained. These studies need to be extended also to anti-EA reactions observed occasionally in the sera of patients with other malignant or nonmalignant diseases or of apparently healthy donors.

INADVERTANT TRANSMISSION OF EBV BY BLOOD

Experimental transmission of EBV to susceptible healthy volunteers appears indefensible because of the often prolonged course of IM and the suspected oncogenic potential of the virus, which it might exert on rare occasions, most probably in conjunction with other, undefined factors. Thus, the third of the Henle-Koch postulates remains unfulfilled, a not unusual omission when dealing with potentially dangerous agents. However, since EBV is often present in circulating leukocytes (see above) its transmission by blood transfusion cannot be avoided. Indeed, antibody conversion has been observed following heart surgery with extracorporeal circulation (Gerber et al., 1969; Henle, W. et al., 1970b) which, in one of Gerber's cases, was accompanied by the IM-like postperfusion syndrome. Cytomegalovirus, the usual cause of this disease (Kääriäinen, Klemola & Paloheimo, 1966), was excluded serologically in this patient. The infrequency of EBV-induced postperfusion syndromes may be due to the fact that donors of EBV-containing blood provide at the same time antibodies that might abort or prevent the disease.

CONCLUDING REMARKS

It has been suggested that EBV is unrelated to IM and merely activated by this disease, which is caused by another, as yet unidentified, agent (Glade, Hirschhorn & Douglas, 1969). This hypothesis assumes that EBV is either transmitted vertically from mother to child, evoking no antibody response, or that antibodies formed after early primary infections decline subsequently to undetectable levels. As already pointed out, neither vertical transmission of EBV nor complete loss of antibodies to VCA has been demonstrated. In fact, anti-VCA persists at nearly constant levels for years, if not for life (Niederman et al., 1968, 1970; Henle, G. & Henle, W., 1970; Henle, W. et al., 1971b). While anti-VCA nevertheless might occasionally decline to undetectable levels one could hardly expect that such a rare

event would have occurred in every IM patient. One would expect, however, that other viruses causing lymphoproliferative responses would activate latent EBV infections. This has not been observed in rubella (Sawyer et al., 1971), infectious lymphocytosis (Blacklow & Kapikian, 1970), hepatitis (Henle, G. et al., 1968), cytomegalovirus (Klemola et al., 1970), coxsackie, adeno- or myxovirus infections (Niederman et al., 1968).

The fact that the anti-VCA titres seen in IM are at times matched or exceeded in other diseases, namely BL (Henle, G. et al., 1970), NPC (Henle, W. et al., 1971a), other malignancies (Johansson et al., 1970), sarcoidosis (Hirshaut et al., 1970) and systemic lupus erythematosus (Evans et al., 1971), has hindered acceptance of the aetiological rôle of EBV in IM. If it were based solely on the regular presence of usually high antibody levels in the acute stage of IM there would be cause for doubt, but one has to consider all the evidence: (a) IM occurs only in antibody-negative individuals and, in turn, presence of antibodies denotes immunity; (b) rises and falls in antibody titres in the course of IM are now frequently noted with the anti-EA test; (c) all individuals with documented histories of IM have anti-VCA; (d) the sero-epidemiology of EBV conforms with predictions based on the epidemiology of IM (Evans, 1960): the causative agent must be readily transmitted early in life under low socio-economic conditions, resulting in asymptomatic or mild, but immunizing infections, whereas under high socio-economic conditions exposure is postponed to an age at which classical IM may ensue. Indeed, anti-VCA is acquired by children from poor surroundings at higher frequency early in life than by children from middle or upper middle class families (Henle, G. & Henle, W. 1967; Porter, Wimberly & Benyesh-Melnick, 1969); (e) IM is a lymphoproliferative disease and EBV stimulates growth of cultured lymphocytes in vitro; and (f) EBV can cause the IM-like postperfusion syndrome. With these facts, the causal relation of EBV to IM can hardly be questioned even though the last of the Henle-Koch postulates has not been truly fulfilled.

The association of EBV with other diseases is not as well supported and, indeed, when based entirely on high antibody titres in a higher proportion of patients than in controls, the "association" might not be causal but incidental. Hypergammaglobulinaemia might shift titres into the "high" category or lymph-node involvement in certain diseases may activate an EBV carrier state following earlier primary infections. The association of EBV with BL and NPC does not rest solely on high antibody titres. EB viral nucleic acid has regularly been detected in considerable quantities in tumour biopsies (zur Hausen et al., 1971). This does not necessarily exclude activation of a passenger virus, but an explanation must then be found for the absence of EBV-DNA in other, related tumours of anti-VCA positive patients, likely to be carriers of EBV, and for the failure of these patients to show similarly high levels of antibodies to EBV-related antigens. Indeed, the presence of anti-EA, or its increase or decrease in BL patients, has prognostic implications (Henle, G. et al., 1971) and loss of antibodies to EBV-determined cell-membrane antigens in long-term survivors, free of detectable tumours, has presaged the imminence of relapses (Klein et al., 1969). It would hardly seem plausible that antibodies to a passenger virus would be of prognostic significance. Nevertheless, the present evidence affords insufficient proof of an aetiological relation of EBV to BL, and even less so to NPC and certain other malignancies. While these relationships remain unexplained they provide the incentive for further intensive studies.

ACKNOWLEDGEMENTS

The work of the authors was supported by grant CA 04568 and contract PH-43-66-477 within the Special Virus Cancer Program, from the National Cancer Institute, US Public Health Service, and by the US Army Medical Research and Development Command, Department of the Army, contract DA-49-193-MD-2474, under the sponsorship of the Commission on Viral Infections, Armed Forces Epidemiological Board.

REFERENCES

Banatvala, J. E. & Gryllis, S. G. (1969) Serological studies in infectious mononucleosis. Brit. med. J., 3, 444-446

Blacklow, N. R. & Kapikian, A. Z. (1970) Serological studies with EB virus in infectious lymphocytosis. Nature (Lond.), 326, 647

Dameshek, W. (1966) Immunocytes and immunoproliferative disorders. In: Wolstenholme, G. E. W. & Porter, R., eds, The thymus: experimental and clinical studies, London, Churchill (CIBA Foundation Symposium)

Diehl, V., Henle, G., Henle, W. & Kohn, G. (1968) Demonstration of a herpes group virus in cultures of peripheral leukocytes from patients with infectious mononucleosis. J. Virol., 2, 663-669

Diehl, V., Henle, G., Henle, W. & Kohn, G. (1969) Effect of a herpes group virus (EBV) on growth of peripheral leukocyte cultures. In Vitro, 4, 92-99

Epstein, M. A., Henle, G., Achong, B. G., & Barr, Y. M. (1965) Morphological and biological studies on a virus in cultured lymphoblasts from Burkitt's lymphoma. J. exp. Med., 121, 761-770

Evans, A. S. (1960) Infectious mononucleosis in University of Wisconsin students. Amer. J. Hyg., 71, 342-346

Evans, A. S., Niederman, J. C. & McCollum, R. W. (1968) Seroepidemiologic studies on infectious mononucleosis with EB virus. New Engl. J. Med., 279, 1121-1127

Evans, A. S., Rothfield, N. F. & Niederman, J. C. (1971) Raised antibody titers to EB virus in systemic lupus erythematosus. Lancet, i, 167-168

Gerber, P., Hamre, D., Moy, R. A. & Rosenblum, E. N. (1968) Infectious mononucleosis: Complement-fixing antibodies to herpes-like virus associated with Burkitt's lymphoma. Science, 161, 173-175

Gerber, P., Walsh, G. H., Rosenblum, E. N. & Purcell, R. H. (1969) Association of EB-virus with the post-perfusion syndrome. Lancet, i, 593-596

Gerber, P., Whang-Peng, J. & Monroe, J. H. (1969) Transformation and chromosome changes induced by Epstein-Barr virus in normal human leukocyte cultures. Proc. nat. Acad. Sci. (Wash.), 63, 740-747

Gergely, L., Klein, G., & Ernberg, I. (1971) Appearance of EBV associated antigens in infected Raji cells. Virology (in press)

Glade, P. R., Karel, J. A., Morse, H. L., Whang-Peng, J., Hoffman, P. E., Kammermeyer, J. K. & Chessin, L. N. (1968) Infectious mononucleosis continuous suspension cultures of peripheral leukocytes. Nature (Lond.), 217, 564-565

Glade, P. R., Hirschhorn, K. & Douglas, S. D. (1969) Herpes-like virus. Lancet, i, 1049-1050

Henle, G. & Henle, W. (1967) Immunofluorescence, interference and complement fixation techniques in the detection of the herpes type virus in Burkitt tumor cell lines. Cancer Res., 27, 2442-2446

Henle, G. & Henle, W. (1970) Observations on childhood infections with the Epstein-Barr virus. J. infect. Dis., 121, 303-310

Henle, G., Henle, W., & Diehl, V. (1968) Relation of Burkitt's tumor-associated herpes-type virus to infectious mononucleosis. Proc. nat. Acad. Sci., (Wash.) 59, 94-101

Henle, G., Henle, W., Clifford, P., Diehl, V., Kafuko, G. W., Kirya, B. G., Klein, G., Morrow, R. H., Munube, G. M. R., Pike, M., Tukei, P. M. & Ziegler, J. L. (1970) Antibodies to Epstein-Barr virus in Burkitt's lymphoma and control groups. J. nat. Cancer Inst., 43, 1147-1157

Henle, G., Henle, W., Klein, G., Gunven, P., Clifford, P., Morrow, R. H., & Ziegler, J. L. (1971) Antibodies to early Epstein-Barr virus-induced antigens in Burkitt's lymphoma. *J. nat. Cancer Inst.*, **46**, 861-871

Henle, W. & Henle, G. (1970) Evidence for a relation of Epstein-Barr virus to Burkitt's lymphoma and nasopharyngeal carcinoma. In: Dutcher, R. M., ed., *Comparative leukemia research 1969*, Basel, München & Paris, Karger, pp. 706-713

Henle, W., Hummeler, K. & Henle, G. (1966) Antibody coating and agglutination of virus particles separated from the EB3 line of Burkitt lymphoma cells. *J. Bact.*, **92**, 269-271

Henle, W., Diehl, V., Kohn, G., zur Hausen, H. & Henle, G. (1967) Herpes-type virus and chromosome marker in normal leukocytes after growth with irradiated Burkitt cells. *Science*, **157**, 1064-1065

Henle, W., Henle, G., Zajac, B. A., Pearson, G., Waubke, R. & Scriba, M. (1970a) Differential reactivity of human serums with early antigens induced by Epstein-Barr virus. *Science*, **169**, 188-190

Henle, W., Henle, G., Scriba, M., Joyner, C. R., Harrison, F. S., Jr, von Essen, R., Paloheimo, J. & Klemola, R. (1970b) Antibody responses to the Epstein-Barr virus and cytomegalovirus after open-heart surgery. *New Engl. J. Med.*, **282**, 1068-1074

Henle, W., Henle, G., Ho, H. C., Burtin, P., Cachin, Y., Clifford, P., Schryver, A., de-Thé, G., Diehl, V. & Klein, G. (1971a) Antibodies to Epstein-Barr virus in nasopharyngeal carcinoma, other head and neck neoplasms, and control groups. *J. nat. Cancer Inst.*, **44**, 225-231

Henle, W., Henle, G., Niederman, J. C., Klemola, E. & Haltia, K. (1971b). Antibodies to Epstein-Barr virus-induced early antigens in infectious mononucleosis. *J. infect. Dis.*, **124**, 58-67

Hirshaut, Y., Glade, P., Moses, H., Manaker, R. & Chesin, L. (1969) Association of herpes-like virus infection with infectious mononucleosis. *Amer. J. Med.*, **47**, 520-527

Hirshaut, Y., Glade, P., Octavio, L., Vieira, B. D., Ainbender, E., Dvorak, B. & Siltzbach, L. E. (1970) Sarcoidosis—serologic evidence for herpes-like virus infection. *New Engl. J. Med.*, **283**, 502-505

Johansson, B., Klein, G., Henle, W. & Henle, G. (1970) Epstein-Barr virus (EBV)-associated antibody patterns in malignant lymphomas and leukemia. I. Hodgkin's disease. *Int. J. Cancer*, **6**, 450-462

Kääriäinen, L., Klemola, E. & Paloheimo, J. (1966) Rise of cytomegalovirus antibodies in an infectious-mononucleosis like syndrome after transfusion. *Brit. med. J.*, **1**, 1270-1272

Klein, G., Pearson, G., Henle, G., Henle, W., Diehl, V. & Niederman, J. C. (1968) Relation between Epstein-Barr viral and cell membrane immunofluorescence in Burkitt tumor cells. II. Comparison of cells and sera from patients with Burkitt's lymphoma and infectious mononucleosis. *J. exp. Med.*, **128**, 1021-1030

Klein, G., Clifford, P., Henle, G., Henle, W., Geering, G. & Old, L. J. (1969) EBV-associated serological patterns in a Burkitt lymphoma patient during regression and recurrence. *Int. J. Cancer*, **4**, 416-421

Klemola, E. & Kääriäinen, L. (1965) Cytomegalovirus as a possible cause of a disease resembling infectious mononucleosis. *Brit. med. J.*, **2**, 1099-1102

Klemola, E., von Essen, R., Henle, G. & Henle, W. (1970) Infectious-mononucleosis-like disease with negative heterophil agglutination test. Clinical features in relation to Epstein-Barr virus and cytomegalovirus antibodies. *J. infect. Dis.*, **121**, 608-614

Mayyasi, S. D., Schidlovsky, G., Bulfone, L. M. & Buscheck, F. I. (1967) The coating reaction of the herpestype virus isolated from malignant tissues with an antibody present in sera. *Cancer Res.*, **27**, 2020-2024

Niederman, J. C., McCollum, R. W., Henle, G. & Henle, W. (1968) Infectious mononucleosis. Clinical manifestations in relation to EB virus antibodies. *J. Amer. med. Ass.*, **203**, 205-209

Niederman, J. C., Evans, A. S., Subrahmanyan, M. S. & McCollum, R. W. (1970) Prevalence, incidence and persistence of EB virus antibody in young adults. *New Engl. J. Med.*, **282**, 361-365

Pearson, G., Dewey, F., Klein, G., Henle, G. & Henle, W. (1970) Relation between neutralization of Epstein-Barr virus and antibodies to cell membrane antigens induced by the virus. *J. nat. Cancer Inst.*, **45**, 989-995

Pereira, M. S., Blake, J. M. & Macrae, A. D. (1969) EB virus antibody at different ages. *Brit. med. J.*, **4**, 526-527

Pope, J. H. (1967) Establishment of cell lines from peripheral leukocytes in infectious mononucleosis. *Nature (Lond.)*, **216**, 810-811

Pope, J. H., Horne, M. K. & Scott, W. (1968) Transformation of foetal human leukocytes *in vitro* by filtrates of a human leukemic cell line containing herpeslike virus. *Int. J. Cancer*, **3**, 857-866

Porter, D. D., Wimberly, I. & Benyesh-Melnick, M. (1969) Prevalence of antibodies to EB virus and other herpes viruses. *J. Amer. med. Ass.*, **208**, 1675-1679

Sawyer, R. N., Evans, A. S., Niederman, J. C. & McCollum, R. W. (1971) Prospective studies of a group of Yale University freshmen. I. Occurrence of infectious mononucleosis. *J. infect. Dis.*, **123**, 263-270

Wahren, B., Lantorp, K., Sterner, G., & Espmark, A. (1970) EBV antibodies in family contacts of patients with infectious mononucleosis. *Proc. Soc. exp. Biol. (N.Y.)*, **133**, 934-939

zur Hausen, H. & Schulte-Holthausen, H. (1970) Presence of EB virus nucleic acid homology in a "virus-free" line of Burkitt tumor cells. *Nature (Lond.)*, **227**, 245-248

zur Hausen, H., Schulte-Holthausen, H., Klein, G., Henle, W., Henle, G., Clifford, P. & Santesson, L. (1970) EB-virus DNA in biopsies of Burkitt tumors and anaplastic carcinomas of the nasopharynx. *Nature (Lond.)*, **228**, 1056-1058

Virology and Immunology of Nasopharyngeal Carcinoma: Present Situation and Outlook—A Review

G. DE-THÉ [1]

INTRODUCTION

Variations in the incidence of the different human tumours as between various ethnic groups or geographical areas are well known. The study of such variations by the integration of field and laboratory studies, and by collaboration between investigators in different fields and countries should add a new dimension to human tumour research.

This is exemplified by recent developments in nasopharyngeal carcinoma (NPC). From successive epidemiological studies (Ho, 1967; Muir & Shanmugaratnam, 1967; Yeh, 1967), it became obvious that the Southern Chinese constituted a high-risk group from the point of view of the development of NPC while the incidence of this tumour was low in other ethnic groups and geographical areas (see Muir, 1972[2]). Studies on migrant Chinese populations (Djojopranoto, 1960; Zippin et al., 1962; Scott & Atkinson, 1967) suggested that genetic factors might play an important rôle in the development of NPC. If a comparison is made with the experimental murine system, the relationship between an ethnic group and a particular tumour calls to mind the high incidence of leukaemia and mammary tumour in certain inbred strains of mice, in which a genetically determined susceptibility to the development of virus-induced tumours has been established (Lilly, 1966). Our NPC studies were therefore designed to investigate the possibility of a high susceptibility on the part of the Southern Chinese to an oncogenic agent. As we shall see, an association has now been established between NPC and a herpesvirus closely related to the Epstein-Barr virus (EBV).

The aim of this paper is to review briefly the virological and immunological results, and to discuss the possibility of establishing the nature of the association between the herpesvirus and NPC.

MATERIALS AND METHODS

Specimens (biopsies and sera) were collected by Dr H. C. Ho, at the Medical and Health Department, Institute of Radiology, Queen Elizabeth Hospital, Kowloon, Hong Kong, by Mr P. Clifford, at the Department of Head and Neck Surgery, Kenyatta National Hospital, Nairobi, Kenya, and recently by Dr Mourali, at the Institut national de Cancérologie, Tunis. In addition, sera were collected at the Institute of Radiology by Dr Ho, both before treatment and at different intervals after radiotherapy, and the relevant clinical data recorded. A pathological slide for each Chinese tumour was kindly provided by Dr C. C. Lin, of the Department of Pathology, Queen Elizabeth Hospital, Kowloon, Hong Kong. The techniques used in our laboratory for tissue culture and serological studies have been described in detail elsewhere (de-Thé et al., 1970; Henle & Henle, 1966).

RESULTS AND DISCUSSION

1. Establishment of long-term cultures containing a herpesvirus

Since May 1968, we have been attempting to study the behaviour of NPC biopsies in tissue culture. Table 1 shows the types of cultures obtained during the last 15 months. As it is generally accepted that the various histopathological types of NPC represent variants of epithelioid carcinoma (Shanmugaratnam

[1] Unit of Biological Carcinogenesis, International Agency for Research on Cancer, Lyon, France.
[2] See p. 367 of this publication.

— 275 —

Table 1.   Types of cultures obtained from NPC biopsies between Jan. 1970 and April 1971

Origin of the specimens	No. of biopsies cultured	Epithelial growth	Early lymphocytic production	Fibroblastic cultures	Long-term lympho-blastoid cultures
Hong-Kong	28	5 (18 %)	20	23	13 (46 %)
Kampala	3	0	3	3	1
Nairobi	13	0	11	13	3
Tunis	6	6 (100 %)	5	6	1
Total	50	11	39	45	18 (36 %)

& Muir, 1967), our main concern was to establish epithelial cultures. Short-term (2 to 4 weeks) cultures of epithelial cells (Fig. 1) were obtained when the interval between the time of biopsy and that of culture was short (not more than 1 or 2 days), but so far we have not succeeded in establishing long-term cultures of such cells. The presence of cytoplasmic bundles of keratin fibrils and of desmosomes detected by electron microscopy (Gazzolo *et al.*, 1972) in these cells grown *in vitro* established their epithelial nature. In some cases, these epithelial cells were elongated and took on a fibroblastic appearance, and desmosomes were present, suggesting that change from an epithelial to a fibroblastic morphology was possible. Early lymphocytic production was noted in nearly 80% of the cultures (Table 1), corresponding to the multiplication of lymphoid elements present in the tumour (de-Thé *et al.*, 1970). Such early lymphocytic production was radically different from the lymphoblastoid cultures obtained later on. In fact, 30 to 50% of the fibroblastic cultures emerging from the epithelial outgrowths or growing as primary cultures from the explants gave rise, after 30 to 120 days, to round, free-floating cells that, in most cases, developed to give permanent cultures of lymphoblastoid cells (Fig. 2) (de-Thé *et al.*, 1969, 1970).

Origin of the "lymphoblastoid transformation". This phenomenon, also noted by Sugano *et al.* (1970) and referred to as spontaneous lymphoblastoid

transformation (Benyesh-Melnick, Fernbach & Lewis, 1963) is not well understood. The origin of the lymphoblastoid cells in the NPC cultures is not established, but it is logical to assume that they originate from the lymphoid elements present in the original biopsy specimens. Morphologically, however, fibroblastlike cells may convert into round free-floating cells and vice-versa.

As most of these lymphoblastoid cultures (we obtained more than 50 from material from different sources) contained a herpes-type virus (Figs. 3, 4 & 5), we have suggested that the presence of this virus in the original material was the cause of the "spontaneous lymphoblastoid transformation" and of the subsequent establishment of the cultures (de-Thé *et al.*, 1970). The indirect immunofluorescence test and electron microscopy showed, in these NPC-derived lines, that 0.1 to 10% of the cultured cells were producing virions at any given time. Variations in the proportion of virus-producing cells, and transformation of virus-producing lines into non-virus-producing cultures, were occasionally observed; the reverse of the latter situation was also met, although very rarely.

Electron microscopy. The synthesis of herpesvirus in these cultures was achieved by two apparently different mechanisms: (1) by cell death, where empty capsids and nucleocapsids were seen in nuclear cell debris (Fig. 3); or (2) by the budding of nucleocapsids through the nuclear membrane and their

---

Fig. 1.   Epithelial growth from NPC explant.   Note the presence of polykaryons (arrows).   × 405.

Fig. 2.   NPC-derived, mixed lymphoblastoid and fibroblastic culture with free-floating cells growing in clumps on top of fibroblastic cells attached to the surface of the flask.   × 150.

Fig. 3.   Nuclear cell debris with presence of numerous nucleocapsids and a few empty capsids, but with no enveloped virion.   × 51 000.

Fig. 4.   Migration of herpes virions through cytoplasmic channels.   × 55 000.

Fig. 5.   Extracellular complete herpes virions, with fuzzy outer envelope.   × 66 000.

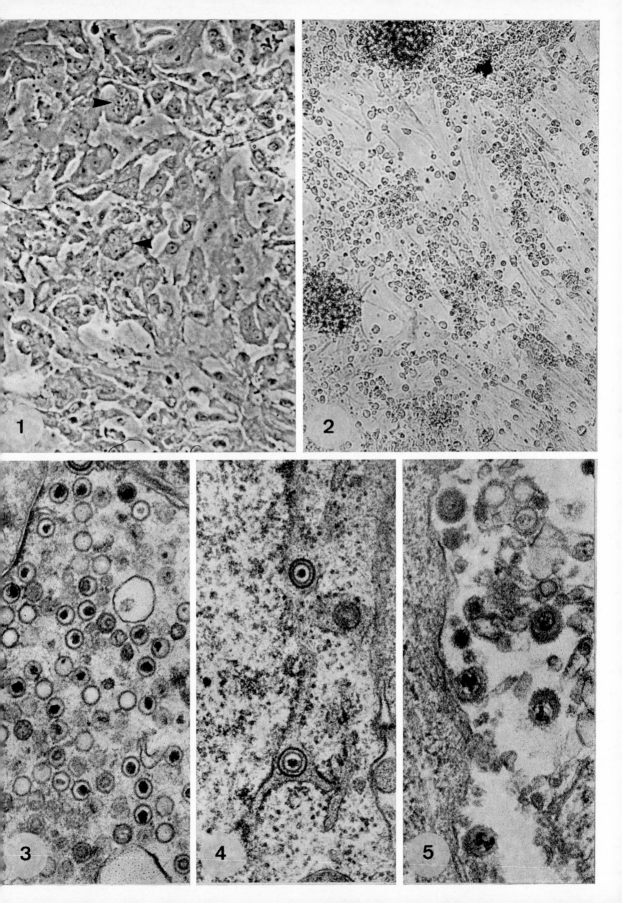

migration through cytoplasmic channels (Fig. 4) in ultrastructurally preserved cells, and liberation of mature virions in the extracellular spaces (Fig. 5).[1]

Cultures from control specimens. Lymphoblastoid cultures of similar appearance and virus content were obtained from biopsies of tumours other than NPC in Chinese, as well as from non-neoplastic specimens from Chinese, such as biopsies of tonsillitis and apparently normal nasopharynx (de-Thé *et al.*, 1970). Nilsson, Pontén & Philipson (1968) have also shown that non-neoplastic human lymphoid tissues could give rise to long-term lymphoblastoid cultures. This does not preclude the herpesvirus present in these various cultures from being a potential carcinogenic agent, as it is well known that oncogenic viruses can replicate in non-neoplastic cells and often better than in cells that they have transformed.

2. Serological studies

Immunoprecipitation test. In 1966, Old *et al.* (using sera from NPC patients as "control" sera in their studies on Burkitt's lymphoma) found that these NPC sera had precipitating antibodies against an antigen extracted from the Jijoye cell line very similar to those of BL patients' sera. In a subsequent publication, Oettgen *et al.* (1967) surveyed the sera from various groups of tumour- and non-tumour-bearing patients, and found that 83% of the NPC patients, 30% of the lymphosarcomas, 20% of the acute leukaemias and 5 to 13% of the nontumorous patients showed positive reactions. As a rule, a positive correlation was found between high titres in the immunofluorescence Henle test and the immunoprecipitation test.

Antibodies against VCA and EA. Henle *et al.* (1970a), when testing NPC sera from Hong Kong

and East Africa, found that they all contained antibody activity against viral capsid antigen(s) (VCA), 84% of them having high titre ($\geqslant$ 1:160) with a geometric mean titre (GMT) of 1:348. Chinese and East African patients with carcinomas other than NPC showed a much lower activity, 13% of them having a VCA antibody titre $\geqslant$ 1:160, with a GMT of 1:36. Ito *et al.* (1969) similarly found high anti-EBV-type reactivities in NPC sera from Formosa. Recently, Henle *et al.* (1970b) described the induction of "early antigen(s)" (EA) in Raji cells when challenged with semi-purified EBV (according to zur Hausen *et al.* (1970), the Raji cell-line carried the EBV genome in an integrated state). Henle *et al.* (1970b) reported the presence of antibodies against EA in NPC patients' sera as well as in BL patients' sera, where high titres were of bad prognostic significance.

Antibodies against membrane-associated antigens. Antibodies against membrane-associated antigens were detected in NPC patients' sera by de Schryver *et al.* (1969, 1970). These authors showed that NPC sera had a high blocking activity against two reference sera, namely Mutua (an African BL patient in long-term remission) and Kipkoech (an African NPC patient), when both a BL-derived line (Jijoye) or a NPC-derived line (HKLY-1) were used as antigens. An increase in membrane reactive antibody titres was noted in African NPC patients after radiotherapy of the tumour (Einhorn, Klein & Clifford, 1970).

Postradiotherapy sera. The increase in membrane antibody activity in sera after radiotherapy noted above differs from the behaviour of the VCA-type antibody titres (IF test) studied in sera of Chinese NPC patients at regular intervals after radiotherapy.[2] As seen from Table 2, the majority (67%) of such

---

[1] Unpublished data.

[2] Unpublished data.

Table 2. Change ($>$ 1 dilution) in VCA antibody titre of NPC sera 8 to 15 months after radiotherapy

Decrease		No change		Increase		Total
No.	%	No.	%	No.	%	
35	24.6	95	66.9	12	8.4	142
Stage I 1	10	Stage I 9	90			10
Stage II 3	16	Stage II 15	80	Stage II 1	4	19
Stage III 26	27	Stage III 60	63	Stage III 9	10	95
Stage IV 4	31	Stage IV 8	61	Stage IV 1	8	13
		Stage V 1				
Stage unknown 1		Stage unknown 2		Stage unknown 1		

patients showed no change in their VCA antibody titres over a period of 8 to 15 months after completion of radiotherapy, while 25% showed a decrease in their titres of more than one dilution. Comparative analysis of these serological results with the clinical data at each bleeding did not reveal any direct relationship between clinical and serological (VCA-type antibodies) behaviour. The distribution of antibody titres at T_o (before radiotherapy) and at T_{1y} (one year after completion of radiotherapy) (see Fig. 6), shows a downward trend. For purposes of comparison, we have included in Fig. 6 the distribution curves of the titres obtained from the sera of untreated Chinese patients with tumours other than NPC, and from normal Chinese individuals as controls. The distribution curves of the titres obtained with sera of patients with tumours other than NPC showed a peak intermediate between that for the normal controls and that for the NPC.

The general downward trend of the initially high titre of NPC sera (Fig. 6) was similar to that observed over a period of 12 to 18 months in the high titres found in the general population surveyed in the West Nile district of Uganda by Kafuko et al., 1972.

Complement-fixing antibodies. Complement-fixing reactions were recently carried out by Sohier & de-Thé (1971) on NPC sera with a "soluble" antigen extracted from the QMIR-WILL line developed by Pope (1968) from a leukaemic patient. Preliminary results show high reactivities of NPC sera.

Indirect radiolabelled antibody (IRA) test. Greenland et al.[1] developed the IRA test, which is essentially identical to the indirect fluorescent antibody (IFA) technique, except that a paired-labelled mixture of ^{125}I-labelled rabbit antihuman gamma globulin and ^{131}I-labelled normal rabbit gamma globulin is used to reveal the uptake of antibody by the antigen from the test serum. The ratio of ^{125}I to ^{131}I for the sample decreases with increasing dilution of the test serum, and the dilution at which this ratio falls to that measured on antigen untreated with any human serum is taken as the IRA titre of that serum. Over 100 sera have been titrated by means of this technique, on two different cell lines (Jijoye and HKLY28 derived from NPC biopsy). The same sera have also been titrated on the same two

[1] Unpublished data.

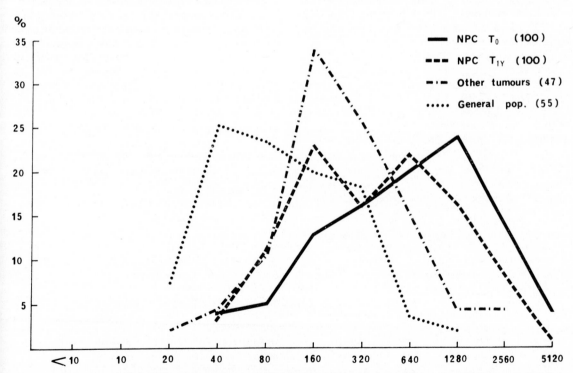

Fig. 6. Distribution of IF antibody titres (Henle test) in NPC sera before treatment (T_o) and 1 year after radiotherapy (T_{1y}) and two control groups (other tumours and general population).

cell lines using the IFA technique. In each case, there is a very good correlation between the titre of the same serum on the two lines—though by the IFA test all titres were uniformly reduced when one of the lines was used, whereas the IRA technique gave essentially the same titres for the two lines. The correlation between the IFA and the IRA titres for both the cell lines was close, but not so good as the internal correlation between the two lines for a single technique. This may mean that, although the majority of the antibodies measured by the two methods are the same, there are supplementary antibody activities in some of the sera that are recognized only by one of the two techniques.

3. *Comparative studies of lymphoblastoid lines derived from different sources (BL, NPC, IM, etc.)*

The phenomenon referred to as "spontaneous lymphoblastoid transformation" occurs when certain human lymphoid tissues are grown *in vitro*, whether they are tumorous or not (Nilsson *et al.*, 1968; de-Thé *et al.*, 1970). If such is the case, what is the significance of the establishment of permanent lymphoblastoid lines *in vitro* from so many different sources (BL, IM, NPC, normal individuals), and are these cells identical in all these different cell lines? In order to answer this question, comparative analytical investigations of these various lines were carried out.

The serological and tissue-culture results obtained with NPC sera and tumour biopsies were analyzed in the light of the histopathology of the corresponding original tumour. Details of this analysis will be published elsewhere, but Table 3 shows that the amount of lymphoid elements present in the original NPC tumour affects the frequency of early lymphocytic production but has no effect on the frequency of establishment of long-term lymphoblastoid cultures. Such lymphoid infiltration of the tumour biopsies does not seem either to be correlated with

the anti-VCA antibody titre of the corresponding patients' sera. We suggested that the establishment of long-term lymphoblastoid cultures was caused by the presence, in the original material, of a herpesvirus having "transforming activities" (de-Thé *et al.*, 1970). In BL, the lymphoblastoid cells are believed to represent the progeny of the tumorous lymphoid cells probably carrying the EBV *in vivo*, but in NPC, it is not possible, from present tissue-culture data, to state the exact origin of the lymphoblastoid cells growing in culture. Zur Hausen & Schulte-Holthausen (1970) have shown that EB-type viral genomes are present in NPC biopsies in quantities very similar to those found in BL biopsies, but it is not known whether the virus is present in the epithelial tumour cells or in lymphoid cells or in both.

Nishioka *et al.* (1971) analyzed the cell membrane receptor complex by immuno-adherence, and found that NPC-derived cells had IA-type receptors reacting with EA (IgM) C43, while BL-derived cells showed reaction with EA (IgG). It is not clear, however, whether these differences reflect cellular or viral-induced properties of the cell-membrane.

Sohier & de-Thé (1971), using complement fixation with a soluble antigen, extracted from the QMIR-WILL line of Pope (1968), recently reported that NPC sera regularly had high CF antibody titres whereas BL sera reacted very poorly or negatively.

Greenland *et al.*[1], used the uptake of radio-iodine-labelled globulins by different cell lines to show that these differed in detail from each other, and that NPC-derived cell lines appeared to differ from either IM- or BL-derived cell lines. The significance of these findings has yet to be clarified. The rôle of temperature in the synthesis of the herpes-type virus in various NPC, BL and IM lines was studied by Ambrosioni & de-Thé.[2] Statistically

[1] See p. 302 of this publication.
[2] See p. 318 of this publication.

Table 3. Relationships between degree of lymphoplasmocytic infiltration of NPC and the frequency of various events in T.C. and serology

Tumour biopsies		Tissue culture		Serology	
Lympho-plasmocytic infiltration	No. of cases	Early lymphocytic production	No. of lymphoblastoid cultures	Low Henle titre $10 < x < 160$	High Henle titre $x \geqslant 160$
Moderate	21	14 (66%)	7 (33%)	2 (10%)	19 (90%)
Marked	14	13 (93%)	4 (28%)	2 (14%)	12 (86%)
Total	35	27 (77%)	11 (31%)	4 (11%)	31 (89%)

significant differences were observed, the optimal temperatures being 35°C and 39°C for NPC, 33°C for BL, and 37°C for IM cultures, respectively.

It is not possible at present to ascertain the origin (cellular or viral) of the differences described above between NPC-derived lines and BL- or IM-derived lines. Some of these results may reflect cell-type differences, or the BL-associated virus (EBV) and the NPC-associated virus may not be identical, although closely related. Cross-infection and neutralization experiments should clarify this problem.

4. Outlook

What are the possibilities of establishing the nature of the association described above between a herpes-virus and NPC? Two hypotheses can be put forward: (1) the associated virus is a "passenger", without any aetiological rôle; or (2) it is a causative agent (a "driver" or a "co-driver") playing an aetiological rôle in the development of the tumour.

With regard to the first alternative, the herpes-virus may be present in the lymphoid elements infiltrating the epithelial tumour cells, and may be stimulated, in an unknown but specific fashion, by the presence of neighbouring tumour cells. This hypothesis cannot be ruled out, but does not explain why all NPC cases from different parts of the world regularly have antibodies against this type of virus, most of the time at a very high titre, and why sera from patients with other tumours of the ear, nose and throat, such as the carcinomas arising from the tonsils and the back of the tongue, which are frequent in India, and have been studied by de Schryver et al. (1969), do not show similar serological properties, even if the tumours contain abundant lymphoid elements.

The testing of the second hypothesis is not an easy task, as we cannot conduct, in man, experiments of the type used in experimental viral oncology. There is therefore a need for a new approach to the problem and for new tools. Long-term multiphasic programmes, integrating field and laboratory studies, and a close collaboration between interested laboratories of different backgrounds and nationalities may constitute such new tools. Such an approach, because of the size of the population to be studied, and because of the complexity of the relationship between the various groups involved in the field and in the research institutes, requires both careful planning and monitoring and large-scale financial support. Table 4 gives an example of a multiphasic field and laboratory integrated programme for the study of NPC and BL. Phase I is historical for both BL and NPC. Phase II is now being implemented for NPC in Hong Kong and Singapore, where a sero-epidemiological study of the natural history of the herpesvirus is being carried out in populations at high risk (Chinese) and at low risk (Indo-Pakistanis) for NPC. A knowledge of the age-specific prevalence, the incidence of infection and

Table 4. Multiphasic integrated field and laboratory programme for human tumour

	PHASE I *Establishment* of an *association* between a *virus* and a *tumour*	PHASE II Studies on the *natural history* of the virus	PHASE III *Test the proposed hypotheses* and establish the *nature of the association*	PHASE IV *Controlled* trial on the *disease*
Field studies	1. Investigate the epidemiological characteristics suggesting a viral aetiology; 2. Collect tumour specimens and corresponding sera with proper controls.	1. Sero-epidemiology aimed at establishing: prevalence ⎫ of incidence ⎬ infection mode ⎭ in populations at *high* and *low* risk; 2. Search for cofactors.	1. Prospective sero-epidemiological studies—establishment of *sequential events prior to tumour development*; 2. Specific studies on suspected cofactors.	1. Vaccine trials; 2. Sero-epidemiological studies of natural variants or mutant of the causative virus.
Laboratory studies	1. Investigate regular presence of viral infection in tumours; 2. Evidence of specific antibody response in patients' sera.	1. Testing of sera for *neutralizing* antibodies; 2. Other types of antibodies against —capsid antigens, —early antigens; 3. Study oncogenic and immunological properties of the virus.	1. Find experimental models corresponding to situation in man; 2. Study possible variants, mutant virus or subviral components for their immunological and biological properties.	1. Development of vaccine with natural or artificial mutant or variant virus, or with viral subcomponents; 2. Development of protective sera.
Decision point	If a strong association exists, Phase II should be implemented.	If sero-epidemiological results suggest testable hypotheses, Phase III should be implemented.	If succession of events supports or establishes *causal relationship*, Phase IV should be implemented	If trial is successful, mass vaccination can be implemented.

hopefully the mode of transmission of the virus should lead to the formulation of hypotheses to be tested in Phase III. A problem met with in such studies is that of selecting the type of antibodies to test in such large population groups. The VCA test of Henle was selected, but it is realized that other antibodies, such as the IF membrane reactive antibodies or EA antibodies, or those detected by CF or IRA tests, might represent the relevant marker for tumour development. The collected sera are therefore being kept in liquid nitrogen for future studies.

In the case of BL, Phase III is now being implemented (Geser & de-Thé[1]). This will consist of a follow-up study of a child population of 35 000, aimed at establishing the immunological sequence of events between viral infection and tumour development.

The herpesvirus associated with NPC might not be the one and only "driver", it could also act as a cocarcinogen, or a "co-driver", helping a specific chemical carcinogen or a biological oncogenic agent other than the herpesvirus to induce NPC. To investigate this possibility, a general epidemiological approach is at present being planned as an extension of the sero-epidemiological studies now being carried out in Hong Kong and Singapore. Furthermore, a relatively high incidence of NPC (intermediate between that in Chinese and that in Caucasians) was recently found in Tunisia, and it is intended to carry out parallel epidemiological studies in that country.

Such integrated field and laboratory studies should be able to assist in answering the basic questions, namely whether herpesviruses are oncogenic for man, whether a single EB-type herpesvirus induces NPC, BL, IM and possibly certain forms of Hodgkin's disease, depending upon the ethnic groups concerned, or whether there is a family of closely related, but biologically different viruses, each having an aetiopathological potential for inducing a specific disease in particular groups of human beings.

---

[1] See p. 372 of this publication.

SUMMARY

Long-term lymphoblastoid cultures could be obtained with high frequency (46%) from Chinese nasopharyngeal carcinoma (NPC), but also from other sources, such as Chinese inflamed tonsils and apparently normal nasopharyngeal mucosa. These permanent cultures contained a herpesvirus sharing structural antigens with the Epstein-Barr virus (EBV) and it is suggested that the presence of such a virus in the original material was the cause of the establishment of such cultures.

A serological association has been established between this herpesvirus and NPC, as NPC patients' sera regularly showed high immunoprecipitating (IP), immunofluorescent (IF), and complement-fixing (CF) antibodies against this type of virus. Radiotherapy did not change substantially the IF antibody titres of the NPC patients' sera over periods up to 15 months, although a slight tendency was observed for the high titres to fall.

When compared with similar permanent lymphoblastoid cultures derived from Burkitt's lymphoma (BL) biopsies and from infectious mononucleosis (IM), the NPC-derived cultures showed peculiarities in immunoadherence surface receptors (IA), in temperature sensitivity with regard to virus production, and in the type of antigen(s) detected by radio-iodine labelled antisera. These differences might reflect cellular or viral properties.

The possibility of establishing the nature of the association between this herpesvirus and NPC and/or BL is discussed. Multiphasic programmes integrating field and laboratory studies, together with close international collaboration, should provide a new tool which could help to prove or disprove the oncogenic rôle of herpesviruses in human populations.

ACKNOWLEDGEMENTS

Most of the results reported here represent a team effort by the staff of the Unit of Biological Carcinogenesis at IARC, Lyon (Dr A. Geser, Dr D. Simkovic, Dr N. Muñoz, Mr T. Greenland and Mr Ambrosioni, Miss Favre, Miss Desgranges and Mrs Lavoué), and of the collaborating laboratories in Lyon (Professor R. Sohier), in Hong Kong (Dr H. C. Ho, Dr B. Bard), in Singapore (Professor K. Shanmugaratnam and Dr M. Simons), in Melbourne (Mr I. Jack), and in Tunis (Professor Mourali, Dr Voegt-Hoerner).

This investigation was supported by Contract No. NIH-70-2076, within the Special Virus-Cancer Program of the National Institutes of Health, Department of Health, Education, and Welfare, USA.

REFERENCES

Benyesh-Melnick, M., Fernbach, D. J. & Lewis, R. T. (1963) Studies on human leukemia. I. Spontaneous lymphoblastoid transformation of fibroblastid bone marrow cultures derived from leukemic and non-leukemic children. *J. nat. Cancer Inst.,* **31**, 1311-1325

de Schryver, A., Friberg, S., Jr, Klein, G., Henle, W., Henle, G., de-Thé, G., Clifford, P. & Ho, H. C. (1969) Epstein-Barr virus-associated antibody patterns in carcinoma of the post-nasal space. *Clin. exp. Immunol.,* **5**, 443-459

de Schryver, A., Klein, G. & de-Thé, G. (1970) Surface antigens on lymphoblastoid cells derived from nasopharyngeal carcinoma. *Clin. exp. Immunol.,* **7**, 161-171

de-Thé, G., Ambrosioni, J. C., Ho, H. C. & Kwan, H. C. (1969) Lymphoblastoid transformation and presence of herpes-type viral particles in a Chinese nasopharyngeal tumour culture *in vitro. Nature,* **221**, 770

de-Thé, G., Ho, H. C., Kwan, H. C., Desgranges, C. & Favre, M. C. (1970) Nasopharyngeal carcinoma (NPC). I. Types of cultures derived from tumour biopsies and non-tumorous tissues of Chinese patients with special reference to lymphoblastoid transformation. *Int. J. Cancer,* **6**, 189-206

Djojopranoto, M. (1960) *Beberapa segi patologi tumour ganas nasopharynx di-Diawa-Timur* (Thesis, Surabaja, Gita Karya)

Einhorn, N., Klein, G. & Clifford, P. (1970) Increase in antibody titer against the EBV-associated membrane antigen complex in Burkitt's lymphoma and nasopharyngeal carcinoma after local irradiation. *Cancer,* **26**, 1013-1022

Gazzolo, L., de-Thé, G., Vuillaume, M. & Ho, H. C. (1972) Nasopharyngeal carcinoma. II. Ultrastructure of normal mucosa, tumor biopsies, and subsequent epithetial growth *in vitro. J. nat. Cancer Inst.* (in press)

Henle, G. & Henle, W. (1966) Immunofluorescence in cells derived from Burkitt's lymphoma. *J. Bact.,* **91**, 1248-1256

Henle, W., Henle, G., Burtin, P., Cachin, Y., Clifford, P., de Schryver, A., de-Thé, G., Diehl, V., Ho, H. C. & Klein, G. (1970a) Antibodies to Epstein-Barr virus in nasopharyngeal carcinoma, other head and neck neoplasms, and control groups. *J. nat. Cancer Inst.,* **44**, 225-231

Henle, W., Henle, G., Zajac, B. A., Pearson, G., Waubke R. & Scriba, M. (1970b) Differential reactivity of human serums with early antigens induced by Epstein-Barr virus. *Science,* **169**, 188-190

Ho, H. C. (1967) Nasopharyngeal carcinoma in Hong Kong. In: Muir, C. S. & Shanmugaratnam, K., ed., *Cancer of the nasopharynx,* Munksgaard, Copenhagen, pp. 58-63 (*UICC Monograph Series,* Vol. 1)

Ito, Y., Takahshi, T., Kawamura, A., Jr & Tu, S. M. (1969) High anti-EB virus titer in sera of patients with nasopharyngeal carcinoma: a small-scale seroepidemiological study. *Gann,* **60**, 335-340

Kafuko, G. W., Day, N. E., Henderson, B. E., Henle, G., Henle, W., Kirya, G., Munube, G., Morrow, R. H., Pike, M. C., Smith, P. G., Tukei, P. & Williams, E. H. (1972) Epstein-Barr virus antibody levels in children from the West Nile district of Uganda; results of a field study. *J. nat. Cancer Inst.* (in press)

Lilly, F. (1966) The histocompatibility-2 locus and susceptibility of tumour induction. *Nat. Cancer Inst. Monogr.,* **22**, 631-642

Muir, C. S. & Shanmugaratnam, K. (1967) The incidence of nasopharyngeal carcinoma in Singapore. In: Muir, C. S. & Shanmugaratnam, K., ed., *Cancer of the nasopharynx,* Munksgaard, Copenhagen, pp. 47-53 (*UICC Monograph Series,* Vol. 1)

Nilsson, K., Pontén, J. & Philipson, L. (1968) Development of immunocytes and immunoglobulin production in long-term cultures from normal and malignant human lymph nodes. *Int. J. Cancer,* **3**, 183-190

Nishioka, K., Tachibana, Y., Hirayama, T., de-Thé, G., Klein, G., Takada, M. & Kawamura, A., Jr (1971) Immunological studies on the cell membrane receptors of cultured cells derived from nasopharyngeal cancer, Burkitt's lymphoma and infectious mononucleosis. In: Nakahara, W., Nishioka, K., Hirayama, T. & Ito, Y., ed., *Recent advances in human tumor virology and immunology,* University of Tokyo Press, pp. 401-420

Oettgen, H. F., Aoki, T., Geering, G., Boyse, E. A. & Old, L. J. (1967) Definition of an antigenic system associated with Burkitt lymphoma. *Cancer Res.,* **27**, 2532-2534

Old, L. J., Boyse, E. A., Oettgen, H. F., de Harven, E., Geering, G., Williamson, B. & Clifford, P. (1966) Precipitating antibody in human serum to an antigen present in cultured Burkitt's lymphoma cells. *Proc. nat. Acad. Sci. (Wash.),* **56**, 1699-1704

Pope, J. H. (1968) Establishment of cell lines from Australian leukemic patients. Presence of herpes-like virus. *Austr. J. exp. biol. med. Sci.,* **46**, 643-645

Scott, G. C. & Atkinson, L. (1967) Demographic features of the Chinese population in Australia and the relative prevalence of nasopharyngeal cancer among Caucasians and Chinese. In: Muir, C. S. & Shanmugaratnam, K., ed., *Cancer of the nasopharynx,* Munksgaard, Copenhagen, pp. 64-72 (*UICC Monograph Series,* Vol. 1)

Shanmugaratnam, K. & Muir, C. S. (1967) Nasopharyngeal carcinoma: origin and structure. In: Muir, C. S. & Shanmugaratnam, K., ed., *Cancer of the nasopharynx,* Munksgaard, Copenhagen pp. 153-162 (*UICC Monograph Series,* Vol. 1)

Sohier, R. & de-Thé, G. (1971) Fixation du complément avec un antigène soluble: différences d'activité importantes entre les sérums de lymphome de Burkitt, de cancer du rhino-pharynx et de mononucléose infectieuse. *C. R. Acad. Sci. (Paris),* **273**, 121-124

Sugano, H., Takada, M., Chen, H. C. & Tu, S. M. (1970) Presence of herpes-type virus in the culture cell line from a nasopharyngeal carcinoma in Taiwan. *Proc. Japan. Acad., 46*, 453-457

Yeh, S. (1967) The relative frequency of cancer of the nasopharynx and accessory sinuses in Chinese in Taiwan. In: Muir, C. S. & Shanmugaratnam, K., ed., *Cancer of the nasopharynx,* Munksgaard, Copenhagen, pp. 54-57 (*UICC Monograph Series,* Vol. 1)

Zippin, C., Tekawa, I. S., Bragg, K. U., Watson, D. & Linden, G. (1962) Studies on heredity and environment in cancer of the nasopharynx. *J. nat. Cancer Inst., 29* 483-490

zur Hausen, H. & Schulte-Holthausen, H. (1970) Presence of EB virus nucleic acid homology in a "virus-free" line of Burkitt tumour cells. *Nature (Lond.), 227,* 245-248

The Rôle of EBV in the Establishment of Lymphoblastoid Cell Lines from Adult and Foetal Lymphoid Tissue

K. NILSSON,[1] G. KLEIN,[2] G. HENLE[3] & W. HENLE[3]

INTRODUCTION

During the last few years a great many continuous human lymphoblastoid cell lines (LL) have been reported from different laboratories. The frequency of establishment has varied depending on the source of lymphoid cells and the culture technique employed (for review, see Nilsson, 1971a).

Peripheral blood lines can regularly be established only when the donors suffer from acute infectious mononucleosis (IM) – a disease in which Epstein-Barr virus (EBV) (Epstein, Achong & Barr, 1964) has been shown to be the causative agent (Henle, Henle & Diehl, 1968; Niederman et al., 1968).

A relationship between EBV and the establishment of LL has been suggested by studies on the lymphoproliferative effect in vitro of the virus. The first indication of the growth-promoting role of EBV was presented by Henle et al. (1967), who found that X-irradiated EBV-carrying Burkitt's lymphoma (BL) cells promoted the establishment of LL from leukocyte cultures of normal donors. These results were extended by Gerber, Whang-Peng & Monroe (1969), Miller et al. (1969) and Henle & Henle (1970), who observed lymphoblastoid transformation in normal buffy coat cells when exposed to EBV concentrate. Pope, Horne & Scott (1968) and Pope et al. (1971) inoculated foetal lymphoid cells with EBV-containing cell-free filtrates and noted the same phenomenon. None of the groups reported

the establishment of LL when the cultures were exposed to filtrates or concentrates of EBV-free lines.

With peripheral blood from non-IM patients, a high success rate (20–40%) was reported by Moore & Minowada (1969) only when large volumes from each donor (1 litre) were used for the initiation of multiple cultures that were pooled during the subsequent long-term cultivation. The growth of LL from blood from 6 out of 8 normal donors was reported by Gerber & Monroe (1968), who employed the same technique. Most of these lines were found to be EBV carriers.

Nilsson, Pontén & Philipson (1968) reported that lymph nodes from normal individuals and patients with lymphomas yielded lines at a frequency of 50% when cultivated by the lens-paper grid technique (Pontén, 1967). This indicated that solid lymphoid tissue might be a better source of cells with the potential of infinite replication in vitro than buffy coat cells. This grid culture method has been improved by the addition of gelatin foam (Spongostan) to the grid. By means of this technique, it has recently been shown that LL can be established from the lymphoid tissue of almost any adult individual without a recent history of IM (Nilsson, 1971b). It was important therefore to investigate whether EBV was also involved in these seemingly "spontaneous" lymphoblastoid transformations.

The high efficiency of the Spongostan culture made it ideal for tests to determine whether foetal lymphoid tissue could give rise to LL. If so, and if the lines contained EBV-related antigens or EBV-DNA, EBV must be considered to be transmitted vertically.

This paper reports the results of a search for EBV and membrane antigens in LL established by the

[1] The Wallenberg Laboratory, Uppsala University, Uppsala, Sweden.
[2] Department of Tumor Biology, Karolinska Institutet, Stockholm, Sweden.
[3] Virus Laboratories, The Children's Hospital, Philadelphia, Pennsylvania, USA.

Spongostan culture technique, and for the presence of antibodies to EBV in the donor sera. The results of the long-term cultivation of human foetal lymphoid and haematopoietic tissues, and the induction of LL by cell-free filtrates from an EBV-antigen-carrying LL of normal origin, are also reported.

MATERIAL AND METHODS

Lines were established by the grid organ culture technique described previously (Pontén, 1967; Nilsson et al., 1968; Nilsson, 1971b). Briefly fragmented lymphoid tissue was placed on top of a stainless steel grid covered either by lens paper alone or by lens paper and gelatin foam (Spongostan, Ferrosan, Malmö, Sweden). Peripheral lymphocytes were separated from peripheral blood (10–20 ml) and resuspended together with trypsinized allogeneic adult skin fibroblasts in 0.5 ml of medium. The cell mixture was gently pipetted into the Spongestan. LL usually became established within 3 months. Grid cultures were maintained in nutrient mixture F-10 (Ham, 1963; Grand Island Biological Company, New York) supplemented with 10% postnatal calf serum, 100 IU/ml penicillin, 50μg/ml streptomycin and 1.25 μg/ml amphotericin B. Established LL were propagated in Erlenmeyer flasks as suspension cultures without stirring and fed with the above medium.

All lines were tested for the presence of EBV-associated antigens at least once within two months of establishment.

PREPARATION OF FILTRATES OF LYMPHOBLASTOID CELL LINES

Cell-free filtrates were prepared from one EBV-carrying line derived from the lymph node of an elderly patient operated on for cholecystopathy and from one "EBV-negative" line originating from the peripheral blood of a myeloma patient. The method described by Pope et al. (1968) was used with slight modification.

FOETAL CULTURES

Spleen, thymus and bone marrow from 16 foetuses (13–20 weeks gestation) were cultivated by the Spongostan technique already described (Nilsson, 1971b). The tissue fragments of four other foetuses were exposed to fresh filtrate once within 3–6 hours of the initiation of the cultures.

IMMUNOFLUORESCENCE TESTS

The EBV immunofluorescence test to detect EBV in acetone-fixed cells, and the test for the assay of antibodies to EBV capsid antigens (anti-VCA) in the donor's sera have been described by Henle & Henle (1966) and Henle G. et al. (1969).

The direct membrane immunofluorescence test (Klein et al., 1969) was performed with living cells as described by Nilsson et al. (1971).

RESULTS

The results of the immunofluorescence tests are shown in the Table, and are discussed below.

Presence of EBV capsid antigens in the cells

All lines except one derived from a patient with multiple myeloma (282 Pl) were EBV-positive. In five lines (214 L, 288 L, 303 L, 115 C and 279 C) 2–3% of stained cells were counted while in the others <1% were positive. Two lines (199 Sp and 296 L) contained cells (5% and 0.3% respectively) that were stained with the anti-EBV-negative control serum. These lines contained IgG-producing cells. They were kindly examined by Dr G. de-Thé by electron microscopy. Line 199 Sp contained EBV particles, while in line 296 L no such particles were detected by this means.

Detection of EBV-dependent membrane antigens

Only line 282 Pl was consistently negative in the direct membrane immunofluorescence test. For lines 211 L, 298 L and 297 Bm, the reactivity was weak but positive. Reactivity in the EBV-antigen and membrane immunofluorescence tests were in agreement in all lines on all occasions. During the course of repeated testing, most lines gradually lost EBV, but the agreement between the two tests remained high.

The change in the number of membrane- and EBV-antigen-containing cells during prolonged *in vitro* cultivation will be reported elsewhere (Nilsson et al., 1971). Briefly, only 8 out of 21 lines showed no significant change in antigens, while in the remainder a complete or partial loss of antigens was noted during the four months of observation.

Anti-EBV titres in donor sera

Sera from 16 of the 20 donors were available. All these sera except three (61, 115 and 282) were

Detection of EBV-determined antigens in human lymphoblastoid cell lines, and anti-VCA antibodies in donors' sera

Cell line	Donor			EBV antigen identified by:	
	Diagnosis	Age (years)	Anti-VCA titre in serum	VCA tests	Membrane antigen tests
185 L	Cholecystopathy	25	1 : 40	+	+
211 L		87	1 : 20	(+)	(+)
214 L		79	1 : 80	+	+
288 L		81	n.t.[1]	+	+
293 L		55	1 : 40	+	+
296 L		24	1 : 5–1 : 10	+	+
296 PI				+	+
298 L		32	n.t.[1]	(+)	(+)
300 L		36	1 : 10	+	+
303 L		63	1 : 40	+	+
275 L	Duodenal ulcer	58	1 : 40	+	+
292 L		27	1 : 40	+	+
199 Sp	Haemolytic anaemia	22	1 : 40	+	+
202 L	Disseminated lupus erythematosus	34	1 : 320	+	+
255 Bm	Myeloma	53	1 : 40–1 : 80	+	+
282 PI		60	<1 : 10	−	−
297 Bm		57	n.t.[1]	(+)	(+)
61 M	Hodgkin's disease	45	<1 : 5 [2]	+	+
115 C	Stomach cancer	62	<1 : 10 (1 : 5)	+	+
279 C	Chronic lymphatic leukaemia	58	n.t.[1]	+	+

[1] Not tested.
[2] Serum is positive for anti-EBV antibodies by CF test.

anti-VCA-positive in dilutions > 1 : 10. The three exceptional sera came from elderly patients with various malignancies. The serum of patient 61 was negative in a dilution of 1 : 5, while serum 115 gave a faint positive reaction. Serum 61 was kindly titrated by Dr V. Dunkel[1] by the CF test with concentrated EBV antigen and found positive. Sera from patients 115 and 282 were no longer available for test.

Sera from 9 out of 20 foetuses were tested and found negative in all instances except one. The gestational age of this foetus was 18–20 weeks.

Culture of foetal lymphoid tissue

Gradually diminishing numbers of small lymphocytes were observed in the grid cultures of 16 foetuses during the initial 3–4 weeks. In parallel with this fall in cell production, macrophage-like cells that were attached to the bottom of the Petri dish became dislodged from the grids. These eventually died and were replaced by fibroblastoid cells. The latter were noticed throughout the period of observation (5–7 months). No spontaneous lymphoblastoid transformation was noted.

Long-term culture of foetal lymphoid tissue exposed to filtrate

The growth pattern in all noninfected control cultures and in cultures exposed to filtrates derived from the EBV-antigen-negative line was similar to that observed in the other 16 long-term cultured foetuses. In the case of foetuses inoculated with filtrate from the EBV-antigen-positive line, three cultures underwent lymphoblastoid transformation 25–34 days after explantation. The morphology of these lines and their tissue culture characteristics were indistinguishable from those of LL established from adult lymphoid tissue or peripheral blood.

The three lines, derived from two spleen cultures and one liver culture, were examined by the immunofluorescence test after four months of culture. Two lines were found to contain EBV and membrane antigens. The third line, derived from the liver, was negative in both tests.

[1] Personal communication.

DISCUSSION

EBV seems so far to be exclusively lymphotropic, since it has not been found in other cell types either *in vitro* or *in vivo*. It has been shown to be the cause of at least one lymphoproliferative diseases—infectious mononucleosis (Henle *et al.*, 1968; Niederman *et al.*, 1968). In cell lines derived from patients with acute IM, EBV-related antigens are consistently found if lines are examined shortly after establishment. During the course of prolonged *in vitro* cultivation, the lines seem to lose viral and membrane antigens more rapidly (Klein *et al.*, 1968) than BL lines.

In a review of human haematopoietic cell lines, Moore & Minowada (1969) reported the establishment of about 500 LL. The characteristics of all these lines except one, which was derived from a myeloma patient and shown to be a myeloma line, were indistinguishable from those of the IM lines. Although it was not stated how many of these lines were examined, the proportion carrying EBV in this enormous material was claimed to be 80%. However, since many of these non-IM lines were tested for virus after several months *in vitro*, the true frequency of EBV at the time of establishment is not known.

The LL examined in this study seem to be unique in that they were derived from unselected adult patients without histories of IM with a frequency close to 100%, and all lines were tested for EBV within two months of establishment. The finding of EBV in virtually all these lines strongly suggests that truly EBV-free lines are very rare or probably nonexistent, especially since lines without viral-dependent antigens may still contain the viral genome (zur Hausen & Schulte-Holthausen, 1970). As in the IM lines, the proportion of cells containing membrane and viral-capsid antigens diminished in the majority of the lines during observation periods of four months.

The presence of EBV-antigens in the cell lines was correlated with the finding of antibodies to EBV in the donor sera. In two cases, namely 61 M and 115 C, the sera were negative at a dilution of 1:10. EBV antibodies were shown, however, by the CF test, to be present in serum 61 (Dr V. Dunkel[1]), and in serum 115 by a faint positive reaction at a dilution of 1:5. The very low levels of anti-EBV in the sera may be explained by the age and the clinical status of the donors. Both had advanced malignancy

---

[1] Personal communication.

that might have caused nonspecific impairment of antibody formation.

The cell line of donor 282 and his serum were exceptional. The line was consistently free of virus-antigen-containing cells on repeated tests and the serum was negative for anti-EBV antibodies even at a dilution of 1:5. The 60-year-old patient was in the terminal stage of multiple myeloma. The explanation for the absence of anti-EBV titre in his serum could therefore well have been the same as for the negative reactions of the sera of patients 61 and 115, i.e., old age and advanced malignancy resulting in a decline in antibody titres.

The negativity of cell line 282 Pl is perhaps more difficult to explain. The two human myeloma cell lines described by Moore & Kitamura (1968) and Nilsson *et al.* (1970) did not contain EBV-antigen-carrying cells. However, line 282 was not derived from myeloma cells since the immunoglobulin produced *in vitro* was unrelated to the patient's myeloma protein. Because of the general presence of EBV in all LL, it will be of importance to examine this line for EBV-DNA before it can be said to be negative.

In view of the high efficiency of the technique employed with adult lymphoid cells, the absence of "spontaneous" lymphoblastoid transformation in foetal tissue cultures is of particular interest. This observation indicates that EBV is not vertically transmitted. Similar results have been presented by Pope *et al.* (1968, 1971). The establishment of LL only in foetal cultures treated with filtrate from an EBV-positive, but not from an EBV-negative line, strongly suggests a growth-promoting rôle for EBV. Since unpurified material was used and no serial passage was carried out, alternative explanations cannot be excluded. The frequency of establishment of continuous cultures after exposure to EBV-containing filtrate was low as compared to the results reported by Pope *et al.* (1968, 1971). The reason for this discrepancy is not known.

The absence of anti-EBV in the majority of foetal sera most probably reflects the inability of the IgG of the mother to cross the placenta during the early gestational period (Martin, 1954).

SUMMARY AND CONCLUSIONS

The EBV-carrier status of 21 permanent human lymphoblastoid cell lines has been examined by assessing EBV-associated membrane antigens and EBV capsid antigens by immunofluorescence tests. These LL were established by a grid culture technique with a frequency close to 100% for lymph nodes from "normal" adults, and with one of about 60% for the lymphatic tissue of patients with malignancy. All lines except one contained EBV- and membrane-antigen-positive cells at the time of the initial test, performed within two months of their establishment. A reduction in the percentage of antigen-containing cells was noted in all but 8 lines during an observation period of four months. Agreement between the results of the membrane and the EBV capsid-antigen tests was regularly observed.

All donors except one, a multiple myeloma patient who yielded an EBV-negative line, revealed serological evidence of previous EBV infections.

In contrast to the results obtained with adult lymph nodes, no lines could be derived under identical culture conditions from the lymphoid tissue of 16 foetuses aged 13–20 weeks. However, when the lymphoid tissue of four other foetuses was exposed to EBV-containing, but not when inoculated with EBV-free, filtrates, three lines were established from two of the foetuses.

The data presented suggest that EBV infection may be a prerequisite for the permanent growth of human lymphoblastoid cells *in vitro*. The absence of "spontaneous" lymphoblastoid transformation of foetal lymphoid tissue indicates that, as a rule, EBV is not vertically transmitted.

ACKNOWLEDGEMENTS

This work was supported by grant No. 70:23 from the Kigg Gustav V Jubilee Fund, by research grant CA 04568 and contracts Nos. PIL-43-66-477 and NIH-69-2005 within the Special Virus Cancer Program, National Cancer Institute, National Institutes of Health, US Public Health Service. We thank Dr Pontén, Dr Dunkel and Dr de-Thé for valuable help and advice. The skilful technical assistance by Mrs B. Karlsson, Mrs M. Adams and Miss E. Hutkin is gratefully acknowledged.

REFERENCES

Epstein, M. A., Achong, B. G. & Barr, Y. M. (1964) Virus particles in cultured lymphoblasts from Burkitt's lymphoma. *Lancet,* i, 702-703

Gerber, P. & Monroe, J. H. (1968) Studies on leukocytes growing in continuous culture derived from normal human donors. *J. nat. Cancer Inst.,* 40, 855-866

Gerber, P., Whang-Peng, J. & Monroe, J. H. (1969) Transformation and chromosome changes induced by Epstein-Barr virus in normal human leucocyte cultures. *Proc. nat. Acad. Sci. (Wash.),* 63, 740-747

Ham, R. G. (1963) An improved nutrient solution for diploid chinese hamster and human cell lines. *Exp. cell Res.,* 29, 515-526

Henle, G. & Henle, W. (1966) Immunofluorescence in cells derived from Burkitt's lymphoma. *J. Bact.,* 91, 1248-1256

Henle, G., Henle, W., Clifford, P., Diehl, V., Kafuko, G. W., Kirya, R. G., Klein, G., Morrow, R. H., Manube, G. M. R., Pike, P., Tukei, P. M. & Ziegler, J. L. (1969) Antibodies to Epstein-Barr virus in Burkitt's lymphoma and control groups. *J. nat. Cancer Inst.,* 43, 1147-1158

Henle, G., Henle, W. & Diehl, V. (1968) Relation of Burkitt's tumor-associated herpes-type virus to infectious mononucleosis. *Proc. nat. Acad. Sci. (Wash.),* 59, 94-101

Henle, W., Diehl, V., Kohn, G., zur Hausen, H. & Henle, G. (1967) Herpes-type virus and chromosome marker in normal leukocytes after growth with irradiated Burkitt cells. *Science,* 157, 1064-1065

Henle, W. & Henle, G. (1970) Evidence for a relation of Epstein-Barr virus to Burkitt's lymphoma and nasopharyngeal carcinoma. In: Dutcher, R. M., ed., *Comparative leukemia research,* Basel, Karger, pp. 706-713

Klein, G., Pearson, G., Henle, G., Henle, W., Diehl, V. & Niederman, J. C. (1968) Relation between Epstein-Barr viral and cell membrane immunofluorescence in Burkitt's lymphoma and infectious mononucleosis. *J. exp. Med.,* 128, 1021-1030

Klein, G., Pearson, G., Henle, G., Henle, W., Goldstein, G. & Clifford, P. (1969) Relation between Epstein-Barr viral and cell membrane immunofluorescence in Burkitt tumor cells. III. Comparison of blocking of direct membrane immunofluorescence and anti-EBV reactivities of different sera. *J. exp. Med.,* 129, 697-705

Martin, N. H. (1954) Agammaglobulinemia—a congenital defect. *Lancet,* ii, 1094

Miller, G., Enders, J. F., Lisco, H. & Kohn, H. I. (1969) Establishment of lines from normal human leucocytes by co-cultivation with a leucocyte line derived from a leukemic child. *Proc. Soc. exp. Biol. (N.Y.),* 132, 247-252

Moore, G. E. & Kitamura, H. (1968) Cell line derived from patient with myeloma. *N.Y. State J. Med., 68,* 2054-2060

Moore, G. E. & Minowada, J. (1969) Studies of human hematopoietic cells. Hemic cells in vitro. *In Vitro, 4,* 100-184

Niederman, J. C., McCollum, R. W., Henle, G. & Henle, W. (1968) Infectious mononucleosis. Clinical manifestations in relation to EB virus antibodies. *J. Amer. med. Ass., 203,* 205-209

Nilsson, K. (1971a) *Human hematopoietic cells in continuous culture.* In: *Abstracts of Uppsala dissertations from the Faculty of Medicine,* Vol. 107, Stockholm, Almqvist and Wiksell

Nilsson, K. (1971b) High frequency establishment of human immunoglobulin-producing lymphoblastoid lines from normal and malignant lymphoid tissue and peripheral blood. *Int. J. Cancer, 8,* 432-442.

Nilsson, K., Bennich, H., Johansson, S. G. O. & Pontén, J. (1970) Established immunoglobulin producing myeloma (IgE) and lymphoblastoid (IgG) cell lines from an IgE myeloma patient. *Clin. exp. Immunol., 7,* 477-489

Nilsson, K., Pontén, J. & Philipson, L. (1968) Development of immunocytes and immunoglobulin production in long term cultures from normal and malignant human lymph nodes. *Int. J. Cancer, 3,* 183-190

Pontén, J. (1967) Spontaneous lymphoblastoid transformation of long term cell cultures from human malignant lymphoma. *Int. J. Cancer, 2,* 311-325

Pope, J. H., Horne, M. K. & Scott, W. (1968) Transformation of foetal human leukocytes *in vitro* by filtrates of a human leukaemic cell line containing herpes-like virus. *Int. J. Cancer, 3,* 857-866

Pope, J. H., Scott, W., Reedman, B. M. & Walters, M. K. (1971) In: Nishioka, K., ed., *First International Symposium of the Princess Takamatsu Cancer Research Fund,* Tokyo (in press)

zur Hausen, H. & Schulte-Holthausen, H. (1970) Presence of EB virus nucleic acid homology in a "virus-free" line of Burkitt tumour cells. *Nature (Lond.), 228,* 1056-1058

A Comparative Study of the Ultrastructure of Various Cell Lines Harbouring Herpes-type Virus

I. KIMURA[1] & Y. ITO[1]

INTRODUCTION

The present study is concerned with the cellular morphology of THE-2, THE-3 and NPC-204 cells, and with the unique inclusions observed in the cells of these lines and in those of the P3J Burkitt line.

THE-2 and THE-3 are transformed human embryonic cell lines, established from foci of morphologically altered cells derived from embryonic cell cultures exposed to the cell-free supernatant of human leukaemic culture fluid *in vitro* (Osato & Ito, 1967, 1968). NPC-204 is a cell line cultured *in vitro* from a biopsy specimen from a patient with nasopharyngeal carcinoma (Sugano *et al.*, 1970). Ultrastructural studies have revealed the presence of herpes-type virus (HTV) particles in all these cell lines (Kimura *et al.*, 1968; Kimura & Ito, 1969; Sugano *et al.*, 1970) (Figs. 1 & 2).

GENERAL ULTRASTRUCTURAL MORPHOLOGY OF PREDOMINANT CELLS IN THE-2, THE-3, NPC-204 AND P3J LINES

The predominant cells in the THE-3 line were 12 to 17 μ in diameter with smooth contours. The nucleus was usually oval, with a prominent nucleolus located in the centre of the nucleus. The cyctoplasm contained scattered endoplasmic reticulum (ER), a relatively well-developed Golgi apparatus, a few mitochondria and osmiophilic granules surrounded by a single membrane. Thus the cells in THE-3 line closely resemble those of the P3J line or so-called lymphoblastoid cells.

THE-2 cells were irregular in contour, due to the presence of numerous cytoplasmic projections. The

---

[1] Laboratory of Viral Oncology, Aichi Cancer Centre Research Institute, Nagoya, Japan.

nucleus was irregularly indented and multilobulated, with a prominent nucleolus attached to the nuclear membrane. The pattern of the ER also showed different characteristics from those seen in THE-3 and P3J cells.

The main cellular element in the NPC-204 line was relatively large and elongated. The cells contained numerous mitochondria, small lysosomal bodies and phagosomes. These cytoplasmic elements were often seen clustered at one side of the nucleus. Scattered or often moderate amounts of lamellae in the ER were observed in the cytoplasm on the other side of the nucleus. Thus the cells in the NPC-204 line resemble those in the reticulum cells or the macrophage stage (Moulton, Krauss & Malmquist, 1971).

ULTRASTRUCTURE OF CELLS HARBOURING HTV

Although the cells differed considerably from each other in morphology, cells harbouring HTV in all these lines showed similar ultrastructural changes in the organelles or inclusions. The findings can be listed as follows:

(1) Particles of approximately 65 nm in diameter present in the nucleus (Fig. 3) and sometimes in the cytoplasmic vacuoles (Fig. 1).

(2) Particles of approximately 25 nm in diameter outside the cell membrane close to the cell surface and in the cytoplasmic vacuoles (Fig. 4).

(3) An electron-dense tubular structure with a diameter of 35 nm (Fig. 5).

(4) Unusual margination of the nuclear substance.

(5) Duplication of the nuclear membrane (Fig. 3) and the membrane of the ER.

(6) Fusion of the mitochondria and unusual alterations of the inner structure of the mitochondrion (Fig. 5).

UNUSUAL INCLUSIONS OBSERVED IN THE-2, THE-3, NPC-204 AND P3J CELLS

In addition to the changes in the organelles and the inclusions observed in the cells harbouring HTV, the following unique inclusions were also observed:

(1) Bundles of fibres or filamentous structures, which have been described by Chandra, Moore & Brandt (1968) (in cells of the THE-3, P3J, and NPC-204 lines).

(2) Undulating tubules (Chandra, 1968) (in cells of the THE-2, THE-3, NPC-204 and P3J lines) (Fig. 6).

(3) Annulate lamellae (Kassel, 1968) (in cells of the THE-3 and P3J lines) (Fig. 7).

(4) A ribosome-membrane complex (Kimura & Ito, 1969) (in cells of THE-2, THE-3 and P3J lines) (Fig. 7).

DISCUSSION AND SUMMARY

The predominant cells in the THE-2, THE-3, NPC-204 and P3J lines differed considerably from each other in their morphology, as described above. Nevertheless, unique cytoplasmic inclusions, such as annulate lamellae, undulating tubules, a ribosome-membrane complex and bundles of fibres, were commonly observed in these cells. These inclusions or similar structures have also been observed in several kinds of cells which are presumably not infected with HTV. Although the functions of the inclusions have not been clarified, it is possible,

therefore, that they may merely represent the morphological expression of some unique metabolic stage of special kinds of cells adapted to certain conditions of cultivation.

The problem of whether the viruses that appear in the cell lines studied belong to the same "species" still remains to be solved. However, if the viruses in these cell lines are essentially the same, it may reasonably be assumed that HTV can develop in different kinds of cells derived from patients with different diseases. If so, it is possible that HTV is capable of inducing a number of different diseases in various circumstances when some "unknown co-factor(s)" may also be present.

Some of the unique ultrastructural changes in the organelles or the inclusions seen in cells harbouring HTV may represent only the unique degenerative process taking place in HTV-infected cells. Thus the particles of 25 nm in diameter and the electron-dense tubular structure may bear some relationship to a particular developmental stage of the virus, because virus particles were frequently observed near such inclusions.

Particles of 65 nm in diameter or similar particles 50 nm in diameter, as described above, have also been reported to appear in the cells of renal adeno-carcinoma in frogs (Lunger, Darlington & Granoff, 1965), in cells of Marek's disease in chickens (Cook & Sears, 1970) and in herpes-simplex-virus-infected cells (Nii, Morgan & Rose, 1968). Lunger et al. (1965) have suggested that such particles may represent the inner core of the herpesvirus. It is possible, therefore, that these findings may be related to the morphogenesis of the viruses of the herpes group as a whole, including HTV.

Fig. 1. Double-membraned herpes-type virus particles and budding particles observed in a THE-3 cell.   × 41 500.

Fig. 2. Herpes-type virus particle observed in a degenerating NPC-204 cell. Single-membraned virus particles are seen on the left of the photograph and double-membraned virus particles are seen surrounded by membraneous structures.   × 41 500.

Fig. 3. Electron micrograph of a cell of the P3J line harbouring herpes-type virus. Aggregates of small spherical particles of approximately 65 nm in diameter are seen in the nucleus and duplication of the nuclear membrane is also observed.   × 34 500.

Fig. 4. Portion of a cell of the THE-3 line harbouring herpes-type virus. Aggregates of small particles of 25 nm in diameter are seen at the cell surface and in the cytoplasmic vacuoles.   × 18 500.

Fig. 5. Portion of a cell of the THE-3 line harbouring herpes-type virus. Mitochondrial changes and an electron-dense tubular structure with a diameter of 35 nm are seen.   × 15 000.

Fig. 6. Undulating tubules observed in a THE-2 cell.   × 34 500.

Fig. 7. Portion of a THE-3 cell. Annulate lamellae, ribosome-membrane complex, and pseudo-inclusion body in the nucleus are seen in the photograph.   × 15 000.

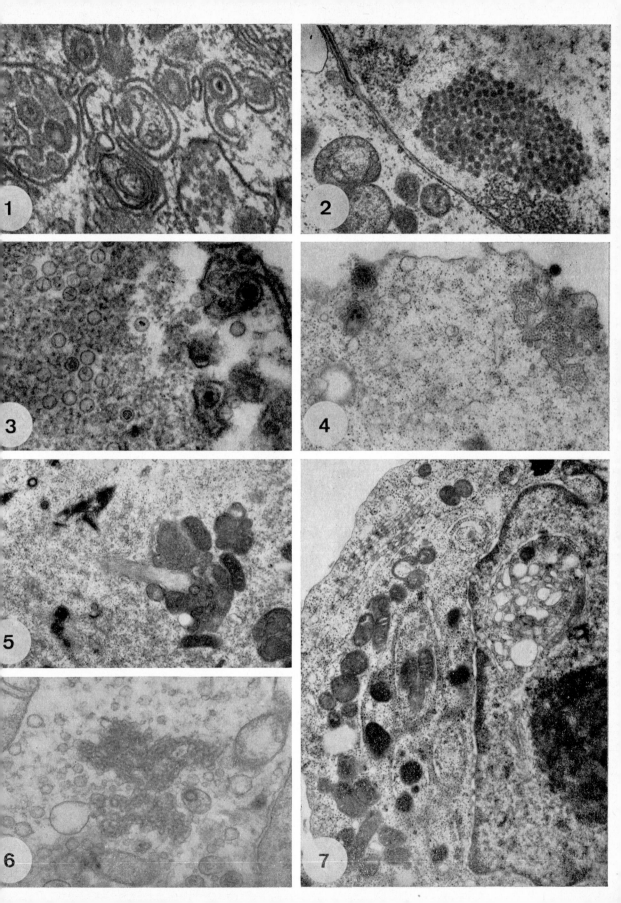

ACKNOWLEDGEMENTS

This study was supported by contract NIH-69-96 of the National Cancer Institute under the Special Virus Cancer Program.

REFERENCES

Chandra, S. (1968) Undulating tubules associated with endoplasmic reticulum in pathologic tissues. *Lab. Invest., 18,* 422-426

Chandra, S., Moore, G. E. & Brandt, P. M. (1968) Similarity between leukocyte cultures from cancerous and noncancerous human subjects: An electron microscopic study. *Cancer Res., 28,* 1982-1989

Cook, M. K. & Sears, J. F. (1970) Preparation of infectious cell-free herpes-type virus associated with Marek's disease. *J. Virol., 5,* 258-261

Kassel, R. G. (1968) Annulate lamellae. *J. Ultrastruct., 24,* 5-82

Kimura, I., Osato, T., Nagano, T. & Ito, Y. (1968) Transformation *in vitro* of human embryo cells by human leukemic culture fluid. II. Demonstration of herpes-like virus particles in transformed cells. *Proc. Japan Acad., 44,* 95-98

Kimura, I. & Ito, Y. (1969) Ultrastructural studies on the transformed human embryonic cell line, THE-2 and THE-3. *Gann Monogr., 7,* 115-153

Lunger, P. D., Darlington, R. W. & Granoff, A. (1965) Cell-virus relationship in the Lucké renal adenocarcinoma: An ultrastructure study. *Ann. N.Y. Acad. Sci., 126,* 289-314

Moulton, J. E., Krauss, H. H. & Malmquist, W. A. (1971) Transformation of reticulum cells to lymphoblasts in cultures of bovine spleen infected with *Theileria parva. Lab. Invest., 24,* 187-196

Nii, S., Morgan, C. & Rose, W. M. (1968) Electron microscopy of herpes simplex virus. II. Sequence of development. *J. Virol., 2,* 517-536

Osato, T. & Ito, Y. (1967) Morphological transformation of human embryo cells *in vitro.* An effect of human leukemic culture fluid. In: *Subviral carcinogenesis; Monograph of the 1st International Symposium on Tumor Viruses,* pp. 414-418

Osato, T. & Ito, Y. (1968) Transformation *in vitro* of human embryo cells by human leukemic culture fluid. I. Isolation and establishment of transformed cells. *Proc. Japan Acad., 44,* 89-94

Sugano, H., Takada, M., Chen, H. C. & Tu, S. H. (1970) Presence of herpes-type virus in the culture cell line from nasopharyngeal carcinoma in Taiwan. *Proc. Japan Acad., 46,* 453-457

EBV- associated Membrane Antigens

G. KLEIN[1]

A few years ago, when it became obvious that virus-induced animal tumours carry common, group-specific, membrane-associated antigens, detectable by graft rejection, membrane immunofluorescence, cytotoxicity, mixed haemadsorption, and other methods, we asked whether it would be possible to search for group-specific membrane antigens in tumours of unknown aetiology in order to obtain clues as to possible tumour-associated viruses. Our attention was turned to Burkitt's lymphoma (BL), due to its postulated viral aetiology and the clinical indications that host defences may play a rôle in the eventual outcome of the disease (Burkitt, 1963; Clifford, 1966; Burkitt, 1967; Ngu, 1965).

Membrane-reactive antibodies were identified in the sera of Burkitt's lymphoma patients, competent to react with BL biopsies, but not with biopsies from a number of other lymphomas or with normal lymph-node or bone-marrow cells (Klein, G. et al., 1966; Klein, E. et al., 1967; Klein, G. et al., 1967). Isoantibodies could not provide the explanation of this phenomenon, because tumour cells reacted whereas normal bone-marrow cells from the same Burkitt's lymphoma donor failed to react. In a number of cases, the reaction could be demonstrated, furthermore, with the autochthonous serum-tumour cell combinations.

A survey of established lymphoblastoid tissue culture lines, derived from Burkitt's lymphomas, indicated that the Burkitt-associated membrane antigen (MA) may be determined by the Epstein-Barr virus (EBV), a herpes-like agent first discovered by Epstein, Achong & Barr (1964) by electron microscopy of a BL-derived cell culture. Virus capsid antigens (VCA) determined by this agent have been identified in a small proportion of cells in carrier cultures by an immunofluorescence test described

by Henle & Henle (1966). A relationship between MA and EBV was suggested by the fact that MA-positive (MA+) cells were found only in cultures that carried a relatively high "load" of EBV (between 1% and 5% VCA-positive cells), while lines with no demonstrable capsid antigens showed no MA reactivity. The MA reaction was negative in lines that carried a relatively low "load" of EBV (less than 1% VCA-positive cells). These findings, originally made on randomly collected lines (Klein, G. et al., 1967), were confirmed in a prospective study on newly established lines (Klein et al., 1968). A relationship between MA and VCA was further indicated by the finding of a good correlation between the anti-MA and anti-VCA reactions in 80% of 279 human sera compared (Klein et al., 1969b), although the lack of correlation in the remaining 20% indicated that the anti-MA and anti-VCA antibodies are directed against different antigenic specificities. A final proof that MA is actually induced by EBV has been obtained by superinfecting established, EBV-antigen-free lymphoblastoid lines with the virus and demonstrating the appearance of the membrane antigen (Horosziewicz et al., 1971; Henle & Henle 1971a; Gergely, Klein & Ernberg, 1971b).

The nonidentity of the VCA and MA antigen complexes has been shown by the ten-fold difference in the frequency of reactive cells in certain lines (Klein et al., 1968), by the existence of the discordant sera already mentioned, by the fact that it is possible to absorb membrane-reactive antibodies with viable, MA-positive cells, leaving the anti-VCA titre virtually intact (Pearson et al., 1969), and by direct two-colour fluorescence tests (Klein, Gergely & Goldstein, 1971).

The virus-neutralizing activity of MA-VCA concordant and discordant sera was studied in relation to their anti-MA and anti-VCA titres. In concordant sera, the two reactivities were correlated with each

---

[1] Department of Tumor Biology, Karolinska Institutet, Stockholm, Sweden.

other and also with the virus-neutralizing titres. In discordant sera with one predominant activity, virus neutralization was related to the anti-MA but not to the anti-VCA levels (Pearson et al., 1970). Carrier cultures with a high frequency of MA-positive cells absorbed virus neutralizing antibody better than cultures with a low frequency of MA-positive cells, when both contained approximately equal numbers of VCA-positive cells (Gergely et al., 1971a). This indicates that MA may represent viral envelope components, inserted in the outer plasma membrane, perhaps by analogy with the new membrane glyco-proteins that appear in the membrane of herpes-simplex-infected cells (Roizman & Spring, 1967). Immunoferritin staining of EBV-carrier cells also showed that MA is present on the outer envelope of the virus, but not on naked virus particles (Sil-vestre et al., 1971). The Table summarizes the evidence obtained with representative sera, showing either concordant or discordant anti-MA and anti-VCA reactivities.

The relationship of MA to the virus cycle and to other EBV-associated antigens was studied recent-ly in abortively infected Raji-cells (Gergely et al., 1971b, c). MA and the recently described (Henle et al., 1970b) intracellular "early antigen" (EA) were both early products of the viral genome. They were fully expressed in the presence of DNA inhibi-tors, such as cytosine arabinoside (Ara-C) and iododeoxyuridine (IUDR) (Gergely et al., 1971b). VCA, on the other hand, depended on DNA syn-thesis (Gergely et al., 1971a). In carrier cultures treated with Ara-C, VCA-positive cells disappeared whereas EA-positive cells accumulated, presumably

as a result of the blockade of DNA synthesis and the absence of late viral products. Removal of DNA inhibition by deoxycytidine led to a rapid accumulation of VCA-positive cells and a decrease in the excess of EA+VCA– cells.

The question whether the appearance of MA, EA and VCA influences host-cell macromolecular synthesis in infected cells and in carrier cultures was studied by a combination of autoradiography and immunofluorescence (Gergely et al., 1971c). MA+ EA–VCA– cells continued to make RNA and protein like the antigen-negative controls. Their DNA synthesis showed certain differences compared to antigen-negative controls, but these may be attributable to the fact that membrane antigens are maximally expressed during the G1 phase of the cell cycle (Cikes, 1971). MA+ cells and antigen-negative cells may therefore represent population samples that are at least partially out of phase with one another. EA+ cells showed a progressive inhibition of DNA, RNA and protein synthesis, increasing with time after infection. This indicates that the appearance of EA signals the entry of the cell into a lytic cycle. Macromolecular synthesis was also reduced in the VCA+ cells of the carrier cultures, but these cells still made DNA, although at a reduced level, in comparison with antigen-negative cells.

The relationship of the three antigens to each other was studied by two-colour fluorescence (Klein et al., 1971). All VCA+ cells contained MA and EA as well. The regular presence of MA is in line with the nature of MA as an envelope antigen, already mentioned. It is not known, however, whether the

EBV-associated immunoferritin reactivity of cell membranes, viral envelopes and naked particles after the application of human sera with concordant and discordant anti-MA and anti-VCA reactivities [1]

Serum	Diagnosis	Anti-MA		Anti-VCA	Anti-HLA	Virus neutraliza-tion	Immunoferritin coating of:		
		BI[2]	BT[3]				cell mem branes	viral enve-lopes	naked par-ticles
Agnes	BL	0.76	64–128	1280–2560	0	+++	+++	+++	+++
Ciorimberia	NPC	0.77	2–4	640–1280	0	++	++	++	++
Vincent	Healthy BL-relative	0.55	1	20	0	+	+	+	–
Nathan	NPC	0.17	0	640–1280	1:2	–	±	±	+++
Bibia	NPC	0.24	0	1280	0	–	–	–	+++
Labolle	Multitransfused serum donor	0	0	20	1:16	n.t.	+++	±	–
Vacher	Acute leukaemia	0	0	0	0	n.t.	–	–	–

[1] Silvestre et al. (1971).
[2] Obtained with undiluted test serum, against FITC-Mutua conjugate, as described by Klein et al. (1969).
[3] With 0.5 blocking index (BI) endpoint against FITC-Mutua conjugate, as described by Gunvén & Klein (1971).

early MA is completely identical with the late MA, from the point of view of antigenic specificity. The regular presence of EA in VCA+ cells indicated that this antigen is still required at a late stage of virus formation. It is not known whether it becomes incorporated into the virion.

EA+VCA− cells can be MA+ or MA−. The relative proportion of EA+MA− cells increased with time after infection, presumably due to the disappearance of MA from the membrane of EA+ cells. This may reflect a progressive inhibition of protein synthesis in EA+ cells. Puromycin reduced MA expression in carrier cultures within 6–12 hours. The expression of the membrane antigen is thus dependent on continuous protein synthesis. This is probably related to the rapid turnover of the outer cell membrane in growing cells (Warren & Glick, 1968).

Biopsies taken from Burkitt tumours *in vivo* contained MA+EA−VCA− cells, but EA+ and VCA+ cells were absent, as a rule. The presence of MA+ cells in the growing tumour is in line with the finding that host-cell macromolecular synthesis is not inhibited in MA+ cells. The absence of EA+ and VCA+ cells from the biopsy may mean either that the virus cycle is suppressed in the tumour *in vivo* or, alternatively, that virus-activated cells that enter the lytic cycle are rapidly eliminated from the tumour, or at least from the viable free-cell suspensions prepared for study. Macrophage action may be responsible. Complete suppression of the viral cycle is unlikely, since the presence and level of anti-EA antobodies was shown to be related to the tumour burden in BL patients (Henle *et al.*, 1971b).

The expression of MA in EBV-carrier cultures is subject to considerable physiological fluctuations and can be changed by altering the culture conditions or adding certain inhibitors of macromolecular synthesis (Yata & Klein, 1969; Yata *et al.*, 1970). In two lines, crowding of the cells, or the addition of very small doses of mitomycin, actinomycin or methotrexate, or small doses of X-irradiation, led to a conspicuous and sometimes dramatic increase in the proportion of MA+ cells. Larger doses of actinomycin had no effect. Puromycin reduced MA expression, and prevented its induction by the agents mentioned. Ara C did not induce MA, although it increased the proportion of EA+ cells, as already mentioned. This showed that inhibition of cell multiplication was not sufficient *per se* to induce a high proportion of MA+ cells.

Antibody formation against the different EBV-associated antigens shows different disease-related patterns. Anti-EA levels appear to be related to the tumour burden and, in regression patients, to the residual tumour (Henle *et al.*, 1971b). Immunoprecipitating antibody may show a similar pattern (Klein *et al.*, 1969a; Klein *et al.*, 1970), but the antibody specificities involved are clearly different (Henle *et al.*, 1971b). Membrane-reactive antibodies are related to the disease in a more complex way. In long-term regression patients, their level is often high (Klein *et al.*, 1966; Yata *et al.*, 1970). Before or around the time of tumour recurrence, they often decrease; it is not clear whether this is a cause or a consequence of recurrence (Gunvén *et al.*;[1] Klein *et al.*, 1970). During continued progressive growth of the tumour, they may rise again, obviously as a result of continued antigenic stimulation. Anti-VCA antibodies show only limited disease-related fluctuations.

DISEASE-ASSOCIATED SEROLOGICAL PATTERNS AND THE AETIOLOGICAL DILEMMA

Mean anti-EBV (VCA) and anti-MA titres are much higher in groups of Burkitt's lymphoma patients than in healthy African controls, or African patients with neoplastic diseases other than Burkitt's lymphoma and nasopharyngeal carcinoma (Gunvén *et al.*, 1970; Henle *et al.*, 1969). The same is true for immunoprecipitating antibodies against a soluble antigen extracted from an EBV-carrying Burkitt line (Old *et al.*, 1968). Nasopharyngeal carcinoma was unique among the head and neck tumours so far investigated in showing high EBV-associated antibody levels in the anti-MA, anti-VCA and immunoprecipitin tests as well (Old *et al.*, 1968; de Schryver *et al.*, 1969; Henle *et al.*, 1970a). It was particularly remarkable that hypopharyngeal and oropharyngeal carcinomas were quite different serologically; their EBV-associated antibody pattern resembled that of the controls (de Schryver *et al.*, 1969). The evidence concerning sarcoidosis is less conclusive, and it is not yet clear whether this is also generally a high anti-EBV-associated disease (Hirshaut *et al.*, 1970; Wahren *et al.*[1]).

Interesting findings have been obtained in Hodgkin's disease (Johansson *et al.*, 1970). The prognostically unfavourable, lymphocyte-poor, sarcomatous form gave high anti-VCA and anti-MA levels, comparable to those of Burkitt's lymphoma

---

[1] Unpublished data.

and nasopharyngeal carcinoma. The lymphocyte-rich, paragranulomatous form having a better prognosis was characterized, in contrast, by low anti-VCA and anti-MA titres, similar to those of the controls. The granulomatous form was intermediate, from both the histological and serological points of view.

POSSIBLE IMPLICATIONS

There is strong evidence that EBV is involved in the aetiology of infectious mononucleosis (Henle, Henle & Diehl, 1968). Prospective studies showed that anti-EBV seropositive young adults are protected from mononucleosis, while a substantial proportion of seronegatives acquire the disease in the same environment (Niederman et al., 1970). It is also clear, however, that EBV is widespread and ubiquitous. Infection in early childhood apparently causes seroconversion only, but does not appear to be associated with any recognized clinical syndrome.

If it is accepted that EBV can cause infectious mononucleosis, its relationship to Burkitt's lymphoma can be considered from the point of view of three main hypotheses, namely the cofactor hypothesis, the multiple virus hypothesis and the passenger hypothesis.

The cofactor hypothesis implies that EBV is the causative agent of both infectious mononucleosis and Burkitt's lymphoma, but that malignant proliferation is also due to the action of some cofactor together with the virus. One such hypothesis has been formulated in very concrete terms by Burkitt himself (Burkitt, 1969), who suggested that chronic holoendemic malaria could turn an otherwise limited lymphoproliferative disease into a malignancy. His reasoning was mainly based on the observation that a number of geographical areas, located within the holoendemic region but where malaria control has been practised for some time, were apparently free of the disease. Preliminary observations, not yet statistically significant, suggest, furthermore, that BL is less frequent in children with the sickling trait, known to give substantial protection against severe falciparum malaria, than in normal controls (Pike et al., 1970).

There are, of course, many animal models demonstrating that agents capable of stimulating cell proliferation in a given target tissue, but without causing malignant tumours, promote the oncogenic action of chemical or viral carcinogens, administered at subthreshold levels (Southam et al., 1969); some may even act synergistically with other stimuli that are not oncogenic by themselves. The postulated cocarcinogenic action of malaria has not been investigated extensively, but recent reports indicate that the mouse malaria agent, *Plasmodium berghei*, may increase the incidence of "spontaneous" or virally induced lymphoma in mice (Jerusalem, 1968; Weddenburn, 1970).

The multiple virus hypothesis implies that different variants of EBV exist in nature. Some of these may induce benign, and others malignant, disease. Attempts to show antigenic differences between EBV associated with infectious mononucleosis, Burkitt's lymphoma and nasopharyngeal carcinoma, have failed to reveal any differences so far, but very few such attempts seem to have been made and the methodology is still very crude and probably quite unable to reveal minor differences in type. It may be recalled that the avian-leukosis-sarcoma complex was believed to be due to a single viral agent not very long ago. Distinction between different viral subgroups became possible only after the interference test had been discovered (Rubin, 1960). This led to the realization that there are a large number of different subgroups of virus with common group-specific antigens and identical appearance under the electron microscope, but with widely differing oncogenic potential. Types exist capable of inducing erythromyeloid leukaemia, lymphomatosis, various kinds of sarcomas, or no disease at all. There is also considerable variation in the oncogenic potency of the same agent in relation to different host genotypes.

A very similar situation has developed in the field of murine leukaemia viruses. C-type viruses, with identical morphological appearance under the electron microscope and common group-specific antigens, can give rise to very different conditions, including thymic lymphoma (e.g., the Gross and Moloney agents), reticulum cell neoplasia (Rauscher or Friend virus), myeloid leukaemia (Graffi agent) and sarcoma (MSV). At least one agent, the so-called L-cell virion, could not be shown to induce any disease at all. If transmission experiments could not have been performed and serological methods had alone been available to trace the distribution of these agents, by analogy with the situation in man, it would have been very difficult to determine which agent was related to which disease.

The multiple virus hypothesis is thus perfectly possible, but there is, of course, no direct evidence for or against it as far as EBV is concerned.

The passenger hypothesis is based on the fact that EBV is an inhabitant of human lymphoid tissues and is both widespread and ubiquitous. When lymphomas or other lymphocyte-rich tumours arise and proliferate, due to causes quite unconnected with EBV, it would be carried along, with a corresponding increase in antigen load and antibody response. This is also perfectly possible, and the passenger hypothesis cannot be excluded at present. Its likelihood has been somewhat reduced, however, by the following facts:

(*a*) Although 10–15% of the children in the holo-endemic areas and within the age-groups at risk are seronegative, all African, histologically confirmed cases of Burkitt's lymphoma are seropositive and most of them have high antibody titres. If the causation of the disease is entirely unrelated to EBV, a few seronegative cases, at least, would be expected.

(*b*) Other lymphomas and other malignant diseases of the lymphoreticular system would be expected to show a high serological reactivity similar to that of Burkitt's lymphoma, but this is by no means the case (Old *et al.*, 1968; Henle *et al.*, 1969; de Schryver *et al.*, 1969; Johansson *et al.*, 1970). Hodgkin's disease is interesting in this respect; as already mentioned, the sarcomatous type, poorest in lymphoid elements, shows a high EBV-associated serological reactivity, whereas the lymphocyte-rich paragranuloma resembles the controls (Johansson *et al.*, 1970).

(*c*) In the case of nasopharyngeal carcinoma, it is hard to see why other tumours of the same or closely adjacent regions do not show a similar serological pattern, if the passenger hypothesis is correct. A variety of tumours localized to the nasopharynx that showed low anti-MA and anti-VCA reactivity turned out, on histological examination by an unbiased observer who was not aware of the serological findings, to have entirely different diagnoses, including reticulum cell sarcoma, Hodgkin's disease, craniopharyngioma, salivary gland tumour, etc. (de Schryver *et al.*, 1969). Hypo- and oropharyngeal carcinomas did not show a high EBV-associated serological pattern either (de Schryver *et al.*, 1969; Henle *et al.*, 1970a). In contrast, Chinese, American, Swedish, African and French cases of nasopharyngeal carcinoma showed identical serological patterns, namely high EBV-associated reactions in the anti-VCA, anti-MA and immunoprecipitin tests (Old *et al.*, 1968; de Schryver *et al.*, 1969; Henle *et al.*, 1970a). Serological uniformity, unaffected by geographical location, is one of the few criteria that may be usefully applied in trying to distinguish between chance passenger agents and viruses more intimately associated with certain tumours.

A difference in the intimacy of the EBV-tumour association in BL and NPC, in contrast to other neoplasms arising in EBV-seropositive patients, is also indicated by the recent nucleic acid hybridization studies of zur Hausen and co-workers (1970). BL and NPC biopsies contained DNA that hybridized specifically with purified EBV-DNA, whereas tumours of other kinds, arising in EBV-seropositive patients, contained no detectable hybridizable DNA. In 13 Burkitt biopsies, the approximate number of EBV-genome equivalents per cell varied between 2 and 26. In 10 NPC biopsies, the number of EBV-genome equivalents varied between 1 and 19. Interestingly enough, three BL patients from whom double biopsies were taken showed closely similar genome equivalent values in anatomically distant tumours (2–2, 7–8 and 21–25, in the three cases). Since Burkitt's lymphoma is a monoclonal disease (Fialkow *et al.*, 1970), this would mean that the EBV"load" is fairly constant and characteristic for a given clone. This is reminiscent of the different and characteristic numbers of SV40 genome copies in different clones of SV40-transformed cells (Westphal & Dulbecco, 1968).

While the nucleic acid hybridization studies merely confirm that EBV is more closely associated with BL and NPC than with a number of other tumours, this evidence is at least consistent with the behaviour of known oncogenic DNA virus systems.

ACKNOWLEDGEMENTS

The studies of the author and his associates were conducted under USPHS Contract No. NIH-69-2005 within the Special Virus Cancer Program of the National Cancer Institute, National Institutes of Health. They were also supported by grants from the Swedish Cancer Society.

REFERENCES

Burkitt, D. (1963) A lymphoma syndrome in tropical Africa. *Int. Rev. exp. Path.*, **2**, 67-138

Burkitt, D. (1967) Chemotherapy of jaw tumors. In: Burchenal, J. M., ed., *Treatment of Burkitt's tumor*, Heidelberg, Springer Verlag, pp. 94-101 *(UICC Monograph Series)*

Burkitt, D. (1969) Etiology of Burkitt's lymphoma—an alternative hypothesis to a vectored virus. *J. nat. Cancer Inst.*, **42**, 19-28

Cikes, M. (1971) Variation in expression of surface antigens on cultured cells. *Ann. N.Y. Acad. Sci.* **177**, 190-200

Clifford, P. (1966) Further studies in the treatment of Burkitt's lymphoma. *E. Afr. med. J.*, **43**, 179-199

de Schryver, A., Friberg, S., Klein, G,. Henle, W., Henle, G., de-Thé, G., Clifford, P. & Ho, H. C. (1969) Epstein-Barr virus associated antibody patterns in carcinoma of the post-nasal space. *Clin. exp. Immunol.*, **5**, 443-459

Epstein, M. A., Achong, B. G. & Barr, Y. M. (1964) Virus particles in cultured lymphoblasts from Burkitt's lymphoma. *Lancet*, **i**, 702-703

Fialkow, P. J., Klein, G., Gartler, S. M. & Clifford, P. (1970) Clonal origin for individual Burkitt tumors. *Lancet*, **i**, 384-386

Gergely, L., Klein, G. & Ernberg, I. (1971a) The action of DNA antagonists on Epstein-Barr virus (EBV)-associated early antigen (EA) in Burkitt lymphoma lines. *Int. J. Cancer*, **7**, 293-302

Gergely, L., Klein, G. & Ernberg, I. (1971b) Appearance of EBV-associated antigens in infected Raji-cells. *Virology* (in press)

Gergely, L., Klein, G. & Ernberg, I. (1971c) Host cell macromolecular synthesis in cells containing EBV induced early antigens, studied by combined immunofluorescence and radioautography. *Virology* (in press)

Gunvén, P., Klein, G., Henle, G., Henle, W. & Clifford, P. (1970) Antibodies to EBV-associated membrane and viral capsid antigens in Burkitt lymphoma patients. *Nature (Lond.)*, **228**, 1053-1056

Henle, G. & Henle, W. (1966) Immunofluorescence in cells derived from Burkitt's lymphoma. *J. Bact.*, **91**, 1248-1256

Henle, G. & Henle, W. (1971a) Evidence for a relation of Epstein-Barr virus to Burkitt's lymphoma and nasopharyngeal carcinoma. In: *International Symposium of Comparative Leukemia Research*, Basel, Karger, pp. 706-713

Henle, G., Henle, W., Clifford, P., Diehl, V., Kafuko, G. W., Kirya, B. G., Klein, G., Morrow, R. H., Munube, G. M. R., Pike, P., Tukei, P. M. & Ziegler, J. L. (1969) Antibodies to Epstein-Barr virus in Burkitt's lymphoma and control groups. *J. nat. Cancer Inst.*, **43**, 1147-1158

Henle, G., Henle, W. & Diehl, V. (1968) Relation of Burkitt's tumor-associated herpes-type virus to infectious mononucleosis. *Proc. nat. Acad. Sci. (Wash.)*, **59**, 94-101

Henle, G., Henle, W., Klein, G., Gunvén, P., Clifford, P., Morrow, R. H. & Ziegler, J. L. (1971b) Antibodies to early Epstein-Barr virus-induced antigens in Burkitt's lymphoma. *J. nat. Cancer Inst.*, **46**, 861-871

Henle, W., Henle, G., Ho, H. C., Burtin, P., Cachin, Y., Clifford, P., de Schryver, A., de-Thé, G., Diehl, V. & Klein, G. (1970a) Antibodies to Epstein-Brar virus in nasopharyngeal carcinoma, other head and neck neoplasms, and control groups. *J. nat. Cancer Inst.*, **44**, 225-231

Henle, W., Henle, G., Zajec, B. A., Pearson, G., Waubke, R. & Scriba, M. (1970b) Differential reactivity of human serums with early antigens induced by Epstein-Barr virus. *Science*, **169**, 188-190

Hirshaut, Y., Glade, P., Octavio, L., Viera, B. D., Ainbender, E., Dvorak, B. & Silzbach, L. E. (1970) Sarcoidosis, another disease associated with serologic evidence for herpes-like virus infection. *New Engl. J. Med.*, **283**, 502-506

Horoszewicz, J. S., Dunkel, V. C., Avila, L. & Grace, J. T. (1971) EB-virus infection and propagation in human hematopoietic cells. In: *International Symposium of Comparative Leukemia Research*, Basel, Karger, pp. 722-728

Jerusalem, C. (1968) Active immunization against malaria *(Plasmodium berghei)*. I. Definition of antimalarial immunity. *Trop. Med. Parasitol.*, **19**, 171-181

Johansson, B., Klein, G., Henle, W. & Henle, G. (1970) Epstein-Barr virus (EBV)-associated antibody patterns in malignant lymphoma and leukemia. I. Hodgkin's disease. *Int. J. Cancer*, **6**, 450-462

Klein, E., Clifford, P., Klein, G. & Hamberger, C. A. (1967) Further studies on the membrane immunofluorescence reaction of Burkitt lymphoma cells. *Int. J. Cancer*, **2**, 27-36

Klein, G., Clifford, P., Henle, G., Henle, W., Geering, G. & Old, L. J. (1969a) EBV-associated serological patterns in a Burkitt lymphoma patient during regression and recurrence. *Int. J. Cancer*, **4**, 416-421

Klein, G., Clifford, P., Klein, E., Smith, R. T., Minowada, J., Kourilsky, F. M. & Burchenal, J. H. (1967) Membrane immunofluorescence reaction of Burkitt lymphoma cells from biopsy specimens and tissue cultures. *J. nat. Cancer Inst.*, **39**, 1027-1044

Klein, G., Clifford, P., Klein, E. & Stjernswärd, J. (1966) Search for tumor specific immune reactions in Burkitt lymphoma patients by the membrane immunofluorescence reaction. *Proc. nat. Acad. Sci.*, **55**, 1628-1635

Klein, G., Geering, G., Old, L. J., Henle, G., Henle, W. & Clifford, P. (1970) Comparison of the anti-EBV titer and the EBV-associated membrane reactive and precipitating antibody levels in the sera of Burkitt lymphoma and nasopharyngeal carcinoma patients and controls. *Int. J. Cancer*, **5**, 185-194

Klein, G., Gergely, L. & Goldstein, G. (1971) Two-colour immunofluorescence studies on EBV-determined antigens. *Clin. exp. Immunol.*, **8**, 593-602

Klein, G., Pearson, G., Henle, G., Henle, W., Goldstein, G. & Clifford, P. (1969b) Relation between Epstein-Barr viral and cell membrane in immunofluorescence in Burkitt tumor cells. III. Comparison of blocking of direct membrane immunofluorescence and anti-EBV reactivities of different sera. *J. exp. Med., 129,* 697-706

Klein, G., Pearson, G., Nadkarni, J. S., Nadkarni, J. J., Klein, E., Henle, G., Henle, W. & Clifford, P. (1968) Relation between Epstein-Barr viral and cell membrane immunofluorescence of Burkitt tumor cells. I. Dependence of cell membrane immunofluorescence on presence of EB virus. *J. exp. Med., 128,* 1011-1020

Ngu, V. A. (1965) The African lymphoma (Burkitt tumors): Survivals exceeding two years. *Brit. J. Cancer,* 19, 101-107

Niederman, J. C., Evans, A. S., Subrahmanyan, L. & McCollum, R. W. (1970) Prevalence, incidence and persistence of EB-virus antibodies in young adults. *New Engl. J. Med.,* 282, 361-365

Old, L. J., Boyse, E. A., Geering, G. & Oettgen, H. F. (1968) Serological approaches to the study of cancer in animals and in man. *Cancer Res.,* 28, 1288-1299

Pearson, G., Dewey, F., Klein, G., Henle, G. & Henle, W. (1970) Correlation between antibodies to Epstein-Barr virus (EBV)-induced membrane antigens and neutralization of EBV infectivity. *J. nat. Cancer Inst.,* 45, 989-997

Pearson, G., Klein, G., Henle, G., Henle, W. & Clifford, P. (1969) Relation between Epstein-Barr viral and cell membrane immunofluorescence in Burkitt tumor cells. IV. Differentiation between antibodies responsible for membrane and viral immunofluorescence. *J. exp. Med.,* 129, 707-718

Pike, M. C., Morrow, R. H., Kisuule, A. & Mafigiri, J. (1970) Burkitt's Lymphoma and sickle cell trait. *Brit. J. prev. soc. Med.,* 24, 39-41

Roizman, B. & Spring, S. B. (1967) Alteration in immunologic specificity of cells infected with cytolytic viruses.

In: Trentin, J. J., ed., *Proceedings of the Conference on Cross Reacting Antigens and Neo-antigens,* Baltimore, Williams & Wilkins Co., pp. 85-197

Rubin, H. (1960) A virus in chick embryos which induces resistance *in vitro* to infection with Rous sarcoma virus. *Proc. nat. Acad. Sci. (Wash.),* 46, 1105-1119

Silvestre, D., Kourilsky, F. M., Klein, G., Yata, J., Neauport-Sautes, C. & Levy, J. P. (1971) Relationship between the EBV-associated membrane antigen on Burkitt lymphoma cells and the viral envelope, demonstrated by immunofenitin labelling. *Int. J. Cancer,* 8, 222-233

Southam, C. M., Tanaka, S., Arata, T., Simkovic, D., Miura, M. & Peptiopules, S. F. (1969) Enhancement to responses to chemical carcinogens by nononcogenic viruses and anti-metabolites. *Progr. exp. Tumor Res. (Basel),* 11, 194-212

Warren, L. & Glick, M. C. (1968) Membranes of animal cells. II. The metabolism and turnover of the surface membrane. *J. cell Biol.,* 37, 729-746

Weddenburn, N. (1970) Effect of concurrent malarial infection on development of virus-induced lymphomas in Balb/c mice. *Lancet,* ii, 1114-1116

Westphal, H. & Dulbecco, R. (1968) Viral DNA in polyoma- and SV40-transformed cell lines. *Proc. nat. Acad. Sci. (Wash.),* 59, 1158-1165

Yata, J. & Klein, G. (1969) Some factors affecting membrane immunofluorescence reactivity of Burkitt lymphoma tissue culture cell lines. *Int. J. Cancer,* 4, 767-775

Yata, J., Klein, G., Hewetson, J. & Gergely, L. (1970) Effect of metabolic inhibitors on membrane immunofluorescence reactivity of established Burkitt lymphoma cell lines. *Int. J. Cancer,* 5, 394-403

zur Hausen, H., Schulte-Holthausen, H., Klein, G., Henle, W., Henle, G., Clifford, P. & Santesson, L. (1970) EBV-DNA in biopsies of Burkitt tumours and anaplastic carcinomas of the nasopharynx. *Nature (Lond.),* 228, 1056-1058

Detection and Analysis of Antigens in Lymphoblastoid Cell Lines Using Radio-labelled Antisera; Preliminary Results

T. B. GREENLAND,[1] G. DE-THÉ[1] & N. E. DAY[1]

INTRODUCTION

Sera taken from patients with nasopharyngeal carcinoma (NPC), Burkitt's lymphoma (BL), or infectious mononucleosis (IM) usually contain high titres of antibodies that react with several different antigens expressed by the lymphoblastoid cells that may be cultured from biopsies of the tumours. Antibodies that will react with herpes-type viral capsid antigens (zur Hausen et al., 1967; Epstein & Achong, 1967), virally induced membrane antigens (Klein et al., 1968), or "early antigens" (Henle et al., 1970) have been detected by the use of immuno-fluorescence techniques. Soluble antigens may be extracted from the cells and their reaction with antibody may be shown by immunoprecipitation in gel (Old et al., 1966), where some three different reactivities may be observed, or by complement fixation, which can detect antigen not only from cell lines positive in the immunofluorescence test, but also from negative lines (Pope et al., 1969).

We have developed a technique whereby we compare the uptake of total antibody activity from a panel of sera by acetone-fixed cell smears made from different cell lines. The uptake of the antibodies is measured by labelling the gamma-globulin-enriched fraction of each serum with radio-iodine, and counting the radioactivity remaining on the cell smears after treatment with the labelled fractions and washing. In order to compare the activities of the sera under identical conditions, each smear is treated with a mixture of two sera, each of which has been labelled with a different isotope of radio-iodine. By com-parison of the relative strengths of the same sera on different cell lines, and of the weight of globulin taken up from the different sera (calculated from the number of counts per specimen, and a known sample of the labelled globulin), it is possible to estimate the minimum number of antigenic activities, and of corresponding antibody activities, necessary to explain the observed pattern of uptake.

MATERIALS

Cells were obtained from cultures originating from nasopharyngeal carcinomas, Burkitt's lymphomas, or infectious mononucleosis. The origins of the different lines used are reported in Table 1, with the exception of HKLY 28, which is a line derived from a Chinese NPC. The cells were grown in RPMI 1640 medium (Grand Island Biological Co., Grand Island, N.Y.) supplemented with 20% foetal calf serum (FCS) at 37°C, and were then held at 35°C for four days before being used as antigen, unless otherwise stated.

Sera were obtained from African patients with suspected NPC, Chinese patients with confirmed NPC, African patients with confirmed BL, and from healthy Caucasian controls. In most cases a crude gamma-globulin-enriched fraction was prepared from these sera by precipitation with 50% saturated ammonium sulphate, the precipitate being redis-solved and exhaustively dialyzed against phosphate-buffered saline (PBS) at pH 7.8 before use. In a few cases a more highly purified gamma-globulin-fraction, namely that excluded from a diethylaminoethyl (DEAE) cellulose column at pH 8.0 and ionic strength 0.005 M (phosphate), was used for labelling.

[1] Units of Biological Carcinogenesis and Epidemiology and Statistics, International Agency for Research on Cancer, Lyon, France.

Table 1. Origins and antigenicities of 12 lymphoblastoid cell lines

Cell line	Origin [1]	"Antigens" [2]							IFA [3] behaviour
		A	B	C	D	E	F_a	F_b	
HKLY 11	NPC Ch	+	−	+	+	−	−	−	+
HKLY 17	NPC Ch	+	+	+	−	−	−	−	+
HKLY 37	NPC Ch	+	+	+	+	−	−	−	+
HKLY 1	NPC Ch	−	−	+	−	+	+	+	−
HKLY 2	NPC Ch	−	+	+	−	−	+	+	−
HKLY 38	NPC Ch	−	−	+	−	−	+	+	−
LY 39	NPC Af	−	+	+	−	+	+	+	−
Jijoye	BL Af	+	+	−	−	−	−	−	+
Silfere	BL Af	+	−	−	+	+	(+	−)	+
Ester	BL Af	+	−	−	+	+	(+	−)	+
Raji	BL Af	−	−	−	+	+	+	+	+
IM 71	IM Ca	+	−	−	+	+	(+	−)	+

[1] NPC = nasopharyngeal carcinoma.   Ch = Chinese.
 BL = Burkitt's lymphoma.   Af = African.
 IM = infectious mononucleosis.   Ca = Caucasian.
[2] As determined by the radiolabelled antibody method.
[3] + denotes the presence of fluorescent cells in the indirect test after treatment with patients serum; − denotes no fluorescent cells under the same conditions.

METHODS

Preparation of the cell smears

The cultured cells were washed in PBS, and resuspended at a concentration of 2×10^6 cells/ml. 0.05-ml aliquots of this suspension were smeared on to glass cover slips, air dried, and fixed in anhydrous acetone for 10 minutes at room temperature. After fixing, the smears were stored at −70°C until use.

Labelling of the sera

The gamma-globulin-enriched fractions from the different sera were adjusted to a concentration corresponding to an optical density at 280 nm (OD^{280}) of 1.32. This would be a concentration of 1 mg/ml if they were composed of pure gamma globulin. 1-ml aliquots of these solutions were labelled with 1 mCi of either ^{131}I or ^{125}I by the chloramine T method of Greenwood, Hunter & Glover (1963). The labelled proteins were then mixed with 0.1 ml of FCS as a carrier and passed through 10 cm × 1.3 cm columns of Sephadex G 25 (Pharmacia, Uppsala) to remove unreacted isotope. The eluates were then dialyzed overnight against PBS, filtered through 0.2-μ filters and made up to a volume of 10 ml in 10% FCS in PBS.

Preparation of the paired labelled mixtures (PLMs)

After removal of a small quantity of each labelled serum for dilution as a known reference standard, the remainder was divided into as many portions as there were sera in the panel. One portion of each labelled serum was then combined with one portion of every serum labelled with the other isotope. A small portion of each PLM was then taken as a reference standard.

Treatment of the cell smears

The cell smears from all the cell lines were incubated with 0.1 ml of each PLM for 30 minutes at 37°C, after which the unreacted PLM was rinsed off with PBS, and the smears washed three times for 10 minutes in PBS. After washing, the smears were allowed to air dry, then were put into plastic tubes for counting.

Counting

The samples were counted in a Packard "Auto gamma" three-channel gamma-ray spectrometer, model 5312. The ^{125}I was measured by using its emission peak at 25–37 KeV, and the ^{131}I by using that at 364 KeV. Specimens were counted for 2 minutes each, and the background was automatically subtracted.

Calculations

The counts measured in the ^{125}I channel must be corrected for the contribution due to the emission from the ^{131}I at that energy. This is done by using a correction factor (CF) established by counting a pure sample of ^{131}I where:

$$CF = \frac{\text{Counts measured in the } ^{125}I \text{ channel}}{\text{Counts measured in the } ^{131}I \text{ channel}}$$

The true ^{125}I counts for a specimen in which ^{131}I is present are then calculated as follows:

True ^{125}I
$= ^{125}$I channel counts $- $ (CF $\times\ ^{131}$I counts).

The ratio of the ^{125}I and ^{131}I uptakes (R) is then calculated:

$$R = \frac{\text{True }^{125}\text{I counts.}}{^{131}\text{I counts}}$$

This ratio needs still to be corrected for the isotopic composition of the PLM used to treat the specimen. The isotopic ratio in the PLM (R_0) is then calculated as above for the PLM standard sample, and the relative uptake of the two labelled sera by the cells is expressed as an uptake quotient (Q), where:

$$Q = \frac{R}{R_0} - 1.$$

Weight uptakes are determined by taking the average of the number of counts due to a given labelled serum fixed on to the cell line under consideration and dividing by the number of counts measured in a standard dilution of the labelled serum. These are then corrected for loss of specimen during washing by measuring the protein content of the counted smears, using a modification of the Lowry method, as described by Tanigaki *et al.* (1967); the weight fixed per 1 OD^{750} unit is used as the corrected value.

RESULTS

The cell lines listed in Table 1 have all been analyzed at least twice, and some of them up to five times, in some 10 separate experiments, the same panel of sera being used in each case. The antigenic contents were found not to vary appreciably as between different measurements. The findings for the HKLY 28 line are the result of a single experiment for which a different panel of five sera was used. Results from a number of these experiments have been chosen to illustrate certain points of interest. Fig. 1 shows the uptake quotients (Q) measured for the HKLY 28 cell line, using a panel of five sera arranged in a checkerboard pattern. It can be seen that the top left to bottom right diagonal, which corresponds to like serum pairs, contains values close to zero—in other words there is no appreciable difference in the uptake of the same serum labelled

^{125}I - labelled sera:

	1	2	3	4	5
A	0.06	-.74	-.52	-.86	-.18
B	2.84	-.03	1.16	-.53	1.69
C	0.80	-.50	0.09	-.80	0.26
D	4.90	0.58	2.70	-.17	3.39
E	0.27	-.53	-.17	-.74	0.10

^{131}I - labelled sera:

Fig. 1. Uptake quotients for the 25 paired labelled mixtures (PLMs) (made by combination of the five labelled sera) and HKLY 28 cell smears.

with the different isotopes. The values of Q can be arranged in decreasing order of size horizontally, or in increasing order of size vertically, to give a relative order of strength of the sera; this can be done for each of the five rows and five columns to give 10 classifications of the sera into their order of strength for each line. The sera are finally classified by plotting the uptake of each serum by every cell line in the form of histograms of the position in the order of strength versus the number of times the serum appears in that position. An example of such histograms for two sera (taken from a panel of six in a different experiment from the above) and two different cell lines is shown in Fig. 2. It can be seen that clear changes in the relative strengths of certain sera can be observed when different lines are used as antigen. This suggests that lines may vary in their antigenic content, and that different sera may contain qualitatively different antibodies. By means of a computer, it has been possible to determine the minimum number of antigen/antibody pairs needed to explain the observed differences in the relative order of

strengths of the same sera for different cell lines. An alternative approach is to estimate the weight of labelled globulin fixed from each serum on to a standardized quantity of antigen, as described above. Table 2 shows an example of such corrected weight uptakes by a cell line (HKLY 28) cultured at three different temperatures for 7 days before being used as antigen. It can be seen that, whereas the fixing of some sera (especially No. 3) remains invariant, other sera attach much more strongly to cells cultured at one temperature than at either of the other two. Serum No. 1 is taken up more strongly

Table 2. Weights of labelled protein taken up from five labelled sera by smears of HKLY 28 cells cultured at three different temperatures

Temperature	Weight taken up from Serum No:				
	1	2	3	4	5
35°C	9.04	2.25	6.63	0.00	6.00
37°C	6.91	2.92	6.78	0.00	5.98
39°C	5.85	3.57	6.82	0.00	9.28

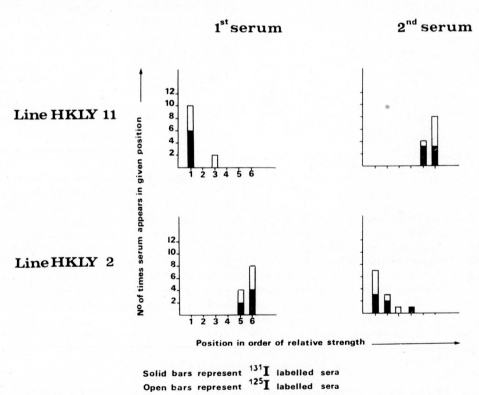

Fig. 2. Example of the change in relative order of strength of two sera when tested on two different cell-lines.

by the cells cultured at 35°C than by cells cultured at either 37°C or 39°C, whereas serum No. 5 fixes preferentially on to the cells cultured at 39°C. This behaviour can be explained by postulating three different antigens, as shown in Table 3A, and three corresponding antibodies, as shown in Table 3B.

When more cell lines are examined, some 7 antibody/antigen pairs are found to be necessary to explain the observed uptakes. The antigenic composition of 12 lines, as determined by this method, is shown in Table 1.

Table 3A.   Antigens expressed by HKLY 28
at three different culture temperatures

Temperature	"Antigens"		
	α	β	γ
35°C	+	+	−
37°C	+	−	−
39°C	+	−	+

Table 3B.   Origin and antibody contents of the five sera
used in the HKLY 28 experiment

Serum	Origin [1]	"Antibodies"			IFA [2]
		α	β	γ	
1	NPC	+ +	+	−	1280
2	BL	+	−	±	640
3	NPC	+ +	−	−	2560
4	EC	−	−	−	<10
5	BL	+ +	−	+	5120

[1] NPC = nasopharyngeal carcinoma.
BL   = Burkitt's lymphoma.
EC   = European control.
[2] Titre in the indirect fluorescein-labelled antibody test.

DISCUSSION

Several correlations between the antigenic patterns of the different cell lines, as determined by this method, and other properties of these lines can be observed. The antigen referred to as A is always expressed by the lines containing a detectable percentage of cells positive by the immunofluorescence test, and never by the negative lines. Whether this reactivity corresponds to the viral capsid antigenicity, the membrane antigenicity, or to a combination of both, is at present not clear. Another antigen—referred to as F—is also correlated with the immunofluorescent behaviour of the cells, but this time negatively. This reactivity of the immuno-fluorescence-negative lines seems to disappear when the highly purified IgG fraction of the sera, rather than the crude 50% saturated ammonium sulphate precipitated fraction, is used. This could either mean that the antibodies to this antigen are not of the IgG class, or that the cells have a receptor site for some serum protein present in the crude fraction. Both A and F reactivities are easily detected whether computer analysis or weight uptake analysis is used.

A very interesting reactivity seen when weight uptake analysis is used is that referred to as C. This reactivity has been found in all NPC-derived cell lines, regardless of their immunofluorescent behaviour, and has never been seen in any lines derived either from BL or IM. Computer analysis has not confirmed this reactivity, but has been unable to rule it out completely. Other antigenicities are required by either form of analysis to explain all the details of the observed fixations, but these have not as yet been found to correlate with specific characteristics of the cells.

An interesting result is that illustrated in Tables 2, 3A, & 3B. Ambrosioni & de-Thé [1] have shown that lines of different origin differ in the modification of their immunofluorescent behaviour on incubation at different temperatures. NPC-derived lines, of which HKLY 28 is one, show a higher percentage of fluorescent cells on cultivation at both 35°C and 39°C than they do if they are cultivated at 37°C. By the radiolabelled antibody technique, it appears that an antigenicity present at 35°C is absent at higher temperatures, and that another present at 39°C is absent at lower temperatures. The 35°C cells and the 39°C cells both have supplementary determinants, different from one another, that are absent from the 37°C cells. A further interesting finding is that antibodies to one of these supplementary determinants were seen only in the BL sera, while antibodies to the other were present only in one of the NPC sera; only two sera each from NPC and BL were examined, however, in this experiment.

[1] See p. 318 of this publication.

ACKNOWLEDGEMENTS

This investigation was sponsored by Contract No. NIH-70-2076, within the Special Virus Cancer Program, with the National Institutes of Health, Department of Health, Education, and Welfare, USA.

REFERENCES

Epstein, M. A. & Achong, B. G. (1967) Formal discussion: Immunologic relationships of the herpes-like EB virus of cultured Burkitt lymphoblasts. *Cancer Res.,* **27,** 2489-2493

Greenwood, F. C., Hunter, W. M. & Glover, J. S. (1963) The preparation of I-131-labelled human growth hormone of high specific radioactivity. *Biochem. J.,* **89,** 114-123

Henle, W., Henle, G., Zajac, B. A., Pearson, G., Waubke, R. & Scriba, M. (1970) Differential reactivity of human sera with EBV-induced "early antigens". *Science,* **169,** 188-190

Klein, G., Pearson, G., Nadkarni, J. S., Nadkarni, J. J., Klein, E., Henle, G., Henle, W. & Clifford, P. (1968) Relation between Epstein-Barr viral and cell membrane immunofluorescence of Burkitt tumor cells. I. Dependance of cell membrane immunofluorescence on presence of EB virus. *J. exp. Med.,* **128,** 1011-1020

Old, L. J., Boyse, E. A., Oettgen, H. F., Deharven, E., Geering, G., Williamson, B. & Clifford, P. (1966) Precipitating antibody in human serum to an antigen present in cultured Burkitt's lymphoma cells. *Proc. nat. Acad. Sci. (Wash.),* **56,** 1699-1704

Pope, J. H., Horne, M. K. & Wetters, E. J. (1969) Significance of a complement-fixing antigen associated with herpes-like virus and detected in the Raji cell line. *Nature (Lond.),* **222,** 186-187

Tanigaki, N., Yagi, Y. & Pressman, D. (1967) Application of the paired label antibody technique to tissue sections and cell smears. *J. Immunol.,* **98,** 274-280

zur Hausen, H., Henle, W., Hummeler, K., Diehl, V. & Henle, G. (1967) Comparative study of cultured Burkitt tumor cells by immunofluorescence, autoradiography, and electron microscopy. *J. Virol.,* **1,** 830-837

Analysis of EBV-carrying Burkitt's Lymphoma Cell Line Using Density Gradients of Gum Acacia

T. OSATO,[1] K. SUGAWARA [1] & F. MIZUNO [1]

INTRODUCTION

It has been shown by comparative immuno-fluorescence and electron microscopy investigations that a small fraction of the cells in most continuous cultures derived from Burkitt's lymphomas harbour the Epstein-Barr virus (EBV) (Epstein, Achong & Barr, 1964; Rauscher, 1968). Clonal analyses of these carrier cultures, however, have subsequently suggested that the EBV genome may be present in all the cells (Hinuma & Grace, 1968; Zajac & Kohn, 1970). This viral genome could be activated under appropriate environmental conditions, since viral synthesis is enhanced by maintaining cultures at lower temperatures or by incubating with an arginine-deficient medium (Hinuma et al., 1967; Osato & Ito, 1968; Henle & Henle, 1968).

The present work was undertaken to determine the virus-cell relationship in EBV-carrier cultures by separating, if possible, cells infected with EBV from uninfected cells. A gum acacia density-gradient method can be used for this purpose, and a virological analysis of a cloned Burkitt's lymphoma cell line has been made in this way. The preliminary data obtained have been reported (Sugawara, Mizuno & Osato, 1971).

MATERIALS AND METHODS

Cells

The P3HR-1 cell line (Hinuma et al., 1967), which was derived from Burkitt P3J as a cloned subline, was used. The cells were maintained in Eagle's minimum essential medium supplemented with 10% foetal calf serum and 10% tryptose phosphate broth.

---

[1] Department of Virology, Cancer Institute, Hokkaido University School of Medicine, Sapporo, Japan.

Density-gradient centrifugation

Gum acacia was used to prepare solutions of different densities, as described by Kimura, Suzuki & Kinoshita (1960), but with a slight modification. A discontinuous gradient with densities of 1.030, 1.040, 1.050, and 1.060 g/ml was loaded with approximately 10^7 P3HR-1 cells resuspended in a small amount of phosphate-buffered saline (PBS) containing 0.002 M disodium ethylenediamine tetraacetate (EDTA), and centrifuged at 1000 g for 25 minutes. Several zones of opalescence that formed at the density interfaces were collected, and the cells in each zone were used in the present studies.

Cell counts and cytological staining

Cells were counted in a haemocytometer chamber, using 0.1% trypan blue. For cytological studies, they were stained with Giemsa solution.

Immunofluorescence techniques

The indirect staining method was used as described by Henle & Henle (1966) for intracellular fluorescence and by Klein et al. (1966) for membrane fluorescence, respectively. The human sera used in these investigations were from a Japanese patient with nasopharyngeal carcinoma and from a normal Japanese subject. The first contained antibodies to membrane antigen (Klein et al., 1966), capsid antigen (Henle & Henle, 1966), and to early antigen (Henle et al., 1970), whereas the second contained anti-membrane and anticapsid antibodies, but not anti-early antibody. These antigens are known to be associated with EBV. Antihuman γ-globulin rabbit serum conjugated with fluorescein isothiocyanate (FITC) was used as labelled antibody. The stained preparations were examined with an Olympus fluorescence microscope with a BG 12 exciter filter. The light source was an Osram HBO 200 lamp.

The specificity of the positive reactions was established by the following tests: (1) NC-37 cells, an EBV-negative lymphoblastoid cell line derived from a human donor without evidence of malignant disease, were treated with selected serum followed by treatment with FITC-conjugated rabbit antihuman γ-globulin antibody; (2) P3HR-1 cells were treated with immunofluorescence-negative human serum or PBS followed by the conjugated antibody; (3) tests for isoantibody reactivity in the sera used; (4) confirmation of the reactivities of the sera used with EB virions and P3HR-1 cell surface by the immuno-ferritin technique (Sugawara & Osato, 1970).

RESULTS

Differences in the buoyant densities of cloned P3HR-1 cells on discontinuous gum acacia gradients

When the cloned Burkitt P3HR-1 cells were centrifuged on discontinuous gum acacia gradients, several opalescent zones were clearly evident at the density interfaces. Different percentages of cells showing intracellular fluorescence were found in the different zones. A large number of cells at the top of the tubes (zone I at the EDTA-PBS-1.030 g/ml interface) were intensely positive (Fig. 1), and considerable numbers of cells in zone II at the 1.030–1.040 g/ml interface fluoresced strongly as well. In contrast, in the middle of the tubes (zone III at the 1.040–1.050 g/ml interface), specific intracellular fluorescence appeared only in a small number of cells (Fig. 2). Lower down, zone IV (at the 1.050–1.060 g/ml interface) again contained numbers of positive cells and a number of stained cells were also found in zone V (at the bottom of the tubes). These results are summarized in Table 1.

Table 1. Distribution of P3HR-1 cells showing intracellular immunofluorescence in gum acacia gradients [1]

Zone	IIF+ cells [2] (%)	Viability (%)
0 [3]	5–20	70–95
I	50–70	80–95
II	20–40	>95
III	<2	>95
IV	10–20	30–50
V	5–10	<10

[1] Summarized data from several experiments.
[2] Immunofluorescence-positive cells in acetone-fixed preparations (intracellular immunofluorescence-positive cells); stained with NPC serum.
[3] Initial cell sample.

Cell morphology was also found to be different in the different zones. Distinct cytopathological changes, characterized by marked enlargement and multinucleated giant cell formation, were prominent in zone I (Fig. 3). Such changes were also frequently observed in zone II. The majority of the cells in the lower zones were destroyed and stained with trypan blue, and large numbers of dead cells sedimented out at the bottom of tubes. Cells in zone III, in contrast, appeared to be quite healthy. They were round in shape and uniform in size (Fig. 4).

Density changes of P3HR-1 cells with continued incubation

Density changes in the P3HR-1 cells were then studied during continued incubation. Fresh P3HR-1 cultures were kept at the ordinary incubation temperature of 36°C for 10 days. With continued incubation, as cultures became more crowded, increasing numbers of cells were stained specifically. The proportion of cells showing intracellular fluorescence amounted to approximately 4% at 4 days, 10% at 6 days, and about 20% at 8 days after incubation. Centrifugation of the P3HR-1 cells on gum acacia gradients was carried out at 6 and 8 days of incubation, and total numbers of cells and numbers of immunofluorescence-positive cells in each zone were calculated. The majority of the cells were recovered from zone III at 6 days, whereas at 8 days after incubation more cells were obtained from the remaining zones and there was a considerable decrease in the numbers in zone III. In the older cultures, fluorescence-positive cells clearly increased in number in zone I, zone II and the lower zones. These results are shown in Table 2.

Density distribution of P3HR-1 cells showing different types of immunofluorescence

The P3HR-1 cells separated on gum acacia density gradients were submitted to both intracellular and membrane immunofluorescence tests. The percentages of membrane-fluorescence-positive cells and of intracellular-fluorescence-positive cells calculated for each zone, were usually in agreement in zones I, II, IV, and V. In contrast, in zone III, a lack of agreement between these two immunofluorescence tests was clearly observed. Many more membrane-fluorescence-positive cells were seen. The difference was most striking in the early stages of incubation, and the percentage of cells showing intracellular

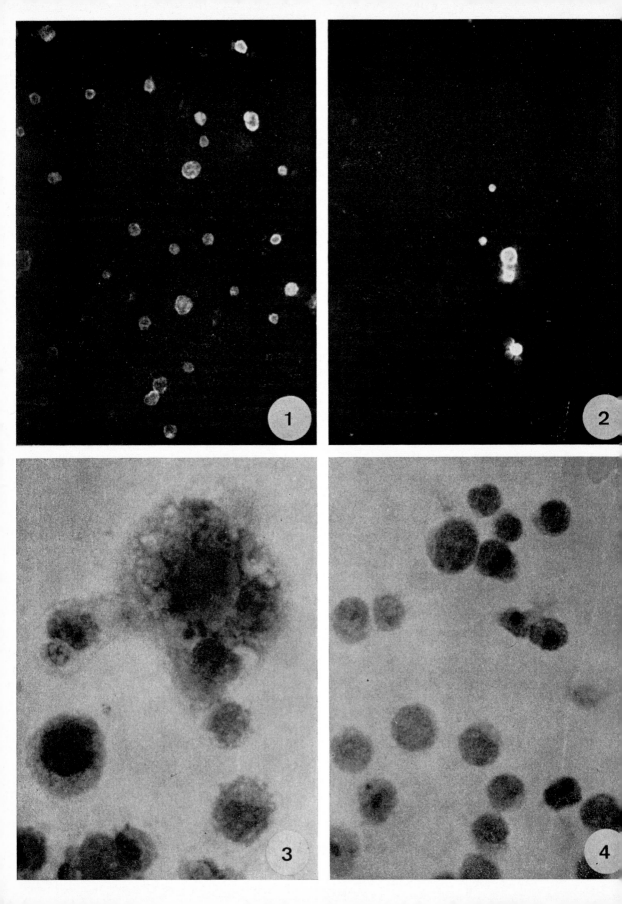

fluorescence rather rapidly reached the level for membrane immunofluorescence as incubation continued. These results are shown in Table 3.

Another difference was noted in intracellular reactivities. Again in zone III, in the early stages of incubation, intracellular fluorescence was produced only by NPC serum but not by normal serum. No significant difference was found between the reactivity with NPC serum and that with normal serum in the remaining 4 zones at any time. These results are shown in Table 4.

DISCUSSION

Methods utilizing differences in the buoyant densities of cells have been extensively used for separating one cell type from another (Cutts, 1970). We have observed certain pleomorphic characteristics of cells in EBV-carrying cell lines (Osato & Ito, 1968) that might suggest differences in the specific gravities of these cells.

The findings of the present study of a clonal Burkitt line seem to indicate that the cells may

Table 2. Total numbers of cells and number of intracellular-fluorescent cells recovered from various zones after density-gradient centrifugation for different incubation periods

No. of cells ($\times 10^5$)	Period of incubation (days)	No. of cells ($\times 10^5$)				
		Zone I	Zone II	Zone III	Zone IV	Zone V
Total	6	1.8	9.7	63.2	10.1	50.2
	8	3.4	34.8	36.0	13.9	70.1
Intracellular-fluorescent [1]	6	1.0	2.9	2.5	0.8	3.2
	8	2.0	7.9	3.6	2.6	5.3

[1] Calculated from the percentage of immunofluorescence-positive cells and the total number of cells in each zone.

Table 3. Distribution patterns of intracellular- and membrane-immunofluorescence-positive cells in gradients

Zone	IIF+ cells [1] (%)			MIF+ cells [2] (%)			MIF/IIF		
	Incubation period (days)			Incubation period (days)			Incubation period (days)		
	1	6	10	1	6	10	1	6	10
0 [3]	2.4	6.1	16.9	5.6	8.9	20.9	2.3	1.5	1.2
I	58.3	56.3	67.9	80.2	85.6	92.0	1.4	1.5	1.4
II	20.7	29.5	24.6	26.1	32.6	34.7	1.3	1.1	1.4
III	0.3	1.5	4.8	5.2	4.7	4.9	17.3	3.1	1.0
IV	16.6	12.8	20.8	18.4	14.1	30.4	1.1	1.1	1.5
V	5.3	7.0	8.6	6.2	7.7	6.1	1.2	1.1	0.7

[1] Immunofluorescence-positive cells in acetone-fixed preparations (intracellular-immunofluorescence-positive cells); stained with NPC serum.
[2] Membrane-immunofluorescence-positive cells; stained with NPC serum.
[3] Initial cell sample.

Fig. 1. Intracellular immunofluorescence of P3HR-1 cells in zone I. The majority of the cells are positive. ×90.

Fig. 2. Intracellular immunofluorescence of P3HR-1 cells in zone III. Only a small number of positive cells are seen, but large numbers of negative cells. ×90.

Fig. 3. Giemsa staining of P3HR-1 cells in zone I. Multinucleated giant cells are seen in large numbers. ×360.

Fig. 4. Giemsa staining of P3HR-1 cells in zone III. Cells are healthy and normal in morphology. ×360.

Table 4. Intracellular immunofluorescence
with different kinds of human sera

Zone	1 day after incubation		6 days after incubation	
	NPC serum	Normal serum	NPC serum	Normal serum
0 [1]	2.4 (1.7) [2]	0.9 (0.5) [2]	6.1	5.7
I	58.3	57.1	56.3	51.8
II	20.7 (16.2) [2]	15.2 (18.0) [2]	29.5	31.3
III	0.3 (1.0) [2]	0.01 (0.02) [2]	1.5	0.6
IV	16.6	14.9	12.8	10.1
V	5.3	5.1	7.0	7.2

[1] Initial cell sample.
[2] Data from another experiment.

initially have lower buoyant densities as EBV synthesis takes place, since a large proportion of cells showing immunofluorescence were present in the two upper zones; these cells also showed distinct cytopathological changes. A similar result has recently been reported in a study of infection with several different viruses, in which Ficoll gradients were used (Sykes *et al.*, 1970). Cells in the two lower zones also fluoresced considerably but the majority of the cells were dead. This suggests that the EBV-producing cells, initially of lower buoyant density, may increase in density as cellular degeneration proceeds. They may finally be sedimented out when dead. This hypothesis is supported by the fact that the number of cells in zone III, in which fluorescence-negative P3HR-1 cells were predominant, decreased markedly with continued incubation, whereas the numbers of cells in the other four zones increased continuously under the same conditions, with a concomitant increase in the number of fluorescence-positive cells. It is probable that the cells in EBV-carrier cultures may eventually die once viral replication occurs.

In the present investigations, zone III contained large numbers of morphologically intact fluorescence-negative cells and a very small number of fluorescence-positive cells. This suggests that cells initiating viral synthesis in EBV-carrier cultures may possibly move to zone III together with uninfected cells. Intracellular fluorescence in this zone in the early stages of incubation appeared to represent the early antigen recently reported by Henle *et al.* (1970), as judged from the reactivities with the sera used. The difference noted between the percentage of cells showing intracellular fluorescence and that of membrane-positive cells in the early stages of incubation suggests that the surface antigen may be synthesized prior to the early antigen in EBV replication.

It has been strongly suggested that the viral genome may be present in all cells of EBV-carrier cultures (Hinuma & Grace, 1968; Zajac & Kohn, 1970). Activation of this genome may occur as the cultures become crowded as well as at lower temperatures (Hinuma *et al.*, 1967; Osato & Ito, 1968) and with arginine deficiency (Henle & Henle, 1968); this may result in the synthesis in succession of membrane antigen, early antigen, and capsid antigen, followed by any cellular changes that may occur. These changes may be responsible for the density differences found in P3HR-1 cells. The cytopathological changes found in this study were characterized by marked cell enlargement and multinucleated giant cell formation, and were consistent with those reported recently by Durr *et al.* (1970) in human lymphoblastoid cells acutely infected with EBV.

It is known that cells producing EBV are apparently limited in number in the virus-carrying Burkitt lines so far established (Rauscher, 1968). The present gum acacia gradient method may be used not only in collecting the cells infected with EBV but also in separating cells carrying different antigens associated with this particular virus.

SUMMARY

Cells of EBV-carrying P3HR-1, a clonal line of Burkitt P3J, were centrifuged in discontinuous gum acacia gradients. Several zones of opalescence appeared at the density interfaces and the cells in each zone were studied by means of immunofluorescence techniques with selected human sera. It was striking that the majority of the cells in zone I at the medium-1.030 g/ml interface were intensely fluorescent inside the cells and at the cell surface, and showed distinct cytopathological changes. Cells in zone II at the 1.030–1.040 g/ml interface also fluoresced considerably. In contrast, in zone III at the 1.040–1.050 g/ml interface, only a small number of cells showed intracellular fluorescence whereas many more membrane-fluorescence-positive cells were seen. This intracellular fluorescence appeared to represent early antigen and the cells in this zone were of normal morphology. Cells in zone IV at the 1.050–1.060 g/ml interface were positive to a large extent, but the majority were stained with trypan blue. Dead cells usually sedimented out at the bottom of the tubes. The implications of these findings and the possible applications of the present density-gradient method are discussed.

ACKNOWLEDGEMENTS

We thank Dr Y. Hinuma, Tohoku University, and Dr E. M. Jensen, Pfizer Inc., Maywood, for the generous supply of the P3HR-1 cells and the NC-37 cells. This work was supported in part by grants from the Ministry of Education, Ministry of Health and Welfare, Naito Research Foundation, the Eisai Company, and the Chiyoda Insurance Company.

REFERENCES

Cutts, J. H. (1970) *Cell separation, methods in hematology,* New York & London, Academic Press, p. 114

Durr, F. E., Monroe, J. H., Schmitter, R., Traul, K. A. & Hirshaut, Y. (1970) Studies on the infectivity and cytopathology of Epstein-Barr virus in human lymphoblastoid cells. *Int. J. Cancer,* 6, 436-449

Epstein, M. A., Achong, B. G. & Barr, Y. M. (1964) Virus particles in cultured lymphoblasts from Burkitt's lymphoma. *Lancet,* i, 702-703

Henle, G. & Henle, W. (1966) Immunofluorescence in cells derived from Burkitt's lymphoma. *J. Bact.,* 91, 1248-1256

Henle, W. & Henle, G. (1968) Effect of arginine-deficient media on the herpes-type virus associated with cultured Burkitt tumor cells. *J. Virol.,* 2, 182-191

Henle, W., Henle, G., Zajac, B. A., Pearson, G., Waubke, R. & Scriba, M. (1970) Differential reactivity of human sera with EBV-induced "early antigens". *Science,* 169, 188-190

Hinuma, Y., Konn, M., Yamaguchi, J., Wudarski, D. J., Blakeslee, J. R., Jr. & Grace, J. T., Jr (1967) Immunofluorescence and herpes-type virus particles in the P3HR-1 Burkitt lymphoma cell line. *J. Virol.,* 1, 1045-1051

Hinuma, Y. & Grace, J. T., Jr (1968) Cloning of Burkitt lymphoma cells cultured *in vitro. Cancer,* 22, 1089-1095

Klein, G., Clifford, P., Klein, E. & Stjernswärd, J. (1966) Search for tumor-specific immune reactions in Burkitt lymphoma patients by the membrane immunofluorescence reaction. *Proc. nat. Acad. Sci. (Wash.),* 55, 1628-1635

Kimura, E., Suzuki, T. & Kinoshita, Y. (1960) Separation of reticulocytes by means of multi-layer centrifugation. *Nature (Lond.),* 188, 1201-1202

Osato, T. & Ito, Y. (1968) Transformation *in vitro* of human embryo cells by human leukemic culture fluid. I. Isolation and establishment of transformed cells. *Proc. Japan Acad.,* 44, 89-94

Rauscher, F. J., Jr (1968) Virologic studies in human leukemia and lymphoma: The herpes-type virus. *Cancer Res.,* 28, 1311-1318

Sugawara, K. & Osato, T. (1970) An immunoferritin study of a Burkitt lymphoma cell line harboring EB virus particles. *Gann,* 61, 279-281

Sugawara, K., Mizuno, F. & Osato, T. (1971) Density changes of cultured Burkitt's lymphoma cells following EB viral synthesis. *Nature New Biol. (Lond.),* 233, 106-107

Sykes, J. A., Whitescarver, J., Briggs, L. & Anson, J. H. (1970) Separation of tumor cells from fibroblasts with use of discontinuous density gradients. *J. nat. Cancer Inst.,* 44, 855-864

Zajac, B. A. & Kohn, G. (1970) Epstein-Barr virus antigens, marker chromosome, and interferon production in clones derived from cultured Burkitt tumor cells. *J. nat. Cancer Inst.,* 45, 399-406

A Precipitating Antigen in Human Milk:
Its Relation to Herpesvirus Antigens*

W. M. GALLMEIER, G. GEERING, C. HERTENSTEIN & E. TITZSCHKAU

Milk has been shown by Nowinski *et al.* (1967, 1968) to be a convenient source of precipitating antigen related to tumour or leukaemia viruses in the mouse. Since the Epstein-Barr virus (EBV), a herpes-type virus, is the first virus associated with certain malignancies in man, namely Burkitt's lymphoma (Epstein, Achong & Barr, 1964; Henle & Henle, 1969) and carcinoma of the nasopharynx (Henle & Henle, 1970), we tried to study the occurrence of precipitating antigen related to this virus in human milk. We used a human serum with known precipitating activity for the antigen derived from the Burkitt tissue culture cell line P3J (Old, 1966).

The precipitation reactions were performed in Ouchterlony plates of commercial origin (Hyland patterns C and D, 2% agar) as well as in slides containing 0.7% agar in 0.9% NaCl containing 7.5% glycine and 0.1% sodiumazide. The antiserum used was a serum from a normal blood donor, blood group O Rh+, the EBV titre of which in the Henle fluorescence test (Henle & Henle, 1967) was 1 : 20. This antiserum formed one line of precipitation when plated against the P3J cell line derived soluble antigen, which gave a line of identity with the standard precipitating antibody of Old.

The milk was provided by the local milk bank. Specimens were collected under sterile conditions by trained nurses from healthy donors previously screened clinically. All samples were tested for bacterial growth and with minor exceptions were found to be sterile. The milk samples were then stored at −70°C for up to 3 years but were otherwise unprocessed.

We tested 187 samples from different donors, of which 68 specimens showed a clear precipitation reaction between 6 and 10 days after plating. This represents an incidence of 37%. Generally only one line of precipitation was seen, but in rare instances double lines were observed. All reactions gave lines of identity with each other indicating that the same precipitating antigen was being detected. Of particular interest was the very strong reaction obtained with a milk specimen from a woman with Hodgkin's disease.

Many attempts to concentrate the antigen were made in order to intensify the reactions and to shorten the reaction time. The best reactions were seen when concentrated skim milk fractions, with or without ultracentrifugation, were used (Fig. 1). Precipitation occurred as early as 12 hours after plating. Here again splitting of the lines or double lines were sometimes observed.

The following investigations were carried out with concentrated milk from three different milk donors showing lines of identity with each other. The women had 1, 2 and 3 children respectively. There was no history of serious illness. The grandmother of one donor had breast cancer.

Preliminary studies of the physicochemical properties of the milk antigen showed the following. The antigen was heat-resistant since it retained its precipitating activity after 30 minutes exposure to 56°C.

Reactivity was still present in the supernatant after 90 minutes centrifugation at 80 000 g. There was, however, some precipitating activity in the pellet. Twelve hours centrifugation at 150 000 g, however, deleted its precipitating reactivity. Treatment with ether or acetone neither increased nor decreased its reactivity. The antigen showed a better migration in 0.7% agar than in 2% agar,

---

* From the Innere Klinik und Poliklinik (Tumorforschung) Essen, der Ruhruniversität Bochum, Federal Republic of Germany and Division of Immunology, Sloan-Kettering Institute for Cancer Research, New York, N.Y., USA.

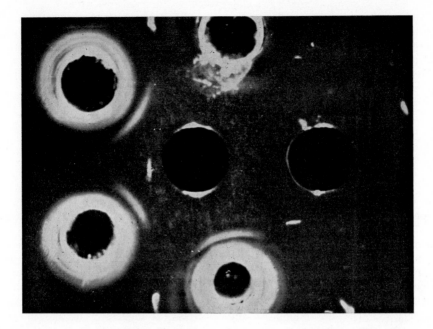

Fig. 1.   Milk reactions in 2% agar.   Centre well: positive human serum.   Outer wells: milk samples.

where it remained close to the antigen well. On Sephadex column chromatography, it appeared to migrate with the macromolecules.

Antibody to the milk antigen (s) was often found in the serum of patients with precipitating antibody to the Burkitt herpesvirus (EBV) (see Table 1). Preliminary surveys showed that the occurrence of antibodies to the milk antigen(s) appeared to be unrelated to the presence of precipitating antibody to type-specific antigens of herpes simplex or cytomegalovirus. Two of 6 reference herpesvirus group-specific antisera (see Kirkwood, Geering & Old, 1971 [1]) reacted with positive milk specimens. These two reference group-specific antisera, namely rabbit anti-herpes-simplex nucleocapsids and rat anti-herpesvirus Lucké nucleocapsids, gave a reaction of nonidentity with milk antigen and group-specific herpesvirus antigen, which was confirmed by absorption studies.   Thus the milk antigen is not the group-specific herpesvirus antigen.

[1] See p. 479 of this publication.

Table 1.   Comparison of serum reactivity against the soluble Burkitt antigen (P3J), milk antigen and a soluble herpes simplex antigen [1]

Diagnosis	Soluble Burkitt antigen	Human milk antigen	Soluble herpes antigen
Nasopharyngeal cancer	6/6	2 [2]/6	3/3 [3]
Nasopharyngeal cancer [4]	10/13	2 [2]/13	N.D.
Chronic lymphatic leukaemia	8/10	5(3) [2]/10	3/3 [2]
Cancer of the cervix	24/28	7(5) [2]/28	19/19 [3]
Lung cancer	5/15	0/15	14/14
Breast cancer	1/9	1 [2]/9	N.D.
Cancer of the ovary	6/14	3 [2]/21	N.D.

[1] Numbers positive/numbers tested.   N.D. = not done.
[2] Same serum positive for Burkitt antigen and milk antigen.
[3] Same serum positive for Burkitt antigen.
[4] Hong Kong.

The major milk precipitin band gives a line of non-identity with the major type-specific precipitin line of Burkitt herpesvirus when plated against a human type-specific antiserum precipitating the soluble Burkitt herpesvirus antigen. However, one preparation of Burkitt herpesvirus (P3J) cell antigen, but not other batches of antigen, shared a common antigen.

There was no relation between the major type-specific components of herpes simplex and the milk antigens, since we recorded lines of nonidentity. Furthermore, absorption of rabbit anti-herpesvirus nucleocapsid antiserum with the milk antigen removed the milk reaction but not the reaction against the soluble type-specific herpesvirus antigen. The rabbit anti-herpesvirus nucleocapsid antiserum, after absorption with normal human tonsils, lost its activity against a cytomegalovirus-soluble antigen but not against milk.

It thus appears that the milk antigen is not one of the common known major human herpesvirus type-specific antigens. The distribution of antibodies against the milk antigen(s) is shown in Table 2.

There is no clear-cut pattern discernible in the distribution of positive sera among patients with

Table 2.   Reactivity of sera against the milk antigen

Diagnosis	Numbers positive/ Numbers tested	%
Normal	4/83	5
Breast cancer	4/30	13
Cancer of the cervix	16/58	28
Acute leukaemia	8/30	27
Chronic leukaemia [1]	6/30	20
Hodgkin's disease	7/30	23

[1] Myelogenous and lymphatic.

different malignant diseases. Normals seem to have a distinctly lower frequency than patients with malignant disease. It is however noteworthy that the strongest precipitating reactions against milk antigens were seen in the group of patients with carcinoma of the cervix, where the highest incidence of reaction was also found (Fig. 2).

In summary, we here report a new precipitating antigen in human milk having some relation to herpesvirus antigens. This antigen can be detected with human sera. The frequency of the antigen in normal human milk is about 37%. The frequency of antibodies is 5% in normal blood donors and between 20–30% in patients with malignant disease.

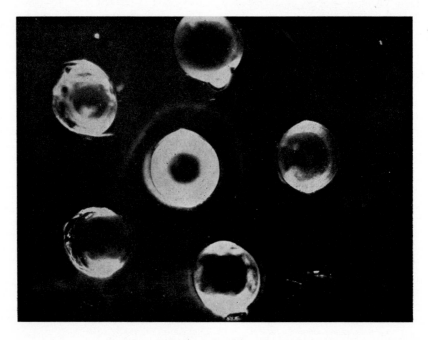

Fig. 2.   Reactions of sera from patients with carcinoma of the cervix (0.7% agar).   Centre well: milk antigen.   Outer wells: milk samples.

The antigen, although detected by antisera with group- and type-specific reactivity for herpesviruses so far does not appear to be one of the known major type-specific or group-specific antigens of the class of human herpesviruses. The milk antigen may therefore represent a new and previously undescribed type-specific herpesvirus antigen. Further investigations are needed to analyze this antigenic system thoroughly and to define its relation to benign or malignant disease in man.

ACKNOWLEDGEMENTS

We thank Dr B. Roizman, Departments of Microbiology and Biophysics, University of Chicago, for the provision of herpesvirus antigens, Dr A. Dortmann, Kinderklinik, Klinikum Essen, and Dr W. Lubold, Blutbank am Klinikum Essen, for co-operation in providing milk samples and normal blood specimens respectively. This work was supported by Deutsche Forschungsgemeinschaft and Landesamt für Forschung, NRW, and in part by NCI grant CA 08748 and a grant from the John A. Hartford Foundation, Inc.

REFERENCES

Epstein, M. A., Achong, B. G. & Barr, Y. M. (1964) Virus particles in cultured lymphoblasts from Burkitt's lymphoma. *Lancet,* **i**, 702-703

Henle, G. & Henle, W. (1969) Present status of herpes-group virus associated with cultures of the hematopoietic system. *Perspect. Virol.,* **6**, 105

Henle, W. & Henle, G. (1970) Evidence for a relation of Epstein-Barr virus to Burkitt's lymphoma and nasopharyngeal carcinoma. In: Dutcher, R. M., ed., *Comparative leukemia research,* Basel, Karger, pp. 706-713

Henle, G. & Henle, W. (1967) Immunofluorescence in cells derived from Burkitt's lymphoma, *J. Bact.,* **91**, 1248-1256

Nowinski, R. C., Old, L. J., Moore, D. H. & Geering, G. (1967) A soluble antigen of the mammary tumor virus. *Virology,* **31**, 1-14

Nowinski, R. C., Old, L. J., Boyse, E. A., de Harven, E. & Geering, G. (1968) Group-specific viral antigens in the milk and tissues of mice naturally infected with mammary tumor virus or Gross leukemia virus. *Virology,* **34**, 617-629

Influence of Temperature on the Percentage of Virus-producing Cells in Various Lymphoblastoid Cultures

J. C. AMBROSIONI [1] & G. DE-THÉ [1]

INTRODUCTION

The rôle of temperature in the synthesis of the herpesvirus present in the various lymphoblastoid cultures obtained from tumour biopsies from patients with nasopharyngeal carcinoma (NPC) and Burkitt's lymphoma (BL), or from the peripheral blood of patients with infectious mononucleosis (IM) was investigated. The percentage of fluorescent cells, detected by the indirect immunofluorescence method on fixed cells (Henle & Henle, 1966), was used as an indicator of the percentage of virus-producing cells (Henle & Henle, 1967).

MATERIAL AND METHODS

The following established lines were used in the present study:

1. NPC-derived lines: HKLY-11, HKLY-12, HKLY-28, HKLY-34;

2. BL-derived lines: Jijoye, Esther, Silfere;

3. IM-derived lines: IM-63, IM-71, Scheizer, Kaplan.

Cells (5×10^5 per ml) in RPMI 1640 medium (Grand Island Biological Co., Grand Island, N. Y.) supplemented with 1 or 5% foetal calf serum were incubated for seven to fourteen days at four temperatures (33°C, 35°C, 37°C and 39°C). The percentage of fluorescent cells was determined on duplicate coded smears and the reading was made on approximately 2000 cells for each line at each temperature.

[1] Unit of Biological Carcinogenesis, International Agency for Research on Cancer, Lyon, France.

RESULTS

1. NPC cultures: As shown in the Figure (A & B), the percentage of fluorescent cells was maximal when the cells were incubated at either 35°C or 39°C. These results were observed only in smears, reacted with sera from NPC or IM patients. If the smears were incubated with a BL patient's serum, only a single peak was observed at 37°C. (Ambrosioni & de-Thé [2]).

2. BL cultures: The results obtained are shown in the Figure (C & D) (Jijoye and Esther cultures). The highest percentage of fluorescent cells was obtained when the cells were incubated either at 33°C or 35°C. The percentage of fluorescent cells decreased with increasing temperatures to a minimum at 39°C. No difference was noted when the smears were treated with either NPC, BL or IM patients' sera.

3. IM cultures: The results obtained are given in the Figure (E & F) (Scheizer and Kaplan cultures): a single optimum was observed when the cultures were incubated at 37°C. Similar results were obtained in a culture derived from a nonmalignant nasopharyngeal mucosa. No modifications were observed when the smears were treated with NPC or IM patients' sera, but when the smears were treated with a BL serum, very few or no fluorescent cells were detected.

DISCUSSION

Although variations in the percentage of fluorescent cells for the different temperatures are not very great, a specific pattern emerged for all NPC, BL, or IM cultures. Some cultures were studied

[2] Unpublished data.

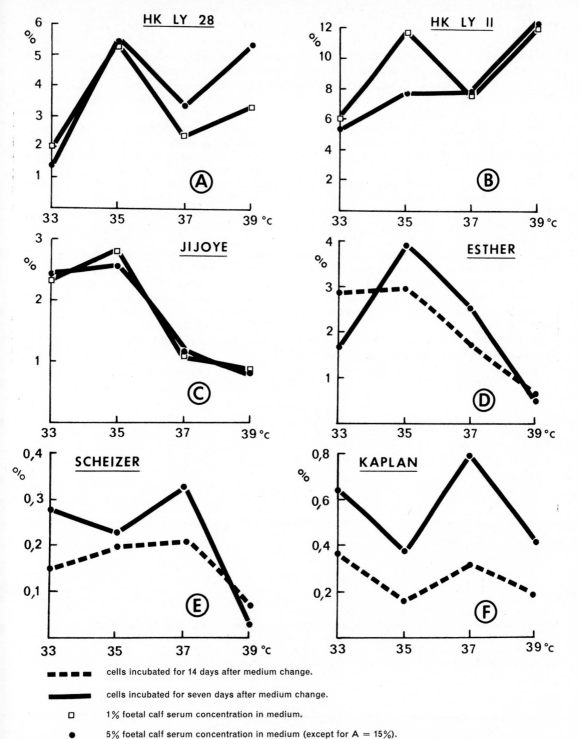

Fluctuations in the percentage of fluorescent cells (ordinate) in different lymphoblastoid cultures grown at 33°C, 35°C, 37°C and 39°C (abcissa).

A and B: NPC-derived cultures (HKLY-28 and HKLY-11). C and D: BL-derived cultures (Jijoye and Esther). E and F: IM-derived cultures (Scheizer and Kaplan).

repeatedly, and regularly showed the same behaviour. Statistical analysis of the results by Dr N. Day (Unit of Biostatistics, IARC, Lyon) showed that the differences between NPC, BL and IM patients' sera were statistically significant; the differences between both NPC and IM, and BL and IM were significant at the 0.1% level, and that between NPC and BL at the 0.5% level. The differences are independent of the ethnic source of the cells, since NPC-derived cultures from Chinese and African patients behave similarly. Furthermore, preliminary results on the effect of exposure to high temperature (45°C) for a short period (30 minutes) on subsequent virus yield (following the technique of Vonka, presented at the European Tumour Virus Group Meeting in Bad Wimpfen, 1971) showed that the proportion of IF-positive cells increased four-fold in Jijoye culture (BL-derived line) but remained unchanged in HKLY-28 (NPC-derived line).

The differences reported here may be related to the properties of either the cell type or the virus present in the various cultures. For example, the effect of temperature on the multiplication of poliovirus appears to depend on intracellular lysosomal enzymes, which mediate the degradation of viral nucleic acid (Lwoff, 1969). Thus the differences we report may be due to the presence of different cell types in the cultures derived from NPC, BL or IM. Temperature-sensitive mutants of many viruses have, however, been isolated and their optimal temperature for replication seems to be a viral property. The effect of temperature on the production of Lucké tumour herpesvirus is well known (see review by Granoff [1]) and the polyhedral cytoplasmic deoxyribo virus (PCDV) has an optimal temperature for replication that is the same in fish-, bird- or mammal-derived cultures. The effect of temperature on virus synthesis in NPC, BL or IM cultures may therefore indicate that these diseases are associated with distinct, although closely related, viruses. This would be in line with the situation in chickens and mice, where related RNA viruses induce different types of tumour or leukaemia.

---

[1] See p. 171 of this publication.

SUMMARY

Two temperatures, 35°C and 39°C, were found to be optimal for virus production in NPC-derived cultures, whereas in BL-derived and IM-derived cultures, the optimal temperatures were 35°C and 37°C respectively. These findings may reflect differences either in cell types or in virus strains.

ACKNOWLEDGEMENTS

This investigation was supported by Contract No. NIH-70-2076 within the Special Virus Cancer Program, National Institutes of Health, Department of Health, Education, and Welfare, USA.

REFERENCES

Gravell, M. & Granoff, A. (1970) Viruses and renal carcinoma of *Rana pipiens*. IX. Influence of temperature and host cell on replication of frog polyhedral cytoplasmic deoxyribovirus (PCDV). *Virology,* **41**, 596-602

Henle, G. & Henle, W. (1966) Immunofluorescence in cells derived from Burkitt's lymphoma. *J. Bact.,* **91**, 1248-1256

Henle, G. & Henle, W. (1967) Immunofluorescence, interference, and complement fixation technics in the detection of herpes-type virus in Burkitt tumor cell lines. *Cancer Res.,* **27**, 2442-2446

Lwoff, A. (1969) Death and transfiguration of a problem. *Bact. Rev.,* **33**, 390-403

Detection of Epstein-Barr Viral Genomes in Human Tumour Cells by Nucleic Acid Hybridization

H. zur HAUSEN [1] & H. SCHULTE-HOLTHAUSEN [1]

In previous experiments we tried to demonstrate Epstein-Barr virus (EBV) nucleic acid in the "virus-free" Raji line of Burkitt tumour origin by nucleic acid hybridization (zur Hausen & Schulte-Holthausen, 1970). By annealing Raji-cell-DNA with purified fragmented radioactive EBV-DNA, we showed that the viral DNA hybridized specifically with Raji-DNA, in contrast to DNA controls of other human origin. Hybridization of tritium-labelled cellular RNA with nonlabelled EBV-DNA resulted in greater binding of Raji RNA to viral DNA, as compared to other human RNAs. These results were taken as evidence for the persistence of EBV in Burkitt tumour cells in a masked form, by exerting at least some genetic activity.

These experiments were extended to biopsies derived from Burkitt tumours and anaplastic carcinomas of the nasopharynx (zur Hausen et al., 1970). Both types of tumours were reported to be associated with EBV infection, as shown by serological and tissue culture studies. Hybridization of radioactive EBV-DNA with DNA derived from these tumours, as well as with DNA from other human tumours of the head and neck, gave greater hybridization of the former, as compared with the various controls.

The test, however, suffered from certain disadvantages. The main difficulty could be attributed to the low yields of viral DNA from relatively large volumes of EBV-synthesizing cell cultures. This was in part due to the extensive purification procedures necessary for the isolation of viral nucleic acids. The maintenance of virus-synthesizing cells in the same medium for prolonged periods of time in order to obtain optimal viral harvests posed another problem: labelling with ^{3}H-thymidine at low isotope concentrations was necessary for cell survival. This, in turn, led to viral DNA preparations of low specific radioactivity, thus limiting the resolving power of the hybridization tests. In addition, variations in the specific radioactivities of the viral DNA used in different experiments made comparisons between the results difficult.

To avoid these difficulties and to improve the resolution, purified EBV-DNA was transcribed by Escherichia coli RNA polymerase in the presence of radioactive nucleosidetriphosphates (zur Hausen et al.[2]). The resulting complementary RNA (cRNA) was purified after DNase treatment and phenol extraction by Sephadex G-50 column chromatography, and characterized by sucrose velocity sedimentation. Nonradioactive cellular RNA was used as a sedimentation marker. The cRNA banded at approximately 9–10 S. After subjecting this cRNA to Cs_2SO_4 equilibrium centrifugation, two peaks were observed: a light peak banding at a density of about 1.6 gm/cm³ and an irregularly shaped heavy peak at a density of approximately 1.66 gm/cm³. Boiling of the cRNA for 10 minutes followed by rapid cooling prior to the density-gradient centrifugation abolished banding of one of the peaks. This result suggested that part of the transcribed RNA was self-complementary and double-stranded. Treatment of the synthetic RNA with RNase provided further support for this interpretation. About 30% of the RNA proved to be RNase-resistant. This percentage was reduced to 4% by boiling of the RNA for 10 minutes prior to RNase digestion.

Hybridizations with untreated and with heat-denatured cRNA demonstrated that denaturation increased the hybridization efficiency by about 30%. About 30% of the cRNA was therefore double-

---

[1] Institut für Virologie der Universität Würzburg, Würzburg, Federal Republic of Germany.

[2] Unpublished data.

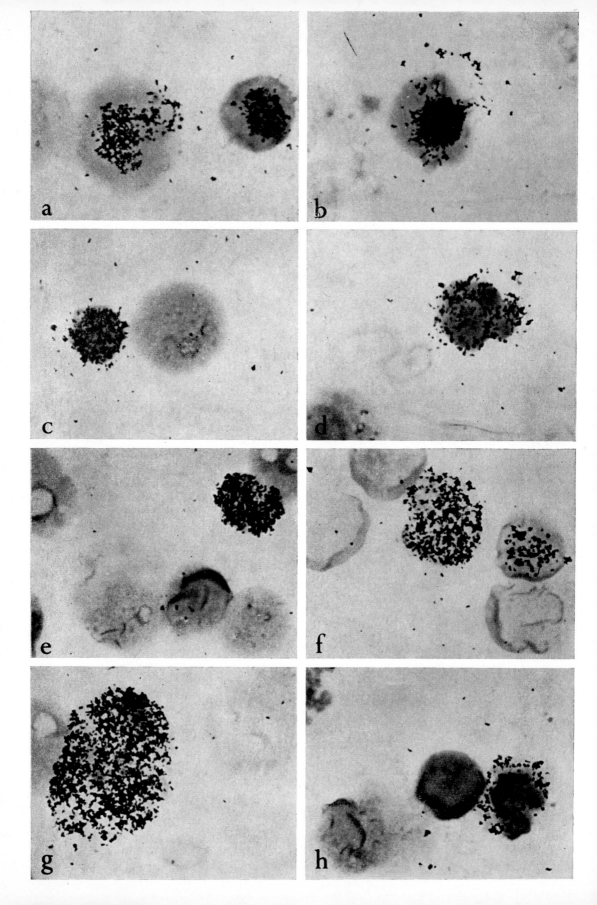

stranded. The reason for this amount of double-stranded transcription of EBV-DNA *in vitro* is not known.

All the following experiments were performed with heat-denatured cRNA. The resolving power of hybridizations with this RNA was tested in reconstruction experiments. Artificial mixtures of 25 μg human KB-cell-DNA and various amounts of EBV-DNA were annealed with constant inputs of EBV-cRNA. Hybridization increased linearly with increasing concentrations of viral DNA in the range 1–30 ng. At 80 ng, the curve levelled off slightly. These results indicated that hybridization with EBV-cRNA permitted the detection of as little as 1–2 ng of EBV-DNA. At an input of 25μg of cellular DNA, this would correspond to 2–3 viral genome equivalents per cell, provided that complete genomes of the virus were present within these cells. With this reservation, we are therefore able to calculate concentrations of viral DNA from hybridized counts annealed under identical conditions.

Various amounts of DNA derived from EBV-synthesizing P3HR-1 cells, "EBV-free" Raji cells, both of Burkitt tumour origin, as well as DNA from human and hamster cell controls, were annealed with equal amounts of EBV-cRNA. An almost linear increase in hybridized counts of EBV-cRNA was observed up to approximately 25 μg of DNA from P3HR-1 and Raji cells. Thereafter, the curves levelled off, reaching a plateau between 50 and 100 μg in the case of Raji DNA. At an input of 100 μg, the P3HR-1-DNA annealed at lower counts (8000 counts/min) than at an input of 50 μg (11 000 counts/min). This reduced hybridization was not due to quenching of radioactivity. Unavoidable renaturation of virus-specific and cellular DNA at these DNA concentrations according to a zipper mechanism is a possible explanation. Human and hamster cell controls hybridized slightly above the values given by blank filters.

These results clearly confirmed the data previously presented and obtained by DNA-DNA hybridization for the "virus-free" Raji cells (zur Hausen & Schulte-Holthausen, 1971). They showed, however, that the original calculation of viral-genome equivalents per Raji cell, previously estimated to be about 6, was much too low. This was due to the high DNA inputs used in those experiments. The number of viral-genome equivalents per Raji cell should be in the range 45–55. In view of the high molecular weight of EBV-DNA, this would indicate that about 0.1 % of the total Raji-DNA is virus-specific. This value exceeds those reported for viral DNA in animal-virus-induced tumour cells.

EBV-synthesizing P3HR-1 cells, as well as virus-particle-free Raji cells, have been studied by *in situ* hybridization according to the technique of Pardue & Gall (1969) and Jones (1970). Human KB and hamster H-A12-7 cells served as controls. The DNA of cytological preparations was denatured and annealed with cRNA under cover slips. The slides were then washed, treated with RNase, and submitted to autoradiography. By 3 days after exposure, EBV-synthesizing P3HR-1 cells were heavily labelled and easily distinguishable from nonvirus-producing nuclei (Fig. 1). The percentage of cells showing immunofluorescence for viral capsid antigens correlated well with the number of heavily labelled nuclei. In some of the nuclei, the label was localized in clusters, suggesting the beginning of viral DNA replication. This was similar to what was seen in the autoradiographic pictures that we obtained four years ago by pulse-labelling of the same cells with tritiated thymidine (zur Hausen *et al.*, 1967). Some metaphases of P3HR-1 cells showed labelling after *in situ* hybridization. In this case the chromosomes were usually extremely pyknotic. Apparently EBV-DNA replication is not inhibited during mitosis; this parallels the results obtained with herpes simplex virus (Waubke, zur Hausen & Henle, 1968). Human adenoviruses, in contrast, do not replicate DNA late in G-2 and in mitosis (zur Hausen, 1967).

By exposing Raji cells, after *in situ* hybridization, for prolonged periods to autoradiography, interphase nuclei as well as metaphase chromosomes showed an increasing uptake of label. Numerous grains were seen in association with metaphase chromosomes 30 days after exposure to the emulsion (Fig. 2). No grains were observed in control preparations of human KB and hamster H-A12-7 cells subjected to the same procedure. Raji cell metaphase chromosomes frequently revealed iso-chromatid labelling (Fig. 2, arrows) which suggests the firm association of virus-specific DNA sequences with specific chromosomal sites. In addition, these figures demonstrate that virus-specific DNA is localized in the chromosomes of all size groups.

Fig. 1. *In situ* hybridization of EBV-cRNA with cytological preparations of P3HR-1 cells. EBV-synthesizing cells are heavily labelled.

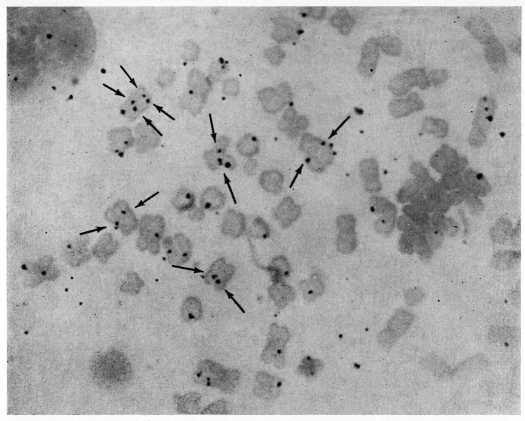

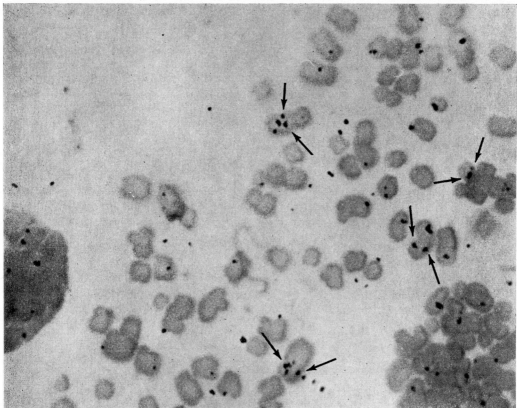

Fig. 2. *In situ* hybridization of EBV-cRNA with chromosomal preparations of Raji cells. After 30 days of exposure to the emulsion, numerous grains are found in association with the metaphase chromosomes. The arrows point to isochromatid labelling, which suggests a firm association of viral DNA with specific sites in the host-cell chromosomes.

ACKNOWLEDGEMENTS

This work was supported by the Deutsche Forschungsgemeinschaft, Bad Godesberg.

REFERENCES

Jones, K. W. (1970) Chromosomal and nuclear location of mouse satellite DNA in individual cells. *Nature (Lond.)*, **225**, 912-915

Pardue, M. L. & Gall, J. G. (1969) Molecular hybridization of radioactive DNA to the DNA of cytological preparations. *Proc. nat. Acad. Sci. (Wash.)*, **64**, 600-604

Waubke, R., zur Hausen, H. & Henle, W. (1968) Chromosomal and autoradiographic studies fo cells infected with herpes simplex virus. *J. Virol.*, **2**, 1047-1054

zur Hausen, H., Henle, W., Hummeler, K., Diehl, V. & Henle, G. (1967) Comparative study of cultured Burkitt tumor cells by immunofluorescence, autoradiography and electron microscopy. *J. Virol.*, **1**, 830-837

zur Hausen, H. (1967) Induction of specific chromosomal aberrations by adenovirus type 12 in human embryonic kidney cells. *J. Virol.*, **1**, 1174-1185

zur Hausen, H. & Schulte-Holthausen, H. (1970) Presence of EB virus nucleic acid in a "virus-free" line of Burkitt tumor cells. *Nature (Lond.)* **227**, 245-248

zur Hausen, H., Schulte-Holthausen, H., Klein, G., Henle, W., Henle, G., Clifford, P. & Santesson, L. (1970) EBV DNA in biopsies of Burkitt tumors and anaplastic carcinomas of the nasopharynx. *Nature (Lond.)*, **228**, 1056-1058

Molecular Events in the Biosynthesis of Epstein-Barr Virus in Burkitt Lymphoblasts

Y. BECKER [1] & A. WEINBERG [1]

INTRODUCTION

Burkitt lymphoblasts, isolated from human lymphoma patients (Epstein & Barr, 1964), were established *in vitro* as continuous cell lines. Under normal growth conditions, only a small number of the lymphoblasts contain herpes-type Epstein-Barr (EB) virions (Epstein, Achong & Barr, 1964) and virus-specific antigens (Henle & Henle, 1966; Epstein & Achong, 1968). A marked increase in the number of cells with EBV antigen was observed in lymphoblasts cultured in an arginine-deficient medium (Henle & Henle, 1968). From such cells, EB virions were isolated and characterized (Weinberg & Becker, 1969). The EB virions were found to contain one DNA genome each, of about 100×10^6 Daltons molecular weight, and an enveloped nucleocapsid made up of eight structural proteins (Weinberg & Becker, 1969). The EB virions resemble herpes simplex virions in their DNA (Becker, Dym & Sarov, 1968) and proteins (Olshevsky & Becker, 1970).

The ability of the human Burkitt lymphoblasts to synthesize EB virions when grown *in vitro* stems from the presence of EB viral genomes in the tumour cells, prior to *in vitro* establishment (zur Hausen *et al.*, 1970). These viral DNA genomes exist in the cells in a masked form but, under suitable conditions, can code for the synthesis of several viral proteins. The replication of the viral DNA molecules was found to be induced in EB3 cells within the initial 30 hours after arginine deprivation. The coating of these viral molecules is a slow process and occurs during a period of several days (Weinberg & Becker, 1969; Becker & Weinberg, 1971). At the time when

EB virions are detected, all the viral structural proteins can be found, by means of the acrylamide gel electrophoresis technique, in the cytoplasm of the lymphoblast. The presence of the viral proteins in the lymphoblasts was detected with the aid of antibodies present in Burkitt's lymphoma patients' sera. With this technique, several EBV-specific antigens were revealed: (1) soluble complement-fixing antigen (Old *et al.*, 1968); (2) "early" antigen (Henle *et al.*, 1970); (3) membrane antigen (Klein *et al.*, 1966; Klein *et al.*, 1968); and (4) capsid antigen (Henle & Henle, 1966; zur Hausen *et al.*, 1967). These viral antigens might be the structural and nonstructural EBV proteins, synthesized in the Burkitt lymphoblasts according to the genetic information in the viral genomes (Klein *et al.*, 1968). Previous studies (Weinberg & Becker, 1970; Becker & Weinberg, 1971) demonstrated that arginine deprivation resulted in the inhibition of cellular RNA and protein synthesis in the lymphoblasts and in the synthesis of EBV proteins. This effect of arginine made possible the study on the time course of EBV protein synthesis. The present study deals with: (*a*) the induction of EBV-DNA synthesis; (*b*) the sequence of "early" and "late" EBV protein synthesis; and (*c*) inhibition of EBV replication by the antibiotic distamycin A.

RESULTS

Synthesis of EBV-DNA genomes

Effect of arginine deprivation on EBV-DNA synthesis. Analysis by CsCl density-gradient centrifugation (Fig. 1) of the DNA molecules synthesized in arginine-deprived cells demonstrated the presence of two DNA species: (1) cellular DNA molecules with a density of 1.696 g/ml; and (2) EBV-DNA molecules with a density of 1.719 ± 0.001 g/ml

---

[1] Department of Virology, Hebrew University/Hadassah Medical School, Jerusalem, Israel.

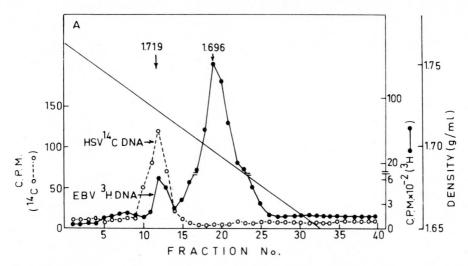

Fig. 1. Induction of EBV-DNA synthesis in arginine-deprived lymphoblasts. EB3 cells were resuspended in Eagle's medium without arginine (10^7 cells/10 ml). The cells were labelled with 10 μc of ³H-thymidine (specific activity 16 400 mCi/mM, obtained from the Radiochemical Centre, Amersham, England) and incubated for 30 hours at 37°C. The cells were harvested, the nuclei were isolated and resuspended in 0.04 M tris HCl buffer, pH 8.7, containing 0.04 M EDTA and treated with 1% (w/v) sodium deoxycholate (B.D.H., England) and 1 mg/ml pronase (Calbiochem, USA, B grade), and incubated overnight at 37°C. Herpes simplex virus DNA labelled with ¹⁴C-thymidine (specific activity 410 mC/mM) was used as a density marker. CsCl crystals were added to give the initial density of 1.69 g/ml. The gradient was centrifuged for 48 hours at 40 000 rpm at 10°C in the SW 65 rotor of the Beckman model 65 B ultracentrifuge. The tubes were punctured and the gradients collected drop-wise. The TCA precipitable radioactivity was collected on Millipore filters (0.45 μ) and counted in a Packard liquid scintillation counter. The density was determined by weighing 50 λ aliquots in a Mettler analytical balance.

(Becker & Weinberg, 1971). EBV-DNA molecules were not detected in CsCl density gradients of DNA molecules synthesized in EB3 lymphoblasts grown in complete medium.

To clarify the mechanism of induction of EBV-DNA synthesis in arginine-deprived lymphoblasts, the rôle of protein synthesis in this process was studied. Puromycin (100 μg/10^6 cells) was added to the deprived lymphoblasts at different time intervals, the cells were labelled with ³H-thymidine for 30 hours, and the DNA was analyzed in CsCl gradients. It was found (Table 1) that inhibition of protein synthesis at the time of arginine deprivation prevented EBV-DNA synthesis and also affected cell DNA synthesis. Inhibition of protein synthesis after arginine deprivation had only a partial inhibitory effect (Table 1). These results suggested that the replication of EBV-DNA genomes in the arginine-deprived lymphoblasts was dependent on the synthesis of new proteins, and that these were synthesized immediately after the cells were incubated in an arginine-deficient medium. The nature of these proteins, and their rôle in the induction of EBV-DNA, are currently being investigated.

Table 1. Effect of puromycin on EBV-DNA synthesis[1]

Time of puromycin addition (hours post arginine deprivation)	Radioactivity of EBV-DNA (counts/min)	Radioactivity of cell DNA (counts/min)
0	0	850
10	265	2400
20	640	2000

[1] Three cultures of Burkitt lymphoblasts were incubated in Eagle's medium without arginine. Puromycin (100 μg/ml) was added to one culture at the time of incubation (0 hours), to the second at 10 hours, and to the third at 20 hours after arginine deprivation. The cells were labelled with ³H-thymidine for 30 hours, and treated as described in the legend to Fig. 1; the EBV-DNA was obtained by centrifugation in CsCl gradients. The radioactivity in the viral and cellular DNA bands was determined.

Induction of EBV-DNA synthesis by mitomycin C. The effect of mitomycin C (Iyer & Szybalski, 1963, 1964) on the induction of EBV-DNA synthesis was studied in EB3 Burkitt lymphoblasts. The cells, in complete medium, were treated with mitomycin C (5 μμg/10^6 cells) (Yata *et al.*, 1970) in the presence of ³H-thymidine and incubated for a 30-hour period. Arginine-deprived, untreated, and actinomycin D

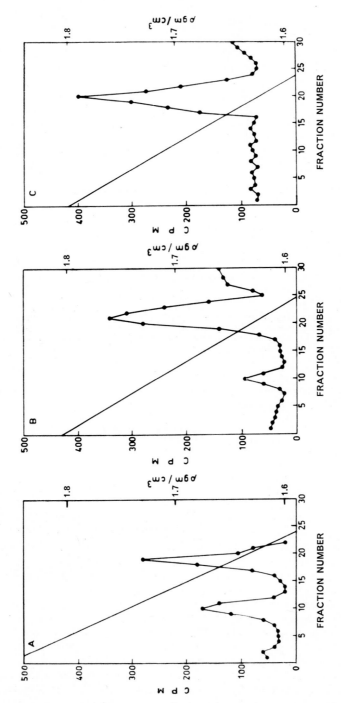

Fig. 2.   Induction of EBV-DNA synthesis by mitomycin C.   EB3 lymphoblasts were divided into 3 cultures (5 × 10⁶ cells/culture).   One culture was resuspended in arginine-deficient medium (A), the second in Eagle's medium containing 5 μμg/ml mitomycin C (Nutritional Biochemical Co., USA) (B), and the third culture in a medium containing 5 μμg of actinomycin D (Merck, Sharpe & Dohme, USA).   The cells were labelled with ³H-thymidine (10 μC/culture) and incubated at 37°C for 30 hours.   The nuclei were isolated and resuspended in 0.001 M sodium phosphate, pH 6.4.   The DNA was isolated from the nuclei by treatment with 1% (w/v) sodium deoxycholate, 1% (w/v) sodium dodecyl sulphate (Matheson, Coleman and Bell, USA) and 500 μg/ml pronase.   The DNA was centrifuged in CsCl gradients, as described in Fig. 1.   The cell DNA obtained by this technique has a lower (1.64 g/ml) density than the density of cell DNA (1.696 g/ml), isolated and analyzed in CsCl gradients described in Fig. 1, due to incomplete purification of cell DNA by the present technique.

treated (5 μμg/10⁶ cells) (Yata *et al.*, 1970) lympho-blasts, which served as controls, were also labelled. The DNA was isolated from the three cell samples and analyzed by centrifugation in CsCl density gradients. It was found (Fig. 2) that treatment of the lymphoblasts with mitomycin C induced the syn-thesis of EBV-like DNA (Fig. 2B), but to a lesser extent than in the arginine-deprived cells (Fig. 2A). Treatment of lymphoblasts with actinomycin D did not result in the synthesis of EBV-DNA (Fig. 2C). These results suggested that treatment with mito-mycin C induced the synthesis of EBV-DNA.

Lack of EB virion synthesis in mitomycin C treated lymphoblasts. To determine whether the EBV-like DNA molecules synthesized in mitomycin C treated cells were also incorporated into EB virions, the formation of the latter was studied. It was found that no complete EB virions were synthesized in the mitomycin C treated cells (Fig. 3B), while EB virions were synthesized in the arginine-deprived cells (Fig. 3A). This suggested that mitomycin C in-duced the replication of either fragmented EBV-DNA genomes, or prevented the late transcription of the viral genomes. Arginine deprivation, how-ever, made possible the normal replication of the viral DNA genomes and the synthesis of all the viral structural proteins (Becker & Weinberg, 1971).

Sequential synthesis of EBV structural proteins

Time course of EBV protein synthesis. In a previous study (Becker & Weinberg, 1971), it was found that only about 20% of the EBV-DNA was synthesized during the initial 10 hours after arginine deprivation; most of the EBV-DNA was synthesized during the subsequent 20 hours. It might thus be possible to characterize the viral proteins synthesized during and after EBV-DNA synthesis. The argi-nine-deprived lymphoblasts were labelled with radioactive leucine at various time intervals, and the cytoplasmic fractions were studied by electrophoresis in acrylamide gels. It was found (Fig. 4) that, dur-ing the initial five hours of arginine deprivation, only two proteins (designated VII and VIII) were synthesized. Protein VII was identified as the core protein of the herpes virion (Olshevsky & Becker, 1970), while the nature of protein VIII is not yet known. This protein was synthesized only during the initial five hours (Fig. 4), and might therefore be connected with the induction of EBV-DNA syn-thesis. During the period from 6 to 22 hours after arginine deprivation, the synthesis of protein VII continued, but concomitantly three additional viral

proteins (designated VI, V, and IV) were synthesized. These might be the glycoproteins that constitute the herpes virion's inner envelope (Fig. 4B). The viral protein II (and also protein IIa, which might repre-sent a precursor of protein II) and the glyco-protein III (present in the outer envelope of the

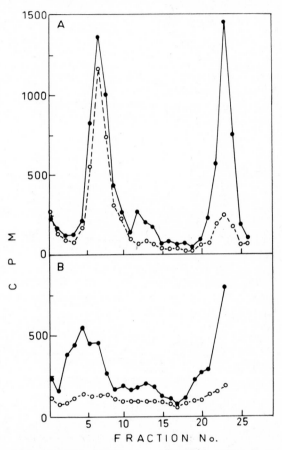

Fig. 3. Absence of EB virion synthesis in mitomycin C treated lymphoblasts. Two cultures of EB3 cells (5 × 10⁶ cells/culture) were resuspended in arginine-deficient medium (A) and in complete medium containing 5 μμg/ml mitomycin C (B) and labelled for 5 days with ³H-thymidine (10 μCi/culture). The cells were harvested and resuspended in 0.001 M phosphate buffer, pH 6.4, frozen and thawed 3 times at −70°C, and centrifuged for 5 min at 2000 rpm in a refrigerated PR2 centrifuge. The supernatant was centrifuged in sucrose gradients (30–60% w/w sucrose in phosphate buffer) in the SW 25.1 rotor for 40 min at 24 000 rpm. The gradients were collected and each fraction was divided into two equiv-alent portions: one was treated with a 30 μg/sample of deoxyribonuclease (Worthington, USA) prior to TCA treatment, and the other with TCA, and the radioactivity in each fraction was determined.

●———● radioactivity;

○— — — —○ DNase-resistant DNA.

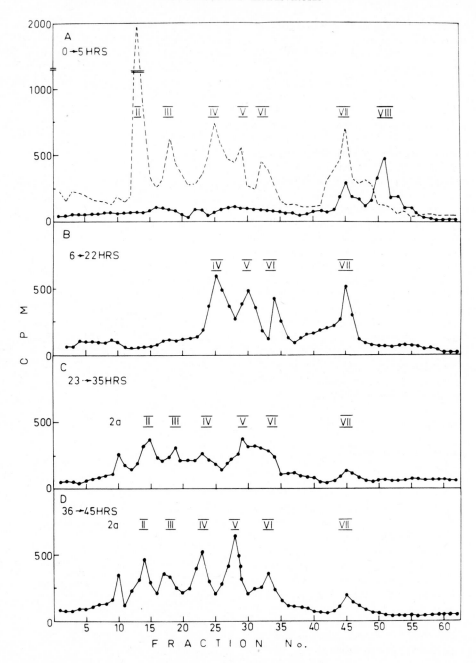

Fig. 4. Time course of EBV protein synthesis. Four cultures of EB3 cells (10⁷ cells/culture) were resuspended in arginine-deficient medium containing 1/100 of the normal leucine concentration. The cells were incubated at 37°C and at various time intervals a different culture was labelled with 5 μc of ¹⁴C-leucine (specific activity 312 mCi/mM) for a period of 12 hours. At the end of the labelling period, the cells were centrifuged and resuspended in a phosphate buffer containing 0.1% (v/v) Nonidet P-40, and the nuclei were centrifuged at 2000 rpm for 5 min. The cytoplasmic fractions were isolated, treated with sodium dodecyl sulfate, urea and 2-mercaptoethanol, and analyzed in acrylamide gels (Olshevsky & Becker, 1970). Purified herpes virions, labelled with ³H-leucine, were used as a marker.

– – –   – – – ³H leucine in herpes simplex virions; ————— ¹⁴C-leucine labelled cells.

herpes virion; Olshevsky & Becker, 1970), did not appear in the cytoplasm of the arginine-deprived lymphoblasts until 22 hours after arginine deprivation (Fig. 4C). The synthesis of all the structural proteins continued in the arginine-deprived lymphoblasts labelled during the period from 36 to 45 hours (Fig. 4D). These results suggested that viral proteins were synthesized in sequence, some being synthesized early and some late in the virus growth cycle. In order to define the "early" and "late" viral functions, the synthesis of proteins was studied in lymphoblasts in which EBV-DNA synthesis was prevented.

EBV proteins synthesized in the absence of viral DNA synthesis. The relation of the various viral proteins to the virus growth cycle was investigated in cytosine arabinoside (Pizer & Cohen, 1960) treated lymphoblasts (5 μg/10⁶ cells) (Gergely, Klein & Ernberg, 1971). The cells were labelled with radioactive leucine at various time intervals, and analyzed by electrophoresis in acrylamide gels (Fig. 5). It was found that, in the absence of EBV-DNA synthesis, proteins II and III were not synthesized, while proteins VII, VI, V and IV were synthesized. These results demonstrated that the last-mentioned pro-

teins were synthesized in the arginine-deprived lymphoblasts in the absence of viral DNA synthesis, and can therefore be regarded as "early" viral proteins. Proteins II and III, whose synthesis was dependent on viral DNA synthesis, represent "late" viral proteins.

The antibiotic distamycin A—an inhibitor of EBV synthesis

Effect of the antibiotic on the formation of EB virions. The antibiotic distamycin A (Fig. 6) (Cassazza et al., 1966) was found to interact with double- and single-stranded DNA molecules, and to prevent their transcription by RNA polymerase (Puschendorf & Grunicke, 1969). It was therefore of interest to determine whether distamycin A could interfere with the synthesis of EBV-DNA in arginine-deprived cells and prevent the formation of EB virions. Distamycin A and its analogue, compound II (which lacks the side-chain [R = COHNN] present in distamycin A) were dissolved in dimethylformamide (final concentration 0.4% in Eagle's medium) and added (50 μg/ml) in the arginine-deficient medium to the EB3 lymphoblasts. The untreated and distamycin A treated cells were

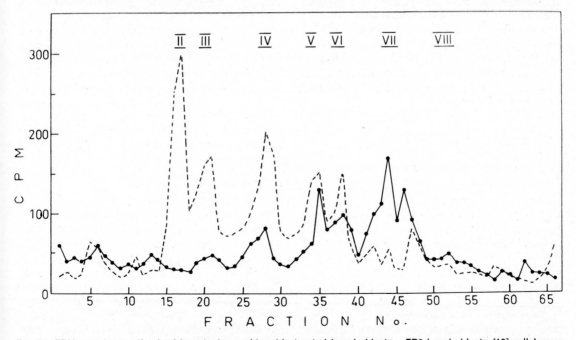

Fig. 5. EBV proteins synthesized in cytosine arabinoside treated lymphoblasts. EB3 lymphoblasts (10⁷ cells) were resuspended in arginine deficient medium containing 1/100 leucine and 5 μg/ml cytosine arabinoside (Sigma, USA). Labelling with ¹⁴C-leucine (312 mCi/mM) was initiated at 30 hours after arginine deprivation and continued for a 20-hour period. The proteins synthesized in the cells were analyzed by electrophoresis in acrylamide gels.

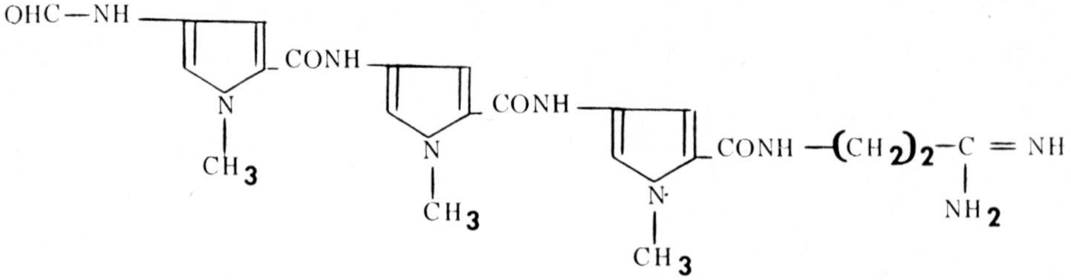

Fig. 6. Distamycin A.

labelled with ^{3}H-thymidine, incubated at 37°C, and the virions isolated in sucrose gradients. It was found that, in the untreated arginine-deprived cells (Fig. 7A), as well as in cells treated with compound II (Fig. 7B), EB virions were synthesized and could be detected in sucrose gradients. However, in the distamycin A treated lymphoblasts, EB virions were not found (Fig. 7C). This suggested that distamycin A, due to its side-chain, is capable of inhibiting the formation of EB virions in the arginine-deprived lymphoblasts.

Mode of antiviral activity of distamycin A. Analyses of the DNA species synthesized in distamycin A treated (50 μg/ml) arginine-deprived cells demonstrated that cell DNA alone, but no EBV-DNA molecules, were synthesized. Analysis of the proteins synthesized in distamycin A treated arginine-deprived cells demonstrated that only proteins VII and VIII ("early" viral proteins) were synthesized in the deprived cells. These findings support the observation (Fig. 7) that distamycin A inhibits the formation of EB virions.

DISCUSSION

State of the EBV genomes in the lymphoblasts

Zur Hausen et al. (1970) demonstrated by the nucleic acid hybridization technique (zur Hausen & Schulte-Holthausen, 1970) that the Burkitt lymphoblasts obtained from human Burkitt tumours contain up to 26 viral DNA genomes per cell.. It is not known, however, whether the viral DNA genomes are integrated into the cell DNA. However, these genomes were activated when the cells were incubated in an arginine-deficient medium, as determined by the synthesis of the EBV capsid antigen in about 75% of the cells (Henle & Henle, 1968). Although the mechanism of this process is not yet known, it was possible to conclude that arginine deprivation of the lymphoblasts: (a) affected the synthesis of

cellular macromolecules differently from that of other amino acids (Weinberg & Becker, 1970); (b) induced the synthesis of EBV-DNA (Becker & Weinberg, 1971); and (c) stimulated the synthesis of viral proteins. The fact that protein synthesis is necessary for the induction of EBV-DNA synthesis was demonstrated by the sensitivity of this process to puromycin treatment. The synthesis of a viral protein (VIII) at the time of induction of EBV-DNA synthesis may suggest that a change in the regulatory processes of the lymphoblasts is necessary to make EBV-DNA synthesis possible. The ability of mitomycin C to induce the replication of EBV-DNA may indicate that replication of EBV followed the formation of cross-linkages in the cell DNA molecules (Iyer & Szybalski, 1963, 1964). It is possible that, due to the effect of mitomycin C on specific regions of the cell's DNA, EBV-DNA molecules are released from cellular control. However, the EBV-DNA released under these conditions might be incomplete, as virions were not synthesized in the mitomycin C treated cells. Further work is still needed to elucidate the mechanism that induces the synthesis of EBV-DNA in the lymphoblasts and on the cellular mechanism that controls the EBV-DNA genomes in the lymphoblasts.

EBV protein and antigens

Burkitt lymphoblasts were found to contain four different EB viral antigens. The viral capsid (Henle & Henle, 1966) and membrane (Klein et al., 1968; Klein et al., 1969) antigens, were found to be absent from lymphoblasts that lack the viral genomes (Yata & Klein, 1969). Since it was possible to demonstrate that EBV structural proteins which were synthesized in arginine-deprived lymphoblasts (Becker & Weinberg, 1971), resembled the structural proteins of herpes simplex virions (Olshevsky & Becker, 1970), the time course of their synthesis was studied. It was concluded from the analyses of EBV proteins synthesized during various stages of the virus growth

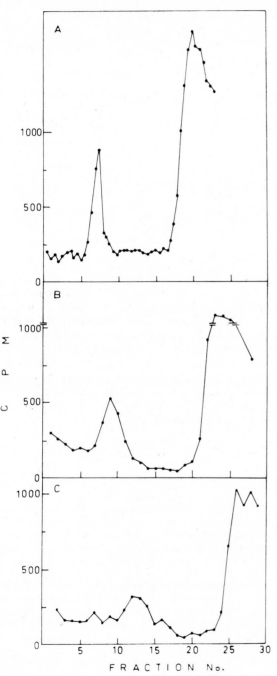

Fig. 7. Effect of distamycin A on the synthesis of EB virions. Three cultures of EB3 cells were set up (10[7] cells/culture) in arginine-deficient medium. One culture served as control (A); to the second, 50 μg/ml of compound II (an analogue of distamycin A; Farmitalia, Milan, Italy) (B), and to the third, distamycin A (Farmitalia, Italy)

cycle in the arginine-deprived cells, that the core protein (VII) and three viral envelope proteins (IV, V and VI) were synthesized prior to, or during the synthesis of EBV-DNA and could therefore be regarded as "early" viral proteins. Since proteins VII and VIII were the first to be synthesized, both in the presence and the absence of DNA synthesis, these proteins could be regarded as "early-early" viral proteins. Proteins IV, V, and VI, which are also synthesized in the absence of DNA synthesis, could be regarded as "early-late" viral proteins (Table 2). The capsid protein II and the glycoprotein III were synthesized after EBV-DNA synthesis and could therefore be regarded as "late" viral proteins. Since protein III is synthesized prior to protein II, the former was regarded as a "late-early" viral protein while the capsid protein II was regarded as a "late-late" viral protein (Table 2). This time course of EBV protein synthesis was taken to indicate that: (a) the newly synthesized proteins are functional viral components; and (b) specific early and late mRNA species are transcribed from the viral DNA genomes in arginine-deprived EB3 lymphoblasts. The synthesis of viral mRNA species in Burkitt cells is currently under investigation. The above-mentioned sequence of EBV protein synthesis is in agreement with the results of studies on the viral antigens in EBV-infected lymphoblasts, and in lymphoblasts treated with metabolic inhibitors (Yata & Klein, 1969; Yata et al., 1970). A correlation between the various EBV antigens and structural proteins is presented in Table 2. Further studies demonstrated that the sequence of EBV protein synthesis in the arginine-deprived Onesmas cell line of Burkitt lymphoblasts (Yata et al., 1970) resembled that of viral protein synthesis in arginine-deprived EB3 cells.

Distamycin A—an antiviral antibiotic

EBV has been shown to be associated with Burkitt's lymphoma (Henle & Henle, 1966), nasopharyngeal carcinoma (de Schryver et al., 1969; de-Thé et al., 1969) and infectious mononucleosis (Henle, Henle & Diehl, 1968). The synthesis of

(50 μg/ml) (C) were added. The antibiotics were dissolved in dimethylformamide and diluted to 1 mg/ml in the medium. The cells were labelled with ³H-thymidine (10 μCi/culture, incubated for 5 days at 37°C), harvested and treated as described in the legend to Fig. 3. EB virions were isolated in sucrose gradients (12–52% w/w made in tris-HCl-saline buffer, pH 7.4). The gradients were collected and the radioactivity of each fraction determined.

Table 2. Sequence of EBV protein synthesis in arginine-deprived Burkitt lymphoblasts

Viral proteins [1]		Time of synthesis (hours)			Proteins in relation to virus cycle	Correlation with viral antigens
Designation	Function	0 to 5	6 to 22	23 to 35		
VIII		VIII				
VII	Core protein	VII (VIIa)	VII	(VII) ⎱	Early-early	Soluble antigen [2] (CF)
VI	Glycoprotein ⎰ in the		VI	VI ⎱		Early antigen [3] (EA)
V	Glycoprotein ⎰ inner		V	V ⎰	Early-late	Membrane
IV	Glycoprotein ⎰ envelopes		IV	IV ⎰		antigen [4] (MA)
	⎰ in the					
III	Glycoprotein ⎰ outer			III	Late-early	Late membrane antigen [5]
	⎰ envelope					
II	Capsid			II		
IIa	Capsid precursor?			IIa	Late-late	Capsid antigen [6] (VCA)

[1] Olshevsky & Becker (1970); Weinberg & Becker (1969).
[2] Old et al. (1968).
[3] Henle et al. (1970).
[4] Klein et al. (1969).
[5] Not yet known.
[6] Henle & Henle (1966).

specific antibodies to EBV antigens in human cancer patients indicated that the replication of EBV might accompany the disease processes. The ability of distamycin A to inhibit the replication of EBV might therefore be of interest. Distamycin A is a basic oligopeptide antibiotic, isolated from cultures of *Streptomyces distallicus* and its structure was determined by chemical analysis (Arcamone *et al.*, 1964). The antibiotic molecules interact with native DNA (Puschendorf & Grunicke, 1969; Chandra, Zimmer & Thrum, 1970) and prevent its use as a template for the transcription of RNA and for DNA synthesis. In the present study it was found that distamycin A inhibited (at 50 μg/10^6 cells) the formation of EB virions in arginine-deprived lymphoblasts. At this concentration, the antibiotic had only a slight inhibitory effect on the macromolecular processes of HeLa cells. Since the replication of EBV in lymphoblasts is a slow process, and distamycin A inhibits EBV formation, it is possible that the antibiotic might be of value in the treatment of EBV infections, if it is not toxic to man.

SUMMARY

Incubation of Burkitt lymphoblasts (EB3 cells) in an arginine-deficient medium resulted in the induction of EBV-DNA replication during the period of cellular DNA synthesis. The replication of EBV-DNA, but not cellular DNA, was dependent on *de novo* protein synthesis and was sensitive to puromycin treatment. Treatment of the undeprived lymphoblasts with mitomycin C also resulted in the induction of EBV-DNA replication, but contrary to the effect of arginine deprivation, EB virions were not synthesized. The EBV structural proteins were synthesized in sequence in the arginine-deprived lymphoblasts. The viral protein VII (core protein), and proteins VI, V, and IV (viral envelope glycoproteins) were synthesized during the initial 24 hours. These proteins are regarded as the "early" viral proteins. Subsequently, protein III (the glycoprotein of the outer viral envelope), and protein II (capsid protein), were synthesized. The latter proteins were not synthesized in the absence of EBV-DNA synthesis in cytosine arabinoside treated cells, and were therefore regarded as "late" viral proteins. The formation of EB virions in arginine-deprived cells was inhibited by distamycin A, an oligopeptide antibiotic, which interferes with the synthesis of EBV-DNA.

ACKNOWLEDGEMENTS

We wish to thank Dr M. Ghione, Farmitalia, Milan, Italy, for the supply of distamycin A and compound II. The help of Mr Udy Olshevsky is greatly appreciated. This work was supported by grants from the Leukemia Research Foundation, Inc., Chicago, Illinois, and from the Israel Cancer Association.

REFERENCES

Arcamone, F., Penco, S., Nicolella, V., Orezzi, P. & Pirelli, A. M. (1964) Structure and synthesis of distamycin A. *Nature (Lond.)*, **203**, 1064-1065

Becker, Y., Dym, H. & Sarov, I. (1968) Herpes simplex virus DNA. *Virology*, **36**, 185-192

Becker, Y. & Weinberg, A. (1971) Burkitt lymphoblasts and their Epstein-Barr virus: synthesis of viral DNA and proteins in arginine deprived cells. *Israel J. med. Sci.*, **7**, 561-567

Cassazza, A. M., Fioretti, A., Ghione, M., Soldati, M. & Verini, M. A. (1966) Distamycin A, a new antiviral antibiotic. In: *Antimicrobial agents and chemotherapy-1965*, pp. 593-598

Chandra, P., Zimmer, C. & Thrum, H. (1970) Effect of distamycin A on the structure and template activity of DNA in RNA-polymerase system. *FEBS Letters*, **7**, 90-94

de Schryver, A., Friberg, S., Klein, G., Henle, W., Henle, G., de-Thé, G., Clifford, P. & Ho, H. C. (1969) Epstein-Barr virus-associated antibody patterns in carcinoma of the post-nasal space. *Clin. exp. Immunol.*, **5**, 443-459

de-Thé, G., Ambrosini, J. C., Ho, H. C. & Kwan, H. C. (1969) Lymphoblastoid transformation and presence of herpes-type viral particles in a Chinese nasopharyngeal tumor cultured *in vitro*. *Nature (Lond.)*, **221**, 770-771

Epstein, M. A., Achong, B. G. & Barr, Y. M. (1964) Virus particles in cultured lymphoblasts from Burkitt's lymphoma. *Lancet*, **i**, 702-703

Epstein, M. A. & Barr, Y. M. (1964) Cultivation *in vitro* of human lymphoblasts from Burkitt's malignant lymphoma. *Lancet*, **i**, 252-253

Epstein, M. A. & Achong, B. G. (1968) Specific immunofluorescence test for the herpes-type EB virus of Burkitt lymphoblasts, authenticated by electron microscopy. *J. nat. Cancer Inst.*, **40**, 593-607

Gergely, L., Klein, G. & Ernberg, I. (1971) The action of DNA antagonists on Epstein-Barr virus (EBV)-associated early antigen (EA) in Burkitt lymphoma lines. *Int. J. Cancer*, **7**, 293-302

Henle, G. & Henle, W. (1966) Immunofluorescence in cells derived from Burkitt's lymphoma. *J. Bact.*, **91**, 1248-1256

Henle, G. & Henle, W. (1968) Effect of arginine deficient medium on herpes-type virus associated with cultured Burkitt tumor cells. *J. Virol.*, **2**, 182-191

Henle, G., Henle, W. & Diehl, V. (1968) Relation of Burkitt's tumor associated herpes-type virus to infectious mononucleosis. *Proc. nat. Acad. Sci., (Wash.)*, **59**, 94-101

Henle, W., Henle, G., Zajac, B. A., Pearson, G., Waubke, R. & Scriba, M. (1970) Differential reactivity of human sera with EBV-induced "early antigens". *Science*, **169**, 188-190

Iyer, V. N. & Szybalski, W. (1963) A molecular mechanism of mitomycin action: linking of complementary DNA strands. *Proc. nat. Acad. Sci., (Wash.)*, **50**, 355-362

Iyer, V. N. & Szybalski, W. (1964) Mitomycins and porfiromycin: chemical mechanism of activation and cross-linking of DNA. *Science*, **145**, 55-58

Klein, G., Clifford, P., Klein, E. & Stjernswärd, J. (1966) Search for tumor-specific immune reactions in Burkitt lymphoma patients by the membrane immunofluorescence reaction. *Proc. nat. Acad. Sci., (Wash.)*, **55**, 1628-1635

Klein, G., Pearson, G., Nadkarni, J. S., Nadkarni, J. J., Klein, E., Henle, G., Henle, W. & Clifford, P. (1968) Relation between Epstein-Barr viral and cell membrane immunofluorescence of Burkitt cells. I. Dependence of cell membrane immunofluorescence on presence of EB virus. *J. exp. Med.*, **128**, 1011-1020

Klein, G., Pearson, G., Henle, G., Henle, W., Goldstein, G. & Clifford, P. (1969) Relation between Epstein-Barr viral and cell membrane immunofluorescence in Burkitt tumor cells. III. Comparison of blocking of direct membrane immunofluorescence and anti-EBV reactivities of different sera. *J. exp. Med.*, **129**, 697-706

Old, L. J., Boyse, E. A., Geering, G. & Oettgen, H. F. (1968) Serologic approaches to the study of cancer in animals and in man. *Cancer Res.*, **28**, 1288-1299

Olshevsky, U. & Becker, Y. (1970) Herpes simplex virus coat proteins. *Virology*, **40**, 948-960

Pizer, L. I. & Cohen, S. S. (1960) Metabolism of pyrimidine arabinonucleosides and cyclonucleosides in Escherichia coli. *J. biol. Chem.*, **235**, 2387-2392

Puschendorf, B. & Grunicke, H. (1969) Effect of distamycin A on the template activity of DNA in RNA polymerase system. *FEBS Letters*, **4**, 355-357

Weinberg, A. & Becker, Y. (1969) Studies on EB virus of Burkitt's lymphoblasts. *Virology*, **39**, 312-321

Weinberg, A. & Becker, Y. (1970) Effect of arginine deprivation on macromolecular processes in Burkitt's lymphoblasts. *Exp. cell. Res.*, **60**, 470-474

Yata, J. & Klein, G. (1969) Some factors affecting membrane immunofluorescence reactivity of Burkitt lymphoma tissue culture cell lines. *Int. J. Cancer*, **4**, 767-775

Yata, J., Klein, G., Hewetson, J. & Gergely, L. (1970) Effect of metabolic inhibitors on membrane immunofluorescence reactivity of established Burkitt lymphoma cell lines. *Int. J. Cancer*, **5**, 394-403

zur Hausen, H., Henle, W., Hummeler, K., Diehl, V. & Henle, G. (1967) Comparative study of cultured Burkitt tumor cells by immunofluorescence auto-radiography and electron microscopy. *J. Virol.*, **1**, 830-837

zur Hausen, H. & Schulte-Holthausen, H. (1970) Presence of EB virus nucleic acid homology in a "virus-free" line of Burkitt tumor cells. *Nature (Lond.)*, **227**, 245-248

zur Hausen, H., Schulte-Holthausen, H., Klein, G., Henle, W., Henle, G., Clifford, P. & Santesson, L. (1970) EBV DNA in biopsies of Burkitt tumors and anaplastic carcinomas of the nasopharynx. *Nature (Lond.)*, **228**, 1056-1058

Methylation Pattern of the DNA of Burkitt's Lymphoma Cells

E. D. RUBERY [1]

Methylation of DNA is generally considered to occur by the transfer of the methyl group from S-adenosyl methionine to a particular base in the polymer. 5-methylcytosine and 6-methylamino-purine are the only two minor methylated bases known to occur in DNA, and of these only 5-methylcytosine has been found in mammalian DNA. No minor methylated bases have been detected in the three animal viral DNAs so far studied, namely polyoma, herpes simplex and pseudorabies (Kaye & Winocour, 1967; Low, Hay & Keir, 1969; Low [2]).

Burkitt's lymphoma cells (EB3) were maintained in suspension culture in Eagle's minimal essential medium (Eagle, 1955) supplemented with 10% foetal calf serum, nonessential amino-acids and sodium pyruvate (Epstein & Barr, 1965). They were subcultured by the addition of an equal volume of fresh medium at intervals of about three days. Incubation with ³H-methylmethionine, (20 µc/ml) was carried out in a low methionine medium produced by the addition of an equal volume of methionine-deficient Eagle's medium plus supplements to cells ready for subculturing. Under these conditions, it was shown that DNA and RNA synthesis proceeded at the same rate as in control cultures for a period of 72 hours.

DNA was isolated from cells incubated with ³H-methylmethionine by the method of Burdon & Adams (1969). The DNA was hydrolyzed to bases in formic acid (Vischer & Chargaff, 1948), and the bases separated by chromatography on Whatman's P81 phosphocellulose paper using isopropanol/concentrated HCl/water (Rubery & Newton, 1971). 85% of the radioactivity in the bases was found in 5-methylcytosine, and the remaining 15% in thymine

[1] Department of Biochemistry, Cambridge, UK.
[2] Personal communication.

(Fig. 1). No radioactivity was detected in the position of a 6-methylaminopurine marker. The 5-methylcytosine and thymine spots were both eluted and rechromatographed in n-butanol/ammonia/water (Markham, 1955). Each base migrated as a homogeneous peak and both were therefore considered pure.

A further 50 µg of DNA labelled with ³H-methylmethionine was treated with 0.1 N H_2SO_4 for 60 minutes (Shapiro & Chargaff, 1960). This treatment removes purines from the DNA, leaving an apurinic acid. When the pH of the solution is raised above 10 by the addition of ammonia, the apurinic acid undergoes β-elimination reactions at the sites of purine loss, causing breakage of the deoxyribosephosphate backbone and release of pyrimidine oligonucleotides of the general structure $Py_nP_{(n+1)}$. These can be separated on Whatman's No. 1 paper in isobutyric acid/0.5N NH_3 (Shapiro & Chargaff, 1963). Under these conditions, the shorter oligonucleotides travel nearest to the fluid front, while the longer oligonucleotides remain closer to the origin. Fig. 2 shows the pattern obtained when the EB3-DNA oligonucleotides, labelled with ³H-methylmethionine, were treated in this manner, and compares the pattern to that obtained from the same DNA labelled with ³H-5-methylthymidine, which passes into the thymine of the DNA alone.

It is clear that the distribution of the thymine bases in the DNA is preferentially within long pyrimidine sequences, while that of 5-methylcytosine occurs frequently as short pyrimidine sequences within purine regions. This difference is even more marked when it is remembered that 15% of the label from the methionine passes into the thymine of the DNA, and the peak at the origin in the methionine-labelled DNA probably contains the majority of this thymine.

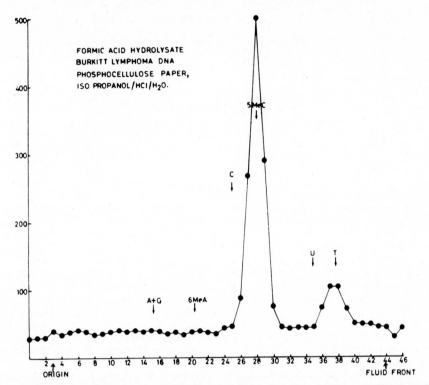

Fig. 1. Chromatography of a formic acid hydrolysate of Burkitt's lymphoma DNA on phosphocellulose paper. Markers for the seven bases indicated were added (0.5% in 0.1 N HCl; 10 μl sample). The position of the markers was detected by UV contact photography, (Markham, 1955); the paper was shredded into strips 1 cm × 3 cm, and each strip was placed in scintillant (3.5 g of 2 : 5 diphenyl oxazole and 50 mg of 1 : 4 2-(5-diphenyl oxazolyl) benzene in 1 litre of sulphur-free toluene) and counted in a Beckman L150 scintillation counter. Abbreviations: A = adenine; G = guanine; 6MeA = 6-methylaminopurine; C = cytosine; 5MeC = 5-methylcytosine; U = uracil; T = thymine.

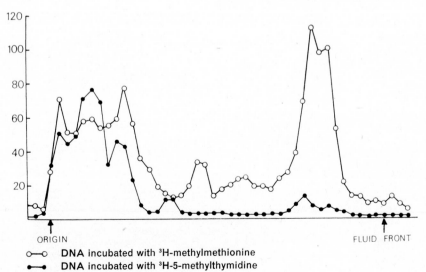

○—○   DNA incubated with ³H-methylmethionine
●—●   DNA incubated with ³H-5-methylthymidine

Fig. 2. 60-minute hydrolyses of Burkitt's lymphoma DNA in 50 μl of 0.1 N HCl.

SUMMARY

The DNA of Burkitt's lymphoma cells (EB3 line) has been shown to contain 5-methylcytosine as the only minor methylated base. The pattern of methylation of the DNA is nonrandom and shows that this minor methylated base tends to occur within short pyrimidine sequences more frequently than is the case with the major base thymine, which preferentially occurs within longer pyrimidine runs.

ACKNOWLEDGEMENTS

The author wishes to thank the Medical Research Council for a Junior Research Fellowship held while this work was in progress. Financial assistance towards the purchase of isotopes was also obtained from a grant from the Cancer Research Campaign to Dr A. A. Newton, Department of Biochemistry, Cambridge.

REFERENCES

Burdon, R. H. & Adams, R. L. P. (1969) The in vivo methylation of DNA in mouse fibroblasts. *Biochim. biophys. Acta*, **174**, 322-329

Eagle, W. R. (1955) Nutritional needs of mammalian cells in tissue culture. *Science*, **122**, 501-504

Epstein, M. A. & Barr, Y. M. (1965) The characteristics and mode of growth of a tissue culture strain (EB1) of human lymphoblasts from Burkitt's lymphoma. *J. nat. Cancer Inst.*, **34**, 231-240.

Kaye, A. M. & Winocour, E. (1967) On the 5-methyl-cytosine found in the DNA extracted from polyoma virus. *J. molec. Biol.*, **24**, 475-478

Low, M., Hay, J. & Keir, H. M. (1969) DNA of Herpes Simplex virus is not a substrate for methylation in vivo. *J. molec. Biol.*, **46**, 205-207

Markham, R. (1955) *Modern methods of plant analysis*, Vol. 4, Berlin, Springer-Verlag, pp. 246-304

Rubery, E. D. & Newton, A. A. (1971) A simple method for the separation of methylated adenines and cytosine from the major bases in nucleic acids. *Analyt. Biochem*, **42**, 149-154

Shapiro, H. S. & Chargaff, E. (1960) Studies on the nucleotide arrangement in DNA iv: patterns of nucleotide sequence in the DNA of rye germ and its fractions. *Biochim. biophys. Acta*, **39**, 68-82

Shapiro, H. S. & Chargaff, E. (1963) Studies on the nucleotide arrangement in DNA vii: direct estimation of pyrimidine nucleotide runs. *Biochim. biophys. Acta*, **76**, 1-8

Vischer, E. & Chargaff, E. (1948) The composition of the pentose nucleic acids of yeast and pancreas. *J. biol. Chem.*, **176**, 715-734

Immunofluorescent Studies of the Infection of Fibroblastoid Cells with Human Cytomegaloviruses

I. JACK [1, 2] & M. C. WARK [1]

Post-transfusion mononucleosis (Foster, 1966) has been associated with two herpesviruses, cytomegalovirus (CMV) (Foster & Jack, 1968), and Epstein-Barr-type virus (EBTV) (Gerber *et al.*, 1969). CMV infection may be a good model for understanding the cell-virus relationship with EBTV, and has the advantage of being easier to study *in vitro*. We have therefore studied the development of CMV antigens after infection of human diploid fibroblasts (Fb) with CMV.

METHODS

CMV strains RCH-234 and RCH-455 at passage levels of 170 and 70 were inoculated at high titre (approximate multiplicity 0.005 to 0.05) to coverslip cultures of well-spread, semiconfluent monolayers of local strain diploid Fb. After 60 minutes' adsorption of virus the cells were rinsed with saline and incubated at 36°C in basal medium Eagle containing 2% foetal calf serum, and were harvested at 12-hour intervals over a period of 96 hours. After having been washed free of medium, the cells were fixed in cold acetone (4°C) for 10 minutes and then examined by indirect immunofluorescence (IF) using convalescent sera diluted 1 : 4 and goat anti-human IgG (Hyland) diluted 1:8. The human sera were selected on the basis of positive complement-fixing antibody (CF) activity. In some later work we also used a rabbit anti-CMV serum prepared by the method of Mäntyjärvi (1968) and goat anti-rabbit globulin. All fluorescein-conjugated sera were absorbed with acetone-dried rabbit liver powder and gave no background staining of uninfected Fb, nor of infected Fb in the absence of

specific CMV reactive sera. Fluorescence was scored for intensity (+ to +++), with the + level being taken as that just discernible when the photomicrography prism was inserted in the light path. Microscopical examination was with a Leitz Ortholux illuminated by a Philips CS 150 mercury lamp fitted with a UG 1 2-mm exciter filter and a Leitz K 430 barrier filter. Darkground illumination was obtained with a Leitz 1.20 N.A. condenser oiled to the lower surface of the slide. This gave an incomplete field coverage for the × 40 objective, but a good intensity of ultraviolet illumination.

RESULTS AND CONCLUSIONS

The development of fluorescence in CMV-infected Fb was generally seen to commence in the nucleus. In contrast to the reports of others (McAllister *et al.*, 1963; Rapp, Rasmussen & Benyesh-Melnick, 1963) and either because of the serum used in this series (Mat.) or the virus strains used, antigen(s) were detected as early as 6 hours after initiation of infection. The nuclear fluorescence was at first diffuse, but later showed additional intensely fluorescent bodies about the size of the nucleoli (12–24 hours). It is not thought that these bodies are actually the nucleoli. By 48 hours, the small bodies had fused into the large nuclear inclusion body (NIB) characteristically seen in standard histological staining. Secondary involvement of adjacent Fb cells was also seen at this time.

Cytoplasmic staining of infected cells was first seen as a diffuse staining at 24 hours with serum Mat. and by 48 hours the cytoplasmic inclusion (CIB) was found to contain antigen. This CIB was present earlier, as judged by standard stains and as inferred from the eccentric position of the nucleus shown by

[1] Royal Children's Hospital, Parkville, Victoria, Australia.
[2] WHO Immunology Centre, University of Singapore.

IF. One serum (Hos.) of low CF titre consistently showed the CIB at an earlier stage (12 hours). In this case the CIB was seen as a finely granular structure which progressed to give a more solid fluorescence by 24 hours. At this time serum Hos. also became reactive with the NIB, but less intensely so than serum Mat.

In using several CF-negative human sera as negative controls, it was found that most were reactive with CIB but not with NIB antigen(s). A group of 49 sera from adults and children was examined for CF, IF-NIB and IF-CIB. Of these, 15 were positive in CF, 17 in IF-NIB, while all 49 were positive in IF-CIB. Thus two CF-negative sera were found positive by IF-NIB, suggesting an increased sensitivity of the latter test at this location. The 100% IF-CIB positivity was suspect.

No difference in the nature of the IF-CIB was detected between CF-positive and CF-negative sera with the exception of Hos., as above. In both groups the first few serum dilutions tended to show a $++$ IF-CIB but this fell to a $+$ level toward the end-points, which ranged from 1/32 to 1/1028. The IF-CIB was also detected when nonimmune rabbit serum was used in conjunction with goat anti-rabbit serum. It may therefore reflect a non-immunological uptake of serum globulins by the late-phase CIB. Fluorescence of this CIB can be blocked by the use of unlabelled anti-human IgG or anti-rabbit globulins before the finalization of the indirect IF with the respective labelled reagents.

Three sets of unwashed, acetone-fixed smears of peripheral blood of CMV-viraemic positive patients were available for examination after 2 years' storage at $-70°C$. To avoid the use of indirect IF with human sera, rabbit serum with specific IF reactivity at the NIB was conjugated with fluorescein and used directly. Prior to conjugation, this serum was found capable of specifically blocking the indirect IF reaction of human serum at the NIB. The labelled rabbit serum was reactive at the NIB to a titre of 1 : 64 and at the CIB to a titre of 1 : 128. As it appeared specific at the NIB site, it was used to examine the stored human blood smears. Only a few leukocytes were found to be fluorescent but the nucleus was negative and the cytoplasmic fluorescence diffuse. It is not known whether this represents nonspecific fluorescence or whether the serum is incapable of detecting the full range of nuclear antigens, at least as expressed in leukocytes. Further tests must await direct infection of leukocytes by CMV and collection of freshly harvested leukocyte smears from post-transfusion mononucleosis (PTM) patients.

It may be that the technique of using well-spread cells in the CMV-Fb system will be applicable to the IF reactions of EBTV-infected lymphoblasts, which frequently tend to be seen as shrunken cells in the usual smears. Initial studies suggest that a 1-minute exposure of such cell suspensions to a 1 : 3 dilution of saline produces a nondisruptive hypotonic swelling of the lymphoblasts.

REFERENCES

Foster, K. M. (1966) Post-transfusion mononucleosis. *Aust. Ann. Med.*, **15**, 305-310

Foster, K. M. & Jack, I. (1968) Isolation of cytomegalovirus from blood leucocytes of a patient with post-transfusion mononucleosis. *Aust. Ann. Med.*, **17**, 135-140

Gerber, P., Walsh, J. H., Rosenblum, E. N. & Purcell, R. H. (1969) Association of EB-virus infection with the post-perfusion syndrome. *Lancet*, i, 593-596

Mäntyjärvi, R. (1968) Preparation of cytomegalovirus immune serum in rabbits with alkaline-extracted virus. *Acta path. microbiol. scand.*, **72**, 345-346

McAllister, R. M., Straw, R. M., Filbert, J. E. & Goodheart, C. R. (1963) Human cytomegalovirus: cytochemical observations of the intracellular lesion development correlated with viral synthesis and release. *Virology*, **19**, 521-531

Rapp, F., Rasmussen, E., & Benyesh-Melnick, M. (1963) The immunofluorescent focus technique for studying the replication of cytomegalovirus. *J. Immunol.*, **91**, 709-179

Discussion Summary

P. GERBER [1]

Results were presented of studies with soluble (S) antigens derived from lymphoid cell cultures. This antigen is present in cell extracts of both EBV-positive and EBV-negative lymphoblastoid cell lines. Complement-fixing (CF) antibodies to S-antigen can be detected in 55–80% of sera from normal adult subjects, but is found only rarely among infants 7–12 months of age. Since only sera with antibodies to the viral EBV antigens react with S-antigen, it is assumed that infection by EBV is required for the appearance of S-antibodies.

Sera from patients with infectious mononucleosis were tested at various periods after illness for antibodies to viral (V) and soluble (S) antigens. It was found that all of the 50 sera tested in the first month after onset of illness were positive for V-antibodies but none had detectable S-antibodies at a 1 : 10 serum dilution. From the 8th month onwards, about 50% of the sera also reacted with S-antigen, and this incidence increased to 80% in sera obtained 6–8 years after illness. These results were interpreted as follows: S-antibody develops much later than V-antibody following EBV infection or, alternatively, repeated EBV infection may be necessary for the appearance of S-antibody.

Data were presented which indicated that EBV can stimulate cellular DNA synthesis in leukocytes obtained from normal donors without previous EBV infection. This reaction could be inhibited by inactivation of viral infectivity. Since only oncogenic DNA viruses can stimulate host DNA synthesis, it was suggested that EBV possesses oncogenic properties. This was further supported by the fact that lymphocytes transformed *in vitro* by EBV formed massive, metastatic tumours when injected into immunosuppressed newborn mice, while the peripheral leukocytes from the blood donors in question failed to form tumours in these animals.

Evidence was given indicating that lymphocytes sensitized to EBV as a result of previous infections of the donors, were stimulated *in vitro* when exposed to UV-inactivated EBV preparations. It was suggested that this could be developed into a test to measure cellular immunity to EBV in patients with infectious mononucleosis or Burkitt's lymphoma.

During the general discussion, the following questions were raised: What is the inter-relationship between EBV as a possible aetiological agent in NPC and the effect of high EBV antibody levels in patients? Are there any animal models known where antiviral antibody increases with tumour growth? The possible aetiological rôle of EBV in infectious mononucleosis, Burkitt's lymphoma and NPC was discussed on the basis of three possible hypotheses: (1) the cofactor hypothesis, in which the same virus causes all these diseases but with different cofactors for Burkitt's lymphoma and NPC; (2) the multiple virus hypothesis, in which different, but serologically closely related viruses are postulated as being responsible for these diseases; (3) the hypothesis that EBV causes mononucleosis, but is merely a passenger in the malignant diseases, and as the tumours grow, an increase in viral antigen and antibody occurs. Arguments were advanced that weakened the case for a passenger rôle of EBV.

A question was raised concerning the possible relationship between cell-mediated immunity and EBV antibody levels. The example of leprosy was cited, where cell-mediated immune deficiency allows antigen to escape from the cells, resulting in increased antibody formation. Is there a similar situation in patients with Burkitt's lymphoma and NPC?

Studies to obtain answers to these questions are currently in progress in a number of laboratories.

Are lymphoblastoid cell lines obtained from biopsies of nasopharyngeal tumours of malignant origin? Since these tumours are considered to be carcinomatous, it was suggested that the lymphoid

[1] Division of Biologics Standards, National Institutes of Health, Bethesda, Maryland, USA.

cell lines derived from these tissues should not be referred to as "NPC cells", but as "NPC-derived cultures". The significance of EBV present in such NPC-derived lymphoblastoid cell lines with regard to a possible aetiological rôle remains to be determined.

The present state of uncertainty regarding the rôle of EBV in malignancy was summed up as follows: "If one wishes to rule out the possible aetiological rôle of EBV for these particular tumours (Burkitt's lymphoma and NPC) one is hard put to find a better candidate at this time."

EPIDEMIOLOGY OF BURKITT'S LYMPHOMA, INFECTIOUS MONONUCLEOSIS AND NASOPHARYNGEAL CARCINOMA

Chairman — W. Henle

Rapporteur — R. H. Morrow, Jr

The Trail to a Virus—A Review

D. P. BURKITT [1]

[1] Medical Research Council External Staff, London, W.1, UK.

INTRODUCTION

I have been asked to outline the events leading up to the present belief that Burkitt's lymphoma might turn out to be the first form of human cancer shown to be due, at least in part, to a virus.

In the present climate of medical research, simple clinical observations tend to be overshadowed by sophisticated techniques, elaborate equipment and experimental work. It is therefore gratifying to simple clinicians like myself to remember that it was initially a bedside observation that triggered off the vast volume of research into the tumour which has become known as Burkitt's lymphoma (BL), an eponym that I should like to consider as primarily reflecting the generosity of my colleagues, who first suggested it, rather than as singling out the work of one individual.

The first step was the realization that a number of tumours occurring in children in different anatomical sites tended to be related to one another in individual patients (Burkitt, 1958). The simultaneous occurrence of tumours in different locations such as the maxilla, mandible, thyroid, ovaries, liver and kidneys, demanded *explanation* and suggested a common origin. The clinical distribution appeared to preclude a primary tumour with bizarre metastases, and the alternative of a multifocal tumour seemed to be more acceptable.

Subsequently it was shown that the different clinical manifestations tended not only to be associated in individual patients but were also associated geographically (Burkitt & O'Conor, 1961; Burkitt & Davies, 1961). This double association, which proved to be such a useful guide in the investigation of BL, has subsequently been used to throw light on the cause of other diseases (Burkitt, 1970a, b).

The sharing of a common and unusual age-distribution was an additional factor linking the tumours presenting in different anatomical sites.

Shortly after the clinical syndrome became apparent, O'Conor & Davies (1960) and O'Conor (1961) identified the tumour as a malignant lymphoma, and confirmed that the differing clinical manifestations were, in fact, all part of a single neoplastic condition. Wright took the investigation a step further when he identified the tumour as a specific type of lymphoma distinguishable on histological and cytological criteria from other lymphomas (Wright, 1963, 1964). This view was subsequently confirmed at a meeting convened by WHO in 1967, although it was agreed that pathological classification could not in all cases be settled on histological grounds.

At the same time the geographical distribution in Africa was further clarified and the relationship of tumour frequency to temperature and rainfall was shown (Burkitt, 1962a, b).

The climatic factors which determined tumour distribution in Africa were found also to determine distribution in Papua-New Guinea, the only other part of the world in which the condition was then known to be common (Booth *et al.*, 1967).

THE SEARCH FOR A VIRUS

The demonstrated relationship of BL to climatic factors suggested to several workers independently the possibility that some biological organism, possibly an insect vector, might play a part in the causation of the tumour. The question then arose as to what an insect could carry that might cause cancer. Viruses seemed to be the obvious answer, for certain virus diseases were known to be transmitted by insects, and some animal cancers were known to be virus-induced. The distribution patterns of yellow fever and O'nyong-nyong fever in Africa were found to bear a close resemblance to that of BL, and both are caused by mosquitovectored viruses.

These observations initiated an intense search for viruses in the tumour by teams of workers from different countries which resulted in the discovery by Epstein and Barr of the eponymous Epstein-Barr virus (EBV) in cultured cells from an African child (Epstein & Barr, 1964) and also led to important studies on other viruses (Bell, 1967) and mycoplasma (Dalldorf et al., 1966), although these now seem less likely to be causally related.

A vast amount of serological and other evidence has since been accumulated to strengthen the original theory of a virus aetiology, but this will be discussed by other participants in the Symposium. The hypothesis has certainly been strengthened by the evidence of strong host-defence mechanisms, possibly of an immunological nature, shown not only by the high cure rate following chemotherapy (Burkitt, 1967; Clifford, 1967; Ziegler et al., 1969 [1]) but also by demonstrated spontaneous remissions (Burkitt & Kyalwazi, 1967; Ngu, 1967), a situation which parallels that of virus-induced tumours in animals. Furthermore, the recognition of the clustering, both in time and space, of patients in some areas (Pike, Williams & Wright, 1967) indicates the presence of an infective agent.

THE IMPLICATION OF MALARIA

As further investigations by the Kleins and Henles and others (Klein, 1970) made EBV the most likely suspect, it became increasingly evident that it would not alone explain the geographical distribution of the tumour. Not only is this virus not known to be insect-vectored, but serological tests have shown that infection is no more common in areas where the tumour is endemic than in regions where it is rare (Levy & Henle, 1966). Moreover, although there are no very cold or dry areas where the tumour is common, there are many warm moist areas where it is rare or unknown.

These observations led to a search for an alternative explanation to that of a vectored virus that might account for the climatic dependence of the tumour. The suggestion first made by Dalldorf (1962) and subsequently by O'Conor (1961) that there might be a causative relationship between the

---

[1] Unpublished data.

tumour and chronic malaria, was re-examined. When this suggestion was first made, the extreme current interest in the possible viral aetiology of some forms of cancer prevented its being given the attention it deserved, and the possibility of both factors playing a part was not at that time considered.

Epidemiological studies on a worldwide scale and certain aspects of the clinical features strongly suggest that hyperendemic and holoendemic malaria are indeed causally related to the tumour and responsible for its geographical distribution (Burkitt, 1969).

This does not, however, in any way detract from the evidence that this may be a virus-induced tumour. In fact the knowledge that BL can occur as a rare tumour anywhere in the world but is only common in highly malarious regions, suggests that an ubiquitous virus, such as EBV, may well be the primary cause, and that intense chronic malaria renders the lymphoreticular system more liable to malignant change in the presence of a virus.

The close relationship between endemic tumour distribution and the existence of holoendemic and hyperendemic malaria has been described in some detail (Kafuko & Burkitt, 1970); suffice it to say that no area is known where the tumour is endemic and malaria is not hyperendemic or the reverse. Moreover, the relationship with malaria is further strengthened by the association between the distribution of BL and other conditions now considered to be caused by malaria, such as big spleen disease (Hamilton et al., 1965) and the nephrotic syndrome of the tropics (Kibukamusoke, Hutt & Wilks, 1967).

The way in which even distantly related diseases can throw light on the aetiology of one another is exemplified by the fact that patients with BL appear to have a significantly lower incidence of sickle cell trait than do controls (Pike et al., 1969). This would be expected if the tumour were causally related to malaria, since this haemoglobinopathy confers some protection against malarial infection.

CURRENT CONCEPT OF CAUSATION

The theory of the causation of BL that best fits all the available epidemiological and other evidence is that the profound changes known to occur in the lymphoreticular system as a result of repeated, prolonged and severe infection with malaria, or possibly as a result of the action of other stimulants, renders the lymphoid cells particularly liable to malignant change in the presence of EBV or of some other virus (see Figure).

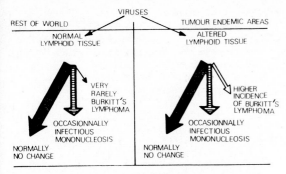

VIRUSES

REST OF WORLD | TUMOUR ENDEMIC AREAS

NORMAL LYMPHOID TISSUE | ALTERED LYMPHOID TISSUE

VERY RARELY BURKITT'S LYMPHOMA | HIGHER INCIDENCE OF BURKITT'S LYMPHOMA

OCCASIONNALLY INFECTIOUS MONONUCLEOSIS | OCCASIONNALLY INFECTIOUS MONONUCLEOSIS

NORMALLY NO CHANGE | NORMALLY NO CHANGE

Theory of causation of Burkitt's lymphoma.

Since it is now known that EBV can lead to non-malignant lymphoid proliferation (infectious mononucleosis) in man (Henle, Henle & Diehl, 1968) and

that a morphologically similar virus causes Marek's disease, a malignant lymphoid proliferation in fowl (Churchill & Biggs, 1967) it seems not unreasonable to suspect that in certain circumstances EBV may be responsible for malignant change in the lympho-reticular system in man.

The situation might well be analogous to that of poliomyelitis, tumour formation corresponding to the occurrence of paralysis. In both conditions, infection is probably almost universal, but usually only slightly pathogenic. In both, more severe but usually not dangerous symptoms are common, and again in both diseases, serious pathological changes occur in a small proportion of those infected. In the case of infection with one of the poliomyelitis viruses the result is paralysis, and in the case of EBV infection possibly a tumour.

REFERENCES

Bell, T. (1967) Viruses associated with Burkitt's tumor. *Progr. med. Virol.*, **9**, 1-34

Booth, K., Burkitt, D. P., Bassett, D. J., Cooke, R. A. & Biddulph, J. (1967) Burkitt lymphoma in Papua, New Guinea. *Brit. J. Cancer*, **21**, 657-664

Bull. Wld Hlth Org. (1969) Histopathological definition of Burkitt's tumour, **40**, 601-607

Burkitt, D. P. (1958) A sarcoma involving the jaws in African children. *Brit. J. Surg.*, **46**, 218-223

Burkitt, D. P. (1962a) A "Tumour Safari" in East and Central Africa. *Brit. J. Cancer*, **16**, 379-386

Burkitt, D. P. (1962b) Determining the climate limitations of a children's cancer common in Africa. *Brit. med. J.*, **2**, 1019-1023

Burkitt, D. P. (1967) Chemotherapy of jaw tumours. In: Burchenal, J. H. & Burkitt, D. P., eds, *Treatment of Burkitt's tumour*, Heidelberg, Springer-Verlag, pp. 94-101

Burkitt, D. P. (1969) Etiology of Burkitt's lymphoma—an alternative hypothesis to a vectored virus. *J. nat. Cancer Inst.*, **42**, 19-28

Burkitt, D. P. (1970a) Relationship as a clue to causation. *Lancet*, **ii**, 1237-1240

Burkitt, D. P. (1970b) Relationship as a guide to etiology of disease. *Int. Path.*, **2**, 3-6

Burkitt, D. P. & Davies, J. N. P. (1961) Lymphoma syndrome in Uganda and tropical Africa. *Med. Press*, **145**, 367-369.

Burkitt, D. P. & O'Conor, G. T. (1961) Malignant lymphoma in African children. I. A clinical syndrome. *Cancer*, **14**, 258-269

Burkitt, D. P. & Kyalwazi, S. K. (1967) Spontaneous remission of African lymphoma. *Brit. J. Cancer*, **21**, 14-16

Churchill, A. E. & Biggs, P. M. (1967) Agent of Marek's Disease in tissue culture. *Nature (Lond.)*, **215**, 528-530

Clifford, P. (1967) Observations on the treatment of Burkitt's lymphoma. In: Burchenal, J. H. & Burkitt, D. P., eds, *Treatment of Burkitt's tumour*, Heidelberg, Springer-Verlag, pp. 77-93

Clifford, P. (1970) Treatment—response to particular chemotherapeutic and other agents and treatment of CNS involvement. In: Burkitt, D. P. & Wright, D. H., eds, *Burkitt's lymphoma*, Edinburgh, E. & S. Livingstone, pp. 52-63

Dalldorf, G. (1962) Lymphoma of African children with different forms of environmental influences. *J. Amer. med. Ass.*, **181**, 1026-1028

Dalldorf, G., Linsell, C. A., Barnhart, F. E. & Martyn, R. (1966) An epidemiological approach to the lymphomas of African children and Burkitt's sarcoma of the jaws. *Perspect. Biol. Med.*, **7**, 435-449

Epstein, M. A. & Barr, Y. M. (1964) Cultivation *in vitro* of human lymphoblasts from Burkitt's malignant lymphoma. *Lancet*, **i**, 252-253

Hamilton, P. J. S., Hutt, M. S. R., Wilks, N. E., Olweny, C., Ndawula, R. L. & Mwamje, L. (1965) Idiopathic splenomegaly in Uganda. Part 2, Geographical aspects. *E. Afr. med. J.*, **42**, 196-202

Henle, G., Henle, W. & Diehl, V. (1968) Relation of Burkitt's tumor-associated herpes-type virus to infectious mononucleosis. *Proc. nat. Acad. Sci. (Wash.)*, **50**, 94-101.

Kafuko, G. W. & Burkitt, D. P. (1970) Burkitt's lymphoma and malaria. *Int. J. Cancer*, **6**, 1-9

Kibukamusoke, J. W., Hutt, M. S. R. & Wilks, N. E. (1967) The nephrotic syndrome in Uganda and its association with quartan malaria. *Quart. J. Med.*, **36**, 393-408

Klein, G. (1970) Some immunological studies. In: Burkitt, D. P. & Wright, D. H., eds, *Burkitt's lymphoma*, Edinburgh, E. & S. Livingstone, pp. 172-185

Levy, J. A. & Henle, G. (1966) Indirect immunofluorescence tests with sera from African children and cultured Burkitt lymphoma cells. *J. Bact.,* **92**, 275-276

Ngu, V. A. (1967) Clinical evidence of host defences in Burkitt's tumour. In: Burchenal, J. H. & Burkitt, D. P., eds, *Treatment of Burkitt's tumour*, Heidelberg, Springer-Verlag, pp. 204-208

O'Conor, G. T. (1961) Malignant lymphoma in African children. II. A pathological entity. *Cancer,* **14**, 270-283

O'Conor, G. T. & Davies, J. (1960) Malignant tumours in African children. *J. Ped.,* **56**, 526-535

Pike, M. C., Williams, E. H. & Wright, B. (1967) Burkitt's tumour in the West Nile District of Uganda, 1961-65. *Brit. med. J.,* **2**, 395-399

Pike, M. C., Morrow, R. H., Kisuule, A. & Mafigiri, J. (1970) Burkitt's lymphoma and sickle cell trait. *Brit. J. prev. soc. Med.,* **24**, 39-41

Wright, D. H. (1963) Cytology and histochemistry of the Burkitt's lymphoma. *Brit. J. Cancer,* **17**, 50-55

Wright, D. H. (1964) Cytology and histochemistry of the malignant lymphomas seen in Uganda. In: Roulet, F. C., ed., *The lymphoreticular tumours in Africa*, Basel, Karger, pp. 291-303

Some Epidemiological Problems with "EBV + Malaria gives BL"—A Review

M. C. PIKE [1] & R. H. MORROW [2]

Burkitt has discussed the evidence for a relationship between Burkitt's lymphoma (BL) and falciparum malaria, and we have been repeatedly reminded that BL patients have high Epstein-Barr virus (EBV) antibody titres, as shown by the Henle test for anti-viral-capsid antigen (VCA).

There is some question that a very small number of histologically diagnosed BL cases have no detectable VCA antibodies, but due to the extreme difficulties occasioned by having to ask pathologists to say what is a BL rather than being able to ask God, this question of the existence of a "true" BL with no VCA antibodies is likely to take a long time to clear up. BL patients outside the endemic areas may help in this.

This is so since the amount of EBV infection in Europe and the USA, as measured by antibody levels, is considerably less than in Africa, and continued association of EBV and BL in patients in Europe and the USA would thus be most informative.

In tropical Africa there is unfortunately the possibility that virtually all persons are infected with EBV very early in life, but some have antibody levels, at least after a few years, below the threshold of detection by the available test system.

The evidence for this comes from two sources. (1) To the best of our knowledge every one of a small number of Ugandan patients with either Hodgkin's disease or reticulum cell sarcoma and examined for EBV antibodies were positive. (2) The results of a serological survey in the West Nile District of Uganda conducted by Dr Kafuko, Director of the East African Virus Research Institute, Dr Munube, Dr Henderson and others, and analyzed by the

[1] University of Oxford Department of the Regius Professor of Medicine, Radcliffe Infirmary, Oxford, UK.
[2] Department of Tropical Public Health, Harvard School of Public Health, Boston, Massachusetts, USA.

Henles showed that the proportion of persons with positive ($\geqslant 1/10$) titres reached a peak of 90% around age 3 to 4 years and then *declined* gradually to about 70% by age 15 years. This could of course be a cohort effect with a differentially greater transmission to the younger children. In other studies, done mainly in the USA and involving small numbers of people, VCA antibodies once positive have remained so. However, the decline with age could mean that persons *under circumstances like those obtaining in the West Nile* have gradually declining antibodies to EBV, and that in some persons after a certain time they are not detectable.

If the latter is true then there is no evidence to disprove the contention that virtually everyone has met EBV at a very early age. All that would then distinguish the BL patients is the level of their antibody titres. One should however point out here that it would be very difficult to establish that "everyone" had met the virus, since in the BL context of a rare disease "everyone" must mean somewhat greater than 99.9%, if the possibility of sufficient true negatives being around to become the BL patients is to be eliminated.

There are a number of other epidemiological facts that should again be brought to your attention. (1) BL is a childhood disease; it starts around age 3, has a peak about age 7 and is rare after puberty. It is as uncommon as childhood leukaemia is in Europe and the USA: about 40 cases per 1 000 000 per year up to age 15 or about 1 in 2000 children born will get BL by age 16. (2) Late teenage and rare adult BL cases in the endemic BL areas tend to be immigrants from nonendemic areas. (3) The disease displays time-space clustering in certain localities. The most recent and most dramatic example of this was in Bwamba County in Uganda. Seven cases of BL have to date been diagnosed in this remote and isolated part of the country. No case had ever been reported before October 1966. Then there were seven cases with clinical onset of the tumour in the 27 months from October 1966 to December 1968, five of them in the last six months of this period. There has been no case since. Two were brother and sister living in the same house. (4) We now have details of three pairs of sibling cases of BL. They give in a striking form highly pertinent information as to the relative rôle of environmental as opposed to host factors. The Bwamba pair were aged 3 and 9 years, with clinical onset of the tumour only 5 months apart. This points to something major having happened in the environment.

What sort of "malaria plus EBV" hypothesis do you need to cope with these observations?

A simple "malaria plus EBV gives BL" is clearly nonsense—the incidence of BL is much too low. Moreover, the EBV antibody curve does not fit the BL incidence curve unless we postulate a long and highly variable latent period, and that makes it very difficult to explain clustering.

Maybe the person who gets BL is "peculiar". There is no good evidence from case histories for this; patients have no gross immunological deficiencies or other differences by history from matched controls, and individual peculiarities would not help explain clustering. There is clearly no great tendency for cases to occur in families, and the Bwamba sibs argue strongly in favour of an environmental effect.

A critical time in the host is possibly needed. This could, for example, be in relation to the development of full immunological competence, to child development, say in the manner of paralytic polio, or to severe physical stress during the incubation of the virus. No hypothesis along these lines has been suggested that accounts for the known facts—and in particular for the adult cases occurring mostly in immigrants.

The most favoured hypotheses are either along the lines of: (a) primary infection with EBV occurring at a critical stage of malaria response; or (b) EBV infection followed by a critical reaction to malaria. (Field testing of (a)-type hypotheses will be discussed later by Dr Geser.[1]) These hypotheses need some form of epidemic pattern of EBV or malaria. However, there is no good evidence for this. It could be that under special circumstances EBV is transmitted by an unusual route (the usual route is in fact unknown) in an epidemic pattern. We just don't know.

It therefore appears to us at the moment that: (1) the evidence for a critical rôle for malaria is good (but could do with tightening by further case-control studies for sickle cell trait and possibly other factors that alter malaria infection such as G-6PD); (2) clinching evidence for a necessary rôle for EBV is not yet at hand; and (3) possibly a third, as yet totally unknown, factor is needed to produce the observed epidemic nature of the disease. The proposed study in the West Nile District of Uganda to be discussed by Dr Geser[1] would appear to provide the only hope of obtaining a direct answer.

[1] See p. 372 of this publication.

Epidemiology of Infectious Mononucleosis
A Review*

A. S. EVANS & J. C. NIEDERMAN

This discussion of the epidemiology of infectious mononucleosis will be founded on three basic concepts, of which the first is that Epstein-Barr virus (EBV) is the cause of infectious mononucleosis. The evidence for this has already been presented by Henle.[1] The most convincing immunological evidence is the absence of antibody prior to illness, its regular appearance during illness, and its persistence for years thereafter. This has been well documented in prospective sero-epidemiological surveys carried out in our laboratory; these now include over 2000 young adults followed carefully from 9 months for up to 8 years after determination of their initial antibody status (Niederman et al., 1968; Evans, Niederman & McCollum, 1968; Niederman et al., 1970; Sawyer et al., 1971) (Table 1). In 977 young adults whose serum lacked antibody to EBV at 1 : 5

or 1 : 10 dilution by the indirect immunofluorescence test (Henle & Henle, 1966) on initial screening, 100 subsequently developed clinical infectious mononucleosis, a rate of 10.2 per 100 susceptibles. The absence of antibody in many of these sera has been confirmed further by examining undiluted serum samples (Hallee et al.[2]). In 1141 students whose serum possessed EBV antibody on initial screening, not a single recognized case of infectious mononucleosis occurred over the next 9 months to 8 years. This work has been confirmed by Tischendorf et al., (1970) and Pereira, Blake & Macrae, (1969), and recently by Hirshaut, Christenson & Perlmutter, (1971) in a prospective study of 800 Cornell Medical students over a 2-year period.

The second basic concept is that the occurrence of clinical infectious mononucleosis in a country or in an individual is inversely related to the age at which EBV infection occurs. In countries or individuals in which infection occurs early in life, clinical infectious mononucleosis will be rare. In countries or individuals in which the occurrence of infection is delayed until late childhood or young adult life, infectious mononucleosis will be a common and important clinical disease. This concept was repeatedly advanced, prior to the discovery of EBV, as much as 10 years ago (Evans, 1960; Evans, 1961–62; Evans, 1968) and has been confirmed in recent sero-epidemiologic studies with EBV (Evans et al., 1968; Henle & Henle, 1970; Niederman et al., 1970).

The third concept is that EBV produces a spectrum of host responses, some of which we regard as primary and others as secondary or delayed. The primary responses include asymptomatic EBV infection in early life, mild pharyngitis and tonsillitis of childhood, infectious mononucleosis in the young

Table 1. Relation between EBV antibody status and subsequent development of infectious mononucleosis in 2118 subjects at Yale University and West Point Military Academy

Initial EBV antibody status [1]	No. of subjects [2]	Cases of infectious mononucleosis during the next 9 months to 8 years	
		No.	%
Negative	977	100	10.2
Positive	1141	0	0
Totals	2118	100	4.7

[1] On entrance to college.
[2] Age: 17–19 years.

* WHO Serum Reference Bank and Section of International Epidemiology, Department of Epidemiology and Public Health, Yale University School of Medicine, New Haven, Connecticut, USA.

[1] See p. 269 of this publication.

[2] Unpublished data.

adult, and EBV transfusion mononucleosis in susceptible recipients of EBV-positive blood at any age. EBV causes both heterophile positive mononucleosis cases and some heterophile negative cases (Evans *et al.*, 1968). Other cases with features of infectious mononucleosis, but without sore throat, lymphadenopathy, or elevated heterophile antibody, are due to cytomegalic infection (Klemola & Kääriäinen, 1965; Klemola *et al.*, 1970) and other viruses may also be involved in other similar mononucleosis syndromes (Evans, 1968). The secondary or delayed group of EBV infections include Burkitt's lymphoma (Burkitt & Wright, 1970), nasopharyngeal cancer (Henle *et al.*, 1971; Kawamura *et al.*, 1970), Hodgkin's disease (Johansson *et al.*, 1970; Carvalho *et al.*, 1971), sarcoidosis (Hirshaut *et al.*, 1970), leprosy (Papageorgiou *et al.*, 1971) and systemic lupus erythematosus (Dalldorf *et al.*, 1969; Carvalho *et al.*, 1970; Evans, Rothfield & Niederman, 1971). The rôle of EBV in these chronic diseases is not clear, may vary from one to another, and may range from passenger to cause. As a working hypothesis we propose that a defect in cellular immunity, leading to endogenous reactivation of EBV multiplication, may play an important rôle in the pathogenesis of these diverse states (Evans, 1971).

These three basic concepts mean that a discussion of infectious mononucleosis must include the factors affecting the occurrence of clinical disease as well as those affecting the total pattern of primary EBV infection. In this context, the presence of EBV antibody can be taken to represent either clinical or subclinical infectious mononucleosis; the situation therefore resembles that found with poliomyelitis and other common infections. The chief factors affecting the occurrence of clinical infectious mononucleosis are summarized below.

Factors tending to increase the occurrence of clinical disease	Factors tending to decrease the occurrence of clinical disease
Temperate climate	Tropical climate
Developed economy	Developing economy
High socio-economic level:	Low socio-economic level
Private school	Public school
Private hospital	Public clinic
Good housing (little crowding)	Poor housing (crowding)
High income	Low income
High educational level	Low educational level
High standards of hygiene	Low standards of hygiene

Whatever factors tend to delay the occurrence of clinical infection until older childhood and young adult life, will tend to increase the incidence of clinical infectious mononucleosis.

INCIDENCE

EBV infection is worldwide; it even occurs in remote Indian or Eskimo tribes (Black *et al.*, 1970; Tischendorf *et al.*, 1970), whereas clinical infectious mononucleosis is largely confined to the so-called developed countries. Antibody surveys for EBV infection have shown an initially high prevalence rate in the first six months of life, a gradual acquisition of antibody through the years, with 50–80% of people being EBV antibody-positive by young adult life (Evans, 1970; Henle & Henle 1970; Niederman *et al.*, 1970; Porter, Wimberly & Benyesh-Melnick, 1969; Svedmyr & Demissie, 1968; Deinhart *et al.*, 1969). In a typical middle-income American community, such as Danbury, Connecticut, where 1701 healthy children aged 1–15 were tested, antibody acquisition proceeded in an irregular stepwise fashion with an average increase of 4.1% per year during the first 10 years of life; typical prevalence rates were at age 2—13.1%; age 5—27.8%; age 9—45.8% (Nimmannitya, 1969).

The incidence of *clinical* infectious mononucleosis is of the order of 50 per 100 000 per year in the general population in the USA (Christine, 1968; Evans, 1968). In five colleges the incidence rates for hospitalized cases have ranged from 316–1449 per 100 000 students (average 749) (Evans *et al.*, 1968). The incidence of the *disease* in both the civilian and military populations places it among the first 5 common communicable diseases (the other four being streptococcal sore throat, rubella, measles and hepatitis) (Evans, 1969). Prospective studies of about 1500 college students, where both data on EBV antibody and clinical data were used, have indicated an average incidence rate of clinical infectious mononucleosis of 5460 per 100 000 *susceptible* students per year (Evans *et al.*, 1968; Niederman *et al.*, 1970; Hallee *et al.*, 1971 [1]). The total infection rate (clinical plus subclinical) averaged 11 500 per 100 000 susceptibles yearly. While this suggests a clinical : subclinical ratio of about 1 : 2, we feel that almost all EBV infections are associated with clinical illness in young adults if they are carefully enough followed (Sawyer *et al.*, 1970).

---

[1] Unpublished data.

GEOGRAPHICAL DISTRIBUTION

As already indicated, infection with EBV occurs earlier in developing countries and at lower socioeconomic levels than in developed countries and at higher socio-economic levels. Prevalence rates have been 95 to 100% in Argentine and Colombian recruits (Evans, Jeffrey & Niederman, 1971; Evans et al.[1]); in such a population no clinical infectious mononucleosis would be expected to occur in military life as all recruits are already immune on entry. In University of the Philippines students, an EBV antibody prevalence rate of 82% was found, and during five years of observation not a single clinical case of infectious mononucleosis was diagnosed in over 5000 infirmary admissions (Evans & Campos, 1971). Prevalence rates within a single country may also vary geographically. Thus in 426 cadets entering the US Military Academy who had resided in the same state for 12 or more years, an EBV antibody prevalence rate of 75.6% was found in those from the warm South Central States as compared to a 38.0% prevalence in the colder West North Central States (P<0.01) (Hallee et al. [1]) (see Table 2).

Table 2. Frequency of EBV antibody in the serum of 426 freshman cadets at West Point who had lived in one geographical area for 12 years or more

Area of USA in which resident for previous 12 years	Proportion positive at 1:10 for EBV antibody (%)
South central	75.6
East north central	58.9
Atlantic	58.1
Mountain	54.5
Pacific	54.3
New England	46.2
West north central	38.0
Total	56.7

Note: $\chi^2 = 21.03$; P < 0.01.

SOCIO-ECONOMIC STATUS

As in poliomyelitis, this is an important factor in EBV infections. In both conditions, infections occur earlier in life in lower socio-economic settings, where crowding and poor personal hygiene are important determinants. This effect is hard to separate from climate because socio-economic levels are generally lower in many tropical settings. In West Point

Cadets an analysis of yearly income of the parents revealed a 77% prevalence for incomes under $6000 and a 55.5% prevalence for those over $30000 (Hallee et al.[2]).

AGE

This is the most critical factor in determining the host response. The frequency of *clinical* infectious mononucleosis with elevated heterophile antibody depends on the number of individuals who escape infection until age 15 to 25; in these young adults almost all primary EBV infections are clinically manifested as classical infectious mononucleosis. This peak in late childhood and young adult life has been found in all developed countries, such as the United States (Hoagland, 1969; Evans, 1968), Great Britain (Davidson, 1970; Newall, 1957), and Scandinavia (Strom, 1960; Thomsen, 1942). A search for cases of heterophile-positive infectious mononucleosis in Brazil has shown the clinical disease to be rare but to occur in the age group 5–10; presumably all persons older than this are immune (Carvalho et al., 1971 [2]). Heterophile-negative cases in the USA tend to occur earlier in life than heterophile-positive cases, with many of these under age 10 (Evans et al., 1968; Davidson, 1970). There is some evidence of a bimodal distribution of EBV *infection* with high peaks under age 6 and over age 12 but a relative hiatus between (Henle & Henle, 1970; Deinhart et al., 1969; Evans et al.[2]).

SEX

There is little difference in incidence between the sexes except that females tend to develop clinical disease and EBV infection at a somewhat earlier age (Davidson, 1970). Recently freshman women entering Yale University showed 55.3% EBV antibody prevalence as compared to 51.0% in freshman men (Evans, A. S. & Niederman, J. C.[2]).

CONTAGIOUSNESS

The earlier observations that infectious mononucleosis has a low contagiousness, even in roommates and in families (Hoagland, 1969; Evans & Robinton, 1950) has been confirmed in recent work with EBV in which the susceptibility of those exposed and the occurrence of subclinical infection can both be identified (see Table 3). In a family setting, Henle & Henle (1970) found evidence of secondary trans-

---

[1] Unpublished data.

[2] Unpublished data.

Table 3.   Infection rates for EB virus in susceptible persons or family members

Place where study carried out	Group	No. susceptible [1]	No. infected	Rate (%)	Reference
Yale University	Freshmen	175	23	13.1	Sawyer *et al.* (1971)
	Room-mates of cases	18	1	5.5	
West Point	Freshmen	437	54	12.4	Unpublished data
	Room-mates of cases	24	2	8.3	
Canada	Family contacts	67	7	10.4	Joncas & Mitnyan (1970)
Sweden	Family contacts	21	3	14.2	Wahren *et al.* (1970)
Cleveland, Ohio	Family members	8 families	3 families [2]	42.8	Henle & Henle (1970)

[1] Lacking EBV antibody.
[2] There was evidence of multiple infection in 3 of 8 families, each with some susceptible children.

mission in 3 of 8 families (37.5%), Wahren *et al.* (1970) in 7 of 21 families (33.3%), and Joncas & Mitnyan (1970) in 7 of 67 susceptible family contacts (10.4%).   In a study of students in three dormitories at Yale University we have not found attack rates to be higher in susceptible room-mates (1 out of 18, or 5.5%) than in the total group of susceptibles (13.1%) (Sawyer *et al.*, 1971). Among 175 susceptible students in three dormitories, infection rates were 13.1, 18.7 and 21.2% respectively, with some evidence of clustering in social groups served by one entry-way.   This low efficiency of transmission in family and college settings is in contrast to the apparently high efficiency of spread in a day nursery setting in England (Pereira *et al.*, 1969) and in an orphanage in the United States (Tischendorf *et al.*, 1970).

METHOD OF TRANSMISSION

Oral/faecal spread in young children and close intimate contact, including kissing, in young adults are the most likely methods of transmission (Hoag-

land, 1955; Evans, 1960). The poor personal hygiene of very young children and the propensity to oscultatory activities in young adults may account for a bimodal age distribution.   As EBV is cell-bound, transmission may actually require the transfer of infected cells from person to person; this might be termed "cellular kissing".   The demonstration of EBV in the throat and stool remains an important challenge to investigators.

INCUBATION PERIOD

The early estimates of about 33 to 49 days in adults made by Hoagland (1955) and by Evans (1960) still remain valid.   For example, in one susceptible Yale student, definite illness with an elevated heterophile antibody titre developed 41 days after recognition of illness in his room-mate (Sawyer *et al.*, 1971).   If there is an earlier identifiable prodromal period, as in children, the apparent incubation period will be shorter.

SUMMARY

The work with EBV has shown that it causes a wide spectrum of primary infections, including asymptomatic infection, mild pharyngitis and tonsillitis, transfusion mononucleosis, and clinical infectious mononucleosis. The age at which EBV infection occurs is the major determinant of the host response. When this is early in life, as in tropical and developing countries or in low socioeconomic settings, mild and asymptomatic infection will be predominant and infectious mononucleosis may be rare or even unrecognized as a clinical entity. In settings where infection with EBV is delayed until older childhood or young adult life, classical infectious mononucleosis will occur as a common disease. The contagiousness of infectious mononucleosis is low, the incubation period is long, and immunity is permanent.

ACKNOWLEDGEMENTS

The work on which this paper was based was supported by grants from the Public Health Service, National Institutes of Health (AI 08731), and the National Communicable Diseases Center (CC-00242). It was carried out, in part, under the sponsorship of the Commission on Viral Infections of the Armed Forces Epidemiological Board, through US Army Medical Research and Development Command, research contract No. DADA17-69-C-9172, and also received financial assistance from the World Health Organization.

REFERENCES

Black, F. L., Woodall, J. P., Evans, A. S., Lebhaber, H. & Henle, G. (1970) Prevalence of antibody against viruses in the Tiriyo, an isolated Amazon tribe. *Amer. J. Epidem.*, **91**, 430-438

Burkitt, D. P. & Wright, D. H. (1970) *Burkitt's lymphoma*, Edinburgh & London, E. & S. Livingstone

Carvalho, R. P. S., Frost, P., Dalldorf, G., Jamra, M. & Evans, A. (1971) Pesquisa de anticorpos precipitantes para antígeno do linfoma de Burkitt (células Jijoye) em soros colhidos no Brasil. *Rev. Bras. Pes. med. biol.*, **4**, 75-82

Christine, B. (1968) Infectious mononucleosis. *Conn. Hlth Bull.*, **82**, 115-119

Dalldorf, G., Carvalho, R. P. S., Jamra, M., Frost, P., Erlich, D. & Marigo, C. (1969) The lymphomas of Brazilian children. *J. Amer. med. Ass.*, **28**, 1365-1368

Davidson, R. J. L. (1970) A survey of infectious mononucleosis in the North-East Regional Hospital Board area of Scotland, 1960-69. *J. Hyg. (Camb.)*, **68**, 393-400

Deinhardt, F., Tischendorf, P., Shramek, J., Maynard, J. & Noble, G. (1969) Distribution of antibodies to EB virus (EBV) in various American population groups. *Bact. Proc.*, **1**, 178-179

Evans, A. S. (1960) Infectious mononucleosis in University of Wisconsin students. *Amer. J. Hyg.*, **71**, 342-346

Evans, A. S. (1961-62) Infectious mononucleosis. Observation from a public health laboratory. *Yale J. Biol. Med.*, **34**, 261-276

Evans, A. S. (1968) Epidemiology and pathogenesis of infectious mononucleosis. In: *Proceedings of the International Infectious Mononucleosis Symposium*, Washington, American College Health Association, pp. 40-55

Evans, A. S. (1969) Infectious mononucleosis—recent developments. *GP*, **60**, 127-134

Evans, A. S. (1970) Infectious mononucleosis in the armed forces. *Military Med.*, **135**, 300-304

Evans, A. S. (1971) The spectrum of infections with Epstein-Barr virus: a hypothesis. *J. infect. Dis.*, **124**, 330-337.

Evans, A. S. & Campos, L. E. (1971) Acute respiratory diseases in University of the Philippines students, 1964-1969. *Bull. Wld Hlth Org.* (in press)

Evans, A. S., Jeffrey, C. & Niederman, J. C. (1971) The risk of acute respiratory infection in two groups of young adults in Colombia, South America. *Amer. J. Epidem.*, **93**, 463-471

Evans, A. S., Niederman, J. C. & McCollum, R. W. (1968) Seroepidemiologic studies of infectious mononucleosis with EB viruses. *New Engl. J. Med.*, **274**, 1121-1127

Evans, A. S. & Robinton, E. D. (1950) The epidemiologic study of infectious mononucleosis in a New England college. *New Engl. J. Med.*, **242**, 492-496

Evans, A. S., Rothfield, N. F. & Niederman, J. C. (1971) Raised antibody levels to EB virus in systemic lupus erythematosus. *Lancet*, **i**, 167-168

Henle, G. & Henle, W. (1966) Immunofluorescence in cells derived from Burkitt's lymphoma. *J. Bact.*, **91**, 1248-1258

Henle, W. & Henle, G. (1970) Observations on childhood infections with the Epstein-Barr Virus. *J. infect. Dis.*, **121**, 303-310

Henle, W., Henle, G., Ho, H. C., Burtin, P., Cachin, V., Clifford, P., de Schryver, A., de-Thé, G., Diehl, V., & Klein, G. (1971) Antibodies to Epstein-Barr Virus in nasopharyngeal carcinoma, other head and neck neoplasms, and control groups. *J. nat. Cancer Inst.*, **44**, 225-231

Hirshaut, Y., Glade, H., Viera, B. D., Ainbender, E., Dvorak, B. & Siltsbach, L. E. (1970) Sarcoidosis, another disease with serologic evidence for herpes-like virus infection. *New Engl. J. Med.*, **283**, 502-505

Hirshaut, Y., Christenson, W. N. & Perlmutter, J. C. (1971) Prospective study of herpes-like virus. Role in infectious mononucleosis. (Abstract). *Clin. Res.*, **19**, 43a

Hoagland, R. J. (1955) The transmission of infectious mononucleosis. *Amer. J. med. Sci.*, **229**, 262-272

Hoagland, R. J. (1964) The incubation period of infectious mononucleosis. *Amer. J. publ. Hlth*, **54**, 1699-1705

Hoagland, R. J. (1967) *Infectious mononucleosis.* Grune & Stratton

Johansson, B., Klein, G., Henle, W. & Henle, G. (1970) Epstein-Barr virus (EBV)—associated antibody patterns in malignant lymphoma and leukemia. I. Hodgkin's disease. *Int. J. Cancer*, **6**, 450-462

Joncas, J. & Mitnyan, C. (1970) Serological response of the EBV antibodies in pediatric cases of infectious mononucleosis and in their contacts. *Canad. med. Ass. J.*, **6**, 1260-1263

Kawamura, A., Takada, M., Goto, H. A., Hamajima, K., Sanpe, T., Murata, M., Ito, Y., Takahashi, T., Yoshioa, T., Hirayama, T., Tu, S.-M., Liu, C.-H., Yang, C.-S. & Wang, C.-H. (1970) Seroepidemiological studies on nasopharyngeal cancer by fluorescent antibody techniques with cultured Burkitt's lymphoma cell. *Gann*, **61**, 55-71

Klemola, E. & Kääriäinen, L. (1965) Cytomegalovirus as a possible cause of a disease resembling infectious mononucleosis. *Brit. med. J.*, **2**, 1099-1102

Klemola, E., von Essen, R., Henle, G. & Henle, W. (1970) Infectious mononucleosis-like disease with negative heterophile agglutination test. Clinical features in relation to Epstein-Barr Virus and cytomegalovirus antibodies. *J. infect. Dis.*, **121**, 608-614

Newall, K. W. (1957) The reported incidence of infectious mononucleosis, an analysis of a report of the Public Health Service. *J. clin. Path.*, **10**, 20

Niederman, J. C., McCollum, R. W., Henle, G. & Henle W. (1968) Infectious mononucleosis. Clinical manifestation in relation to EB virus antibodies. *J. Amer. med. Ass.*, **203**, 205-229

Niederman, J. C., Evans, A. S., Subrahmanyan, M. S. & McCollum, R. W. (1970) Prevalence, incidence and persistence of EB virus antibody in young adults. *New Engl. J. Med.*, **282**, 361-365

Nimmannitya, S. (1969) *A survey for EB virus antibodies in healthy children.* (Thesis, Yale University).

Papageorgiou, P. S., Sorokin, C., Kouzoutzaglou, K. & Glade, P. R. (1971) Herpes-like Epstein-Barr virus in leprosy. *Nature (Lond.)*, **231**, 47-49.

Pereira, M. S., Blake, J. M. & Macrae, A. D. (1969) EB virus antibody at different ages. *Brit. med. J.*, **2**, 526-527

Porter, D. D., Wimberly, I. & Benyesh-Melnick, M. (1969) Prevalence of antibodies to EB virus and other herpesviruses. *J. Amer. med. Ass.*, **208**, 1675-1679

Sawyer, R. N., Evans, A. S., Niederman, J. C. & McCollum, R. W. (1971) Prospective studies of a group of Yale University freshmen. I. Occurrence of infectious mononucleosis. *J. infect. Dis.*, **123**, 263-270

Strom, J. (1960) Infectious mononucleosis. Is the re-incidence increasing? *Acta med. scand.*, **168**, 35

Svedmyr, A. & Demissie, A. (1968) Age distribution of antibodies to Burkitt cells. *Acta path. microbiol. scand.*, **73**, 653-654

Thomsen, S. (1942) *Studier over mononucleosis infectiosa*, Munksgaard, Copenhagen

Tischendorf, P., Shramek, R. C., Balagtas, Deinhardt, F., Knopse, W. H., Noble, G. R. & Maynard, J. E. (1970) Development and persistence of immunity to Epstein-Barr virus in man. *J. infect. Dis.*, **122**, 401-409

Wahren, B. K., Lantorp, K., Stermer, G. & Epmark, A. (1970) EBV antibodies in family contacts of patients with infectious mononucleosis. *Proc. Soc. exp. Biol. (N.Y.)*, **133**, 934-939

Current Knowledge of the Epidemiology
of Nasopharyngeal Carcinoma—A Review

H. C. HO [1]

INTRODUCTION

Nasopharyngeal carcinoma (NPC) has for a long time intrigued epidemiologists because it is a rare tumour in most people but Chinese are unusually susceptible to it even if they have migrated to a distant country. Recently, interest in the tumour has extended to oncovirologists and immunologists because it has been found to have an association with an infection by a herpes-type virus immunologically very similar to the Epstein-Barr virus (EBV), a virus suspected to be oncogenic. Lilly (1966) has shown that in inbred strains of mice the incidence of lymphomas and leukaemias is linked to a genetically determined susceptibility to tumour induction by oncogenic viruses. In NPC we may have a similar situation at the human level. This discovery has therefore aroused much interest and led to a multinational and multidisciplinary collaborative investigation on the nature of the association which, it is hoped, may shed some light on whether viruses cause cancer in man.

HISTOPATHOLOGICAL DISTRIBUTION OF MALIGNANT TUMOURS OF THE NASOPHARYNX

In high-risk population groups, e.g., Chinese, at least 99% of the malignant tumours arising from the nasopharynx are carcinomas of the squamous type showing in the majority of cases poor or no differentiation on light microscopy. On electron microscopy, however, many of the undifferentiated and anaplastic cells show features of squamous differentiation. In low-risk population groups, the ratio

of the number of carcinomas (including the so-called "lympho-epitheliomas") to that of all types of malignant lymphomas (including reticulosarcomas) is significantly lower (Ho, 1971). Before we proceed to find an explanation for this difference, it is perhaps necessary to determine how much of it is actually due to a difference in diagnostic criteria that might be eliminated by a wider use of electron microscopy.

HISTOGENESIS

Liang *et al.* (1962), Ch'en (1964), and Shanmugaratnam & Muir (1967a) found that both the classical squamous and the undifferentiated carcinomas could arise from squamous, transitional or respiratory epithelium. It would appear, therefore, that squamous metaplasia is not a prerequisite.

EARLY CASES

It is important to investigate whether NPC has afflicted man since early times or is a relatively new disease. Since NPC often produces quite typical erosions in the base of the skull, examination of ancient skull specimens might give some information. A few such specimens have been reported to bear marks of damage probably caused by NPC. In one of them, Skull No. 236 kept at Duckworth Laboratory, Cambridge, reported by Wells (1963), the damage was more likely to have been due to myeloma or carcinoma of the left maxillary alveolus or floor of the maxillary sinus than to NPC (Ho, 1971). This specimen was thought to have been derived from an inhabitant of North East Africa and the Middle East from the period 3500–3000 BC. A "strongly probable one" (Wells, 1964) from Tepe

[1] Medical and Health Department Institute of Radiology, Queen Elizabeth Hospital, Kowloon, Hong Kong.

Hissar, Iran, c. 3000 BC, was described by Krogman (1940), who thought the condition might be primarily due to sinus infection brought about by dental disease. This specimen shows the centre of bone destruction to have been in the left facial bones. Smith & Dawson (1924), in their studies of Egyptian mummies, observed that no evidence of true cancer could be discovered until mummies from comparatively recent (Byzantine) times were examined, when two cases were found, one with lesions involving the base of skull and the other the rectum. The first of these was a male pre-Christian Nubian (c. 4th–6th century AD), first described by Derry (1909) as having suffered extensive destruction at the cranial base, from the cribiform plate in front to the basiocciput behind, and almost reaching the foramen magnum. This was either a case of NPC or sphenoidal sinus carcinoma. Even in patients the differential diagnosis between the two can be very difficult and sometimes impossible.

In China, the tumour has been known as "Kwang-tung tumour". A search of old Chinese medical writings has so far revealed that a fatal disease called "shih ying" or "shih jung", described in The Encyclopaedia of Chinese Medical Terms edited by Wu (1921), fits very closely the clinical picture of nasopharyngeal carcinoma with cervical nodal metastases (Ho, 1971). It is, however, not known when it was first noted.

RACIAL SUSCEPTIBILITY

Although the Chinese are of mongoloid stock, not all mongoloid people have a high risk of NPC. For instance, the disease is rare among Japanese (Miyaji, 1967). The Malays throughout South East Asia have a risk intermediate between that for the Chinese and the Indians in Singapore; the latter have a very low risk, like other caucasoid people elsewhere in the world (Shanmugaratnam & Muir, 1967b). In China itself the cancer appears to be much less common in the north than in the south (Ho, 1971). The Malays in Malaya, Singapore and Indonesia have had a long period of association, including some racial intermingling, with the Chinese, mainly those from the south, dating back to the early Ming dynasty in the 14th century. The Japanese have had an even longer association with the Chinese, dating back to the Ch'in dynasty (221–227 BC), but mainly with the northern Chinese. A comparatively high incidence of NPC is also found both in Chinese and the indigenous mongoloid people in Sarawak (Muir & Oakley, 1967; Arulambalam, 1968) and in Sabah (Muir, Evans & Roche, 1968), but there has been practically no racial intermingling between the latter and the Chinese, and the living habits of the various racial groups in Sarawak and Sabah are widely different. The disease is, however, very rare in Australian New Guinea, where the population is largely of Melanesian stock (Booth et al., 1968).

EFFECT OF MIGRATION ON THE RISK

The Cape Malays in South Africa came originally from Java. Rossall Sealy (1970) [1] estimates that the incidence of NPC in this group is over twice that in the "Coloured" population in the Magisterial districts of Cape Town, Wynberg, Bellville and Simonstown. Four Cape Malay and 11 Cape "Coloured" patients with the disease were seen in his department of the Groote Schuur Hospital over a period of 15 years. The population of the two communities have been roughly estimated as 60 000 and 360 000 respectively.

In migrant Chinese it has been established that those originating from different parts of China and living in the same city of adoption largely retain their original relative risks of NPC. Mekie & Lawley (1954) found that, in Singapore, Teochews (people from Chiu Chau) and Hokkinese (people from Fukien) have significantly lower NPC frequencies than Cantonese and people from other parts of Kwangtung. These findings are in agreement with those reported by Ho (1967a), which were subsequently confirmed by another study (Ho, 1971). In addition, Ho found that people originating from provinces outside Kwangtung and Kwangsi have significantly lower incidence rates than people from Kwangtung and Kwangsi, with Chiu Chau and Hong Kong excluded ($P < 0.01$ for males and $P = 0.025$ for females). People from Chiu Chau are ethnically more closely related to people from Fukien than to people from other parts of Kwangtung. It would seem, therefore, that geographical factors have little or no influence on the risk, which is determined largely by ethnic origin.

In Hong Kong and Singapore, Chinese constitute the majority of the population and consequently many of them still follow their ancestral living habits to a large extent, making it difficult to assess the

---

[1] Personal communication.

relative importance of the two factors involved—genetic and environmental. To carry the investigation a stage further, one should study the frequency of NPC in migrant and local-born Chinese who have settled in places such as Australia and the United States where the disease is rare, and the Chinese, constituting only a very small proportion of the population, are less likely by force of circumstances to follow closely their ancestral living habits. In Australia, Scott & Atkinson (1967), in an analysis of 227 (200 Caucasian and 27 Chinese) cases of NPC reported by the Radiotherapy Departments and the major private radiotherapeutic practices throughout Australia during the 11-year period 1953–63, found that it apparently made little difference to the risk of suffering from NPC whether a person of Chinese descent was born inside or outside Australia. Six of the 27 patients were born in Australia and the rest outside.

In Hawaii, Quisenberry & Reimann-Jasinski (1967) analyzed a series of 14 NPC cases of Chinese descent (9 born in the USA and 5 born elsewhere) reported to the Hawaiian Tumor Registry during 1960–62, and found the average annual incidence rate for those born elsewhere to be 54.2 per 100 000, which is over 6 times higher than the rate of 8.3 for those born in the USA. A rate of 54.2 is also about 3.5 times higher than the corresponding rate in Hong Kong. It makes one wonder whether some NPC patients who came from Hong Kong or South East Asia for treatment might have been included in the born-elsewhere group or whether this group might have included a very high proportion of middle-aged people, who have the highest risk of NPC. If patients of both these types were not excluded, no valid comparison can be made.

In California, Zippin et al. (1962) investigated the place of birth of 31 Chinese males reported to the Californian Tumor Registry during the 16-year period 1942–57, and found the ratio of the number of observed to expected cases (O/E) by age-group to be more than 8 times higher in Chinese under the age of 55 born outside the USA than in those born inside the country. However, no difference was found between the two groups above this age.

Buell (1965), in contrast, investigated the records of deaths from NPC in 67 men and 13 women of Chinese descent during the 14-year period 1949 to 1962. He found the risk of NPC for the immigrant group to be higher than that for the local-born (men and women combined and all age-groups included) by a factor of 1.5 to 2.

None of the above-mentioned reports have convincingly shown that Chinese born in the USA have a lower risk of NPC than Chinese immigrants. They have, however, established that both immigrant and local-born Chinese have a significantly higher risk of NPC than Caucasians living in the same area. The subject was reviewed by Ho (1971).

RISK IN PEOPLE WITH PART-CHINESE ANCESTRY

Ho (1971) produced evidence suggesting that "Macaonese", largely descendants of Portuguese who intermarried with Chinese from Kwangtung in Macao, had a much higher frequency of NPC than the rest of the non-Chinese population in Hong Kong. In Thailand, Garnjana-Goochorn and Chantarakul (1967) roughly estimated that the ratios of relative frequencies of NPC in Chinese, Chinese with part-Thai ancestry and Thais were 3.4 : 2.2 : 1.0.

It is interesting to note that, despite the fact that the Thais are mongoloid people and mostly Buddhists by religion, whereas Portuguese are Caucasians and traditionally Catholics, in both cases the offspring from their intermarriage with Chinese appear to inherit a part of the high risk of their Chinese ancestors.

FAMILIAL AGGREGATION

In a retrospective search of the medical records of cancer cases (sex-determined cancers excluded) seen in 1969 for family history of NPC, Ho (1971) found a significantly higher frequency of NPC in close blood relatives of NPC patients than in those of patients suffering from other cancers. The familial aggregation of NPC was quite random, and appeared to be as likely to occur in the vertical as in the horizontal direction and not to be sex-linked. The results are shown in Tables 1 and 2.

These findings have been confirmed subsequently by a prospective study of similar types of cancer cases during the first three months of 1971 (see Table 3).

It is of interest to note from Table 1 that 3 of the 12 families with aggregation of NPC are "boat" people, a very small, highly inbred ethnic group, that constitutes not more than 2.7% of the total population of Hong Kong.

So far, there has been only one report of NPC occurring in both partners of a pair of dizygotic

Table 1.   Frequency of family history of NPC in patients with NPC and in those with other cancers (OC) diagnosed at Medical and Health Department Institute of Radiology, Hong Kong, in 1969

Cancer	Families with history	Families with no history	Total
NPC	12 [1]	385	397 [2]
OC	2	687	689 [3]
Total	14	1072	1086

$\chi^2 = 14.77785$ (P < 0.001)   t = 3.864234 (P = 0.00011)

[1] 3 of the 12 families are "boat" people.
[2] 72 cases excluded because a family history was unobtainable, unreliable or not obtained.
[3] 515 cases excluded, 230 for the above reasons and 285 because the cancers are sex-determined, e.g., gynaecological, penile, etc.

Table 3.   Frequency of family history of NPC in patients with NPC and in patients with other cancers seen during the first three months of 1971

Cancer	Families with history	Families with no history	Total
NPC	4 [1]	109	113
OC	0	109	109 [2]
Total	4	218	222

$\chi^2 = 3.92920$ $0.025 > P < 0.05$

[1] (1) Paternal side: grandmother, an uncle and an aunt died of NPC (hearsay diagnoses and typical history); (2) youngest sis. has verified NPC; (3) eldest bro. has verified NPC, and (4) mother died of NPC (typical history).
[2] 3 cases excluded because family history unobtainable.

twins of Sze Yap origin (Ho, 1971). The ages at clinical onset were 38 and 49.

Ho (1971) has also reported a pedigree study of a Cantonese family that had settled in Hong Kong since generation II. In this family, members of three successive generations had NPC (see Fig. 1).

GENETIC PROFILE

Clifford (1970), in comparing the ABO blood group distribution in 233 Kenyan patients with NPC and in controls, found a significance level of 3% in the comparison A/O which, he considers, is highly suggestive that in Kenya, group-A persons are "pro-

Table 2.   Familial aggregations of NPC in patients seen in 1969

Families	Vertical	Horizontal	Remarks
(1)	Daughter (II/2M+4F) & father	1 female paternal 1st cousin	
(2)	Son (V/2M+3F) & mother [1]		
(3)		2 sisters (I & II/6M+3F)	"Boat" people
(4)	Son (IV/4M+5F) & mother		
(5)		2 paternal 1st cousins: Male (III/3M+2F) & M(2M)	
(6)	Son (V/4M+3F) & father		
(7)		Male (V/3M+2F) & bro. (II)	
(8)	Daughter (IV/1M+5F) & father [1]	& bro. (I)	
(9)	Son (only child) & father		
(10)		Male (only child) & his 1/2-bro. by same father, but latter's 1/2 bro. by same mother well	"Boat" people
(11)	Mother (II/1M+3F) & daughter		
(12)		Bro. (?/4M+1F) & 1 bro.	"Boat" people
	Total = 7	Total = 7	

[1] Diagnosis based on typical history only.
Note : (II/2M+4F) means propositus is the 2nd child (II) of a family of 2 sons and 4 daughters.

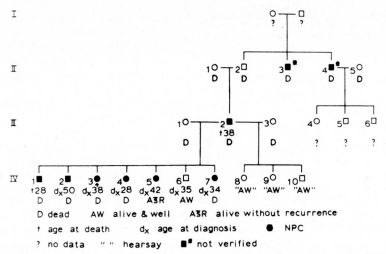

Fig. 1. Pedigree study of a Chinese family of Sze Yap origin with aggregation of NPC in three successive generations.

tected" or at less risk of NPC than persons with other blood groups. In contrast, Ho (1971), in Hong Kong, failed to find any significant difference between the blood group distribution (A/O and B/O) in 1000 consecutive Chinese NPC patients and in controls.

ENVIRONMENTAL FACTORS

NPC occurring in both marital partners

Ho (1971) reported only two verified instances amongst 5070 cases seen in Hong Kong. Although this must be considered the minimum incidence, one would have expected a much higher frequency if those environmental factors that are normally the same for both partners, since they are commonly found in Chinese homes, were of significance in adult life.

Occupation and socio-economic level

Neither in Singapore (Shanmugaratnam & Higginson, 1967) nor in Hong Kong (Ho, 1971) has the risk of NPC been found to be higher in certain occupations than in others. In Singapore, Polunin (1967) found a lack of any clear-cut association between ways of living and NPC, and in Hong Kong, Ho (1971) had the impresssion that a higher or lower risk was not associated with any particular socio-economic level. Andrews & Michaels (1968a, b) reported three cases of NPC in Canadian bush pilots, a very small occupational group. They thought that in these cases NPC could be associated with

frequent and rapid air pressure changes in unpressurized aircraft, and also with cigarette smoke. No subsequent case has been reported in these bush pilots or in other groups of airmen working under somewhat similar conditions.

Inhalants

Ho (1967a) found that the marine population of Hong Kong had a significantly higher incidence rate of NPC than the land population living largely in congested dwellings, and that the rate for "boat" people engaged in fishing as a livelihood was significantly higher than that for the rest of the Chinese population. The latter finding was confirmed by a later study, the results of which are shown in Table 4. These "boat" people live practically all their lives in boats and cook their food in the open. It could be safely assumed that they inhale much less household carcinogens than the land-dwellers, yet they have a higher risk. Furthermore, it is the custom of the "boat" people for their females to do all the cooking, yet NPC affects their males 2.7 times more frequently than their females, much the same as in the case of the land-dwellers. It would appear, therefore, that common household inhalants cannot account for the high incidence of NPC in Chinese.

In Kenya, Clifford & Beecher (1964) suspected that the inhalation of smoke from burning exotic trees in ill-ventilated huts might have some bearing on the tribal distribution and incidence of NPC, but the crude incidence rate for the Nandi tribe,

Table 4. Incidence rates of nasopharyngeal cancer for "boat" people engaged in fishing as main livelihood and for the rest of the Hong Kong Chinese population

Year	No. of cases		
	All Chinese	"Boat" people	Others
1959	261 [1]	12	249
1960	294 [2]	10	284
1961	366	19	347
1962	339	11	328
1963	343	15	328
Total	1603 [4]	67 [3]	1536
1961 population	3 079 901	81 649	2 998 252
Cases/million/annum	104	164	102

P < 0.001

[1] 2 cases of malignant melanoma included.
[2] 1 case of reticulosarcoma included.
[3] 2 cases not histologically proven.
[4] 30 cases not histologically proven.

which has the highest rate, is only slightly higher than that for the Swedes, who live in well-ventilated homes. Booth *et al.* (1968) found living conditions similar to those of Kenyans in more than 1 million people living in the highlands of Australian New Guinea, but NPC is a rarity there.

Chinese incense or joss sticks are burnt not only in Chinese Buddhist and Taoist temples but also on a small scale in many Chinese homes in Hong Kong, South East Asia and the USA even today; this was also formerly the practice in China itself. Many "boat" people in Hong Kong customarily place three burning incense sticks at the bow of their boats in the morning and another three in the evening, weather permitting, but their living quarters are at the rear. Again it is the females who do this. In addition, they worship their patron saint, Tien Hau, in temples at least once a year during the Tien Hau Festival. The amount of incense smoke they could have inhaled from such sources is negligible compared with that inhaled by non-Christian Chinese land-dwellers in Hong Kong. Ho (1967b) found the NPC frequency not to be higher in Buddhist and Taoist religious workers, who spend much of their time in an incense-smoke-laden atmosphere, than in the rest of the Hong Kong population. This is in agreement with the observation by Sturton, Wen & Sturton (1966) that NPC was rare in Hangchow in central China where, in a population of about 800 000 before the Sino-Japanese War, there were about 10 000 Buddhist monks and a smaller number

of Taoist priests. The disease is, on the other hand, not rare in Catholic "Macaonese" and Muslim Malays, neither of whom burn Chinese incense.

According to Aaron E. Freeman and his associates of Microbiological Associates Inc., Bethesda (Freeman, 1970 [1]), the smoke from a burning Chinese incense stick could cause a pronounced transformation of rat embryo cells in tissue culture equivalent to that caused by a potent chemical carcinogen, but he thinks that these studies may not be of significance in revealing the potential rôle of incense sticks in causing human cancer for the following reasons: (1) tissue culture assay systems are, perhaps, too sensitive; (2) *in vitro* transformation may not be related to *in vivo* transformation; (3) rat cells may be less resistant to transformation than human cells; (4) whole animals may have much better detoxifying systems than cell cultures; and (5) some noncarcinogenic chemicals can transform cells in culture.

Of the other inhalants, opium has been excluded as without significance (Ho, 1971). Antimosquito coils, cigarettes, pipe tobacco, snuff, Chinese medicinal inhalants and drugs for intranasal applications are probably of no significance (Shanmugaratnam & Higginson, 1967).

Ingestants

Druckrey *et al.* (1964a) have shown that compounds of the nitrosamine group can act systemically and are organ specific. They have further shown (Druckrey *et al.*, 1964b) that a single dose of such a carcinogen can initiate a train of events that will culminate in cancer development, after a long latent period, without a further dose. Nitrosamines are formed when nitrites are used as food additives or preservatives. Cantonese salt fish is likely to contain *N*-nitrosamines, which are potent carcinogens, and should be investigated because it is a common item of food among southern Chinese, rich or poor, inside or outside China, in South East Asia, Australia or the United States. Many "Macaonese" in Hong Kong and Macao, and even the Malays in Malaya, eat it occasionally. The common method of preparing it is to place the fish in pickle as soon as they are caught. The fish then die of suffocation and are not cleaned, thus allowing the enzymes in the blood and the digestive juices to remain in the fish. This is in marked contrast to the practice of British, Canadian and American fishermen, who clean their fish as soon as they are caught (McCarthy

[1] Personal communication.

& Tausz, 1952). Because of the special method of preparation, which permits a kind of "benign decay" to take place in the fish so as to produce a "gamey" flavour relished by most southern Chinese palates, such salt fish, when exported, goes only to Chinese communities abroad. In 1969, 3262 cwt of salt fish were exported to the USA and 224 cwt to Australia. It would be exceptional to find a southern Chinese who had never eaten salt fish in his life.

INTERNAL FACTORS, INCLUDING VIRUSES

No convincing evidence has been reported that the risk of NPC is associated with malnutrition, avitaminosis, certain hormonal states, chronic upper respiratory infection or vasomotor rhinitis (Ho, 1971).

If a viral aetiology is postulated, the virus must be one that is ubiquitous on a world-wide basis in order to explain the regular high incidence of NPC in people of Chinese descent domiciled in widely scattered parts of the world, and its mode of action must be highly selective to account for the marked differences in incidence in different ethnic groups living in the same geographical locality. The selectivity may be determined by cofactors, intrinsic or extrinsic, that may be present in some and absent in others. The subject has been reviewed by de-Thé [1].

---

[1] See p. 275 of this publication.

COMPARATIVE PATHOLOGY

In animals other than primates there is hardly any postnasal space because the nasal septum extends backwards to the posterior wall of the pharynx. In Hong Kong, Teoh (1971) found tumours in the posterior part of the nasal cavities, which may be considered the analogue of the nasopharynx in man, in 16 Friesian cows. Four of the tumours were epidermoid carcinomas, five were carcinomas and seven were adenocarcinomas. The tumours appeared as fungating ulcerative lesions showing gross invasion of the base of the skull and brain and of the paranasal sinuses, as typically found in the invasive type of NPC in man. No serological studies for EBV-associated antibodies have been done.

PATTERN OF AGE DISTRIBUTION

Fig. 2 shows the age distributions of NPC in the male Chinese populations of Hong Kong and Singapore and in the male population of Sweden. The pattern for Chinese in both Hong Kong and Singapore shows a rapid, almost uninterrupted and fairly regular rise after 20–24 years of age, i.e., earlier than for most epithelial cancers. The incidence rate reaches a high plateau at 45–54 years of age and begins to decrease thereafter. This would suggest that, in Chinese, the disease is unlikely to be due to continued exposure to an external carcinogenic agent

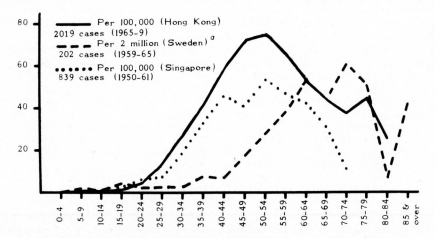

Fig. 2. Male age-specific incidence rates of NPC for Sweden and for Chinese in Hong Kong and in Singapore. For Sweden, only transitional cell, squamous and undifferentiated carcinomas were included.

[a] Only transitional-cell, squamous and undifferentiated carcinomas included.

throughout life, as is postulated for most of the common epithelial cancers, or, alternatively, that susceptibility to the carcinogen is influenced by an internal factor, possibly of hormonal nature, that increases the susceptibility progressively after adolescence, and causes a progressive decrease in susceptibility after 50–54 years of age. Figs. 3 and 4 show the age-specific incidence curves for Hong Kong and Sweden respectively. In both places the curves for the two sexes show a difference in height only but do not differ greatly in shape. However,

when the male curves for the two places are compared it becomes immediately evident that the curves for Sweden show a progressive increase in incidence up to 70–74 years of age, resembling the pattern shown by bronchial carcinoma in cigarette smokers and by most of the common epithelial cancers. This difference would seem to suggest that NPC in the two ethnic groups may not have a common aetiology. The Swedes also have an age-standardized male-to-female incidence ratio of only 1.86 : 1 compared with 2.38 : 1 in Hong Kong Chinese (Ho, 1971).

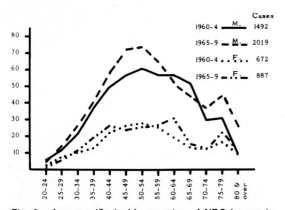

Fig. 3. Age-specific incidence rates of NPC by sex in Hong Kong. Patients who came from outside Hong Kong were excluded.

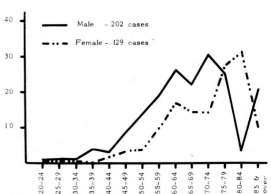

Fig. 4. Age-specific incidence rates of NPC (transitional-cell, squamous and undifferentiated) in Sweden by sex.

SUMMARY

Of the tumours of the nasopharynx, carcinoma of the squamous type, more often undifferentiated than differentiated, occurs with high frequency in southern Chinese, irrespective of their place of domicile, and has a close association with an infection by a herpesvirus immunologically resembling Epstein-Barr virus. Of the ancient pathological skull specimens previously suspected of bearing marks of damage due to nasopharyngeal carcinoma, only that derived from a male pre-Christian Nubian, described by Derry (1909), is considered as possibly bearing such marks. It is not known when the disease was first described in China. The evidence available tends to suggest that the high risk of the disease in southern Chinese is determined by a combination of genes that are not sex-linked or related to the ABO blood

group, but the possibility of other factors also playing an aetiological rôle has not been excluded. No such factor has been found, but southern Chinese salt fish and a viral factor are most deserving of a thorough investigation.

The nature of the close association between nasopharyngeal carcinoma and an EBV-like herpesvirus infection, which is not influenced by geography or ethnic origin, is not known. Carcinomas have been found in Friesian cows in Hong Kong in the posterior part of the nasal cavities, which is probably the analogue of the nasopharynx in man. The difference in the patterns of the age-distribution of nasopharyngeal carcinoma in Hong Kong and Sweden may be due to a difference in aetiology.

ACKNOWLEDGEMENTS

Thanks are due to Dr the Hon. Gerald H. Choa, Director of Medical and Health Services of Hong Kong, for his permission to publish this article, Dr T. B. Teoh and Dr C. C. Lin for histopathological diagnoses, the medical staff of the Medical and Health Department

Institute of Radiology for their care in obtaining family histories, Mr C. M. Lam for statistical assistance, Mrs P. Liu for careful secretarial assistance, Messrs K. Fung, H. K. Tam and Y. S. Lau for data collection, and Mr K. W. Leung for photographic assistance.

The author is indebted to the Swedish Cancer Registry and Professor Nils Ringertz, Scientific Surveyor of the Registry, for the generous supply of Swedish data, Professor J. Mitchell, FRS, and Mr J. A. Fairfax Fozzard of Cambridge University for supplying photographs and radiographs taken of skull No. 236 kept at Duckworth Laboratory, and Dr Sealy Rossall of the Department of Radiotherapy, Groote Schuur Hospital, South Africa, for information relating to the Cape Malays.

REFERENCES

Andrews, P. A. J. & Michaels, L. (1968a) Nasopharyngeal carcinoma in Canadian bush pilots. *Lancet*, **ii**, 85-87

Andrews, P. A. J. & Michaels, L. (1968b) Aviator's cancer. *Lancet*, **ii**, 639

Arulambalam, T. R. (1968) Cancer in Sarawak. *Far East Med. J.*, **4**, 321-325

Booth, K., Cooke, R., Scott, G. & Atkinson, L. (1968) Carcinoma of the nasopharynx and oesophagus in Australian New Guinea 1958-1965. In: Clifford, P., Linsell, C. A. & Timms, G. L., eds, *Cancer in Africa*, Nairobi, East African Publishing House, pp. 319-322

Buell, P. (1965) Nasopharyngeal cancer in Chinese of California. *Brit. J. Cancer*, **19**, 459-470

Ch'en, C. C. (1964) The morphological changes and biochemical reactions of the stroma associated with the process of malignant transformation. *Abstracts of papers of 1964 Cancer Conference of Chung Shan Medical College (Commemorative publication for the opening of the Huanan Cancer Hospital)*, p. 13

Clifford, P. & Beecher, J. L. (1964) Nasopharyngeal cancer in Kenya. Clinical and environmental aspects. *Brit. J. Cancer*, **18**, 25-43

Clifford, P. (1970) A Review on the epidemiology of nasopharyngeal carcinoma. *Int. J. Cancer*, **5**, 287-309

Derry, D. E. (1909) Anatomical report (B). *Archaeological survey of Nubia*, Cairo, Egyptian Ministry of Finance, pp. 40-42 (*Bulletin* No. 3)

Druckrey, H., Ivankovic, S., Mennel, H. D. & Preussmann, R. (1964a) Selektive Erzeugung von Carcinomen der Nasenhöhle bei Ratten durch N,N -Nitrosopiperazin, Nitrosopiperidin, Nitrosomorpholin, Methylallyl, Dimethyl und Methyl vinyl-Nitrosamin. *Z. Krebsforsch.*, **66**, 138-150

Druckrey, H., Steinhoff, D., Preussmann, R. & Ivankovic, S. (1964b) Erzeugung von Krebs durch eine einmalige Dosis von Methylnitroso-Harnstoff und verschiedenen Dialkyl-nitrosaminen an Ratten. *Z. Krebsforsch.*, **66**, 1-10

Garnjana-Goochorn, S. & Chantarakul, N. (1967) Nasopharyngeal cancer at Siriraj Hospital, Dhonburi, Thailand. In: Muir, C. S. and Shanmugaratnam, K., eds, *Cancer of the nasopharynx*, Copenhagen, Munksgaard, pp. 33-37 (*UICC Monograph Series*, No. 1)

Ho, H. C. (1967a) Nasopharyngeal carcinoma in Hong Kong. In: Muir, C. S. and Shanmugaratnam, K., eds, *Cancer of the nasopharynx*, Copenhagen, Munksgaard, pp. 58-63 (*UICC Monograph Series*, No. 1)

Ho, H. C. (1967b) Cancer of the nasopharynx. In: Harris, R. J. C., ed., *Panel II, Ninth International Cancer Congress*, Berlin, Heidelberg, New York, Springer-Verlag, pp. 110-116 (*UICC Monograph Series*, No. 10)

Ho, H. C. (1971) Nasopharyngeal carcinoma (NPC). *Advanc. Cancer Res.* (in press)

Krogman, W. M. (1940) The skeletal and dental pathology of an early Iranian site. *Bull. Hist. Med.*, **8**, 28-48

Liang, P. C., Ch'en, C. C., Chu, C. C., Hu, Y. F., Chu, H. M. & Tsung, Y. S. (1962) The histopathologic classification, biologic characteristics and histogenesis of nasopharyngeal carcinomas. *Chin. med. J.*, **81**, 652

Lilly, F. (1966) The histocompatibility-2 locus and susceptibility of tumour induction. *Nat. Cancer Inst. Monogr.*, **22**, 631-642

McCarthy, J. P., & Tausz, J. (1952) *Salt fish industry in Hong Kong*. Hong Kong, The Governemnt Printer, p.6

Mekie, D. E. C. & Lawley, M. (1954) Nasopharyngeal carcinoma. *Arch. Surg.*, **69**, 842

Miyaji, T. (1967) Cancer of the nasopharynx and related organs in Japan based on mortality, morbidity and autopsy studies. In: Muir, C. S. and Shanmugaratnam, K., eds, *Cancer of the nasopharynx*. Copenhagen, Munksgaard, pp. 29-32 (*UICC Monograph Series*, No. 1)

Muir, C. S. & Oakley, W. F. (1967) Nasopharyngeal cancer in Sarawak (Borneo). *J. Laryng.*, **81**, 197-207

Muir, C. S., Evans, M. D. E. & Roche, P. J. L. (1968) Cancer in Sabah (Borneo), a preliminary survey. *Brit. J. Cancer*, **22**, 637-645

Polunin, I. (1967) The ways of life of peoples with high risk of nasopharyngeal carcinoma. In: Muir, C. S. and Shanmugaratnam, K., eds, *Cancer of the nasopharynx*, Copenhagen, Munksgaard, pp. 106-111 (*UICC Monograph Series*, No. 1)

Quisenberry, W. B. & Reimann-Jasinski, D. (1967) Ethnic differences in nasopharyngeal cancer in Hawaii. In: Muir, C. S. and Shanmugaratnam, K., eds, *Cancer of the nasopharynx*, Copenhagen, Munksgaard, pp. 77-81 (*UICC Monograph Series*, No. 1)

Scott, G. C. & Atkinson, L. (1967) Demographic features of the Chinese population in Australia and the relative prevalence of nasopharyngeal cancer among Caucasians and Chinese. In: Muir, C. S. & Shanmugaratnam, K., eds, *Cancer of the nasopharynx*, Copenhagen, Munksgaard, pp. 64-72 (*UICC Monograph Series*, No. 1)

Shanmugaratnam, K. & Higginson, J. (1967) Aetiology of nasopharyngeal carcinoma. In: Muir, C. S. and Shanmugaratnam, K., eds, *Cancer of the nasopharynx*, Copenhagen, Munksgaard, pp. 130-137 (*UICC Monograph Series*, No. 1)

Shanmugaratnam, K. & Muir, C. S. (1967a) Nasopharyngeal carcinoma and structure. In: Muir, C. S. and Shanmugaratnam, K., eds, *Cancer of the nasopharynx*, Copenhagen, Munksgaard, pp. 153-162 (*UICC Monograph Series*, No. 1)

Shanmugaratnam, K. & Muir, C. S. (1967b) The incidence of nasopharyngeal cancer in Singapore. In: Muir, C. S. and Shanmugaratnam, K., eds, *Cancer of the nasopharynx*, Copenhagen, Munksgaard, pp. 47-53 (*UICC Monograph Series*, No. 1)

Smith, G. E. & Dawson, W. R. (1924) *Egyptian mummies*. London, Allen & Unwin, p. 157

Sturton, S. D., Wen, H. L. & Sturton, O. G. (1966) Etiology of cancer of the nasopharynx. *Cancer*, **19**, 1666-1668

Teoh, T. B. (1971) The pathologist and surgical pathology of head and neck tumours. *J. roy. Coll. Surg. Edinb.*, **16**, 122-123

Wells, C. (1963) Ancient Egyptian pathology. *J. Laryng.*, **77**, 261-265

Wells, C. (1964) Two mediaeval cases of malignant disease. *Brit. med. J.*, 1611

Wu, C. H. & disciples (1921) *The encyclopaedia of Chinese medical terms*, Vol. 1, p. 756. (In Chinese)

Zippin, C., Tekawa, I. S., Bragg, K. U., Watson, D. A. & Linden, G. (1962) Studies in heredity and environment in cancer of the nasopharynx. *J. nat. Cancer Inst.*, **29**, 483-490

Nasopharyngeal Carcinoma in Non-Chinese Populations

C. S. MUIR [1]

While nasopharyngeal carcinoma (NPC) has long been known to be very common in Southern Chinese (Dobson, 1924; Digby, Thomas & Hsiu, 1930), little attention has been given to the occurrence of the neoplasm in other ethnic groups. The occurrence of the disease is examined, largely in those populations in South East Asia and Africa that have levels of risk intermediate between those of the Southern Chinese and the white populations of Europe and North America. Unless otherwise stated, all rates are per 100 000 population per annum and are age-adjusted to the "world" standard population (UICC, 1970). The findings are discussed in the light of the hypothesis that this cancer is due to an unknown environmental factor acting on genetically susceptible populations.

CHINESE

In virtually all occidental countries NPC rates are very low, usually well below unity. The incidence rates given in Table 1 for Chinese vary between 10 and 20 for males and from 5 to 10 for females. There are, of course, differences within China (for references, see Shanmugaratnam (1967)).

NON-CHINESE MONGOLOIDS

NPC is common in the non-Chinese mongoloid populations of South East Asia (Table 2). It is very rare in Japan (Miyaji, 1967) and probably infrequent in Korea (Yun, 1949), Mongolia and North China (Hu & Yang, 1959).

---

[1] Unit of Epidemiology and Biostatistics, International Agency for Research on Cancer, Lyon, France.

Table 1. Age-adjusted morbidity rates per 100 000 per annum for nasopharyngeal cancer in Overseas Chinese, Hawaii and Israel [1]

Population group	Cases		Rates	
	M	F	M	F
Overseas Chinese				
California [2]	5	1	12.3	5.4
Hawaii [3]	12	5	10.4	4.6
Hong Kong [4]	1980	875	24.3	10.2
Singapore [5]	442	202	20.2	9.0
Taiwan [6]	1516		7.1	3.1
Hawaii [3]				
Hawaiians	7	1	7.8	0.7
Caucasians	4	3	1.1	0.9
Chinese	12	5	10.4	4.6
Filipinos	8	0	3.1	0
Japanese	4	2	0.9	0.3
Israel [3]				
Jews born in Israel	6	3	0.7	0.3
Jews born in Africa or Asia	38	23	1.8	1.0
Jews born in Europe or America	26	6	0.6	0.1
Non-Jews	9	4	1.3	0.6

[1] Age-adjusted to the "world" standard population (UICC, 1970). M = Male; F = Female.
[2] State of California (1967).
[3] UICC (1970).
[4] Ho (1972); see p. 357 of this publication.
[5] Muir, Shanmugaratnam & Tan (1970).
[6] Lin et al. (1971).

AFRICA

Sudan and Zanzibar

El Hassan et al. (1968) noted that 5% of all biopsied cancers seen at the Khartoum Civil Hospital (KCH) in 1962–65 were NPC. Saad (1968) found that 7 of 17 cases seen by him at the Ear, Nose and Throat (ENT) department of the KCH were in Nubas. Daoud and El Hassan kindly provided the figures in Table 2, noting that 20% of patients were below 20 years of age.

Table 2. Relative frequency of nasopharyngeal cancer in African and Asian populations

Population group	Cases[1]		Relative frequency %	
	M	F	M	F
Malaysia[2]				
Chinese	150	77	11.5	6.0
Malays	54	20	10.6	4.2
Indians	7	2	1.3	0.4
Singapore[3]				
Chinese	442	202	13.9	8.1
Malays	20	5	11.2	2.8
Indians	2	0	0.5	0
Thailand (Bangkok)[4]				
Chinese		27		15.9
Chinese/Thai		20		10.3
Thais (Chiang Mai)[5]		29		4.6
North Thais	34	19		
			3.7	2.1
Sabah[6]				
Chinese	9	0	8.9	0
Indigenous	11	3	9.2	3.4
Sarawak[7]				
Chinese	6	5	4.8	3.8
Malays	1	1	3.2	4.2
Dayaks	8	4	8.6	4.9
Jawa[8]				
Chinese	31	3	18.2	1.4
Javanese	108	44	10.3	2.9
Formosa[9]	1260	446	23.2	5.2
Filipinos[10] (Philippines)		182		2.9
Filipinos[11] (Hawaii)	8	0	2.5	0
Vietnamese[12]	105	58		
		163		3.7
Algeria[13]				
Algerians		364		5.0
Europeans		83		3.5
Tunisia[14]				
Tunisians	125	25		4.3
Europeans				1.6
Congo[15] (Brazzaville)	1	0	0.4	0
Senegal[16] (Dakar)	1	0	0.1	0
Ivory Coast[17] (Abidjan)	2	0	0.5	0
Morocco[16]	95	46	8.5	3.8
Sudan[18]	27	12	5.6	2.1
Zanzibar[19]				
Negroes		1		0.4
Arabs		1		1.3
Mozambique[20] (Lourenço Marques)	3	0	0.7	0
Nigeria[11] (Ibadan)	5	3	0.7	0.4
South Africa[11] (Cape Province: Coloureds)	4	1	0.5	0.1

[1] No. of cases of nasopharyngeal cancer. These figures should not be used without consulting the text and/or the source material. The relative frequencies for Lourenço Marques, Ibadan and Cape Coloureds are derived from incidence data. M = Male; F = Female.   [2] Marsden (1958).
[3] Muir et al., (1971).   [4] Garnjana-Goonchorn & Chantarakul (1967).   [5] Menakanit et al., (1971).   [6] Muir et al., (1968).
[7] Muir & Oakley (1966).   [8] Djojopranoto & Soesilowati (1967).
[9] Yeh (1967).   [10] Pantangco et al., (1967).   [11] UICC (1970).
[12] Huong et al., (1969).   [13] Bréhant (1967).   [14] Chadli (1966).

Kenya

Clifford (1967) has claimed that NPC is common in certain parts of Kenya and certainly the cancer is of high relative frequency among head and neck malignancies treated in Nairobi. By contrast, Linsell (1968) noted a relative frequency for NPC of 2.3 % in Kenya biopsy material in 1957–61, this being greatest in the Kalenjin tribe (3.2 %) and least in the Coastal Bantu (0.2 %).

When crude incidence rates are computed from Clifford's (1967) material, the maximum is 1.34 in male Nandi. These rates are not age-adjusted and it is likely that there is considerable under-diagnosis. Nevertheless, it is doubtful whether the "true" age-adjusted rates would be greater than 5.

If differential utilization of medical services is not the underlying cause, then the observed tribal variations in crude incidence within Kenya are of great interest. The cancer rarely occurs in the Coastal Bantu and has not been reported in the Hamitic group; the crude incidence in the Nilo-Hamitic and Central Bantu is of the same order. There are important environmental differences, probably related to altitude and climate, between these four major ethnic groups, which Clifford believes lead to the inhalation of wood fuel fumes.

Uganda

Schmauz & Templeton (1970) have examined tribal distribution, finding a higher minimum age-adjusted incidence rate in Nilotics and Para-Nilotics than in the Bantu/Sudanic groups. In general terms, the North of the country showed a markedly higher incidence than did the South, despite the better hospital and transport systems in the South. The authors conclude that it is an open question as to whether this variation is the result of environmental or genetic factors.

Algeria, Morocco, Tunisia

Relative frequency studies from Algeria, Morocco and Tunisia (Table 2) suggest a moderately elevated NPC level. The Moroccan material, from a radiotherapy department, is certainly biased. The suggestion that NPC was less frequent in Europeans living in Algeria and Tunisia is difficult to evaluate. However, in 23 months, 156 nasopharynx cancers (115 males, 41 females) were diagnosed at the Institut

[15] Tuyns & Ravisse (1970).   [16] Tuyns (1971); personal communication.   [17] Duvernet-Battesti (1970).   [18] Daoud & El Hassan (1970); personal communication.   [19] Chopra (1968).
[20] UICC (1966).

National de Carcinologie, Tunis (Zaouche, 1970): as in the Sudan, many patients were very young. It is likely that there is gross under-reporting; nevertheless, if these cancers be related to the entire Tunisian population (4.5 million), the crude minimum incidence rates for males and females are 2.6 and 0.9 respectively (the age-adjusted minimum incidence rates would probably be double). NPC incidence would appear to be raised in Tunisians.

Other African countries

NPC incidence rates, available from several cancer registries in Africa, are generally low (UICC, 1966, 1970).

HAWAII AND ISRAEL

Hawaii and Israel are considered together as both contain migrant groups coming from areas of elevated frequency.

Hawaii

The NPC incidence rates for the ethnic groups of Hawaii are given in Table 1. The rates, based on very small numbers, for Hawaiians are 7.8 and 0.7 for males and females respectively, compared to the Chinese rates of 10.4 and 4.6. Part of the Hawaiian increase may be due to the fact that this group includes part-Hawaiians, some of whom are the offspring of Hawaiian/Chinese marriages. The rates for the other groups also appear raised, an intriguing finding.

Israel

The incidence in the non-Jewish (Arab) population and in Jews born in Africa or Asia (many of whom were born in Morocco or Tunisia) is rather higher than in the Israel-born Jews or in Jews born in America or Europe (Table 1). While North African Jews have intermarried with non-Jews to some extent, these findings, in consonance with the elevation of risk in Tunisia, would suggest the importance of environmental factors in North Africa.

DISCUSSION

In any discussion of the significance of the reported differences, it should be borne in mind that many are based on relative frequency data and small numbers of cases.

With the above caveats in mind concerning the validity of the basic data, two questions merit discussion: firstly, whether the effect on NPC level of intermarriage with Chinese can be assessed from available material and, secondly, whether this cancer is common in certain parts of Africa.

Effect of intermarriage with Chinese

If genetic influence is strong in NPC, then NPC incidence in the offspring of marriages between Chinese and other ethnic groups, preferably residing in the same area and following the same way of life as pure-blood Chinese, would be of interest. Information on this question is available, to some extent, for South East Asia.

Thus, in Thailand, where Chinese migration has occurred over the centuries, one would expect to find Thai rates higher than those for occidentals. Further, one would anticipate that the highest rates would be found in recent migrants from China, the lowest rates in Thais, and intermediate rates in persons of mixed Thai and Chinese blood. In Table 2 it will be seen that, in Bangkok, persons of mixed Chinese/Thai blood have a relative frequency for NPC falling between those for pure-blood Chinese and Thais. Further, in Singapore, Malaya and Indonesia, the Chinese have higher frequencies than the Malays and Indonesians.

In Borneo, the frequency in the Dayaks of Sarawak is higher than that in Chinese, and in the indigenous people of Sabah it is of the same order. While the Borneo series need to be extended and confirmed, they are of great interest as the ways of life of all the groups concerned are still quite different (Muir & Oakley, 1967), and it is difficult to think of a common environmental exposure.

Unfortunately, it is not yet possible to examine NPC incidence in the offspring of marriages between Chinese and groups entirely free from Chinese genetic influence.

In summary, despite certain anomalies, the evidence from South East Asia suggests that the greater the admixture of Chinese blood the more likely is the NPC level to be raised. But the diminishing frequency of NPC in North China and the virtual absence in the Mongoloid Japanese confirm that there is more to this disease than genes.

Nasopharyngeal cancer in Africa

The relative frequency data for Algeria, Tunisia, the Sudan and possibly Morocco strongly suggest a moderate elevation of risk, which, if confirmed, offers for study populations of genetic constitution presumably far removed from that of the Chinese.

There is probably, despite the manifest under-reporting, a *moderate* increase in incidence in certain tribal groups in East Africa.

SUMMARY

Nasopharyngeal cancer (NPC) is very common in Chinese, the age-adjusted incidence rates per 100 000 *per annum* varying between 10 to 20 for males and 5 to 10 for females. In virtually all occidental countries, the rates are generally well below 1.0 per 100 000 *per annum*.

Rare in Japan and India, NPC is fairly common in non-Chinese mongoloid groups in South East Asia.

NPC incidence in Tunisia is raised, and the relative frequency is probably raised in the Sudan and Algeria. NPC is probably not unduly common in Kenya or Uganda, but there may be differences between ethnic groups.

NPC incidence in Hawaii is raised for all ethnic groups, the rate for male Hawaiians approaching that for male Chinese. In Israel, the incidence in the non-Jewish population and in Jews born in Africa or Asia is higher than in Israel-born Jews or in Jews born in America or Europe.

The effect of intermarriage with Chinese is examined. There is a general trend suggesting that NPC is commoner in those South East Asian groups with admixture of Chinese blood.

The demonstration of an intermediate NPC risk level in Tunisia and other parts of Africa should, when confirmed, permit studies of viral and other aetiological hypotheses in populations with little or no Chinese genetic material.

REFERENCES

Bréhant, J. (1967) Premières esquisses de la physionomie du cancer en Algérie. *Rev. méd. Moy. Or.*, **24**, 116-125

Chadli, A. (1966) Etude histologique et fréquence du cancer du cavum en Tunisie. *Méd. Afr. noire*, **13**, 391-392

Chopra, S. A. (1968) A nine-year study of malignancy in the people of Zanzibar and Pemba. In: Clifford, P., Linsell, C. A. & Timms, G. L., eds, *Cancer in Africa*, Nairobi, East African Publishing House, pp. 19-30

Clifford, P. (1967) Malignant diseases of the nasopharynx and paranasal sinuses in Kenya. In: Muir, C. S. & Shanmugaratnam, K., eds, *Cancer of the nasopharynx*, pp. 82-94 (*UICC Monograph Series*, No. 1)

Digby, K. H., Thomas, G. H. & Hsiu, S. T. (1930) Notes on carcinoma of the nasopharynx. *Caduceus*, **9**, 45-68

Djojopranoto, M. & Soesilowati (1967) Nasopharynx cancer in East Java (Indonesia). In: Muir, C. S. & Shanmugaratnam, K., eds, *Cancer of the nasopharynx*, pp. 43-46 (*UICC Monograph Series*, No. 1)

Dobson, W. J. (1924) Cervical lymphosarcoma. *China med. J.*, **38**, 786

Duvernet-Battesti, F. (1970) *Le cancer en Côte d'Ivoire.* (Etude statistique portant sur 816 cas confirmés histologiquement en trois ans.) (Thesis, Abidjan)

El Hassan, A. M., Milosev, B., Daoud, E. H. & Kashan, A. (1968) Malignant diseases of the upper respiratory tract in the Sudan. In: Clifford, P., Linsell, C. A. & Timms, G. L., eds, *Cancer in Africa*, Nairobi, East African Publishing House, pp. 307-314

Garnjana-Goonchorn, S. & Chantarakul, N. (1967) Nasopharyngeal cancer at Siriraj Hospital, Dhonburi, Thailand. In: Muir, C. S. & Shanmugaratnam, K., eds, *Cancer of the nasopharynx*, pp. 33-37 (*UICC Monograph Series*, No. 1)

Hu, C. H. & Yang, C. (1959) A decade of progress in morphologic pathology. *Chin. med. J.*, **79**, 409-422

Huong, B. Q., Buu-Joi, N. P., Duong, P. N., Te, N. H. & Hoang, D. D. (1969) Les cancers du nasopharynx au Vietnam: Epidémiologie, aspects cliniques, facteurs étiologiques possibles. *Ann. oto-rhino-laryng. (Paris)*, **86**, 267-278

Lin, T. M., Hsu, M. M., Chen, K. P., Chiang, T. C., Jung, P. F. & Hirayama, T. (1971) Morbidity and mortality of cancer of the nasopharynx in Taiwan. *Gann Monogr.*, **10**, 137-144

Linsell, C. A. (1968) Cancer in Kenya. In: Clifford, P., Linsell, C. A. & Timms, G. L., eds, *Cancer in Africa*, Nairobi, East African Publishing House, pp. 7-12

Marsden, A. T. H. (1958) The geographical pathology of cancer in Malaya. *Brit. J. Cancer*, **12**, 161-176

Menakanit, W., Muir, C. S. & Jain, D. K. (1971) Cancer in Chiang Mai, North Thailand. A relative frequency study. *Brit. J. Cancer* **25**, 225-236

Miyaji, T. (1967) Carcinoma of the nasopharynx and related organs in Japan based on mortality, morbidity and autopsy studies. In: Muir, C. S. & Shanmugaratnam, K., eds, *Cancer of the nasopharynx*, pp. 29-32 (*UICC Monograph Series*, No. 1)

Muir, C. S., Evans, M. D. E. & Roche, P. J. L. (1968) Cancer in Sabah (Borneo). A preliminary survey. *Brit. J. Cancer*, **22**, 637-645

Muir, C. S. & Oakley, W. F. (1966) Cancer in Sarawak (Borneo). A preliminary survey. *Brit. J. Cancer*, **20**, 217-225

Muir, C. S. & Oakley, W. F. (1967) Nasopharyngeal carcinoma in Sarawak (Borneo). *J. Laryng.*, **81**, 197-207

Muir, C. S. & Shanmugaratnam, K. (1967) The incidence of nasopharyngeal cancer in Singapore. In: Muir, C. S. & Shanmugaratnam, K., eds, *Cancer of the nasopharynx*, pp. 47-53 (*UICC Monograph Series*, No. 1)

Muir, C. S., Shanmugaratnam, K. & Tan, K. K. (1971) Incidence rates for microscopically diagnosed cancer in the Singapore population 1960-1964. *Singapore med. J.* (in press)

Pantangco, E. E., Basa, G. F. & Canlas, M. (1967) A survey of nasopharyngeal cancers among Filipinos: A review of 203 cases. In: Muir, C. S. & Shanmugaratnam, K., eds, *Cancer of the nasopharynx*, pp. 38-42 (*UICC Monograph Series*, No. 1)

Saad, A. (1968) Observations on nasopharyngeal carcinoma in the Sudan. In: Clifford, P., Linsell, C. A. & Timms, G. L., eds, *Cancer in Africa*, Nairobi, East African Publishing House, pp. 281-285

Schmauz, R. & Templeton, A. C. (1970) Nasopharyngeal carcinoma. II. Histological and epidemiological aspects in Uganda 1964-1968 (in press)

Shanmugaratnam, K. (1967) Nasopharyngeal carcinoma in Asia. In: Shivas, A. A., ed., *Racial and geographical factors in tumour incidence*, Edinburgh, Edinburgh University Press, pp. 169-188 (*Pfizer Medical Monographs* 2)

State of California (1967) *Incidence of cancer in Alameda County, California, 1960-1964*. Berkeley, State of California Department of Public Health, pp. 73-75, 93-95

Tuyns, A. J. & Ravisse, P. (1970) Cancer in Brazzaville, the Congo. *J. nat. Cancer Inst.*, **44**, 1121-1127

UICC (1970) *Cancer incidence in five continents, Volume II*, Doll, R., Muir, C. S. & Waterhouse, J., eds, Geneva, UICC

UICC (1966) *Cancer incidence in five continents—A technical report*, Doll, R., Payne, P. & Waterhouse, J., eds, Geneva, UICC

Yeh, S. (1967) The relative frequency of cancer of the nasopharynx and accessory sinuses in Chinese in Taiwan. In: Muir, C. S. & Shanmugaratnam, K., eds, *Cancer of the nasopharynx*, pp. 54-57 (*UICC Monograph Series*, No. 1)

Yun, I. S. (1949) A statistical study of tumors among Koreans. *Cancer Res.*, **9**, 370-371

Zaouche, A. (1970) *Les tumeurs malignes de la sphère ORL en Tunisie*. (A propos de 644 tumeurs des voies aéro-digestives supérieures observées à l'Institut National de Carcinologie de Tunis du 1.10.67 au 15.8.69.) (Thesis, Paris)

Does the Epstein-Barr Virus Play on Aetiological Rôle in Burkitt's Lymphoma?

(The Planning of a Longitudinal Sero-epidemiological Survey in the West Nile District, Uganda)

A. GESER [1] & G. DE-THÉ [1]

BACKGROUND

In the study proposed in this paper, a child population of about 35 000 living in an area where the risk of Burkitt's lymphoma (BL) is high will be followed for five years in order to see whether the development of Burkitt's lymphoma depends upon previous infection with a herpesvirus (Epstein-Barr virus or EBV). The plan is to bleed the children at an early age and then to follow them to observe whether the BL cases that will arise in them do so in particular immunological subgroups or whether they occur independently of previous virus infection.

Laboratory work has established that an association does exist between a herpesvirus (EBV) and Burkitt's lymphoma, and epidemiological investigations (Burkitt, 1962a, b; Williams, Spit & Pike, 1969) have shown that the incidence of BL is high enough in the West Nile District of Uganda to make an aetiological follow-up study feasible there. The proposed study is, to our knowledge, the only one in which the aetiological rôle of a virus in a human cancer will be directly observed. The field work involved in the bleeding and the follow-up of a large child population is considerable, but as no other approach seems available to test the hypothesis of viral oncogenicity in man, the effort appears justified.

TESTABLE HYPOTHESES

The degree of complexity in the relationship between EBV and BL that can be ascertained

[1] Unit of Biological Carcinogenesis, International Agency for Research on Cancer, Lyon, France.

depends upon the size and duration of the prospective study. For the study proposed, the possible relationship between EBV and BL can be described in terms of the following hypotheses:

Hypothesis 1 : There is no causal relationship between EBV and BL. According to this hypothesis, the EBV antibody status of those who subsequently develop BL is in no way different from the EBV antibody status of those who do not. BL patients would have high antibody levels because the tumour is a suitable medium for the multiplication of EBV.

Hypothesis 2 : Primary EBV infection is necessary for the development of BL, which appears clinically after a fairly short and relatively constant latent period (e.g., 18 ± 6 months).

This hypothesis implies that the EBV antibody titre is negative two years or more before the appearance of the tumour.

Hypothesis 3 : BL develops in children who have had a long and heavy exposure to EBV. This implies that BL develops in those who have a continued high antibody titre. An argument in favour of this hypothesis is that BL patients do have high EBV antibody titres.

Hypothesis 4 : There is a causal relationship between EBV and BL of a more complicated nature than that described in hypotheses 2 and 3, involving, for example, a long or variable latent period, or perhaps a special relationship between certain co-factors and EBV infection.

Hypothesis 4 implies that, in a prospective study, the pattern of titres in children who later develop BL would differ from that in the general population, but that these differences may be undetectable in a study of practical size.

The expected results from the proposed study could also give information on the duration of the latent period between virus infection and tumour appearance. Knowledge of the latent period is important for two reasons: first, for the design of an EBV vaccine trial for the prevention of BL; second, for the design of further studies into the rôle of co-factors; in particular, the relationship in time between malaria infection and EBV infection could be crucial (Burkitt, 1969; Burkitt, 1972[1]; Wedderburn, 1970). To investigate such a relationship one would need to know the time interval between EBV infection and tumour development.

The number of cases of BL that would be expected

[1] See p. 345 of this publication.

to arise under hypothesis 2, year by year during a 5-year follow-up period, is shown in Tables 1b and 1c for two different latent periods. In these tables, it is assumed that six new cases of BL develop per year in the study population and that the risk of developing BL is increased ten-fold by being EBV-negative $X + \frac{1}{2}$ year or more previously.

With 10% of the population initially EBV-negative, a ten-fold increase in risk would give equal numbers of BL cases among EBV-positives and EBV-negatives, as shown in Tables 1b and 1c. Table 1 clearly illustrates the information the study will yield, both with regard to hypothesis 2 and to the latent period. The values given are the expected numbers; the actual numbers will have a Poisson distribution around the expected mean.

DESIGN AND PLACE OF THE STUDY

The study is primarily designed to test hypotheses 2 and 3. The aetiological significance of the hypotheses depends upon the increase in risk of BL found in the relevant subgroups. An increase of risk of

Table 1.   Initial serological status of children developing BL in subsequent years of follow-up [1]

(a) *Hypothesis I : no aetiological relationship*

Year of follow-up	1st	2nd	3rd	4th	5th
Initial serological status of BL cases	0.6 −ve 5.4 +ve	0.6 −ve 5.4 +ve	0.6 −ve 5.4 +ve	0.6 −ve 5.4 +ve	0.6 −ve 5.4 +ve

(b) *Hypothesis 2, with latent period : 18 months ± 6 months*

Year of follow-up	1st	2nd	3rd	4th	5th
Initial serological status of BL cases	6 +ve	4½ +ve 1½ −ve	3 +ve 3 −ve	3 +ve 3 −ve	3 +ve 3 −ve

(c) *Hypothesis 2, with latent period : 30 months ± 6 months*

Year of follow-up	1st	2nd	3rd	4th	5th
Initial serological status of BL cases	6 +ve	6 +ve	4½ +ve 1½ −ve	3 +ve 3 −ve	3 +ve 3 −ve

[1] Expected numbers serologically positive (titres ⩾1:10) or negative (titre <1:10), if it is assumed that 6 new cases occur each year.

Table 2. Number of cases of BL found in the West Nile District 1965-1969 by age and by county

County	Pop. in 1969 census	Est. child pop., 4–8 years in 1969	Number of cases of BL in W. Nile District 1965–1969 (age in years)													Age 4–8 years	
			1	2	3	4	5	6	7	8	9	10	11	12	<12 or unknown	No. of cases (5 years)	Annual incidence
Aringa [1]	56 819	9500	−	−	−	1	1	1	2	3	1	1	1	0	2	8	$17/10^5$
Koboko	37 428	6200	−	−	−	−	−	−	−	−	1	−	−	−	1	−	−
Maracha [1]	59 475	10 000	−	−	−	1	4	0	0	2	1	−	−	−	1	7	$14/10^5$
Tereco [1]	57 012	9500	−	1	0	3	7	1	1	1	2	0	1	−	−	13	$27/10^5$
Vurra	35 155	6000	−	−	1	1	0	1	0	1	−	−	−	−	−	3	$10/10^5$
Madi [1]	46 976	7800	−	−	−	0	4	2	1	−	−	−	−	−	−	7	$18/10^5$
Ayivu	76 449	12 600	−	−	−	−	1	0	1	1	0	0	1	−	−	3	$5/10^5$
Jonam	49 531	8200	1	−	−	−	1	2	2	2	−	−	−	−	−	7	$17/10^5$
Okoro	78 186	13 000	−	−	−	−	0	2	−	1	2	−	−	−	1	3	$4/10^5$
Padyere	75 049	12 500	−	−	−	−	−	−	−	−	−	−	−	−	−	0	−
All counties	572 078	95 000	1	1	1	6	18	9	7	11	7	1	3	0	5	51	$11/10^5$
Proposed counties	220 280	36 800	−	1	0	5	16	4	4	6	4	1	3	0	3	35	$19/10^5$

[1] Counties proposed for the prospective study.

less than, say, five-fold would not establish the necessary nature of the rôle of EBV in the development of BL. Thus, the basic requirement for this design is that it should be capable, with high probability, of detecting a five-fold increase in risk in either of two subgroups, each comprising about 10% of the total population. For this purpose, an expected number of approximately 30 cases is required.

The number of cases of BL detected over the last five years in the West Nile District was analyzed by county (Table 2). On the basis of this table, it was possible to select four adjacent counties (Aringa, Maracha, Terego and Madi) that will yield about 35 cases over five years. These counties have a total of approximately 37 000 children between the ages of 4 and 8 years with an average BL incidence of about 19/100 000, resulting in a total of about 7 cases per year.

Table 3. Number of BL cases expected to occur in each successive year of follow-up in each yearly cohort of children

Age of child at start	No. of BL cases found in each year of follow-up					Total
	1st	2nd	3rd	4th	5th	
0	0	0	0	0	2	2
1	0	0	0	2	3	5
2	0	0	2	3	2	7
3	0	2	3	2	1	8
4	2	3	2	1	1	9
5	3	2	1	1	0	7
Total	5	7	8	9	9	38

The proposal is to collect blood samples from all children initially aged 2–5 years inclusive. Those initially under the age of 2 will also be registered in the study, as will those born during it. In order to ensure sufficient cases of BL, those under two years of age at the start will be bled in the first two years of the trial. The number of cases expected in each year of the trial from each yearly cohort of children is shown in Table 3, where the numbers have been increased so as to give an integer in each case.

ORGANIZATION OF SURVEY WORK

Field work

The people in the West Nile District live scattered over the land they cultivate, without forming villages or other definite population clusters. The technicians must walk from house to house to register the survey population and bleed the examinees. Children who subsequently develop BL must be matched with the proper serum specimen. For this purpose, it is essential to ensure that all examinees can be correctly identified. This will be done on the basis of the name of the examinee and that of the parents, and the house in which the child lives. All houses included in the survey will be given serial numbers which will be marked on a sketch map of the area to facilitate subsequent identification.

The essential activity during the follow-up period is to detect cases of BL arising in the surveyed children. Detection will be effected by the existing health services in the West Nile District, assisted by the teams established under the BL research project.

Past experience in the West Nile indicates that a certain proportion of BL cases are, at present, missed in the area. The project personnel will help to increase the probability of discovering BL cases by maintaining a constant surveillance in the area. Medical assistants from the research project will regularly visit all localities included in the survey and inquire about the occurrence of any illness resembling BL. They will also keep close contact with personnel working at peripheral health units in the area and constantly remind them of the importance of diagnosing and reporting all cases of confirmed and suspected BL. Five field units, each comprising five members, will be established in the survey area to collect blood samples and to maintain surveillance of the study population.

Testing of sera

For the sake of testing the hypotheses regarding the virus-tumour relationship, it is not necessary to test all sera but only those of the thirty children who subsequently develop BL and those of properly matched controls.

The presence of antibodies against virus-induced antigens (viral capsid antigens, early antigens) as well as tumour-specific antigens (membrane-bound antigens) will have to be determined in the "pre-sera" by immunofluorescence. Furthermore, any other available method may be used to determine the immunological status of the sera (with respect to EBV), since each serum collected will be preserved in liquid nitrogen.

Malaria and Burkitt's lymphoma

In view of the omnipresence of EBV and the geographical restriction of BL, it is obvious that the virus cannot be the sole cause of the tumour, and one or more cofactors must be assumed to exist. During the survey in the West Nile District, an attempt will be made to identify factors other than EBV that may influence the risk of BL in a population.

Hyperendemic malaria has repeatedly been mentioned (Burkitt, 1972 [1]) as a possible aetiological factor in BL since the geographical distributions of the two conditions practically coincide: BL reaches a high incidence in tropical countries only, and in these countries only up to an altitude that favours malaria transmission. It also appears that even in the tropics, BL is rare in urban areas where malaria has been controlled.

It is thus conceivable that malaria may play an aetiological rôle in BL, and a study of the relationship between malaria infection and Burkitt's lymphoma will be included in the West Nile survey. An attempt will be made to determine whether the children who eventually develop BL have experienced a heavier malaria burden in the past than comparable children living in the same localities who do not get BL.

At a later stage, an experimental approach may be applied to test whether malaria control can reduce the incidence of BL in a high-risk area.

[1] See p. 345 of this publication.

ACKNOWLEDGEMENTS

This investigation was supported by Contract No. NIH-70-2076 within the Special Virus-Cancer Program, National Institutes of Health, Department of Health, Education, and Welfare, USA.

REFERENCES

Burkitt, D. (1962a) A "tumour safari" in East and Central Africa. Brit. J. Cancer, 16, 379-386

Burkitt, D. (1962b) A children's cancer dependent on climatic factors. Nature (Lond.), 194, 232-234

Burkitt, D. (1969) Etiology of Burkitt's lymphoma—an alternative hypothesis to a vectored virus. J. nat. Cancer Inst., 42, 19-28

Wedderburn, N. (1970) Effect of concurrent malarial infection on development of virus-induced lymphoma in Balb-c mice. Lancet, ii, 1114-1116

Williams, E. H., Spit, P. & Pike, M. C. (1969) Further evidence of space-time clustering of Burkitt's lymphoma patients in the West Nile district of Uganda. Brit. J. Cancer, 23, 235-246

The Laboratory Differentiation
of the Infectious-mononucleosis Syndrome
with a Study of Epstein-Barr Virus Antibodies
within the Family Group

R. J. L. DAVIDSON [1] & J. E. BANATVALA [2]

A. LABORATORY DIFFERENTIATION OF THE INFECTIOUS-MONONUCLEOSIS SYNDROME

The glandular-fever or infectious-mononucleosis (IM) syndrome comprises a heterogeneous group of diseases of which IM with heterophile antibody (HA) accounts for only 20–30% of the cases (Hobson, Lawson & Wigfield, 1958; Penman, 1968; Davidson, 1970). Few attempts have been made to establish precise diagnosis in the remainder, which may include cases of IM without HA.

In view of the possible causal rôle of Epstein-Barr virus (EBV) and recent improvements in diagnostic virology, this prospective study was undertaken to establish, if possible, the specific laboratory diagnosis in a series of 65 patients, clinically suspected of having IM. The clinical and laboratory findings with conclusions are briefly presented.

PATIENTS INVESTIGATED

The 65 patients consisted of 27 males with a mean age of 22.5 years (range, 18–58) and 38 females with a mean age of 19.7 years (range, 10–37). All patients underwent a full clinical examination on presentation and after intervals of 10 days and 6 weeks.

LABORATORY INVESTIGATIONS AND METHODS

At each of the three clinical examinations, clotted and sequestrene blood samples and a throat swab were taken.

Heterophile antibodies. All sera were screened with the Monospot test (Ortho Diagnostics Ltd.) and those reacting positively further investigated by a differential absorption test (Davidson, 1967).

Viral antibodies. Complement-fixation tests were carried out by a modification of the method of Bradstreet & Taylor (1962) using three times the minimum haemolytic dose (MHD_{50}) of complement. All sera were screened against the following antigens: adenovirus, influenza A & B, herpes simplex, mumps, Sendai (para-influenza I), respiratory syncytial virus, psittacosis, *Mycoplasma pneumoniae*, *Rickettsia burneti* and cytomegalovirus. Rubella antibodies were determined by the haemagglutination-inhibition test, serum inhibitors being removed according to Plotkin, Bechtel & Sedwick (1968). EB virus antibodies were detected by the method of Henle & Henle (1966), as modified by Banatvala & Grylls (1969). All sera were screened for toxoplasmosis by the Toxoplasma Agglutinotest (Italdiagnostic) and those reacting positively were then investigated for cytoplasm-modifying antibodies according to Sabin & Feldman (1948). Anti-streptolysin O (ASO) titres were determined by means of Bacto-Streptolysin O Reagent (Difco Ltd.). Immunoglobulins were analysed by the single radial immunodiffusion technique (Fahey & McKelvie, 1965) and serum transaminases (SGOT and SGPT) [3] estimated by a modification of the method of Mohun & Cook (1957). Throat swabs were plated on two 10% horse blood agar

[1] Regional Laboratory, City Hospital, Aberdeen, UK.
[2] Clinical Virology Department, St. Thomas's Hospital, London, S.E.1, UK.

[3] Serum glutamic oxalo-acetic-acid-transaminase and serum glutamic pyruvic transaminase.

plates. Peripheral blood examination included total and differential leukocyte counts and scrutiny of a Leishman stained film for atypical mononuclear cells.

RESULTS

In this section, the main clinical and laboratory findings in 31 patients (11 males and 20 females), diagnosed retrospectively as having IM, are compared and contrasted with those of the remaining 34 (16 males and 18 females). The former group includes 26 patients with positive HA reactions and 5, who although consistently HA-negative, had raised EBV antibody titres.

Clinical features. Comparison of the clinical manifestations of the two groups (Table 1), apart from indicating a common presenting triad of pyrexia, sore throat and lymphadenopathy, serves to emphasize the close similarity of the overall clinical picture. Palatal petechiae, exudative pharyngotonsillitis and splenomegaly may however be helpful distinguishing features because of their increased frequency in the IM group.

Haematological findings. At the first examination, 25/31 IM patients had an absolute lymphocytosis, 3 a leukopenia, and 23 numerous atypical mononuclear cells. The 3 leukopenic patients all presented within 48 hours of the onset of their symptoms and only developed an absolute lymphocytosis with atypical mononuclears and positive HA reactions by the time of the second examination. The 5 IM patients who failed to show atypical mononuclears remained HA-negative throughout.

In contrast, only 3/34 non-IM patients had an absolute lymphocytosis but 15 had a neutrophilia, 2 a leukopenia and 3 an eosinophilia. None showed atypical mononuclears in significant number.

SEROLOGICAL RESULTS

Heterophile antibodies. Of the 26 patients with HA, 23 were found to be positive at the time of presentation (range of positivity, 1 : 64–4096) and a further 3 (see above) at the second examination. During surveillance, sera from 10 patients exhibited a rising, and those from 16 patients a falling titre, only 3 of which were completely negative at 6 weeks.

EB virus antibodies. The highest of the 3 serial titres recorded for each of the 62 patients tested are shown in Table 2. Thus all 25 HA-positive patients tested showed rising or elevated antibody titres ($\geqslant$1 : 80), 22/25 having convalescent titres of $\geqslant$1 : 160. In the 5 IM patients without HA, 4 had elevated titres while the remaining patient had a rise in titre from 1 : 10 to 1 : 80. In contrast, none of the 32 non-IM patients tested showed a rising titre and only 1 had a titre of >1 : 80.

Other viral antibodies. Three IM and 3 non-IM patients gave "positive" results. Of the former group, high titres to rubella were obtained in 2 and to mumps in the third. None showed characteristic clinical manifestations of either infection or a rising

Table 1.   Comparison of the main clinical manifestations of IM and non-IM patients

Clinical manifestations	IM patients		Non-IM patients	
	No. of cases	(%)	No. of cases	(%)
Pyrexia	24	(77.4)	18	(52.9)
Headache	9	(29.0)	24	(70.6)
Productive cough	3	(9.7)	6	(17.6)
Jaundice	2	(6.4)	–	–
Skin rash	1	(3.2)	2	(5.9)
Purpura	–	–	–	–
Palatal petechiae	15	(48.4)	9	(26.5)
Sore throat:	27	(87.1)	32	(91.2)
(a) Simple pharyngitis	4	(12.9)	14	(41.2)
(b) Pharyngotonsillitis without exudate	6	(19.3)	9	(26.5)
(c) Pharyngotonsillitis with exudate	17	(54.8)	9	(26.5)
Lymphadenopathy:	31	(100)	32	(94.1)
(a) Cervical	10	(32.2)	12	(35.3)
(b) Cervical and axillary	15	(48.4)	11	(32.2)
(c) Generalized	6	(19.3)	9	(26.5)
Hepatomegaly	1	(3.2)	9	(26.5)
Splenomegaly	7	(22.6)	1	(2.9)
Hepato/splenomegaly	–	–	3	(8.8)
Total no. of cases	31	–	34	–

Table 2. EBV antibody titres (62/65 patients)

Reciprocal antibody titres	IM patients		Non-IM patients
	HA-positive	HA-negative	
<10	–	–	9
10	–	–	1
20	–	–	4
40	–	–	10
80	3	1	7
160	8	2	1
320	10	2	–
640	4	–	–
Totals	25	5	32

antibody titre. These findings remain unexplained but may have resulted from the presence of HA and increased amounts of IgM.

In the non-IM group, one patient showed un-equivocal serological evidence of infection with adenovirus (titres: 1 : 16–1 : 256), one equivocal evidence of infection with cytomegalovirus (titres: 1 : 32 × 3) while the third, who had a streptococcal throat infection, showed a rising titre to herpes simplex.

Throat swab cultures and ASO titres. A normal throat flora was obtained in all 31 IM patients. On the other hand, 8/34 non-IM patients gave a moderate to profuse growth of beta-haemolytic streptococcus from at least 2/3 swabs taken and all had an accompanying neutrophilia and significant rise in ASO titre.

Immunoglobulin patterns. Analyses were performed on serial samples from 8 of the IM (all HA-positive) and 11 of the non-IM patients. In the IM group, 6 had IgM levels in excess of 280 mg/100 ml, while the 2 remaining showed at least a twofold reduction between the first and third samples. In 7 of these 8 patients, the IgM levels parallelled the HA titre. None of the 11 non-IM patients had IgM levels above the normal range.

Serum transaminases. These were strikingly different in the two patient groups. Thus, in the IM patients, an elevated SGPT was found in 23 (74%), 16 of whom also had high SGOT levels. Con-

sistently normal SGPT and SGOT levels were found in all but 6 (18%) of the non-IM patients.

The specific laboratory diagnoses made in the 65 patients are summarized in Table 3; they include one patient with herpes zoster. This exceptional case, a 27-year old student, presented with sore throat and enlarged, tender neck glands. He was found to have pharyngitis, generalized lymphadenopathy, hepatomegaly, elevated transaminases and a relative lymphocytosis. Within 10 days he had developed a typical dorsal root eruption of herpes zoster.

Table 3. Definitive diagnosis in 65 patients with "IM syndrome"

Diagnosis	No. of cases	Proportion (%)
Seropositive IM (typical)	26	40
Seronegative IM (typical)	5	7.7
Streptococcal sore throat	8	12.3
Adenovirus infection	1	} 4.6
Herpes zoster infection	1	
Cytomegalovirus infection	1	
No laboratory diagnosis	23	35.4

CONCLUSIONS

(*a*) Clinical differentiation within the syndrome is unreliable.

(*b*) Some 10% of seropositive cases of IM do not show atypical mononuclears or HA antibody at the time of presentation.

(*c*) The incidence of hepatic involvement is high in seropositive IM but low in other cases of the syndrome.

(*d*) The majority of seropositive cases of IM show an EBV antibody titre of ≥1 : 160, but a rising titre cannot always be demonstrated.

(*e*) The results lend support to the view that HA-negative IM is a genuine entity.

(*f*) Other than seropositive IM, the commonest demonstrable cause of the syndrome is streptococcal sore throat (12.3%).

(*g*) The cases eluding laboratory definition (35.4%) remain a mystery and diagnostic challenge.

B. A STUDY OF EB VIRUS ANTIBODIES WITHIN THE FAMILY GROUP

Infectious mononucleosis is generally regarded as a sporadic disease of low infectivity even within the setting of family or small, closed community groups. Several minor epidemics have been recorded and lathough their validity has been challenged (Hoag-

land, 1967) or accepted with reluctance (Evans, 1968), clinical and haematological evidence of spread to about one third of close contacts has been found in at least one study (Pejme, 1966). Until recently, recognition of contact susceptibility, asymptomatic cases and the carrier state has not been possible but with the candidacy of EBV or a closely-related herpesvirus as the causal agent of IM, some of these problematic aspects of the disease are now open to exploratory investigation and have already been the subject of several reports (Joncas & Mitnyan, 1970; Wahren *et al.*, 1970).

This communication gives a brief account of our findings, including EBV antibody levels, in all members of 11 families each of which contained an individual with unequivocal IM (the index case).

PATIENTS AND METHODS

A total of 55 individuals were investigated.   These consisted of 11 index cases, all HA-positive with typical clinical and haematological features of the disease, and the 22 parents and 22 siblings of these cases.   Each family member was kept under clinical surveillance for at least 2 months following laboratory confirmation of the index case and each had sequestrene and clotted blood samples taken within 10 days of that time and after an interval of 6–10 weeks.

LABORATORY INVESTIGATIONS

Peripheral blood examination included: total and differential leukocyte counts; serum transaminases (SGOT and SGPT); batch testing of paired sera for EBV antibodies after storage at $-20°C$; and screening of all sera for HA by the Monospot test followed by differential absorption tests on those giving a positive slide-test reaction.   The methods employed were as described in A above.

RESULTS

Throughout the period of study all parents and siblings of the index cases remained well and none developed HA.   Recent enquiry has further revealed that, even after intervals varying from 6–14 months, none have developed symptoms suggestive of IM. Of the 44 family contacts, only 1 (member "c" of family 2) was known to have had IM, as confirmed by laboratory tests, and that was 2 years previously.

The EBV antibody titres of the paired serum samples obtained from each individual are recorded in Table 4.   In summary, these reveal that:

(*a*) Of the 11 index cases, whose mean age was 14.2 years (range, 6–23), 3 had a rise and 1 a fall in antibody level; a titre of $\geqslant 1 : 160$ was recorded in 8.

Table 4.   EBV antibody levels in 55 family members

Family	Member	Age	Sex	EBV antibody titres
1	a	52	M	10 – 10
	b	43	F	80 – 160
	c	21	F	10 – 10
	d	17	M	10 – 10
	e [1]	10	M	10 – 80
2	a	41	M	640 – 640
	b	39	F	10 – 10
	c	19	M	20 – 20
	d [1]	17	M	80 – 40
	e	13	M	10 – 10
	f	4	F	10 – 10
3	a	46	M	1280 – 320
	b	44	F	160 – 160
	c	21	M	640 – 1280
	d [1]	16	M	160 – 160
	e	12	F	10 – 10
4	a	46	M	10 – 10
	b	42	F	10 – 10
	c [1]	23	M	160 – 320
5	a	47	M	160 – 160
	b	44	F	80 – 80
	c [1]	14	M	320 – 320
	d	9	M	40 – 40
	e	7	F	40 – 40
6	a	40	M	160 – 160
	b	37	F	320 – 160
	c [1]	12	M	320 – 320
7	a	35	M	40 – 80
	b	32	F	160 – 160
	c	11	M	40 – 40
	d	8	F	40 – 40
	e [1]	6	M	40 – 40
	f	4	M	40 – 40
8	a	48	M	160 – 320
	b	39	F	80 – 80
	c	19	M	20 – 40
	d	17	F	10 – 10
	e [1]	15	F	160 – 160
	f	4	M	10 – 10
9	a	46	M	160 – 160
	b	45	F	80 – 80
	c	18	F	10 – 10
	d	16	F	40 – 80
	e [1]	12	F	640 – 640
10	a	45	M	40 – 40
	b	44	F	80 – 80
	c	17	M	10 – 10
	d [1]	14	F	320 – 320
	e	9	M	10 – 10
	f	9	M	10 – 10
11	a	45	M	80 – 80
	b	43	F	160 – 160
	c	20	F	640 – 640
	d	18	M	80 – 80
	e [1]	17	F	320 – 320

[1] Index case.

(*b*) Of the 22 parents, whose mean age was 42.9 years (range, 32–52), 3 had a rise and 2 a fall in antibody level; a titre of $< 1 : 10$ was recorded in 4 and a titre of $\geqslant 1 : 160$ in 11.

(c) Of the 22 siblings, whose mean age was 13.3 years (range, 4–21), 3 had a rise and none a fall in antibody level; a titre of <1 : 10 was recorded in 10 and a titre of ≽1 : 160 in 2.

Atypical mononuclear cells were not detected in significant number in any of the parents or siblings but an absolute lymphocytosis (4000 lymphocytes/mm³) was initially found in 3 ("b" of family 1, "b" of family 6 and "a" of family 9).

Serum transaminases were elevated in 8/11 index cases, 3/22 parents ("a" of family 5, "a" of family 7 and "a" of family 9) and 2/22 siblings ("e" of family 5 and "d" of family 7).

CONCLUSIONS

None of the 44 family contacts developed clinical features of the disease or a positive HA reaction and only 7 (15.9%) showed what must be regarded as equivocal laboratory evidence of the disease, namely an absolute lymphocytosis and/or slightly elevated serum transaminase levels.

Four parents (18.2%) and 10 siblings (45.4%) had EBV antibody levels of <1 : 10 and thus were possibly susceptible contacts. Eight of the 44 contacts (18.2%) exhibited a rise or fall in antibody levels but as this was never more than one dilution, except in the father of family 3, the changes do not necessarily signify recent infection.

Thirteen contacts (29.5%) were found to have EBV antibody titres in the range 1 : 160–1280. Even the significance of these high titres must be interpreted with caution, as such levels may merely represent persistence of antibody from a past rather than a recent infection (Marker, 1970). If not related to recent infection, one may speculate however that in some of these cases the high levels resulted from the "booster" effect of recent exposure.

This study would appear to indicate that IM and EBV, if causally related, have a low infectivity even within the family setting.

REFERENCES

Banatvala, J. E. & Grylls, S. G. (1969) Serological studies in infectious mononucleosis. *Brit. med. J.*, **3**, 444-446

Bradstreet, C. M. P. & Taylor, C. E. D. (1962) Technique of complement-fixation test applicable to the diagnosis of virus diseases. *Mth. Bull. Minist. Hlth Lab. Ser.*, **21**, 96-104

Davidson, R. J. L. (1967) New slide test for infectious mononucleosis. *J. clin. Path.*, **20**, 643-646

Davidson, R. J. L. (1970) A survey of infectious mononucleosis in the North-East Regional Hospital Board area of Scotland, 1960-9. *J. Hyg. (Camb.)*, **68**, 393-400

Evans, A. S. (1968) Epidemiology and pathogenesis of infectious mononucleosis. In: *Proceedings of the International Infectious Mononucleosis Symposium*, Washington, American College Health Association, pp. 40-55

Fahey, J. L. & McKelvey, E. M. (1965) Quantitative determination of serum immunoglobulins in antibody-agar plates. *J. Immunol.*, **94**, 84-90

Henle, G. & Henle, W. (1966) Immunofluorescence in cells derived from Burkitt's lymphoma. *J. Bact.*, **91**, 1248-1256

Hoagland, R. J. (1967) *Infectious mononucleosis*, New York & London, Grune & Stratton

Hobson, F. G., Lawson, B. & Wigfield, M. (1958) Glandular fever; a field study. *Brit. med. J.*, **1**, 845-852

Joncas, J. & Mitnyan, C. (1970) Serological response of the EBV antibodies in paediatric cases of infectious mononucleosis and in their contacts. *Canad. med. Ass. J.*, **102**, 1260-1263

Marker, O. (1970) Fluorescent EBV antibodies in infectious mononucleosis. *Acta path. microbiol. scand. Sect. B*, **78**, 305-310

Mohun, A. F. & Cook, I. J. Y. (1957) Simple methods for measuring serum levels of the glutamic-oxalacetic and glutamic-pyruvic transaminases in routine laboratories. *J. clin. Path.*, **10**, 394-399

Pejme, J. (1964) Infectious mononucleosis. A clinical and haematological study of patients and contacts and a comparison with healthy subjects. *Acta med. scand. Suppl.* 413, pp. 1-83

Penman, H. G. (1968) Glandular fever-like illness. *J. roy. Coll. Gen. Practit.*, **16**, 275-292

Plotkin, S. A., Bechtel, D. J. & Sedwick, W. D. (1968) A simple method for removal of rubella haemagglutination inhibitors from serum adaptable to finger-tip blood. *Amer. J. Epidem.*, **89**, 232-238

Sabin, A. B. & Feldman, H. A. (1948) Dyes as microchemical indications of a new immunity phenomenon affecting a protozoan parasite (Toxoplasma). *Science*, **108**, 660-663

Wahren, B., Lantorp, K., Sterner, G. & Espmark, A. (1970) EBV antibodies in family contacts of patients with infectious mononucleosis. *Proc. Soc. exp. Biol. (N.Y.)*, **133**, 934-939

Longitudinal Sero-epidemiological Survey of the Development of Antibodies to Epstein-Barr Virus

R. SOHIER,[1] M. HENRY-AYMARD [2] & J. TERRAILLON [2]

A longitudinal sero-epidemiological survey has been carried out to study the variations in antibody titres against Epstein-Barr virus (EBV) in two groups of healthy individuals, and also to investigate the cause of any variations in the antibody titres, when observed.

The first group was composed of 56 normal individuals in the age-group 16–65. The antibody titres, determined by the indirect immunofluorescence

[1] Unité de Virologie, INSERM, Lyon, France.
[2] Université Claude Bernard, Département de Bactériologie-Virologie-Immunologie, Lyon, France.

test, were stable in 41 cases, 11 remaining negative for intervals varying from 1 to 12 years (Table 1). In contrast, the antibody titres changed with time in 15 cases (Table 2).

The second group contains 236 blood donors (18 to 60 years of age), blood specimens being taken at intervals varying from 3 to 24 months. The results are shown in Table 3: in 203 cases (86%), the antibody titre remained stable, while in 33 cases (14%), variations were observed. Antibodies appeared in 2 cases (M.G. and C.D.) who had no detectable antibodies at the time of the first bleeding. The titres

Table 1. Stability of immunofluorescent antibody titres among 41 individuals with sera taken over a 12-year period

Titre	No. of titres remaining stable for:												Total
	1 yr	2 yrs	3 yrs	4 yrs	5 yrs	6 yrs	7 yrs	8 yrs	9 yrs	10 yrs	11 yrs	12 yrs	
<10	2	2	2	1	1		2					1	11
10	3	2											5
20													
40	7		1				1	3	2	1			15
80	2	1	1										4
160	3			1			1						5
320													
640		1											1
Total	17	6	4	2	1		4	3	2	1		1	41

Table 2. Variation in immunofluorescent antibody titres in a 12-year period in 15 individuals

Seroconversion from negative (<10) to positive	Change from initial titre of $\geqslant 10$ [1]			Total
	Increase	Decrease	Fluctuations	
4	1	8	2	15

[1] These changes represent an increase or decrease in antibody titres equal to or greater than 2 dilutions.

Table 3.   Variation in immunofluorescent antibodies during a period
of three to eighteen months in 236 individuals

Age groups (years)	Seroconversion from negative (<10) to positive	Change from initial titre of ⩾10 [1]			Stable cases	Total
		Increase	Decrease	Fluctuations		
18–29	2	3	3	2	40	50
30–39	–	5	3	1	67	76
40–49	–	8	3	–	64	75
50–60 and older	–	2	1	–	32	35
Total	2	18	10	3	203	236

[1] These changes represent an increase or decrease in antibody titres equal to or greater than 2 dilutions.

Table 4.   Antibody titres to EBV during a period of four to twelve months,
measured by the complement-fixation test (soluble antigen)

Unchanged titres		Changed titres		Total
Negative	Positive	Increase	Decrease	
34	43	1 [1]	1	79

[1] This antibody increase coincided with increases in antibody to para influenza viruses type 3 and mycoplasma pneumonia (see Table 5).

Table 5.   Antibody titres to EBV in individuals with increases in antibodies to other pathogens

Patients	Simultaneous change [1]	Indirect immunofluorescence test	Complement-fixation test (soluble) antigen [2]	Influenza A	Influenza B	Para influenza 3	Mumps	Respiratory syncytial virus	Adenovirus	Ornithosis	Mycoplasma pneumonia
D. M.	Yes	1280	8	<2	–	–	–	–	–	–	–	–
		80	8	16	–	–	–	–	–	–	–	–
M. R.	No	160	ND	<2	–	–	–	–	–	–	–	–
		160	ND	16	–	–	–	–	–	–	–	–
S. C.	Yes	160	16	2	–	–	–	–	–	–	–	–
		640	16	16	–	–	–	–	–	–	–	–
O. A.	No	40	ND	<2	–	–	–	–	–	–	–	–
		40	ND	8	–	–	–	–	–	–	–	–
P. L.	Yes	160	ND	<4	–	–	–	–	–	–	–	–
		10	ND	16	–	–	–	–	–	–	–	–
D. J.	No	80	<8	–	<2	–	–	–	–	–	0	–
		40	<8	–	16	–	–	–	–	–	64	–
F. C.	Yes	160	<8	–	–	2	–	–	–	–	–	<8
		80	16	–	–	16	–	–	–	–	–	64
C. E.	No	20	<8	–	–	–	0	–	–	–	–	–
		20	<8	–	–	–	64	–	–	–	–	–
D. J.	No	40	<8	–	–	–	–	2	–	–	–	–
		40	<8	–	–	–	–	8	–	–	–	–
B. M.	No	20	±8	–	–	–	–	2	–	–	–	–
		40	8	–	–	–	–	16	–	–	–	–
F. J.	No	<10	<8	–	–	–	–	–	2	–	–	–
		<10	<8	–	–	–	–	–	16	–	–	–
C. G.	Yes	160	16	–	–	–	–	–	–	2	–	–
		40	16	–	–	–	–	–	–	8	–	–

[1] In antibodies against pathogenic human viruses or mycoplasma pneumonia and anti-EBV antibodies.
[2] ND = not done.

of these 2 donors rose to 40, after 1 year and 4 months respectively. Clinical investigation revealed that the first donor (M.G.) had shown an adenopathy without other symptoms between the two bleedings, and that the second donor (C.D.) had had two attacks of stomatitis without fever. No apparent clinical syndrome was detectable among the individuals whose antibody titre rose from an initial titre > 10.

Sera of 79 individuals were tested by complement fixation (CF) using a soluble antigen prepared from a cell line producing EBV derived from a myeloblastic leukaemia. The CF antibody titres were found to be stable in 77 cases (97.4%) and a significant variation was observed only in 2 cases (one increase and one decrease) (Table 4).

The complement-fixation test was also applied to the sera of 138 individuals with antigens derived from 10 viruses pathogenic for man (influenza A, B; parainfluenza 1, 2, 3; mumps; respiratory syncytial virus; adenovirus; measles; herpes simplex) and from ornithosis and mycoplasma pneumonia. The results are shown in Table 5: a significant rise in antibody titre against one of these viruses was observed in 4 cases (influenza A: 3 cases; ornithosis: 1 case) simultaneously with a change in anti-EBV titre (3 decreases, one increase). In one case, increase of antibodies against parainfluenza virus type 3 and against mycoplasma pneumonia occurred at the same time as a significant increase in CF anti-EBV antibodies.

ACKNOWLEDGEMENTS

The technical assistance of J. Gramusset and P. Giraudon is acknowledged.

Relationship of Epstein-Barr Virus Antibodies to Disease State in Hodgkin's Disease, Chronic Lymphocytic Leukaemia and American Burkitt's Lymphoma *

P. H. LEVINE

Recent studies implicating herpes-type viruses as a necessary factor in the aetiology of animal lymphomas (Churchill & Biggs, 1968; Melendez et al., 1969) have led to a more intensive evaluation of the oncogenic potential of Epstein-Barr virus (EBV), a human herpes-type virus. Since Henle, Henle & Diehl (1968) first implicated this virus as the probable cause of infectious mononucleosis, serological studies have been carried out by many investigators to determine whether other lymphoproliferative diseases could also be caused by EBV. One of the major reasons for suspecting EBV of being a tumour virus has been the report (Henle et al., 1969) of consistently high antibody levels in African patients with Burkitt's lymphoma, a disease appearing to have an infectious aetiology because of its geographical distribution (Burkitt, 1963) and time-space clustering (Pike, Williams & Wright, 1967). The reports of high EBV titres in several nonmalignant diseases, including sarcoidosis (Hirshaut et al., 1970), systemic lupus erythematosus (Evans, Rothfield & Niederman, 1971), and lepromatous leprosy (Papageorgiou et al., 1971), have raised serious questions as to the significance of EBV antibody elevations.

Since direct proof of oncogenesis in humans is not possible, several indirect approaches to the problem of determining the rôle of EBV in human cancer are necessary. These include sero-epidemiology, case control studies, and prospective serological studies measuring EBV-related antibodies in patients, family members, and controls to evaluate the effects of genetic, environmental, and disease factors on the humoral response to EBV.

In each disease that has been studied thus far, the correlation between clinical parameters and EBV antibody titres reveals a distinct pattern of immunity. This report will summarize our findings with the viral capsid antibody (VCA) in American patients with three lymphoproliferative diseases: Hodgkin's disease, chronic lymphocytic leukaemia and American Burkitt's lymphoma. The differences between antibody patterns suggests that the rôle of EBV may not be the same in each disease associated with high antibody titres.

Our studies initially concentrated on Hodgkin's disease (Levine et al., 1971a) because of the clinical and epidemiological features suggesting an infectious aetiology. The initial description by MacMahon in 1966 of the geographical differences in the incidence of Hodgkin's disease in young adults was followed by a report indicating that socio-economic conditions are important factors in the age of onset and histology of this disease. Correa & O'Conor (1971) found that less developed countries have a prominent incidence peak of the disease in childhood, and that improving socio-economic conditions are associated with a gradual shift of this incidence peak from childhood to young adulthood. Most of the childhood Hodgkin's patients have the histological subtypes associated with a poor prognosis (lymphocyte depletion and mixed cellularity) while the subtypes associated with a better prognosis (lymphocyte predominance and nodular sclerosis) are commoner in the more developed countries. These epidemiological studies, which point to an environmental factor

---

* From the Viral Leukemia and Lymphoma Branch of the National Cancer Institute, National Institutes of Health, Bethesda, Maryland, USA.

Table 1.   Viral capsid antibody titres in lymphoproliferative disease

Diagnosis	No. of patients	Negative		Titre >1:640		GMT [1]
		No.	%	No.	%	
Chronic lymphatic leukaemia	34	1	3	10	29	1:444
Hodgkin's disease	68	1	1	24	35	1:426
Burkitt's lymphoma (American)	23	4	17	4	17	1:100
Burkitt's lymphoma (African)	15	0	0	10	67	1:1280
Other lymphomas	49	1	2	6	12	1:160
Nonmalignant diseases	241	32	13	17	7	1:64
Normal controls	125	9	7	8	6	1:100

[1] Geometric mean titre.

as an important aetiological component, are largely responsible for our current emphasis on Hodgkin's disease at the National Cancer Institute. The geometric mean titre (GMT) of our total Hodgkin's groups is not only higher than that of normal controls (Table 1), but there is also a marked difference between histological subgroups in Hodgkin's pa-

tients, as has also been shown by Johansson *et al.* (1970) in an independent study of Swedish Hodgkin's patients. Both reports point to the lymphocyte-depletion or sarcomatous form of Hodgkin's disease as the group with the highest VCA antibody titre to EBV. In the American Hodgkin's study, treated patients in each histological category had higher

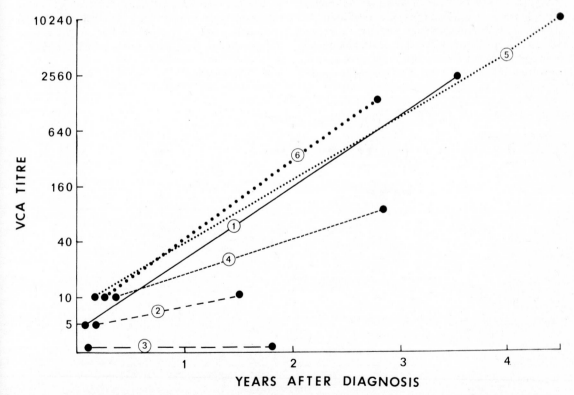

Fig. 1.   Serial EBV titres in Hodgkin's disease.   Serial serum samples were obtained from six of seven Hodgkin's patients who were found to have low EBV titres in an initial survey (Levine *et al.*, 1971a).   All have remained free of clinical disease and are being carefully followed serologically because of the evidence that high titres correlate with poor prognosis in Hodgkin's disease.

EBV titres than untreated patients, and the finding of a high VCA titre correlated with a relatively poor prognosis. Seven patients with low EBV titres were identified in our initial study and each of these patients has remained free of disease for the past 3 years. The titres in three of these patients have risen to abnormally high levels (>1 : 640) (Fig. 1) and we are carefully following them to determine whether the rising titres predict early relapse.

Patients with chronic lymphocytic leukaemia (CLL) were also of interest because of the possibility that the elevated antibody levels in the lympho-proliferative diseases were due to the accumulation of abnormal lymphocytes, a condition favorable for virus growth. If this were indeed the reason for high EBV titres in lymphomas, one would expect that CLL patients with high white blood counts would have higher EBV titres than CLL patients with low blood counts. Since this disease is not treated until the later stages, it was relatively easy to follow a group of patients over a long period of time. In a study of 34 CLL patients who were followed at the University of Colorado between 1964 and 1969 (Levine et al., 1971b), 91 serum samples were tested for viral capsid antibody and immunoglobulins measured. Excluded from the study were patients presenting initially with massive lymphadenopathy or splenomegaly, thus eliminating lymphosarcoma patients with a late leukaemia phase and leaving a more homogeneous population of CLL patients. All patients accepted into the study had a progressive lymphocytosis of at least one year's duration and minimal or no peripheral lymphadenopathy at the time of diagnosis. Some initial samples were obtained from patients with presenting white blood counts as low as 20 000 per mm³. In contrast to the rising titres noted in several of our Hodgkin's patients followed over a 3-year period of time, the titres of all CLL patients were remarkably stable. In addition, there was no relationship between VCA titre and the absolute lymphocyte count or the white blood count (Fig. 2), and there was no correlation with prognosis in this group. Therapy did not appear to affect the EBV titres in any of the patients. The high VCA titres seen in this group in the initial stages of the disease and prior to the appearance of any immunological abnormalities, as shown by measurements of immunoglobulins and skin tests, suggested to us that the elevated titres reflected an event occurring prior to the onset of disease or in the very early stages. We believe this study excludes the possibility that EBV titres in CLL reflect a nonspecific rise paralleling the increase in total body lymphocytes.

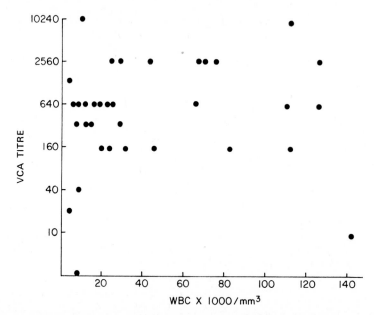

Fig. 2. Correlation of white blood count and EBV titre in chronic lymphocytic leukaemia. A comparison of white blood count and EBV titre in 34 patients with chronic lymphocytic leukaemia demonstrated no correlation between the two, suggesting that the elevated EBV titres in this disease were not solely a reflection of lymphoid mass.

Studies in families with multiple cases of chronic lymphocytic leukaemia have offered an opportunity to evaluate the significance of EBV titres in a different fashion. Fraumeni (1969) has summarized the evidence for a genetic influence on the incidence of cancer, and virological markers such as SV_{40} transformation (Todaro & Takemoto, 1969; Snyder et al., 1970) have been reported as a possible indicator of susceptibility to human leukaemia. We are currently studying families with multiple cases of lymphoma and leukaemia to determine whether EBV titres can detect susceptibility to certain tumours in a similar fashion.

In one particular family, under the care of Dr Khalil Hatoum at the Veterans Administration Center Hampton, Virginia, high antibody titres to the viral capsid antibody as well as antibody to the EBV early antigen (Henle et al., 1971) have been observed in normal family members as well as in two siblings with chronic lymphocytic leukaemia (Fig. 3).

The continued follow-up of these and other families with multiple cases of CLL may provide us with an opportunity to learn more about the immune reactivity to EBV in these normal individuals who appear to be at high risk to CLL.

In our most recent studies (Levine, O'Conor & Berard, 1971c), which have concentrated on childhood lymphomas, we have compared the antibodies to EBV in American Burkitt patients to age-matched patients with acute lymphocytic leukaemia, African Burkitt's lymphoma, and a variety of controls. Criteria for the diagnosis of Burkitt's lymphoma were based on the histological criteria established by the World Health Organization, which has classified Burkitt's lymphoma as a malignant lymphoma composed of lymphoreticular cells or stem cells which are generally PAS-negative, oil red O-positive, and pyroninophilic.

Our studies indicated that the pattern of antibody to VCA in American Burkitt's lymphoma was

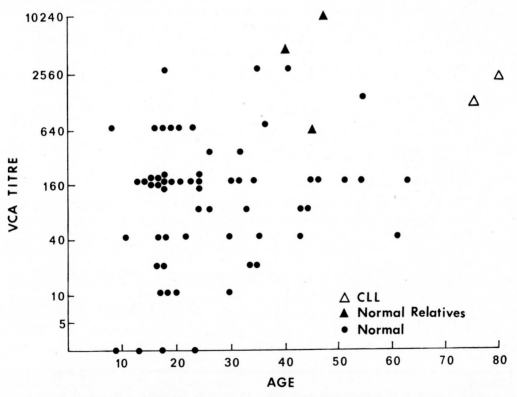

Fig. 3. Familial chronic lymphocytic leukaemia. The possible rôle of elevated EBV titres as a marker of susceptibility to lymphoma is being evaluated by studying families with multiple cases of lymphomas. The high titres seen in normal members of one such family, shown here, suggest that high titres may be related to the aetiology of this disease.

distinct from that found in Hodgkin's disease and CLL. The titres, as in CLL, were stable throughout the course of the disease and were unaffected by treatment or stage of disease. However, age and prognosis did correlate with antibody levels. Young American Burkitt patients, namely those under the age of eight, had higher titres (GMT = 1 : 640) than age-matched leukaemia patients (GMT = 1 : 10) and normal controls (GMT = 1 : 8), and although they did not reach the GMT of the age-matched African Burkitt patients (GMT = 1 : 3162) they did form a distinctive group. In contrast with the Hodgkin's patients, a high VCA titre suggested a good prognosis in our American Burkitt group. Five of the seven patients with titres greater than 1 : 320 are still alive as compared with less than half (7 out of 17) with titres of less than 1 : 320. Four American Burkitt patients were found whose serum had no detectable VCA antibody.

It is of great importance to study Burkitt patients in a nonendemic area since it may be possible to identify cofactors that will give further clues to its aetiology. For this purpose we have established an American Burkitt registry at the National Cancer Institute where we hope to obtain sufficient cases to allow a careful epidemiological and virological evaluation. The clustering of cases around major medical centres suggests that Burkitt's lymphoma is generally an under-reported disease in the Western hemisphere and therefore it is hoped that additional cases in the next few years will allow meaningful comparisons between these patients and their African counterparts.

In summary, we have attempted to follow patients with several lymphoproliferative diseases throughout the course of their illness, comparing antibody levels to EBV and clinical parameters in an attempt to determine whether this virus is aetiologically related to any of these diseases. The dissimilar patterns of antibody in each of these diseases (Table 2) indicate that the rôle of EBV may be different in each lymphoma. With the development of new serological approaches as well as sophisticated tests of cellular immunity to EBV, it is possible that continued studies of patients, families and groups having a high risk of developing tumours will provide more direct evidence as to the precise nature of the rôle of EBV in each of these lymphoproliferative diseases.

Table 2.   Antibody patterns in lymphoproliferative disease

Diagnosis	Correlates with histological subgroups	Stable VCA titres throughout course of disease	Correlates with survival [1]	Frequently EA+
Hodgkin's disease	+	−	+ (IC)	+
Chronic lymphatic leukaemia	+	+	−	+
Burkitt's lymphoma (United States)	−	+	+ (DC)	+
Burkitt's lymphoma (Africa)	−	+	+ (IC) [2]	+
Acute lymphatic leukaemia	−	−	+ (DC) [3]	−

[1] IC = inverse correlation; DC = direct correlation.
[2] Henle et al., 1971.
[3] Stevens et al., 1971.

ACKNOWLEDGEMENTS

The author gratefully acknowledges the assistance of Dr Deward Waggoner, Mrs Shirley Norris and Mrs Wilma Varrato for their assistance in data compilation and calculations.

REFERENCES

Burkitt, D. (1963) A lymphoma syndrome in tropical Africa. *Int. Rev. exp. Path.*, **2**, 69-136

Churchill, A. E. & Biggs, P. M. (1968) Herpes-type virus isolated in cell culture from tumor of chickens with Marek's disease. *J. nat. Cancer Inst.*, **41**, 1371-1375

Correa, P. & O'Conor, T. (1971) Epidemiologic patterns of Hodgkin's disease. *Int. J. Cancer*, **8**, 192-201.

Evans, A. S., Rothfield, N. F. & Niederman, J. C. (1971) Raised antibody titres to EB virus in systemic lupus erythematosus. *Lancet*, **i**, 167-168

Fraumeni, J. F. (1969) Clinical epidemiology of leukemia. *Seminars Hemat.*, **6**, 250-260

Henle, G., Henle, W. & Diehl, V. (1968) Relation of Burkitt's tumor associated herpes-type virus to infectious mononucleosis. *Proc. nat. Acad. Sci. (Wash.)*, **58**, 94-101

Henle, G., Henle, W., Clifford, P., Diehl, V., Kafuko, G. W., Kirya, G., Klein, G., Morrow, R. H., Munube, G. M. R., Pike, P., Tukei, P. M. & Ziegler, J. L. (1969) Antibodies to Epstein-Barr virus in Burkitt's lymphoma and control groups. *J. nat. Cancer Inst.*, **43**, 1147-1157

Henle, G., Henle, W., Klein, G., Gunvén, P., Clifford, P., Morrow, R. H. & Ziegler, J. L. (1971) Antibodies to early Epstein-Barr virus induced antigens in Burkitt's lymphoma. *J. nat. Cancer Inst.*, **46**, 861-871

Hirshaut, Y., Glade, P., Octavio, L. Vieira, B. D., Ainbender, E., Dvorak, B. & Siltzbach, L. E. (1970) Sarcoidosis: Another disease associated with serological evidence for herpes-like virus infection. *New Engl. J. Med.*, **283**, 502-505

Johansson, B., Klein, G., Henle, W. & Henle, G. (1970) Epstein-Barr virus (EBV)-associated antibody patterns in malignant lymphoma and leukemia. I. Hodgkin's disease. *Int. J. Cancer*, **6**, 450-462

Levine, P. H., Ablashi, D. V., Berard, C. W., Carbone, P. P., Waggoner, D. E. & Malan, L. (1971a) Elevated antibody titers to Epstein-Barr virus in Hodgkin's disease. *Cancer*, **27**, 416-421

Levine, P. H., Merrill, D. A., Bethlenfalvay, N. C., Dabich, L., Stevens, D. A. & Waggoner, D. E. (1971b) A longitudinal comparison of antibodies to Epstein-Barr virus and clinical parameters in chronic lymphocytic leukemia and chronic myelocytic leukemia. *Blood*, **38**, 479-484

Levine, P. H., O'Conor, G. T. & Berard, C. W. (1971c) Antibodies to Epstein-Barr virus (EBV) in American patients with Burkitt's lymphoma. *Proc. Amer. Ass. Cancer Res.*, **12**, 59

MacMahon, B. (1966) Epidemiology of Hodgkin's disease. *Cancer Res.*, **26**, 1189-1200

Melendez, L. V., Hunt, R. D., Daniel, M. D., Garcia, F. G. & Fraser, C. E. O. (1969) Herpesvirus Saimiri. II. Experimentally induced malignant lymphoma in primates. *Lab. Anim. Care*, **19**, 378-386

Papageorgiou, P. S., Sorokin, C., Kouzoutzakoglou, K. & Glade, P. R. (1971) Herpes-like Epstein-Barr virus in leprosy. *Nature (Lond.)*, **231**, 47-48

Pike, N. C., Williams, E. H. & Wright, B. (1967) Burkitt's tumor in the West Nile District of Uganda 1961-5. *Brit. Med. J.*, **2**, 395-399

Snyder, A. L., Li, F. P., Henderson, E. S. & Todaro, G. J. (1970) Possible inherited leukemogenic factors in familial acute myelogenous leukemia. *Lancet*, **i**, 586-589

Stevens, D. A., Levine, P. H., Lee, S. K., Sonley, M. J. & Waggoner, D. E. (1971) Concurrent infectious mononucleosis and acute leukemia. *Amer. J. Med.*, **50**, 208-217

Todaro, G. J. & Takemoto, K. K. (1969) "Rescued" SV40: Increased transforming efficiency in mouse and human cells. *Proc. nat. Acad. Sci.*, (Wash.), **62**, 1031

The Spectrum of Epstein-Barr Virus Infections in Brazil

R. P. S. CARVALHO,[1] P. FROST,[2] G. DALLDORF,[2] M. JAMRA,[3] A. S. EVANS[4] & M. E. CAMARGO[1]

This investigation is concerned with the prevalence of Epstein-Barr virus (EBV) precipitating and immunofluorescent antibodies in Brazil, and is a continuation of studies previously reported (Dalldorf *et al.*, 1969; Carvalho *et al.*, 1971).

MATERIALS AND METHODS

Serum samples

Sera from 1171 individuals, made up of 1135 Brazilians and 36 Africans were studied. Of the Brazilian sera, 582 were from normal subjects, 337 from subjects suffering from various diseases, 130 from patients with nasopharyngeal carcinoma, Burkitt's lymphoma, Hodgkin's disease, or various lymphomas and leukaemias, 62 from patients with systemic lupus erythematosus, rheumatoid arthritis or dermatomyositis, and 24 from patients with infectious mononucleosis. Of the African sera, 7 were from patients with nasopharyngeal carcinoma, 13 from patients with Burkitt's lymphoma, and 16 from patients with various diseases.

The major part of these sera came from the State of São Paulo, but many were collected from other Brazilian regions, from adults and children, of both sexes and covering a wide spectrum of ages. Most of the sera were collected during the last three years but many, both from normal subjects and from patients with various diseases, have been collected since 1956. Most of the Indian sera are from adults from an isolated Indian population living in the Upper-Xingu River area of Central Brazil.

The sera were collected aseptically and stored at —20°C.

Precipitin test

The immunodiffusion (Ouchterlony) test with type C plates from Hyland Laboratories, 2% agar and 5 mm distance between wells, and at room temperature, was used. A soluble antigen from the Jijoye line of Burkitt tumour cells was used, prepared at the Sloan-Kettering Institute for Cancer Research, N.Y., according to the method of Old *et al.* (1966). As controls, standard sera and antigens sent by Dr Lloyd J. Old were used.

Indirect immunofluorescent test

The technique of Henle, Henle & Diehl (1968) was used. Smears on glass slides were prepared with aged Jijoye cells grown in RPMI 1640. The sera were screened at 1/10 dilution, as unknowns, with positive and negative sera and saline controls. Additional tests for immunofluorescent antibody have also been carried out with EB3 Burkitt-tumour cells, in a large number of sera.

RESULTS

Tables 1 and 2 show the results of the precipitin test, while the results of the indirect immunofluorescence test are given in Tables 3 and 4.[5]

Table 5 shows the agreement between the precipitin and immunofluorescence tests. Overall agreement

[1] Instituto de Medicina Tropical da USP, São Paulo, Brazil.
[2] Research Corporation, New York, USA.
[3] Faculdade de Medicina da USP, São Paulo, Brazil.
[4] Department of Epidemiology, Yale University School of Medicine, New Haven, Connecticut, USA.

[5] The immunofluorescence tests with the Jijoye cells were carried out at the Instituto de Medicina Tropical de São Paulo, Brazil; a large number of the same sera have also been tested in the Department of Epidemiology, Yale University, USA, with EB3 cells. There was very good agreement between the results of the two series of tests, the titres with EB3 cells being usually one dilution lower.

Table 1.   Results of precipitin test (Ouchterlony) with Jijoye antigen
(normal subjects and various diseases) [1]

Sera	No. of positive sera/no. tested	Percentage of positive sera	Totals
Sera from normal subjects			
Children	16/349	5	
Umbilical cord	0/12	0	5%
Adults	4/120	3	(29/582)
Xingu indians (adults and children)	9/93	10	
Multiparous females	0/8	0	
Sera from patients with various diseases			
Toxoplasmosis	2/32	6	
Tegumentary leishmaniasis	2/11	18	
South american blastomycosis	3/37	8	
Leprosy	0/10	0	
Diphtheria	0/7	0	
Syphilis	1/24	4	
Smallpox	0/10	0	
Infectious hepatitis or serum hepatitis	1/30	3	
Poliomyelitis	2/31	6	7%
Thyroid disease	2/20	10	(22/337)
Schistosomiasis	0/5	0	
Nephritis	1/10	10	
Nephrosis	1/5	20	
Scleroderma	0/6	0	
Sjögren's syndrome	0/6	0	
Miscellaneous diseases	5/57	9	
Malignant diseases [2]	1/15	7	
Repeated blood transfusion	1/21	5	

[1] Data from Carvalho et al. (1971).
[2] Excluding lymphomas, leukaemias and nasopharyngeal cancer.

Table 2.   Results of precipitin test (Ouchterlony) with Jijoye antigen
(lymphatic neoplasms and autoimmune diseases) [1]

Disease	No. of patients tested	No. positive	Percentage positive
Brazilian sera			
Nasopharyngeal cancer	7	6	86
Burkitt's lymphoma	10	6	60
Hodgkin's disease	41	14	34 31%
Lymphomas	16	3	19
Leukaemias	47	9	19
Reticulosarcoma	9	0	0
Lupus erythematosus	29	10	34
Rheumatoid arthritis	23	6	26 31%
Dermatomyositis	10	3	30
Infectious mononucleosis	24	2	8
Normal subjects	582	29	5
African sera			
Nasopharyngeal cancer	7	6	86
African Burkitt's lymphoma [2]	13	11	85
Various diseases	16	3	19

[1] Data from Carvalho et al. (1971).
[2] Previously selected sera.

Table 3.   Distribution of EBV immunofluorescent antibody titres
(fixed Jijoye cells) in normal subjects

Age (years)	No. of sera tested	EBV antibody titre					
		≥ 1/10		≥ 1/160		≥ 1/320	
		No. positive	% positive	No. positive	% positive	No. positive	% positive
Under 10	120	71	59.1	14	11.6	3	2.5
Over 10	32	25	78.1	8	25.0	3	9.3
Total [1]	152	96	63.1	22	14.4	6	3.9

[1] Excluding 12 umbilical cord sera.

Table 4.   Indirect immunofluorescence test with fixed Jijoye cells

Disease	No. of sera tested	EBV antibody titre ≥ 1/320	
		No. positive	% positive
Brazilian sera			
Nasopharyngeal cancer	4	4	100.0
Burkitt's lymphoma	9	5	55.5
Hodgkin's disease	39	14	35.8
Lymphomas	15	3	20.0
Leukaemias	42	14	33.3
Lupus erythematosus	14	7	50.0
Rheumatoid arthritis	11	2	18.1
Dermatomyositis	8	3	37.5
Infectious mononucleosis	24	7	29.1
Normal subjects	164	6	3.6
African sera			
Nasopharyngeal cancer	6	6	100.0
African Burkitt's lymphoma	5	3	60.0

Table 5.   Agreement between precipitin (Ouchterlony) and indirect
immunofluorescence (Henle) tests with Jijoye antigen
(immunofluorescence-positive titres ≥ 1/320)

Disease	No. of sera in agreement (+ or —)/No. tested [1]	Percentage agreement
Brazilian sera		
Normal subjects	151/154	98
Nasopharyngeal cancer	4/4	100
Burkitt's lymphoma	13/19	68
Hodgkin's disease	42/59	71
Lymphomas	13/15	87
Leukaemias	51/69	74
Lupus erythematosus-rheumatoid arthritis-dermatomyositis	23/34	68
Infectious mononucleosis	17/24	71
African sera		
Burkitt's lymphoma-nasopharyngeal cancer	7/11	64
Total	321/389	83

[1] With certain patients, several serum samples were tested.

between elevated titres of 1/320 or more by the immunofluorescence test (Jijoye cells) and the precipitin test has been obtained in 83% of the sera (321/389).

DISCUSSION

The results of the precipitin test with Brazilian sera were similar to those reported by Old *et al.* (1966, 1968) for African and North American sera. The percentage of positive tests among normal subjects or in patients with various non-neoplastic diseases or nonlymphatic malignant diseases, was very low (6%). In lymphatic neoplasms and certain autoimmune diseases, not only was the percentage of positive sera higher but the reactions were more intense and rapid. Of the patients with systemic lupus erythematosus, rheumatoid arthritis and dermatomyositis, 31% gave positive reactions. These sera often produced more than one precipita-

tion line, together with a reaction of identity with known positive Burkitt's lymphoma sera.

The indirect immunofluorescence test with Jijoye cells yielded 59.1% positives at the 1/10 dilution in normal children under 10 years of age, and 78.1% in the other normal subjects. Various proportions of positive tests have been reported for normal children (Demissie & Svedmyr, 1969; Pereira, Blake & Macrae, 1969; Candeias & Pereira, 1970). As only 3.9% of the normal sera reacted at the $\geqslant 1/320$ dilution, such titres were considered as abnormal. Table 4 is based on this criterion; it shows that the percentages found for the lymphatic neoplasms and autoimmune diseases were very similar to those found by the precipitin test (Table 2); agreement between the two tests is very close (Table 5).

A high frequency of EBV antibody in lupus erythematosus, rheumatoid arthritis and dermatomyositis has been reported by Dalldorf *et al.* (1969), Evans, Rothfield & Niederman (1971) and Carvalho *et al.* (1971).

ACKNOWLEDGEMENTS

We are indebted to Drs Celeste Fara Netto, Rubens Guimarães Ferri, Wilson Cossermelli and Carlos Marigo for many of the serum specimens. This work was supported by a research grant from the Brown-Hazen Fund-Research Corporation, New York, USA, and IARC Collaborative Research Agreement No. RA/70/016.

REFERENCES

Candeias, J. A. N. & Pereira, M. S. (1970) Pesquisa de anticorpos para o virus EB em adultos e crianças. *Rev. Inst. Med. trop. São Paulo*, **12**, 333-338

Carvalho, R. P. S., Frost, P., Dalldorf, G., Jamra, M. & Evans, A. S. (1971) Pesquisa de anticorpos precipitantes para antígeno do Linfoma de Burkitt (células Jijoye) em soros colhidos no Brasil. *Rev. Bras. Pes. med. biol.*, **4**, 75-82

Dalldorf, G., Carvalho, R. P. S., Jamra, M., Frost, P., Erlich, D. & Marigo, C. (1969) The lymphomas of Brazilian children. *J. Amer. med. Ass.*, **208**, 1365-1368

Demissie, A. & Svedmyr, A. (1969) Age distribution of antibodies to EB virus in Swedish females as studied by indirect immunofluorescence on Burkitt cells. *Acta path. microbiol. scand.*, **75**, 457-465

Evans, A. S., Rothfield, N. F. & Niederman, J. C. (1971)

Raised antibody levels to EB virus in systemic lupus erythematosus. *Lancet*, **i**, 167-168

Henle, G., Henle, W. & Diehl, V. (1968) Relation of Burkitt's tumor associated herpes-type virus to infectious mononucleosis. *Proc. nat. Acad. Sci. (Wash.)*, **59**, 94-101

Old, L. J., Boyse, E. A., Geering, G. & Oettgen, H. F. (1968) Serologic approaches to the study of cancer in animals and in man. *Cancer Res.*, **28**, 1288-1299

Old, L. J., Boyse, E. A., Oettgen, H. F., De Harven, E., Geering, G., Williamson, B. & Clifford, P. (1966) Precipitating antibody in human serum to an antigen present in cultured Burkitt's lymphoma cells. *Proc. nat. Acad. Sci. (Wash.)*, **56**, 1699-1704

Pereira, M. S., Blake, J. M. & Macrae, A. D. (1969) EB virus antibody at different ages. *Brit. med. J.*, **4**, 526-527

Relation of Epstein-Barr Virus and Malaria Antibody Titres in Two Population Groups *

A. S. EVANS, P. HESELTINE, I. KAGAN & J. C. NIEDERMAN

Malaria was postulated as a possible cofactor with Epstein-Barr virus (EBV) in the causation of Burkitt's lymphoma by Dalldorf *et al.* (1964). Burkitt and others have endorsed this concept and it has received much additional attention (Burkitt, 1969; *Lancet*, 1970). The main evidence in its favour is the occurrence of this lymphoma in children residing in areas in which EBV and severe malaria infections are common and occur early in life (O'Conor, 1970). Additional support has come from the recent experimental demonstration that concurrent infection with a murine plasmodium and Moloney virus increased the incidence of malignant lymphoma in adult Balb/C mice (Wedderburn, 1970). The mechanism by which malaria may enhance susceptibility is obscure, but may involve a selective effect on immune mechanisms. High EBV antibody titres are a regular feature of patients with Burkitt's lymphoma (Henle, G. *et al.*, 1969). It is possible, therefore, that malaria infection interferes with immunity at a cellular level, permitting multiplication of latent and potentially oncogenic EBV with a subsequent boost of antibody levels, and that, in some patients, this immunological defect may result in tumour formation. If this concept is correct, higher EBV antibody titres would be expected in persons with malaria than in those without. The observations to be reported explore this concept.

MATERIALS AND METHODS

Serum samples obtained from two WHO epidemiological surveys for yaws were studied. One of these surveys was conducted in rural areas of the Philippines in 1963 and the other in eastern Nigeria during 1964. EBV antibody titres were determined by the indirect immunofluorescence test using the EB3 cell line (Henle, G. & Henle, W., 1966). Readings were made independently by two observers and positive and negative controls were included in each test run. In this system a titre of 1 : 160 or greater is regarded as elevated. Malaria antibody was measured by a passive haemagglutination test (PHA) in which *Plasmodium knowlesi* is absorbed on to human group 0 cells, pretreated with tannic acid (Rogers, Fried & Kagan, 1968).

RESULTS

Sera from 62 Philippine subjects lacking malaria antibody and from 57 whose sera possessed malaria PHA antibody at levels of 1 : 128 or greater were selected for testing for EBV antibody levels. Their ages ranged from 2 to 60 years. All but one of these 119 sera (99%) had EBV antibody detectable at a 1 : 5 serum dilution or higher, while 41% had anti-EBV levels equal to or greater than 1 : 160.

Of the 62 Philippine sera lacking malaria antibody, 26% had elevated EBV titres of 1 : 160 or higher; of 57 specimens possessing malaria antibody at titres equal or greater than 1 : 128, 58% had elevated EBV antibody levels. This difference between 26 and 58% with elevated titres was statistically significant with a value of P less than 0.01. By age-group, this difference was significant only for those 68 subjects under the age of 20.

Eighty serum samples from Nigerian children under the age of 10 were tested without preselection as to their malaria antibody titre. Of these, 98% had detectable EBV antibody at 1 : 5 dilution or greater and 34% had high titres at levels of 1 : 160 or above. Malaria antibody was present in sera

* From the WHO Serum Reference Bank, Department of Epidemiology and Public Health, Yale University School of Medicine, and the Parasitology Laboratory, National Center for Disease Control, Atlanta, Georgia, USA.

from 63 of these 80 children; elevated malaria titres were found in 13 subjects. As shown in the Table, no correlation existed between elevated EBV and elevated malaria titres; indeed, the reverse seemed to be true, but the numbers are small. In addition, no correlation existed between the exact malaria and exact EBV antibody titres in sera from these Nigerian children. EBV antibody levels tended to be lower rather than higher in the presence of elevated malaria antibody.

Relation of high EBV to high malaria titres

Malaria antibody titre	EBV antibody titres			
	Philippine sera		Nigerian sera	
	No. tested	% elevated [1]	No. tested	% elevated [1]
Negative	62	25.8	17	41.2
Elevated [2]	57	57.9	13	15.4
Total	119	41.2	30	30.0

[1] Levels ⩾ 1 : 160.
[2] Levels of 1 : 128, 1 : 160 or higher.

DISCUSSION

This study has demonstrated the high frequency and early acquisition of EBV antibody in two populations from the Philippines and Nigeria. All but 5 (98.4%) of the total 199 sera tested had demonstrable antibody at a 1 : 5 dilution. Antibody levels of 1 : 160 or greater were found in 41.2% of Philippine and 33.8% of Nigerian sera. These rates are much higher than the 10–15% of normal persons whose sera have titres at these levels.

A statistically higher frequency of elevated EBV titres was found in Philippine subjects with high malaria antibody than in those lacking malaria antibodies. However, this correlation was not seen in Nigerian children. In addition, EBV antibody titres did not correlate directly with malaria antibody levels in sera from the Nigerian children.

The hypothesis that high EBV and high malaria titres may go hand in hand was supported in the Philippine but not in the Nigerian sera. The reasons for this are not clear, and certainly larger numbers must be tested before any final conclusions can be reached.

ACKNOWLEDGEMENTS

This study was supported by a grant from the National Institutes of Allergy and Infectious Diseases (AI 08731). Appreciation is expressed to Dr Wilbur Downs, Yale Arbovirus Research Laboratory, who provided the sera from the WHO Nigerian Survey and Dr Karel Zacek, WHO Serum Reference Bank, Prague, Czechoslovakia, for the WHO Philippine collection. The statistical assistance of Miss Virginia Richards is acknowledged.

REFERENCES

Burkitt, D. P. (1969) Etiology of Burkitt's lymphoma—an alternative hypothesis to a vectored virus. *J. nat. Cancer Inst.*, **42**, 19-28

Dalldorf, G., Linsell, C. A., Barnhart, F. E. & Martyn, R. (1964) An epidemiological approach to the lymphomas of African children and Burkitt's sarcoma of the jaws. *Perspect. Biol. Med.*, **7**, 435-449

Henle, G. & Henle, W. (1966) Immunofluorescence in cells derived from Burkitt's lymphoma. *J. Bact.*, **91**, 1248-1256

Henle, G., Henle, W., Clifford, P., Diehl, V., Kafako, C. W., Kinya, B. G., Klein, G., Morrow, R. H., Munube, G. M. R., Pike, P., Tukey, P. M. & Ziegler, J. L. (1970) Antibodies to Epstein-Barr virus in Burkitt's lymphoma and control groups. *J. nat. Cancer Inst.*, **43**, 1147-1157

Lancet (1970) Burkitt lymphoma and malaria, i, 300-301

O'Conor, G. T. (1970) Persistent immunologic stimulation as a factor in oncogenesis with special reference to Burkitt's tumor. *Amer. J. Med.*, **48**, 279-285

Rogers, W. A., Fried, J. A. & Kagan, I. G. (1968) A modified indirect microhemagglutination test for malaria. *Amer. J. trop. Med. Hyg.*, **17**, 804-809

Wedderburn, N. (1970) Effect of concurrent malarial infection on development of virus induced lymphoma in Balb/C mice. *Lancet*, **ii**, 1114-1116

Discussion Summary

R. H. MORROW, JR [1]

Criteria for definition of Burkitt's lymphoma

Several aspects of the problem of definition of Burkitt's lymphoma (BL) were discussed. The histopathological criteria still provide the single best definition of this entity. Much intensive work with interchange of coded slides and case histories amongst pathologists have led to the establishment of fairly widely accepted criteria which have been published by WHO.[2] Of the patients listed in this paper who were accepted as having typical BL on the basis of histopathology, two non-African patients were negative for antibodies to Epstein-Barr virus (EBV). Thus in time it should be possible to obtain a group of typical BL patients from outside Africa. It would be particularly valuable to look at this group of patients for factors other than malaria or EBV that might be involved in the aetiology.

However, the acceptance of any single criterion, even that of histopathology, as absolute creates logical problems in particular situations. Thus although all African lymphoma patients fulfilling both clinical and histopathological criteria for BL have been positive for EBV antibodies thus far, there are a few non-African lymphoma patients who have fulfilled at least the histopathological criteria for BL but who have been negative for EBV antibodies. Until we have a definition based on aetiological criteria (if we ever do), we cannot say whether these patients represent cases of Burkitt's lymphoma that are negative for EBV or whether they have a different kind of disease. Different definitions might be useful for different purposes. It is interesting that this particular dilemma apparently does not exist for nasopharyngeal carcinoma (NPC) thus far. G. Klein reported that all patients who have fulfilled the histopathological criteria for NPC, whether from Africa, Asia, or Europe, have also had antibodies to EBV.

Epidemic nature of Burkitt's lymphoma

The importance of clustering theory is that it represents a challenge to the simple idea that a direct combination of malaria plus EBV produces Burkitt's lymphoma. However, it may not be necessary to introduce a third factor to explain the clustering. It might be a matter of the timing of the infection with EBV in relation to the malaria. The tendency has been to consider malaria as the background to the EBV infection. However, some of the epidemiological information may indicate that the process proceeds in the opposite direction, i.e., malarial stimulation of EBV-infected lymphocytes may initiate the malignant change. Although it is possible that the relationship could be reciprocal, it would seem more plausible if it were in one direction only. The importance of cluster theory epidemiologically is that we must look for some form of micro-epidemic behaviour in the malaria or in the EBV in the areas where Burkitt's lymphoma behaves as an epidemic disease.

The fact that, in Africa, EBV infection is highly prevalent, and in fact almost universal, does not necessarily rule out the possibility of epidemics of EBV. Parallels were drawn with paralytic poliomyelitis and with measles to indicate that highly prevalent, universal infections could nevertheless be epidemic in nature. Similarly there may be important shifts in the intensity of transmission of even holoendemic malaria from season to season and place to place. However, very little work has been done to determine whether such epidemics of EBV, or shifts in malarial intensity to parallel the time-space clusters of BL, really occur.

[1] Department of Tropical Public Health, Harvard School of Public Health Boston, Massachusetts, USA.
[2] Histopathological definition of Burkitt's tumour. *Bull. Wld Hlth Org.*, 1969, **40**, 601-607.

EBV cohort studies and false negative EBV titres

"Negative" EBV antibodies have been shown to be meaningful in the studies done on infectious mononucleosis, and since the same serological methods have been used in the African serological surveys, it would seem that these "negatives" should not be readily discounted.

With regard to the population surveys, it should be understood that a serum dilution of 1 to 10 was made for convenience and thus those with anti-bodies of less than 10 did indeed include at least some positives with low titres of 1 in 5 or even 1 in 2.

Morrow reported that in the West Nile, where serial specimens from populations were obtained, it would appear that about 20% of those with titres of less than 10 had been previously positive and hence this figure, though based on small numbers, could be considered a minimum "false negative" rate for this population.

Even if there is a larger proportion of false negatives than seems apparent, only a very small number of susceptibles (true negatives) in the population would be required, since BL has such a low incidence. Possibly a delayed primary EBV infection superimposed upon malaria with its intense stimulation of the reticulo-endothelial system (RES) could explain the observed incidence of BL.

The point must be made, however, that in a country such as the USA, where the proportion of the population with positive sera continues to rise with age, no serious problem is involved in assuming that the negatives, as measured by the test, contain a fair proportion of true negatives. However, in a population in which the proportion of positives declines with age, as in the West Nile, and in which nearly all are positive at the age of 2–4, it is much harder to believe that a very large proportion of the titres of less than 10 are not really positive. The only way to determine the rôle of EBV in BL is through a study such as the IARC is going to conduct in the West Nile. However, it is important to appreciate that, although the study will test one of the most critically important hypotheses, namely that EBV is an agent that necessarily leads to the induction of BL within a limited latent period, it cannot test several other potentially very important hypotheses, such as the possibility that infection with EBV must take place within a few weeks or months of birth, with induction occurring as the result of a subsequent independent event, leading to a long and variable latent period. To test such

a possibility would require a population size at least five times as great as that envisaged.

An important point concerning the possible rôle of EBV in BL is that virally induced animal tumours, which are generally highly antigenic, must grow very fast because otherwise they will be rejected. This factor makes it unlikely that, if EBV induces BL, the latent period would be very long, and it would tend to make it less likely that individuals with high titres to EBV could remain for long periods at continuing risk of developing BL. Thus the alternative possibility that the tumours would be expected to have a relatively short latency period is more likely.

Pathogenesis and transmission of EBV infection

The many major gaps in our understanding of the pathogenesis of EBV infection became evident during the discussions. Very little is known about the method of transmission, portal of entry, method of dissemination, duration of infectivity, or capacity for reinfection. This is largely due to the fact that there is still no method of readily isolating the virus by transfer to a susceptible cell so that, at present, only indirect epidemiological approaches are available.

In infectious mononucleosis (IM), which provides at least a clinical marker for infection, it is unusual to be able to identify the source of infection. In one documented instance, the period between an isolated contact and onset of symptoms was 40 days. There is some information as to the possibility of, or symptoms associated with, recurrent infection, but little attention has been paid to these aspects. Evans reported that, in general, there is a very good correlation between antibody levels of 1 : 10 or above and immunity to clinical disease, but there is at least one documented case of recurrence of IM associated with a rise in EBV antibody levels. Exacerbation, rather than true recurrence, is commoner, but this generally occurs within 6 weeks of the acute illness.

There is little information on reactivation of EBV in immunosuppressed or transplant patients, such as occurs with herpes simplex virus or cytomegalovirus. There is also little or no information concerning other possible parallels with the herpes and cytomegalovirus group, such as perinatal and transplacental transmission.

The spread of EBV infection within the families of patients with IM is as difficult to demonstrate as the infectivity of this virus for other cells in the laboratory. In an epidemiological study, which was the subject of a recent preliminary report,[1] a very low proportion (less than 10%) of susceptible family contacts converted from negative to positive. In addition, in those who converted, the "latent period" between contact and conversion varied from 1 week to 2 years, so that it is doubtful whether these periods were true "incubation" periods. It may be that IM is really an early or late postinfection syndrome of an essentially immunological nature. Possibly the disease process results from a massive immune rejection by cytotoxic antibodies or from a cell-mediated immune survey mechanism against rapidly proliferating lymphoid cells carrying EBV-specific cell-surface antigen. IM is certainly an unusual infection in many respects, and it is difficult to regard it as a primary infection of a cytolytic type with a relatively short incubation period, similar to most other common viral infections.

The idea that IM might be a rejection syndrome of EBV-converted lymphoid cells was attractive to several participants, who considered the various possible ways in which clinical symptoms might be explained and how experimental procedures, such as transplanting EBV-infected lymphoid cells, might be accomplished.

Immunological reactions to EBV

It was suggested that an EBV-specific IgM that is detected by indirect immunofluorescence and persists for only a few months and then declines may turn out to be useful in the diagnosis of infectious mononucleosis.

The dissociation described by Sohier between the complement-fixing (CF) antibody and the immunofluorescence (IF) antibody might well be expected. CF antibodies generally fall into the IgG class of immunoglobulin, but the nature of the indirect IF antibodies is apparently not yet clear. It should now be possible to determine this.

There is also a considerable difference between the different antigens involved. The antigen used in Sohier's tests was derived from a cell line that does not express viral antigens, but only CF antigens.

The work reported by Henle on the dissociation of early antigen into the diffuse form, which appears and disappears with infectious mononucleosis, and the restricted form, which may be specifically associated with BL, may have important aetiological significance.

G. Klein said that in the African population some of the relatives of BL patients were remarkable in having high membrane-reactive antibody levels but nevertheless low anti-VCA levels. It was interesting that there seemed to be high activity in relatives of chronic lymphatic leukaemia patients as well as in BL.

Little work has been done on the relatives of American BL patients. However, Levine said that in a case reported by Dr D. Stevens, an 8-year-old boy developed classical BL, and a month later his sister had developed a clinical course of acute leukaemia in which there were circulating cells, primarily in the peripheral blood, that seemed indistinguishable from Burkitt's cells. This is now under investigation.

Geographical coincidence of Burkitt's lymphoma and malaria

A number of questions concerning the geographical coincidence of BL and malaria were raised. In Malaysia, where malaria is hyperendemic, a number of cases of BL have been reported, but in areas of less intense seasonal malaria, BL is not found. Hyperendemic malaria no longer exists in Ceylon or India. Unfortunately, in the areas of South and Central America, where malaria is still hyper- or holoendemic, very little information is available concerning BL. It appears that the incidence of BL has declined in those parts of New Guinea and Malaysia where malaria has been controlled.

Only two documented cases of BL have occurred in Asians living in Uganda and none in Kenya. This is less than would be expected for the Asian population in these areas, but at least a partial explanation is that a large proportion live in cities, especially Nairobi, where there is little or no malaria and most either use mosquito netting or take antimalarials regularly.

Immunological reactions to malaria

The mechanisms whereby malaria may play a part in the aetiology of BL were discussed in general terms. It was recognized that the effect of malaria on the immunological system of the host was exceedingly complex and was now the subject of a great deal of study. Malaria, both quantitatively and qualitatively, is an enormous challenge to both humoral and cellular defence mechanisms. The

---

[1] J. H. Joncas (1970) *Can. med. Assoc. J.*, **102**, 1260-1263.

effect of malaria-host interaction on any particular antibody response has been shown in some situations to result in immunosuppression and in other situations in enhancement. There is no doubt that, morphologically, malaria has an intense effect on the RES, stimulating it markedly and thus causing splenomegaly, hepatomegaly and general lymphoidal hyperplasia. There is as yet no evidence for an impaired immunity induced by severe malarial infection in patients with BL, but it seems likely that the tests used so far for evaluation of cellular immunity may not be sufficiently sensitive or selective.

In relation to its interaction with EBV, although intense malaria infection might serve as a specific immunosuppressive, it might also act as a promoter in a two-stage carcinogenesis model, of which there are a number of examples, and its rôle in BL need not have a purely immunological explanation.

Possible models for Burkitt's lymphoma

There was some discussion of the potential value of other herpesviruses, particularly Marek's disease virus and herpesvirus saimiri, as models for EBV in its relation to BL, and also of the possibility of developing certain experimental models that might provide parallels to BL. In Marek's disease, in the natural situation, the study of the antibodies to the virus would not have been helpful in understanding the aetiology of the disease. Most chickens carry antibodies to the virus, and those that develop Marek's disease do not have higher antibody levels than those that do not. It may be that there are a number of different strains of virus of varying pathogenicity. A second interesting point in Marek's disease is that, within quite closely genetically related groups of chickens challenged with exactly the same dose of virus at the same age at the same time, the incubation period may vary from a few weeks to several months. Finally, there is a very interesting relationship between an intestinal protozoal condition in chickens called coccidiosis, and Marek's disease. From field observations it appeared that coccidiosis was primary and that it produced some kind of suitable environment in which Marek's disease could develop. Biggs reported that under experimental conditions, however, this did not prove to be the case. It turned out that not only was coccidiosis not necessary for the development of Marek's disease, but rather the reverse, i.e., those chickens developing Marek's disease were apparently

more susceptible to prolonged infection with coccidiosis.

Several participants urged the need for more experimental model work with both Marek's disease virus and herpesvirus saimiri, and also with these viruses in conjunction with malaria, in order to observe the effect on tumour formation.

Epidemiology of nasopharyngeal carcinoma

It is interesting that the incidence rate for nasopharyngeal carcinoma (NPC) in the indigenous people of Sarawak and Sabah is nearly the same as that of the Chinese. Although these are Mongoloid people, Ho pointed out that their living habits are entirely different from those of the Chinese. There has been little or no intermarriage recently, but information from an archeologist indicates that there is evidence in China that Mongoloid features appeared first in southern China rather than in the north, and that this may perhaps have some bearing on the high incidence of NPC in southern Chinese.

Muir said that these incidence figures are based on quite small numbers, but the biases are probably such that the above conclusions are basically justified. The Chinese in both Sarawak and Sabah are generally much more prosperous than the rest of the population, and are more likely to seek treatment than the indigenous peoples who never move far from the coast, live in houses on stilts, and make their livelihood from fishing. The Dayaks are particularly reluctant to use medical treatment facilities, and the higher frequency for the indigenous people of these areas probably does indicate a genuine higher risk.

EBV and other diseases

The nature of the relationship between EBV and various other diseases may be quite different in each case. The similarity between the epidemiology of IM and that of Hodgkin's disease, in that they both occur in young adults with an age-peak difference of 5 to 10 years and that the occurrence of both is greater for the higher socioeconomic groups and in developed countries, as compared to developing countries, may be important. The interesting suggestion that, in lymphocyte-depleted Hodgkin's disease, the increased antibody levels were related to the progress of the disease requires further confirmation in longitudinal studies. Levine reported that, in 7 low titred patients with Hodgkin's disease who are now alive and free of the disease three years later and who have been followed closely, the titres of three have risen whereas those of three others have remained absolutely stable. There has so far been no change in clinical status to which these serological data can be related.

Evans commented that in lupus erythematosus, the best information at the moment is that EBV antibody titres are generally elevated, or at least that fluorescent cells are more common. Though this is probably true, the possibility remains that other antinuclear components may interfere. It is important to realize that, in lupus, there are elevated titres to rubella, measles, para-influenza, and to other viruses as well. The titres to these viruses are not all elevated in the serum from the same patient, however; there is usually only one antibody, so that it might be pos-

tulated that selective immunological defects exist that permit modification of one or the other of these viruses in different circumstances. Just how such hypothetical viral modification occurs is, of course, a point of critical importance, and such modification may have a causal relationship with the syndrome with which it is associated. Thus, on the one hand, there may be production of additional tumour cells, as in BL, and on the other hand, there may be a common defect that both occurs in the condition and permits viral multiplication. Possibly in lupus, viral modification might participate in antigen-antibody complexes that deposit and cause tissue damage. In any case, it is clear that more attention should be devoted to the question of the immunological deficiencies that may be present in these conditions. They have so far been well documented only in advanced Hodgkin's disease, and may, of course, be effect rather than cause.

OTHER HERPESVIRUSES ASSOCIATED WITH NEOPLASMS

Chairman — G. de-Thé

Rapporteur — R. J. C. Harris

Genital Herpes and Cervical Cancer—Can a Causal Relation be Proven?—A Review*

A. J. NAHMIAS, Z. M. NAIB & W. E. JOSEY

In 1964, we noted an association between genital herpes simplex virus (HSV) infection and cervical neoplasia (Naib, Nahmias & Josey, 1966). Subsequent work by our group (Josey, Nahmias & Naib, 1968; Naib et al., 1969; Nahmias et al., 1970a; Nahmias, Naib & Josey, 1971), and others (Rawls, Tompkins & Melnick, 1969; Royston & Aurelian, 1970a; Sprecher-Goldberger et al., 1970; Catalano & Johnson, 1971), substantiated this association and raised the question of a possible cause-and-effect relationship. From the onset of our investigations, we have struggled with the same problems as those confronting other virologists seeking proof of viral causation of human cancers. The problems are similar to those experienced by other scientists who, over a 30-year span, have attempted to demonstrate a causal relation between cigarette smoking and lung cancer.

In order to place in perspective the current status of the genital-herpes-cervical-cancer question, it seemed appropriate to adopt for this discussion the criteria used to uphold a causal relation between cigarette smoking and lung cancer. In the US Surgeon General's report on smoking and health (1964), the association between cigarette smoking and lung cancer was discussed in terms of its: (1) coherence; (2) consistency; (3) strength; (4) specificity; and (5) temporal relationship. An alternative to the causative hypothesis, in the case of the cigarette smoking-lung cancer association, is that of a genetic predisposition to both entities (the "constitutional hypothesis"). In the case of the genital herpes-cervical cancer association, an analogous problem is the necessity to disprove the hypothesis that certain women with increased sexual activity are predisposed to the development of both conditions.

COHERENCE OF THE ASSOCIATION

This criterion requires that there must be coherence with known facts in the natural history and biology of the disease. It should be realized that in 1964 there was considerable doubt regarding the oncogenic potential of any herpesvirus. Our present meeting, only 7 years later, is an answer to such doubts.

The hypothesis would gain coherence if it were found that the herpesvirus infecting the female genital tract was different from that infecting other body sites. When we began to explore this question, we expected to find only small differences; however, we found that the differences between genital and nongenital herpesviruses were larger than anticipated Dowdle et al., 1967; Nahmias & Dowdle, 1968). Not only is the genital virus (type 2) most often different antigenically from the nongenital virus (type 1), but the two viruses also differ in a large variety of biological characteristics (Nahmias & Dowdle, 1968; Nahmias, Naib & Josey, 1971). The microtubules observed by electron microscopy with type 2 and not with type 1 HSV (Couch & Nahmias, 1969) are of interest, since similar structures have been observed with some of the oncogenic animal herpesviruses.

Crucial to the coherence of the association was the need to establish that genital herpes affected the cervix, since it had generally been thought that the virus affected only the external genitalia. Not only was the cervix shown to be a common site of HSV involvement, but it was found to be the most common site, with the majority of cervical infections being asymptomatic (Josey, Nahmias & Naib, 1968; Nahmias, Naib & Josey, 1971).

For coherence, it was also necessary to show that the epidemiological characteristics of genital herpes

* From the Departments of Pediatrics, Preventive Medicine, Pathology, Gynecology and Obstetrics, Emory University School of Medicine, Atlanta, Georgia, USA.

were similar to those of cervical cancer. Here again, the prevalence of both genital herpes and cervical cancer has been found to be the same in various groups of women, and both entities have been shown to have the epidemiological characteristics of venereal diseases (Josey, Nahmias & Naib, 1968; Nahmias *et al.*, 1969).

Results of other studies, still mostly unconfirmed, provide further support for the coherence of the association. After inoculation of newborn hamsters with either live or ultraviolet-inactivated type 2 HSV (Nahmias *et al.*, 1970b), eleven sarcomas were found (a rate of approximately 1.5% of newborn hamsters that survive viral inoculation). Of five of these tumours, transplantable either in other hamsters or in tissue culture, two were shown to have C-type particles on electron-microscopic examination, and the same two were also found to contain C-type particle antigens (as studied by Dr R. Huebner). We have been unable to detect antibodies which react with herpes simplex viruses in the serum of hamsters with tumours. About 90% of sera from individuals with cervical cancer and/or HSV type 2 antibodies react with soluble antigens prepared from the hamster tumours; however, 48% of control sera with no antibodies or HSV type 1 antibodies have also been found to react with the soluble antigens.

In mice genitally inoculated with HSV (Nahmias, Naib & Josey, 1971), two *in situ* cervical cancers were observed, whereas none of the control non-inoculated mice developed cervical cancer. It is of interest to note that Munoz (1971)[1] is reporting at this Symposium the development of cervical cancer in two mice inoculated genitally with type 2 HSV, one of which had also received oestrogen. We have also established a model of genital HSV type 2 infection in monkeys (Nahmias *et al.*, 1971a). Of three monkey species studied, infection could not be established in Rhesus or spider monkeys. However, we found over two years ago that we could establish a genital infection in female Cebus monkeys that mimics closely infection in humans. Up to the present, we have not observed any definite cervical cytological changes in the Cebus monkeys inoculated over 2 years ago.

Watkins (1964) has described a form of transformation in cells infected with HSV. The membranes of the cells were found to be changed by the virus, even though viral synthesis was prevented by the presence of DNA inhibitors. We have con-

firmed these observations with the use of membrane-immunofluorescent techniques. However, the best evidence for *in vitro* transformation by HSV type 2 has been obtained by Rapp & Duff (1971)[2] who are reporting their findings in this Symposium.

Several attempts have been made to find antigens common to both cervical cancer and HSV type 2. Royston & Aurelian (1970b) have reported that exfoliated cells from women with cervical neoplasia reacted with anti-HSV type 2 serum by immunofluorescent techniques. We have also obtained similar results using *in vitro* cervical cancer cells. McKenna, Sanderson & Blakemore (1962) reported earlier a common precipitin line, demonstrated by immunodiffusion, between HeLa cells, originating from cervical cancer, and herpesviruses (type undetermined). More recently, Hollingshead (1971)[3] has also obtained evidence of common antigens between cervical cancer cells and HSV type 2, using complement-fixation tests.

We have also noted that cervical cancer cells are less susceptible to herpesviruses *in vitro* (Nahmias, Naib & Josey, 1971). The extent of nucleic acid hybridization between HSV type 2 and human or hamster tumour cells is currently under investigation. It must be noted, however, in connection with all the studies noted above that, even if some antigens or nucleic acids are found to be shared by cervical cancer cells and HSV type 2, we shall still be faced with an alternate, noncausal, hypothesis, namely that the antigens or nucleic acids originate from viruses that were latent in the cancer cells.

CONSISTENCY OF THE ASSOCIATION

So far, results of the retrospective studies of our group (Nahmias *et al.*, 1970a; Nahmias *et al.*, 1971[3]) and five others (Rawls, Tompkins & Melnick, 1969; Royston & Aurelian, 1970a; Sprecher-Goldberger *et al.*, 1970; Catalano & Johnson, 1971; Wildy, 1972[4]) have demonstrated a consistency in the association between cervical neoplasia and genital HSV type 2. These studies have used a variety of serological tests to detect HSV type 2 antibodies in women with cancer and in "control" groups. It appears that the more recent results of Rawls, Adams & Melnick (1972)[5] disagree with the earlier

---

[1] See p. 443 of this publication.

[2] See p. 447 of this publication.
[3] Unpublished data.
[4] See p. 409 of this publication.
[5] See p. 424 of this publication.

Cumulative percentage risk of carcinoma *in situ* and cervical dysplasia
in women with or without genital herpes detected cytologically

Group	Cumulative percentage risk at end of 5 years [1]		
	Carcinoma *in situ*	Cervical dysplasia	Total
Herpes group			
Herpes detected intra- or postpartum	6.2 %	22.2 %	28.4 %
Herpes detected in nonpregnant women	1.3 %	16.6 %	17.9 %
No herpes group	0.6 %	11.9 %	12.5 %

[1] Determined by life table method.

findings of that group (Rawls, Tompkins & Melnick, 1969).

In a recent prospective study, we have preliminary information showing that the cumulative percent risk for a woman with genital herpes to develop carcinoma *in situ* is 2 to 10 times higher than in a control group (see Table), depending on her pregnancy status at the time of herpes detection. We shall come back to the possible influence of pregnancy as a cofactor with genital herpes in cervical carcinogenesis.

STRENGTH OF THE ASSOCIATION

In the retrospective or prospective studies that have shown a consistent association, the ratio of cervical neoplasia rates for women with genital herpes, as compared to those without genital herpes, has varied with the presence of dysplasia, or *in situ* or invasive cancer. For dysplasia, the ratio is 2–3 : 1; for *in situ* carcinoma, 1.5–10 : 1 and for invasive cancer, 1.5–3 : 1. The ratios are also greater for the younger women, since the prevalence of HSV type 2 antibodies is high in the older women in the control populations.

The possibility of a dose effect might be considered, as it has in the lung cancer–cigarette smoking parallel: the more cigarettes smoked, up to a point, the higher the risk of lung cancer. In order to examine this question, we have analyzed cases of cervical neoplasia according to whether the herpetic infection was primary or recurrent. A higher frequency of cervical neoplasia was noted in the recurrent than in the primary cases (Nahmias *et al.*, 1971 [1]), suggesting that transformation may occur

during a recurrence as well, and that it may take longer for the development of neoplasia after a primary than a recurrent infection.

SPECIFICITY OF THE ASSOCIATION

The question arises as to whether there may be more than one cause of cervical cancer—one must remember that approximately 20% of lung cancers occur in individuals who have never smoked cigarettes (Report of the Advisory Committee to the Surgeon General of the Public Health Service, 1964). The differences among pathologists in their histological interpretation of what constitutes cervical dysplasia or *in situ* cancer increase the problems involved in assessing specificity. A similar problem exists with regard to attempts to demonstrate the specificity of the association, in connection with the serological procedures used to detect type 2 HSV antibodies. Five such tests are in current use. Of the neutralization tests, the first is the micro-neutralization (MN) test with neutralizing potency analysis, developed by Nahmias *et al.* (1970c) and adapted with slight modification by Rawls *et al.* (1970); the second is the kinetic neutralization test, originally used by the Baylor group (Rawls, Tompkins & Melnick, 1969) and later by Belgian and British investigators (Sprecher-Goldberger *et al.*, 1970; Wildy, 1972 [2]); the third is the multiplicity analysis method of Roizman, adapted by Royston & Aurelian (1970a). A micropassive haemagglutination test, first reported by Fuccillo *et al.* (1970) has also been used, and we have recently developed an inhibition passive-haemagglutination (InPHA)

---

[1] Unpublished data.

[2] See p. 409 of this publication.

test to differentiate HSV antibodies (Schneweis & Nahmias, 1971). Our comparison of the InPHA and microneutralization tests gave 86% agreement (Nahmias *et al.*, 1971 [1]).

The reproducibility of any of the serological tests to ascertain type 2 HSV antibodies when done within the same laboratory is no better than 80–90%. Furthermore, the results obtained with some of these tests have differed by as much as 50% when comparative serum samples were submitted to different laboratories. The need for improved serological tests is therefore obvious, and for this purpose, in collaboration with Dr K. E. Schneweis of Bonn, Federal Republic of Germany we have initiated studies designed to obtain a clearer understanding of the antigens of the two HSV types. Since the cross-reaction between HSV type 1 and type 2 and their homologous sera is more like a one-way cross, the question arose as to whether it would be possible to demonstrate specific antigens for each of the two HSV types. However, by application of the immuno-diffusion technique, we were able to demonstrate distinct precipitin lines for type 1 and type 2 HSV, as well as common lines (Schneweis & Nahmias, 1971). We developed a working formula for the two HSV types: type 1 = AC, type 2 = BC, where A = specific type 1 antigens, B = specific type 2 antigens, and C = common antigens. The formula was tested by two methods. The first involved the use of an indirect passive haemagglutination (InPHA) test; it confirmed the expectation that sera containing anti-1 or anti-2 antibodies would be more markedly inhibited by the homologous virus when tested with tanned sheep red blood cells sensitized with the homologous antigens. The second method, membrane fluorescence (Nahmias *et al.*, 1971b), also gave confirmatory results, in that it was possible to remove the common surface antigens with heterologous viral-infected cells. Since the serum could also be made type-specific for its neutralizing activity, this technique, incidentally, should be a good way of detecting similarities and differences in neutralizing antigens of different animal and human herpesviruses.

Associated causative factors might play a rôle in permitting cervical neoplasia to develop after HSV infection. As mentioned earlier, there appeared to be an increased incidence of cervical neoplasia when HSV was detected during pregnancy or the puerperium. In a small series of women who

developed *in situ* cervical cancer prior to 20 years of age, pregnancy in association with type 2 HSV infection appeared to be correlated with such an early development of cervical neoplasia (Nahmias *et al.*, 1971 [2]). In a larger group of women, the only differences found between those patients without cancer and those with cervical dysplasia or *in situ* cancer were the presence of type 2 HSV antibodies and age of first pregnancy. Since the variables could have been coindependent, the frequency of type 2 HSV antibodies in women who became pregnant at an early age was examined; the results indicated that the two variables were not independent of each other.

TEMPORAL RELATIONSHIP OF THE ASSOCIATION

The median age for genital herpes, detected by cytological or virological means, is 5–30 years earlier than that for cervical dysplasia and for *in situ* and invasive cancer (Naib *et al.*, 1969; Nahmias *et al.*, 1969). Similar conclusions have also been reached by analysis of serological data on the incidence of acquisition of HSV type 2 antibodies by age (Rawls, Gardner & Kaufman, 1970; Nahmias *et al.*, 1971 [2]).

CONSTITUTIONAL HYPOTHESIS

Increased promiscuity of women has been found to correlate with both cervical cancer and genital herpes. The question has therefore been raised as to whether the association between genital herpes and cervical cancer is the consequence of such promiscuity. In a recent study in which several factors, including those related to sexual behaviour, attendance at venereal disease clinics and frequency of syphilis antibodies, were considered, it was found that control groups and patients with cervical neoplasia differed with respect only to two factors, namely frequency of HSV type 2 antibodies and age of first pregnancy (Nahmias *et al.*, 1971 [2]). Royston & Aurelian (1970a) have found a greater frequency of HSV type 2 antibodies in women with cancer than in controls, although two indices of promiscuity—presence of *Trichomonas* and of syphilis antibodies—were similar in both control and cancer groups. Rawls, Tompkins & Melnick (1969), in an earlier study, also reported that the frequency of antibodies in women with invasive cervical cancer was higher

---

[1] Unpublished data.

[2] Unpublished data.

than that found in prostitutes. However, more recently, the same group (Rawls, Adam & Melnick, 1971 [1]) has observed that once the patients with cancer were matched as regards various sexual characteristics, the differences in frequency of HSV type 2 antibodies between cancer and control groups decreased, if not disappeared completely.

DISCUSSION AND CONCLUSIONS

In conclusion, the difficulties that we have encountered in attempting to show a causal relation between genital herpes and cervical cancer are those that have been found, and will continue to be found, whenever an association between a virus and a human cancer, e.g., Epstein-Barr virus and Burkitt's lymphoma and nasopharyngeal carcinoma, is suggested. Although the demonstration of causation with animal herpesviruses is less difficult, it is worth noting that there is still some doubt regarding the causative rôle of the frog herpesviruses and Lucké's renal adenocarcinoma, despite the dozen or more years that that system has been studied.

Our present concept is that the herpesvirus is most likely to be carcinogenic if it is present in the cervix of a woman who is pregnant. Coppleson (1969) has suggested that benign squamous metaplasia of the

---

[1] See p. 424 of this publication.

cervix, which occurs most frequently during the first pregnancy, is a prerequisite for the possible neoplastic transformation of cervical cells under the influence of a venereally-transmitted mutagen, possibly a virus. This theory is particularly attractive, since it might help to explain the low frequency of penile cancer in the face of a relatively high frequency of genital herpes in males. We are therefore working at present on the effect of female sex hormones on the virus-cancer relationship.

Since the most conclusive evidence might be the demonstration that women protected from developing HSV type 2 infection would, over the years, have a lower frequency of cervical neoplasia than an unprotected comparable gioup, we are also actively engaged in the immunological aspects of herpetic infection. The studies described here have opened up many areas of interest—maternal genital herpes as a source of severe infection in the newborn and as a possible cause of abortions; genital herpes as the second most common venereal disease (after gonorrhoea) in women; the concept of a hormonal effect on the potential oncogenicity of herpesviruses. Finally, we believe that, based on the criteria used to establish a causal association between cigarette smoking and lung cancer, the genital herpes-cervical cancer hypothesis stands up quite well to these criteria, enough at least to suggest the need for further studies.

ACKNOWLEDGEMENTS

This work was supported by grants from the American Cancer Society and the National Institutes of Health. A. Nahmias is the recipient of a Research Career Development Award IK-AI-18637 from the National Institutes of Health.

REFERENCES

Catalano, L. W., Jr & Johnson, L. D. (1971) *Herpesvirus hominis* antibody in relation to carcinoma in situ of the uterine cervix. *J. Amer. med. Ass.*, **217**, 447-450

Coppleson, M. (1969) Carcinoma of the cervix—epidemiology and aetiology. *Brit. J. Hosp. Med.*, **2**, 961-980

Couch, E. & Nahmias, A. (1969) Filamentous structures of type 2 *Herpesvirus hominis* infection of the chorioallantoic membrane. *J. Virol.*, **3**, 228-232

Dowdle, W., Nahmias, A. J., Harwell, R. & Pauls, F. (1967) Association of antigenic type of Herpesvirus hominis to site of viral recovery. *J. Immunol.*, **99**, 974-980

Fuccillo, F., Moder, L., Catalano, L., Jr, Vincent, M. & Sever, J. (1970) Herpesvirus hominis types I and II: a specific microindirect hemagglutination test. *Proc. Soc. exp. Biol. (N.Y.)*, **133**, 735-739

Josey, W., Nahmias, A. & Naib, Z. (1968) Genital herpes simplex infection: Present knowledge and possible relationship to cervical cancer. *Amer. J. Obstet. Gynec.*, **101**, 718-729

McKenna, J. M., Sanderson, R. P. & Blakemore, W. S. (1962) Extraction of distinctive antigens from neoplastic tissue. *Science*, **135**, 370-371

Nahmias, A. & Dowdle, W. (1968) Antigenic and biologic differences in *Herpesvirus hominis*. *Progr. med. Virol.*, **10**, 110-159

Nahmias, A., Dowdle, W., Naib, Z., Josey, W., McClone, D. & Domescik, G. (1969) Genital infection with type 2 *Herpesvirus hominis*—a commonly occurring venereal disease. *Brit. J. vener. Dis.*, **45**, 294-298

Nahmias, A. J., Josey, W. E., Naib, Z. M., Luce, C. F. & Guest, B. A. (1970a) Antibodies to *Herpesvirus hominis* types 1 and 2 in humans. II. Women with cervical cancer. *Amer. J. Epidem.*, **91**, 547-552

Nahmias, A. J., Naib, Z. M., Josey, W. E., Murphy, F. A. & Luce, C. F. (1970b) Sarcomas after inoculation of newborn hamsters with *Herpesvirus hominis* type 2 strains. *Proc. Soc. exp. Biol. (N.Y.)*, **134**, 1065-1069

Nahmias, A., Josey, W., Naib, Z., Luce, C. & Duffey, C. (1970c) Antibodies to *Herpesvirus hominis* types 1 and 2 in humans. I. Patients with genital herpetic infection. *Amer. J. Epidem.*, **91**, 539-546

Nahmias, A., Naib, Z. & Josey, W. (1971) *Herpesvirus hominis* type 2 infection—association with cervical cancer and perinatal disease. *Perspect. Virol.*, **7**, 73-89

Nahmias, A., London, W., Catalano, L., Fuccillo, D., Sever, J. & Graham, C. (1971a) Genital *Herpesvirus hominis* type 2 infection—an experimental model in Cebus monkeys. *Science*, **171**, 297-298

Nahmias, A., delBuono, I., Schneweis, K., Gordon, D. & Thies, D. (1971b) Surface antigens of herpes simplex virus type 1 and 2 as detected by membrane fluorescent technics. *Proc. Soc. exp. Biol. (N.Y.)*, **138**, 21-27

Naib, Z. M., Nahmias, A. J. & Josey, W. E. (1966) Cytology and histopathology of cervical herpes simplex. *Cancer*, **19**, 1026-1030

Naib, Z. M., Nahmias, A. J., Josey, W. E. & Kramer, J. H. (1969) Genital herpetic infection—association with cervical dysplasia and cancer. *Cancer*, **23**, 940-945

Rawls, W. E., Tompkins, W. A. F. & Melnick, J. L. (1969) The association of herpesvirus type 2 and carcinoma of the uterine cervix. *Amer. J. Epidem.*, **89**, 547-554

Rawls, W., Iwamoto, K., Adam, E. & Melnick, J. (1970) Measurement of antibodies to herpesvirus types 1 and 2 in human sera. *J. Immunol.*, **104**, 599-606

Rawls, W. E., Gardner, H. L. & Kaufman, R. L. (1970) Antibodies to genital herpesvirus in patients with carcinoma of the cervix. *Amer. J. Obstet. Gynec.*, **7**, 710-716

Report of the Advisory Committee to the Surgeon General of the Public Health Service (1964) *Smoking and cancer*, Washington, US Government Printing Office (*Public Health Service Publication* No. 1103)

Royston, I. & Aurelian, L. (1970a) The association of genital herpesvirus with cervical atypia and carcinoma in situ. *Amer. J. Epidem.*, **91**, 531-538

Royston, I. & Aurelian, L. (1970b) Immunofluorescent detection of herpesvirus antigens in exfoliated cells from human cervical carcinoma. *Proc. nat. Acad. Sci. (Wash.)*, **67**, 204-212

Schneweis, K. E. & Nahmias, A. J. (1971) Antigens of Herpes simplex virus types 1 and 2—immunodiffusion and inhibition passive-hemagglutination studies. *Z. ImmunForsch. exp. Ther.*, **141**, 479-487

Sprecher-Goldberger, S., Thiry, L., Catoor, J. P., Hooghe, R. & Pestian, J. (1970) Herpesvirus type 2 infection and carcinoma of the cervix. *Lancet*, **ii**, 266

Watkins, J. F. (1964) Adsorption of sensitized sheep erythrocytes to HeLa cells infected with herpes simplex virus. *Nature (Lond.)*, **202**, 1364-1365

Herpesvirus and Antigens

P. WILDY [1]

THE HERPESVIRUS GROUP

In 1953 the herpes group consisted of four viruses: herpes simplex, B virus, pseudorabies and virus III. The introduction of the negative stain technique added a new dimension to virology and has resulted in the growth of the group so that it now includes some fifty members, all of which have a similar morphological pattern. It so happens that these viruses are also characterized by other common features that have been used by the International Committee on Nomenclature of Viruses as descriptive of the group.

Some herpesviruses have, in the past, been shown by a variety of tests to possess common antigens. Sometimes these have occurred in surprising combinations (Burrows, 1970; Nahmias & Dowdle, 1968). Interest is increasing in these antigens and more are being reported at this Symposium. This happens at a time when the events occurring in infected cells are being intensively investigated by biochemical methods, and we are naturally interested to know how these virus-specified antigens fit into the picture. I shall deal almost entirely with the antigens of herpes simplex virus and shall draw in fair measure on work now in progress in our own laboratory.

HERPES SIMPLEX VIRUS

Biological differences between virus types 1 and 2

Nahmias & Dowdle (1968) presented a comprehensive review of the properties of herpes simplex viruses. The picture they drew was of two fairly well-defined groups that differed from each other in mode of natural infection, clinical picture, and pathogenicity in animals and eggs. Since that time,

there have been several reports of other differences; for example, type 1 viruses infect chick embryo cells abortively (Figueroa & Rawls, 1969; Lowry, Melnick & Rawls, 1971; Lowry, Bronson & Rawls, 1971). Type 2 virus will not grow above 40°C (Ratcliffe, 1971). The thymidine kinase induced by type 2 virus is heat labile, in contrast to that induced by type 1 (Thouless & Skinner, 1971). The last is of interest in view of a probable serological difference between these enzymes (see later). These studies involved a number of strains and lead to the conclusion that there may be two distinct groups of viruses. However, there are also some indications that this may not be so (Terni & Roizman, 1970; Ejercito, Kieff & Roizman, 1968). To decide the matter, a full comparative study of a large number of strains would have to be carried out by a single research organization.

Serological discrimination between the herpes simplex viruses using neutralization tests

The serological discrimination between herpesviruses of oral and genital origin was put on a firm footing by Schneweis (1962), Plummer (1964), Dowdle et al. (1967) and Pauls & Dowdle (1967), all of whom used neutralization tests for this purpose. Altogether, a considerable number of strains from various sources have been examined and the work has led to the conclusion that there are both common antigenic determinants available to participate in the neutralization reaction and some type-specific determinants that can be used broadly to divide strains into two major types according to their source. However, Ejercito et al. (1968) found differences in the serological behaviour of four strains, suggesting that the pattern may not be simple or clear-cut. It is important, therefore, not only to ascertain whether two main groupings exist or not, but also to determine the degree of relatedness between strains and between the main groups. This

---

[1] Department of Virology, The Medical School, University of Birmingham, UK.

is also important from a practical point of view since reciprocal neutralization tests form the basis of sero-epidemiological studies.

We have recently carried out a comparison of 44 randomly-collected strains of herpes simplex virus using reciprocal kinetic neutralization tests (Skinner, Thouless & Gibbs[1]). Antisera were prepared, one for each strain, by injecting groups of 10 mice intra-peritoneally with each virus strain. Mice each received 10^8 PFU of virus (formalin inactivated) intraperitoneally, followed by five intraperitoneal injections of 10^6 PFU at monthly intervals. The mice were bled out 14 days after the last injection and the sera of each group pooled. Inactivated sera were tested against each strain by a kinetic neutralization test (Skinner, Thouless & Jordan, 1971). To date, we have tested only 18 of these sera but the data are sufficient to justify a preliminary report. The results were analyzed by the principal coordinates method (Gower, 1966). Fig. 1 illustrates the distribution of strains on the first and third vectors. If

the second vector is taken into account, the pattern is but little altered. Clearly, the oral isolates and genital isolates cluster predominantly at opposite ends of the diagram. However, there is some over-lapping of the clusters; the nearest genital and oral isolates are closer than the most distant isolates within each group.

We can conclude, on the basis of a single character, that the strains fall into two contiguous overlapping groups and that some oral and genital strains have greater affinity than do some of the strains within each group.

The following further conclusions were also drawn from the experiment: (1) For any one serum where the characteristics are fully known, strains of oral and genital origin can be distinguished with a probability of about 80%. (2) Oral strains are, in general, more neutralizable than genital strains, whatever immunizing virus was used to raise the antiserum. This is analogous to the results of Watson *et al.* (1967), who found that B virus anti-serum would neutralize herpes simplex virus type 1 more readily than homologous virus. (3) The sera

---

[1] Unpublished data.

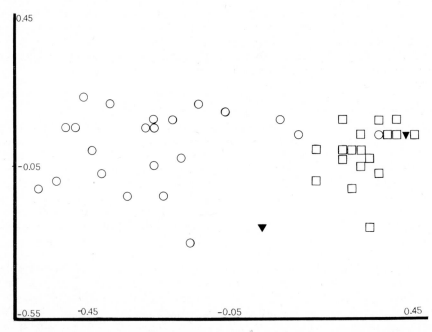

Fig. 1.   Distribution of herpes simplex virus isolates determined by the principal co-ordinates method of Gower (1966), using data from reciprocal neutralization tests.

    □   genital strains
    ○   oral strains (including strains isolated from pharynx, brain and whitlows)
    ▼   strains isolated from neonatal skin

that discriminate best between types 1 and 2 virus, in general, have low k-values against both viruses. (4) Seventeen out of 18 antisera were useful in discriminating between 38/41 virus strains correctly. (5) On the basis of the crude data, the prospect of predicting the immunizing virus type for any particular serum was extremely poor, and even worse when two arbitrarily chosen strains were used, as is usually done in sero-epidemiological studies. However, this may be improved if the data are normalized. It might be concluded from these data that all the sero-epidemiological evidence so far published may be suspect. However, we should remember that these antisera were prepared in an artificial manner in an animal that was not the natural host for these viruses.

Antigens of type 1 and 2 herpes simplex virus

In view of the size of the herpesvirus genome, we should expect it to be capable of specifying a large number of polypeptides. The protein molecules containing these polypeptides certainly fulfil two functions (acting both as structural elements of the virion and as enzymes concerned with DNA synthesis) and many fulfil others. We can recognize some of these proteins by conventional serological tests, such as the precipitin or complement-fixation tests, or by the specific neutralization of functions such as infectivity or enzyme activity. It would be logical to consider these proteins in terms of the rôles they play in the life history of the virus, but only in a few cases can this be done. I shall therefore first deal with them in descriptive terms as diffusible antigens, surface antigens, etc.

Diffusible antigens. It has been known for some time that herpes simplex virus specifies complement-fixing antigens of much smaller size than the virion (Hayward, 1949; Wildy & Holden, 1954). This has been confirmed, using gel-diffusion tests, and it is clear that there are many species of diffusible antigens (Tokumaru, 1965a, b; Watson et al., 1966). We have counted up to 12 separate lines in gel-diffusion tests where extracts of herpesvirus-infected cells are tested against "general antiserum". This serum is prepared by a prolonged course of immunization with broken virus-infected RK13 cells (Watson et al., 1966). Thouless has examined the cross-reactions between type 1 and 2 herpes virus in immunodiffusion tests.[1] When extracts of cells infected with type 1 or type 2 virus were tested reciprocally using

general antisera against prototype strains, up to eight precipitin lines were found showing reactions of identity and indicating that types 1 and 2 herpesvirus share at least this number of diffusible antigens. There were also lines that appeared not to be shared. These were revealed more clearly when the antisera were absorbed with excess heterologous infected cell extract. The absorbed sera then reacted specifically to homologous virus extracts, three (or four) lines being visible with type 1 virus and one or two with type 2.

Antigens involved in virus neutralization tests. Herpes simplex and pseudorabies virus particles probably contain at least eight polypeptides in the nucleocapsid and at least four in the envelope (Spear & Roizman, 1968; Spear, Keller & Roizman, 1970; Olshevsky & Becker, 1970; Shimono, Ben-Porat & Kaplan, 1969; Kaplan & Ben-Porat, 1970; Robinson & Watson, 1971). Some of these will be incorporated into the surface of the virion and so will be detectable by neutralization tests. It is important to recognize that in discussing diffusible antigens (see above), we were thinking in terms of antigenic determinants present on single molecules (even if they comprise more than one polypeptide). With neutralization tests, we are dealing with antigenic determinants that may represent the sum of the exposed antigenic determinants recognizable also on some diffusible antigens, or new determinants that arise as a result of the juxtaposition of subunits.

In an earlier section I discussed cross-neutralization reactions between type 1 and 2 herpes simplex viruses. It has been established (Watson & Sim, 1972; Geder & Skinner, 1971) that type 1 and 2 herpes simplex viruses have distinct antigenic determinants, some of which are type-specific and others shared in common. Type-specific neutralization sera for both types 1 and 2 virus were obtained by:

(*a*) Reacting hyperimmune rabbit sera with excess heterologous virus-infected cell extract.

(*b*) Reacting sera with an excess of whole cells infected with heterologous virus.

(*c*) Reacting sera with excess heterologous electrofiltrate (diffusible antigens obtained by preparative polyacrylamide gel electrophoresis) (see Watson, 1969).

Antigens on the cell surface. O'Dea & Dineen (1957), using immunofluorescence, noted that the surface of herpesvirus-infected cells developed altered antigenic specificity. The nature of the alteration and its relevance to the social behaviour of infected

---

[1] Unpublished data.

cells has been extensively studied by Roizman and his co-workers (Roizman, 1962, 1969). The property of fusion with uninfected cells is associated with the appearance of virus-specific antigens detectable by immunofluorescence or by immune adherence (Peterknecht, Bitter-Sauerman & Falke, 1968) and "nonspecific" avidity for sensitized sheep erythrocytes (Watkins, 1964). Geder & Skinner (1971) have recently examined the appearance of virus-specific antigens on the surface of BHK cells infected with herpes simplex virus. Using indirect immunofluorescence, they found that all cells infected at high multiplicity gave strong fluorescence by 5½ hours after infection. By absorbing general antisera to type 1 and 2 virus with whole BHK cells infected with heterologous virus, they obtained type-specific reactions. The technique thus appears to be useful for discriminatory purposes, since it gave unequivocal results, not only with prototype strains, but also with eight recent isolates from various sources. More recently, these workers have extended their observations, making use of surface immunofluorescence and an immune cytolytic test (Klein & Klein, 1964). The latter was developed following the observation (Roizman, 1962) that the surface anti-

gens of herpesvirus-infected cells can be used to prevent infective centre formation in the presence of antibody and complement. Though the surface immunofluorescence technique is the more sensitive of the two there is excellent agreement between the results obtained. Figs. 2 and 3 show the development of various virus antigens on the surface of BHK cells infected with type 1 and type 2 virus.

Antigens and enzyme activities. The activities of two enzymes involved in DNA metabolism (DNA polymerase and thymidine kinase) are vastly enhanced after herpesvirus infection of BHK cells. These activities are inhibited by virus-specific antisera (Keir *et al.*, 1966; Klemperer *et al.*, 1967). The thymidine kinases of herpes simplex type 1 and pseudorabies show antigenic specificity (Buchan & Watson, 1969) in reciprocal enzyme-inhibition tests. Furthermore, type 1 antiserum does not inhibit type 2 thymidine kinase (Thouless & Skinner, 1971).

Summary. The results indicate that there are virus-specific antigenic determinants both common to type 1 and type 2 virus and type-specific. These

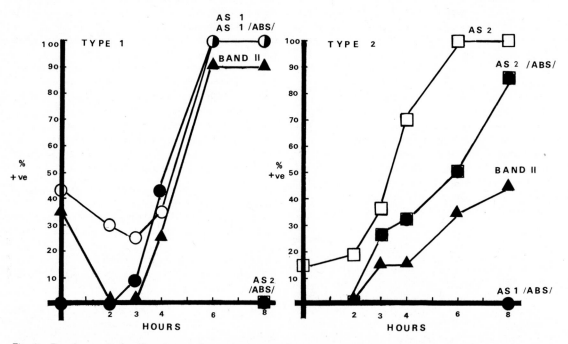

Fig. 2. Development of cell surface antigens by immunofluorescence. AS 1: general antiserum to type 1 herpes simplex virus; AS 2: general antiserum to type 2 virus; Band II: using anti-Band II; ABS: absorbed with cells infected with heterologous type virus.

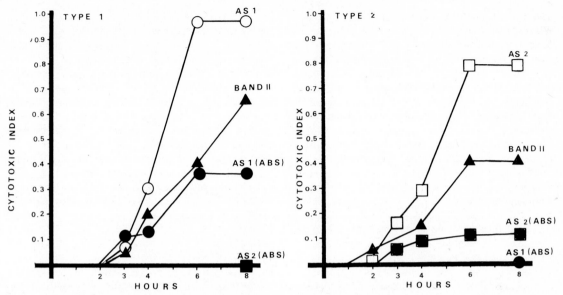

Fig. 3.   Development of cell surface antigens by cytolytic test.   For key, see Fig. 2.

are located on diffusible particles of molecular size, at the surface of the virion and at the cell surface (Table 1). This is in general agreement with the findings of other workers (Roizman, 1969; Nahmias *et al.*, 1971).

Properties of individual antigens

We, in Birmingham, have been interested in separating herpes antigens and in defining their rôle in the growth processes of the virus. The approach has depended upon the production of antisera that react uniquely with the antigen under examination. These antisera can then be used to detect and identify the antigen in question by a variety of tests. At present we have information on four antigens.

Band II antigen. The proteins in infected cell extracts were separated by polyacrylamide gel electrophoresis (Watson, 1969) and fractions obtained that then gave single precipitin lines when reacted with general antisera in immunodiffusion tests. One of these lines was located in a particular region of the electropherogram, which we call Band II. The agar containing these single lines in Band II was removed, washed and used to immunize rabbits by lymph-node injection (Watson & Wildy, 1969). The sera of these rabbits react with extracts of infected cells, giving only one precipitin line, and we refer to these "monoprecipitin" sera as anti-Band II.

When infected cells are extracted and fractionated in the usual manner (Watson, 1969), Band II appears reproducibly as diffusible antigen. Because anti-Band II sera prepared against type 1 antigen neu-

Table 1.   Type 1 and type 2 herpes simplex virus antigens

Type of antigen	Diffusible	Involved in virus neu-tralization	Detected by immuno-fluorescence (IF) tests	Detected at cell surface	
				By IF	By cytolysis
Common antigens	8–10	+ (Band II)	+	+	+
Type-specific antigens:					
Type 1	3–4	+	+	+	+
Type 2	1–2	+	+	+	+

tralize both types of herpes simplex (Watson & Wildy, 1969; Skinner [1]) we can infer: (a) that Band II is a structural antigen; (b) that it is represented at the surface of the virion; and (c) that it is at least one of the common antigens participating in the neutralization test. We can now go farther. Watson & Sim (1972) have blocked the neutralizing activity of general antisera against type 1 and type 2 virus with excess of type 1 virus Band II antigen. The resulting sera react specifically in neutralization tests with homologous virus. This indicates that Band II determinants at the surface of the virion must account for virtually all the common antigen component. The procedure is one that might well be useful in diagnostic serological tests. It might be supposed that the Band II determinants are on the surface of the envelope since there is a good deal of evidence that this is a necessary constituent of the virion. However, it is clear that anti-Band II agglutinates naked particles detected by electron microscopy. The kinetics of the formation of Band II have been followed by Barton et al. (1971),[2] using anti-Band II serum. Early (4–6 hours) in infection there was massive accumulation of Band II reacting material in the cytoplasm of the cell. Later (7 hours onwards) faint fluorescence was detected in the nucleus. Immunodiffusion studies confirmed that antigen accumulated in the cytoplasm but failed to detect its presence in the nucleus. Geder [2] has demonstrated the antigen on the surface of cells infected with both types of herpes simplex virus, both by immunofluorescence and cytolytic tests, from the third hour after infection (Figs. 2 & 3).

All these findings agree, in general, with the notion that viral structural proteins are synthesized in the cytoplasm, pass into the nucleus and become incorporated into membranes at the surface of the cell (cf. Ross, Watson & Wildy, 1968; Roizman, Spring & Roane, 1967; Spear & Roizman, 1968; Fujiwara & Kaplan, 1967; Ben-Porat, Shimono & Kaplan, 1969; Roizman & Spear, 1971).

To summarize, Band II antigen can be extracted from virus-infected cells as a diffusible antigen. It is a structural antigen, whose determinants are represented on the surface of both type 1 and type 2 viruses. It appears to represent the active common element participating in virus neutralization. It accumulates in the cytoplasm early in infection and has been detected later in the nucleus. It appears early on the surface of cells infected with both types of virus. It will agglutinate naked virus particles (Table 2).

PRC antigen. Watson et al. (1967) showed that extracts of cells infected with pseudorabies virus gave a single precipitin band with herpes simplex antiserum. Such a precipitate was used to immunize a rabbit, giving an antiserum that gave a single line with herpesvirus but several lines with pseudorabies virus infected cell extracts (Watson & Wildy, 1969). The single antigen in herpesvirus is called PRC. We know that this antigen exists as a diffusible entity, but it is not involved in virus neutralization and does not agglutinate naked virus particles (Watson & Wildy, 1969). It would seem that this antigen also appears in type 2 virus infected cell extracts.

Capsid antigen. Robinson & Watson (1971) extensively purified herpes simplex virus. The final product was inactivated with formalin and used to immunize a rabbit by the lymph-node technique. The resulting antiserum fortuitously reacts to give a single precipitin line when tested by gel diffusion against extracts of herpes infected cells (Watson [3]).

---

[1] Personal communication.
[2] Unpublished data.

[3] Personal communication.

Table 2. Main properties of some herpes simplex virus antigens

Antigen	Diffusible in cell extracts	Structural antigen	Involved in virus neutralization	Associated with thymidine kinase	Present at cell surface	Type
Band II	+	+	+	−	+	Common
PRC	+	NK[1]	−	−	NK[1]	Common
Capsid	+	(+)	−	−	NK[1]	Common
TdKA[2]	+	NK[1]	−	+	NK[1]	Specific

[1] Not known.
[2] Thymidine kinase antigen.

We call this antigen capsid antigen. It can be inferred that this is a structural antigen, since the anti-capsid antigen level was raised by using highly purified virus. Anti-capsid antigen does not neutralize type 1 virus. It reacts in immunodiffusion tests against type 2 virus infected extracts. We therefore presume that it is a common antigen.

Thymidine kinase antigen (TdKA). Dubbs & Kit (1964) isolated a mutant of herpesvirus deficient in thymidine kinase. Extracts of cells infected with the mutant have been used to absorb herpes-reacting antibodies. The absorbed sera react in gel diffusion to give only one line of precipitate with extracts of cells infected with the wild type of virus; they no longer neutralize infective virus but retain their ability to neutralize herpesvirus-induced thymidine kinase activity (Buchan, Luff & Wallis, 1970). We call the antigen TdKA. We can infer that this antigen is associated with thymidine kinase and is probably the enzyme itself. Interestingly, this antigen is type-specific (Thouless[1]). Type 1 "general" antisera, after absorption with excess type 2 virus infected extract, react specifically with type 1 virus infected cell extracts in immunodiffusion tests. When the same antiserum is absorbed with excess extract of type 1 TdKA mutant virus-infected cells, the result is a reaction of identity with one of the type 1 specific lines (Fig. 4).

Table 2 summarizes the main properties of these four antigens.

SERO-EPIDEMIOLOGICAL EVIDENCE FOR THE ASSOCIATION BETWEEN TYPE 2 HERPES SIMPLEX AND CARCINOMA OF THE CERVIX UTERI

During the past four years, several sero-epidemiological surveys have been carried out by four laboratories, and the results obtained have indicated an association between herpes simplex type 2 infection and carcinoma of the cervix. More recently, there have been disquieting reports that the serological findings seem to vary for different communities and classes (see, for example, Rawls, Adam & Melnick, 1971 [2]). The first major difficulty in assessing the probability of infection by type 1 or type 2 virus is that each has both common and specific antigenic determinants so that it is impossible to tell whether a particular antibody level actually represents infection with one type of virus or whether there has been a double infection. In previous studies, the method used has been to calculate the ratio of type 2 to type 1 serological values, and this has been justified by various means by different authors. However, the use of this method is suspect. Firstly, the ratio will vary in the same direction if the anti-type 2 values are high or the anti-type 1 values are low. Secondly, the method

[1] Personal communication.

[2] See p. 424 of this publication.

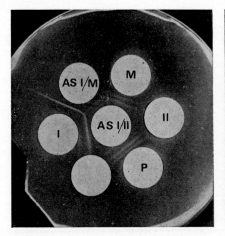 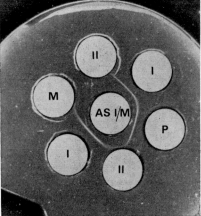

Fig. 4. The type-specific TdK antigen of herpes simplex virus. Immunodiffusion tests with absorbed antisera and crude virus-infected cell extracts. AS 1/M: type 1 general antiserum absorbed with thymidine-kinase-deficient mutant; AS 1/II: same antiserum absorbed with type 2 virus-infected cell extract; I: type 1 extract; II: type 2 extract; M: type 1 TdK⁻ mutant extract; P: type 1 TdK⁺ parent extract.

of expressing the properties of groups of subjects with type 2 antibodies is clearly dependent on the sensitivity of the serological tests used. Thirdly, the ratio may sometimes reflect a dual infection, which can certainly occur (Terni & Roizman, 1970). Finally, the ratios may be misleading, as predicted from the results given earlier.

Yet another survey has just been completed in Birmingham (Skinner, Thouless & Jordan, 1971). In this survey, antisera were tested against a prototype virus of each type of virus (HFEM and LOVE-LACE) and the neutralization constant k measured. The mean k-value for each virus type was calculated for each clinical group and the results assessed independently. In this way it was hoped to minimize the effects of the first three difficulties. The clinical material was derived from 276 patients attending the Birmingham & Midlands Hospital for Women and the Venereal Disease Clinic. Patients were grouped according to age and cytological and/or histological condition of the cervix. Of the patients, 74% belonged to social class 3 and most were white. Half the patients showed no abnormality on cytological examination while the remainder had abnor-

mal cytology and/or histology, ranging from "suspicious" smear to invasive carcinoma. With all groups of patients (Fig. 5), the mean anti-type 1 k-values increased with age. Analysis of variance showed no significant difference between clinical groups at any age. On the other hand, the mean anti-type 2 k-values remained more or less constant in the control groups and increased with age only in the group with abnormal cytology. Analysis of variance showed that the mean anti-type 2 k-values of cytologically and histologically abnormal patients differed significantly from those for controls at all ages ($P < 0.05$ or < 0.01). The results, indeed, demonstrate an association between anti-type 2 k-values and cervical abnormality. Since we can be sure that this is independent of the behaviour of type 1 antibody, it is interesting to look at the ratio of the mean k-values (type 1/type 2; see Fig. 6). It is quite clear that all abnormal groups have a very different ratio of means at all ages and at all stages of disease. It is impossible to tell from any of the serological studies, however, what the association means. If, indeed, type 2 herpesvirus can cause carcinoma, we urgently require critical evidence of a quite different kind.

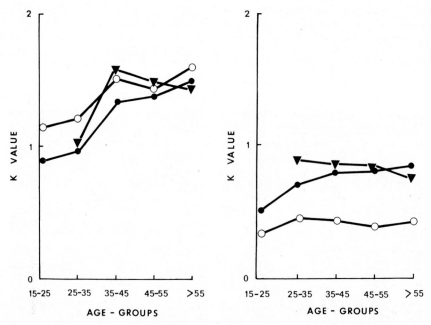

Fig. 5. Mean anti-type 1 k-values (left) and anti-type 2 k-values (right) for herpes simplex virus, for women patients by age-group.
   ○ negative cytology
   ● abnormal cytology
   ▼ histological evidence of cervical disease

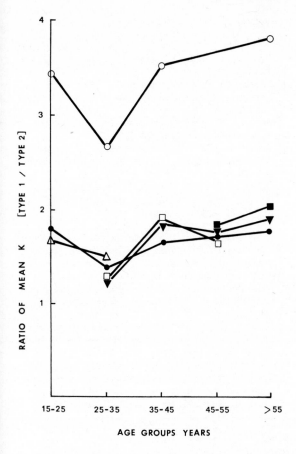

Fig. 6. Ratio of means of k-values for women patients by age-group (type 1/type 2).

○ negative cytology
● abnormal cytology
▼ histological evidence of cervical disease
□ carcinoma-*in-situ*
△ suspicious cytology
■ invasive carcinoma

ACKNOWLEDGEMENTS

I am grateful to Dr D. H. Watson, Dr A. Buchan, Dr L. Geder, Dr G. R. B. Skinner, Dr K. Cameron, Mrs M. E. Thouless and Mrs C. Sim for helpful discussions and for allowing me to present their unpublished results.

REFERENCES

Ben-Porat, T., Shimono, H. & Kaplan, A. S. (1969) Synthesis of proteins in cells infected with herpesvirus. II. Flow of structural viral proteins from cytoplasm to nucleus. *Virology*, **37**, 56-61

Buchan, A., Luff, S. & Wallis, C. (1970) Failure to demonstrate interaction of subunits of thymidine kinase in cells simultaneously infected with herpes virus and a kinaseless mutant. *J. gen. Virol.*, **9**, 239-242

Buchan, A. & Watson, D. H. (1969) The immunological specificity of thymidine kinases in cells infected by viruses of the herpes group. *J. gen. Virol.*, **4**, 461-463

Burrows, R. (1970) The general virology of the herpesvirus group. In: *Proc. 2nd Int. Conf. Equine Infect. Dis., Paris 1969*, Basel, Karger, pp. 1-12

Dowdle, W. R., Nahmias, A. J., Harwell, R. W. & Pauls, F. P. (1967) Association of antigenic type of herpesvirus hominis with site of viral recovery. *J. Immunol.*, **99**, 974-980

Dubbs, D. & Kit, S. (1964) Mutant strains of herpes simplex deficient in thymidine kinase-inducing activity. *Virology*, **22**, 493-502

Ejercito, P. M., Kieff, E. D. & Roizman, B. (1968) Characterization of herpes simplex virus strains differing in their effects on social behaviour of infected cells. *J. gen. Virol.*, **2**, 357-364

Figueroa, M. E. & Rawls, W. E. (1969) Biological markers for differentiation of herpesvirus strains of oral and genital origin. *J. gen. Virol.*, **4**, 259-267

Fujiwara, S. & Kaplan, A. S. (1967) Site of protein synthesis in cells infected with pseudorabies virus. *Virology*, **32**, 60-68

Geder, L. & Skinner, G. R. B. (1971) Differentiation between type 1 and type 2 strains of herpes simplex virus by an indirect immunofluorescent technique. *J. gen. Virol.*, **12**, 179-182

Gower, J. C. (1966) Some distance properties of latent root and vector methods used in multivariate analysis. *Biometrika*, **53**, 325-338

Hayward, M. E. (1949) Serological studies with herpes simplex virus. *Brit. J. exp. Path.*, **30**, 250-254

Kaplan, A. S. & Ben-Porat, T. (1970) Synthesis of proteins in cells infected with herpesvirus. VI. Characterization of the proteins of the viral membrane. *Proc. nat. Acad. Sci. (Wash.)*, **66**, 799-806

Keir, H. M., Subak-Sharpe, J. H., Shedden, W. I. H., Watson, D. H. & Wildy, P. (1966) Immunological evidence for a specific DNA polymerase produced after infection by herpes virus. *Virology*, **30**, 154-157

Klein, E. & Klein, G. (1964) Antigenic properties of lymphomas induced by the Moloney agent. *J. nat. Cancer Inst.*, **32**, 547-568

Klemperer, H. G., Haynes, G. R., Shedden, W. I. H. & Watson, D. H. (1967) A virus-specific thymidine kinase in BHK 21 cells infected with herpes simplex virus. *Virology*, **31**, 120-128

Lowry, S. P., Bronson, D. L. & Rawls, W. E. (1971) Characterization of the abortive infection of chick embryo cells by herpesvirus type 1. *J. gen. Virol.*, **11**, 47-51

Lowry, S. P., Melnick, J. L. & Rawls, W. E. (1971) Investigation of plaque formation in chick embryo cells as a biological marker for distinguishing herpesvirus type 2 from type 1. *J. gen. Virol.*, **10**, 1-10

Nahmias, A. J., DelBuono, B. S., Schneweis, K. E., Gordon, D. S. & Thies, D. (1971) Type-specific surface antigens of cells infected with herpes simplex virus (1 and 2). *Proc. Soc. exp. Biol. (N.Y.)* **138**, 21-27

Nahmias, A. J. & Dowdle, W. R. (1968) Antigenic and biologic differences in herpesvirus hominis. *Progr. med. Virol.*, **10**, 110-159

O'Dea, J. F. & Dineen, J. K. (1957) Fluorescent antibody studies with herpes simplex virus in unfixed preparations of trypsinized tissue cultures. *J. gen. Microbiol.*, **17**, 19-24

Olshevsky, U. & Becker, Y. (1970) Synthesis of herpes simplex virus structural proteins in arginine depressed cells. *Nature (Lond.)*, **226**, 851-853

Pauls, F. P. & Dowdle, W. R. (1967) A serologic study of herpesvirus hominis strains by microneutralization tests. *J. Immunol.*, **98**, 941-947

Peterknecht, W., Bitter-Sauermann, D. & Falke, D. (1968) Immunadhärenz zum Nachweis virusspezifischer Antikörper und Antigene. II. Immunadhärenz-Hämadsorption zum Nachweis zell- und virusspezifischer Antigene auf Kulturzellen. *Z. med. Mikrobiol. Immunol.*, **154**, 234-244

Plummer, G. (1964) Serological comparisons of the herpes virus. *Brit. J. exp. Path.*, **45**, 135-141

Ratcliffe, H. (1971) The differentiation of herpes simplex virus types 1 and 2 by temperature markers. *J. gen. Virol.*, **13**, 181-183

Robinson, D. J. & Watson, D. H. (1971) Structural proteins of herpes simplex virus. *J. gen. Virol.*, **10**, 163-171

Roizman, B. (1962) Polykaryocytosis. *Cold Spring Harb. Symp. quant. Biol.*, **27**, 327-342

Roizman, B. (1969) The herpesviruses. A biochemical definition of the group. *Curr. Top. Microbiol. Immunol.*, **49**, 1-64

Roizman, B. & Spear, P. G. (1971) Herpesvirus antigens on cell membranes detected by centrifugation of membrane antibody complexes. *Science*, **171**, 298-300

Roizman, B., Spring, S. B. & Roane, P. R. (1967) Cellular compartmentalization of herpesvirus antigens during viral replication. *J. Virol.*, **1**, 181-192

Ross, L. J. N., Watson, D. H. & Wildy, P. (1968) Development and localization of virus-specific antigens during the multiplication of herpes simplex virus in BHK 21 cells. *J. gen. Virol.*, **2**, 115-122

Schneweis, K. E. (1962) Serologische Untersuchungen zur Typendifferenzierung des Herpesvirus hominis. *Z. ImmunForsch. exp. Ther.*, **124**, 24-48

Shimono, H., Ben-Porat, T. & Kaplan, A. S. (1969) Synthesis of proteins in cells infected with herpesvirus. I. Structural viral proteins. *Virology*, **37**, 49-55

Skinner, G. R. B., Thouless, M. E. & Jordan, J. (1971) Antibodies to type 1 and type 2 herpes virus in women with abnormal cervical cytology. *J. Obstet. Gynaec. Brit. Cwlth*, **78**, 1031-1038

Spear, P. G., Keller, J. M. & Roizman, B. (1970) Proteins specified by herpes virus. II. Viral glycoproteins associated with cellular membranes. *J. Virol.*, **5**, 123-131

Spear, P. G. & Roizman, B. (1968) The proteins specified by herpes simplex virus. I. Time of synthesis, transfer into nuclei and properties of proteins made in productively infected cells. *Virology*, **36**, 545-555

Terni, M. & Roizman, B. (1970) Variability of herpes simplex virus: isolation of two variants from simultaneous eruptions at different sites. *J. infect. Dis.*, **121**, 212-216

Thouless, M. E. & Skinner, G. R. B. (1971) Differences in the properties of thymidine kinase produced in cells infected with type 1 and type 2 herpes virus. *J. gen. Virol.*, **12**, 195-197

Tokumaru, T. (1965a) Studies of herpes simplex virus by the gel diffusion technique. I. Distribution of precipitating antibodies among human sera. *J. Immunol.*, **95**, 181-188

Tokumaru, T. (1965b) Studies of herpes simplex virus by the gel diffusion technique. II. The characterization of viral and soluble precipitating antigens. *J. Immunol.*, **95**, 189-195

Watkins, J. F. (1964) Adsorption of sensitized sheep erythrocytes to HeLa cells infected with herpes simplex virus. *Nature (Lond.)*, **202**, 1364-1365

Watson, D. H. (1969) The separation of herpes virus-specific antigens by polyacrylamide gel electrophoresis. *J. gen. Virol.*, **4**, 151-161

Watson, D. H., Shedden, W. I. H., Elliot, A., Tetsuka, T., Wildy, P., Bourgaux-Ramoisy, D. & Gold, E. (1966) Virus-specific antigens in mammalian cells infected with herpes simplex virus. *Immunology*, **11**, 399-408

Watson, D. H. & Wildy, P. (1969) The preparation of "monoprecipitin" antisera to herpes virus-specific antigens. *J. gen. Virol.*, **4**, 163-168

Watson, D. H., Wildy, P., Harvey, B. A. M. & Shedden, W. I. H. (1967) Serological relationships among viruses of the herpes group. *J. gen. Virol.*, **1**, 139-141

Watson, D. H. & Sim, C. (1972) In: Herpes virus workshop, International Virology II (in press)

Wildy, P. & Holden, H. F. (1954) The complement-fixing antigen of herpes simplex virus. *Aust. J. exp. Biol. med. Sci.*, **32**, 621-632.

Localization of Structural Viral Peptides in the Herpes Simplex Virion

Y. BECKER [1] & U. OLSHEVSKY [1]

The herpes simplex virus replicates in the nuclei of infected cells in which empty capsids, nucleocapsids and enveloped virions are formed. The analysis of the mature virions demonstrated the presence in each virion of: (a) a DNA genome of about 100×10^6 Daltons in molecular weight (Becker, Dym & Sarov, 1968); (b) a capsid made up of 162 hollow cylindrical capsomeres (Wildy, Russell & Horne, 1960); and (c) a lipid-containing envelope (Asher, Heller & Becker, 1968). The isolation of viral capsids and virions, labelled with radioactive amino acids, by zone centrifugation in sucrose gradients (Levitt & Becker, 1967) made possible the study of the viral peptides by electrophoresis in acrylamide gels after detergent treatment (Shapiro, Viñuela & Maizel, 1967). It was demonstrated that 9 peptides are present in the virion of herpes simplex virus (Olshevsky & Becker, 1970) and Epstein-Barr virus (Weinberg & Becker, 1969). In the present communication we describe: (a) the peptide composition of the virion as determined by staining of the viral proteins; and (b) the localization of the major viral peptides within the herpes virion (Becker & Olshevsky, 1971).

Disruption of infected nuclei, labelled with ^{3}H-leucine, and centrifugation of the homogenates in sucrose gradients (12–52% w/w) led to the isolation of empty capsids and nucleocapsids as well as enveloped virions. The acrylamide gel electrophoretic analysis of the empty capsids, after dissolution in sodium dodecyl sulphate (SDS), urea and 2-mercaptoethanol, disclosed that the viral capsid is composed mainly of one peptide (designated as II) (Olshevsky & Becker, 1970). It was therefore concluded that each capsomere is made up of several

(about 9) molecules of peptide II. As the molecular weight of peptide II is 110 000 Daltons (see Table), it is assumed that the molecular weight of one capsomere is about 1×10^6 Daltons, and that the molecular weight of the capsid is 162×10^6 Daltons.

Acrylamide gel electropherograms of labelled nucleocapsids (Fig. 1) demonstrate that, in addition to peptide II, the major component of the viral particle, additional peptides are present. These peptides are designated as VIII (32 000 Daltons molecular weight), VII (40 000 Daltons), VI (60 000 Daltons) and V (69 000 Daltons). Peptide VII was found to be rich in arginine (Olshevsky & Becker, 1970) and it is therefore assumed that it serves as an internal protein within the nucleocapsid. The rôle of peptides VIII and VI is not yet known. Peptide V is a virus-specific glycoprotein, and is probably the first to attach to the nucleocapsid during the process of virion envelopment.

The electrophoretic analysis of the peptides present in enveloped virions (Fig. 1) demonstrated that three additional peptides (designated as III, IV and V) are added to the viral nucleocapsid during its envelopment. Labelling of the infected cells with glucosamine demonstrated that these three peptides were glycosylated and are the viral glycopeptides. From the study of the effect of the nonionogenic detergent, Nonidet P-40, on enveloped virions it was concluded that peptide III is situated in the outer layer of the viral envelope, while peptides IV and V are closer to the viral nucleocapsid. Peptide V, which was also found in the nucleocapsids (Fig. 1), is in close proximity to the viral capsomeres.

A similar composition for the peptides in herpes virions is also suggested by the quantitative analysis of the Coomassie blue stained peptides (determinations of the optical density of the stained gels) (Fig. 2). The molecular weight estimations for the

[1] Department of Virology, Hebrew University/Hadassah Medical School, Jerusalem, Israel.

— 420 —

Content and molecular weight estimation of the structural peptides present
in herpes simplex enveloped virions

No.	Designation of peptide	Molecular weight ($\times 10^{-3}$ Daltons)	Relative protein content (%) [1]		
			Virions from BSC₁ cells		Virions from HeLa cells
			Nuclear	Cytoplasmic	
—	top	—	12.4	8.7	8.4
1	I	125	0.6	1.7	8.5
2	Ia	123	4.3	5.4	2.9
3	II	110	24.8	25.7	18.9
4	IIIa	108	2.0	2.7	1.46
5	III	106	9.5	12.2	13.4
6	IVa	99	4.9	5.4	4.75
7	IV	83	13.6	11.3	10.7
8	V	69	11.2	11.0	8.2
9	VI	60	9.0	4.9	12.8
10	VIIa	47	N.D.	2.3	N.D.
11	VII	40	4.4	5.6	6.75
12	VIII	32	1.9	2.2	1.3
13	IX	28	1.6	1.7	2.0

[1] The area underneath the various bands (Fig. 2) in the optical density tracings of Coomassie blue stained gels was determined and calculated as percentage of the total area under all the viral protein bands. N.D. = not possible to determine.

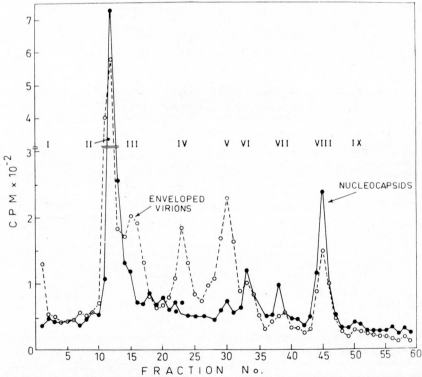

Fig. 1. Acrylamide gel electropherograms of purified herpes simplex virus (HSV) nucleocapsids (●———●) and enveloped virions (O— – – –O) labelled with ³H-leucine.   BSC₁ cells were infected with the HF strain of HSV and labelled with 1 μc/ml of ³H-L-leucine.   The cultures were incubated at 37°C for 18 hours and then harvested and sonicated. The cell sonicates were layered on 12–52% (w/w) sucrose gradients and the various virus structures were isolated and removed with syringes.   The nucleocapsids were obtained, treated with a solution of sodium deoxycholate (1% w/v) and rebanded in a similar sucrose gradient.   The treated nucleocapsids and the enveloped virions were sedimented, dissolved in a solution of sodium dodecyl sulphate (SDS), urea and mercaptoethanol and analyzed by electrophoresis on two separate polyacrylamide gels.

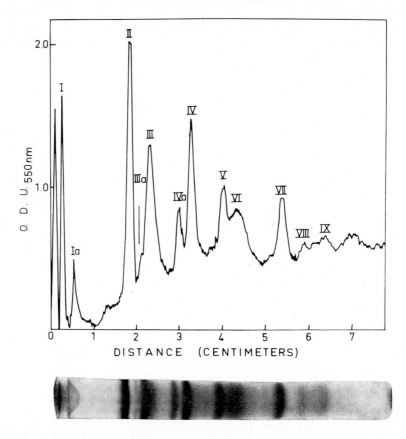

Fig. 2. Structural peptides of enveloped herpes simplex virions. The enveloped herpes virions were obtained from the sucrose gradient, diluted in buffer and centrifuged for 60 min at 100 000 g in the Beckman L-2 preparative ultracentrifuge. The pellets were dissolved in a solution containing sodium dodecyl sulphate (SDS, 0.5% w/v), urea (0.5 M) and 2-mercaptoethanol (0.1% v/v) made in 0.1 M phosphate buffer, pH 7.2. The preparations were left at room temperature for 10 hours and 100 λ samples, containing sucrose (10% w/v) and bromphenol blue (0.01% w/v), were layered on top of the acrylamide gels. The latter consisted of recrystallized acrylamide (7.5% w/v; Eastman Organic Chemical, New York), ethylene diacrylate (0.192% v/v), N, N, N', N'-tetramethylethylenediamine (TEMED) (0.05% v/v), urea (0.5 M) and SDS (0.1% w/v) prepared in 0.1 M sodium phosphate buffer, pH 7.2. The gels were placed between two chambers filled with 0.1 M sodium phosphate buffer, pH 7.2, containing 0.1% (w/v) SDS and electrophoresed for 6 hours at 6 mA per gel. The gels were then removed from the tubes and stained for 2 hours in a solution of 0.25% Coomassie blue (Edward Gurr Ltd., London, England), which was dissolved in a solution of methanol : acetic acid : water at a ratio of 5 : 1 : 5 (Burgess et al., 1969). The gels were destained in a solution containing 7.5% acetic acid and 5% methanol and scanned in a Gilford gel scanner at a wavelength of 550 nm and the optical density recorded. The molecular weight estimation of the viral protein was carried out by the procedure reported by Shapiro, Viñuela & Maizel (1967) using bovine albumin (M.W. 68 000) chemotrypsinogen A (M. W. 25 000) and trypsin (M.W. 23 000) as markers. The various proteins were electrophoresed on separate acrylamide gels, stained and the migration distance was measured.

viral peptides present in herpes simplex enveloped virions, which were propagated in BSC_1 and HeLa cells (Neuman & Becker[1]) (see Table), resemble those obtained previously from the analyses of radioactively labelled peptides (Olshevsky & Becker, 1970). The relative content of each peptide in the virion (grown either in BSC_1 or HeLa cells) was also the same. Nevertheless, several peptides (designated as Ia, IIIa, IVa and VIIa) were detectable in the stained gels but not in the labelled electropherograms. It is possible that: (a) the staining technique is more sensitive than slicing the gels for radioactivity determinations; (b) the amount of the radioactive label in these peptides is very low (possibly due to a low content of leucine); or (c) these peptides are of cellular origin and are present in the viral envelopes

[1] Unpublished data.

as cellular components within the envelopes. Further experiments are needed to determine the nature of these peptides.

From the analyses of the relative content of the various viral peptides in the enveloped herpes simplex virions (see Table) it is concluded that seven peptides (designated as II, III, IV, V, VI, VII and VIII) constitute about 76% of the virion-associated peptides. These peptides are regarded as the major viral peptides, while the rest are the minor constituents. The organization of the major peptides within the herpes virions is shown schematically in Fig. 3. The localization and function of the other viral peptides is under investigation.

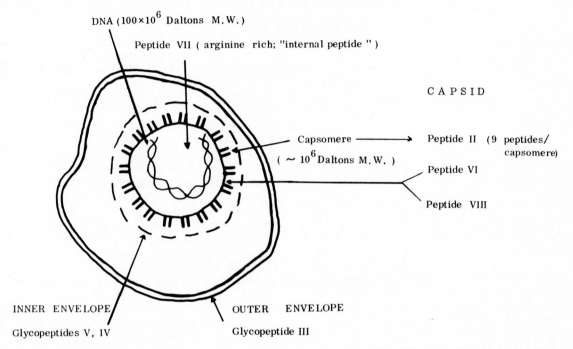

Fig. 3. Schematic diagram suggesting the localization of the seven major viral peptides in the herpes simplex virion.

REFERENCES

Asher, Y., Heller, M. & Becker, Y. (1969) Incorporation of lipids into herpes simplex virus particles. *J. gen. Virol.*, **4**, 65-76

Becker, Y., Dym, H. & Sarov, I. (1968) Herpes simplex virus DNA. *Virology*, **36**, 184-192

Becker, Y. & Olshevsky, U. (1971) The molecular composition of herpes simplex virions. In: Sanders, M., ed., *Viruses affecting man and animals*, St Louis, Warren H. Green, Inc., pp. 73-77

Burgess, R. R., Travers, A. A., Dunn, J. J. & Bautz, E. R. F. (1969) Factor stimulating transcription by RNA polymerase, *Nature (Lond.)*, **221**, 43-46

Levitt, J. & Becker, Y. (1967) The effect of cytosine arabinoside on the replication of herpes simplex virus. *Virology*, **31**, 129-134

Olshevsky, U. & Becker, Y. (1970) Herpes simplex virus structural proteins. *Virology*, **40**, 948-960

Shapiro, A. L., Viñuela, E. & Maizel, J. V. (1967) Molecular weight estimation of polypeptide chains by electrophoresis in SDS-polyacrylamide gels. *Biochem. Biophys. Res. Commun.*, **28**, 815-820

Weinberg, A. & Becker, Y. (1969) Studies on EB virus of Burkitt's lymphoblasts. *Virology*, **39**, 312-321

Wildy, P., Russell, W. C. & Horne, R. W. (1960) The morphology of herpes virus. *Virology*, **12**, 204-222

Geographical Variation in the Association of Antibodies to Herpesvirus Type 2 and Carcinoma of the Cervix

W. E. RAWLS,[1] E. ADAM [1] & J. L. MELNICK [1]

INTRODUCTION

Herpesvirus strains producing genital lesions can be distinguished antigenically and biologically from herpesvirus strains producing lesions at other sites of the body. The virus isolated from genital sites, type 2, is transmitted primarily by venereal means. Previously, it was reported that there was a much greater occurrence of antibodies to herpesvirus type 2 among women with cervical cancer than among control women matched for race, age, and social level (Rawls, Tompkins & Melnick, 1969; Royston & Aurelian, 1970; Nahmias et al., 1970). These studies were carried out primarily among Negro women of lower socio-economic urban populations.

The association between antibodies to the virus and the neoplastic disease could represent an aetiological one or the virus infection and the malignancy could be covariables of sexual activity. If the virus infection was a necessary prerequisite of the malignancy, the association between antibodies to the virus and cervical carcinoma should be found in all population studies. The present paper reports the results of studies of the occurrence of herpesvirus type 2 antibodies among women from different geographical areas.

MATERIALS AND METHODS

Women with histologically proven squamous-cell carcinomas of the uterine cervix were selected from hospital wards, outpatient clinics or the practices of private physicians. Information regarding age, race and socio-economic level were obtained from all patients. In some areas, additional information was obtained that included age at first intercourse, present marital status, age at first marriage, age at first pregnancy, number of live births, number of sex partners in a lifetime and, where applicable, tribal origin. Women to serve as controls were selected from similar sources and the same personal data as those obtained from the women with cervical cancer were also obtained from control women. Blood samples were obtained from the women. The blood was allowed to clot and the serum was removed and stored at —20°C. The presence of antibodies to herpesvirus type 2 was determined by a micro-neutralization test previously described (Rawls et al., 1970).

RESULTS

The occurrence of antibodies to herpesvirus type 2 among females residing in different geographical areas is shown in Table 1. Antibodies to herpesvirus type 2 were not found in the population until the age when heterosexual activity began. For example, in Uganda only 5 of 61 (8%) females aged 14 years or less had antibodies to the virus and 4 of the 5 were aged 12 years or over. The occurrence of antibodies to the virus increased rather rapidly during the late teens and remained relatively constant after 30 years of age. Antibodies were found among 38% of females in Uganda between 15 and 19 years of age; 60% of women of the next decade had antibodies to the virus and 77% of women over age 30 were found to have herpesvirus type 2 antibodies. A similar pattern of the development of antibodies to the virus

[1] Department of Virology and Epidemiology, Baylor College of Medicine, Houston, Texas, USA.

Table 1. Occurrence of antibodies to herpesvirus type 2 among females residing in different geographical areas

Population	Race [1]	Antibodies to herpesvirus type 2 at ages [2]				
		≤14	15–19	20–29	30–39	≥40
Uganda	Negro	5/61 (8)	5/13 (38)	9/15 (60)	27/35 (77)	36/47 (77)
Jamaica	Negro	—	—	6/10 (60)	6/12 (50)	5/5
Texas, U.S.A.						
Rural	Negro	—	—	—	5/6	23/28 (82)
Urban	Negro	—	—	27/89 (30)	41/98 (42)	126/257 (49)
Urban	Caucas. 1	—	—	6/26 (23)	3/19 (16)	19/66 (29)
	Caucas. 2	—	—	0/6	2/16 (13)	2/22 (9)
West Virginia, U.S.A.	Caucas.	—	—	5/18 (28)	3/26 (12)	19/72 (26)
New Zealand	Caucas.	—	—	1/11 (9)	4/14 (29)	8/28 (29)
Taiwan	Oriental	—	—	—	3/7	12/23 (52)
Colombia	Mestizo	2/60 (3)	20/100 (20)	36/100 (36)	26/78 (33)	40/110 (36)

[1] Predominant race in sample; 1 = lower socio-economic class; 2 = upper socio-economic class.
[2] Number with antibody to herpesvirus type 2 over number tested. Number in parentheses represents percentage with antibody to herpesvirus type 2.

was observed among women in Colombia, but only 33–36% of adult women had evidence of a past infection with the virus.

There were marked differences in the occurrence of antibodies to herpesvirus type 2 among the different populations sampled. Considerably greater occurrences of antibodies to the virus were found among Negro populations than among populations composed predominantly of other races. Among Texas Negroes, a higher occurrence of antibodies was found among rural inhabitants than among urban inhabitants. Antibodies were found among 23 of 28 (82%) women over 40 years of age from a small rural community, as compared to 126 of 257 (49%) of women of comparable age residing in metropolitan Houston. In the same age-group of the urban population, only 9% of Caucasian women of the higher socio-economic class had antibodies to the

virus, while 29% of Caucasian women from the lower socio-economic class had such antibodies.

Considerable variation was also found in the occurrence of antibodies to herpesvirus type 2 among women with cervical cancer from different geographical areas (Table 2). Among Negro women from both Uganda and the urban Texas population, the occurrence of antibodies ranged from 68 to 96%. However, among non-Negro populations the occurrence of antibodies was considerably lower. Only about 60% of Caucasian women in Houston belonging to the lower socioeconomic class had antibodies to the virus; about 50% of women from rural West Virginia and only about one-third of women from New Zealand were found to possess such antibodies. Similar low occurrences were observed among women with cervical cancer from Taiwan and Colombia.

Table 2. Occurrence of antibodies to herpesvirus type 2 among women
with cervical cancer

Population	Race [1]	Antibodies to herpesvirus type 2 at ages [2]		
		≤39	40–49	≥50
Uganda	Negro	23/24 (96)	14/20 (70)	16/21 (76)
Texas, U.S.A.	Negro	13/19 (68)	14/17 (82)	20/23 (87)
	Caucas.	1/4	6/10 (60)	7/12 (58)
West Virginia, U.S.A.	Caucas.	5/11 (45)	9/18 (50)	13/23 (57)
New Zealand	Caucas.	3/5	3/11 (27)	12/36 (33)
Taiwan	Oriental	2/4	6/11 (55)	4/10 (40)
Colombia	Mestizo	10/25 (40)	10/31 (32)	10/26 (38)

[1] Predominant race in sample.
[2] Number with antibody to herpesvirus type 2 over number tested. Number in parentheses represents percentage with antibody to herpesvirus type 2.

As reported previously, significant differences were observed in the occurrence of antibodies to herpesvirus type 2 among Negro women with cervical cancer and women matched for age and socio-economic level. For example, in Houston, the differences between the occurrence of antibodies to the virus among the control women (Table 1) and the women with cervical cancer (Table 2) was highly significant ($\chi^2 = 21.7$). However, when the control women were selected according to age at first intercourse, age at first marriage, age at first pregnancy and number of live births, in addition to race, age and socio-economic level, the differences in the occurrence of antibodies to the virus among cases and controls was markedly reduced. This is illustrated in Table 3, where the occurrence of antibodies to

Table 3. Occurrence of antibodies to herpesvirus type 2 among women with cervical cancer and control women matched for multiple attributes

Population	Antibodies to herpesvirus type 2 [1]		χ^2
	Patients with cervical cancer	Matched control women	
Uganda	27/32 (84)	22/32 (69)	1.9
Houston, Texas	48/60 (80)	57/84 (69)	2.05
Colombia	13/36 (36)	12/36 (33)	0.05

[1] Number with antibodies to herpesvirus type 2 over number tested. Number in parentheses means percentage.

herpesvirus type 2 among cases and matched controls is shown. The value of χ^2, as determined by the method of Mantel & Haenszel (1959), is not significant in any of the three areas studied.

DISCUSSION

In three different laboratories, antibodies to herpesvirus type 2 were found in 83, 83, and 100% of Negro women with cervical cancer in Houston, Atlanta and Baltimore, respectively, while the occurrence of antibodies among control women was found to be 22, 35 and 67% for the same three respective locations. Considerable variation in the occurrence of antibodies to the virus was reported among women with cervical dysplasia and carcinoma *in situ* (Rawls & Kaufman, 1970). These differences could represent differences in antibody assay techniques or differences in population sampling.

The results of the present study clearly show that, using the same antibody assay techniques, there are significant differences in the occurrence of antibodies to herpesvirus type 2 in different population segments. The distribution of antibodies is age-dependent. At similar social levels, antibodies to the virus were consistently found in greater proportions among Negro women than among women of other races. A lower occurrence of antibodies to the virus was found among urban Negro women than among a small sample of Negro women obtained from a rural community, and a lower occurrence of antibodies was also found among Caucasian women of the higher socio-economic classes than among women of the lower socio-economic classes.

The occurrence of antibodies to herpesvirus type 2 among women with cervical cancer varied from 27 to 96%. A high occurrence—68 to 96%—was observed among Negro women and a lower occurrence, 27 to 60%, was found among women of other races. A greater occurrence of antiviral antibodies was found among women with cervical cancer than among control women in all populations sampled, where the controls were matched only for race, age, and socio-economic level. If the controls were also matched for age at first intercourse, age at first marriage, age at first pregnancy and number of live births, the differences in the occurrence of antibodies to the virus between women with cervical cancer and control women dropped to insignificant levels. However, those members of the closely matched control population possessing type 2 antibodies might be in their incubation period and their examination a few years hence might show them to belong in the cervical cancer group.

The occurrence in populations of herpesvirus type 2 infections, a venereally transmitted agent, depends upon the sexual mores and practices of the social groups within the population. It is not surprising that variations in the occurrence of antibodies to the virus were found among groups of persons sampled in different areas with diverse racial, cultural and socio-economic backgrounds. Whether the difference in occurrence of antibodies among Negro women and women of other races is due entirely to these factors and not related to a biological difference in the antibody response to herpesvirus types 1 and 2 remains to be determined. Assuming that a biological difference in the antibody response to the herpesvirus related to race does not exist, herpesvirus type 2 would have to play an aetiological rôle of different proportions in cervical cancer cases in different populations. A more likely explanation of the observed association between the virus and the malignancy is that infection with herpesvirus type 2 is associated with the same sexual and reproductive factors that are associated with an increased risk of developing cervical cancer. More data will be required to clarify the observed geographical differences in the occurrences of antibodies to herpesvirus type 2 and the association between the virus and cervical cancer.

SUMMARY

The occurrence of antibodies to herpesvirus type 2 was determined in selected samples of women from different geographical areas. Both women with cervical cancer and women without cervical cancer were studied. A considerable variation in the occurrence of antibodies to the virus was found between the areas studied. Generally, Negro women were found to have a higher occurrence of antibodies to the virus than women from other races. Among women with cervical cancer, 27 to 96% were found to have antibodies to the virus; a higher occurrence was found among Negro women with cervical cancer than among women of other races with the malignancy. In all areas, women with cervical cancer had a greater occurrence of antibodies to the virus than control women matched only for age, race and socio-economic level. When controls were selected by race, age, socio-economic and sexual and reproductive factors, the differences in the occurrence of antibodies between women with cervical cancer and control women fell to insignificant levels. Assuming no racial differences in the antibody responses to the herpesviruses, the data suggest that the association between herpesvirus type 2 and cervical carcinoma represents one of covariability with sexual activity.

ACKNOWLEDGEMENTS

The data reported were collected with the support of Contract PH 43-68-678 within the Special Virus Cancer Program, National Cancer Institute; The Ruth Estrin Goldberg Memorial for Cancer Research; Clinical Investigation Grant No. CI23 from the American Cancer Society, Inc.; and Career Development Award 5-K3-AI25943 from the National Institute of Allergy and Infectious Diseases.

REFERENCES

Mantel, N. & Haenszel, W. (1959) Statistical aspects of the analysis of data from retrospective studies of disease. *J. nat. Cancer Inst.*, **22**, 719-748

Nahmias, A. J., Josey, W. E., Naib, Z. M., Luce, C. F. & Guest, B. A. (1970) Antibodies to *Herpesvirus hominis* types 1 and 2 in humans. II. Women with cervical cancer. *Amer. J. Epidem.*, **91**, 547-552

Rawls, W. E., Iwamoto, K., Adam, E. & Melnick, J. L. (1970) Measurement of antibodies to herpesvirus types 1 and 2 in human sera. *J. Immunol.*, **104**, 599-606

Rawls, W. E. & Kaufman, R. H. (1970) Herpesvirus and other factors related to the genesis of cervical cancer. *Clin. Obstet. Gynec.*, **13**, 857-872

Rawls, W. E., Tompkins, W. A. F. & Melnick, J. L. (1969) The association of herpesvirus type 2 and carcinoma of the cervix. *Amer. J. Epidem.*, **89**, 547-554

Royston, I. & Aurelian, L. (1970) Association of genital herpesvirus with cervical atypia and carcinoma *in situ*. *Amer. J. Epidem.*, **91**, 531-538

Herpes Simplex Type 1 and Type 2 Antibody Levels in Patients with Carcinoma of the Cervix or Larynx

K. E. K. ROWSON [1] & H. M. JONES [1]

The microneutralization test (Pauls & Dowdle, 1967), kinetic neutralization test (Rawls *et al.*, 1968) and plaque reduction test (Aurelian, Royston & Davis, 1970) have been used successfully to measure antibody levels to two types of herpes simplex virus, but these methods are expensive in time and materials. The indirect haemagglutination test described by Fuccillo *et al.* (1970) would be more suitable for routine use if it could be shown to give results comparable with those of the other tests.

We have used the indirect haemagglutination test as described by Fuccillo *et al.* (1970) except that our virus stocks have been grown in BHK 21 cells instead of MA-196 human cell line, and we have used formalinized sheep erythrocytes instead of fresh sheep erythrocytes.

MATERIALS AND METHODS

Sera have been obtained from: (*a*) patients with carcinoma of the cervix; (*b*) patients with carcinoma of the larynx; (*c*) patients in the same wards as the cases of carcinoma of the larynx but with other conditions; (*d*) antenatal patients; and (*e*) cases of suspected genital herpes.

Strains of herpes simplex virus (the HFEM strain (type 1) and the NAM strain (type 2)) were obtained from Dr D. H. Watson, The Medical School, University of Birmingham, and have been grown in BHK 21 cells. Virus for antigen production was prepared by infecting confluent sheets of BHK 21 cells grown in 4-oz medical flats. The cell culture growth medium consisted of Eagles BHK medium (Wellcome Laboratories, Beckenham, Kent) with 10% bovine serum (B.D.H. Chemicals Ltd., Poole), 10% tryptose phosphate broth, penicillin (200 units/ml) and streptomycin (200 μg/ml). After removal of growth medium, the cultures were inoculated with 1 ml of type 1 or type 2 virus (titre of 10^5–10^7 PFU/ml on BHK 21 cells). Virus absorption was allowed to occur for 60 minutes at 37°C, fresh medium was added and the bottles incubated at 37°C. Total destruction of the cell sheet normally took place in 24 to 48 hours. The cells were then freed from the glass by shaking and the suspension centrifuged at 600 g for 5 minutes. The supernatant was poured off and the cells resuspended in 0.85% sodium chloride (pH 6.4) to give a 10% suspension. The resuspended cells were disrupted by 2 minutes sonication in a Dawe sonicleaner (Dawe Instruments Ltd., Acton, London, W.3). Cell debris was removed by centrifugation at 600 g for 5 minutes and the supernatant fluid used as antigen.

Sensitized erythrocytes were prepared according to the method of Fuccillo *et al.* (1970), except that formalinized cells (Difco Laboratories, East Molesey, England) were used instead of fresh cells. A 3% suspension of formalinized cells was washed three times in 0.85% sodium chloride and then mixed with an equal volume of freshly prepared tannic acid solution (50 mg per 1000 ml) (B.D.H. Chemicals Ltd., Poole, England). The mixture was incubated in a water bath at 37°C for 15 minutes, the cells spun down, washed once and resuspended to 3% concentration in saline. Cells were sensitized with a previously determined dilution of antigen by mixing equal volumes of cells and antigen and allowing the mixture to stand at room temperature for 15 minutes. The cells were spun down, washed twice in 1% heat inactivated normal rabbit serum (NRS), and resuspended in 1% NRS to give a concentration of 1%, which was used in the haemagglutination test.

[1] Institute of Laryngology and Ontology, London, W.C.1, UK.

The dilution of antigen used to sensitize the tanned cells was determined by a preliminary test in which various dilutions of antigen were used to sensitize erythrocytes. Table 1 shows the haemagglutination titres obtained in such a preliminary test. It can be seen that type 1 antigen gave an optimum activity when used as prepared whereas the type 2 antigen gave an optimum activity when the preparation was diluted tenfold.

Table 1. Haemagglutination titres obtained in preliminary test

Serum	Antigen	Reciprocal of haemagglutinin titre at dilution of:		
		10^0	10^{-1}	10^{-2}
1	Type 1	256	128	<8
2	Type 2	512	1024	64

The indirect haemagglutination test was carried out using a micromethod (Flow Laboratories Ltd., Irvine, Scotland). Test sera were heat-inactivated at 56°C for 30 minutes, then absorbed with tanned erythrocytes for 15 minutes at room temperature (0.1 ml serum and 0.7 ml unsensitized 3% tanned cells). Beginning with the 1 : 8 dilution of serum, doubling dilutions were prepared in 2% NRS. Each well on the plate received three 0.025-ml volumes: one of test serum, one of 2% NRS and one of 1% sensitized erythrocytes. Plates were shaken, sealed with cellophane tape and read after incubation at room temperature for 1½ hours. Each serum was tested against erythrocytes sensitized with type 1, type 2 and uninfected BHK 21 cells. A standard antiserum was included in all tests. Titres were read as the highest dilution of serum giving a 3+ agglutination on a scale of 0 to 4+ and were expressed as the reciprocal of the serum dilution.

Immune rabbit sera were prepared by injecting rabbits subcutaneously with either type 1 or type 2 virus. The viruses were grown on the chorioallantoic membranes of 11-day-old chick embryos. Each rabbit received two injections at 14-day intervals of the homogenized tissue from two membranes. The animals were bled after the second injection.

RESULTS

To test the specificity of the indirect haemagglutination test rabbits were immunized with either the HFEM (type 1) or the NAM (type 2) strain and

their sera tested against erythrocytes sensitized with type 1 or type 2 virus or an extract of uninfected BHK 21 cells. Table 2 shows the result of these tests. At a dilution of 1 : 8 or greater the rabbit sera show no cross-reaction, and none of the sera agglutinated the control cells. Human sera obtained

Table 2. Antibody response in rabbits injected with herpes simplex virus

Virus used for immuniza-tion	Reciprocal of haemagglutinin titre			
	Before immunization		After immunization	
	Type 1	Type 2	Type 1	Type 2
Type 1	<8	<8	64	<8
Type 2	<8	<8	<8	128

from various sources were tested against erythrocytes sensitized with the same antigens. Fig. 1 shows the haemagglutinin titre for each serum against type 1 and type 2 antigen. There is a marked correlation between the type 1 and type 2 titres but there are a few patients with antibodies to type 2 only. However, there is only one serum that reacts to type 1 and is negative for type 2.

Fig. 2 shows the haemagglutinin titres obtained against type 2 antigen in single serum samples from 21 patients with carcinoma of the cervix and 20 patients with carcinoma of the larynx. All the patients with carcinoma of the larynx and all but four of the patients with carcinoma of the cervix had antibodies to type 2 virus. Similar antibody levels were obtained in 9 patients suspected on clinical grounds of having genital herpes. In contrast to these results are those obtained in 10 antenatal patients and 11 patients in the same wards as the patients with carcinoma of the larynx but with other conditions.

DISCUSSION

In contrast to the expensive and time-consuming plaque reduction and microneutralization tests, the indirect haemagglutination test appears to offer a rapid and easily performed method of measuring type-specific herpesvirus antibodies in sera. The results obtained when immune rabbit sera raised against herpes simplex virus types 1 and 2 were tested against red cells sensitized with each virus (Table 2) suggest that the test is type-specific. Similar results were obtained by Fuccillo *et al.*, (1970) but Bernstein & Stewart (1971) failed to

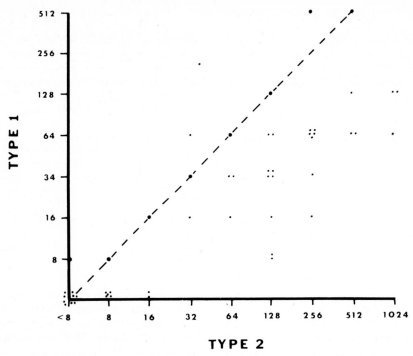

Fig. 1. Correlation between antibody levels to herpes simplex types 1 and 2

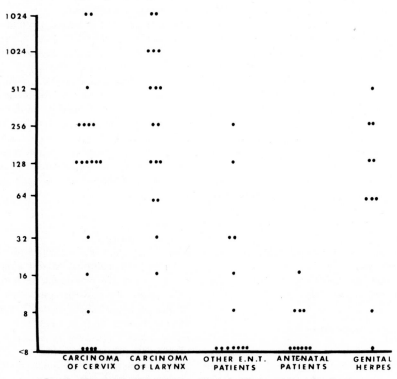

Fig. 2. Herpes simplex type 2 antibody levels in groups of patients

obtain type-specific rabbit sera. However, on absorption of the sera with virus, the indirect haemagglutination test gave good results.

In the human sera that we tested, there appears to be some correlation between the results obtained for type 1 and type 2 antibody levels (Fig. 1). This may be due to a lack of type-specificity in the test, but in view of the results obtained with the rabbit sera it seems likely that it is due to the production of cross-reacting antibodies. Very few sera showed a higher titre to type 1 than to type 2 and this may be because the test for type 2 antibodies was less sensitive than that for type 1 antibodies. An optimal antigen concentration was not obtained (Table 1) with the type 1 virus and this may explain the lower levels obtained and the absence of sera reacting to type 1 only.

The serological results obtained in different clinical groups of patients (Fig. 2) suggest that herpes simplex type 2 virus infection is common in patients with carcinoma of the cervix or larynx. The patients clinically diagnosed as having genital herpes

also showed a high incidence of type 2 antibodies. The results obtained in the patients with carcinoma of the cervix are similar to those obtained by others (Josey, Nahmias & Naib, 1968; Rawls et al., 1968, 1969) and antibodies to type 2 virus are to be expected in patients with genital herpes. However, the high incidence of type 2 antibodies in the patients with carcinoma of the larynx is surprising. That there is some direct relationship between carcinoma of the larynx and the presence of herpes type 2 antibodies seems likely as other patients with ear, nose and throat conditions and antenatal patients showed no signs of a high incidence of these antibodies.

Epidemiologically, carcinoma of the cervix and genital herpes have an elevated incidence in women with multiple sexual contacts, and so the relationship between these two conditions may be purely coincidental. The relationship between carcinoma of the larynx and herpes type 2 infection is more difficult to explain. It could be that the virus is causally related to the tumour or equally well that the tumour predisposes to infection with the virus.

SUMMARY

The indirect haemagglutination test has been used to test sera from patients with carcinoma of the larynx or cervix, genital herpes, other ear, nose and throat conditions and sera from antenatal patients for antibodies to herpes simplex types 1 and 2. High type 2 antibody levels were found in sera from patients with carcinoma of the cervix, larynx, and genital herpes, whereas the sera from the antenatal patients and from patients with ear, nose and throat conditions other than those mentioned gave relatively low type 2 antibody titres.

ACKNOWLEDGEMENTS

This work was supported by a grant from the Cancer Research Campaign.

REFERENCES

Aurelian, L., Royston, I. & Davis, H. J. (1970) Antibody to genital herpes simplex virus: association with cervical atypia and carcinoma in situ. J. nat. Cancer Inst., 45, 455-464

Bernstein, M. T. & Stewart, J. A. (1971) Method for typing antisera to Herpesvirus hominis by indirect haemagglutination inhibition. Appl. Microbiol., 21, 680-684

Fuccillo, D. A., Moder, F. L., Catalano, L. W. Jnr., Vincent, M. M. & Sever, J. L. (1970), Herpesvirus hominis types 1 and 2: A specific microindirect haemagglutination test. Proc. Soc. exp. Biol. (N.Y.), 133, 735-739

Josey, W. E., Nahmias, A. J. & Naib, Z. M. (1968) Genital infection with type 2 Herpesvirus hominis. Present knowledge and possible relation to cervical cancer. Amer. J. Obstet. Gynec., 101, 718-729

Pauls, F. P. & Dowdle, F. R. (1967) A serologic study of Herpesvirus hominis strains by microneutralization tests. J. Immunol., 98, 941-947

Rawls, W. E., Tompkins, W. A. F., Figueroa, M. & Melnick, J. L. (1968) Herpesvirus type 2: Association with carcinoma of the cervix. Science, 161, 1255-1256

Wheeler, C. E., Jr., Briggaman, R. A. & Henderson, R. R. (1969) Discrimination between two strains (types) of herpes simplex virus by various modifications of the neutralization test. J. Immunol., 102, 1179-1192

Biochemical Comparisons of Type 1 and Type 2 Herpes Simplex Viruses

I. W. HALLIBURTON [1]

Herpes simplex viruses have been divided into two types. Despite the contention by Roizman *et al.* (1970) that some of the evidence may be better interpreted as indicating the occurrence of a spectrum of herpes simplex viruses, the serological data from a number of laboratories have clearly indicated, with many different viral strains and various serological tests, the presence of two major groups (Schneweis, 1962; Plummer, 1964; Pauls & Dowdle, 1967; Nahmias *et al.*, 1969; Fuccillo *et al.*, 1970). In addition, numerous biological comparisons have provided supportive evidence for the classification (Nahmias & Dowdle, 1968; Figueroa & Rawls, 1969; Plummer *et al.*, 1970). Apart from the interesting observation by Goodheart, Plummer & Waner (1968) that the DNAs of types 1 and 2 differ in density, biochemical evidence of differences between the two strains is, however, sadly lacking. The present project was undertaken with two major objectives in view: first, to verify and extend the findings of Goodheart *et al.* (1968) regarding possible differences in the DNAs of types 1 and 2; and second, to look for differences in respect of specific proteins between types 1 and 2, which might eventually be correlated with the antigenic differences.

The virus strains employed are both established laboratory strains and some Scottish field isolates. Of the type 1 strains used, MP 17, isolated from a facial lesion and "McDonald" from a herpetic infection of the cornea are local field isolates. Strain α is a plaque isolate of the HFEM strain. Dr Peutherer, University of Edinburgh, supplied three type 2 strains: HG 52, an isolate from herpetic anal lesions; and MS and Dawson, both established type 2 strains (Peutherer, 1970). Dr Ross, Ruchill

Hospital, Glasgow, supplied HG 48, a locally obtained field isolate from a genital lesion.

Although these strains could be tentatively classified on the basis of their site of origin, this was checked by kinetic neutralization. As shown in Fig. 1, the strains listed above as types 1 or 2 fall clearly into the appropriate category. This neutralization test has since been performed with strain Dawson, confirming its classification as a type 2 strain.

Before the DNAs and proteins of types 1 and 2 were compared, one member of each group was selected and examined in detail. The observed differences were then sought in the other strains. The strains selected for the initial detailed study were MP 17 and HG 52 because both are under intensive genetic study by my colleagues in the Institute of Virology. The one-step growth curves of these two viruses are very similar, maximal viral yields being attained about 10 hours post infection, although, as has previously been reported by others (Figueroa & Rawls, 1969; Plummer *et al.*, 1970), the type 2 strain yield reached a plateau about 1–1½ logs lower than that attained by the type 1 strain. As the lag periods and the time taken to attain maximal viral yields were the same for both strains, comparative pulse-labelling experiments can be performed in which the duration and timing of the pulse is the same for both strains.

The rate of viral DNA synthesis was then examined in BHK 21/C13 cells by labelling with ^{3}H-thymidine during successive 2-hour intervals after infection. The viral DNA was separated from the cell DNA by isopyknic centrifugation in caesium chloride and the total radioactive counts under each peak determined. As shown in Fig. 2, viral DNA synthesis for both strains was detectable within the first two hours and the rate of synthesis reached a maximum 4–6 hours after the end of the one-hour

[1] Medical Research Council Virology Unit, Institute of Virology, Glasgow, W.1, UK.

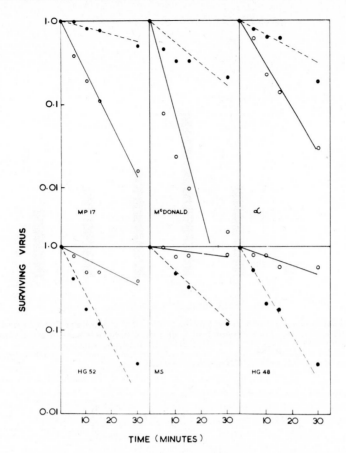

Fig. 1. Kinetic curves of neutralization of herpes simplex viruses with antiserum prepared against: (a) type 1 virus (McDonald) at a dilution of 1 : 100 (O———O); and (b) type 2 virus (HG 52) at a dilution of 1 : 30 (●– – –●).

absorption period; the rate decreased thereafter. Although cell DNA synthesis progressively decreased after infection, it was still detectable at 10–12 hours with both virus strains. Despite the lower titres attained by the type 2 virus, the incorporation of ³H-thymidine into the type 2 DNA was in fact greater than that into the type 1 DNA. Several other experiments suggest that this surprising result is not atypical.

To determine whether density differences exist between the DNAs of the two strains, ³H- and ¹⁴C-labelled DNA was cocentrifuged in fully orthogonal experiments, as shown for the comparison between MP 17 and HG 52 in Fig. 3. Similarly compared were MS × MP 17; MS × HG 52; α × MP 17; α × HG 52; HG 48 × MP 17; Dawson × MP 17; McDonald × HG 52. The results show unambiguously that type 2 DNA is denser than type 1

DNA. To determine the magnitude of the density difference, the densities of the DNAs of MP 17 and HG 52 were determined on the model E ultracentrifuge. Values of 1.7254 ± 0.0018 (mean ± standard error for four estimations) and 1.7275 + 0.00006 were obtained respectively. The guanine + cytosine (G + C) contents of the DNAs determined from these density values by the method of Schildkraut, Marmur & Doty (1962) were 66.7 and 68.9% respectively. Goodheart *et al.* (1968) have reported G + C contents of 68.3 and 70.4% for type 1 and type 2 strains respectively, also reflecting a density difference of 0.002. Despite the difference in base composition of type 1 and 2 DNAs, nearest neighbour analysis of the DNAs (Halliburton, Hill & Russell[1]) disclosed no obvious differences in

[1] Unpublished data.

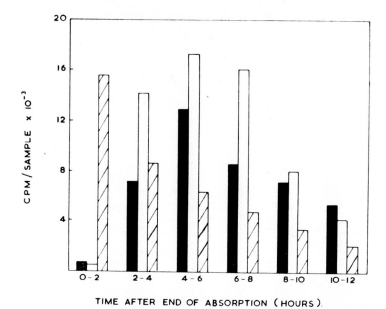

Fig. 2.   The amount of thymidine (methyl-H3) incorporated into viral and cellular DNA during successive two-hour intervals after the end of absorption.   For each point, 40 × 10⁶ BHK 21/C13 cells were infected with the appropriate virus at a multiplicity of infection (MOI) of about 10, the virus being absorbed by shaking at 37°C for 1 hour.   No check was made that all of the cells were infected; this could account for the continued cellular DNA synthesis.   Type 1 DNA, black bars; type 2 DNA, open bars; cellular DNA, cross-hatched bars.

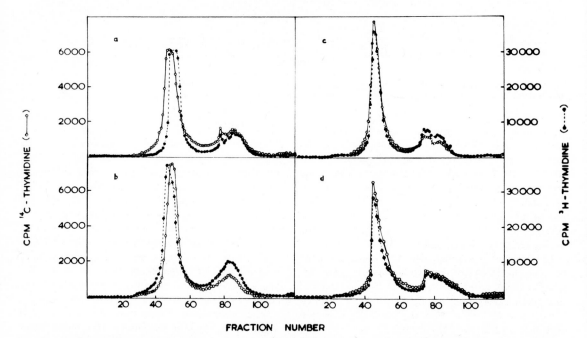

Fig. 3.   Cocentrifugation of type 1 (MP17) and type 2 (HG 52) DNA on caesium chloride gradients.   25 × 10⁶ cells were infected on monolayers at a MOI of about 5 and labelled with thymidine (methyl-H3) or thymidine-2-C14 from 2.5–18 hours post infection.   (a) ³H — MP17 + ¹⁴C — HG 52; (b) ³H — HG 52 + ¹⁴C — MP 17; (c) ³H — MP 17 + ¹⁴C — MP 17; (d) ³H — HG 52 + ¹⁴C — HG 52.

doublet patterns or General Design (Subak-Sharpe, 1969).

The 2.2% difference in G + C content, however, allows about 6600 base differences between the DNAs of types 1 and 2, assuming the DNAs of both types to have molecular weights of 100×10^6 Daltons, as has been reported for type 1 (Becker, Dym & Sarov, 1968). If these base differences are in stretches of DNA that are transcribed and translated, then differences should exist in the virus-specified proteins. With the recognized antigenic differences, it seemed distinctly possible that protein differences could be demonstrated by polyacrylamide gel electrophoresis. Nuclear or cytoplasmic fractions of infected cells, however, contain so many viral and cell proteins that adequate resolution by existing techniques is almost impossible. We therefore decided to fractionate infected cells further so as to resolve individual polypeptides. The approach used has been to select basic proteins by extracting nuclei of infected cells with dilute hydrochloric acid. The basic proteins (which include the cell histones) were then resolved on polyacrylamide gels. Following infection with either type 1 or type 2 viruses,

arrest of histone synthesis is complete by 6 hours post infection, whereas host cell DNA synthesis can still be detected after 13 hours.

In addition to the cell histones, a number of virus-induced proteins appear in the acid extracts. Since all of these polypeptides incorporate tryptophan, they are not "classic" histones. Comparison of separately obtained acrylamide gel profiles of acid extracts of the nuclei of type 1 and type 2 infected cells reveals general similarities. Orthogonal experiments involving coelectrophoresis of acid extracts of type 1 and 2 infected cells in sodium dodecyl sulphate (SDS) acrylamide gels however, have indicated a number of differences. Fig. 4 shows the results of such an experiment on acid extracts of cells infected with MP 17 or HG 52 and labelled with ^{3}H-arginine or ^{14}C-arginine between 4.5 and 6.5 hours post infection. Arrows numbered 1, 2 and 3 indicate the polypeptides in which the two types differ. Arrow 1 indicates that in type 1 infected extracts there is a major polypeptide of molecular weight about 40 700 Daltons (estimated according to the method of Shapiro, Viñuela & Maizel, 1967) whereas in type 2 infected extracts the "corresponding"

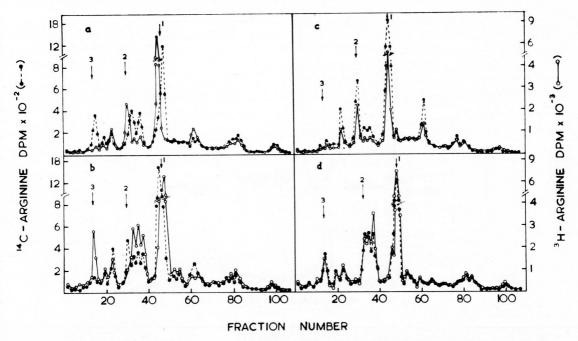

Fig. 4. SDS acrylamide gel electrophoresis of acid-soluble proteins from the nuclei of cells infected with type 1 (MP 17) or type 2 (HG 52) herpes simplex viruses. Proteins labelled with ^{3}H- or ^{14}C-arginine from 4.5–6.5 hours post infection. (a) ^{3}H — MP 17 + ^{14}C — HG 52; (b) ^{3}H — HG 52 + ^{14}C — MP 17; (c) ^{3}H — MP 17 + ^{14}C — MP 17; (d) ^{3}H — HG 52 + ^{14}C — HG 52.

polypeptide appears to have a molecular weight of about 39 900 Daltons. This apparent difference could arise in at least four ways. First, it could reflect simply quantitative differences: there might be a type 2 polypeptide of molecular weight 40 700 Daltons and a type 1 polypeptide of 39 900 Daltons but both present in such small quantities relative to the polypeptides of the other virus type as to be virtually undetectable. Second, the type 1 polypeptide might contain the same sequence of amino acids as the type 2 plus approximately 7 amino acids. Third, *in vivo*, the two virus types might contain a polypeptide of the same length but the type 2 polypeptide might lose a small piece during the extraction procedure. Fourth, the polypeptides from the two strains may in fact be of totally different amino acid content but, by chance, of almost identical molecular weight. We are currently trying to distinguish between these possibilities by homology studies on extracted polypeptides. It is also relevant that the amount of the type 2 polypeptide synthesized soon after infection is relatively much less than that of its corresponding type 1 polypeptide (Fig. 5).

A further difference between type 1 and type 2 polypeptides is shown by arrow 2. Arrow 3 clearly indicates a region with a polypeptide present in the acid extract from type 2 but probably not from type 1 infected cells. This difference is complicated, however, by the fact that a polypeptide migrating at this position is observed both in extracts of nuclei of type 1 infected cells and in acid extracts of nuclei labelled during the first three hours after the end of absorption. Fig. 5 for example, in which the labelling was from 2–3 hours after infection, clearly shows the presence of both type 1 and type 2 polypeptides at this position. At times later than three hours, no polypeptide is found at this position. Nonidentity of a fourth polypeptide can be demonstrated if mixed acid extracts are applied to a gel in the absence of SDS. Under these conditions, separation is principally on the basis of charge whereas with SDS gels it is based on size. Since this fourth difference does not occur with SDS gels, both type 1 and type 2 strains apparently contain a polypeptide of the same size but of slightly different charge.

Fig. 5 shows coelectrophoresis of acid extracts of the two types labelled soon after infection, from 1–2 hours after the end of absorption. In addition, Fig. 5 shows uninfected cell extracts labelled at a corresponding time after mock infection. The

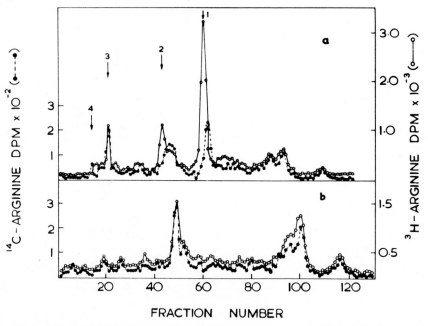

Fig. 5. SDS acrylamide gel electrophoresis of acid-soluble proteins from the nuclei of cells infected with type 1 (MP 17) or type 2 (HG 52) herpes simplex viruses labelled at an early time (2–3 hours) after infection. (a) Infected cell extracts. O———O type 1 polypeptides; ●– – –● type 2 polypeptides. (b) Mixed uninfected cell extracts labelled with ³H- or ¹⁴C-arginine at 2–3 hours after mock infection with used medium.

arrows 1, 2 and 3 apply to the same polypeptides as in Fig. 4. Differences 1 and 2 are clearly demonstrable at this early time, but as mentioned above, the situation under arrow 3 is altered in as much as both types synthesize a polypeptide at this time. In addition, there is a fourth region where the polypeptides do not coelectrophorese on SDS gels (arrow 4). The quantities of the polypeptide involved are very small but the difference has been found to be reproducible, and can be demonstrated only during the first two hours after the end of absorption; thereafter, these polypeptides do not seem to be synthesized.

Having confirmed these polypeptide differences with several strains of each type, it is now important to find which proteins are represented by the radioactive peaks and to define their functions. Acid extracts have been coelectrophoresed with purified virus to determine whether the proteins referred to in Figs. 4 and 5 are structural or nonstructural. Virus was prepared from infected cells labelled with ³H-arginine for 48 hours. The cells were sonicated and the virus banded on a 10–65% w/v sucrose gradient by centrifugation at 50 000 g for 40 minutes. The virus band was rebanded on a similar gradient, dialyzed to remove sucrose, and the virus pelleted by centrifugation. The viral proteins were then dissolved in a buffer containing SDS, urea and β-mercaptoethanol, mixed with an acid extract of the nuclei of cells infected with the same virus labelled with ¹⁴C-arginine and coelectrophoresed on SDS gels. Fig. 6 shows the results of such an experiment with MP 17. As can be seen, the proteins of the acid extract arrowed 1 and 2 in Figs. 4, 5 and 6 are structural proteins. The nature of protein 3 is uncertain but protein 4 is not a structural entity.

In conclusion, differences between types 1 and 2 herpes simplex viruses have been demonstrated in the density and base composition of their DNAs, in at least two structural, one nonstructural and two proteins of undetermined origin. In addition, current studies with the same virus strains show that the DNA polymerases of types 1 and 2 can be distinguished from one another on the basis of response to varying ammonium ion concentration and thermal stability. These results clearly strengthen the case for the division of herpes simplex viruses into two types.

It is also of interest that genetic studies with temperature-restricted mutants of types 1 and 2 show that the two types complement each other (Timbury & Subak-Sharpe [1]).

---

[1] Personal communication.

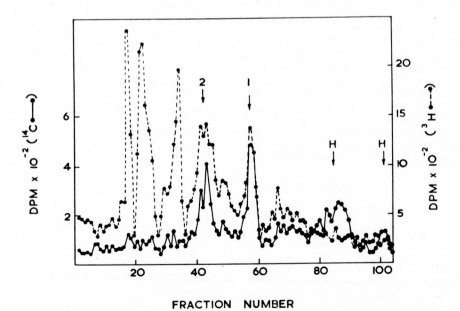

Fig. 6. Coelectrophoresis of MP 17 structural viral proteins (●– – –●) and acid-soluble proteins from the nuclei of infected cells (●———●). Peaks marked H represent cell histones.

SUMMARY

Comparison of the viral DNAs and proteins in type 1 and type 2 herpes simplex virus infected BHK21/C13 cells has revealed differences in two general areas. First, the density of the DNA of type 2 strains is 0.002 g/ml greater than that of type 1 strains, corresponding to a 2% difference in guanine + cytosine content. Second, six qualitative and/or quantitative differences have been demonstrated in the viral proteins by polyacrylamide gel electrophoresis of dilute acid extracts of the nuclei of infected cells.

ACKNOWLEDGEMENTS

I am grateful to Mrs E. A. Hill for skilled technical assistance. My thanks are due to Dr M. C. Timbury for doing the neutralization tests, Mr G. Russell for assistance with the DNA studies, Dr J. Hay who collaborated in the enzyme assays, Professor J. H. Subak-Sharpe and Dr C. R. Madeley for discussion of the manuscript and to Professor Subak-Sharpe for general help and encouragement.

REFERENCES

Becker, Y., Dym, H. & Sarov, I. (1968) Herpes simplex virus DNA. *Virology*, **36**, 184-192

Figueroa, M. E. & Rawls, W. E. (1969) Biological markers for differentiation of Herpes-virus strains of oral and genital origin. *J. gen. Virol.*, **4**, 259-267

Fuccillo, D. A., Moder, F. L., Catalano, L. W., Vincent, M. M. & Sever, J. L. (1970) Herpesvirus hominis types 1 and 2: A specific microindirect haemagglutination test. *Proc. Soc. exp. Biol. (N.Y.)*, **133**, 735-739

Goodheart, C. R., Plummer, G. & Waner, J. L. (1968) Density difference of DNA of human Herpes simplex viruses types 1 and 2. *Virology*, **35**, 473-475

Nahmias, A. J., Chiang, W. T., Del Buono, I. & Duffey, A. (1969) Typing of Herpesvirus hominis strains by a direct immunofluorescent technique. *Proc. Soc. exp. Biol. (N.Y.)*, **132**, 386-390

Nahmias, A. J. & Dowdle, W. R. (1968) Antigenic and biologic differences in Herpesvirus hominis. *Progr. med. Virol.*, **10**, 110-159

Pauls, F. P. & Dowdle, W. R. (1967) A serologic study of Herpesvirus hominis strains by microneutralisation tests. *J. Immunol.*, **98**, 941-947

Peutherer, J. F. (1970) The specificity of rabbit antisera to Herpes virus hominis and its dependence on the dose of virus inoculated, *J. med. Microbiol.*, **3**, 267-272

Plummer, G. (1964) Serological comparison of the Herpes viruses. *Brit. J. exp. Path.*, **45**, 135-141

Plummer, G., Waner, J. L., Phuangsab, A. & Goodheart, C. R. (1970) Type 1 and type 2 Herpes simplex viruses: Serological and biological differences. *J. Virol.*, **5**, 51-59

Roizman, B., Keller, J. M., Spear, P. G., Terni, M., Nahmias, A. & Dowdle, W. (1970) Variability, structural glycoproteins, and classification of Herpes simplex viruses. *Nature (Lond.)*, **227**, 1253-1254

Schildkraut, C. L., Marmur, J. & Doty, P. (1962) Determination of the base composition of deoxyribonucleic acid from its buoyant density in CsCl. *J. molec. Biol.*, **4**, 430-443

Schneweis, K. E. (1962) Serologische Untersuchungen zur Typendifferenzierung des Herpesvirus hominis. *Z. ImmunForsch. exp. Ther.*, **124**, 24-48

Shapiro, A. L., Viñuela, E. & Maizel, J. V. (1967) Molecular weight estimation of polypeptide chains by electrophoresis in SDS-polyacrylamide gels. *Biochem. biophys. Res. Commun.*, **28**, 815-820

Subak-Sharpe, J. H. (1969) The doublet pattern of the nucleic acid in relation to the origin of viruses. In: Lima de Faria, A., ed., *Handbook of molecular cytology*, Amsterdam & London, North Holland Publishing, pp. 67-87

The Different Cellular Alterations Induced by Various Herpesvirus Hominis Strains as Evidenced by Electron Microscopy

K. MUNK [1] & C. WALDECK [1]

In previous work we were concerned with differences in the properties of various herpesvirus hominis (HVH) strains (Munk & Donner, 1963). We are able to correlate the site of origin of the particular strains with the biological properties of plaque morphology and cytopathic effect. The genital strains produce, in the plaque assay of Dulbecco & Vogt (1954), large plaques in addition to small plaques whereas the nongenital strains produce small plaques only. After plaque cloning, the clones derived from large plaques always produce both large plaques and small plaques, whereas the clones derived from small plaques of either the genital strains or the nongenital strains produce small plaques only (Munk & Ludwig, 1972). The large plaque clones form membranous syncytia, whereas the small plaque clones of the genital strains form small syncytia. The small-plaque-producing nongenital strains do not form syncytia but rather a rounding up of the infected cells.

During the course of this previous study we also found differences in the pattern of the HVH-specific antigens in the infected cell, as detected by the immunofluorescence method (Munk & Fischer, 1965). This apparent difference in cell response led us to investigate further the interactions between the various herpes strains and the host cell using the electron microscope. We searched for possible differences in either the morphological appearance and/or cell response of cells infected with the various HVH strains. This particular study was also considered necessary because the descriptions of the electron microscopical appearance of herpesvirus production in the cell have varied a great deal.

Reports on the electron microscopy of the strains of HVH have been published by Epstein (1962), Siminoff & Menefee (1966), Nii, Morgan & Rose (1968), Couch & Nahmias (1969), and Schwartz & Roizman (1969a). Schwartz & Roizman (1969b) have recently described specific differences in the ultrastructural pattern of genital and nongenital strains.

MATERIALS AND METHODS

Virus. The herpesvirus hominis strains reported here were isolated from diagnosed efflorescences of patients at the Dermatological University Hospital, Heidelberg. The genital strains were: HOF (23–35 pass.), ANG (8–11 pass.), BER (8–12 pass.) and LEO (14 pass.); the nongenital strains were: WAL (12–28 pass., and a new isolate of the same patient after 6 years), MUN (2–4 pass.), ETT (1 pass.). The strains were isolated and passaged in HeLa cells. The titre was determined by plaque assay.

Cells. HeLa-S3 cells were grown in Hanks solution with 0.5 % lactalbumin, 5 % calf serum and antibiotics and incubated at 37°C. The cells were infected with virus usually at a multiplicity of 1 to 10, and were examined at 12, 24, 48 and 72 hours post infection. For purposes of the comparison of the strains used, the cells were examined 24 hours post infection.

Electron microscope method. The infected cells were fixed in glutaraldehyde, postfixed in osmium tetroxyde and embedded in Araldite. The Siemens electron microscope "Elmiskop I" was used.

RESULTS AND DISCUSSION

We have observed most of the intracellular phenomena that have been described by the various

[1] Institut für Virusforschung, Deutsches Krebsforschungszentrum, Heidelberg, Federal Republic of Germany.

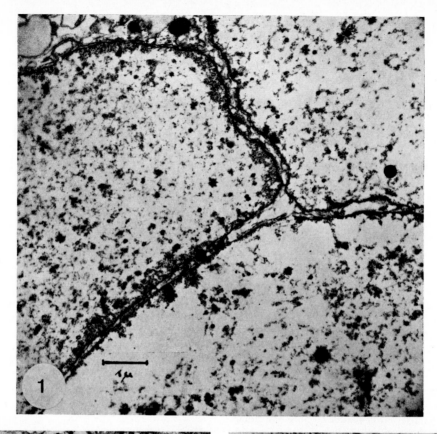

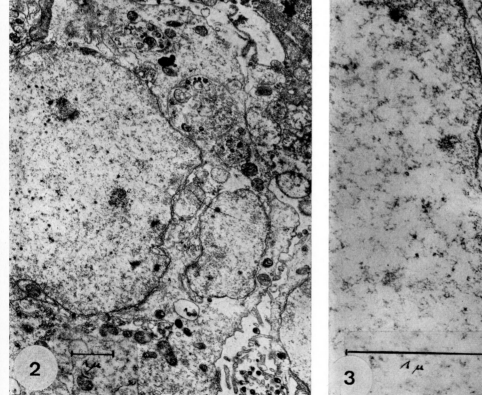

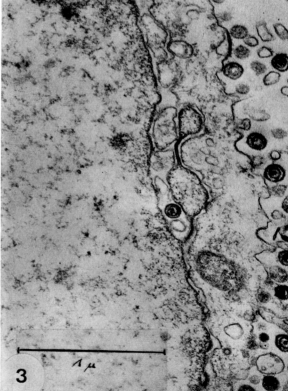

authors, such as the different forms of viral particles within the nucleus, granules similar to those described by Schwartz & Roizman (1969), and disintegrating virions in the cytoplasm (Nii *et al.*, 1968; Schwartz & Roizman, 1969). These phenomena were observed in all the virus strains investigated so far. We have found intranuclear viral crystals only with the Armstrong 1116 strain, at 24 hours and later times post infection. We were not able to demonstrate the crystals in other strains regardless of their origin, the number of passages, or the time after infection.

We have never seen in infected HeLa cells the intranuclear lattice structures described by Couch & Nahmias (1969) and Schwartz & Roizman (1969). We have found, however, the intranuclear filament structures in HeLa cells infected with genital strains, similar to those reported by Couch & Nahmias (1969).

One special phenomenon was seen in the nuclei of HeLa cells infected with the different HVH strains at 24 hours and later post infection. In syncytia-forming strains, the nuclei of the infected cells are loosely structured (Fig. 1). Due to the fixation procedure, the inner part of the nucleus is retracted to a large extent. The membrane is somewhat fuzzy but regular. In one of the syncytia-forming strains, we generally found small nuclear pockets and small nuclear loops along the nuclear membrane. There was no virus within the pockets. In contrast, the nuclei of HeLa cells infected with non-syncytia-forming strains were highly lobated (Fig. 2). They showed numerous projections (Fig. 3) and loops of

various sizes. These projections are not known in normal HeLa cells. We have repeatedly examined our uninfected HeLa cells for this phenomenon with negative results. This phenomenon is seen only in normal lymphoid cells and in cells from human leukaemia (McDuffie, 1967). It has, however, been described in cells of fish melanoma (Vielkind & Vielkind, 1970) and in human dermatofibrosarcoma cells (Mollo, Canese & Stramignoni, 1969). In particular it has been observed by Achong & Epstein (1966) in Burkitt's tumour lymphoblasts. The fact that similar cellular alterations are found in the Burkitt's cell and in the herpes cell may point perhaps to similar mechanisms of action. Nuclear projections are generally considered to be an indication of an actively proliferating type of cell. That we have found these nuclear projection in cells infected with nongenital strains may be a sign that these types of HVH also possess the ability to activate the nucleus initially even though the cells eventually undergo cytopathic changes and disintegrate.

The nuclear projections were regularly seen in contact with a disintegrating virion. This observation leads to the speculative hypothesis that virions re-enter the nucleus and in so doing start a new cycle of virus multiplication (Munk & Waldeck, 1969). This could explain the long-term excretion of HVH from infected cells.

In conclusion, we have found specific nuclear alterations in herpesvirus-infected cells. These alterations resemble those found predominantly in actively proliferating cells.

REFERENCES

Achong, B. G. & Epstein, M. A. (1966) Fine structure of the Burkitt tumor. *J. nat. Cancer Inst.*, **36**, 877-897

Couch, E. F. & Nahmias, A. J. (1969) Filamentous structures of type 2 herpesvirus hominis infection of the chorioallantoic membrane. *J. Virol.*, **3**, 228-232

Dulbecco, R. & Vogt, M. (1954) Plaque formation and isolation of pure lines with poliomyelitis viruses. *J. exp. Med.*, **99**, 167-182

Epstein, M. A. (1962) Observations on the mode of release of herpesvirus from infected HeLa cells. *J. Cell Biol.*, **12**, 589-597

McDuffie, N. G. (1967) Nuclear blebs in human leukaemic cells. *Nature (Lond.)*, **214**, 1341-1342

Mollo, F., Canese, M. G. & Stramignoni, A. (1969) Nuclear sheets in epithelial and connective tissue cells. *Nature (Lond.)*, **221**, 869-870

Munk, K. & Donner, D. (1963) Cytopathischer Effekt und Plaque-Morphologie verschiedener Herpes-simplex-Virus-Stämme. *Arch. ges. Virusforsch.*, **13**, 529-540

Munk, K. & Fischer, H. (1965) Fluoreszenzimmunologische Unterschiede bei Herpes-simplex-Virus-Stämmen. *Arch. ges. Virusforsch.*, **15**, 539-548

Fig. 1. Herpesvirus-induced HeLa cell syncytium 24 hours post infection; nuclei loosely structured, membranes regular.

Fig. 2. Herpesvirus-infected HeLa cell 24 hours post infection; nucleus lobated, nuclear pockets.

Fig. 3. Herpesvirus-infected HeLa cell 24 hours post infection; nuclear projections.

Munk, K. & Ludwig, G. (1972) Properties of plaque variants of herpes virus hominis strains of genital origin. *Arch. ges. Virusforsch.* (in press)

Munk, K. & Waldeck, C. (1969) Kernrandveränderungen bei Herpes-simplex-Virus-infizierten HeLa-Zellen. *Naturwissenschaften,* **56**, 567-568

Nii, S., Morgan, C. & Rose, H. M. (1968) Electron microscopy of herpes simplex virus. II. Sequence of development. *J. Virol.,* **2**, 517-536

Schwartz, J. & Roizman, B. (1969) Similarities and differences in the development of laboratory strains and freshly isolated strains of herpes simplex virus in HEp-2 cells: electron microscopy. *J. Virol.,* **4**, 879-889

Siminoff, P. & Menefee, M. G. (1966) Normal and 5-bromodeoxyuridine-inhibited development of herpes simplex virus. An electron microscopy study. *Exp. Cell Res.,* **44**, 241-255

Vielkind, U. & Vielkind, J. (1970) Nuclear pockets and projections in fish melanoma. *Nature (Lond.),* **226**, 655-656

Effect of Hormonal Imbalance and Herpesvirus Type 2 on the Uterine Cervix of the Mouse

N. MUÑOZ [1]

INTRODUCTION

Epidemiological studies have shown that carcinoma of the cervix is related to sexual behaviour: the disease is practically nonexistent in women who have no sexual experience and is more common among married women. A consistent finding has been the higher incidence among women who have experienced sexual intercourse early in life and who have had multiple sexual partners (Christopherson & Parker, 1965; Martin, 1967; Pereyra, 1961; Terris & Oalman, 1960; Terris et al., 1967; Towne, 1955). From these facts it has been concluded that cancer of the cervix behaves epidemiologically like a venereal disease. An association between carcinoma of the cervix and genital herpesvirus has been proposed. The evidence for this association has been reviewed by Nahmias, Naib & Josey,[2] and Rawls, Adam & Melnick.[3] On the other hand, there also are some data that suggest that hormonal factors might play a rôle on the aetiology of cervical carcinoma: (1) The age-distribution curve of the cancer of the cervix: in most countries the incidence increases regularly with age, reaches a peak at the fourth or fifth decades, and then decreases (Doll, Muir & Waterhouse, 1970); (2) It is known that oestrogens are carcinogenic for several laboratory animals (Allen & Gardner, 1941; Bertolin, 1964); (3) Although the evidence is controversial, it has been suggested that oral contraceptives increase the risk of cervical carcinoma (Dunn, 1969; Melamed et al., 1969).

To test the hypothesis of a combined effect of genital herpesvirus and hormonal imbalances on the aetiology of carcinoma of the cervix, the following experiment is being carried out. Preliminary results only will be described.

MATERIALS AND METHODS

Twelve groups of 20 BALB/c female mice, 2–3 months old, were treated with different combinations of hormones and human genital herpesvirus (Benfield strain). Three groups were treated with the virus after treatment with oestrogens, progesterone and Enovid (norethynodrel and metranol) respectively. Three groups were treated with oestrogens alone, progesterone alone, and Enovid alone, respectively, and another group with the virus alone. Two control groups were used, one having a normal diet and the other the liquid diet used to administer the Enovid (see Table). The groups treated with the virus were previously immunized with genital herpesvirus inactivated with ultraviolet light. Viral inoculation was carried out by introducing a cotton pellet soaked in an undiluted virus solution (titre $10^{4.3}$) in the vagina. Oestrogens and progesterone were administered as subcutaneous pellets of a hormone-cholesterol mixture and the oral contraceptive was administered in a liquid diet (Metrecal).

Vaginal smears were taken at intervals to check the level of viral infection and to detect any early neoplastic change. The virus was administered twice, namely at the beginning of the experiment and then again 10 months after the first inoculation.

RESULTS

Characteristic cytological changes of herpesvirus infection were observed in 40% of 140 mice treated with the virus. Multinucleated giant cells were first observed two days after viral inoculation, with a

[1] Unit of Biological Carcinogenesis, International Agency for Research on Cancer, Lyon, France.
[2] See p. 403 of this publication.
[3] See p. 424 of this publication.

Treatment of mice with hormones and human genital herpesvirus (HVH) [1]

Group	Initial no. of mice	Treatment	No. of mice at risk [2]	Tumours No.	Tumours %
I	40 [3]	Oestrogens + HVH-2	36	1	2.8
II	40 [3]	Progesterone + HVH-2	39	0	—
III	40 [3]	Enovid + HVH-2	34	0	—
IV	20	Oestrogens	17	0	—
V	20	Progesterone	20	0	—
VI	20	Enovid	18	0	—
VII	20	HVH-2	20	1	5.0
VIII	20	Control (liquid diet)	17	0	—
IX	20	Control (solid diet)	19	0	—

[1] The mice used were BALB/c females, 2–3 months old.
[2] Number of mice alive 6 months after the initiation of the experiment, when the first tumour was detected by cytology.
[3] 20 mice were treated with the virus one week after hormonal treatment and another 20 mice three weeks after hormonal treatment.

maximum number at the 6th day and a decline ten days after viral inoculation (Figs. 1 & 2). Margination and clumping of the chromatin and a very small number of viral inclusions were also observed.

No deaths occurred after the first inoculation with herpesvirus with previous immunization, but a 10% mortality occurred after the second inoculation without previous immunization.

A malignant tumour was cytologically diagnosed 7 months after the first viral infection in the group treated with oestrogens and herpesvirus (Figs. 3 & 4). Four months later, the tumour was 20 × 15 × 15 mm, surrounding the vagina and vulva and compressing the urinary bladder and rectum. At the autopsy it was possible to establish its origin in the cervix. The vaginal epithelium was atrophic in some areas and moderately hyperplastic in others. The tumour was a well-differentiated squamous-cell carcinoma infiltrating the vaginal walls, bladder, rectum and mammary gland tissue (Fig. 5). No metastasis was observed. The tumour was transplanted subcutaneously into BALB/c female mice for 7 generations and conserved the well-differentiated pattern (Fig. 6). In one case metastasis to the lung was observed. It was also placed in tissue culture; although growth of epithelial cells has been observed, a permanent cell line has not been established yet.

A second tumour appeared in the group treated with genital herpesvirus alone. Atypical cytological changes were detected 9 months after the first inoculation of herpesvirus, and 6 months later a tumour of 15 × 10 × 10 mm was surrounding the vagina and vulva. Smears taken from the tumour showed clusters of very bizarre cells with keratin formation, which led to the diagnosis of squamous-cell carcinoma (see Table), but after the *post mortem* examination it was shown, in fact, to be a mixed tumour containing both squamous- and sebaceous-cell carcinoma.

The sacrifice of the survivor mice has just been started, 18 months after the beginning of the experiment, so that final results are not yet available. Serum from all the mice will be collected. Immunological studies to detect antibodies in the sera of mice bearing transplanted tumours and to detect herpesvirus in these tumours will be carried out.

DISCUSSION

Spontaneous carcinoma of the uterine cervix is very rare in all animals, whether domestic, laboratory or wild (Cotchin, 1964).

Figs. 1 & 2. Giant multinucleated cells. One of them has an intranuclear inclusion (arrow) (Papanicolaou, ×380).

Fig. 3. Malignant epithelial cell with coarse granular pattern of the chromatin (Papanicolaou, ×950).

Fig. 4. Malignant tadpole cell. Note inflammatory reaction (Papanicolaou, ×950).

Fig. 5. Well-differentiated squamous-cell carcinoma of the cervix infiltrating the vaginal wall. Note the vaginal epithelium, which is atrophic in some areas and hyperplastic in others (Haematoxylin-eosin, ×95).

Fig. 6. Generation 3. Transplanted squamous-cell carcinoma conserving the same characteristics as the primary tumour. The overlying skin is normal (Haematoxylin-eosin, ×95).

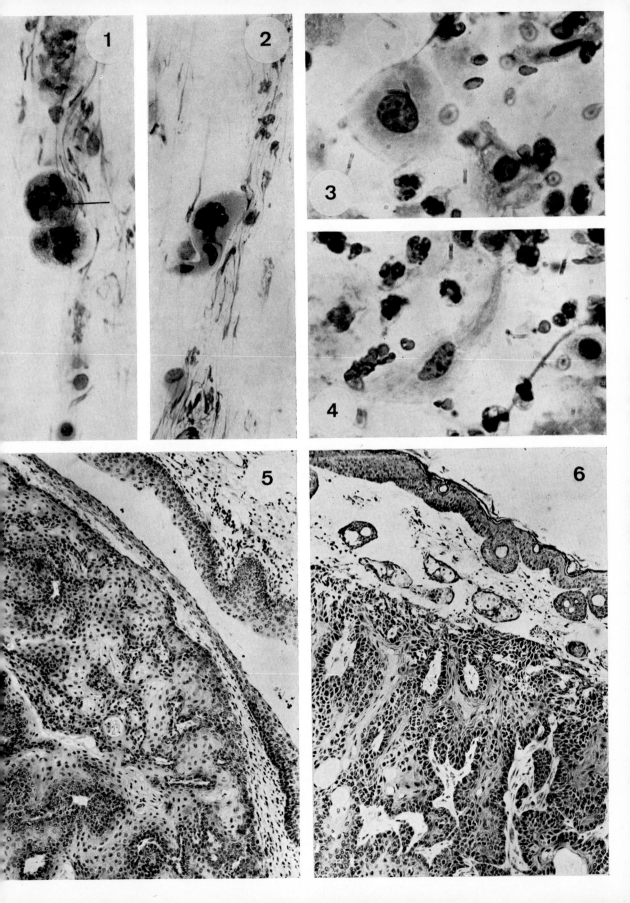

This tumour could be considered as almost non-existent in mice. Only one glandular carcinoma (which may have arisen in the cervix or the vagina) was found among 39 000 mice by Slye, Holmes & Wells, 1923. One stock of mice has been described in which malignant tumours of the uterus and vagina were frequent, but the stock was lost because of infertility (Gardner & Pan, 1948).

Two carcinomas in mice are described in the present study, one a squamous-cell carcinoma of the uterine cervix, and the other a mixed squamous- and sebaceous-cell carcinoma of the vagina. The former arose in the group treated with oestrogens and genital herpesvirus hominis, and the latter in the group treated with herpesvirus alone. Although virological and immunological studies have not been carried out yet to demonstrate the presence of virus or viral antigens in the tumour or in the serum of mice bearing transplanted tumours, these preliminary results suggest a possible oncogenic effect of genital herpesvirus hominis in mice.

SUMMARY

BALB/c mice were treated with different combinations of hormones (oestrogens, progesterone and Enovid) and genital herpesvirus hominis. Characteristic cytological changes of herpesvirus infection were observed in 40% of 140 mice treated with the virus; these had previously been immunized with genital herpesvirus inactivated with ultraviolet light. Two squamous-cell carcinomas of the cervix have been observed: one in the group treated with oestrogens and herpesvirus and the other in the group treated with herpesvirus alone. These preliminary results suggest a possible oncogenic effect of genital herpesvirus hominis in mice.

REFERENCES

Allen, E. & Gardner, W. U. (1941) Cancer of the cervix of the uterus in hybrid mice following long-continued administration of estrogen. *Cancer Res.*, **1**, 359-366

Bertolin, A. (1964) Estrogeni e tumorigenesi epiteliale uterina: aspetti sperimentali. *Attual. Ost. Gin.*, **10**, 994-1022

Christophersen, W. M. & Parker, J. E. (1965) Relation of cervical cancer to early marriage and childbearing. *New Engl. J. Med.*, **273**, 235-239

Cotchin, E. (1964) Spontaneous uterine cancer in animals. *Brit. J. Cancer*, **18**, 209-277

Doll, R., Muir, C. S. & Waterhouse, J., eds, (1970) *Cancer incidence in five continents*, Berlin, Springer-Verlag (*UICC Monograph Series*, No. 2)

Dunn, T. (1969) Cancer of the uterine cervix in mice fed with a liquid diet containing an antifertility drug. *J. nat. Cancer Inst.*, **43**, 671-692

Gardner, W. U. & Pan, S. C. (1948) Malignant tumours of the uterus and vagina in untreated mice of the PM stock. *Cancer Res.*, **8**, 241-256

Martin, C. E. (1967) Epidemiology of cancer of the cervix. Marital and coital factors in cervical cancer. *Amer. J. publ. Hlth*, **57**, 803-814

Melamed, M., Koss, L., Flehinger, B. J., Kelinsky, R. P. & Dubrow, H. (1969) Prevalence rates of uterine cervical carcinoma in situ for women using the diaphragm or contraceptive oral steroids. *Brit. med. J.*, **3**, 195-200

Pereyra, A. J. (1961) The relationship of sexual activity to cervical cancer. Cancer of the cervix in a prison population. *Obstet. and Gynec.*, **17**, 154-159

Slye, M., Holmes, H. F. & Wells, H. G. (1924) Primary spontaneous tumours of the uterus in mice. *J. Cancer Res.*, **8**, 96-118

Terris, M. & Oalman, M. (1960) Cancer of the cervix. An epidemiologic study. *J. Amer. med. Ass.*, **174**, 1847-1851

Terris, M., Wilson, F., Smith, H., Spring, E. & Nelson, J. H. Jr (1967) The relationship of coitus to carcinoma of the cervix. *Amer. J. publ. Hlth*, **57**, 840-847

Towne, J. E. (1955) Carcinoma of the cervix in multiparous and celibate women. *Amer. J. Obstet. Gynec.*, **69**, 606-611

Transformation of Hamster Cells after Infection by Inactivated Herpes Simplex Virus Type 2 *

F. RAPP & R. DUFF

Herpes simplex viruses (HSV) have been extensively investigated in efforts to link these viruses to the aetiology of human cancer. Much of the evidence for the association of HSV type 2 (HSV-2) with human neoplasms has been obtained by epidemiological methods (Naib, Nahmias & Josey, 1966; Naib et al., 1969; Rawls, Tompkins & Melnick, 1969). These results have indicated a possible relationship between HSV-2 infection and cervical carcinoma. In addition, it has recently been reported that exfoliated human cervical tumour cells contain HSV-2 antigens whereas normal cells from the same individual do not (Aurelian, Royston & Davis, 1970). However, the direct demonstration of in vivo oncogenicity or of cell transformation in vitro by HSV-2 has been difficult. When HSV-2 is injected into newborn mice or hamsters, most animals die as a result of the infection and the surviving animals rarely develop tumours (Rapp & Falk, 1964; Nahmias et al., 1970). Cells from the few hamster tumours that have developed did not contain demonstrable HSV-2 antigens and the aetiology of those tumours is therefore obscure. Similarly, the cytopathic effect of the virus has prevented attempts to demonstrate in vitro transformation of cells. For these reasons, HSV-2 was inactivated by ultraviolet irradiation prior to exposure of hamster embryo fibroblasts (HEF).

Primary HEF were infected with HSV-2 that had been inactivated with ultraviolet light for 2, 4, 6, or 8 minutes. Most cultures that were exposed to virus inactivated for 2 or 4 minutes developed typical HSV cytopathic effects and, as a result, no transformed foci developed. No transformation or cytopathic effects were observed when the HSV was

irradiated for 6 minutes. However, transformed foci were observed in cultures that had received HSV-2 previously irradiated for 8 minutes. Cells from one transformed focus were isolated and grown into a cell line. During passage 14, this cell line (333–8–9) was injected into weanling (21 to 28 days old) Syrian hamsters; no tumours developed in these animals. The same cell line was then injected into newborn hamsters during passage 21. At 10 to 18 weeks after injection, tumours developed in 11 of 31 hamsters. Cells from several of these tumours were cultured and re-injected into weanling hamsters. In contrast to the original cell line, these cells were highly oncogenic in weanling hamsters and induced tumours in 100% of the animals injected with 10^6 cells. The latent period was less than 3 weeks.

The tumours induced in hamsters by the original line were of two basic types. The first type was classified as an undifferentiated mesenchymal tumour. The second type was composed of interlacing bundles of fibroblasts. In addition, tumours with areas of both cellular types were observed. Both types of tumours were found to be invasive, and the invasion of tumour cells into muscle, bone, and nerve areas was observed.

Three types of cells were observed in the original in vitro transformed cell line. The first type was classified as essentially undifferentiated (Fig. 1). This cell type was usually predominant and 90% of the cells in a culture were of this type. The second cell type was fibroblastic and the third was a giant cell, which was usually multinucleated.

Cells from the tumours and from the primary cell line were examined by electron microscopy by Dr Ronald Glaser. These cells exhibited morphological damage usually associated with lytic infection by herpesvirus. The chromatin material was marginated at the nuclear membrane and the nucleoli

* From the Department of Microbiology, College of Medicine, The Milton S. Hershey Medical Center, Pennsylvania State University, Hershey, Pennsylvania, USA.

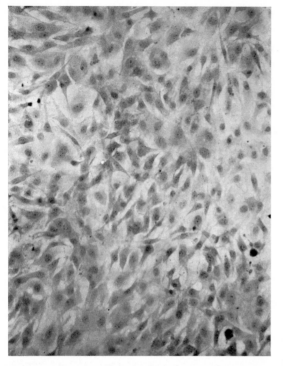

Fig. 1. Photomicrograph of cell culture following transformation of hamster embryo fibroblasts *in vitro* after exposure to irradiated herpes simplex virus type 2. Stained with haematoxylin and eosin. × 95.

were characteristic of herpesvirus-infected cells. Herpesvirus particles were also observed in a few degenerating tumour cells.

The presence of HSV-2 in the transformed cells was also investigated by immunofluorescence techniques. HSV-2 antigens were detected in the cytoplasm of 5% to 20% of the cells transformed *in vitro* (Fig. 2). This antigen has persisted through 50 cell passages. Furthermore, the cells isolated and grown *in vitro* from hamster tumours also contained cytoplasmic HSV antigens in a small percentage of the cells. In addition to the cytoplasmic antigens, some cells contained what appeared to be specific nuclear virus antigens, but the exact relationship of these antigens to the original HSV-2 infection has not yet been determined.

The sera from tumour-bearing animals were also examined for neutralizing antibodies against HSV-2 and HSV type 1 (HSV-1). As shown in Fig. 3, sera from these animals effectively neutralized HSV-2 but not HSV-1. Sera from normal control animals did not neutralize either virus type and an antiserum

prepared by the injection of HSV-2 into hamsters effectively neutralized both HSV-1 and HSV-2.

Attempts to rescue infectious HSV-2 from transformed and tumour cells have been unsuccessful. Perhaps the virus, which was inactivated for 8 minutes before virus-cell interaction, had been rendered so defective that rescue of the virus is not possible.

These results demonstrate the continued presence of the human HSV-2 in a transformed and oncogenic hamster cell line. Whether the virus genome induced the transformation event or is merely a passenger in a spontaneously transformed cell line has not yet been determined. However, the cellular transformation has now been repeated and the resulting cell type is nearly identical to the original transformed cell line. In addition, the newly transformed cell lines also contain HSV-2 antigens. The close similarities of both cell types strongly suggest that the HSV-2 genome played an important rôle in the induction of the transformed state of these cells.

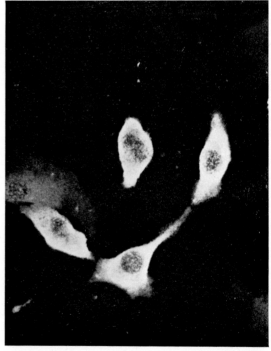

Fig. 2. Immunofluorescence photomicrograph of hamster cells transformed *in vitro* after exposure to irradiated herpes simplex virus type 2 (HSV-2). The cells were exposed to anti-HSV-2 serum prepared in hamsters and to anti-hamster globulin labelled with fluorescein isothiocyanate. × 380.

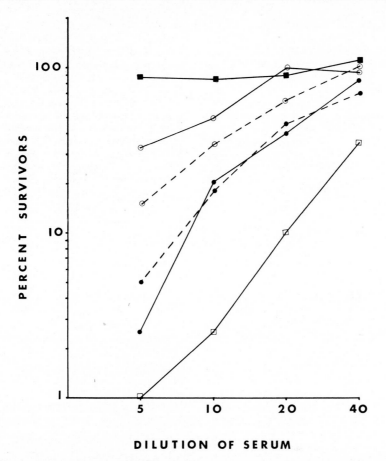

DILUTION OF SERUM

Fig. 3. Neutralization tests with herpes simplex virus type 1 (HSV-1) and herpes simplex type 2 (HSV-2) to detect antibodies in hamsters bearing tumours induced by cells transformed *in vitro* following exposure to irradiated HSV-2. The viruses were reacted with various serum dilutions for 40 minutes at 37°C and the mixtures then tested for plaque-forming ability in rabbit kidney cells. From Duff & Rapp (1971).

——■—— normal hamster serum against HSV-2.

——□—— anti-HSV-2 hamster serum against HSV-2.

——●— - tumour-bearing hamster serum no. 1 against HSV-2.

- -●- - tumour-bearing hamster serum no. 2 against HSV-2.

——O—— tumour-bearing hamster serum no. 1 against HSV-1.

- -O- - tumour-bearing hamster serum no. 2 against HSV-2.

ACKNOWLEDGEMENTS

The research upon which this publication is based was supported by Contract No. NIH-70-2024 from the National Institutes of Health, Department of Health, Education, and Welfare.

REFERENCES

Aurelian, L., Royston, I. & Davis, H. J. (1970) Antibody to genital herpes simplex virus: Association with cervical atypia and carcinoma *in situ*. *J. nat. Cancer Inst.*, **45**, 455-464

Duff, R. & Rapp, F. (1971) Oncogenic transformation of hamster cells after exposure to herpes simplex virus type 2. *Nature (Lond.)*, **233**, 48-50.

Nahmias, A. J., Naib, Z. M., Josey, W. E., Murphy, F. A. & Luce, C. F. (1970) Sarcomas after inoculation of newborn hamsters with *Herpesvirus hominis* type 2 strains. *Proc. Soc. exp. Biol. (N.Y.)*, **134**, 1065–1069.

Naib, Z. M., Nahmias, A. J. & Josey, W. E. (1966) Cytology and histopathology of cervical herpes simplex infection. *Cancer*, **19**, 1026-1031

Naib, Z. M., Nahmias, A. J., Josey, W. E. & Kramer, J. H. (1969) Genital herpetic infection. Association with cervical dysplasia and carcinoma. *Cancer*, **23**, 940-945

Rapp, F. & Falk, L. A. (1964) Study of virulence and tumorigenicity of variants of herpes simplex virus. *Proc. Soc. exp. Biol. (N.Y.)*, **116**, 361-365

Rawls, W. E., Tompkins, W. A. & Melnick, J. L. (1969) The association of herpesvirus type 2 and carcinoma of the uterine cervix. *Amer. J. Epidem.*, **89**, 547-554

Lymphoma Viruses of Monkeys: Herpesvirus Saimiri and Herpesvirus Ateles, the First Oncogenic Herpesviruses of Primates * — A Review

L. V. MELENDEZ, R. D. HUNT, M. D. DANIEL, C. E. O. FRASER, H. H. BARAHONA, F. G. GARCIA & N. W. KING

INTRODUCTION

The purpose of this presentation is: (1) to review briefly the known information on herpesvirus saimiri (HVS), the first herpesvirus of primates with oncogenic capacity (Melendez et al., 1969a; Hunt et al., 1970); (2) to mention new studies done with this agent; (3) to describe briefly the isolation, characterization and in vivo properties of the second lymphoma virus of monkeys: herpesvirus ateles; and (4) to discuss briefly other herpesviruses of animals associated with malignancies: jaagsiekte in sheep (Smith & MacKay, 1969), cotton-tail rabbit lymphoma (Hinze, 1969), and guinea pig leukaemia (Hsiung & Kaplow, 1969).

HERPESVIRUS SAIMIRI: THE FIRST LYMPHOMA VIRUS OF MONKEYS

This virus was isolated in 1968 (Melendez et al., 1968; Melendez et al., 1969b). To date, no similar virus isolate has been obtained by other research groups. The virus in use in a number of laboratories is our original strain S-295C.

HVS was isolated from spontaneously degenerating primary squirrel monkey kidney culture. This virus is carried as a latent agent by the Saimiri species. The incidence of this virus in the wild must be high, since 50–70 % of squirrel monkeys brought directly from their natural environment to our Center have antibodies for this agent.

HVS has been isolated so far only from squirrel monkeys, and mainly from kidney cultures of these animals.

HVS CYTOPATHOGENICITY

HVS multiplies well in cultures derived from the following animal species: owl monkey (kidney, cornea and iris), squirrel monkey (kidney, heart, lung and intestine), and marmoset monkey (kidney). It also grows in BSC-1, Vero, and LLC-MK4 cells, and poorly in whole human embryo, human embryo skin and muscle and human embryonic lung cell cultures (Daniel et al., 1970a). A varying susceptibility of owl monkey kidney (OMK) cultures for this virus has been observed. A titre of 7.5 $TCID_{50}/1.0$ ml was obtained in primary OMK culture, but in a 15th transfer of the same culture a 2 $\log_{10}$ reduction of the viral titre was observed. This was perhaps due to the development of a less susceptible cell population during serial transfer. HVS did not produce cytopathic effects (CPE) in the following cell cultures: whole mouse embryo, mouse kidney, dog kidney, cebus monkey kidney and mouse monocytes (Melendez et al., 1969b).

We have recently found that the virus can multiply in dog foetus lung cell cultures after several blind passages following defined procedures that have led to the attenuation of the virulence of this agent.

EVIDENCE INDICATING THAT HVS BELONGS TO THE HERPESVIRUS GROUP

This agent produced type A intranuclear inclusions in various in vitro cell cultures, as well as polykaryocytes. Its capacity to produce CPE was suppressed

* From the Harvard Medical School, New England Regional Primate Research Center, Southborough, Massachusetts, USA.

or inhibited by heating it at 56°C for a period of 30 minutes. The same result was observed when the agent was treated with bromodeoxyuridine or 20% ether v/v overnight (Melendez *et al.*, 1968).

That this agent is also a DNA virus was indicated by pyrimidine analogue treatment, acridine orange staining and specific studies of its DNA content. Coincidentally, the density of the DNA of this agent was similar to that of human herpesvirus simplex type 2. This latter virus is thought to be involved in human cervical carcinoma (Goodheart, 1970).

Morphological confirmation of the herpes nature of this oncogenic virus has also been provided by electron microscopy studies (Morgan *et al.*, 1970; Melendez *et al.*, 1970a). Typical herpesvirus particles have been observed in sections of infected *in vitro* cultures. Immature hexagonal particles, either empty or with central ring-shaped or dense nucleoids were seen in the nucleus and cytoplasm. These particles were about 108 nm in diameter. The mature particles with an additional enveloping membrane measured 140 nm across.

HVS PLAQUE DEVELOPMENT

Several materials, namely methylcellulose, agar and starch have been employed to overlay a number of different infected cell cultures used in plaque assays (Daniel *et al.*, 1971). These cultures were: squirrel monkey heart (SMH), squirrel monkey lung (SML), squirrel monkey intestine (SMI) and marmoset monkey kidney (MMK). The most sensitive culture for plaquing studies was found to be SMH, followed by SML, SMI and MMK. The highest number of plaques were found with methylcellulose, but for practical working purposes the most suitable medium was that prepared with agar.

Plaque cloning of this agent consistently produced populations of heterogenous plaque size, and since no selection was made for antigenicity, density and pathogenicity, it was not possible to determine whether the plaquing procedures employed could be useful in obtaining viral variants.

MULTIPLICATION AND CYTOPATHOGENICITY OF HVS AND ITS LIKELY VARIANT SQUIRREL MONKEY HEART ISOLATE IN HUMAN CELL CULTURES

The virulence of HVS and its variant squirrel monkey heart isolate (SMHI) were tested in human cell cultures (Daniel *et al.*, 1970b).

SMHI produced in whole human embryonic cell cultures an ill-defined cell layer alteration 12 days post inoculation. In stained preparations intranuclear inclusions were seen. Although, later, no CPE were observed, viable virus was recovered up to 112 days. Some of these cultures have been transferred 23 times within one year. Electron microscopic examination of a sixth transfer showed no viral particles. Co-cultivation with susceptible cells also failed to reveal viable virus.

Human embryonic lung cell cultures inoculated with HVS showed the presence of CPE 19 days after inoculation and viable virus was recovered up to 45 days. These cultures have been transferred 15 times in one year. The virus failed to multiply in human embryo skin and muscle.

These results indicate that these oncogenic viruses are capable of infecting human cell cultures. However, it is not clear what is the fate of the virus in the infected transferred cultures. Perhaps the virus is integrated into the cell genome.

HVS ANTIGENICITY

To assess the behaviour of HVS in its natural host, the squirrel monkey, frozen, heated and fresh virus was inoculated into this animal species. No clinical evidence of disease was observed, even though multiple inoculation and large doses of virus were employed. The procedures followed in this inoculation have been described in detail elsewhere (Melendez *et al.*, 1969b). It was remarkable to observe that the level of antibodies developed was never higher than that found in noninoculated animals. In these latter animals the neutralization index (NI)[1] of the antisera is about 2.0–2.5.

The inability to obtain antibodies of high titre made it difficult to characterize the several indigenous agents that behaved similarly to HSV. Hamster, mouse and rabbit were also inoculated to prepare antibodies. The first two animal species developed neither antibodies nor disease after HSV inoculation, while the rabbits developed a neoplastic disease but no antibodies (Daniel *et al.*, 1970b).

The search for a suitable species to prepare specific high titred antibodies against HSV ended when it was found that the goat is an excellent animal for this purpose (Fraser *et al.*, 1971).

Both fluorescent and neutralizing antibodies developed in the inoculated goats, and the antibody

---

[1] NI = $\log_{10}$ (titre of control serum) — $\log_{10}$ (titre of test serum).

titres were much highert han those developed in the other species tested. It was also found that the fluorescent antibodies were not identical to the neutralizing antibodies, since significant levels of neutralizing antibodies were present in the absence of fluorescent antibodies.

Synthetic polynucleotides (Poly I : C), inoculated together with the virus seem to affect the two types of antibodies differently. They consistently suppressed the production of fluorescent antibodies, and may suppress the production of neutralizing antibodies initially but no suppressive effect was observed after the booster inoculations.

SEROLOGICAL DIFFERENTIATION OF HSV

The CPE produced by this agent in cell cultures were not neutralized by antisera to the following herpesviruses: herpes simplex, herpesvirus B, herpesvirus T, the virus of infectious bovine rhinotracheitis, herpesvirus suis, sand rat nuclear inclusion agent, herpesvirus aotus (Melendez et al., 1970b), herpesvirus ateles,[1] herpesvirus saguinus (Melendez et al., 1970b), AT-46,[1] spider monkey herpesvirus,[1] and ground squirrel agent.[2] This fact provided evidence that HSV is a distinct member of the herpesvirus group. In our hands serum neutralization, expressed in terms of the NI, has been the most specific serological procedure for differentiation between these herpesviruses.

HSV DISEASE IN NONHUMAN PRIMATES
AND IN RABBITS

Cotton-top marmosets

Our first studies with these marmosets *(Saguinus oedipus)* indicated that of 23 animals inoculated intramuscularly with various virus dilutions (undiluted to 10^{-3}) of an inoculum having a titre of $10^{5.5}$ $TCID_{50}/1.0$ ml in OMK cultures, all succumbed to malignant lymphoma within 18 to 48 days.

The lymphoma that developed in this animal species was characterized by marked enlargement of the organs of the reticuloendothelial system (liver, lymph nodes, spleen, and thymus). This enlargement was due to an invasion and replacement of the normal cyto-architecture of the affected tissues by reticulum cells. The invasion of these neo-elements into other tissues such as kidney, lung, testes and choroid plexus, gave this disease the features of a neoplastic disease (Hunt et al., 1970).

These findings have been confirmed recently by other investigators employing HSV strain S-295C. Inoculated animals never survived longer than 35 days. White-lipped tamarins were also inoculated, but were found to be less susceptible (Falk, Wolfe & Deinhardt, 1971).

We have also demonstrated that HSV can be recovered from cell cultures prepared from kidney tissues affected with malignant lymphoma (Melendez et al., 1970a). This was also confirmed, one year later, by another group working independently (Falk et al., 1971).

Owl monkeys

Our earlier studies (Melendez et al., 1969a; Hunt et al., 1970a) reported that of 10 owl monkeys *(Aotus trivirgatus)* inoculated with HSV, all developed malignant lymphoma within 29 days. Though peripheral blood studies were not done, the histopathological features were similar to those in marmosets (Melendez et al., 1970a). The animals employed in these early studies were inoculated by the intramuscular route.

Recently we have determined that the *Aotus* species inoculated with approximately 3.16×10^5 $TCID_{50}$ by various routes can develop malignant lymphoma and leukaemia within 11 to 36 weeks. We have demonstrated that lymphoma can be induced by intravenous, subcutaneous and intradermal inoculation of HSV, as well as by the intramuscular route. We have also demonstrated that lymphocytic leukaemia develops in this species in association with malignant lymphoma (Melendez et al., 1971).

The incidence of the neoplasm in this latter study was 50% (six of twelve animals), as compared with 100% in owl monkeys inoculated in earlier studies. What could be the explanation for this lower incidence? Perhaps variation in the original HSV strain, but thus far we have no *in vitro* evidence for this assumption. A more likely explanation is that monkeys are a poorly-defined laboratory animal.

Cinnamon ringtail monkeys

Four monkeys *(Cebus albifrons)* were inoculated intramuscularly with approximately 1.58 $\times 10^6$ $TCID_{50}/1.0$ ml of HSV. All these animals died between 18 and 20 days after inoculation with

---

[1] Unpublished data.
[2] This is a new herpesvirus isolated by Dr S. Katz from ground squirrel kidney culture; detailed information on this agent will be published elsewhere.

a disease very similar to that developed in the *Aotus* and *Saguinus* species. Each of the four animals presented an extensive infiltration of reticulum cells in several organs (lymph node, thymus, spleen, pancreas, liver, kidney and lung). The infiltrate closely resembled reticulum cell sarcoma or Hodgkin's sarcoma (Melendez *et al.*, 1970c). The disease in this animal species was predominantly characterized by a sharply delineated collection of reticulum cells associated with reticular fibres. Lesser numbers of eosinophils and lymphocytes were also present. The disease can be best considered to be a peculiar reticuloproliferative disorder resembling reticulum cell sarcoma or Hodgkin's sarcoma.

African green monkeys

Six monkeys *(Cercopithecus aetiops)* were inoculated as described for *Cebus albifrons*. Two of the six monkeys died approximately 40 days later, and the other 4 animals were sacrificed at the end of five months. The morphological picture of the

disease observed in the two dead animals closely resembled that described for the Cebus monkey (Melendez *et al.*, 1970b).

Black spider monkey

Four spider monkeys *(Ateles geofroyii)* were inoculated intramuscularly (i.m.) and two subcutaneously (s.c.) each with approximately 3.16 × 10^5 TCID$_{50}$. One animal inoculated s.c. (No. 2623) died within 77 days with a condition termed lymphocytosis. Another (No. 2691) inoculated i.m. died on day 129 with typical lesions of malignant lymphoma. Most lymph nodes in this animal were entirely replaced by lymphoblasts with obliteration of any normal cortical or medullary architecture (Figs. 1 & 2).

Two animals were killed 179 days after i.m. inoculation, and 27 days after a second inoculation. In one animal (No. 2764), gross and microscopic lesions of malignant lymphoma were observed. There were solid, grey tissue masses surrounding the oesophagus,

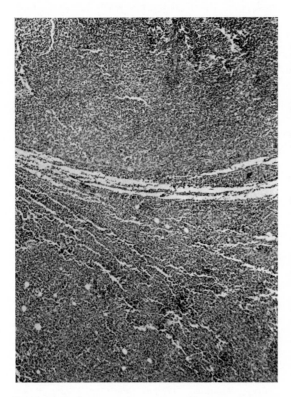

Fig. 1. Lymphoma produced by herpesvirus saimiri in spider monkey lymph node. The entire node as well as surrounding tumour are replaced by a diffuse sheet of lymphoblasts. ×30.

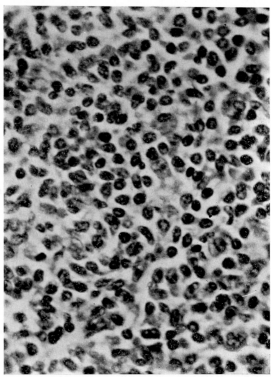

Fig. 2. Lymphoma produced by herpesvirus saimiri in spider monkey lymph node. A higher magnification than Fig. 1 to illustrate the uniform character of the neoplastic lymphoblasts. ×350.

adjacent to and invading the lesser and greater curvature of the stomach, within the pancreas and mysentery, in the myocardium and diaphragm and in the wall of the duodenum and jejunum (Fig. 3). These solid tumour masses were composed of dense sheets of small to large lymphoblasts which invaded and replaced adjacent structures (Fig. 4). A detailed description of this disease in the *Ateles* species will be presented elsewhere.

The pattern of malignant lymphoma induced by HSV in this species is comparable to that seen in owl and marmoset monkeys.

The spider monkey is the first animal species to develop solid tumours after HSV inoculation.

Macaca monkeys

Weaned monkeys. Two stumptail monkeys *(Macaca arctoides)* and two rhesus monkeys *(M. mulatta)* were inoculated i.m. No sign of disease was observed during a 5-month period, at the end of which time the animals were sacrificed.

Newborn monkeys. Nine rhesus, two bonnet and five cynomolgus monkeys were inoculated i.m. each with approximately 3.16×10^5 TCID$_{50}$. No detectable clinical sign of disease has been observed during one year. These animals will be observed for a total of 16 months, at which time the animals will be sacrificed.

Baboon

No signs of disease have been observed in animals inoculated i.m. with HVS during a one-year period.

Chimpanzee

No detectable clinical symptoms have been observed in animals inoculated i.m. with HVS during a one-year period.

Rabbits

HVS and its variant, the squirrel monkey heart isolate (SMHI), were inoculated in 4-week and 4-month-old male albino rabbits for antibody preparation. Equal parts of HVS and Freund's adjuvant (FA) were inoculated subcutaneously followed by virus only in the 4-month-old rabbits. The 4-week-old rabbits received one injection of live virus intravenously. These animals became ill and died within 17 to 30 days. Histopathological studies revealed a reticuloproliferative lesion in the liver, kidney, thymus, lymph nodes, choroid plexus, spleen and adrenal. SMHI was inoculated into 4-month-old rabbits. Group 1 received live virus followed by virus plus FA, group 2, virus plus FA, followed by live virus. In all instances the animals became ill and died or were sacrificed within 42 to 75 days. Malignant lymphoma of the lymphocytic type was

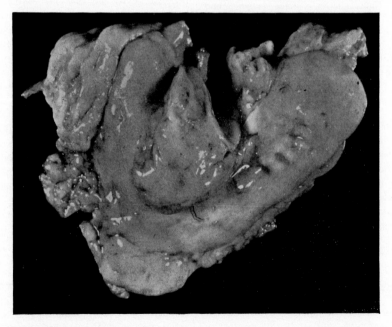

Fig. 3. Lymphoma produced by herpesvirus saimiri in spider monkey stomach. Note tumour masses at the greater and lesser curvature of the stomach.

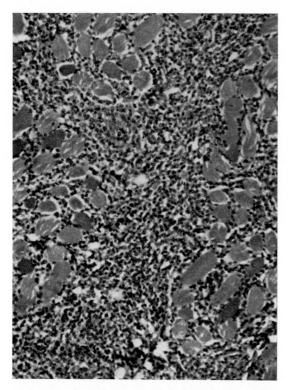

Fig. 4. Lymphoma produced by herpesvirus saimiri in spider monkey. Neoplastic cells have invaded and replaced the skeletal muscle of the diaphragm.   ×75.

seen in rabbits from each group. The cells were identified as lymphocytes by electron microscopy (Daniel *et al.*, 1970b).

HVS, INTERFERON AND SYNTHETIC POLYNUCLEOTIDES

Several studies have demonstrated that interferon and interferon inducers can partially inhibit the replication of some herpesviruses. It has also been reported that they can delay the development of leukaemia and the growth of some tumours in animals. These studies led us to investigate the use of interferon and interferon inducers on the replication and oncogenic capacity of HVS (Barahona & Melendez, 1971).

We have demonstrated that tissue culture monolayers inoculated with HVS can be protected when the medium containing interferon is added after inoculation, that microgram amounts of Poly I : C induced effective resistance against a subsequent challenge with the virus, and that this protective action of Poly I : C can be enhanced with DEAE-

dextran. These studies suggest possibilities for increasing the resistance of susceptible non-human primates to the malignancy produced by HVS.

HERPESVIRUS ATELES, STRAIN 810: THE SECOND MONKEY LYMPHOMA VIRUS

Herpesvirus ateles (HVA) is one of three different herpesviruses isolated from black spider monkeys.

The first known spider monkey herpesvirus (SMHV) was that isolated by Lennette and characterized by Hull. The other two are Guatemala isolate AT-46 and HVA. A detailed description of each of these agents will be presented elsewhere, and they are mentioned here only to provide points of comparison with HVA.

HVA ISOLATION

A primary kidney culture was prepared on 17 June 1970 from the renal cortex of black spider monkey No. 810-69, housed in our laboratories (Southborough) for at least one year. The procedures for culturing this kidney tissue were similar to those employed for Guatemala isolate AT-46. A well-grown cell layer developed in 6 to 8 days. This cell layer developed a large number of bizarre cells and polykaryocytes on day 11. Most of these polykaryocytes shed from the surface of the plastic flasks. On day 16 the cell layer was scraped and together with the culture fluids was collected as isolate 810 and stored at 4°C, —86°C and at —176°C.

Isolate 810, hereafter referred to as HVA, was inoculated undiluted into squirrel monkey foetus lung (SMFL) monolayers. This virus stimulated the development of large spindle-shaped cells in greater numbers than did isolate AT-46. However, in the OMK cell line, the CPE were characterized by scattered foci of swollen and rounded cells. In both cultures the development of intranuclear inclusions was observed in preparations stained with haematoxylin and eosin (HE), but these inclusions were better defined in OMK cell cultures.

CYTOPATHOGENICITY OF HVA

HVA strain 810 reached the following titres ($TCID_{50}$/ml): $10^{4.5}$ in OMK, 10^4 in SMFL, $10^{3.5}$ in HEL and $10^{5.0}$ in RKL 17, 14, 23 and 30 days respectively after inoculation. The type of CPE produced in SMFL varied depending upon the number of passages of the virus in OMK or RKL cell cultures. After three passages in OMK cultures, the virus caused the development of larger nuclear inclusions, more prominent polykaryocytes and was also more virulent for SMFL than the virus passed in RKL continuous cultures.

The following *in vitro* cell cultures were inoculated with Guatemala isolate AT-46, HVA, and SMHV: whole human embryo, human embryonic lung, hamster heart, goat synovial bursa, goat bursa capsule, OMK 210, SMFL and RKL continuous cultures (whole human embryo and goat cultures were not tested with SMHV). These herpesviruses produced CPE in all these cultures, and differed only in the titres reached and in the time it took to develop CPE. Their virulence decreased in the order SMHV to HVA to AT-46. These results are summarized in Table 1. The same range of virulence was observed when these three herpesviruses were inoculated into spider monkey kidney cell cultures (Table 2).

CHARACTERIZATION OF HVA IN THE HERPESVIRUS GROUP

Physicochemical treatments

HVA was treated by: heat at 56°C for 30 minutes, overnight treatment with 20% ether at 4°C, and filtration by 220 nm and 100 nm filters (Millipore). These treatments followed standard procedures.

Table 1. Cytopathogenicity of spider monkey herpesviruses

Cell cultures	Viruses		
	HVA	AT-46	SMHV
Whole human embryo	+	+	N.T.[1]
Hamster heart	+	+	+
Goat synovial bursa	+	+	N.T.[1]
Goat bursa capsule	+	+	N.T.[1]
Owl monkey kidney 210	+	+	+
Squirrel monkey foetus lung	+	+	+
Rabbit kidney	+	+	+
Human embryonic lung	+	+	+

[1] Not tested.

Table 2. Comparative titrations of spider monkey herpesviruses in homologous kidney cultures

SMHV	HVA	AT-46
6.0[1]	4.0	1.5

[1] Titre: log 10/1.0 ml.

The capacity of this virus to produce CPE in various *in vitro* cultures was destroyed after it had been treated by heat and ether. CPE were observed only in cultures inoculated with 220-nm filtrates, but not in those inoculated with 100-nm filtrates. These results are summarized in Table 3.

Table 3. Spider monkey herpesviruses: cytopathogenicity in RKL[1] after physicochemical treatments

Treatments	Viruses		
	SMHV	HVA	AT-46
Heat	−	−	−
Ether	−	−	−
Filtration[2]	+	+	+
Filtration[3]	−	−	−

[1] Rabbit kidney continuous culture.
[2] Through 220 nm filter.
[3] Through 100 nm filter.

Electron microscopy studies

HVA met all the ultrastructural requirements for placement in the herpesvirus group (Fig. 5). A detailed description of these ultrastructural studies will be presented elsewhere.

HVA plaque development

HVA was inoculated into the following tissue culture lines: rabbit kidney (RKL), OMK, hamster heart (HH) and squirrel monkey foetus lung (SMFL). This agent developed plaques in SMFL and HH cultures within 30 to 50 days post inoculation. A detailed description of this work will be presented elsewhere.

HVA antiserum preparation

Goats were inoculated with SMHV, HVA and AT-46 spider monkey herpesviruses, and New Zealand white rabbits with HVA and AT-46.

The immunization procedures for each animal group as well as the results obtained will be described elsewhere.

SMHV produced good levels of antibodies in goats. HVA developed antibodies only in rabbits, and AT-46 produced antibodies in both species of animals. The antisera prepared against these viruses neutralized the homologous virus only. These cross-neutralization tests indicated that the three spider monkey herpesviruses differed from each other. These data are summarized in Table 4. In addition,

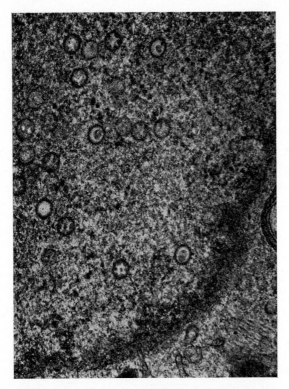

Fig. 5. Herpesvirus ateles inoculated into owl monkey kidney cell cultures. Thin section. Note intranuclear viral particles. Morphological types measuring 101 to 107 nm: empty capsids, capsids with double-shelled cores, capsids with single dense core and some bizarre forms with core material distributed in a clock-face pattern inside capsid membrane. ×48 350.

Table 4. Reciprocal neutralization indices
of spider monkey herpesviruses

Viruses	Antisera		
	HVA	AT-46	SMHV
HVA	3.5	0.0	0.0
AT-46	0.0	2.5	0.0
SMHV	0.0	0.0	4.0

the CPE of HVA were not neutralized by the following antisera: herpesvirus simplex, herpesvirus T, the virus of infectious bovine rhinotracheitis, sand rat nuclear inclusion agent, herpesvirus suis, ground squirrel agent, herpesvirus saimiri, spider monkey herpesvirus, and herpesvirus aotus. These data support the conclusion that HVA is another new member of the herpesvirus group.

INOCULATION OF NON-HUMAN PRIMATES WITH HVA

Two owl monkeys and three cotton-top marmosets were inoculated i.m. with approximately 1.6 × 10^3 $TCID_{50}$ of HVA. Blood was collected weekly. Two of the marmosets died 28 days after inoculation and one was sacrificed in a moribund condition 40 days after inoculation. All the marmosets had malignant lymphoma of a marked similarity to that produced by HVS. Grossly, there was generalized enlargement of lymph nodes, splenomegaly and a reticular pattern in the capsular and cut surfaces of the liver. Microscopically, a cellular infiltrate was seen in tissue sections of spleen, lymph node, liver, kidney, myocardium, adrenal, pancreas, thyroid, and choroid plexus of the lateral ventricles (Fig. 6). The infiltrating cells had round to oval nuclei and prominent nucleoli surrounded by a small to moderate amount of lightly eosinophilic cytoplasm (Fig. 7). The infiltrate invaded and replaced pre-existing structures in all affected organs. The sinuses of lymph nodes were obscured and the infiltrate invaded through the capsules and entered perinodal tissues.

Peripheral blood changes were noted in the last weekly samples taken 1 day before death or sacrifice. Total white blood cell counts were 10 700 (No. 505) 11 700 (No. 497), and 27 200 (No. 501), with 29%, 39% and 51% lymphocytes and 4%, 11%, and 31% lymphoblasts respectively.

HERPESVIRUSES AND MALIGNANCIES IN SHEEP,
RABBITS AND GUINEA PIGS

Sheep pulmonary adenomatosis or jaagsiekte

Evidence has been provided (Smith & Mackay, 1969) that a herpesvirus is present in cultures of alveolar macrophages obtained from the lungs of sheep suffering from pulmonary adenomatosis. Typical herpesvirus particles were observed in thin sections of macrophage cultures. However, so far no direct evidence has been provided to establish this herpesvirus isolate as the aetiological agent of adenomatosis in sheep.

Cotton-tail rabbit lymphoma

A new herpesvirus named herpesvirus sylvilagus was isolated by Hinze from cell cultures prepared with pooled kidney tissue of three weanling cotton-tail rabbits *(Sylvilagus floridanus)* (Hinze, 1969). This herpesvirus produced a lymphoma in six to eight weeks in young cotton-tail rabbits, with in-

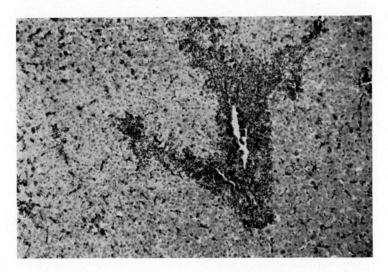

Fig. 6. Lymphoma produced by herpesvirus ateles in cotton-top marmoset monkey liver. Note thick collars of invading cells in the periportal tissues and numerous cells in hepatic sinusoids. ×75.

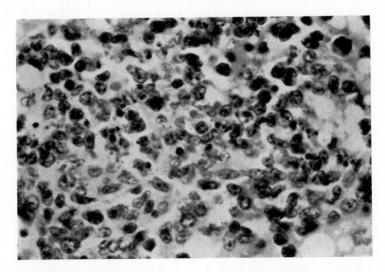

Fig. 7. Lymphoma produced by herpesvirus ateles in cotton-top marmoset monkey liver. The invading cells are lymphoblasts with large leptochromatic nuclei, prominent nucleoli and a moderate amount of cytoplasm. ×350.

vasion of immature lymphoid cells into various organs. The disease has been transmitted by cell-free virus. New Zealand white rabbits *(Oryctolagus cuniculus)* were inoculated with this virus, but were not susceptible to it. So far, only cotton-tail rabbits have been susceptible.

Herpesviruses associated with guinea-pig leukaemia

Hsiung & Kaplow (1969) isolated a herpesvirus from a leukaemia-susceptible guinea-pig *(Cavia cobaya)* strain. This virus has not been isolated from other guinea-pig strains not susceptible to leukaemia. The virus grew in rabbit kidney culture and guinea-pig kidney. So far this virus has not been established as the aetiological agent of guinea-pig leukaemia.

CONCLUSIONS

An important finding from these studies was the discovery of a second lymphoma virus of monkeys:

herpesvirus ateles. This new herpesvirus proved to be oncogenic in marmoset monkeys, which developed malignant lymphoma with terminal leukaemia after virus inoculation.

Herpesvirus saimiri, the first oncogenic herpesvirus of primates, was shown to induce malignant lymphoma and leukaemia in several non-human primate species and in rabbits. The recent demonstration that the spider monkey develops malignant lymphoma after inoculation with the virus increases the list of susceptible primates to five: marmoset, owl, and spider monkey develop malignant lymphoma, and Cebus and African green monkeys develop reticuloproliferative diseases. This striking oncogenic range in non-human primates has not been demonstrated so far for any other leukaemia virus.

This is of great importance since Epstein-Barr virus (EBV), another herpesvirus isolated from human tissues, is considered to be the likely aetiological agent of Burkitt's lymphoma, although further studies are required to prove this. The fact that two herpesviruses, namely HVS and HVA, can induce oncogenic changes in non-human primates provides primate model systems for studying malignant lymphoma and leukaemia.

The oncogenicity of these two viruses in primates suggest that great caution should be used when working with these agents as well as with those species that are their natural reservoirs.

These studies clearly establish two main points: herpesviruses are the aetiological agents of malignant lymphoma in primates and rabbits, and South American monkeys are valuable animal models for studying viral oncogenesis in primates. The use of these primate models may prove of great value in understanding similar disease processes in man.

ACKNOWLEDGEMENTS

This work was supported by the National Institutes of Health, USPHS Grant No. 00168-09.

REFERENCES

Barahona, H. H. & Melendez, L. V. (1971) *Herpesvirus saimiri. In-vitro* sensitivity to virus induced interferon and to polyriboinosinic acid: polyribocytidylic acid. *Proc. Soc. exp. Biol. (N.Y.)*, **136**, 1163-1167

Daniel, M. D., Melendez, L. V., Hunt, R. D. & Trum, B. F. (1970a) The herpes virus group. In: Fiennes, R. N., ed., *Primate pathology*, Basel, Karger (in press)

Daniel, M. D., Melendez, L. V., Hunt, R. D., King, N. W. & Williamson, M. E. (1970b) Malignant lymphoma induced in rabbits by *Herpesvirus saimiri* strains. *Bact. Proc.*, 195

Daniel, M. D., Rabin, H., Barahona, H. H. & Melendez, L. V. (1971) *Herpesvirus saimiri* III. Studies on plaque formation under multi agar overlays, starch and methycellulose in various primate cell cultures. *Proc. Soc. exp. Biol. (N.Y.)*, **136**, 1192-1196

Falk, L. A., Wolfe, L. G. & Deinhardt, F. (1971) Oncogenesis of *Herpes saimiri* in marmosets. *Fedn. Proc. Fedn. Am. socs exp. Biol.*, Abstract No. 1781

Fraser, C. E. O., Melendez, L. V., Barahona, H. H. & Daniel, M. D. (1971) The effect of Poly I: C on fluorescent and neutralizing antibodies to *Herpesvirus saimiri* in goats. *Int. J. Cancer*, **7**, 397-402

Goodheart, C. R. (1970) Herpesviruses and cancer. *J. Amer. med. Assoc.*, **211**, 91-96

Hinze, H. C. (1969) Rabbit lymphoma induced by a new herpesvirus. *Bact. Proc.*, **149**, 157

Hsiung, G. D. & Kaplow, L. S. (1969) Herpeslike virus isolated from spontaneously degenerated tissue culture derived from leukemia-susceptible guinea pigs. *J. Virol.*, **3**, 335-357

Hunt, R. D., Melendez, L. V., King, N. W., Gilmore, C. E., Daniel, M. D., Williamson, M. E. & Jones, T. C. (1970) Morphology of a disease with features of malignant lymphoma in marmosets and owl monkeys inoculated with *Herpesvirus saimiri. J. nat. Cancer Inst.*, **44**, 447-465

Melendez, L. V., Daniel, M. D., Hunt, R. D. & Garcia, F. G. (1968) An apparently new Herpesvirus from primary kidney cultures of the squirrel monkey *(Saimiri sciureus)*. *Lab. Anim. Care*, **18**, 374-381

Melendez, L. V., Hunt, R. D., Daniel, M. D., Garcia, F. G. & Fraser C. E. O. (1969a) *Herpes saimiri*. II. An experimentally induced malignant lymphoma in primates. *Lab. Anim. Care*, **19**, 378-386

Melendez, L. V., Daniel, M. D., Garcia, F. G., Fraser, C. E. O., Hunt, R. D. & King, N. W. (1969b) *Herpes saimiri*. I. Further characterization studies of a new virus from the squirrel monkey. *Lab. Anim. Care*, **19**, 372-377

Melendez, L. V., Daniel, M. D., Hunt, R. D., Fraser, C. E. O., Garcia, F. G., King, N. W. & Williamson, M. E. (1970a) *Herpesvirus saimiri*. V. Further evidence to consider this virus as the etiological agent of malignant lymphoma in primates. *J. nat. Cancer Inst.*, **44**, 1175-1181

Melendez, L. V., Hunt, R. D., Daniel, M. D., & Trum, B. F. New world monkeys, herpesviruses and cancer (1970b) In: Balner, H. & Beveridge, W. J. B., eds, *Infections and immunosuppressions in sub-human primates*, Copenhagen, Munksgaard, pp. 111-117

Melendez, L. V., Hunt, R. D., Daniel, M. D., Fraser, C. E. O., Garcia, F. G. & Williamson, M. E. (1970c) Lethal reticuloproliferative disease induced in *Cebus albifrons* monkeys by *Herpesvirus saimiri*. *Int. J. Cancer*, **6**, 431-435

Melendez, L. V., Hunt, R. D., Daniel, M. D., Blake, J. B. & Garcia, F. G. (1971) Acute lymphocytic leukemia in owl monkeys inoculated with *Herpesvirus saimiri*. *Science*, **171**, 1161-1163

Morgan, D. G., Epstein, M. A., Achong, B. G. & Melendez, L. V. (1970) Morphological confirmation of the Herpes nature of a carcinogenic virus of primates *(Herpes saimiri)*. *Nature (Lond.)*, **228**, 170-172

Smith, W. & Mackay, J. M. K. (1969) Morphological observations on a virus associated with sheep pulmonary adenomatosis (Jaagsiekte). *J. comp. Path.*, **79**, 421-424

Structural Differentiation in the Nucleoid of a Herpesvirus (Herpesvirus Saimiri)

D. G. MORGAN [1] & M. A. EPSTEIN [1]

It has been known for more than a decade that the immature unenveloped particles of herpesviruses seem in thin sections either to contain central ring-shaped or dense spherical nucleoids or to be "empty," lacking such structures and appearing therefore to have uniformly electron lucent centres (Morgan *et al.*, 1959; Epstein, 1962). The form with the dense nucleoid is probably the ultimate stage in the development of the immature particle because only this form undergoes maturation by budding to produce the mature enveloped virus (Falke, Siegert & Vogell, 1959; Epstein, 1962a) and in such enveloped particles the DNA lies within the dense nucleoid (Epstein, 1962b). As regards the other forms, it has been suggested that "empty" particles and those with a ring-shaped nucleoid are either intermediate developmental stages (Roizman, 1969), or particles whose assembly is defective (Smith & Rasmussen, 1963). In any event, these two morphological variants must have some break in the continuity of their capsids since in whole-mount negative-contrast preparations they are readily penetrated by electron-opaque reagents such as phosphotungstic acid (PTA) (Wildy, Russell & Horne, 1960; Watson, Russell & Wildy, 1963; Toplin & Schidlovsky, 1966; Hummeler, Henle & Henle, 1966; Epstein *et al.*, 1968; Mizell, Toplin & Isaacs, 1969); on the other hand, the inner morphology of particles with a central dense nucleoid has not been visualized unequivocally in whole-mount preparations, indicating that their capsids form a more resistant barrier to such penetration.

In three dimensions, all forms of immature herpes particles consist of a capsid shell surrounding an inner volume that some workers (Wildy *et al.*, 1960)

have designated the "core". Particles with ring-shaped nucleoids presumably contain a hollow spherical structure at the centre of the core which appears as the ring when sectioned or filled by PTA. Particles with dense nucleoids must possess a solid spherical central structure since in thin sections the nucleoid appears uniformly electron-opaque; in both cases, thin sections show the nucleoid to be surrounded by electron-lucent material lying below the bases of the capsomeres.

Biochemical analyses have demonstrated at least six different proteins in the immature particles of several herpesviruses (Roizman, 1969; Olshevsky & Becker, 1970; Kaplan & Ben-Porat, 1970; Robinson & Watson, 1971), and it seems likely that some of these components must be located inside the outer protein capsid in association with the nucleoid and its surroundings.

In recent experiments with a carcinogenic herpesvirus of monkeys (herpesvirus saimiri) (Melendez *et al.*, 1969; Morgan, Epstein & Achong, 1970) it was noted that this particular virus permitted new and unusually comprehensive details of fine structure to be visualized. These findings seem of relevance to the wider problem of herpesvirus structure in general and the present paper reports the observations that have been made.

For the experiments, sufficient virus was inoculated into confluent owl monkey kidney cell cultures to give moderate widespread cytopathic change after incubation for 12 days at 37°C. Infected cells were harvested by gently scraping from the glass into suspension in 1 ml of culture medium and were then either made into whole-mount negative-contrast preparations in PTA by the method of Parsons (1963) or were fixed in glutaraldehyde followed by osmium, dehydrated, and embedded in epoxy resin for thin sectioning by methods described elsewhere (Epstein

---

[1] Department of Pathology, Medical School, University of Bristol, UK.

& Achong, 1965). The material was examined in a Philips EM 300 electron microscope.

In thin sections, immature particles with or without nucleoids were clearly seen to be bounded by a regular array of radially arranged rod-shaped capsomeres (Figs. 1 & 2), corresponding to the structure that in most sectioned preparations has presented an appearance described hitherto as a "membrane". It should be noted, however, that with permanganate fixation and embedding in a water soluble epoxy resin, such a "membrane" could not be visualized (Epstein, 1962b). It is of interest that the capsomeres were also visible in thin sections of mature enveloped particles (Fig. 3).

After treatment with PTA in negative-contrast preparations, "empty" immature particles showed the usual appearance for such forms of herpesviruses, with the electron-opaque reagent filling the entire inner volume within the bases of the capsomeres (Fig. 4). Similar filling was observed with immature particles containing a central ring-shaped nucleoid, where, in addition, the PTA also penetrated into the centre of the ring showing that such nucleoids must represent a hollow sphere (Fig. 5).

Besides these familiar findings, many immature particles with dense nucleoids revealed new features in this type of preparation. Unexpectedly, such particles were also often penetrated by PTA, which then likewise filled the subcapsomeric zone, thus surrounding and outlining the central nucleoid (Fig. 6). In such particles the nucleoid itself seemed to be covered by regularly arranged subunits divided into segments by geometrical planes of symmetry (Fig. 7). Groups of immature particles showing examples of all these structural forms in close proximity were sometimes observed (Fig. 8).

Mature enveloped particles were seen only with damaged membranes that allowed PTA to penetrate and reveal the capsid within (Fig. 10). Where preparations were made with PTA on the alkaline side of neutral (up to pH 7.5), the proportion of mature enveloped particles was considerably increased, suggesting that these conditions helped to preserve recognisable fragments of the outer envelope. The relative frequency with which other morphological forms of the virus were found, including some that remained permeable to PTA (Fig. 9), was not affected by using PTA at various pH-values over the range 6 to 7.5. Preservation of the outer envelope by alkaline PTA has been reported previously in the case of herpes simplex virus (Wildy et al., 1960).

The clear resolution of capsomeres in thin sections of herpes saimiri (Fig. 1, 2 & 3) has allowed measurements to be made on this agent in such preparations, whilst the penetration by PTA of particles with dense central nucleoids (Figs. 6, 7 & 8) has enabled the morphology of the latter structures to be studied in a herpesvirus by the negative-contrast technique.

Because of these special attributes it has been possible accurately to compare the size of all visible viral components as measured in two quite different types of preparation. The results obtained are given in the Table and show a close correlation between the

Size of immature particles and components
in two kinds of preparation

Dimension	Size in thin sections [1] (nm)	Size in negative-contrast preparations [1] (nm)
Overall diameter	108	115
Diameter within bases of capsomeres	82	90
Length of capsomeres	13	12.5
Diameter of dense nucleoid	60	60
Diameter of ring-shaped nucleoid	60	60
Width of subcapsomeric zone	11	15

[1] Mean of measurements on at least 30 particles.

dimensions of individual viral components visualized in thin sections and PTA. They also show that a herpesvirus particle with a dense nucleoid can be seen after treatment with PTA to possess a space within the capsid occupied by some material that surrounds the dense nucleoid and holds it at the centre of the particle (Figs. 6, 7 & 8). This confirms, in negative-contrast preparations (Figs. 6 & 7), the reality of the appearance presented by the electron-lucent zone in thin sections (Figs. 1 & 3). It is of further interest that the diameter of the zone within the bases of the capsomeres when delineated by PTA (90 nm—see Table) corresponds exactly with the measurements of this region made earlier on herpes simplex virus in thin sections prepared without dehydrating agents using a water-miscible embedding medium (Epstein, 1962b).

In addition, penetration by PTA of immature particles with dense nucleoids (Figs. 6, 7 & 8) has made it possible to visualize details of structure (Fig. 7) in this type of nucleoid; such nucleoids must be of special significance since only particles with them enter buds to form the final mature enveloped virion. The morphology at the centre of herpes

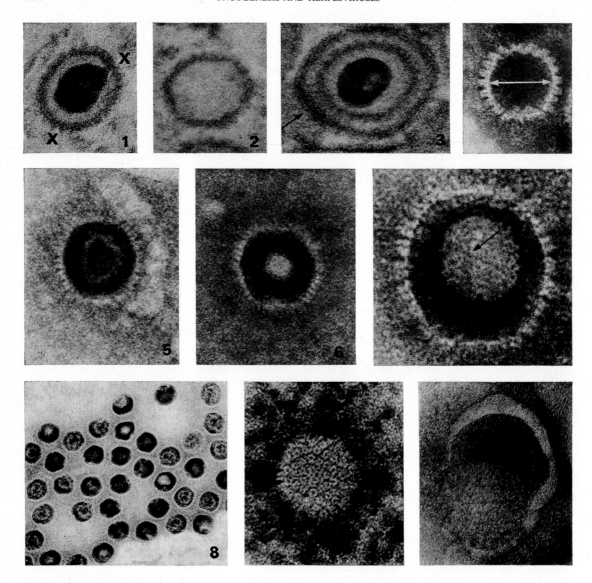

Fig. 1.   Hexagonal immature herpes saimiri virus particle with central dense nucleoid.   Radially arranged rod-shaped capsomeres surround the particle and are well seen at X; they measure 13 nm in length.   An electron-lucent internal zone lying below the bases of the capsomeres surrounds the nucleoid.   Preparation: thin section of material fixed in glutaraldehyde followed by osmium, dehydrated, embedded in epoxy resin and stained in the section with uranyl acetate. $\times$ 200 000.

Fig. 2.   Hexagonal empty immature particle also limited by radiating capsomeres.   The entire inner volume below the bases of the capsomeres is occupied by electron-lucent material.   Preparation: see Fig. 1. $\times$ 200 000.

Fig. 3.   Mature, enveloped, particle; the triple-layered structure of the cell-derived outer envelope can be seen *(arrow)*. The hexagonal inner component is limited by radiating capsomeres enclosing an electron-lucent zone which surrounds the dense central nucleoid.   Preparation: see Fig. 1. $\times$ 200 000.

Fig. 4.   Hexagonal empty immature particle limited by hollow tubular capsomeres 12.5 nm in length (c.f. Figs. 1, 2 & 3).   The inner volume within the bases of the capsomeres *(arrow)* is filled by the electron-opaque reagent.   Whole-mount negative-contrast preparation in PTA. $\times$ 200 000.

Fig. 5.   Hexagonal immature particle containing a ring-shaped nucleoid.   The centre of the ring and the zone below the bases of the capsomeres are filled by the electron-opaque reagent.   Preparation: see Fig. 4. $\times$ 200 000.

saimiri particles reported here reveals a possible site for numerous protein components both in the nucleoid and in the zone between it and the bases of the capsomeres. These internal arrangements may well exist in other members of the herpes family whose internal structure has not yet been clearly seen since the dense nucleoid is much more difficult to demonstrate in PTA preparations.

SUMMARY

Herpes saimiri particles with dense nucleoids have shown unusual penetrability by PTA. This has revealed the subunit structure of the nucleoids, which were also divided into segments by planes of symmetry.

REFERENCES

Epstein, M. A. (1962a) Observations on the mode of release of herpes virus from infected HeLa cells. *J. Cell Biol.*, **12**, 589-597

Epstein, M. A. (1962b) Observations on the fine structure of mature herpes simplex virus and on the composition of its nucleoid. *J. exp. Med.*, **115**, 1-12

Epstein, M. A. & Achong, B. G. (1965) Fine structural organization of human lymphoblasts of a tissue culture strain (EB1) from Burkitt's lymphoma. *J. nat. Cancer Inst.*, **34**, 241-253

Epstein, M. A., Achong, B. G Churchill A. E. & Biggs, P. M. (1968) Structure and development of the herpes type virus of Marek's Disease. *J. nat. Cancer Inst.*, **41**, 805-820

Falke, D., Siegert, R. & Vogell, W. (1959) Electronen-mikroskopische Befunde zur Frage der Doppelmembranbildung des Herpes-Simplex-Virus. *Arch. ges. Virusforsch.*, **9**, 484-496

Hummeler, K., Henle, G. & Henle, W. (1966) Fine structure of a virus in cultured lymphoblasts from Burkitt lymphoma. *J. Bact.*, **91**, 1366-1368

Kaplan, A. S. & Ben-Porat, T. (1970) Synthesis of proteins in cells infected with herpesvirus. VI. Characterization of proteins of the viral membrane. *Proc. nat. Acad. Sci. (Wash.)*, **66**, 799-806

Melendez, L. V., Hunt, R. D., Daniel, M. D., Garcia, F. G. & Fraser, C. E. O. (1969) Herpesvirus saimiri. II. An experimentally induced primate disease resembling reticulum cell sarcoma. *Lab. Anim. Care*, **19**, 378-386

Mizell, M., Toplin, I. & Isaacs, J. J. (1969) Tumour induction in developing frog kidneys by a zonal centrifuged purified fraction of the frog herpes-type virus. *Science*, **165**, 1134-1137

Morgan, C., Rose, H. M., Holden, M. & Jones, E. P. (1959) Electron microscopic observations on the development of herpes simplex virus. *J. exp. Med.*, **110**, 643-656

Morgan, D. G., Epstein, M. A. & Achong, B. G. (1970) Morphological confirmation of the herpes nature of a carcinogenic virus of primates (Herpes Saimiri). *Nature (Lond.)*, **228**, 170-172

Olshevsky, U. & Becker, Y. (1970) Synthesis of herpes simplex virus structural proteins in arginine deprived cells. *Nature (Lond.)*, **226**, 851-853

Parsons, D. I. (1963) Negative staining of thinly spread cells and associated virus. *J. Cell Biol.*, **16**, 620-626

Robinson, D. J. & Watson, D. H. (1971) Structural proteins of herpes simplex virus. *J. gen. Virol.*, **10**, 163-171

Fig 6. Hexagonal immature particle containing a central dense spherical nucleoid outlined by the phosphotungstic acid which fills the remainder of the particle within the bases of the capsomeres (c.f. the electron-lucent zone around the nucleoid in Figs. 1 & 3). Preparation: see Fig. 4. × 200 000.

Fig. 7. Detail of hexagonal immature particle containing a central dense spherical nucleoid. Penetration by phosphotungstic acid reveals that the nucleoid is covered by regularly arranged subunits apparently divided into segments by geometrical planes of symmetry; the point of convergence of four such axes of symmetry is indicated by an *arrow*. Phosphotungstic acid fills the remainder of the subcapsomeric zone and the structure of the hollow tubular surface capsomeres is particularly well seen in side view. Preparation: see Fig. 4. × 400 000.

Fig. 8. Group of hexagonal immature particles showing several morphological variants close together. The particles are either empty or contain nucleoids; besides the familiar ring-shaped form of nucleoid, many particles with a central dense nucleoid can also be seen to have been penetrated by phosphotungstic acid. This property is unusual with other herpesviruses. Preparation: see Fig. 4. × 64 000.

Fig. 9. Intact hexagonal immature particle with hollow tubular surface capsomeres in end-on view arranged in triangular facets. Electron-opaque reagent has failed to penetrate the particle which is presumed to contain a central dense nucleoid. Preparation: see Fig. 4. × 200 000.

Fig. 10. Mature particle with damaged outer envelope allowing phosphotungstic acid to enter and reveal the hexagonal component within. Preparation: see Fig. 4. × 200 000.

Roizman, B. (1969) The herpes viruses—a biochemical definition of the group. *Curr. Top. Microbiol. Immunol.,* **49**, 1-79

Smith, K. O. & Rasmussen, L. (1963) Morphology of cytomegalovirus (salivary gland virus). *J. Bact.,* **85**, 1319-1325

Toplin, I. & Schidlovsky, G. (1966) Partial purification and electron microscopy of the virus in the EB3 cell line derived from a Burkitt lymphoma. *Science,* **152**, 1084-1085

Watson, D. H., Russell, W. C. & Wildy, P. (1963) Electron microscopic particle counts on herpes virus using the phosphotungstate negative staining technique. *Virology,* **19**, 250-260

Wildy, P., Russell, W. C. & Horne, R. W. (1960) The morphology of herpes virus. *Virology,* **12**, 204-222

Pathogenicity Tests in Lambs with an Ovine Herpesvirus

J. M. K. MACKAY [1] & D. I. NISBET [2]

Mackay (1969a, b) reported the occurrence of a herpeslike virus in direct cultures of macrophages from jaagsiekte (sheep pulmonary adenomatosis) lung lesions. Various attempts to reproduce the disease with this virus are reported here.

MATERIALS AND METHODS

Except where otherwise stated, the materials were injected in 1.0-ml volumes intratracheally or 3.0-ml volumes intrathoracically into lambs in a series of experiments, the results of which are summarized in the accompanying Table. The lambs designated "conventional" were born naturally but reared in an environment free from helminths. The hysterectomy-derived lambs were reared in isolation accommodation.[3] All these animals were injected within the first month of life.

The experiments have been grouped according to the nature of the inoculum.

Experiment A. A lung lesion, part of which had yielded virus when directly cultured, was stored intact at —70°C. A 5% w/v suspension of this material in Eagle's medium was homogenized and the crude supernatant was inoculated within 1 hour of preparation.

Experiment B. An 8-day tissue culture of jaagsiekte macrophages showing the characteristic cytopathic effect was subjected to five cycles of freeze-thawing. After centrifugation, the supernatant was filtered through a 0.22 μ Millipore membrane and the filtrate constituted the inoculum.

[1] Moredun Research Institute, Edinburgh, UK. Present address: Bacteriology Department, University of Edinburgh.
[2] Moredun Research Institute, Edinburgh, UK.

[3] Unpublished data.

Results of transmission experiments in lambs

Expt.	Route [1]	No. of lambs [2]	Typical lesions	Early? lesions [3]	Vesicles	Duration (months)	Virus isolates [4]
A	ITR	9 C	5 (3)		0	6–11	2
B	ITR	9 C	1		0	11–13	1
C	ITH	7 C	0	ND	0	2–6	0
C Control	ITH	7 C	0	ND	0	6–12	0
C1	ITR	6 HD	0	4	0	4–6	ND
C1 Control	ITR	11 HD	0	0	0	4–6	ND
C2	ITH	12 HD	0	0	2	7	ND
C3	ITR	14 HD	0	2	4	7	ND
C4	corneal	IHD	0	0	0	7	ND
C5	IC	IHD	0	0	0	7	ND
D	ITR	8 HD	0	0	1	3	ND
D1	IV	8 HD	0	0	3	3	ND
E	ITR	8 C	2 (1)		0	7–8	ND
E1	ITR	6 C	2 (1)		0	10–11	ND

[1] ITR = intratracheal; ITH = intrathoracic; IC = intracerebral; IV = intravenous.
[2] C = conventional; HD = hysterectomy derived.
[3] Number in parenthesis indicates clinical signs.
[4] ND = not done.

Experiments C to C5. All the inocula in this group were from tissue culture passages of the herpesvirus in normal macrophages. The inocula were prepared as previously described (Mackay, 1969b), except that the cultures were not centrifuged after sonication. C and C1 consisted of the 13th serial subculture of pool A10, i.e., a dilution of at least 10^{-14} of the original jaagsiekte lung cultures from which the passage was initiated. C and C1 controls were injected with the normal tissue culture substrates, i.e., a blind passage of aliquots of all of the uninfected macrophage cultures used for C and C1. C2, C3, C4 and C5 were injected with the 18th subculture of pool A10.

Experiments D and D1. Inocula were viable tissue culture cells of ovine kidney in which system the virus was present in cell-associated form.[1]

Experiments E and E1. These inocula were prepared as for Experiment A above from clinical cases of jaagsiekte prior to our demonstration of the herpesvirus and have been included simply for comparison.

RESULTS

Only a proportion of the lambs proved susceptible, even when they were inoculated with the diseased lung tissue (Experiments A, E and E1) or the initial tissue culture material (Experiment B). This agrees with previously published data (Tustin, 1969). It was possible to recover the herpesvirus from lesions of three of the lambs (Experiments A and B) 6–11 months after inoculation, whereas no recoveries were made from the apparently normal lungs of lambs in Experiment C. Several lambs injected with virus in Experiments C2, C3, D and D1 developed perineal vesicles 3–6 weeks after inoculation but no virus recoveries were made from this site. These lesions were not seen in the control groups. The single lamb in Experiment C5, to which the virus was given by corneal scarification, developed an

---

[1] Unpublished data.

herpetic corneal ulceration 10 days after inoculation and typical intranuclear inclusions were seen in smears of this material.

No unequivocal lesions of jaagsiekte were seen in any of the Experiment C lambs, but in two of these experiments, C1 and C3, a total of six animals showed histological evidence of lung lesions. These were suggestive of the early development of the adenomatous lesion. Early adenomatous changes are illustrated (Figs. 1–4).

It should be noted, in connection with Fig. 1, that in naturally-occurring cases of jaagsiekte, the spread of lesions in the lung occurs mainly by peripheral extension. Thus at the periphery of histologically characteristic lesions alveoli can be found in which small epithelial proliferations consisting of a few cuboidal cells only are present, arising from the alveolar surface. In some cases these proliferations occur on the free extremities of alveolar septa which project into the alveolar ducts. Similar small proliferations are also found arising from the epithelial surface of terminal bronchioles in a proportion of cases. The small size of these proliferations would not permit their identification as jaagsiekte lesions if they were considered in isolation, but in conjunction with characteristic adenomatous foci in the same section there is no doubt that they are the earliest recognizable evidence of the disease process. We have found proliferations similar to these consisting of a few cuboidal epithelial cells in 6 of our experimental lambs, but as mentioned above, in the absence of characteristic adenomatous foci these cannot be unequivocally identified as jaagsiekte, so we have termed them early? lesions.

These results, although inconclusive, suggest that further similar transmission experiments are justified and it would be desirable that the test lambs should be maintained for longer periods than those reported here. This type of experiment, although by its nature time-consuming and expensive, is in our view necessary because of the wide comparative interest of this disease.

Fig. 1. Natural case of jaagsiekte. Small proliferative lesions in alveoli and alveolar ducts. Haematoxylin and eosin ($\times$165).

Fig. 2. Natural case of jaagsiekte. Small cuboidal cell proliferations in an alveolus, an alveolar duct and a bronchiole. Haematoxylin and eosin ($\times$165).

Fig. 3. Experimental lamb 120, 4 months post-inoculation. Proliferations of cuboidal epithelial cells in alveoli and alveolar ducts. Haematoxylin and eosin ($\times$165).

Fig. 4. Experimental lamb 119, 4 months post-inoculation. Bronchiole containing papilliform proliferation of cuboidal cells originating from the bronchiolar epithelium. Haematoxylin and eosin ($\times$165).

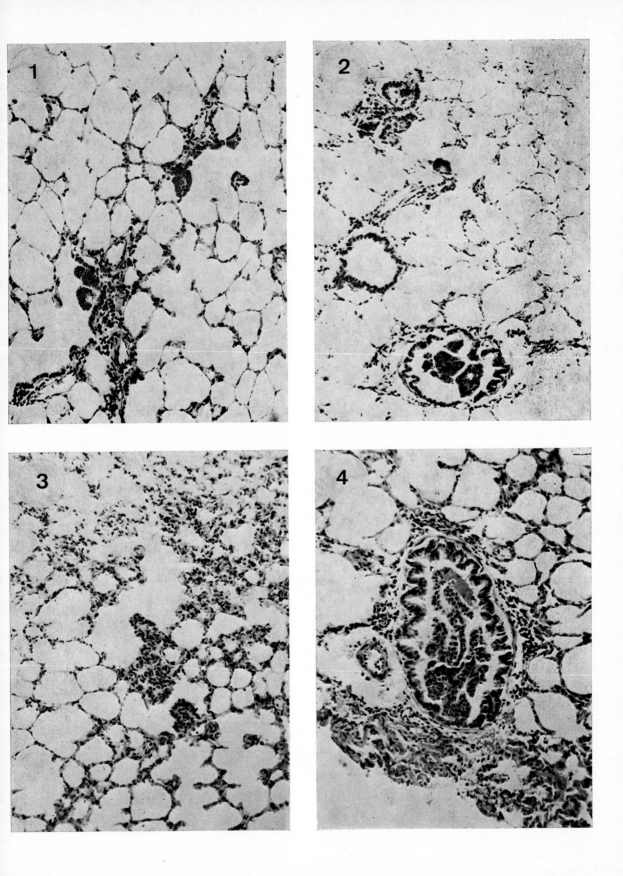

REFERENCES

Mackay, J. M. K. (1969a) Tissue culture studies of sheep pulmonary adenomatosis (jaagsiekte).   I. Direct cultures of affected lungs. *J. comp. Path.*, **79**, 141-146

Mackay, J. M. K. (1969b) Tissue culture studies of sheep pulmonary adenomatosis (jaagsiekte).   II. Transmission of cytopathic effects to normal cultures. *J. comp. Path.*, **79**, 147-154

Tustin, R. C. (1969) Ovine jaagsiekte. *J. S. Afr. vet. med. Ass.*, **40**, 3-23

A New Malignant Lymphoma Induced in Syrian Hamsters Following Their Inoculation with Hamster Embryo Cells Infected with Equine Herpes 3 Virus*

A. KARPAS & A. SAMSO

This report deals with the study of a malignant lymphoma that appeared in two Syrian hamsters after inoculation with embryonic hamster (HE) cells cultured *in vitro* and infected with equine herpes 3 (EH3) virus (Karpas & Samso, 1971).

DEVELOPMENT AND BIOLOGY OF THE TUMOUR

The initial observations that led to the present investigations were made on cultured HE cells infected with EH3 virus at an approximate ratio of one viral particle per ten cells. Initially only a few foci of cytopathogenic effect developed. Following the change of medium, the foci of cell degeneration disappeared and a confluent cell sheet appeared. Examination of stained cultures revealed that the morphology of many of the infected cells was changed (Fig. 2) when compared with noninfected cultures (Fig. 1). Intranuclear inclusion bodies were rarely found, but a large range of nuclear abnormalities was observed. Large nuclei and nucleoli were very common as well as areas of dense growth of cells with small rounded nuclei (Fig. 3). The noninfected cells were of uniform morphology with elongated nuclei, but the cell sheet was a multi-layer (Fig. 1). In spite of the cellular changes that occurred in the infected cultures, the cells continued to grow following trypsinization and passage. A viral-induced transformation was therefore suspected, and several litters of newborn Syrian hamsters were inoculated subcutaneously (s/c) with these cells. In the eighth month it was noticed that one of the hamsters from the litter of ten inoculated with the

second passage of EH3-virus-infected HE cells (of which all ten animals survived) had developed a s/c growth. The growth continued to increase in size in the following weeks and histological examination revealed that the tumour was a lymphosarcoma (Fig. 4). Of the 21 hamsters (of 3 litters) inoculated with the third passage of EH3-virus-infected HE cells, only 11 animals survived the immediate post-inoculation period. A growth identical to the lymphoma that developed in the hamster previously mentioned was detected in one of these 11 animals on the 22nd month following inoculation. None of the other inoculated animals developed a malignant growth, nor did any of the hamsters inoculated with the noninfected HE cells or those inoculated s/c as newborns with 10^4 EH3 virus per animal. Hamsters of various ages inoculated s/c or intraperitoneally (i/p) with the tumour cell suspension developed lymphomas. Metastasis developed in almost all animals and affected many organs. Of particular interest was the frequency of cardiac metastases (71%). A study of the karyotype of the tumour cells revealed that they contained 39 chromosomes. The tumour cells of a hamster lymphoma studied previously contained 51 chromosomes (Cooper, McKay & Banfield, 1964). The normal karyotype of the Syrian hamster is 44. Preliminary electron microscopic examination of freshly fixed tumour tissues did not reveal any distinguishable viral particles.

VIROLOGICAL STUDIES

EH3 virus could not be recovered from cell-free extracts prepared from tumour tissue, when inoculated on susceptible RK cells, nor did co-cultivation of tumour cells with rabbit kidney (RK) cells yield

---

* From the Department of Medicine, University of Cambridge, UK, and Service des Virus, Institut Pasteur, Paris, France.

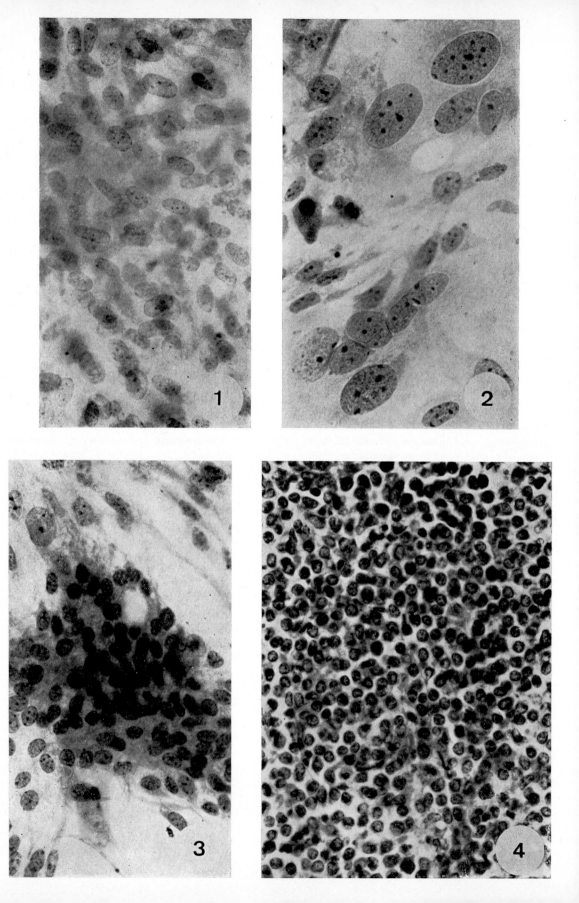

infectious virus. Since it was reported recently that C-type RNA virus is involved in the induction of leukosis in the Syrian hamster (Graffi et al., 1968) attempts were made to find out whether a similar virus was involved in this malignancy. In order to demonstrate the presence of such a virus, the "rescue" technique of Huebner et al. (1966) was used, in which hamster embryo cells were co-cultivated with a hamster cell line (HT-1) carrying the murine sarcoma viral (MSV) genome, and super-infected with the lymphoma extract. An infective MSV could not be recovered, however, from an extract of this mixed cell population, which suggests that a C-type leukaemic virus is not associated with the induction of this tumour.

Development of complement-fixing antibodies
to tumour antigen in adult hamsters
following inoculation with tumour suspension [1]

Hamster No.	Prior to inoculation	Time post inoculation	
		43 days	66 days
2110	< 1 : 5	< 1 : 5	N.D.
2111	< 1 : 5	< 1 : 5	N.D.
2115	< 1 : 5	< 1 : 5	< 1 : 5
2117	< 1 : 5	< 1 : 5	< 1 : 5
2119	< 1 : 5	< 1 : 5	N.D.
2120	< 1 : 5	< 1 : 5	1 : 20
2121	< 1 : 5	< 1 : 5	N.D.
2122	< 1 : 5	< 1 : 5	N.D.
2123	< 1 : 5	< 1 : 5	> 1 : 40
2124	< 1 : 5	< 1 : 5	1 : 40

[1] N.D. = not done.

SEROLOGICAL STUDIES

Sera of tumour-carrying hamsters did not neutralize the infectiousness of EH3 virus, nor could any difference be detected in the amount of fluorescence emitted from RK cells treated with sera from tumour-carrying hamsters, whether the cells had been infected with EH3 virus or not. However, sera of tumour-carrying hamsters reacted with antigen prepared from tumour tissue obtained within two weeks after inoculation into newborn hamsters. This antigen was completely inactivated when heated at 56°C for 30 minutes. Tumour antigen prepared from tissues of adult animals was found to be highly anticomplementary. The titres of the complement-fixing antibody (CF) that developed in hamsters after the inoculation with tumour suspension differed between the individual hamsters and were not always related to the duration of the tumour-carrying period (see Table). As can be seen from the Table, no CF antibodies were detected prior to inoculation and for six weeks thereafter. By the 66th day, three of five tumour-carrying animals showed CF antibodies. Five tumour-carrying hamsters died before the 66th day. Newborn hamsters inoculated with tumour suspensions failed to develop complement-fixing antibodies to the tumour antigen, presumably due to immune tolerance.

SUMMARY AND CONCLUSIONS

The development of a malignant lymphoma in two Syrian hamsters from a colony in which spontaneous lymphomas had not been observed earlier was of great interest; the animals had been inoculated with hamster embryo cells grown in vitro which had undergone morphological transformation after infection with EH3 virus. A direct relationship between EH3 virus and the lymphoma could not be established. The formation of a heat-labile complement-fixing antigen in the lymphoma tissue may have been coded by the incorporation of an incomplete EH3 virus. The tumour-bearing hamsters developed antibodies to this antigen. This may be similar to the early heat-labile antigens induced in hamsters after infection with other "foreign" DNA viruses, such as SV-40 (Ashkenazi & Melnick, 1963) and adenoviruses (Denny & Ginsberg, 1961). Since the growth cycle of these viruses is much longer (20 and 16 hours respectively) than that of EH3 virus, which is only 5 hours (Karpas, 1967), the distinction between early and late antigens is more difficult in the case of EH3 virus.

This tumour differed in its hypoploid karyotype, its high metastatic rate and frequent cardiac involvement, from hamster lymphomas previously studied.

Fig. 1. Noninfected primary hamster embryo cells forming a multilayered sheet of cells. Note the similar shape and size of the nuclei. ×480.

Figs. 2 & 3. Morphological changes in the hamster embryo cell culture (seen in Fig. 1) following infection with EH3 virus. In spite of these changes the cells continued to grow. ×480.

Fig. 4. Cross section of the lymphoma tissue in the kidney.

REFERENCES

Ashkenazi, A. & Melnick, J. L. (1963) Tumorigenicity of simian papova virus SV40 and of virus transformed cells. *J. nat. Cancer Inst.*, **30**, 1227-1265

Cooper, H. L., McKay, C. M. & Banfield, W. G. (1964) Chromosome studies of a contagious reticulum cell sarcoma of the Syrian hamster. *J. nat. Cancer Inst.*, **33**, 691-706

Denny, F. W. & Ginsberg, H. S. (1961) Certain biological characteristics of adenoviruses type 5, 6, 7 and 14. *J. Immunol.*, **68**, 567-574

Graffi, A., Schramm, T., Bender, E., Graffi, I., Horn, K. H. & Bierwolf, D. (1968) Cell free transmissible leukosis in Syrian hamsters, probably of viral aetiology. *Brit. J. Cancer*, **22**, 577-581

Huebner, R. J., Hartley, J. W., Rowe, W. P., Lane, W. T. & Capps, W. I. (1966) Rescue of the defective genome of Moloney sarcoma virus from a non-infectious hamster tumour and the production of pseudo-type sarcoma viruses with various murine leukaemia viruses. *Proc. nat. Acad. Sci. (Wash.)*, **56**, 1164-1169

Karpas, A. (1967) The growth cycle of equine herpes 3 virus as compared with other viruses of the herpes group. *Arch. ges. Virusforsch.*, **22**, 316-323

Karpas, A. & Samso, A. (1971) A new malignant lymphoma induced in the Syrian hamster. *Europ. J. Cancer* (in press)

Discussion Summary

R. J. C. HARRIS [1]

Transformation of hamster cells by herpes simplex virus

In general the transformed cells described by Rapp do not show virus particles. On very rare occasions the old cell shows little clusters that might be called immature herpes particles. These may be degenerating cells and are, perhaps, the same cells as those in which we find intranuclear inclusions. Mycoplasmas have not been found in the cultures.

A possible way of circumventing the difficulty of the cytopathic effect of herpesviruses was to develop temperature-sensitive mutants. Rapp reported that the transformed hamster cells were, in fact, permissive for herpes simplex and it had been hoped that differentiation between the replication of types 1 and 2 might take place. This had been observed in a *different system.*

The question of reactivation was raised. Photoreactivation of herpesviruses had been known for some time but there was now evidence of dark-repair.

The delay before new virus antigen was detectable was about 24 hours and within this time repair was presumably effected. In chick cells, which photoreactivate, this delay can be cut down by light treatment, but BHK cells do not photoreactivate in this way. Since this was a possibility, Rapp had held the input multiplicity to a minimum. Of course the transforming agent could as well have been virus that was initially defective as virus that was damaged by irradiation.

Carcinoma of the cervix and herpesviruses type 2

Nahmias pointed out that an immediate difference between the Burkitt and NP tumours and the cervical carcinomas is that no viruses have been seen in the last-mentioned lesions, nor has the herpesvirus type 2 been isolated routinely except from highly promiscuous women, as reported by Tobin. Cytomegalo-virus has been found, but no other herpesvirus, in studies of 1000 women from ordinary gynaecological or antenatal clinics. As far as antibody levels are concerned, Rawls reported that only two of a group of forty-three nuns were seropositive and these had very low titres. For prostitutes the incidence rose with age to 100%. This represents the two extremes in the occurrence of cervical carcinoma—virtually unknown in nuns and with the highest incidence in prostitutes.

According to Muir in Colombia and South America the incidence rate for carcinoma of the cervix is $100/10^5$. In Israel it is about $10/10^5$. Are there the same differences in antibody status between patients and controls in the two areas? Rawls reported that in Colombia, cases had an incidence of antibodies of 35% and matched controls were similar to general controls at 30–40%. In Israel, cases and matched controls again had a similar incidence but in the general population the incidence of antibodies was considerably less (10%) than in South America. It was agreed that if this difference was real it would suggest that herpesvirus type 2 was not causally related to this carcinoma but was a covariable with promiscuity and other sexual factors.

Lymphoma viruses of monkeys

There was some discussion about the relationship of these viruses to the actual tumours and the problem of virus activation from latency. Melendez reported that the tumours certainly did not contain a separable virus-like material propagable *in vitro.* However, when the kidneys of the tumour-bearing marmosets were cultured a virus appeared. Whether the viral genetic material is incorporated in the cell genome was not clear, but neither virus nor viral nucleic acid was demonstrable. It was possible, of course, that only a very few of the kidney cells were productively infected, which is probably what happens in Marek's disease. In this case the kidneys

[1] Microbiological Research Establishment, Porton Down, Salisbury, Wilts., UK.

— 475 —

look normal enough although a very diligent search might turn up the occasional antigen-containing cell.

It was pointed out that isolation *in vitro* does not necessarily involve the activation of a latent virus.

In the normal course of events, *in vivo*, a cell that is maturing virus may well be destroyed immunologically but cells not revealing surface antigen changes may escape. The fact the kidney acts as a reservoir for herpesvirus may mean nothing more than that tissue cultures can be more readily prepared from kidneys so that the chances of recovering virus from kidneys are that much greater.

Kalter pointed out that experimentally-infected animals, especially monkeys, often react in the same way. Many viruses disappear and are not recoverable for weeks after inoculation, but tissue explants, such as the brain, lungs, spleen and even the intestinal tract, eventually begin to liberate them.

Additional contribution prepared by Dr S. S. Kalter [1]

Kalter reported that he and Nahmias had expanded the original series of nonhuman primates (rhesus, cebus, squirrel) to include the baboon, cebus (as a "control") and marmoset. The same strain of herpesvirus type 2 (Curtis) was employed and the same technique used—packing of virus on cotton plugs around the cervix.

---

[1] Division of Microbiology and Infectious Diseases, Southwest Foundation for Research and Education, San Antonio, Texas, USA.

The results indicated marked species differences, as follows:

Group 1	Baboon	Infection with virus Shedding No gross lesions Seroconversion
Group 2	Cebus	Infection with virus Shedding Cervical and vaginal lesions with purulent discharge Seroconversion
Group 3	Marmoset	Infection with virus Shedding Marked cervical and vaginal lesions with purulent discharge, terminating in death Seroconversion

He suggested that the results reported by Melendez should be clarified. The impression was given that herpesvirus saimiri produced leukaemia and lymphoproliferative disease in all nonhuman primates inoculated. There are exceptions—in collaboration with Melendez, he had demonstrated that the baboon is resistant to infection with this virus. There had not been any evidence of infection in these animals over a period of two years.

RELATIONSHIPS AMONGST ONCOGENIC AND OTHER HERPESVIRUSES

Chairman — W. Henle

Rapporteur — L. J. N. Ross

Demonstration of Group- and Type-specific Antigens of Herpesviruses [1]

J. KIRKWOOD, [2,3] G. GEERING [2] & L. J. OLD [2]

Herpesviruses (HV) cause the Lucké renal adeno-carcinoma of the leopard frog and Marek's disease (neurolymphomatosis) of the chicken, and have been related to three human malignancies, Burkitt's lymphoma, nasopharyngeal carcinoma, and cervical carcinoma. We have analyzed these and other HV by micro-immunodiffusion, and found two classes of antigen—group-specific and type-specific.

Group-specific (gs) antigen is detected by rabbit antisera to nucleocapsid preparations of HV (Lucké), HV (Burkitt), or herpes simplex. Nucleocapsid preparations of HV (Lucké), HV (Burkitt), HV (Marek), herpes simplex, and cytomegalovirus gave reactions of identity (0.7% agarose), confirmed by cross-absorptions. The gs antigen sediments at 80 000 g (1 hour) and this is compatible with its exclusion from 2% agar. Natural antibody of the same specificity is found in 10% or less of normal humans, frogs, rabbits, dogs, cats and cows.

Type-specific (ts) antigens are demonstrable in concentrated extracts of HV-infected cells. They are nonsedimentable at 80 000 g (1 hour), and HV (Burkitt), HV (Marek), herpes simplex (Type 1—strain F, and Type 2—strain G) and cytomegalovirus have more than one ts antigen. Unlike naturally-occurring antibody to gs antigen, incidence and titres of natural ts antibody are related to infection or malignancy. For example, precipitating antibody to ts antigens of HV (Burkitt) is most frequent in patients with Burkitt's lymphoma (71%) and naso-pharyngeal carcinoma (87%). Antibody is usually restricted to the natural host; one exception to this is the high incidence of ts antibody to HV (Burkitt) in chimpanzees (more than 75%).

One ts antigen was shared by HV (Burkitt) and herpes simplex. No ts antigens were found associated with HV (Lucké).

Thus, there exists a group-specific antigen in nucleocapsid preparations of six different HV, to which a low incidence (5–10%) of natural antibody appears in the sera of a wide variety of vertebrates. The presence of this antigen, distinct from previously described ts antigens, may serve as a criterion for the inclusion of further viruses into this group; its utility as a marker for causative viral agents in tumours of unknown origin remains to be demonstrated.

[1] Abstract only.
[2] Sloan-Kettering Institute for Cancer Research, New York, USA.
[3] Present address: Yale University School of Medicine, New Haven, Connecticut, USA.

An Antigen Common to Some Avian and Mammalian Herpesviruses

L. J. N. ROSS,[1] J. A. FRAZIER [1] & P. M. BIGGS [1]

INTRODUCTION

Watson *et al.* (1967) showed that herpes simplex virus, pseudorabies virus and B virus are serologically related. Recent studies suggest that the herpesviruses associated with Lucké adenocarcinoma, Burkitt's lymphoma and Marek's disease are also serologically related (Fink, King & Mizell, 1968; Kirkwood *et al.*, 1969; Naito *et al.*, 1970; Ono *et al.*, 1970). Our intention was to examine the serological relationship between Marek's disease virus, Epstein-Barr virus, herpes simplex and pseudorabies viruses. We have used indirect immunofluorescence and electron microscopy of virus-antibody interactions.

METHODS

Cells

Chick embryo cells were derived from 12-day-old Rhode Island Red embryos (HPRS-RIR) free from leukosis and Marek's disease viruses. Chicken kidney cells were obtained from 20 to 30-day-old HPRS-RIR chickens reared in isolators. Duck embryo cells were derived from 14-day-old embryos obtained commercially (Cherry Valley, England). All cells were grown in 199 medium containing 5% tryptose phosphate broth and 4% calf serum.

Virus

The following herpesviruses grown in BHK cells were provided by Professor P. Wildy: Herpes simplex type 1 (HFEM), herpes simplex type 2 (Lovelace), pseudorabies virus (Dekking). They were propagated in this laboratory in chick embryo cells. The Marek's disease viruses (MDV) were the HPRS-16 and HPRS-16 attenuated (HPRS-16/att) strains.

Turkey herpesvirus (HVT) strain FC126 (Witter *et al., 1970) was obtained from the Wellcome Research Laboratories (England). Burkitt's lymphoblasts EB3 (Epstein, 1965) was given to us by Dr Alison Newton.

Antiserum

Hyperimmune antisera to herpes simplex 1 and 2 and pseudorabies viruses were provided by Professor P. Wildy and had been prepared by inoculating rabbits with extracts of virus-infected rabbit cells, as described by Watson *et al.* (1966). They were highly reactive for virus-specific antigens and produced 10 to 12 precipitin lines when tested against homologous antigen in gel diffusion tests. Antiserum to MDV was obtained from HPRS-RIR cockerels inoculated with extracts of HPRS-16 strain of MDV-infected chicken kidney cells grown in chicken (HPRS-RIR) serum. The antiserum was virus-specific, but produced only 3 precipitin lines against MD antigen in immunodiffusion tests. Goat anti-rabbit globulin and rabbit anti-chicken globulin conjugated to fluorescein isothiocyanate were obtained commercially (Nordic Diagnostics, England).

Indirect immunofluorescence

Preparation of coverslip cultures for infection, and cytological procedures were essentially as described previously (Ross, Watson & Wildy, 1968). Duck embryo fibroblast monolayers were infected at low multiplicity when confluent. Fixation was carried out in methanol at —70°C overnight. Burkitt's lymphoblasts were thoroughly washed in phosphate-buffered saline (PBS) and smeared on glass slides. They were air-dried, fixed and processed as described above.

Virus agglutination

Naked virus particle preparations were used at final concentrations of 0.5 to 1.0 × 10⁹ per ml, and

[1] Houghton Poultry Research Station, Houghton, Huntingdon, UK.

were obtained by extracting homogenates of heavily infected cultures twice with Arcton 113 (ICI). In some cases virus was further concentrated by sedimentation followed by resuspension in PBS. The procedure adopted was to incubate 0.1 ml of virus with 0.1 ml of antiserum in PBS at 37°C for 1 hour. One drop of the mixture was then placed on a Formvar-covered specimen grid, stained *in situ* with potassium phosphotungstate at pH 7.0 and scanned at a magnification of 30 000. At least 250 particles were examined for each preparation and both the number of particles clumped and the total number of particles examined were recorded. A clump was defined as a group of 4 or more particles (Fig. 1). The number of particles in each clump varied from 4 to 10 and rarely exceeded 15. Virus treated with serum from nonimmune rabbits served as controls. The significance of the difference in agglutination produced by immune serum compared with nonimmune serum was tested by χ^2 analysis.

RESULTS AND DISCUSSION

The immunofluorescence results are shown in Table 1. At low dilutions all three hyperimmune rabbit antisera reacted with antigens in the cytoplasm and nucleus of MDV- and HVT-infected cells.

Differences in reactivity were apparent, however, and these were best seen at the end points, where it was obvious that herpes simplex 2 and pseudorabies antisera reacted predominantly with intranuclear antigens (Fig. 2). In contrast, herpes simplex 1 antiserum had greater affinity for antigens in the cytoplasm. We are uncertain of the cross-reactions observed with Burkitt's lymphoblasts because of the lack of adequate controls for these cells. The results obtained with MDV antiserum suggest that MDV is more closely related to HVT than to the mammalian herpesviruses. It is probable that the lack of reactivity of MDV antiserum for cells infected with herpes simplex virus was due to its low potency.

The results of the agglutination experiments are given in Table 2. Because of variations in concentration of virus, in age of virus preparation and in conditions of incubation of virus and antiserum, comparison of agglutinating capacity of antisera is only valid within each experiment. Pseudorabies antiserum reacted consistently with antigens on the surface of MDV and HVT particles. Herpes simplex 2 antiserum was less reactive and herpes simplex 1 antiserum did not react. The surface of pseudorabies and of herpes simplex (HFEM) viruses thus appear to be antigenically distinct. This agrees

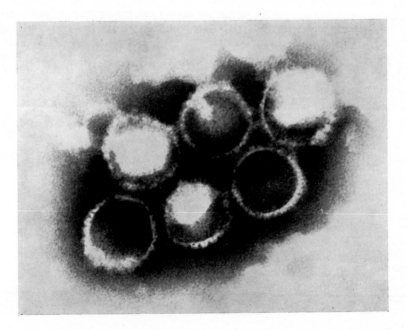

Fig. 1. Clump of Marek's disease virus (HPRS-16) particles agglutinated by pseudorabies antiserum. Note that neighbouring particles are approximately 20 nm apart and are linked presumably by antibody bridges.   ×200 000.

with the findings of Watson [1] and is in accordance with the inability of pseudorabies antiserum to neutralize herpes simplex virus (Watson *et al.*, 1967).

These preliminary observations suggest that the avian and mammalian herpesviruses cross-react antigenically but that there are gradations in the cross-reactions. Differences were observed in the pattern of immunofluorescence and in the antigens on the surface of naked particles. It appears that

---

[1] Personal communication.

there is an association between the capacity of antiserum to agglutinate heterologous virus and its affinity for intranuclear antigens in immunofluorescence tests. This may, however, be fortuitous.

The cross-reactions that we have observed appear to be specific. In immunofluorescence tests, immune sera did not react with uninfected cells or with cells infected with vaccinia virus. Serum from non-immune rabbits did not react with herpes-infected cells. The specificity of virus agglutination by antibody was evident from the fact that herpes simplex 1

Table 1.   Results of indirect immunofluorescence tests

Cell	Virus	Antiserum [1]			
		Herpes simplex I	Herpes simplex 2	Pseudorabies	Marek's disease (HPRS-16)
DEF [2]	MD (HPRS-16)	20 (C)	20 (N)	40 (N)	80 (C)
DEF [2]	HVT (FC126)	20 (C)	40 (N)	40 (N)	40 (N)
DEF [2]	Herpes simplex I	⩾200	200 (N) (C)	80 (C)	<10
DEF [2]	Herpes simplex 2	⩾200	200 (N)	40 (N)	<10
DEF [2]	Pseudorabies	80 (C)	40 (N)	⩾200	10 (C)
EB3 line	EBV	10 (S)	10 (S)	10 (S)	<10

[1] Figures show the reciprocal of the highest dilution giving a positive result; letters in brackets denote the site of fluorescence at the end point, as follows: C = cytoplasm; N = nucleus; S = surface.
[2] Duck embryo fibroblasts.

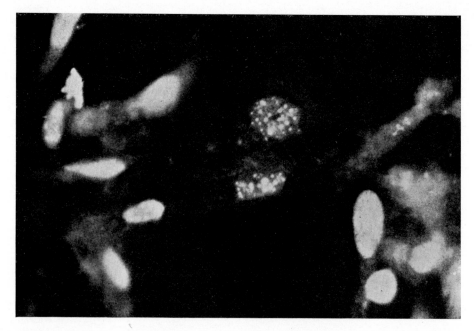

Fig. 2.   Duck embryo fibroblasts infected with Marek's disease virus (HPRS-16).   Indirect immunofluorescence using pseudorabies antiserum.   Note predominantly intranuclear cross-reacting antigens.   × 700.

Table 2. Results of agglutination experiments

Virus	Dilution	% of particles clumped [1]				Experiment no.
		Herpes simplex I serum	Herpes simplex 2 serum	Pseudorabies serum	Control serum	
MD (HPRS16)	1/20	2	9 *	38 *	4.8	1
MD (HPRS16) [2]	1/20	N.T.	N.T.	30 *	4	2
MD (HPRS16/att)	1/10	N.T.	26 *	N.T.	10	3
	1/20	N.T.	13 *	N.T.	5	4
MD (HPRS16/att) [3]	1/10	13	N.T.	72 *	25	5
	1/20	10	N.T.	44 *	19	6
HVT (FC126) [3]	1/20	N.T.	N.T.	92 *	64	7
	1/40	N.T.	N.T.	88 *	46	7
Herpes simplex I	1/10	39 *	N.T.	17	18	8
	1/20	40 *	N.T.	5	10	9
	1/40	45 *	N.T.	N.T.	8	9
Pseudorabies	1/20	4	N.T.	50 *	6	10
	1/40	3	N.T.	42 *	4	11
	1/60	N.T.	N.T.	11 *	4	11

[1] Asterisk denotes significant clumping (0.1 % level); N.T. = not tested.
[2] Virus was used without Arcton treatment but was partially purified by differential centrifugation.
[3] Virus and antiserum were left overnight at 4°C, after 1 hr at room temperature.

antiserum, for example, agglutinated only herpes simplex 1 particles despite the fact that all the viruses were grown in the same type of cells and were prepared in the same way. Furthermore, although the data are not given here, pseudorabies antiserum agglutinated only MDV particles in a mixture of MDV and vaccinia virus particles.

Nevertheless, the results should be interpreted with care since they could have been caused by cross-contamination or by the presence of antibodies to avian herpesviruses in the rabbits prior to immunization. We cannot exclude this latter possibility, however remote, because we have not used preimmune rabbit antisera as controls in our studies. It is known, however, from previous work that the rabbits were originally free from antibodies to herpes simplex and pseudorabies antigens. Furthermore, antiserum from three nonimmune rabbits failed to cross-react either in immunofluorescence tests or in agglutination tests. Contamination of the rabbits with avian herpesviruses or with immunizing antigen is unlikely since the rabbits were kept in a different laboratory where avian herpesviruses are not propagated. Great care was taken to minimize cross-contamination during the preparation of materials for immunofluorescence and agglutination experiments. The type and time of appearance of the cytopathic effects and the higher degree of reaction observed with homologous reagents supported the view that cross-contamination did not occur.

ACKNOWLEDGEMENTS

We are grateful to Professor P. Wildy, Dr D. H. Watson, Dr G. Skinner and Mrs M. Thouless for supplying antiserum to herpes simplex 1, herpes simplex 2 and pseudorabies viruses. We thank Mr B. Whitby and Mr A. Kidd for able technical assistance.

REFERENCES

Epstein, M. A. (1965) Studies with Burkitt's lymphoma. *Wistar Inst. Sympos. Monogr.*, **4**, 69-82

Fink, M. A., King, G. S. & Mizell, M. (1968) Preliminary note: identity of a herpes virus antigen from Burkitt's lymphoma of man and the Lucké adenocarcinoma of frogs. *J. nat. Cancer Inst.*, **41**, 1477-1478

Kirkwood, J. M., Geering, G., Old, L. J., Mizell, M. & Wallace, J. (1969). A preliminary report on the serology of Lucké and Burkitt herpes-type viruses: a shared antigen. In: Mizell, M., ed., *Biology of amphibian tumours*, Berlin, Heidelberg, New York, Springer-Verlag, pp. 365-367

Naito, M., Ono, K., Tanabe, S., Doi, T. & Kato, S. (1970) Detection in chicken and human sera of antibody against herpes type virus from a chicken with Marek's disease and EB virus demonstrated by the indirect immunofluorescence test. *Biken's J.*, **13**, 205-212

Ono, K., Tanabe, S., Naito, M., Doi, T. & Kato, S. (1970) Antigen common to a herpes type virus from chickens with Marek's disease and EB virus from Burkitt's lymphoma cells. *Biken's J.*, **13**, 213-217

Ross, L. J. N., Watson, D. H. & Wildy, P. (1968) Development and localization of virus-specific antigens during the multiplication of herpes simplex virus in BHK21 cells. *J. gen. Virol.*, **2**, 115-122

Watson, D. H., Shedden, W. I. H., Elliot, A., Tetsuka, T., Wildy, P., Bourgaux-Ramoisy, D. & Gold, E. (1966) Virus specific antigens in mammalian cells infected with herpes simplex virus. *Immunology*, **11**, 399-407

Watson, D. H., Wildy, P., Harvey, B. A. M. & Shedden, W. I. H. (1967) Serological relationships among viruses of the Herpes group. *J. gen. Virol.*, **1**, 139-141

Witter, R. L., Nazerian, K., Purchase, H. G. & Burgoyne, G. H. (1970) Isolation from turkeys of a cell-associated herpes virus antigenically related to Marek's disease virus. *Amer. J. vet. Res.*, **31**, 525-537

Immunological Studies on Marek's Disease Virus and Epstein-Barr Virus *

S. KATO, K. ONO, M. NAITO & S. TANABE

A herpes-type virus was isolated from a chicken with Marek's disease (Kato *et al.*, 1970), and was considered to be a strain of Marek's disease virus (MDV) rendered avirulent by passage through duck embryo fibroblasts (DEF). DEF infected with this strain of MDV were used as antigen of MDV. Some of the human sera, including those of patients with Burkitt's lymphoma and nasopharyngeal carcinoma, were found to show antibody activity against MDV by the indirect fluorescent antibody technique (FAT) (see Table). Sera were obtained from 72 subjects whose work was intimately connected with chickens in poultry farms, and from 72 office workers who did not have much contact with chickens. The antibody levels in the two groups of human sera against the antigens of MDV and Epstein-Barr virus (EBV) were titrated simultaneously by the indirect FAT (Naito *et al.*, 1971). The frequency distributions of the anti-MDV antibody titres of the sera are shown in Fig. 1. The frequency distributions of the antibody titres to MDV and to EBV of sera of workers in poultry farms were higher than those of sera of office workers. However, no exact parallel between antibody titres to MDV and to EBV was found.

Four human sera having high antibody activity against both MDV and EBV were fractionated by agar zone electrophoresis. The antibody activity of each fraction against MDV and EBV was examined by direct FAT. The MDV antigen reacted mainly with the β-globulin fraction of the human sera (Fig. 2), while EBV antigen reacted mainly with the γ1 and γ2 fractions of the sera.

A clear spot in the γ region, which was not clear in the control sera, was recognized by simple electrophoresis in the sera of chickens inoculated with

MDV (Fig. 3). Immunoelectrophoresis also revealed that there were more precipitation lines in the γ region of MD sera than in those of control sera (Fig. 4). The sera of the infected chickens, which have high antibody activity against MDV, were fractionated by agar zone electrophoresis. The antibody activity of each fraction against MDV and EBV was examined by direct FAT. The MDV

Fig. 1. Distributions of anti-MDV and anti-EBV titres in sera of 72 workers in poultry farms and 72 control subjects.

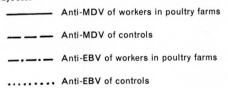

——————— Anti-MDV of workers in poultry farms

— — — Anti-MDV of controls

—·—·— Anti-EBV of workers in poultry farms

········ Anti-EBV of controls

* From the Research Institute for Microbial Diseases, Osaka University, Suita, Osaka, Japan.

Antibody activities to EBV and MDV antigens in human sera by indirect FAT[1]

Serum		Antibody titres			
Origin	Designation	1 : 10	1 : 40	1 : 160	≤ 1 : 640
Patients with Burkitt's lymphoma	KCC-454	+++ / +++	+++ / +++	++ / ++	+ / +
	KCC-766	+++ / ++	+++ / ++	++ / +	+ / −
	9495	+++ / +	+++ / +	++ / −	++ / −
Patients with nasopharyngeal carcinoma	K 245 020471	+++ / +++	+++ / ++	+ / +	+ / −
	955451	+++ / +	+++ / +	++ / −	++ / −
Workers in poultry farm	A-22	+++ / +++	++ / ++	+ / +	+ / +
	I-7	+++ / +++	++ / ++	+ / +	+ / +
	A-1	+ / ++	− / +	− / +	− / −
	A-38	− / +	− / +	− / −	− / −
	S-17	++ / +	+ / −	+ / −	− / −
Office workers	E-61	+++ / −	++ / −	+ / −	+ / −
	E-57	+ / +	+ / −	+ / −	− / −
	E-31	− / −	− / −	− / −	− / −
	E-38	+ / ++	− / +	− / +	− / −
	E-16	++ / +++	++ / ++	+ / +	+ / +

[1] In each category, the first line refers to EBV and the second to MDV.

antigen reacted mainly with $\gamma1$ and $\gamma2$ globulin regions of MD chicken sera (Fig. 5), while EBV antigen did not react with either fraction of the sera. Only 3 out of 42 sera of infected chickens showed antibody titres of more than 1 : 40 by indirect FAT.

The high levels of antibody to MDV in the sera of workers in poultry farms may be due to repeated antigen stimulation from inhaled dust derived from the epithelium of feather follicles of MD chickens.

Fig. 2. Duck embryo fibroblasts infected with MDV were stained with the β globulin fraction of human serum, conjugated with fluorescein isothiocyanate. Granular fluorescent areas were found mainly in the cytoplasm. ×1350.

Fig. 3. Comparison of patterns of simple electrophoresis of sera of three chickens reared in plastic isolators.
Upper: Serum of noninfected chicken, taken at 70 days old.
Middle: Serum of chicken inoculated with MDV (Biken C strain) at 1 day old, taken at 70 days old.
Lower: Serum of chicken inoculated with MDV (virulent strain) at 1 day old, taken at 90 days old.

Fig. 4. Comparison of patterns of immunoelectrophoresis of sera mentioned in Fig. 3 with rabbit anti-MD chicken serum.

Fig. 5. Duck embryo fibroblasts infected with MDV were stained with the $\gamma1$ globulin fraction of the sera of chickens with MD, conjugated with fluorescein-isothiocyanate. Granular and diffuse fluorescent areas were found mainly in the cytoplasm. ×1800.

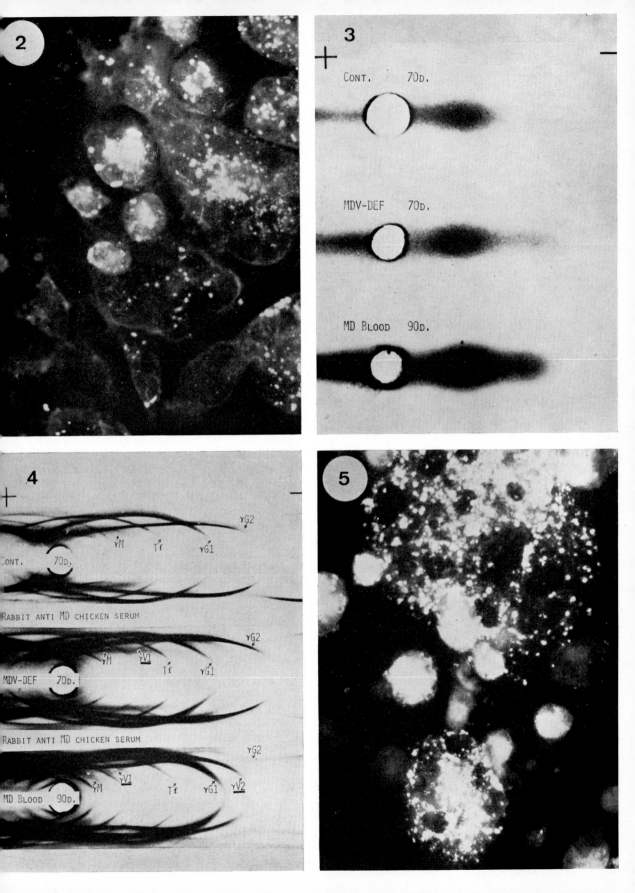

2

3

+   −

Cont.   70D.

MDV-DEF   70D.

MD Blood   90D.

4

+   −

Cont.   70D.   γM   Tf   γG1   γG2

Rabbit anti MD chicken serum

MDV-DEF   70D.   γM   V1   Tf   γG1   γG2

Rabbit anti MD chicken serum

MD Blood   90D.   γM   V1   Tf   γG1   V2   γG2

5

REFERENCES

Kato, S., Ono, K., Naito, M., Doi, T., Iwa, N., Mori, Y. & Onoda, T. (1970) Isolation of herpes type virus from chickens with Marek's disease using duck embryo fibroblast cultures. *Biken's J.*, **13**, 193-203

Naito, M., Ono, K., Doi, T., Kato, S. & Tanabe, S. (1971) Antibodies in human and monkey sera to herpes-type virus from a chicken with Marek's disease and to EB virus detected by the immunofluorescence test. *Biken's J.*, **14**, 161-166

The Involvement of Nuclear Membrane in the Synthesis of Herpes-type Viruses

A. A. NEWTON [1]

In bacterial cells there is much evidence to suggest that the DNA is attached to the cell membrane and that control of DNA synthesis is mediated at this attachment point. The nucleic acid of many DNA-containing bacteriophages is known to be replicated at sites on the host membrane, and it has been suggested that cell death may result from a disturbance of the relationship between host DNA and the membrane. In animal cells it is thought that the nuclear membrane may have a similar function, but its involvement in the synthesis of the nucleic acid of DNA-containing viruses has not been investigated. In this paper a preliminary investigation of the association of DNA with the membrane systems of cells infected by herpes simplex, by an attenuated strain of Marek's disease virus, and in EB3 cells is described.

Use was made, in this investigation, of the sarkosyl method described by Tremblay, Daniels & Schaecter (1969) to isolate membrane-bound DNA; parallel studies using conventional cell fractionation methods have shown that the membrane concerned is the inner nuclear membrane.

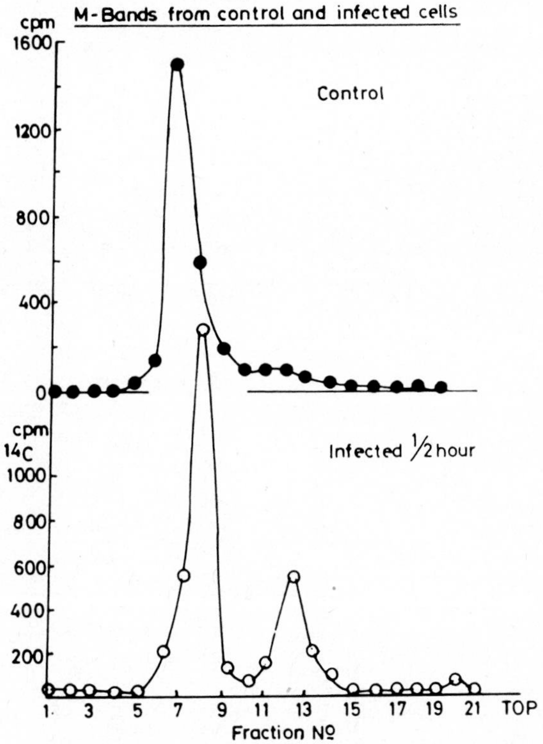

RESULTS

Herpes simplex virus

Pre-incubation of L cells with ^{14}C-labelled thymidine, before infection with a low multiplicity (<10 PFU/cell) of the HFEM strain of herpes simplex virus, showed that within an hour of infection some host cell DNA was displaced from its attachment to the membrane (Fig. 1). This DNA amounted to about 10% of the total and was gradually broken down to low molecular weight material

Fig. 1. DNA of L cells infected with herpes simplex virus. Monolayer cultures of L cells were incubated for 24 hours in Eagle's MEM containing 5% ox serum and 0.1 μC/ml ^{14}C-labelled thymidine (specific activity 5.8 mC/mmole). After washing, they were infected with 8 PFU/cell of HFEM herpes simplex virus for ½ hour. The cells were then washed in buffered saline and removed from the glass. 10[5] cells were layered on top of a 5-ml gradient of 15% sucrose—40% sucrose in 0.01 M tris, pH 7.4, 0.01 M magnesium acetate, 0.1 M KCl, and lysed by the addition of 0.05 ml of 1% sarkosyl NL35. They were centrifuged at 30 000 rpm for 1 hour at 4°C in an SW50 rotor. 5-drop fractions were collected from the bottom of the tube.

---

[1] Department of Biochemistry, Cambridge, UK.

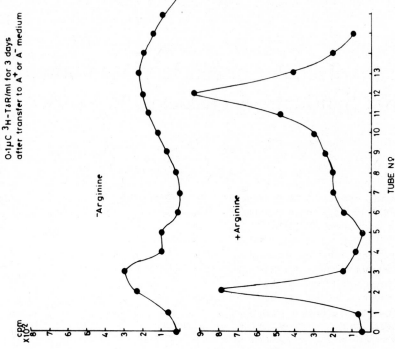

DNA FROM EB3 CELLS

0·1μC ^{3}H-T$_8$R/ml for 3 days
after transfer to A$^+$ or A$^-$ medium

Fig. 3. Cells of the EB3 line of Burkitt's lymphoma were grown in Eagle's MEM supplemented with nonessential amino acids, pyruvate and 10% foetal calf serum. Cells at a density of 10^5/ml were transferred to fresh medium either as above or lacking arginine, each containing 0.1 μC ^{3}H-thymidine/ml (specific activity 5 C/mmole). Cells were harvested after three days' growth and treated as described in Fig. 1.

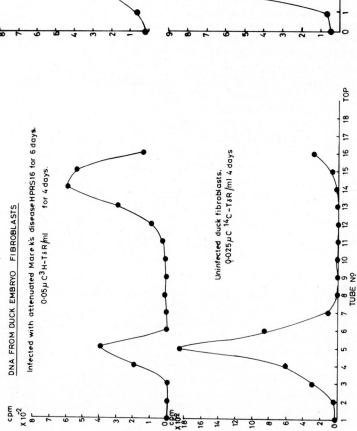

DNA FROM DUCK EMBRYO FIBROBLASTS

Infected with attenuated Marek's disease HPRS16 for 6 days.

0·05 μ C ^{3}H-T$_8$R/ml   for 4 days.

Uninfected duck fibroblasts.
0·025μC ^{14}C-T$_8$R/ml 4 days

Fig. 2. Duck embryo fibroblasts were infected with the attenuated HPRS-16 strain of Marek's disease virus. Two days later, fresh medium (Eagle's MEM containing 2% calf serum) was added containing 0.1 μC/ml ^{3}H-labelled thymidine (specific activity 5 Ci/mmole). Control cultures received similar treatment but received 0.025 μC/ml ^{14}C-labelled thymidine (specific activity 5.8 mC/mmole). Cells were harvested 5 days after infection and analyzed as for Fig. 1.

during the course of infection. These results are very similar to those obtained using virulent T4 bacteriophage infection of *Escherichia coli* (Earheart, 1970).

Pulse labelling was used to find the location of newly synthesized DNA during infection and this was always found in association with the membrane, but label added during viral DNA synthesis accumulated at the top of the gradient. Little DNA was found at the top of the gradient late in infection. These results suggest that viral DNA is synthesized in close association with the cell membrane, and that a disruption of the normal relationship between host cell DNA and the nuclear membrane probably occurs in this system.

Marek's disease

Primary duck embryo fibroblasts were infected with cell-bound preparations of the attenuated HPRS-16 strain of Marek's disease virus, and ³H-labelled thymidine added 2 days after infection; cells were harvested 5–6 days after infection. The pattern of DNA distribution observed was quite different to that obtained with herpes simplex (Fig. 2). A large proportion of the radioactivity was found at the top of the gradient, well separated from the membrane material. This material has a buoyant density of 1.718–1.720 in CsCl at 20°C, as determined by preparative separation, is completely destroyed by digestion with deoxyribonuclease, and may represent viral DNA.

Burkitt's lymphoma

Cultures of EB3 cells were incubated with ^{14}C-labelled thymidine in growth medium and then transferred to fresh medium either with or lacking arginine; ^{3}H-labelled thymidine was then added and incubation continued for another three days. The distribution of DNA in the sarkosyl-sucrose gradients was very similar to that obtained with Marek's disease virus and differed from that obtained with herpesvirus (Fig. 3).

The same pattern of distribution of labelled material was observed whether arginine was present or not; in the absence of arginine less radioactivity was found associated with either the membrane-bound material or the more slowly sedimenting material. It has not so far been possible to tell whether these results indicate a displacement of normal host cell DNA from the membrane or whether the DNA found near the top of the sarkosyl-sucrose gradients represents virus DNA.

CONCLUSIONS

It is clear that there is a difference in the pattern of labelling found in the different systems. It seems possible that in cells infected with more slowly growing viruses, the DNA accumulates at some site separated from nuclear membrane material, whereas in the case of herpes simplex virus, the viral DNA is used more efficiently for assembly of particles, which are then released from the cell.

ACKNOWLEDGEMENTS

This research was supported by a grant from the Cancer Research Campaign.

REFERENCES

Earhart, C. F. (1970) The association of host and phage DNA with the membrane of *Escherichia coli*. *Virology*, **42**, 429-436

Tremblay, G. T., Daniels, M. J. & Schaecter, M. (1969) Isolation of a cell membrane-DNA-nascent RNA complex from bacteria. *J. molec. Biol.*, **40**, 65-76

Discussion Summary

L. J. N. ROSS [1]

Effect of herpesvirus infection on host cell DNA

It was suggested that breakdown of host cell chromosomes could have accounted for the DNA sedimentation profile obtained with lysates of herpes-infected cells in Mg^{++}-sarkosyl gradients. It was argued, however, that this was unlikely since similar patterns were obtained after infecting cells with UV-irradiated virus (survival 10^{-4}), and it was known that the capacity of virus to induce chromosome breaks was reduced by UV with one-hit kinetics. Infection with herpesvirus was reported to inhibit host cell DNA synthesis but Okazaki fragments were synthesized.

Structure and function of herpes simplex

Opinions on the number of structural polypeptides differed. The maximum number reported was 14, of which 6 or 7 were probably part of the envelope. An arginine-rich polypeptide associated with DNA was believed to be an internal structural component. It was agreed that the discrepancies in the number of polypeptides reported could be explained by differences in the method of virus purification and in analytical techniques. Recent studies suggested that the numbers of polypeptides indicated in SDS-polyacrylamide gels were underestimates and that in several instances it was clear that the bands consisted of mixtures of polypeptides. The effect of the host cell on the composition of the virus envelope was emphasized. Some cells were reported to contribute more to the envelope than others, and this could also account for some discrepancies. It was suggested that further studies by means of immunological methods were necessary.

Much information was given on Band II, a structural antigen (or a number of such antigens) of herpes simplex. It was obtained by fractionation of herpes-infected cells by electrophoresis in 7% polyacrylamide gels. It was localized on the surface of virus particles because an antiserum prepared against it neutralized virus infectivity and agglutinated naked virus particles. Band II antigen was also present on the surface of infected cells. It was detected by conventional immunofluorescence in the cytoplasm 3–5 hours after infection, and could be seen later on in the nucleus. These findings illustrated that the same antigenic determinants could be present in different situations and could take part in a number of antigen-antibody reactions.

It was pointed out that the presence of infectious particles in nuclear fractions obtained by cell fractionation was not good evidence that naked particles were infectious since it was possible that intranuclear particles acquired envelopes from invaginations of the nuclear membrane. Although particles were reported to lose infectivity on treatment with non-ionic detergent, this could not be taken as unequivocal evidence that infectivity was a property of enveloped particles only, since it was not known what other effects detergent treatment might have on the particle besides removing the envelope. So far the evidence for the infectivity of naked particles rested on the fact that virus preparations could be obtained in which the number of plaque-forming units exceeded the number of enveloped particles.

Cross-reactions between members of the herpes group

It was reported that antisera which reacted with the group-specific antigen in immunodiffusion tests did not react in immunofluorescence tests and did not neutralize the infectivity of heterologous viruses. Since the evidence for associating the group-specific antigen with nucleocapsids was derived mainly from cross-absorption experiments, it was suggested that further studies, perhaps using labelled virus, were needed to confirm this. Intact nucleocapsids were thought to be too large to take part in immunodif-

[1] Houghton Poultry Research Station, Houghton, Huntingdon, UK.

fusion tests. Degraded virus, in contrast, could have accounted for the reactions observed. However, the presence of virus particles in precipitin lines was mentioned. It was pointed out that in some instances the group-specific antigen could be masked and that partial degradation of virus would be necessary to make it available to antibody.

A large proportion of sera from patients with nasopharyngeal carcinoma reacted with type-specific antigens derived from a line of Burkitt's lymphoblasts. However, only a small proportion of sera from patients with infectious mononucleosis reacted with those antigens. Sera from infectious mononucleosis patients did not react in general with the group-specific antigen either. However, it was reported that previous investigators had found that macroglobulins from infectious mononucleosis patients reacted in immunofluorescence tests with antigens of human cytomegalovirus. Both human and simian cytomegaloviruses were reported to share the group-specific antigen. Sheep and cows had not yet been examined for the presence of antibodies against group- and type-specific antigens.

Alternative explanations for the cross-reactions reported were considered. The possibility of cross-contamination of virus and of animals during experiments was thought to be important and the use of pre-immune antisera as controls was emphasized.

CLASSIFICATION OF HERPESVIRUSES

Chairmen — B. Roizman
G. de-Thé

Rapporteur — M. S. Pereira

Discussion Summary

M. S. PEREIRA [1]

Roizman and de-Thé put forward proposals for the classification and nomenclature of herpesviruses,[2] taking the opportunity of the presence of a large number of workers in the herpesvirus field to ask for suggestions and criticism of these proposals.

Roizman indicated that the proposals and the discussion will be considered by the herpesvirus study group, of which he is chairman, and which was set up by the International Commission for Nomenclature of Viruses to discuss and propose a classification for herpesviruses.

The current nomenclature is unsatisfactory since, in recent years, herpesviruses have been named at random according to the host species they infect, the disease they produce or their discoverers. The practice of naming herpesviruses according to the species they infect — the most desirable of current practices — presents three problems: (1) many species, including man, have more than one herpesvirus; (2) there are too many species with latinized binomial names difficult to remember; (3) some herpesviruses infect naturally more than one species. Roizman and de-Thé also pointed out that the division into A and B groups has no firm scientific basis.

They considered that, although it was probably too early to attempt to classify herpesviruses, there was a real need for a nomenclature as new members of the group were appearing at such a rate that if some plan were not followed in the naming of these new strains confusion was likely to occur.

The classification they proposed was based on a hierarchy of species and subspecies.

Species. The species nomenclature would be in two parts. The first would be a latinized name of the family, super-family, suborder or order containing the smallest known unit that harbours all the host species naturally infected by the virus with the widest host range. The second would be a consecutive numerical designation, one for each species of virus infecting that unit. For example, Herpes simplex would be Herpesvirus primatis 1 and Herpes varicella-zoster would be Herpesvirus primatis 2. Differentiation of species would be on the basis of immunological properties. Both the name and numerical designation would be allocated even if a family, super-family, suborder or order had yielded only one herpesvirus.

Subspecies. Immunologically closely related but biologically different herpesviruses isolated from a given unit of animal classification would be distinguished by means of a consecutive numerical subspecies, in addition to the numerical species, designation. For example, Herpes simplex types 1 and 2 would be Herpesvirus primatis 1 subspecies 1 and 2, and Marek's disease virus and turkey herpesvirus would be Herpesvirus galli 1 subspecies 1 and 2. The absence of a subspecies designation would indicate that none is known.

It was suggested that the latinized form would be used for formal occasions but that for common usage, anglicized forms could be adopted, e.g., primate herpesvirus, poultry herpesvirus, frog herpesvirus, bovine herpesvirus.

Roizman and de-Thé offered this proposal as a basis for discussion in the hope that suggestions could be transmitted to the members of the study group on herpesviruses for them to decide on the sort of classification and nomenclature that would be acceptable, bearing in mind the rules of the International Commission for Nomenclature of Viruses (ICNV).

These rules, which have recently been published in Wildy's book on the classification of herpesviruses, are as follows:

1. The code of bacterial nomenclature shall not be applied to viruses.
2. Nomenclature shall be international.

[1] Virus Reference Laboratory, Central Public Health Laboratory, Colindale Avenue, London, NW9.
[2] Bull. Wld. Hlth. Org. (in press) 1972.

3. Nomenclature shall be universally applied to all viruses.
4. An effort will be made towards a latinized binomial nomenclature.
5. Existing latinized names shall be retained whenever feasible.
6. The law of priority shall not be observed.
7. New sigla shall not be introduced.
8. No person's name shall be used.
9. No nonsense names shall be used.
10. The species is considered to be collections of viruses with like characters.
11. The genus is a group of species sharing certain common characters.
12. The rules of orthography, as listed in Chapter 3 of the proposed international code of nomenclature of names, shall be observed.
13. The ending of the name of a viral genus is -*virus*.
14. To avoid changing accepted usage, numbers, letters or combinations may be accepted for names of species. (This was considered by Wildy to be an important rule as it permits the type of nomenclature proposed by Roizman and de-Thé.)
15. These symbols may be preceded by an agreed abbreviation of the latinized name of a selected host genus, or, if necessary, by the full name. (This rule again would be in accord with the suggested nomenclature.)
16. Should families be required, a specific termination to the name of the family will be recommended.
17. Any family name will end in -*idae*.

Wildy's own views on the proposal were that, if names are to be given, they should be in accordance with a classification; therefore the classification must come first and the nomenclature second. The difficulty is that nomenclature is often needed before it is possible to make a classification. He agreed with Roizman and de-Thé that an enduring classification of herpesviruses is not possible at present, since the fundamental process of classification involves comparing many strains by many characters. This requires a large amount of information that we do not yet possess. The proposed classification would serve as a temporary measure, but he stressed that it would be most unwise to use Latin names as part of a temporary classification as such names tend to become permanent, and this may cause difficulties in the future.

The following points were made from the floor during discussion of the proposal.

Difficulties would be encountered with a classification and nomenclature based on the host because of the difficulty of determining the natural host range of a virus in some cases. The point was also made that numbers were less useful and more difficult to remember and handle than names. In order to ensure that viruses were numbered correctly, it was suggested that an official committee would be required to co-ordinate the numbering of newly isolated viruses.

It was also suggested during the discussion that the division of herpesviruses into groups A and B had little value and also that the terms "herpes-like" and "herpes-type" should be discarded.

Several speakers considered that it was premature to decide on a new nomenclature of herpesviruses because they felt that the next year or two would provide the necessary information on which a classification and nomenclature could logically be based. The authors of the proposal pointed out that these remarks related to classification rather than nomenclature, and that it was nomenclature that urgently required attention.

Some concern was expressed that the various committees dealing with nomenclature might not be in touch with one another and would not be informed of these discussions. Wildy said that the ICNV has taken care to ensure that there should be mutual representation on the appropriate committees. Roizman said that such committees had been asked for their views and suggestions, and in his opinion it would take years before anything was settled.

In concluding the discussion, de-Thé said that the comments and suggestions put forward would be recorded and distributed to members of the herpesvirus study group of ICNV.

CONCLUDING LECTURE

A Summing Up

G. KLEIN [1]

Luis Borges quotes in one of his writings an ancient Chinese classification of animals. According to this, animals fall into the following categories:

1. Those that belong to the Emperor
2. Those that have four legs
3. Wild dogs
4. Those that are likely to break a jar
5. Those that resemble flies, at least from a distance
6. Those that behave in a crazy way
7. Embalmed animals
8. Tame animals
9. Uncountables
10. Those that are drawn with a very fine brush, made of camel hair
11. Mythical beasts
12. Piglets, nursed on milk
13. Etcetera.

I did not come to think of this in relation to the classification of the herpesviruses. It occurred to me when I realized that my task was to consider four species, frog, chicken, monkey and man (with some minor additions), together with the world of herpesviruses. I am expected to use the viruses to link the four species together and to produce an orderly arrangement of their various properties. To this, I will answer, in the same way as everybody else did at the Symposium, whenever a tricky question came up: of course, this is what we are doing, planning to do, and have done, but we cannot yet give the results, since we have not received the latest telegram from the laboratory, etc. We will do it at the next symposium. What I can speak about now, instead, is perhaps a certain complementary effect in the different species and their herpesviruses, when considered in relation to each other. But please remember that all I can do is to feed the information that I was allowed by my physiology to gather during

these five days into my own Chinese system of classification. Others would no doubt do it quite differently. Will you, therefore, bear with me now; attack me later if you wish, and consider and reconsider what we have heard.

I was happy to discover how much is known in the other fields, and even happier to find that so much is not known. That re-established my self-confidence. At the same time, the danger arises of using data gathered from neighbouring fields to support one's own arguments—gathered, that is, with a myopia that hides all the gaps, particularly after listening to some of our most persuasive speakers. Because there are gaps in each system that we dealt with. The main question is whether we can use such data in a mutually helpful way to fill the gaps, or at least whether we can ask the right questions that will stimulate the right kind of experimentation.

What was said?

In his introductory speech, Dr Roizman formulated several important dilemmas. He said that all cells productively infected with herpesviruses die. I have not heard of any exception to this rule during the Symposium. He also said that the ability to establish a state of nonproductive infection was a property of all herpesviruses. This was largely reaffirmed, for several systems, at least in relation to the appropriate target cells. It follows that the oncogenic function must be nonproductive, and possibly a subset of the productive functions. If the UV-irradiated HSV type 2 is really responsible for the transformation reported by Rapp, the latter part of this statement may also be true—the first part must be.

Roizman pointed to the virus-induced membrane changes as the most likely candidates for the key oncogenic function. His own studies on HSV mutants, done long before anyone else suspected that

[1] Department of Tumor Biology, Karolinska Institutet, Stockholm, Sweden.

herpesviruses could be oncogenic, showed not only that viral envelope components were inserted into the plasma membrane of infected cells, but also that this event was accompanied by a change in the "social behaviour" of the target cells, characteristic for each virus mutant. Roizman suggested the possible existence of analogous but nonlytic interactions that might, if compatible with the unlimited proliferation of the target cells, lead to the ultimate model of cellular antisocial behaviour: neoplasia. Meanwhile, virally-induced membrane changes have been found in all virus-induced tumours that have been studied. Since herpesviruses are lytic, it is obvious, however, as Roizman pointed out, that, if they are to play a rôle in neoplastic behaviour, the virus-induced plasma membrane changes must be independent of viral DNA synthesis and of late, cytolytic viral functions. Findings of possible relevance, presented during the meeting, include the EBV-induced membrane antigens carried on the surface of Burkitt's lymphoma cells and, possibly, the antigens on the hamster cells transformed by UV-irradiated, HSV-type 2 virus, described by Rapp. In the latter case, the possible presence of viral envelope components on the surface of the non-productive tumour cells is suggested by the appearance of virus-neutralizing antibodies in the tumour-bearing hamsters. In the former case, EBV-induced membrane antigens were shown to appear in infected cells exposed to various DNA inhibitors. Similar evidence may be forthcoming with regard to the membrane changes induced by HSV types 1 and 2, detected by surface fluorescence and cytotoxicity tests in Wildy's laboratory.

Roizman's dilemma is not fully resolved by these experiments, however, because a nagging doubt remains whether DNA inhibition is really complete in experiments of this type and whether viral DNA may not "sneak through" to a small but sufficient extent, in spite of a complete block of cellular DNA synthesis. Another reason for puzzlement lies in the fact that envelope components are usually thought of as late viral products, since their appearance in cell membranes is presumed to occur in order to facilitate envelopment, the final intracellular stage in the viral strategy. Roizman now reports, however, that HSV-infected cells exposed to the DNA inhibitor, cytosine arabinoside, show an increased glycosylation of membrane proteins. Are the early membrane antigen changes due to the appearance of virally specified and truly early transferases, by analogy with the epsilon conversion in *Salmonella*?

If so, are they adventitious by-products of the viral strategy, not directly aimed at the facilitation of envelopment, but leading nevertheless to the ultimate appearance of virally changed membranes on the envelope, simply because there are no unchanged membranes in the infected cells?

Both Payne and Nazerian have given comprehensive reviews of the epidemiology, virology and immunology of *Marek's disease*. Payne raised the basic question whether the oncogenic virus-cell interaction is of an intrinsic or an extrinsic nature, or, in other words, whether tumour growth is due to a direct, virally induced neoplastic transformation of the target cells or whether it occurs in largely uninfected lymphoid cells, as the expression of a lymphoproliferative response to other, virally infected, perhaps virus-producing cells. Since MD usually has a multifocal origin, and, as far as the multiple cutaneous form is concerned, arises in direct proximity to the productively infected feather follicles, an indirect mechanism is an obvious possibility. On the other hand, Burkitt's lymphoma is also multifocal in a substantial proportion of the cases, and yet there is unequivocal evidence, obtained by the X-linked isozyme marker approach of Fialkow *et al.*, that it is a strictly uniclonal disease. If BL is virally induced, this must mean that the neoplastic transformation or, more probably, the successful development of a potentially neoplastic clone is a relatively rare event. But, as we learned from Nazerian, tumour development is a relatively rare event in Marek's disease as well. It would obviously be important to study the clonality of the MD tumour cell. I know of no suitable X-linked isozyme markers, but perhaps chimeric chickens, of the type used by the Hasek school for the study of immunological tolerance, could be used for this purpose.

We have heard that the neoplastic MD cell is probably not bursa-dependent, since bursectomy had no influence on the development of the disease. However, as Kleven told us, MDHV depresses humoral antibody responses. This indicates that the virus can grow in cells of the bursa-dependent system. There is a potential analogy with the murine leukaemia viruses here. The Moloney virus, for instance, can grow in antibody-producing plaque-forming cells, without turning them into tumour cells. In fact, all known leukaemia viruses can grow in many different cell types that never become neoplastic. Tumour development therefore appears as a relatively rare accident of interaction between cells of the appropriate, transformation-competent type and the virus, and not a regular event that occurs with high probability. If the competent cell is not bursa-dependent in MD, is it thymus-dependent?

The thymectomy experiments, reported by Payne, and, to some extent, the ALG treatment described by Anderson, indicate that the MD cell may be at least partially thymus-dependent. It appears puzzling, at first, that thymectomy increased the incidence of neoplastic lesions in resistant animals, but decreased it in a susceptible strain. In line with the latter finding, ALG had an inhibitory effect in the susceptible strain on which it was tested. There may be an analogy here with virally induced, thymus-derived murine leukaemias. In the Moloney virus-mouse leukaemia system, ALS treatment may lead to inhibition or facilitation of tumour growth, depending on at least three factors: the sensitivity of the neoplastic cell to ALS; its sensitivity to the host immune response; and the strength of the host response. The relative contribution of these factors in the ALS-treated, compared to the untreated, animal determine, in each tumour-host system, whether treatment leads to inhibition or facilitation of tumour growth.

Whether this scheme applies to MD or not, the increased incidence of the neoplastic disease in thymectomized resistant animals indicates that resistance must be due, at least in part, to immunological, rather than receptor-dependent, mechanisms. This would be also in line with Cole's finding that resistance is dominant and probably determined by a small number of genes.

In view of the many advantages of the MD system for experimentation, it is disappointing that there are no MD-derived, established cell lines. Is this a real difference, as compared with EBV, indicating the absence of the corresponding transforming function, or is it due to the species difference and the notorious difficulty of establishing continuous lines of chick cells *in vitro*? Would a start in organ culture, as used by Nilsson for the successful establishment of human blastoid lines from normal donors, help to overcome this?

We were told that some of the MD lesions are self-limiting, just like infectious mononucleosis in man. It is important to understand the immunological basis of the difference between regressive and progressive disease. Which antibodies are most relevant, if any, and at what time? Or does it all depend on cell-mediated immunity? Would all self-limiting lesions proceed to lethal, progressively growing tumours in thymectomized hosts?

The chemotherapy of Marek's disease has not been studied at all for obvious reasons: it is of no interest to the poultry industry. Would this not be of great interest, however, as a possible model for Burkitt's lymphoma and perhaps also for the more general study of the interactions between chemotherapy and host response? Perhaps a natural model of this kind would provide information of greater relevance than the so far unsuccessful attempts to induce tumours by inoculating EBV into various foreign species.

The remarkable maturation of MDHV in the feather follicle, discovered by Calnek, is a fascinating parallel to the previously known behaviour of the Shope papilloma virus, as Becker pointed out. What is the mechanism? Is viral maturation dependent on a highly specialized process of cell differentiation, such as the production of keratin? What would be the molecular basis of such a process, if it occurs? Or could it be simply due to the temperature difference in the outer epidermis? Are the final steps of viral maturation or assembly temperature-dependent? Would it be possible to learn more about this by cell-fusion techniques?

Biggs pointed out that different MDHV isolates may show important differences in the type of disease they produce. Of 9 small-plaque-type isolates, 8 were apathogenic and only one induced classical Marek's disease. The acute form of the disease was obtained only with macroplaque-type (MP) virus (12/16 isolates). Four of 16 MP types induced classical Marek's disease. Hearing this, one regrets the absence of a plaque test for EBV, as a result of the fact that no fully permissive target cells have been found for EBV as yet; all known infectious systems *in vitro* give an abortive cycle.

Purchase has demonstrated the possibility of viral cloning and reported further important viral markers. In addition to the fast growing macroplaque and the slow growing small-plaque types, he also showed that cloned virus strains can differ in certain antigenic components. Host range mutants could be distinguished by their relative chick-duck preference. There was also a difference in the pattern of neoplastic lesions obtained with the different strains. This brings the differences in the distribution of neoplastic lesions in BL patients to mind and raises the question whether such differences are entirely fortuitous, determined by the natural history of the tumour in the individual patient, or may represent some more inherent, essential property of the virus line. Kato added a temperature-sensitive viral strain to the range of MDHV mutants. Such mutants may be helpful in defining the viral functions required for productive infection and/or oncogenic action.

Witter spoke about the natural spread of the virus. In this connection, it must be recalled that MDHV is the only known oncogenic virus with a highly efficient, horizontal transmission in nature. Polyoma virus is also transmitted horizontally, but has no oncogenic effect in normal animals. On the contrary, it protects them against the growth of established polyoma-induced tumour cells. The situation becomes quite different, however, when the virus is allowed to spread among whole-body-irradiated or thymectomized animals. As shown by Law *et al.*, room infection may lead to tumour development in this case. We may recall, in this connection, that MDHV is itself immunosuppressive, as already mentioned. Is this the reason why it is oncogenic when natural contagion occurs?

Witter also spoke about the curious ability of MDHV to remain stable for long periods of time, apparently in large organic materials. Does EBV behave in a similar way? Could this be responsible for its transmission to young children in the lower socio-economic groups? It will be recalled that the mechanism of this transmission is quite unknown so far, and its ease and ubiquity stands in marked contrast to the laborious transmission found in adults, which requires specialized processes, such as "cellular kissing," in Evans' terminology.

Vertical transmission of MDHV was also considered and more or less completely dismissed by Witter. Although the issue is still slightly controversial, there seems to be no doubt that infected flocks can be freed from the virus by isolation rearing. This excludes vertical transmission, at least as a regular phenomenon. This again parallels EBV, where vertical transmission was largely eliminated by Pope and by Nilsson, as reported in Nilsson's contribution.

Witter also mentioned the important fact that anti-MDHV antibodies detectable by fluorescence were not necessarily correlated with precipitating antibodies present during the disease in individual animals. Apparently, the same kinds of antibodies can be detected against MDHV as against EBV. Immunofluorescence on fixed cells probably detects viral capsid antigens and perhaps early antigens as well. The membrane antigen, demonstrated by Chen and Purchase, may be similar to the EBV-induced membrane antigen. It is likely that virus-neutralizing antibodies are directed not only against the viral envelope, but also against the membranes of infected cells, which contain inserted envelope material. The precipitating antibodies that react with soluble antigens may be analogous to the antigens extracted from EBV-carrier cells by Old *et al.* It is clearly important to know more about the relationship between the various antibodies and the course of Marek's disease. In Burkitt's lymphoma, the disease-related patterns of the membrane-reactive antibodies are very different from those of the immunoprecipitating antibodies or the antibodies directed against intracellular early antigens. Chickens might be studied as models, not merely as short-lived commercial products with no other purpose than to be eaten. Such a change of attitude may be also timely in view of the success of the vaccination programme using attenuated MDHV and particularly a HTV vaccine, an antigenically related, nonpathogenic, cell-associated turkey virus. In one group of birds vaccinated with an attenuated MDHV, reported by Biggs, the incidence of the disease fell from 51 to 15%. Good results with vaccination were also reported in various types of infected flocks by Eidson. After this truly remarkable achievement, it may be hoped that our colleagues in the MD field will turn their attention to the possible relevance of their system for other species and particularly for man.

Among the factors that influence the incidence of MD, genetics, age, maternal antibodies, pathogenicity of the virus strain and prior infection are obvious enough, but what is the meaning of social stress if you are a chicken? Is this perhaps the long-desired model, so eagerly sought by the behavioural scientists who postulate a relationship between stress-related factors, immunodepression, and neoplastic disease? One major advantage of MD would lie in the fact that it provides a truly natural model.

Calnek compared the neutralizing and immunoprecipitating antibody status of resistant and susceptible strains of chicken after exposure to virus. Neutralizing antibodies were regularly present in the resistant strain, whereas the susceptible line developed precipitating, but essentially no neutralizing antibodies. This is a very important finding and one that raises a number of questions. Is the absence of neutralizing antibodies in the susceptible line a cause or a consequence of susceptibility? The latter situation could be seen as the result of rapid virus proliferation, leading to the binding of neutralizing antibodies to viral envelopes, particularly if viraemia prevails, and perhaps also to envelope components inserted into the plasma membrane of the infected cells. Precipitating antibodies, directed against internal viral components and/or virally determined soluble cell constituents, would be induced less rapidly and, because of the internal location of the antigen, would be less easily bound or paralyzed. Alternatively, neutralizing antibodies may be more directly involved in bringing about a resistant status. Provided that tumour cells carry viral envelope components in their membranes, as in the EBV system, neutralizing, i.e., presumably envelope reactive antibodies, might exert an anti-tumour effect, in addition to the antiviral effect, contributing to, if not causing, tumour regression. It has been shown recently in several species that responsiveness to defined antigens can vary greatly, depending on the genetic constitution. It is conceivable that susceptible strains may be genetically defective in their response to viral envelope components. This could be true even if the neutralizing antibodies were merely the symptoms, rather than the causes of antitumour immunity, the latter being mediated by the lymphoid cells of the host, as some of the evidence seems to indicate. If cellular immunity was directed against the viral envelope components on the target cell membrane, the appearance of humoral, virus-neutralizing antibodies would be a mere corollary, although an important one.

Surprisingly enough, the reactivity of guinea pigs and mice against some chemically defined antigens is very closely associated with the major histocompatibility locus of the species. Some genes that influence susceptibility to certain virally induced mouse leukaemias are also linked to the major histocompatibility locus. It might be of interest to carry out similar linkage studies in relation to MD susceptibility. The possibility of antigen excess in susceptible lines might be studied by looking for precipitated antigen-antibody complexes in the kidney.

Passive transfer experiments might be helpful in studying the possible antiviral or anticellular rôle of neutralizing antibodies, as well as the rôle of cell-mediated immunity. A purely antiviral mechanism might be expected to take effect prior to, or simultaneously with, virus inoculation. An anticellular effect might act later, during the oncogenic latency period. This seems to have been the case in the ALG experiment already mentioned. Concerning passive transfer experiments on cell-mediated immunity, we were told that these would be difficult, since the

normal lymphoid cells of immunized, ie., infected, chicks can transmit the virus. This may exert a protective effect in itself, by active immunization. Since we have also heard that there is no vertical transmission and that birds reared in isolation can be freed of virus, would it not be worth while to immunize such animals, provided that they belong to a suitably inbred strain, with inactivated virus that is still capable of inducing neutralizing and membrane-reactive antibodies, and therefore, presumably, cell-mediated immunity as well?

Eidson, in his clear and impressive account of the effects of HVT vaccination, pointed to the curious fact that while vaccinated birds have a reduced tumour incidence, they nevertheless continue to shed virus with their feather epithelium. Does this mean that the vaccine has an antitumour, rather than an antiviral, effect? Or could this be spurious, and related to the fact that chicks are never observed for longer periods than those necessary for our own carnivorous species? Perhaps the vaccine merely causes a delay in tumour development, as in certain murine leukaemias; this may not be reflected by such a crude parameter as viral shedding from the surface of a whole animal. There is also a possible analogy with infectious mononucleosis, where virus-neutralizing and membrane-reactive antibodies appear at the time when the proliferative disease goes into reverse and regresses, perhaps as a result of antibody action and/or cell-mediated immunity. Protection against a second infection is then almost complete; Evans has pointed out that recurrent disease is an extreme rarity. Nevertheless, the Henles have shown that it is frequently possible to isolate EBV-carrying cell lines from the peripheral blood of post-IM, anti-EBV-positive donors. This shows that the presence of virus inside cells is in no way incompatible with a state of resistance to proliferative disease. If allowance is made for the fact that no permissive cell has yet been found for EBV, the analogy appears quite close.

We have heard relatively little about the molecular biology of MDHV. Bachenheimer said that no unique GC pattern could be found in the oncogenic or presumably oncogenic herpesviruses, but why should there be, if, as is likely, the neoplastic transformation is an accident of virus-cell-host interaction, and not a property that has contributed to viral evolution? Perhaps more importantly, molecular hybridization experiments gave no support to the hypothesis of a common derivation of MDHV and HSV types 1 and 2.

We have heard from Kaaden that the DNA of HVT was linear and nicked, and Becker reported apparently similar findings for EBV. We have not heard any evidence concerning the important question whether the DNA of MDHV or of the other, presumably oncogenic herpesviruses could enter into covalent linkage with the cellular DNA. In view of the convincing demonstration of the integration of the smaller oncogenic DNA viruses, such as polyoma, SV40 and some human adenoviruses, such studies would be of interest, although perhaps more difficult with the large herpesviruses.

Adldinger studied the stabilization, adsorption and penetration of MDHV in tissue culture systems. Surprisingly, 20 hours after infection, 40–50% of the virus was still susceptible to neutralizing antibody. This may be an important difference from EBV, where irreversible penetration occurs within minutes after infection of, e.g., the Raji cell. Do the two viruses differ in penetrating ability, or is the difference due to the target cells used, namely avian fibroblasts and human blastoid cells?

From chicks we went on to *frogs*, and learned a great deal about how to catch them in Minnesota in the winter. Did we learn anything else?

Rafferty reviewed the remarkable temperature dependence of the *Lucké tumour* and emphasized the inverse relationship between tumour growth and viral maturation. Metastatic growth occurred only at high temperature, and virus maturation only at low temperature. He suggested that the spontaneous regressions seen at low temperature may be related to the winter activation of the virus, resulting in cytopathic effects in the carrier cells.

A system where virus production is temperature-sensitive and malignant transformation occurs at a nonpermissive temperature might provide a remarkable tool for the study of the viral functions required for transformation and productive infection, respectively, if appropriate cell lines could be grown for *in vitro* studies. The laborious search for analogous temperature-sensitive mutants in the more usual (and less natural) systems, such as polyoma and RSV, may be recalled in this connection; in the Lucké system the appropriate biological tools may be available already.

Rafferty also suggested that the localized growth of Lucké tumours in the frog kidney and the absence of contralateral kidney involvement indicated the existence of local immunity. Is this another analogy with Burkitt's lymphoma, with its remarkable absence of lymph-node involvement and the rarity of contralateral jaw tumours or of late relapses in the same site as the original tumour, as pointed out by Burkitt?

Granoff has reaffirmed the significance of a herpesvirus in the frog kidney tumour, but he also raised the question of identity. A herpesvirus, physically separated from HV-containing Lucké tumours, induced tumours in his experiments, but its base composition, hybridization and antigenic properties were different from those of the Rafferty HV. A third, papova-like virus, on the other hand, induced no tumours, either alone or together with the Rafferty HV.

MacKinnell demonstrated that the temperature effect was not a laboratory artefact but occurred in natural frog populations. He went on to discuss his fascinating nuclear transplantation experiments, where a nucleus derived from a triploid frog tumour, implanted into an enucleated egg, was capable of supporting development into a fully triploid tadpole. If the nucleus was really derived from a neoplastic cell and if this cell harboured the HV genome in some form, this would imply either the lack of irreversible nuclear changes and/or the lack of any effect of the changes that have occurred in the tumour nucleus on a very large number of normal tissue differentiation processes. This would again indicate that virally induced tumour formation is an accident of interaction that can occur only in a very restricted range of target cells or only at a very special stage of differentiation. One recalls, in this connection, the occurrence of reversions in SV40 or polyoma transformed cell lines, leading to a phenotypic resemblance to untransformed cells, in spite of the continued presence of the viral genome. Cell hybridization studies have shown, furthermore, that the introduction of a whole normal cell genome into virally transformed or other neoplastic cells may reverse their malignant behaviour. This implies that the viral genome, although perhaps necessary for the neoplastic behaviour of certain target cells, is not sufficient to induce and maintain such behaviour in cells with other types of differentiation. MacKinnell's experiment is obviously extremely important, provided that it can be shown convincingly that the donor cell was a neoplastic cell, rather than a stromal cell. Would it be possible to provide the triploid tumour with a diploid stroma, perhaps by serial passage through diploid frogs, if serial transplantation is possible, and then to repeat the nuclear transplantation experiments? Another approach would be to look for the viral DNA or virally determined products in the tadpole.

Tweedell has presented probably the strongest argument for an aetiological link, since he has shown that only fractions from tumours containing both nuclear inclusions and herpesvirus were oncogenic in frog embryos. The filtrates had a certain "transforming" effect on organ cultures that corresponded, if I understood him correctly, essentially to a cytopathic effect.

Mizell formulated a theory based on an analogy with lysogeny. He is conducting molecular hybridization experiments in collaboration with the Green group that can go a long way towards elucidating this question, provided that cRNA is really what it is supposed to be, i.e., complementary to the viral DNA only, rather than to cellular DNA mixed with it.

If we now turn to *infectious mononucleosis* (IM), Carter has emphasized the variability of the IM cell, as contrasted to the "deadly monotony" of the leukaemia cell, in Dameshek's terminology. It is reasonable to expect, I suppose, that the cellular proliferation in IM is pluriclonal, in contrast to the uniclonal growth of Burkitt tumours. It would be important to establish this also in relation to the question whether the effect of EBV in causing IM is "intrinsic" or "extrinsic", in Payne's terminology, and to the suggestion of Joncas that clinical IM may represent a rejection syndrome against EBV-converted cells.

As Carter so clearly emphasized, the main question in IM is not what makes the cells grow (since, it may be recalled, a proliferation-stimulating effect of EBV vis-a-vis lymphoid cells has now been amply demonstrated), but what stops them. I was amazed by the extensive pathology of the lymphoid system, as shown by him, and impressed by his question whether this was a truly self-limiting neoplastic disease. Are the blast-transformed normal lymphocytes, present in the diseased tissue in addition to the IM cells, reacting against viral or virally induced cellular antigens? Does this lead to rejection, via cell-mediated immunity? Alternatively, are the humoral antibodies responsible, directed against the EBV-induced membrane antigens and known to appear in the serum of mononucleosis patients? The Marek's disease model becomes highly relevant here and leads us back to the question whether the IM-like forms of MD would behave as tumours in immunosuppressed, e.g., thymectomized, animals?

Nilsson confirmed, in very convincing experiments, that EBV is *not* transmitted vertically. This is important, not only because it resembles MDHV, but also because it shows that true EBV-negatives must exist. The prospective studies of Evans *et al.* provide confirmation of this They show that at least 10% of the seronegative (by the <10 criterion) young adults must have been truly negative, at least in the populations studied, since they subsequently contracted mononucleosis, whereas not a single case developed in the large seropositive group.

The regular establishment of EBV-carrying lines from normal, unselected adult donors by Nilsson confirms not only the widespread occurrence of EBV, already well known, but also its probable significance in the unlimited proliferation of lymphoblastoid cells *in vitro*, first postulated by the Henles. The question arises whether truly EBV-negative blastoid lines exist at all? EBV-associated complement-fixing antigens have been found by Henle *et al.*, Pope, Vonka *et al.*, and Gerber *et al.* in lines that contain no viral particles and have no EBV antigens detectable by the various fluorescence methods. Zur Hausen has demonstrated the presence of EBV-DNA in at least one such "EBV-negative" line by molecular hybridization. It would be interesting now to reverse the direction of the search and look for truly EBV-negative lines, if they exist at all. Lines derived from leukaemia or myeloma would be perhaps the most likely candidates, since they might carry other viral genomes; the analogy with the many regularly CoFAL-positive members of the avian leukosis group, as contrasted with the CoFAL negative or only irregularly positive MD, may be recalled as a possible analogy. Zur Hausen's report on the extension of EBV-hybridization from DNA-DNA to cRNA-DNA, with increased sensitivity and the possibility of *in situ* hybridization, may be very useful in this connection, together with improved methods of antigen detection.

Evans stressed the remarkable fact that clinical mononucleosis was inversely related to the age of infection. If Joncas' suggestion is correct, would this imply a more violent rejection syndrome with increasing age? Or are the cells of young adults more competent to transform than the cells of children? Perhaps it would be of interest to study the probability of *in vitro* transformation in relation to the age of the donor, choosing EBV-seronegative donors and adding the virus *in vitro*. Is there perhaps a hormonal cofactor urging the EBV-infected lymphoid system on, in adolescents or young adults?

The session on *Burkitt's lymphoma* was introduced by Burkitt himself. He outlined his hypothesis that hyper- and holoendemic malaria is the most probable cofactor in causing BL, acting together with viral infection. I was alarmed to hear Pike say that the important study on the possible relationship between the distribution of the sickling trait and BL is apparently not being continued. This would be most regrettable, since this is one of the very few available ways to test Burkitt's hypothesis, and one more likely to give meaningful information than tests for malaria antibodies.

There are numerous observations, in addition to the direct serological studies that we have heard a great deal about, indicating that BL is a relatively strongly antigenic tumour. The most impressive is represented by the two cases of spontaneous regression, reported by Burkitt and Kyalwazi some time ago. Another related fact, not discussed at this Symposium, concerns the curious "plateau" in the survival curve of patients treated by chemotherapy, observed at all major clinical centres where substantial number of patients are treated. Very different forms of chemotherapy, and even therapy that was quite inadequate, have regularly led to total tumour regression in 15–25% of the patients, followed by survival for long periods, and perhaps cure. The clinical workers, notably Clifford, Burkitt and Ngu, postulated at an early stage that the factors determining long-term survival, as contrasted with progressive disease, are largely immunological in nature. All that has subsequently been learned about the response of the patients appears to be in line with this concept.

If it is accepted that the host immune response plays a part in determining the clinical course of BL, including its peculiar distribution, with more or less well localized solid tumours and lack of lymph-node involvement, one must ask whether the tumour-associated antigens, responsible for the postulated state of host sensitization, are specified by EBV or by other unrelated factors. Tumour-associated membrane antigens can be regularly demonstrated in BL biopsy materials. Recently, we have also found that IgG eluted from the surfaces of washed BL biopsy cells reacts specifically with the EBV-determined membrane antigen of the *in vitro* carrier lines, as judged by blocking of the direct radioiodine labelled antibody test against viable target cells, and is also capable of neutralizing the virus. This is not proof, of course, but it reinforces the position of EBV as a good candidate for the responsibility for tumour-associated host sensitization.

The diagnosis of BL as a clinical-pathological entity is relatively easy, if done by experienced pathologists in the areas where the tumour is en-demic. It is important to bear in mind that the diagnosis becomes much more difficult outside the endemic areas, however. The clinical and histological picture is then often much less clear, and the position is unlikely to be improved by any pressure on pathologists to reach a consensus of opinion, even if the pressure is exerted by prestigious international organizations. There is, in other words, no guarantee that so-called BL cases in nonendemic areas represent an aetiologically homogeneous group, comparable to the endemic cases.

The uniclonality of African Burkitt's lymphoma has been established unequivocally by Fialkow *et al.*, using X-linked isozyme markers. This raises the question whether neoplastic proliferation may be due to the escape of a single, perhaps relatively immunoresistant, cell clone from restricting host factors that apparently manage to stop the progressive proliferation of the other virally transformed clones that are probably present. Could a cofactor, such as malaria or some other proliferation-promoting agent, increase the probability that such a clone will "sneak through" and reach an irreversible size outside the effective reach of the prevailing host response? The surface receptor studies of Nishioka and the morphological criteria established by Nilsson indicate that the BL cell is a very special kind of lymphoid cell, with characteristic features that differentiate if from other EBV-carrying blastoid lines *in vitro*. The IgG-IA receptor difference between BL lines, on the one hand, and the lines derived from mononucleosis, nasopharyngeal carcinoma or healthy donors, on the other, reported by Nishioka, is particularly impressive. Could the BL cell be more resistant to, e.g., cell-mediated immunity than other blast cells? Could this be a reason why it has been so notoriously difficult to demonstrate an efficient cell-mediated immunity against BL cells in colony inhibition tests *in vitro*? The studies on BL cell immunoglobulin production of the secretory or membrane-bound type, as reported by Eva Klein and by Béchet, respectively, are largely compatible with the clonal derivation of BL, but there are a number of notable exceptions where several immunoglobulin classes are produced by certain BL-derived lines and, what is more surprising, by some biopsies as well. It is important to resolve the contradiction between these results and uniclonality by the isozyme marker approach. It could mean that there was an admixture of normal lymphoid cells, not apparent on the isozyme analysis of large populations, but sufficient to contribute a second immunoglobulin class. Alternatively, a clonal line of lymphoblastoid cells may still be able to produce more than one immunoglobulin class, in contrast to mature lymphocytes or myeloma cells. Van Furth and Eva Klein have recently found that the Ig-product of BL-derived lines is not uniclonal, even in cases where only one Ig-class is produced. Also, Bloom, Choi & Lamb have showed that cloned populations of established lymphoblastoid lines (although not of BL origin) that made both IgM and IgG continued to produce both classes of immunoglobulin in 23 of 25 single-cell derived isolates.

The serological association between EBV and Burkitt's lymphoma has been well reviewed. The consistent, high-titred association shown by 100% of the African cases provides a remarkable contrast to the lower frequency of positives and the much lower titres in healthy African controls and, what is probably more significant, in Africans with other types of lympho- or myeloproliferative diseases as well.

The healthy attack of Pike on the EBV establishment was useful in re-emphasizing that the serological evidence is inconclusive as far as a possible aetiological association is concerned, but I think that Pike was wrong in believing that the laboratory workers in this area have already made up their minds that an aetiological relationship in fact exists or that the earlier conclusion that IM is caused by EBV was reached on analogous grounds. It may be recalled that the clue to a possible association between EBV and IM was found only thanks to a remarkable piece of serendipity on the part of the Henles, when one of their laboratory technicians developed IM and converted from seronegative to positive. The next phase was the retrospective analysis of the Yale sera, in which negativity was consistently found, prior to their illness, in students who developed mononucleosis, and equally consistent seroconversion thereafter. This has not yet led to any definite conclusions, however. It was only the prospective study of Evans and Niedermann, performed together with the Henles, that brought what was accepted by most workers in this field as a decisive proof of an aetiological association. In fact, as we have heard, Dr Henle is still not completely satisfied and is seeking to fulfil Koch's third postulate, by demonstrating EBV-transmission leading to IM in connection with open heart surgery.

As far as BL is concerned, the decisive answer may obviously come from some similar kind of prospective study, and there is every reason to welcome the plan of Geser and de-Thé. We wish them good luck and short latency periods. It is an encouraging thought, in this respect, that BL is a highly antigenic tumour (relatively speaking), and that such tumours have the shortest latency periods in experimental systems, as a rule, because they are rejected unless they grow fast enough to outpace the immune response. One would like to hope that the relatively small number of sera that may become available from children who later develop BL will be tested by several methods, including the most sensitive, to decide whether they contain or lack anti-EBV antibodies. In view of the importance of this study, one also wonders whether 30 000 sera will be enough to give a sample of adequate size.

Pike mentioned that BL has been observed in three sib pairs in Uganda. I may add that three other sib pairs with BL are also known in Kenya. Pike pointed out that the disease occurred at relatively short intervals in the three Uganda families, although the affected children were several years apart in age. As he put it, "there was something very environmental about this" and he argued that EBV was unlikely to account for it, in view of its ubiquity. Is a possible rôle for EBV more unlikely in such a situation, however, than in its documented contact-transmission in relation to IM, which requires two EBV-negative young adults to engage in cell-mediated transfer? Would it be worth while looking for the simultaneous occurrence of two or more EBV-negatives within African families? If this can be shown to occur, it might be rather illuminating and could even serve to contradict the views of Pike on the possible spuriousness of African EBV sero-negatives.

A closer serological and histological study of other lymphoproliferative malignancies inside and outside Africa, as discussed by Levine, is obviously very important. It should be kept in mind, however, that in virally induced experimental leukaemia, serological reactivity follows aetiological and not histopathological boundaries. This can be exemplified by the Gross and Moloney lymphomas, indistinguishable with regard to histology and pathogenesis, but different from the aetiological and serological points of view. On the opposite side, the wide range of lympho- and myeloproliferative diseases induced by, e.g., the Graffi virus or the many different solid tumours induced by polyoma may serve to exemplify cases of aetiological and antigenic identity in spite of widely differing histology.

There have been a number of interesting new findings in relation to EBV and Burkitt's lymphoma.

The morphological transformation of a monolayer culture, found by Epstein after combined EBV infection and cell fusion by inactivated Sendai virus, is very interesting, but its relationship to EBV remains to be proved by antigenic markers or the presence of virus-derived nucleic acid. The Henles demonstrated that the EBV-induced early antigen (EA) can be broken down into two serologically, morphologically, and probably chemically distinct components, designated as R and D. It was particularly interesting that anti-R and anti-D antibody formation showed certain disease-related differences. They had previously found that anti-EA antibodies rarely occur in the sera of healthy anti-EBV-positive, i.e., anti-VCA-positive, individuals, or in BL patients with high anti-VCA levels whose tumours have undergone complete, long-term regression. Positive anti-EA titres in regression patients showed a certain correlation with the risk of recurrence and probably signal the presence of residual tumour. When the anti-EA titres were broken down into anti-R and anti-D, it appeared, surprisingly, that the prognostically unfavourable anti-EA antibody of BL patients was anti-R in the majority of the cases (although not always), whereas the anti-EA-positive sera of acute infectious mononucleosis or of nasopharyngeal carcinoma were usually anti-D, and negative or low-titred for anti-R. Further studies in this area will be important, because they may reflect differences in the virus-cell and cell-host relationships in BL, as contrasted with IM and NPC. They may be informative with regard to the different ways in which viral antigens are activated or stored, and/or presented to the immune system.

In groups with similar anti-VCA titres, ancillary antibodies to other EBV-related antigens show interesting disease-related differences in other test systems as well, e.g., the CF test of Sohier or the S antigen test of Vonka. It would be important to ascertain, by the exchange of suitable reagents, how the various CF antigens are related to each other and to the antigens detected by the different immunofluorescence methods.

Using cRNA-DNA hybridization instead of the previous DNA/DNA hybridization, zur Hausen has now increased the approximate number of 6 genome equivalents per cell in the Raji line to 40–55. It is remarkable that more transcription of the viral genome does not occur under such circumstances than can be detected by the presence of a complement-fixing antigen; none of the antigens detected by immunofluorescence have ever been demonstrated in the Raji cell. Are all copies of the viral genome defective? Or is transcription suppressed as a result of covalent integration, together with a lack of an excision apparatus? Are they any repressors involved? The latter possibility could be tested by fusing the Raji line with a relatively good EBV producer. The introduction of the appropriate drug-resistance, enzyme-deficiency markers required for fusion experiments, into representative blastoid cell lines may turn out to be very interesting for human somatic cell genetics as well, because most lines are diploid or near-diploid, in contrast to the majority of established monolayer lines.

Becker showed that arginine deficiency, a condition known to promote EBV production, may also facilitate the distinction between viral and host proteins in carrier cultures. The contours of a system of early-early, late-early and late viral products begins to emerge here, as in other herpesviruses. It would be important to link these findings with the various EBV-associated antigens that can be demonstrated both in abortively infected and in carrier cells.

Gerber reported the potentially important observation that EBV induces DNA synthesis in infected lymphoid cells. He suggested that a parallel may exist between this finding and the ability of the small oncogenic DNA viruses, SV40 and polyoma, to induce cellular DNA synthesis. Interesting as this would be, it is complicated by the fact that lymphoid cells respond with DNA synthesis and blast transformation when re-exposed to sensitizing antigens or confronted with a number of "nonspecific" stimulating agents. Gerber thinks that antigenic stimulation can be excluded, because inactivated virus did not induce DNA synthesis. It would be interesting to check this observation in relation to viral preparations with known antigenic activity, tested in parallel against the lymphocytes of EBV-immunized (i.e., seropositive) and EBV-negative lymphocyte donors.

Osato presented a new and potentially very useful separation technique, capable of sorting out cells carrying different EBV-associated antigens from carrier cultures. Of particular interest was the fact that cells containing membrane antigen (MA), but no viral capsid antigens (VCA), and also cells containing intracellular early antigens (EA), but no VCA, separated in the same intermediate layer as the antigen-negative cells, whereas VCA+ cells separated out in bands of lower density. From the practical point of view, this may provide a method of isolating VCA+ cells from cell suspensions in which there are very few of them; this is of obvious potential significance, e.g., in relation to studies on biopsy materials or carrier cultures with a very small proportion of VCA+ cells. One may also envisage the separation of the various cell types for biochemical comparisons. It is also of interest that cells that carry only the early products of the viral cycle, namely MA or EA, had a density corresponding to that of the majority of viable cells, whereas the appearance of the late VCA is apparently accompanied by membrane damage, with lower density as a result. It would be interesting to compare the separation pattern in a number of cell lines, differing with regard to the abortiveness of EBV infection, and with accumulation of the early and absence of the late functions in some lines, in contrast to the relatively complete viral cycle characteristic of the P3HR-1 subline used by Osato.

The problems of *nasopharyngeal carcinoma (NPC)* were introduced by the excellent pathological survey of Dr Shanmugaratnam, which gave us a clear picture of the nature of the neoplastic cell in this tumour. One important difference between BL and NPC lines stems from the fact that the pathology of NPC is unambiguous, in contrast to BL. It is therefore striking that NPC cases from Africa, China, Sweden, France and the United States show equally consistent and high EBV-associated serological reactivity in relation to all antigen systems so far investigated (membrane, capsid, early and soluble). Other tumours localized to the nasopharynx, or carcinomas situated in the hypo- or oropharynx do not behave in this way. This must be considered in relation to the passenger hypothesis. In experimental viral oncology, a virally determined agent may be present in the tumour either because the virus has induced the tumour, or superinfected it after it had been induced by other agents. The occasional appearance of polyoma-induced antigens in methylcholanthrene-induced sarcomas can be used as an example of the latter possibility. The two features that differentiate the polyoma antigens due to superinfection from the corresponding antigens in the polyoma-induced tumours (and I am thinking here primarily of the transplantation-type antigen) concern the consistency and the stability of the association. All polyoma tumours, but only some MC-tumours, contain the antigen. All polyoma tumours maintain it, in spite of adverse selection, whereas tumours that have acquired it by superinfection tend to lose it readily. Immunoselection tends to operate

against all antigenic tumours *in vivo* and promotes this loss. The consistent presence of EBV-associated antigens in nasopharyngeal carcinomas, irrespective of geography, does not prove a causative relationship, of course, but it nevertheless casts doubt on any straightforward passenger hypothesis. To maintain this hypothesis, it would be necessary to introduce further qualifications, e.g., by postulating that EBV can grow as a passenger only in a special kind of lymphoid cell that is present in nasopharyngeal carcinoma, but not in the various controls already mentioned. This is possible, but there is no evidence for or against it.

Dr Shanmugaratnam has condemned the use of NPC-derived lymphoblastoid cell lines, since they are not representative of the neoplastic cell. He had a good point as far as direct studies on the neoplastic cell are concerned, but from the EB-viral point of view, the virus associated with the lymphoid cells derived from NPC is of great interest, since it represents a viral strain associated with this disease that may or may not show properties different from the strains associated with BL and IM, respectively.

Since blastoid cells are the only known supporters of EBV multiplication so far, it is obvious that the study of the NPC-associated EBV is possible only in such cells, until methods can be found to demonstrate the virus or virally determined products in other cell types. It is of great interest that de-Thé now reports *in vitro* growth, if only temporary, of the epithelial NPC cell. It would be important to test such cultures for EBV-associated antigens by the most sensitive methods, covering intracellular and surface antigens as well. A modification of the radio-iodine antibody technique, reported by Greenland, might be particularly useful if applied to viable cells. Another sensitive membrane antigen test can be performed by adapting the mixed haemadsorption reaction. Also, it might be possible to look for EBV-DNA by *in situ* hybridization, as reported by zur Hausen, not only at the cytological, i.e., chromosomal, level, but also at the histological level, in order to find out whether the viral nucleic acid is present only in the lymphoid elements, in the carcinoma cells, or in both.

Both Ho and Muir gave us much information about the occurrence of NPC in different geographical regions and ethnic groups. The evidence for the existence of genetic factors influencing susceptibility to NPC are overwhelming, particularly if one considers the first generation offspring from matings between members of susceptible and resistant ethnic groups.

Ho mentioned that husband-wife NPC clusters are extremely rare and he implied that this was an argument against a virus, such as EBV, being involved in the causation of the disease. It may be recalled, however, that most adults are EBV-positive, and the chance of an encounter between two EBV-negatives would be much lower than in children or adolescents. Ho formulated the dilemma very clearly, however, when he asked how a nonethnic virus could lead to an ethnic tumour. If EBV is to play any part at all in NPC, there must either be different forms, with wide differences in oncogenic potential, or genetic and/or environmental cofactors must play the dominant rôle. I have previously argued, in relation to Burkitt's lymphoma, that a long latency period would not be likely to occur with a strongly antigenic tumour. Not too much is known about the antigenicity of NPC, however, and we do not have the same clear indication that it *is* a highly antigenic tumour as in the case of BL. It is interesting that Ho demonstrated a difference between the two-peak age-incidence distribution of NPC in the Chinese, with both a young and an old group being affected, and the Swedish material, where the disease is limited to the older age-group. Could it be that, in certain regions and/or ethnic groups where high susceptibility prevails, latency periods can be short, whereas in groups of lower susceptibility the tumour would only develop after the immune system has undergone an age-dependent involution?

There is a tendency at present to relegate EBV to a passenger rôle in NPC more readily than in BL, but the uniformity of the serology and its independence of geography should warn against too rapid a decision.

Yoshida reported a curiously high incidence of antinuclear antibodies in the sera of NPC patients, not seen in BL and a number of other malignancies. This may obviously introduce complications when distinctive NPC antibodies are sought, and reinforces the need for vigorous specificity controls. Incidentally, is the remarkably high anti-EBV titre found by Carvalho in a number of autoimmune diseases related to Yoshida's finding? Could EBV-induced lysis in certain cells lead to autoimmune reactions?

Cervical carcinoma is the youngest of the possibly herpesvirus-related human diseases. A number of papers dealt with this problem and with related questions concerning the serology of HSV-type 1 and type 2 infections.

Wildy showed that the serological distinction between the envelopes of the two virus types, well known from neutralization tests, also applies to the serology of infected cell membranes, as studied by membrane fluorescence and cytotoxicity. This is obviously related to Roizman's earlier work on the insertion of viral envelope components into the outer plasma membrane. Interestingly enough, intracellular antigens are less rigorously type-specific than membrane antigens. This suggests that in other, less well-known systems, such as EBV, the membrane of the infected cell may be the best place to look for possible type-specific differences.

The comprehensive review of Nahmias affirmed the previously demonstrated relationship between two entities: (*a*) HSV-type 2 infection; and (*b*) carcinoma of the cervix. Since both are related to sexual promiscuity, however, it is hard to know which of the possible relationships is causal. Rawls also stressed the importance of using the same socio-economic and ethnic groups in comparative studies of the frequency of HSV type 2 infection in patients with cervical carcinoma. One very interesting piece of new evidence came from a prospective study of Nahmias, indicating that genital herpes infection during pregnancy was particularly likely to increase the frequency of cervical dysplasia. Is this a case of hormonal-viral co-operation, as known in experimental oncology? The interaction of MTV and oestrogenic hormones in mice is the best known case, but there are many others. An interesting start towards a direct experimental test of HSV-hormonal interaction has been reported by Muñoz.

Obviously, advances in this field will depend on the demonstration of an association between the viral genome and/or virally determined antigens with the carcinoma cells. There are some suggestive findings by Aurelian on exfoliated cells, and by Nahmias on cultured cells, but the evidence is not yet convincing.

It was pointed out that there was no difference between anti-HSV-type 2 antibody levels in patients with cervical carcinoma and in controls, in contrast to the pronounced increase in the anti-EBV titre of BL patients, as compared with controls. Could this be related to a scarcity of virally determined antigens in this case, in contrast to BL, or to the fact that one is looking for antibodies against the wrong (late) antigens? HSV is a very much more cytopathic virus than EBV, of course, and the cell probably cannot afford the production of too many virally determined proteins if it is to multiply. Could the interaction between HSV and the cervical carcinoma cell, if such interaction exists, resemble that of the Raji cell in the case of the Burkitt lines?

Rowson pointed out the curious difference between rabbit anti-HSV-type 1–type 2 sera, in contrast to human sera with the corresponding activities. The rabbit sera were clearly type-specific, whereas the human sera showed considerable cross-reactivity. Could this be related to the fact that the immunocompetent cells of the rabbit are primarily exposed to the surface on intact viral particles, whereas the susceptible species, man, is exposed to the whole viral cycle, with all its antigens? The difference could then be related to the demonstrated fact that intracellular antigens are more cross-reactive than envelope and membrane antigens.

Rowson has also reported a curious relationship between HSV and carcinoma of the larynx. But is the epithelium of the larynx susceptible to herpetic infection? What form could such an infection take?

We all listened with fascination to the remarkable *in vitro* transformation experiment of Rapp. Is the interaction of hamster cells with UV-irradiated HSV-type 2 virus a highly artificial model, such as, e.g., Koprowski's unique SV40 transformed monkey cell, resulting from an accidental encounter between a permissive cell and a defective virus, or does it exemplify a nonproductive interaction that may take place in nature? Since Rapp has demonstrated that the transformed cell was susceptible to lytic HSV, a model situation, rather than a natural nonproductive interaction would appear more likely. Before this question can be settled, it is important to know whether the HSV genome is still present in the transformed cell. This is suggested by the ability of the cells to induce neutralizing antibodies while growing progressively in hamsters. Is the situation similar to Munyon's conversion of TK— cells to TK+, by introducing the virally determined enzyme with UV-irradiated herpes simplex virus? In that case, the viral genome is maintained by selection and it is not known how easily it can be lost. Rapp's case may be quite analogous, since transformed and/or malignant cells are continuously favoured by serial selective passage. A study of reversion frequencies would be most important in establishing the stability of viral-cellular interaction.

If the presence of the HSV genome in the transformed cells could be demonstrated convincingly, Rapp's system will no doubt provide an excellent tool for the study of the relationships between the viral and the cellular genome, together with the various morphological, antigenic and biochemical parameters of transformation.

The oncogenic *monkey herpesvirus* or, now more appropriately, herpesviruses, although only briefly touched upon by Dr Melendez, present perhaps the best models of the human disease. The lymphoma inducing property of herpesvirus saimiri appears to be beyond doubt and, as we heard during the discussion, Rabson *et al.* have succeeded in establishing continuous blastoid lines, resembling the human BL-derived cultures. On the opposite side of the phylogeny, Calnek pointed out the resemblance to MD. Were it not for the practical difficulties of working with primates, and for the possible hazards involved, this would be no doubt the animal system of choice in achieving a better understanding of the human disease.

Very much more than what has been mentioned here was presented at the Symposium and I can only apologize for my inability to fit the rest into the picture. The Table is an attempt to summarize some of the main conclusions, including the major gaps, as I have been able to see them.

It is encouraging that the gaps are in different areas in the different fields and complementary information may be therefore of very great value, in spite of the difficulties involved in all extrapolation. The following are some of the many areas where joint or parallel experimentation might be useful.

It would be of value to investigate the changes in *tumour cell membranes* from several points of view. What is the relationship between the membrane antigen changes seen in tumour cells *in vivo*, as compared to those in the derived tissue culture lines? Do membrane changes always accompany the *in vivo* tumours and is the sensitization of the host against tumour-associated antigens related to the virally induced membrane changes? If that is the case, what is the relationship between the latter and the viral cycle? Is their presence compatible with continued host cell macromolecular synthesis and multiplication? How do the immunologically detectable changes relate to changes in membrane biochemistry, ultrastructure, and behaviour?

Clonality studies of the type performed on BL would be valuable for the other systems as well, particularly for IM and MD, since they could help to distinguish between intrinsic (direct) and extrinsic (indirect) mechanisms of induction. While pluriclonality does not exclude direct induction, uniclonality would be very hard to reconcile with a purely extrinsic (reactive) pathogenesis.

Oncogenesis and herpes-type viruses: Summary of conclusions

Virus	Disease	Causative relation-ship	Intrinsic (direct) action [1]	Extrinsic (indirect) action [1]	Special advantage	Special difficulty	Remarks
MDHV	MD	Yes	?	?	Experimentation Marker approach Viral clonability	No tumour-derived cell lines	—
Lucké virus	Frog kidney cancer	? or (+)	?	?	Temperature-sensitive viral function	No cell lines No *in vitro* tests for virus	—
EBV	IM	Yes	Probable (EBV-carrier line derivation)	Possible	Cell lines Clinical-serological information	No experimentation	
EBV	BL	?	Probable (presence of viral genome, virally determined antigens, uniclonality)	Possible	Clinical-serological information Correlation with tumour status	No experimentation	Common difficulty for all EBV systems: no fully susceptible target cell that supports productive infection so far.
EBV	NPC	?	?	?	Clinical-serological information	No experimentation No tumour-representative cell lines in long-term culture	
HSV-type 2	Cervical carcinoma	?	?	?	Clinical information	No representative cell lines	—
Herpes-virus saimiri	Monkey lymphoma	Yes	?	?	Cell lines Experimentation Primate	Primate	—

[1] If a causative relationship does exist, intrinsic or direct action is defined as the production of neoplastic cells from their normal progenitors by direct action of the viral genome. Extrinsic or indirect action represents the proliferation of lymphoid cells as a reaction to a viral infection in other cells that do not become neoplastic and may undergo lytic changes.

The fact that IM regresses in man, in spite of extensive, tumour-like changes in many internal organs, while some MD lesions regress and some progress, and that most Burkitt's lymphomas progress, although some regress (with or without chemotherapy), urgently requires an immunological explanation. Such an explanation may be of crucial value for the understanding of the immunological behaviour of certain other neoplasms as well. Studies of antibody formation against cell membrane antigens, viral envelopes, viral capsids, soluble or other virally induced, nonstructural antigens are needed in relation to progression and regression, respectively. Equally important is the study of cell-mediated immunity, both alone and in interaction with the various types of human antibodies.

Experiments on comparative chemotherapy would be related, to some extent, to the foregoing. Would neoplasia induced by MDV or some other herpesvirus show an equally favourable response to chemotherapy as a proportion of the BL cases, and could this be due to a synergistic interaction with the proper type of immune response?

The status of viral DNA in the tumour cells and particularly the question whether there is covalent binding between the viral and cellular genome is another important subject where comparative studies can be helpful, particularly as far as the development of suitable methodology is concerned.

The genetics and genetically determined immunology of disease susceptibility and resistance is a field where experiments on the animal species most amenable to genetic study might be of great help in designing appropriate approaches to what may be the corresponding human disease. Is there any linkage between susceptibility and the locus of the major histocompatibility gene? Is resistance related to the presence of a particular kind of antibody or to a definable state of cellular sensitization? Quite apart from the question of genetically determined differences in the immune response, is it possible to demonstrate differences in viral receptors that may be related to susceptibility?

Is a viral genome, such as EBV, necessary for the continuous *in vitro* proliferation of established lymphoblastoid lines? Do all established human blastoid lines have this genome, when studied by the most sensitive techniques, such as nucleic acid hybridization or the detection of complement-fixing antigens? Or can other viruses take the place of EBV in the same rôle? What is the corresponding situation in animal cell lines, established from herpes-virus-induced tumours, such as the lines recently developed from monkey lymphomas, or from apparently normal individuals? Is it possible to overcome the difficulties of establishing MD-derived cell lines by the use of suitable organ cultures? Can different clonal isolates of herpes-type viruses induce neoplastic diseases with different organ distributions, as seems to be the case in MD? Can the approaches of virus cloning and marker isolation be adapted to EBV, if suitable permissive cell types can be found? Is there such a cell type in view and does it correspond to the feather follicle, in relation to MDHV, in the chicken?

Is there a state of local immunity in MD, the Lucké tumour, and BL, that can explain some of the curious tissue distribution patterns and the frequent lack of contralateral involvements or recurrences in the same anatomical site?

Can sero-epidemiological studies on large populations of chickens in relation to MDHV help in designing appropriate sero-epidemiological studies or prospective investigations of the human tumours where EBV is suspected as a possible aetiological agent?

What is the reason for the absence of viral particles and late viral products from *in vivo* tumours in cases such as MD or BL? Why do cells that produce virus particles and late viral antigens promptly appear in culture or, as far as MD is concerned, in a superficial layer of cells where cell death is continuously going on? Do tumour cells not support viral maturation

in vivo? Or are they quickly taken up by macrophages, as soon as they enter the late, i.e., the lytic, cycle? Can virus maturation be suppressed by antibodies? Could this be demonstrated *in vitro*?

Finally, there is one aspect of the studies on lymphomas that has hardly been mentioned, although it may open up a fertile field of study. Lymphoma cells may project highly specialized types of differentiation into large clonal populations. Their study may provide new information about the corresponding normal lymphocyte types, particularly if they are present as a minority, perhaps in a way similar to that in which myeloma cells have been so fruitful for the understanding of the differentiation of normal plasma cells and the various types of immunoglobulins. Several BL-derived lines have now been found that insert molecules with IgM-kappa specificity into their cell membranes, rather than secreting it. They are probably representative of a corresponding normal lymphoid cell, perhaps an antigen-recognition cell. Most of the BL-derived lines also have another receptor, capable of reacting with immunoglobulin-coated antigens of certain types. Are they derived from normal lymphoid cells that respond to antigen-antibody complexes? What is the significance of the IA receptor, present on most blastoid cell lines derived from NPC and IM? Workers interested in the lymphoid system may find a fruitful field of investigation here.

Obviously, there are many other subjects that could be tackled in comparative studies and the most important function of this Symposium, in addition to giving us all a most interesting and enjoyable week in Cambridge, was to reveal the great potentialities inherent in collaborative efforts across species and other barriers.